DIRECTORY OF GRAPHS

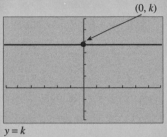

$y = k$
$[-6, 6] \times [-3, 6]$

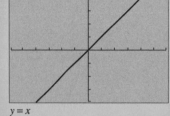

$y = x$
$[-6, 6] \times [-4, 4]$

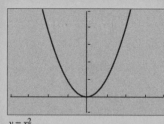

$y = x^2$
$[-4, 4] \times [-1, 5]$

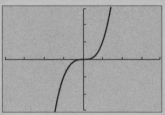

$y = x^3$
$[-4, 4] \times [-3, 3]$

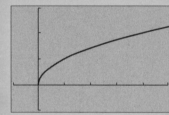

$y = \sqrt{x}$
$[-1, 5] \times [-1, 3]$

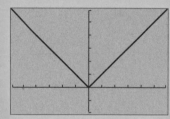

$y = |x|$
$[-6, 6] \times [-2, 6]$

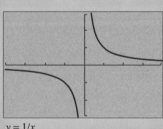

$y = 1/x$
$[-4, 4] \times [-3, 3]$

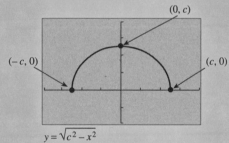

$y = \sqrt{c^2 - x^2}$
$[-4, 4] \times [-2, 4]$

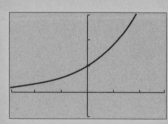

$y = b^x, \; b > 1$
$[-3, 3] \times [-1, 3]$

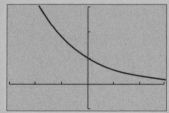

$y = b^x, \; b < 1$
$[-3, 3] \times [-1, 3]$

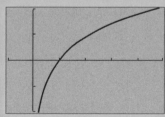

$y = \log_b x, \; b > 1$
$[-1, 5] \times [-2, 2]$

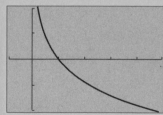

$y = \log_b x, \; b < 1$
$[-1, 5] \times [-2, 2]$

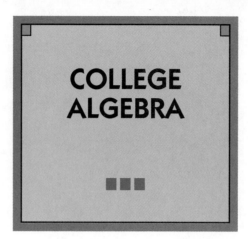

COLLEGE ALGEBRA

COLLEGE
ALGEBRA

Warren L. Ruud   Terry L. Shell

Santa Rosa Junior College

Worth Publishers

COLLEGE ALGEBRA

Library of Congress Catalog Card Number: 96–061160

ISBN: 1–57259–244–3

Printing: 1 2 3 4 5 — 00 99 98 97

Production: Phyllis Niklas

Design: Janet Bollow

Composition: Progressive Information Technologies

Line art: Carl Brown

Printing and binding: R.R. Donnelley & Sons

Cover: *Rainbow Lake, Maine.* Sam Abell, © National Geographic Image Collection

Worth Publishers

33 Irving Place

New York, New York 10003

For Nancy   For Candy

CONTENTS IN BRIEF

CONTENTS

CHAPTER 5

SYSTEMS OF EQUATIONS AND INEQUALITIES 417

APPENDIX A

LINEAR PROGRAMMING   659

As the result of many forces, the traditional college algebra course of years past has evolved into a new course that serves a wider diversity of students, using new technology in its teaching and learning, and providing preparation for many fields of study. In general, this modern college algebra course should provide a good, solid foundation for the successful study of college mathematics: the engineering/science calculus sequence, the applied calculus sequence for social and biological disciplines, finite mathematics, and statistics. It should also prepare students for course work in physical and life sciences (physics, chemistry) as well as social sciences (economics, business, psychology). The course should show this mathematical foundation to be a unified body of knowledge, not a list of skills to be acquired. In particular, it should:

- Stimulate intuition and develop a deeper understanding of the concepts learned by students in previous algebra courses. It should be more than a formal litany of theorems and methods summarizing these courses.

- Use the current technology (such as graphing calculators and computer software) to explore and extend the topics of this course.

- Develop the level of problem-solving ability and mathematical maturity required for success in subsequent college mathematics courses by inviting the student to attempt problems that are honestly different from "Example 3" and require a synthesis of what has been learned.

- Encourage the use of mathematical modeling to apply mathematics to situations similar to those that arise in subsequent courses and real-life situations.

- Use graphs to motivate and explain. The concepts of equations, functions, and graphs should be the primary unifying thread throughout the text. The value of drawing a picture to understand and extend an idea should never be underestimated.

It is toward these ends that this text has been written.

Audience

This text is written for students who will be serious users of mathematics in their subsequent courses and in their lives, and who will be using technology to solve problems. Previous courses in elementary and intermediate algebra are assumed. This text fits three different populations: (a) those using this course along with a trigonometry course to prepare for the traditional calculus sequence; (b) those using this course to prepare for college mathematics courses required by business, the social sciences, and the life sciences; and (c) those for whom this course is the last mathematics course, serving as general education and preparation for other courses that require some mathematical skills.

Features

Use of Technology The content of this text reflects the changes in mathematics education due to the explosion of the use of graphing utilities. The visual approach of the text lends itself to the use of graphics calculators, which are introduced early (Section 1.4). Appropriate use of graphing utilities to explore functions and solve equations is encouraged throughout the book.

Applications A wealth of applications point out to the students the importance of the concepts of this course. Most applications in this text include realistic data. Drawn from such diverse fields as ecology, business, physics, and mathematics, they show students how useful this course will be in their lives. The examples and exercises urge the student to develop mathematical models in a straightforward fashion. Applications in the text are denoted by red example titles and red exercise numbers. Whole sections that discuss applications are also indicated by red titles.

Important Pedagogical Features Many pedagogical threads run through the text. Graphs are used to reinforce and explain algebraic concepts. Students are encouraged to think of functions symbolically, graphically, numerically, and verbally. Most sections of the text contain exercises that ask the student to explain their solutions in writing. Margin notes in the sections and Quick Review notes in the exercise sets assist the student through difficult steps.

Exercise Sets The exercise sets are the backbone of the text, playing a significant part of the learning process. They are designed to reinforce, expand, and synthesize the concepts of each section along with ideas from previous sections. Exercise sets are designed to provide the basic practice required to learn fundamental ideas, to show how mathematics is applied to realistic problems, and to prod the student to think about the "big picture."

Each exercise set is graded into A, B, and C categories. The A exercises deal with the basic concepts and skills of the section. The B exercises are more stimulating and require a synthesis of the concepts of the section and also those of the previous sections. The C exercises take the student beyond the section to think and discover more about mathematics. All three categories include applications throughout.

Supplementary Materials

College Algebra is accompanied by an extensive supplements package.

For the Instructor

Written by John Martin of Santa Rosa Junior College, an **Instructor's Solution Manual** contains worked-out solutions to all exercises in the book.

An **Instructor's Test Manual** includes three different test forms for each chapter of the book. Answers to all test questions are provided.

Test-Generating Software is available in Windows and Macintosh versions. This streamlined program allows instructors to manipulate a large collection of test questions as a basis for building examinations and quizzes.

For the Student

Written by John Martin of Santa Rosa Junior College, the **Student's Solution Manual** offers worked-out solutions to every other odd-numbered exercise in the book.

A new TI-82 graphing calculator **videotape** gradually explores features of the calculator. Demonstrations include: using zoom, math, matrix, and angle menus; creating tables and split screens; and graphing a variety of functions.

Acknowledgments

Many valuable contributions were made by those who reviewed this text as it moved through its various stages. In particular, we thank the following reviewers for their contributions:

Judy Barclay, Cuesta College
Ted Bentley, Capilano College
Richard Bisk, Fitchburg State College
Yungchen Cheng, Southwest Missouri State University
Jess Collins, McLennan Community College
Annalisa Ebanks, Jefferson Community College
Ralph Esparza, Richland College
Larry Friesen, Butler Community College
William Grimes, Central Missouri State University
Gael Mericle, Mankato State University
Audrey Rose, Tulsa Junior College
Steven Rummel, Black Hills State University
Burla Sims, University of Arkansas, Little Rock

Robert Thompson, Lane Community College
Susan Wood, J. Sargent Reynolds Community College
Janet Wyatt, Longview Community College
Joe Yanik, Emporia State University

We also express our appreciation to those students at Santa Rosa Junior College and Tulsa Junior College who used *College Algebra* in its manuscript form. Special thanks go to Abby Tanenbaum, who assisted in proofreading, to Kitty Pellissier and John Martin, who each worked every exercise in the book, and to Janet Bollow, Carl Brown, and Margaret Brown, who worked on the text's design and art. Finally, we would like to express our sincere gratitude to Robert Weinstein, Bill Hoffman, Phyllis Niklas, and the staff at Worth Publishers for their professionalism and their enthusiasm in bringing this manuscript to print.

Warren L. Ruud
Terry L. Shell
Santa Rosa, California

COLLEGE ALGEBRA

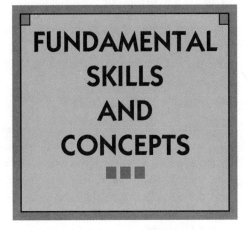

FUNDAMENTAL SKILLS AND CONCEPTS

Algebra is the language of mathematics. By using algebra, we see mathematical connections and patterns more easily, we solve problems, and we reason quantitatively. Instead of performing direct computations with specific numbers as we do in arithmetic, we work with unspecified quantities represented by letters according to a consistent set of rules and properties. These rules and properties are straightforward, much like rules in the game of chess. However, also like the rules of chess, becoming proficient in their use requires diligence and practice.

In this chapter, we reconsider the rules and structure of algebra studied in previous mathematics courses. These topics provide a foundation for the rest of this text.

THE REAL NUMBERS

The set of real numbers is, no doubt, familiar to you. These numbers are used in day-to-day life to measure quantities such as distance, time, and area. In this section, we examine some of the important properties of this set of numbers.

The Real Number System

The best way to visualize the set of *real numbers* is with a *coordinate line* (Figure 1). Each point on the coordinate line corresponds to a real number. Conversely, each real number corresponds to a point on the coordinate line. We say that there exists a *one-to-one correspondence* between the set of real numbers and the set of points on the coordinate line. The point corresponding to 0 is the *origin* of the coordinate line. The *graph* of a particular set of real numbers is the set of points that correspond to those numbers.

Three sets of real numbers are especially important in mathematics. The set of numbers,

$$\dots, -3, -2, -1, 0, 1, 2, 3, \dots$$

is the set of *integers*. The graph of this set is shown in Figure 2.

The ratios of integers form the set of *rational numbers*. Precisely, a rational number can be expressed as a/b, in which a and b are integers ($b \neq 0$). For example,

$$-\frac{1}{2} \qquad \frac{11}{7} \qquad 2.16 = \frac{216}{100}$$

are all rational numbers. Any integer a is a rational number as well, since a can be written as the ratio of two integers:

$$a = \frac{a}{1}$$

The decimal representation of a rational number either repeats in some block of digits or terminates. For instance, the decimal representations of the rational numbers $\frac{2}{7}$, $-\frac{17}{6}$, and $-\frac{7}{4}$ are given by

$$\frac{2}{7} = 0.285\,714\,285\,714\dots = 0.\overline{285\,714}$$
$$-\frac{17}{6} = -2.8333\dots = -2.8\overline{3}$$
$$-\frac{7}{4} = -1.75$$

Some numbers, such as $\sqrt{2}$, cannot be expressed as the ratio of two integers (see Exercise 55). These numbers are called *irrational numbers*. The set of irrational numbers includes, in part, any numerical expression that contains a radical sign when simplified. Another very famous irrational number is π, the ratio of the circumference to the diameter of a circle. The decimal representation of an irrational number neither repeats nor terminates.

Origin

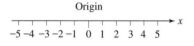

Figure 1

◼ **NOTE:** *The coordinate line is also called the* real number line, *or simply, the* number line.

Figure 2

◼ **NOTE:** *The positive integers* 1, 2, 3, 4, 5, … *are frequently called the* natural numbers, *and the nonnegative integers* 0, 1, 2, 3, 4, 5, … *are often described as the* whole numbers.

◼ *What would the graphs of the set of rational numbers and the set of irrational numbers look like?*

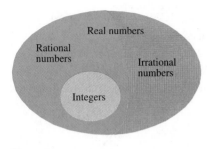

Figure 3

The sets of rational numbers and irrational numbers together form the set of real numbers. Figure 3 shows the relationship between these sets of numbers.

◉ EXAMPLE 1 Identifying Real Numbers

Determine which of the following real numbers are integers, rational numbers, and irrational numbers:

$$-8, \quad \sqrt{5}, \quad 2\pi, \quad -\tfrac{32}{7}$$

Plot the numbers on a coordinate line.

SOLUTION

The first number in the list, -8, is an integer, and it is therefore also a rational number. The second number, $\sqrt{5}$, is an irrational number because 5 is not a perfect square.

The third number, 2π, is also an irrational number. We can prove this claim by first assuming that 2π is a rational number and then showing that this assumption leads to a contradiction. Suppose for a moment that 2π is the ratio of two integers a and b:

$$2\pi = \frac{a}{b}$$

From this equation, it follows that

$$\pi = \frac{a}{2b}$$

implying that π is the ratio of two integers a and $2b$. But we know that π is an irrational number, so 2π must be an irrational number.

The fourth number, $-\tfrac{32}{7}$, is a ratio of two integers, so it is a rational number.

By calculator, we find that $\sqrt{5}$, 2π, and $-\tfrac{32}{7}$ are approximately 2.2, 6.3, and -4.6 (to the nearest 0.1), respectively. The graph of the set is shown in Figure 4. In summary:

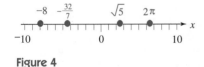

Figure 4

Integers: -8
Rational numbers: $-8, -\tfrac{32}{7}$
Irrational numbers: $\sqrt{5}, 2\pi$

Properties of Real Numbers

Many properties of real numbers are familiar to you from previous algebra courses, but we restate a few of the important ones here. Notice that these properties are expressed using the operations of addition and multiplication.

First, we say that the set of real numbers is *closed* for the operations of addition and multiplication. By this, we mean that for any two real numbers a and b, the sum $a + b$ and the product ab are also real numbers. (For more on this *closure* concept, see Exercise 54.)

The other properties of real numbers that are of interest to us are described in the following box:

Properties of Real Numbers

Suppose that a, b, and c are real numbers.

Commutative properties

Commutative Property of Addition	$a + b = b + a$	Reordering the terms of a sum does not change the value of the sum.
Commutative Property of Multiplication	$ab = ba$	Reordering the factors of a product does not change the value of the product.

Associative properties

Associative Property of Addition	$(a + b) + c = a + (b + c)$	The order in which the terms of a sum are added does not change the value of the sum.
Associative Property of Multiplication	$(ab)c = a(bc)$	The order in which the factors of a product are multiplied does not change the value of the product.

Identities

Additive Identity (Zero)	$a + 0 = a$	Adding 0 to a real number does not change its value.
Multiplicative Identity (One)	$a \cdot 1 = a$	Multiplying a real number by 1 does not change its value.

Inverses

Additive Inverse (Opposite)	$a + (-a) = 0$	The sum of a real number and its opposite (its additive inverse) is 0 (the additive identity).
Multiplicative Inverse (Reciprocal)	$a \cdot \dfrac{1}{a} = 1, \quad a \neq 0$	The product of a nonzero real number and its reciprocal (its multiplicative inverse) is 1 (the multiplicative identity).

Distributive property of multiplication over addition

$a(b + c) = ab + ac$ Multiplying a number by a sum is equivalent to multiplying each term of the sum by the number and adding the resulting products.

The distributive property is the workhorse of algebraic manipulation. This property takes many forms. All the following are examples of the distributive property:

$$4(x + y + z) = 4x + 4y + 4z$$
$$(2a + b)x = 2ax + bx$$
$$-(p + 3q + 2r) = -p - 3q - 2r$$
$$(x + 2)(y + 3) = x(y + 3) + 2(y + 3) = xy + 3x + 2y + 6$$

The process of using the distributive property to change the product $a(b + c)$ to the sum $ab + ac$ is called *simplifying*, or *expanding*, the product. Conversely, changing the sum $ab + ac$ to the product $a(b + c)$ is described as *factoring* the sum.

The other two operations of real numbers, subtraction and division, are defined in terms of addition and multiplication, respectively.

Subtraction and Division

Subtraction of Real Numbers	$a - b = a + (-b)$	Subtracting a number is equivalent to adding its opposite.
Division of Real Numbers	$a \div b = a \cdot \dfrac{1}{b}, \quad b \neq 0$	Dividing by a nonzero number is equivalent to multiplying by its reciprocal.

EXAMPLE 2 Identifying Real Number Properties

Identify the property of real numbers that is exhibited in the statement given.

(a) $2(4 + y) = 2(y + 4)$ **(b)** $(2a)(ac) = 2(aa)c$

(c) $3(x + y - 7) = 3x + 3y - 21$

SOLUTION

(a) The terms in the sum $4 + y$ are reordered as $y + 4$. Thus, the statement demonstrates the commutative property of addition.

(b) The change in grouping symbols reorders the operations, so this statement demonstrates the associative property of multiplication.

(c) A product $3(x + y - 7)$ is changed to a sum by distributing the multiplication by 3 over $x + y - 7$. The statement reflects the distributive property.

Intervals and Inequalities

Algebraic statements that include the symbols $<$, $>$, $\leq$, or $\geq$, are called *inequalities*. Inequalities are often used to describe intervals of real numbers. For example, the set of real numbers greater than or equal to 1 is described by

$$x \geq 1$$

Similarly, the set of real numbers that are both greater than -4 and less than -1 is described by

$$-4 < x < 1$$

An inequality with two inequality symbols, such as $-4 < x < 1$, is a *continued*, or *compound*, *inequality*.

Intervals of real numbers can also be described by *interval notation*. For example, the interval $x \geq 1$ is described by the interval notation

$$[1, +\infty)$$

and the interval $-4 < x < 1$ is described by the interval notation

$$(-4, 1)$$

A parenthesis in interval notation indicates that the corresponding endpoint is not included in the interval, and a bracket is used when the endpoint is included.

The symbol ∞ ("infinity") in the notation $[1, +\infty)$ does not represent a real number. Rather, it is used to show that the interval extends indefinitely without bound in the positive direction. The interval $(-\infty, 1]$ extends indefinitely without bound in the negative direction.

Table 1 (on the facing page) shows the descriptions, graphs, and notations for real number intervals in general.

▣ EXAMPLE 3   Using Inequality and Interval Notation

Sketch the graph of the set described by the inequality notation, and describe it using interval notation.

(a) $-3 \leq x < 2$ **(b)** $x \leq 5$

SOLUTION

(a) The compound inequality $-3 \leq x < 2$ describes a half-open interval with endpoints -3 and 2 (Figure 5). The interval notation for this interval is

$$[-3, 2)$$

(b) The closed infinite interval described by $x \leq 5$ (Figure 6) has interval notation

$$(-\infty, 5]$$

Figure 5

Figure 6

TABLE 1
Inequality Notation and Interval Notation

Description and graph	Inequality notation	Interval notation
Open interval		
$a < x < b$ graph	$a < x < b$	(a, b)
Closed interval		
$a \leq x \leq b$ graph	$a \leq x \leq b$	$[a, b]$
Half-open intervals		
$a \leq x < b$ graph	$a \leq x < b$	$[a, b)$
$a < x \leq b$ graph	$a < x \leq b$	$(a, b]$
Open infinite intervals		
$x < a$ graph	$x < a$	$(-\infty, a)$
$x > a$ graph	$x > a$	$(a, +\infty)$
Closed infinite intervals		
$x \leq a$ graph	$x \leq a$	$(-\infty, a]$
$x \geq a$ graph	$x \geq a$	$[a, +\infty)$

■ **NOTE:** *"Open" implies that the endpoints of the interval, marked by open circles (○), are not included in the interval. "Closed" intervals include the endpoints; they are marked by closed circles (●).*

Absolute Value

The *absolute value* of a real number a, written $|a|$, is the distance from the origin to the point corresponding to a (Figure 7). For example,

$$|-2| = 2 \qquad |2| = 2$$

since -2 and 2 are each 2 units from the origin. The absolute value of a number must be nonnegative because it is the measure of a distance.

In practical terms, the absolute value of a positive real number a is a itself, the absolute value of 0 is 0; and the absolute value of a negative number a is its opposite $-a$.

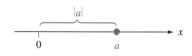

Figure 7

■ **NOTE:** *If $a < 0$, then $-a$ is a positive quantity.*

Absolute Value of a Real Number

The **absolute value** of a real number a is

$$|a| = \begin{cases} a & \text{if } a > 0 \\ 0 & \text{if } a = 0 \\ -a & \text{if } a < 0 \end{cases}$$

EXAMPLE 4   Using Absolute Values

Write each expression without absolute value symbols.

(a) $|\sqrt{11} - 4|$ **(b)** $|x + 6|$, given that $x > -5$

SOLUTION

(a) The problem hinges on whether $\sqrt{11} - 4$ is a positive or negative quantity. The value of $\sqrt{11}$ is approximately 3.3, so $\sqrt{11} - 4$ is negative. Thus, by the definition of absolute value, we get

$$|\sqrt{11} - 4| = -(\sqrt{11} - 4) = 4 - \sqrt{11}$$

(b) The value of x is greater than -5, so $x + 6$ is greater than 1, and is therefore positive. It follows that

$$|x + 6| = x + 6$$

Inequalities involving absolute values arise frequently in science and in fields of mathematics such as calculus. Table 2 (on the facing page) shows a summary of how absolute values are used to describe intervals of real numbers.

EXAMPLE 5   Sketching the Graphs of Absolute Value Inequalities

Sketch each set of numbers described on a coordinate line.

(a) $|x| < 5$ **(b)** $|t| = 4$ **(c)** $|p| \geq 2$

SOLUTION

(a) The inequality $|x| < 5$ is of the form $|u| < a$, where u represents x and a represents 5. By Table 2, we rewrite this absolute value inequality as the compound inequality:

$$-5 < x < 5 \quad \text{See Figure 8}$$

Figure 8

(b) From Table 2, the equation $|t| = 4$ is equivalent to $t = \pm 4$. Thus, $|t| = 4$ describes the points -4 and 4 (Figure 9).

Figure 9

TABLE 2

Absolute Value Equations and Inequalities

	Description	Graph	Set of points
$|u| < a$	All real numbers u that are less than a units from the origin.		Inequality notation: $-a < u < a$ Interval notation: $(-a, a)$
$|u| = a$	All real numbers u that are exactly a units from the origin.		$u = \pm a$
$|u| > a$	All real numbers u that are more than a units from the origin.		Inequality notation: $u < -a$ or $u > a$ Interval notation: $(-\infty, -a)$ or $(a, +\infty)$

Figure 10

(c) The solution to $|p| \geq 2$ is the set of real numbers that are at least 2 units from the origin (Figure 10):

$$p \leq -2 \qquad \text{or} \qquad p \geq 2$$

Using absolute values, we can determine the distance between two points on the coordinate line.

Distance Between Two
Real Numbers

> The **distance** between two real numbers a and b is the absolute value of their difference. That is,
>
> $$d(a, b) = |a - b|$$

The reason we use absolute value in this definition is that the distance between two points is a nonnegative quantity. For example,

$$d(8, 3) = |8 - 3| = 5$$
$$d(2, -7) = |2 - (-7)| = 9$$
$$d(6, 6) = |6 - 6| = 0$$

⬤ EXAMPLE 6   Distance and Absolute Value Inequalities

Sketch each set of numbers described on a coordinate line.

(a) $|x - 2| < 0.5$ **(b)** $|x + 4| \geq 2$

SOLUTION

(a) The expression $|x - 2|$ represents $d(x, 2)$, the distance between x and 2, so the real numbers x that are described by the inequality $|x - 2| < 0.5$ are those within 0.5 unit of 2. This interval is described by $1.5 < x < 2.5$ (Figure 11).

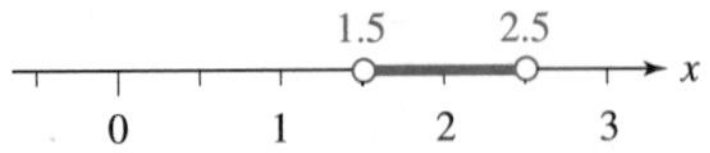

(b) Because $|x + 4| = |x - (-4)|$, this quantity represents $d(x, -4)$, the distance between x and -4. Those numbers x that are 2 or more units from -4 make up the set of real numbers described by $x \leq -6$ or $x \geq -2$ (Figure 12).

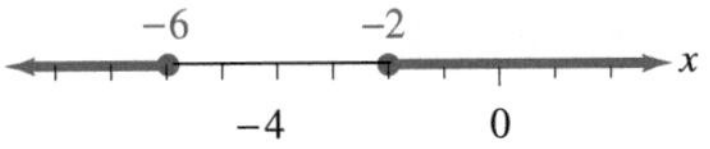

Figure 11

Figure 12

⬤ EXERCISE SET 1

A

In Exercises 1–6, for each set of real numbers, determine which are integers, which are rational numbers, and which are irrational numbers. Plot the numbers on a coordinate line.

1. $-2, 5, 1.34, -\sqrt{5}$

2. $-7, 3, 4.2, \sqrt{30}$

3. $\frac{13}{7}, -\sqrt{8}, \frac{3\pi}{2}, \sqrt{7}$

4. $-\frac{5}{3}, \sqrt{13}, \frac{\pi}{2}, -\sqrt{3}$

5. $\pi, 3.14, \frac{22}{7}, 3$

6. $-\frac{\pi}{2}, -1.57, -\frac{11}{7}, -2$

In Exercises 7–10, identify the property of real numbers that is exhibited in each statement.

7. **(a)** $2t + 1 = 1 + 2t$
    (b) $3(x + t) = 3x + 3t$
    (c) $1 \cdot \frac{3}{5} z = \frac{3}{5} z$

8. **(a)** $ax + 3 = xa + 3$
    (b) $5y + 0 = 5y$
    (c) $\frac{1}{2}(x - 6) = \frac{1}{2}x - 3$

9. **(a)** $-(b - a) = -b + a$
    (b) $3 + x + (-x) = 3$
    (c) $(4t)(px) = 4(tp)x$

10. **(a)** $-b + a = a - b$
    (b) $(x + y) + (6 - n) = x + (y + 6) - n$
    (c) $2 \cdot \frac{1}{2} x = x$

In Exercises 11–18, sketch the graph of each set described by the inequality notation and describe it using interval notation.

11. $x > 4$

12. $x \leq 7$

13. $x \geq \frac{11}{2}$

14. $x < -\frac{5}{2}$

15. $1 < x < 5$

16. $-3 \leq x \leq 0$

17. $-2 \leq x \leq 11$

18. $-7 < x < -1$

In Exercises 19–22, write each expression without absolute value symbols.

19. $|5 + \pi|$ **20.** $|-2 - \pi|$ **21.** $|3 - \sqrt{6}|$

22. $|\sqrt{13} - 4|$

In Exercises 23–28, sketch each set of numbers described by the absolute value equation or inequality on a coordinate line.

23. $|x| = 3$ **24.** $|t| = 6$ **25.** $|n| < 0.3$

26. $|m| < \frac{2}{3}$ **27.** $|p| > \frac{7}{3}$ **28.** $|y| > 0.5$

In Exercises 29–34, sketch each set of real numbers described, and write an absolute value inequality or equation for the set.

29. The real numbers x that are 2 units from the origin.

30. The real numbers x that are 6 units from 3.

31. The real numbers x that are less than 5 units from 4.

32. The real numbers x that are more than $\frac{5}{2}$ units from $-\frac{1}{2}$.

33. The real numbers x that are at least 1 unit from the origin.

34. The real numbers x that are at most 3 units from -3.

B

In Exercises 35–46, for each set of real numbers described, sketch the set on a coordinate line, and describe it using interval notation.

35. $-2 \le x < 4$ **36.** $-5 \le x \le 1$

37. $x < -3$ or $x \ge 4$ **38.** $x \le -1$ or $x > 2$

39. $x \le -4$ or $-1 \le x \le 3$

40. $-4 < x < 0$ or $x \ge 2$

41. $-4 < x \le -2$ or $0 \le x < 4$

42. $-2 < x \le 2$ or $4 \le x < 7$

43. $|x - 5| < 0.5$

44. $|x + 2| < 0.5$

45. $1 < |x| \le 3$

46. $4 \le |x| < 6$

In Exercises 47–52, write each expression without absolute value symbols, given that

$$a > 1 \qquad b \le 3 \qquad 3 < c < 4$$

47. $|a + 2|$ **48.** $|b - 7|$ **49.** $|17 - c^2|$

50. $|17 - 2c^2|$

51. $|a - 1| + |c - 1|$

52. $|a + 6| - |6 - b|$

C

53. Determine the set described by each inequality. Explain your answer.
  (a) $|x - 3| < -2$ (b) $-x^2 > 1$
  (c) $x - \sqrt{x^2 + 4} > 0$

54. In the table below, enter "Yes" if the given set of real numbers is closed for the particular operation, "No" if it is not. For instance, the set of integers is closed for addition, since the sum of any two integers is an integer, so Yes is entered into the table.

Set	Addition	Subtraction	Multiplication	Division
Integers	Yes			
Positive real numbers				
Negative rational numbers				
Odd integers				

55. This exercise provides an outline of the proof that $\sqrt{2}$ is an irrational number.
  (a) Show that if

$$\sqrt{2} = \frac{a}{b}$$

  for two integers a and b ($b \ne 0$), then we can assume that a and b are not both divisible by 2, and that

$$2b^2 = a^2$$

  (b) Explain why the result of part (a) implies that a must be divisible by 2, and that we get

$$2b^2 = (2c)^2$$

  for some integer c.
  (c) Explain why the result of part (b) implies that b must be divisible by 2, and that we can conclude that no such a and b exist. Why does this complete the proof?

56. Show that $\sqrt[3]{2}$ is an irrational number (see Exercise 55).

ALGEBRAIC EXPRESSIONS

An *algebraic expression* is formed by combining constants and variables with the algebraic operations of addition, subtraction, multiplication, division, using exponents, and taking roots. For example, these are all algebraic expressions:

$$2x \qquad 4x^5 - \frac{7}{x} \qquad \sqrt{a^2 + b^2} \qquad \frac{\sqrt{3n} + m}{5n^2 - 4m}$$

In this section (and the three that follow), we review the basic operations and manipulations of algebraic expressions, starting with a discussion of exponents.

Integer Exponents

Natural number exponents allow us to write products with repeated factors. For example,

$$2^5 = 2 \cdot 2 \cdot 2 \cdot 2 \cdot 2 = 32$$
$$(-3)^4 = (-3)(-3)(-3)(-3) = 81$$
$$-7^2 = -(7 \cdot 7) = -49$$

all show how natural number exponents are defined.

Natural Number Exponents

> If a is a real number and n is a natural number, then
>
> $$a^n = \underbrace{a \cdot a \cdot \cdots \cdot a}_{n \text{ factors of } a}$$
>
> In the expression a^n, **the nth power of a,** the number a is the **base** and the number n is the **exponent.**

Multiplying two powers with the same base is straightforward. For instance, consider the product of 2^3 and 2^4:

$$2^3 \times 2^4 = \underbrace{2 \cdot 2 \cdot 2 \times 2 \cdot 2 \cdot 2 \cdot 2}_{7 \text{ factors of } 2} = 2^7$$

Notice that the exponent of the product 2^7 is the sum of the exponents of the factors, 2^3 and 2^4. The generalization of this result, for any real number a and natural numbers m and n, is the *product law of exponents:*

$$a^m a^n = a^{m+n}$$

The definition of natural number exponents can be extended to integer exponents so that the product law of exponents remains consistent. For example, computing $2^3 2^0$ using the product law gives us

$$2^3 2^0 = 2^{3+0} = 2^3$$

This result can be true only if 2^0 is 1, the multiplicative identity. This result suggests the following definition:

Zero Exponent

> If a is a nonzero real number, then
>
> $$a^0 = 1$$
>
> (The expression 0^0 is undefined.)

For example, we have

$$(-126)^0 = 1$$
$$(x^2 + 4)^0 = 1$$
$$-y^0 = -(y^0) = -1 \qquad \text{(if } y \neq 0\text{)}$$

In a similar manner, negative integer exponents can be defined to be consistent with the product law. For example,

$$3^5 3^{-5} = 3^{5+(-5)} = 3^0 = 1$$

leads us to conclude that 3^{-5} is the reciprocal of 3^5, or, equivalently,

$$3^{-5} = \frac{1}{3^5}$$

In general, we have the following definition:

Negative Exponents

> If a is a nonzero real number and n is a natural number, then
>
> $$a^{-n} = \frac{1}{a^n}$$
>
> In particular,
>
> $$a^{-1} = \frac{1}{a}$$
>
> (The expression 0^{-n} is undefined.)

For instance, we have

$$3^{-2} = \frac{1}{3^2} = \frac{1}{9}$$

$$(-5)^{-3} = \frac{1}{(-5)^3} = -\frac{1}{125}$$

$$(x + y)^{-1} = \frac{1}{x + y} \qquad \text{(if } x + y \neq 0)$$

The laws of natural number exponents from basic algebra also apply to integer exponents because of the way we define negative and 0 exponents. These laws are listed here.

Laws of Integer Exponents

If a and b are real numbers, and m and n are integers, then

Product Law	$a^m a^n = a^{m+n}$
Quotient Law	$\dfrac{a^m}{a^n} = a^{m-n}, \quad a \neq 0$
Power Law	$(a^m)^n = a^{mn}$
Distributive Laws	
(Over Multiplication)	$(ab)^n = a^n b^n$
(Over Division)	$\left(\dfrac{a}{b}\right)^n = \dfrac{a^n}{b^n}, \quad b \neq 0$

The proof of the distributive law over multiplication follows. (You are asked to prove the other exponent laws in Exercises 59–62.) We consider the three cases in which n is positive, 0, or negative.

Suppose first that n is a positive integer. Then

$$(ab)^n = \underbrace{(ab)(ab) \cdots (ab)}_{n \text{ factors of } ab} = \underbrace{(aa \cdots a)}_{n \text{ factors of } a} \underbrace{(bb \cdots b)}_{n \text{ factors of } b} = a^n b^n$$

Next, suppose n is 0. Then

$$(ab)^0 = 1$$

and

$$a^0 b^0 = 1 \cdot 1 = 1$$

so

$$(ab)^0 = a^0 b^0$$

Finally, if n is a negative integer, then $n = -k$ for some positive integer k, and

$$(ab)^n = (ab)^{-k} = \frac{1}{(ab)^k} = \frac{1}{a^k b^k} = \frac{1}{a^k} \cdot \frac{1}{b^k} = a^{-k} \cdot b^{-k} = a^n b^n$$

This completes the proof of the distributive law over multiplication for integer exponents.

■ EXAMPLE 1 Using Integer Exponents

Rewrite each expression without negative exponents. (Assume that all variables are nonzero.)

(a) $\dfrac{2}{t^{-4}}$ (b) $\dfrac{x^{-4}}{y^2}$ (c) $\left(\dfrac{2}{n}\right)^{-4}$

SOLUTION

(a) $\dfrac{2}{t^{-4}} = \dfrac{2}{t^{-4}} \cdot \dfrac{t^4}{t^4}$ Multiply the expression by 1.

$\qquad = \dfrac{2t^4}{1}$ The product of t^4 and t^{-4} is 1.

$\qquad = 2t^4$

(b) $\dfrac{x^{-4}}{y^2} = x^{-4} \cdot \dfrac{1}{y^2}$ Use the definition of division.

$\qquad = \dfrac{1}{x^4} \cdot \dfrac{1}{y^2}$ Rewrite x^{-4} using the definition of negative exponents.

$\qquad = \dfrac{1}{x^4 y^2}$

(c) $\left(\dfrac{2}{n}\right)^{-4} = \dfrac{1}{\left(\dfrac{2}{n}\right)^4}$ Use the definition of negative exponents.

$\qquad = \dfrac{1}{\left(\dfrac{2^4}{n^4}\right)}$ Use the distributive property of exponents.

$\qquad = \dfrac{n^4}{2^4} = \dfrac{n^4}{16}$

Example 1 suggests the following properties of negative exponents:

Properties of Negative Exponents

> If a and b are real numbers and n is a natural number, then
>
> **1.** $\dfrac{a^{-n}}{b} = \dfrac{1}{a^{n}b}$ **2.** $\dfrac{a}{b^{-n}} = ab^{n}$ **3.** $\left(\dfrac{a}{b}\right)^{-n} = \left(\dfrac{b}{a}\right)^{n} = \dfrac{b^{n}}{a^{n}}$

Example 2 further illustrates the methods of simplifying algebraic expressions using the laws of exponents.

EXAMPLE 2   Simplifying Expressions Using the Laws of Exponents

Simplify each expression. Express your answer without negative or 0 exponents. (Assume that all variables are nonzero and that n is a natural number.)

(a) $(2xy^{-2})^{-3}$ **(b)** $\left(\dfrac{a^{2}b^{0}}{b^{-5}}\right)^{2}$ **(c)** $\left(\dfrac{x^{3n}}{2}\right)(-x)$

SOLUTION

(a) $(2xy^{-2})^{-3} = 2^{-3}x^{-3}(y^{-2})^{-3}$ Use the distributive law of exponents.

$\qquad\qquad\quad = 2^{-3}x^{-3}y^{6}$ Use the power law of exponents.

$\qquad\qquad\quad = \dfrac{1}{2^{3}} \cdot \dfrac{1}{x^{3}} \cdot y^{6}$ Use the definition of negative exponents.

$\qquad\qquad\quad = \dfrac{y^{6}}{2^{3}x^{3}} = \dfrac{y^{6}}{8x^{3}}$

(b) $\left(\dfrac{a^{2}b^{0}}{b^{-5}}\right)^{2} = \left(\dfrac{a^{2}}{b^{-5}}\right)^{2}$ Use the definition of 0 exponent.

$\qquad\qquad\quad = \dfrac{(a^{2})^{2}}{(b^{-5})^{2}}$ Use the distributive law of exponents.

$\qquad\qquad\quad = \dfrac{a^{4}}{b^{-10}}$ Use the power law of exponents.

$\qquad\qquad\quad = a^{4}b^{10}$ Use property 2 of negative exponents.

(c) $\left(\dfrac{x^{3n}}{2}\right)(-x) = \dfrac{x^{3n}}{2} \cdot \dfrac{(-1)x}{1}$

$\qquad\qquad\quad = \dfrac{(-1)x^{3n} \cdot x^{1}}{2}$

$\qquad\qquad\quad = -\dfrac{x^{3n+1}}{2}$ Use the product law of exponents.

Scientific Notation

Physics, chemistry, astronomy, and other sciences often require computations involving very large numbers and very small numbers. Consider these statements:

The speed of light is 299,800,000 meters per second.

The mass of 1 atom of hydrogen is 0.000 000 000 000 000 000 000 001 67 gram.

The distance from the Earth to the nearest star, *Proxima Centuri,* is 30,900,000,000,000,000 meters.

Handling numbers of such extreme magnitudes is simplified with *scientific notation.*

Scientific Notation

> Every real number can be expressed as a product of a number greater than or equal to 1 but less than 10 and an integer power of 10. That is, for a real number a, we can write
>
> $$a = c \times 10^n$$
>
> for some number c such that $1 \le c < 10$ and some integer n. When a number is written in the form $c \times 10^n$, it is said to be in **scientific notation**.

■ NOTE: *Most scientific calculators are capable of performing computations using scientific notation.*

From our examples above, we can write

$$299,800,000 = 2.998 \times 10^8$$
$$0.000\,000\,000\,000\,000\,000\,000\,001\,67 = 1.67 \times 10^{-24}$$
$$30,900,000,000,000,000 = 3.09 \times 10^{16}$$

Computations involving scientific notation take advantage of the laws of exponents, as we see in Example 3.

● EXAMPLE 3 Computations with Scientific Notation

Evaluate each expression given, and express the answer in scientific notation.

(a) $(2.4 \times 10^4)(3.5 \times 10^7)$ **(b)** $\dfrac{(2.125 \times 10^{-31})(8.12 \times 10^{-14})}{1.13 \times 10^{27}}$

SOLUTION

(a) $(2.4 \times 10^4)(3.5 \times 10^7) = (2.4)(3.5) \times (10^4)(10^7)$
$$= 8.4 \times 10^{11}$$

(b) $\dfrac{(2.125 \times 10^{-31})(8.12 \times 10^{-14})}{1.13 \times 10^{27}} = \dfrac{(2.125)(8.12) \times (10^{-31})(10^{-14})}{1.13 \times 10^{27}}$

$= \dfrac{17.255 \times 10^{-45}}{1.13 \times 10^{27}}$

$= \dfrac{17.255}{1.13} \times \dfrac{10^{-45}}{10^{27}}$

$= 15.3 \times 10^{-72}$

The solution 15.3×10^{-72} is not in scientific notation, because $15.3 > 10$. Thus, we rewrite 15.3×10^{-72} as follows:

$$15.3 \times 10^{-72} = (1.53 \times 10^{1}) \times 10^{-72} = 1.53 \times 10^{-71}$$

Polynomials

An algebraic expression that is a product of a real number constant factor and variable factors is a *monomial*. The following expressions are all monomials:

$$2x \qquad -\tfrac{3}{7}\,ab^2 \qquad 17 \qquad n^3$$

The constant factor is the *numerical coefficient* of the monomial, and the *degree* of the monomial is the number of variable factors of the monomial factors. For the examples above, we say that $2x$ is a monomial in the variable x of degree 1 with numerical coefficient 2; $-\tfrac{3}{7}\,ab^2$ is a monomial in the variables a and b of degree 3 with numerical coefficient $-\tfrac{3}{7}$; 17 is a monomial of degree 0 with numerical coefficient 17; and n^3 is a monomial in the variable n of degree 3 with numerical coefficient 1.

A *polynomial* is an algebraic expression that is the sum of monomials. The *degree* of a polynomial is the greatest degree of its terms. In this text, polynomials in one variable play a large role.

Polynomials

> **A polynomial of degree n in the variable x** is an algebraic expression that can be expressed in the form
>
> $$a_n x^n + a_{n-1}x^{n-1} + a_{n-2}x^{n-2} + \cdots + a_1 x + a_0$$
>
> where n is a nonnegative integer and $a_n, a_{n-1}, a_{n-2}, \ldots, a_1, a_0$ are real numbers, with $a_n \neq 0$.

A polynomial with exactly two terms is a *binomial*, and a polynomial with exactly three terms is a *trinomial*. For example, $a^4 - 2$ is a binomial in the variable a of degree 4, and $x^2 - 2xy + 4y^2$ is a trinomial in the variables x and y of degree 2.

The next two examples review the operations of addition, subtraction, and multiplication of polynomials. (Division of polynomials is discussed in Section 4.)

■ EXAMPLE 4　Adding and Subtracting Polynomials

Perform each operation.

(a) $(3x^2 + 5x + 2) + (4x^3 - 6x)$　　**(b)** $(x^3 - 5x + 11) - (3x^3 - 6x^2 + 7)$

SOLUTION

(a) $(3x^2 + 5x + 2) + (4x^3 - 6x) = 4x^3 + 3x^2 + (5x - 6x) + 2$
$$= 4x^3 + 3x^2 - x + 2$$

(b) $(x^3 - 5x + 11) - (3x^3 - 6x^2 + 7) = x^3 - 5x + 11 - 3x^3 + 6x^2 - 7$
$$= (x^3 - 3x^3) + 6x^2 - 5x + (11 - 7)$$
$$= -2x^3 + 6x^2 - 5x + 4 \qquad ■$$

■ EXAMPLE 5　Multiplying Polynomials

Perform the multiplication:　$(x^2 - 3)(x^2 + 2x - 4)$

SOLUTION

$(x^2 - 3)(x^2 + 2x - 4) = x^2(x^2 + 2x - 4) - 3(x^2 + 2x - 4)$
$$= x^2(x^2) + x^2(2x) + x^2(-4) - 3(x^2) - 3(2x) - 3(-4)$$
$$= x^4 + 2x^3 - 4x^2 - 3x^2 - 6x + 12$$
$$= x^4 + 2x^3 - 7x^2 - 6x + 12 \qquad ■$$

The product of two binomials occurs often enough to warrant special discussion. We perform this type of computation mentally with the so-called *FOIL* ("*First–Outside–Inside–Last*") method from elementary algebra. For example, in simplifying $(x + 5)(2x - 3)$, we get

$$(x + 5)(2x - 3) = 2x^2 + 7x - 15$$

Last term, -15
First term, $2x^2$
First term
Last term
Inside term, $10x$　Sum of inside and outside terms (add these mentally)
Outside term, $-3x$

More examples where the FOIL method can be used follow.

EXAMPLE 6   Using FOIL

Perform each multiplication:

(a) $(x + 2)(x + 5)$ **(b)** $(3t - 1)(t + 4)$ **(c)** $(x^2 - 7)(x^2 - 3)$

(d) $(n - 2)(n + 2)$ **(e)** $(p + 3)^2$

SOLUTION

(a) $(x + 2)(x + 5) = x^2 + 7x + 10$

(b) $(3t - 1)(t + 4) = 3t^2 + 11t - 4$

(c) $(x^2 - 7)(x^2 - 3) = x^4 - 10x^2 + 21$

(d) $(n - 2)(n + 2) = n^2 - 4$

(e) $(p + 3)^2 = (p + 3)(p + 3) = p^2 + 6p + 9$

Parts (d) and (e) of Example 6 hint at the following special products:

Products of Binomials

Perfect Square of a Binomial

$$(u + v)^2 = u^2 + 2uv + v^2$$
$$(u - v)^2 = u^2 - 2uv + v^2$$

Product of Sum and Difference

$$(u - v)(u + v) = u^2 - v^2$$

Simplifying Polynomials

A polynomial in one variable is *simplified* if all terms are expanded, like terms are added together, and the resulting terms are written in order of descending powers of the variable. The next two examples illustrate this process.

EXAMPLE 7   Simplifying a Polynomial

Simplify each polynomial.

(a) $(x + 4)^2 + (x - 2)^2$ **(b)** $(t^2 - 2)(t^2 + 2) - 6t$

SOLUTION

(a) $(x + 4)^2 + (x - 2)^2 = (x^2 + 8x + 16)$ Expand both terms using

$+ (x^2 - 4x + 4)$ perfect square formulas.

$= 2x^2 + 4x + 20$ Collect similar terms and write them in descending powers.

(b) $(t^2 - 2)(t^2 + 2) - 6t = (t^2)^2 - 2^2 - 6t$ Expand first term using sum and difference formula.

$= t^4 - 4 - 6t$

$= t^4 - 6t - 4$ Write terms in decreasing powers.

EXAMPLE 8   Operations with Polynomials

Given that $P = x^2 - 4$ and $Q = 2x + 1$, simplify each expression.

(a) $2P + 3Q$ **(b)** $(P - Q)^2$

SOLUTION

(a) $\begin{aligned} 2P + 3Q &= 2(x^2 - 4) + 3(2x + 1) \\ &= 2x^2 - 8 + 6x + 3 \\ &= 2x^2 + 6x - 5 \end{aligned}$

(b) $\begin{aligned} (P - Q)^2 &= [(x^2 - 4) - (2x + 1)]^2 \\ &= (x^2 - 4 - 2x - 1)^2 \\ &= (x^2 - 2x - 5)^2 \\ &= (x^2 - 2x - 5)(x^2 - 2x - 5) \\ &= x^2(x^2 - 2x - 5) - 2x(x^2 - 2x - 5) - 5(x^2 - 2x - 5) \\ &= x^4 - 2x^3 - 5x^2 - 2x^3 + 4x^2 + 10x - 5x^2 + 10x + 25 \\ &= x^4 - 4x^3 - 6x^2 + 20x + 25 \end{aligned}$

EXERCISE SET 2

A

In Exercises 1–10, rewrite each expression without negative or 0 exponents. (Assume all variables are nonzero.)

1. (a) $\dfrac{4}{a^{-1}}$ **(b)** $\left(\dfrac{2}{3}\right)^{-1}$

2. (a) $\dfrac{1}{3x^{-1}}$ **(b)** $\left(\dfrac{1}{4}\right)^{-1}$

3. (a) $2x^{-1}$ **(b)** $-y^{-2}$

4. (a) $-4b^{-2}$ **(b)** $(5n)^{-1}$

5. (a) $\dfrac{2b}{a^{-3}}$ **(b)** $\dfrac{3x^{-1}}{y^4}$

6. (a) $\dfrac{p^3}{2q^{-5}}$ **(b)** $\dfrac{3a^{-2}}{5b}$

7. (a) $\left(\dfrac{x}{y}\right)^{-2}$ **(b)** $\left(\dfrac{1}{b}\right)^{-4}$

8. (a) $\left(\dfrac{n}{m}\right)^{-3}$ **(b)** $\left(\dfrac{2x}{y^2}\right)^{-1}$

9. (a) $(-3x)^0$ **(b)** $-2m^0$

10. (a) $-q^0$ **(b)** $(-x^2)^0$

In Exercises 11–20, simplify each expression. (Assume all variables are nonzero.)

11. $-4x^{-3}y^7$ **12.** $2^{-1}p^2q^{-5}$

13. $(3a^2b^{-2})^{-1}$ **14.** $(-5a^{-4}b)^{-2}$

15. $(x^{-2}y)(2x^2y^{-2})^2$ **16.** $(-xy^{-4})(x^{-2}y)^3$

17. $\dfrac{4a^2b^{-4}}{2a^{-3}b^8}$ **18.** $\dfrac{3p^{-5}q^4}{6p^3q^{-3}}$

19. $\dfrac{(12x^2y^{-1})(x^0y^{-3})^{-1}}{(2xy^5)^{-2}}$ **20.** $\dfrac{(ab^{-2})^{-3}(10ab^{-2})}{(5a^0b^2)^{-1}}$

In Exercises 21–26, evaluate each expression given, and express the answer in scientific notation.

21. $(1.25 \times 10^3)(4.10 \times 10^{-7})$

22. $(3.6 \times 10^{-13})(2.5 \times 10^{16})$

23. $\dfrac{6.2 \times 10^{32}}{2.8 \times 10^{-15}}$ **24.** $\dfrac{4.32 \times 10^{-21}}{3.72 \times 10^{32}}$

25. $\dfrac{(8.8 \times 10^{45})(4.55 \times 10^{-17})}{2.40 \times 10^{-8}}$

26. $\dfrac{1.36 \times 10^{-12}}{(4.28 \times 10^{-30})(9.12 \times 10^6)}$

In Exercises 27–38, write each polynomial in simplified form.

27. $(2x^2 - 4x) + (x^2 - 2)$

28. $(x^2 - 12) - (4x^2 + 3)$

29. $(2x^2 - 3x - 6) - 2x(x - 5)$

30. $(3x^3 - 1) + 3x(x^2 + x - 2)$

31. $2x(x - 4) + 2x(3x + 6)$

32. $4x(x^2 + 3x) + 2x^2(x + 6)$

33. $(x + 4)(x + 3) - x^2$

34. $(3x - 4)(2x + 3) + 6x^2$

35. $x^2 - (x + 1)^2$

36. $2x + (2x + 1)^2$

37. $(x + 2)(x - 2) - x(x - 4)$

38. $4x(x + 2) - (2x - 1)(2x + 1)$

B

In Exercises 39–46, simplify each expression. Express your answer without negative or 0 exponents. (Assume that all variables are nonzero and that n is a positive integer.)

39. $\dfrac{x^9(x + 1)^{-2}}{x^4(x + 1)^{-10}}$

40. $\dfrac{s^7(s + 2t)^4}{s^5(s + 2t)^{-2}}$

41. $-4(5x^2 + 3)^{-5}(10x)$

42. $-(2x^4 - 1)^{-2}(8x^3)$

43. $(-3a^nb^{n+1})(-4a^2b^{4-n})$

44. $(x^{4-n}y^{2n-3})(7x^{2+n}y^4)$

45. $\dfrac{18x^{4n-3}}{2x^n \cdot 3x^{n-7}}$

46. $\dfrac{24x^{2n}x^{n+4}}{3x^{n-5}}$

In Exercises 47–54, expand and simplify each expression, given that $P = x + 2$, $Q = x^2 - 2x + 4$, and $R = -4x^2$.

47. P^3

48. PQ

49. $Q(R + 2)$

50. Q^2

51. $(P - 2Q)^2$

52. P^2R

53. $(P + 1)(Q - 4)$

54. $(R + 1)(R - 1)$

In Exercises 55 and 56, each statement demonstrates a common error with exponents. Describe the error and its correction.

55. **(a)** $x^3x^4 \not\equiv x^{12}$

  (b) $3x^{-2} \not\equiv \dfrac{1}{3x^2}$

  (c) $(x + y)^{-1} \not\equiv \dfrac{1}{x} + \dfrac{1}{y}$

  (d) $(x^3)^4 \not\equiv x^7$

56. **(a)** $\dfrac{x^{10}}{x^5} \not\equiv x^2$

  (b) $(5x)^2 \not\equiv 5x^2$

  (c) $\dfrac{1}{x^{-1} + y^{-1}} \not\equiv x + y$

  (d) $x^2y^4 \not\equiv (xy)^6$

C

57. Without using a calculator, determine the larger of 3^{2000} and 8^{1000}. Explain your answer.

58. Without using a calculator, determine the larger of 2^{1000} and 10^{300}. Explain your answer.

59. Prove the product law of integer exponents.

60. Prove the quotient law of integer exponents.

61. Prove the power law of integer exponents.

62. Prove the distributive law over division for integer exponents.

● Factoring Polynomials
● Factoring Trinomials
● Special Factoring Formulas
● Factoring Over the Real Numbers

FACTORING

In this section, we reconsider the process of factoring—first as it applies to positive integers and then as it applies to algebraic expressions such as polynomials.

  Factoring a positive integer a means to express a as a product of two or more positive integers. For example, each of these products is a *factorization* of 42:

$$3 \times 14 \qquad 2 \times 21 \qquad 2 \times 3 \times 7$$

Of these factorizations, we say that $2 \times 3 \times 7$ is the *complete factorization over the integers* of 42, since the positive integer factors 2, 3, and 7 cannot be factored themselves (other than as 1×2, 1×3, or 1×7). An integer greater than 1 that has only 1 and itself as factors is said to be a *prime number*. (An integer greater than 1 with other factors besides 1 and itself is said to be a *composite number*.) Thus, a positive integer is *completely factored* if it is expressed as the product of primes.

The other factorizations of 42 shown above, 3×14 and 2×21, are not complete because

$$3 \times 14 = 3 \times (2 \times 7) = 2 \times 3 \times 7$$
$$2 \times 21 = 2 \times (3 \times 7) = 2 \times 3 \times 7$$

Notice that in each case, continuing the factorization of 42 leads to the same representation. This result is true in general; an integer greater than 1 is either prime, or can be represented uniquely as a product of prime factors. For instance, the complete factorizations of 12, 125, and 3549 are

$$12 = 2^2 \times 3 \qquad 125 = 5^3 \qquad 3549 = 3 \times 7 \times 13^2$$

Factoring Polynomials

Many polynomials with integer coefficients can be expressed as products of two or more polynomial factors, also with integer coefficients. For example, the polynomial $x^2 - 4$ can be factored as

$$(x - 2)(x + 2)$$

On the other hand, some polynomials, such as $x^3 + 5$, cannot be factored. These polynomials, as you might suspect, are said to be *prime*.

From here on, to *factor* a polynomial means to factor it completely as a product of two or more polynomials with integer coefficients. As with integers, the factorization of a polynomial is unique.

The most straightforward technique for factoring polynomials is to find a common factor of all the terms of the polynomial and "remove" it by using the distributive property of multiplication over addition:

$$ab + ac = a(b + c)$$

EXAMPLE 1 Removing Common Factors

Factor each polynomial by removing the common factor.

(a) $6m^3 - 2m^2 + 4m$ **(b)** $(y + 4)^3 - 2(y + 4)^2$

SOLUTION

(a) In the polynomial $6m^3 - 2m^2 + 4m$, there are three terms, each with $2m$ as a factor. Using the distributive property gives us the factored form of this polynomial.

$$6m^3 - 2m^2 + 4m = 2m \cdot 3m^2 - 2m \cdot m \qquad 2m \text{ is the common factor.}$$
$$+ 2m \cdot 2$$
$$= 2m(3m^2 - m + 2) \qquad \text{Use the distributive property.}$$

(b) Each of the two terms has $(y + 4)^2$ as a factor. Thus, we get

$$(y + 4)^3 - 2(y + 4)^2 = (y + 4)^2 \cdot (y + 4) \qquad (y + 4)^2 \text{ is the common}$$
$$- (y + 4)^2 \cdot 2 \qquad \text{factor.}$$
$$= (y + 4)^2[(y + 4) - 2] \qquad \text{Use the distributive}$$
$$= (y + 4)^2(y + 2) \qquad \text{property.}$$

Quite often, a polynomial with four or more terms having no common factor can still be factored by the technique of *grouping*. Example 2 illustrates this process.

EXAMPLE 2   Factoring by Grouping

Factor the polynomial: $2x^3 + 6x^2 + 5x + 15$

SOLUTION

There is no common factor for all four terms of the polynomial. However, if we consider the first two terms and the last two terms separately, we see that $2x^2$ is a factor of $2x^3 + 6x^2$, and 5 is a factor of $5x + 15$:

$$2x^3 + 6x^2 + 5x + 15 = 2x^3 + 6x^2 + 5x + 15$$

$$\underbrace{\qquad}_{\substack{\text{Common} \\ \text{factor of } 2x^2}} \quad \underbrace{\qquad}_{\substack{\text{Common} \\ \text{factor of } 5}}$$

$$= 2x^2(x + 3) + 5(x + 3) \qquad \text{Factor each pair of terms.}$$
$$= (x + 3)(2x^2 + 5) \qquad \text{Common factor is } x + 3.$$

Thus, the factorization of $2x^3 + 6x^2 + 5x + 15$ is $(x + 3)(2x^2 + 5)$.

Factoring Trinomials

Many second-degree trinomials may be factored as products of two first-degree binomials by using a trial-and-error method based on the FOIL method of Section 2. We review this method in Example 3.

EXAMPLE 3 Factoring Second-Degree Trinomials

If possible, rewrite each trinomial as the product of two binomials using the FOIL method.

(a) $x^2 + 9x + 14$ **(b)** $12x^2 - 6x - 6$ **(c)** $x^2 + 4x + 2$

SOLUTION

In each case, we suppose that for some integers a, b, c, and d, the factorization of the trinomial is

$$(ax + b)(cx + d) = acx^2 + (ad + bc)x + bd$$

By comparing the coefficients ac, $ad + bc$, and bd with the coefficients of the given trinomial, we can determine the values of these integers. (If no such values exist, then the trinomial is prime.)

(a) Because the coefficient of x^2 is 1, we have

$$x^2 + 9x + 14 = (x + b)(x + d)$$

We can see that bd is 14. A quick check of the possible factors of bd,

$$2 \cdot 7 \qquad 1 \cdot 14 \qquad (-2)(-7) \qquad (-1)(-14)$$

leads us to

$$x^2 + 9x + 14 = (x + 2)(x + 7)$$

(b) The trinomial has a common factor of 6, so we have

$$12x^2 - 6x - 6 = 6(2x^2 - x - 1)$$

Factoring $2x^2 - x - 1$ in a manner similar to part (a) gives us

$$12x^2 - 6x - 6 = 6(2x + 1)(x - 1)$$

(c) Assuming that

$$x^2 + 4x + 2 = (x + b)(x + d)$$

for some integers b and d, leads us to conclude that bd is 2 and $b + d$ is 4. A quick investigation of the candidates ± 1 and ± 2 tells us that no such integers b and d exist. Thus, $x^2 + 4x + 2$ is prime.

Special Factoring Formulas

The following four factoring formulas can be used on many of the polynomials that arise in this and other mathematics courses:

Special Factoring Formulas

Perfect Square:	$u^2 + 2uv + v^2 = (u + v)^2$
	$u^2 - 2uv + v^2 = (u - v)^2$
Difference of Squares:	$u^2 - v^2 = (u - v)(u + v)$
Difference of Cubes:	$u^3 - v^3 = (u - v)(u^2 + uv + v^2)$
Sum of Cubes:	$u^3 + v^3 = (u + v)(u^2 - uv + v^2)$

Each of the special factoring formulas can be proved by expanding the right side and showing it is equal to the left side. (Pause here to verify each of these formulas.) These formulas are demonstrated in the next two examples.

EXAMPLE 4   Factoring with Special Formulas

Factor each polynomial.

(a) $4x^2 + 12x + 9$ **(b)** $t^4 - 16t^2 + 64$
(c) $25n^2 - 1$ **(d)** $(s + t)^2 - s^2t^2$

SOLUTION

(a) $4x^2 + 12x + 9 = (2x)^2 + 2(2x)3 + 3^2$
$= (2x + 3)^2$ Use the perfect square formula.

(b) $t^4 - 16t^2 + 64 = (t^2)^2 - 2(t^2)8 + 8^2$
$= (t^2 - 8)^2$ Use the perfect square formula.

(c) $25n^2 - 1 = (5n)^2 - 1^2$
$= (5n - 1)(5n + 1)$ Use the difference of squares formula.

(d) $(s + t)^2 - s^2t^2 = (s + t)^2 - (st)^2$
$= [(s + t) - st][(s + t) + st]$ Use the difference of squares formula.
$= (s + t - st)(s + t + st)$

EXAMPLE 5   Using the Difference of Cubes Formula

Factor each polynomial.

(a) $x^3 - 27$ **(b)** $a^6 - 8b^9$

SOLUTION

(a) $x^3 - 27 = x^3 - 3^3$ Write the polynomial as the difference of two cubes.
$= (x - 3)(x^2 + 3x + 9)$ Use the difference of cubes formula.

(b) $a^6 - 8b^9 = (a^2)^3 - (2b^3)^3$ Write the polynomial as the difference of two cubes.

$$= (a^2 - 2b^3)[(a^2)^2 + (a^2)(2b^3) + (2b^3)^2]$$ Use the difference of cubes formula.

$$= (a^2 - 2b^3)[a^4 + 2a^2b^3 + 4b^6]$$

In general, higher-degree polynomials present a greater challenge in factoring than do first-degree and second-degree polynomials. However, by using a combination of the techniques discussed so far, we can factor certain higher-degree polynomials.

EXAMPLE 6   Factoring Higher-Degree Polynomials

Factor each polynomial.

(a) $a^4 - 4a^3 - 8a + 32$ **(b)** $t^4 - 11t^2 + 18$

(c) $3(x + 4)^2(2x - 1)^3 + 6(x + 4)^3(2x - 1)^2$

SOLUTION

(a) $a^4 - 4a^3 - 8a + 32 = a^3(a - 4) - 8(a - 4)$ Factor by grouping.

$$= (a - 4)(a^3 - 8)$$ Remove common factor $a - 4$.

$$= (a - 4)(a - 2)(a^2 + 2a + 4)$$ Use the difference of cubes formula.

(b) By replacing t^2 with the variable u, we can express $t^4 - 11t^2 + 18$ as a second-degree polynomial in u:

$$t^4 - 11t^2 + 18 = (t^2)^2 - 11(t^2) + 18$$

$$= u^2 - 11u + 18$$ Let u represent t^2.

$$= (u - 9)(u - 2)$$ Factor using FOIL.

$$= (t^2 - 9)(t^2 - 2)$$ Replace u with t^2.

$$= (t - 3)(t + 3)(t^2 - 2)$$ Factor $t^2 - 9$ as a difference of squares.

Notice that $t^2 - 2$ is not factorable by the difference of squares formula (as $t^2 - 9$ is), since 2 is not the square of an integer.

(c) $3(x + 4)^2(2x - 1)^3 + 6(x + 4)^3(2x - 1)^2$

$$= 3(x + 4)^2(2x - 1)^2[(2x - 1) + 2(x + 4)]$$ Remove common factors.

$$= 3(x + 4)^2(2x - 1)^2(4x + 7)$$

Factoring Over the Real Numbers

In part (b) of Example 6, we noted that the polynomial $t^2 - 2$ cannot be factored because it is not the difference of squares. However, by relaxing the restriction that factors of this polynomial must have integer coefficients, it is factorable:

$$t^2 - 2 = t^2 - (\sqrt{2})^2 = (t - \sqrt{2})(t + \sqrt{2})$$

We say that $t^2 - 2$ is *factorable over the real numbers*. In Example 7, we consider the factorization of a third-degree polynomial over the integers and over the real numbers.

EXAMPLE 7   Factoring Over the Real Numbers

Factor $2x^3 - 12x$:

(a) Over the integers   (b) Over the real numbers

SOLUTION

(a) The terms of the polynomial have the common factor $2x$, so

$$2x^3 - 12x = 2x(x^2 - 6)$$

(b) We rewrite $x^2 - 6$ as the difference of squares to complete the factorization over the real numbers:

$$2x^3 - 12x = 2x(x^2 - 6) = 2x[x^2 - (\sqrt{6})^2] = 2x(x - \sqrt{6})(x + \sqrt{6})$$

EXERCISE SET 3

A

In Exercises 1–6, factor each polynomial by removing the common factors.

1. $3x^3 + 6x$

2. $2t^5 - 6t$

3. $x^3y^2 + 5x^2y^3 + 8xy^4$

4. $4a^3b - 10a^2b^2 + 2ab^3$

5. $5x(x + 3) - 2(x + 3)$

6. $(2a + 1) - 6b(2a + 1)$

In Exercises 7–12, factor each polynomial by grouping.

7. $a^3 - a^2 + 3a - 3$

8. $t^3 + 7t^2 - 7t - 49$

9. $x^6 - 2x^5 - 3x + 6$

10. $y^4 - 3y^3 + 5y - 15$

11. $5n^3 - 15n^2 - 2n + 6$

12. $2p^5 + p^3 - 4p^2 - 2$

In Exercises 13–24, factor each second-degree polynomial, if possible.

13. $x^2 - x - 2$

14. $x^2 - 11x - 12$

15. $3x^2 + 5x + 2$

16. $4x^2 - 5x - 6$

17. $9x^2 + 12x + 4$

18. $16x^2 - 8x + 1$

19. $x^2 - 2x + 9$

20. $x^2 + 6x + 18$

21. $4x^2 - 1$

22. $-x^2 - 16$

23. $x^2 + 9$

24. $25x^2 - 9$

In Exercises 25–30, factor each difference or sum of cubes.

25. $x^3 - 8$

26. $y^3 + 64$

27. $n^3 + m^3$

28. $1000 - t^3$

29. $8x^3 - y^3z^3$

30. $a^3b^3 + 27c^3$

B

In Exercises 31–52, factor each polynomial completely.

31. $6x^2 - 12x + 6$

32. $2ax^2 + 12ax + 18a$

33. $x^3 - 4x$

34. $20x^3y - 5xy$

35. $x^4y - xy^4$

36. $24x^4 - 3x$

37. $x^6 - 5x^5 - 6x^4$

38. $2x^4 - 3x^3 + x^2$

39. $x^4 + 6x^2 + 8$

40. $x^4 + 8x^2y^2 + 16y^4$

41. $16x^4 - y^4$

42. $x^4 - 81$

43. $x^6 - y^3$

44. $8x^6 + 27$

45. $x^6 - 4x^4 + x^2 - 4$

46. $4x^5 - 4x^3 + 32x^2 - 32$

47. $(x - 5)^3 - (x - 5)$

48. $(x + 2)^4 - (x + 2)^2$

49. $10(2x - 3)^4x^2 + 2(2x - 3)^5x$

50. $3(x + 5)^2x^3 + 3(x + 5)^3x^2$

51. $3(x - 3)^2(x^2 + 1)^3 + 6x(x - 3)^3(x^2 + 1)^2$

52. $4(x + 3)^3(x^2 - 2)^2 + 4x(x + 3)^4(x^2 - 2)$

53. Rewrite

$$3x^2 - 4x + \frac{1}{x} - \frac{2}{x^2}$$

as a product of x^{-n} (for some positive integer n) and a fourth-degree polynomial.

54. Rewrite

$$x + 6 - \frac{5}{x^3} + \frac{1}{x^4}$$

as a product of x^{-n} (for some positive integer n) and a fifth-degree polynomial.

In Exercises 55–58:
(a) Factor each polynomial over the integers.
(b) Factor each polynomial over the real numbers.

55. $2x^2 - 12$

56. $x^3 - 8x$

57. $x^4 - 25$

58. $2x^4 - 5x^2 - 3$

C

59. Show that $x^2 + 6x + 7$ can be factored over the real numbers as $(x + 3 + \sqrt{2})(x + 3 - \sqrt{2})$.

60. Show that $x^2 - 4x - 1$ can be factored over the real numbers as $(x - 2 + \sqrt{5})(x - 2 - \sqrt{5})$.

61. Show that

$$x^4 + x^2y^2 + y^4 = (x^2 + y^2)^2 - x^2y^2$$

Use this to factor $x^4 + x^2y^2 + y^4$.

62. Factor $x^4 - 2x^2y^2 + 4y^4$ over the real numbers (see Exercise 61).

SECTION 4

- Simplifying Rational Expressions
- Operations with Rational Expressions
- Polynomial Division

RATIONAL EXPRESSIONS

A *rational expression* is the quotient of two polynomials. Examples of rational expressions are

$$\frac{2}{6x - 5} \qquad \frac{4x^2 + 19x + 6}{2x + 9} \qquad \frac{x}{x^3 - 1}$$

Polynomials are related to rational expressions in the same way that integers are related to rational numbers; just as a rational number is the ratio of two integers, a rational expression is the ratio of two polynomials. Furthermore, rational expressions are added, subtracted, multiplied, divided, and reduced in the same way as with rational numbers. In this section, we review how to perform these operations with rational expressions. Your success in performing these operations will rely on your skills in factoring polynomials.

Simplifying Rational Expressions

The fraction $\frac{6}{15}$ can be reduced to $\frac{2}{5}$ by dividing the numerator and denominator by their common factor, 3. The process can be expressed as follows:

$$\frac{6}{15} = \frac{2 \cdot 3}{5 \cdot 3} = \frac{2}{5}$$

Reducing a fraction in this way is often called *eliminating the common factor*. This principle can be applied to rational expressions.

Fundamental Principle of Rational Expressions

If P, Q, and R are polynomials with $Q \neq 0$ and $R \neq 0$, then

$$\frac{PR}{QR} = \frac{P}{Q}$$

We reduce rational expressions by factoring the numerator and denominator and then eliminating their common factors.

◉ EXAMPLE 1   Reducing a Rational Expression

Reduce each rational expression to lowest terms.

(a) $\dfrac{2x^2 + 2x}{x^2 - 1}$ (b) $\dfrac{6 - 2x}{x^2 + x - 12}$

SOLUTION

Our plan is to factor the numerator and denominator, and then eliminate common factors.

■ **NOTE:** *Be careful to eliminate only common* factors. *For example,*

$$\frac{2x}{x - 1} \neq \frac{2\not{x}}{\not{x} - 1}$$

Since the x in the denominator is not a factor, it cannot be eliminated.

(a) $\dfrac{2x^2 + 2x}{x^2 - 1} = \dfrac{2x(x + 1)}{(x + 1)(x - 1)}$ Factor.

$= \dfrac{2x\cancel{(x + 1)}}{\cancel{(x + 1)}(x - 1)} = \dfrac{2x}{x - 1}$ Eliminate common factors.

(b) $\dfrac{6 - 2x}{x^2 + x - 12} = \dfrac{2(3 - x)}{(x + 4)(x - 3)}$ Factor.

$= \dfrac{-2(x - 3)}{(x + 4)(x - 3)}$ Note that $3 - x = -(x - 3)$.

$= \dfrac{-2\cancel{(x - 3)}}{(x + 4)\cancel{(x - 3)}} = \dfrac{-2}{x + 4}$ Eliminate common factors. ◉

Operations with Rational Expressions

The product of two rational expressions is the product of their numerators over the product of their denominators. It makes sense to write these products in factored form, because we need to reduce the result to lowest terms. Dividing by a rational expression is equivalent to multiplying by its reciprocal.

■ EXAMPLE 2 Multiplying or Dividing Rational Expressions

Perform the indicated operation and simplify.

(a) $\dfrac{x^2 - 4x + 4}{2x^2 + x} \cdot \dfrac{4x^2 - 1}{x - 2}$ (b) $\dfrac{x^2 + 2x - 3}{x^2 - 3x + 9} \div \dfrac{6x - 6}{x^3 + 27}$

SOLUTION

(a) $\dfrac{x^2 - 4x + 4}{2x^2 + x} \cdot \dfrac{4x^2 - 1}{x - 2}$

$= \dfrac{(x - 2)(x - 2)}{x(2x + 1)} \cdot \dfrac{(2x + 1)(2x - 1)}{x - 2}$ Factor.

$= \dfrac{(x - 2)(x - 2)(2x + 1)(2x - 1)}{x(2x + 1)(x - 2)}$

$= \dfrac{(x - 2)(2x - 1)}{x}$ Multiply and eliminate common factors.

(b) $\dfrac{x^2 + 2x - 3}{x^2 - 3x + 9} \div \dfrac{6x - 6}{x^3 + 27}$

$= \dfrac{x^2 + 2x - 3}{x^2 - 3x + 9} \cdot \dfrac{x^3 + 27}{6x - 6}$ Multiply by the reciprocal of the divisor.

$= \dfrac{(x + 3)(x - 1)}{x^2 - 3x + 9} \cdot \dfrac{(x + 3)(x^2 - 3x + 9)}{6(x - 1)}$ Factor.

$= \dfrac{(x + 3)(x - 1)(x + 3)(x^2 - 3x + 9)}{6(x - 1)(x^2 - 3x + 9)}$

$= \dfrac{(x + 3)(x + 3)}{6}$ Multiply and eliminate common factors. ■

A *compound fraction* is a fraction having rational expressions in the numerator or denominator. Compound fractions can be simplified by multiplying the numerator and denominator by the *least common denominator (LCD)* of the rational expressions.

EXAMPLE 3   **Simplifying Compound Fractions**

Simplify each compound fraction.

$$\text{(a)} \quad \frac{\dfrac{4}{5} + 1}{x - \dfrac{3}{2}} \qquad \text{(b)} \quad \frac{\dfrac{1}{x + a}}{\dfrac{1}{x} + \dfrac{1}{a}}$$

SOLUTION

(a) The least common denominator of $\frac{4}{5}$ and $\frac{3}{2}$ is 10. Multiplying the numerator and denominator by 10, we have

$$\frac{\dfrac{4}{5} + 1}{x - \dfrac{3}{2}} = \frac{\left(\dfrac{4}{5} + 1\right) \cdot 10}{\left(x - \dfrac{3}{2}\right) \cdot 10}$$

$$= \frac{\dfrac{4}{5} \cdot 10 + 1 \cdot 10}{x \cdot 10 - \dfrac{3}{2} \cdot 10} \qquad \text{Use the distributive property.}$$

$$= \frac{8 + 10}{10x - 15}$$

$$= \frac{18}{5(2x - 3)}$$

(b) The least common denominator of

$$\frac{1}{x + a}, \quad \frac{1}{x}, \quad \text{and} \quad \frac{1}{a}$$

is $xa(x + a)$. We multiply the numerator and denominator by $xa(x + a)$, and simplify:

$$\frac{\dfrac{1}{x + a}}{\dfrac{1}{x} + \dfrac{1}{a}} = \frac{\dfrac{1}{x + a} \cdot xa(x + a)}{\left(\dfrac{1}{x} + \dfrac{1}{a}\right) \cdot xa(x + a)}$$

$$= \frac{xa}{\dfrac{1}{x} \cdot xa(x + a) + \dfrac{1}{a} \cdot xa(x + a)} \qquad \begin{array}{l} \text{Use the distributive property} \\ \text{in the denominator.} \end{array}$$

$$= \frac{xa}{a(x + a) + x(x + a)}$$

$$= \frac{xa}{(x + a)^2}$$

As with rational numbers, adding or subtracting rational fractions requires expressing each term with a common denominator and adding or subtracting the numerators.

● EXAMPLE 4 Adding or Subtracting Rational Expressions

Perform the indicated operation and simplify.

(a) $\dfrac{2}{3x} + \dfrac{1}{x+4}$ (b) $\dfrac{6}{x^2-36} - \dfrac{x+5}{2x+12}$

SOLUTION

(a) $\dfrac{2}{3x} + \dfrac{1}{x+4}$ Denominators are prime.

$\quad = \dfrac{2(x+4)}{3x(x+4)} + \dfrac{3x \cdot 1}{3x(x+4)}$ The LCD is $3x(x+4)$.

$\quad = \dfrac{2(x+4) + 3x}{3x(x+4)}$ Add numerators.

$\quad = \dfrac{5x+8}{3x(x+4)}$ Simplify and factor the numerator.

(b) $\dfrac{6}{x^2-36} - \dfrac{x+5}{2x+12}$

$\quad = \dfrac{6}{(x+6)(x-6)} - \dfrac{x+5}{2(x+6)}$ Factor denominators.

$\quad = \dfrac{2 \cdot 6}{2(x+6)(x-6)} - \dfrac{(x+5)(x-6)}{2(x+6)(x-6)}$ The LCD is $2(x+6)(x-6)$.

$\quad = \dfrac{12 - (x+5)(x-6)}{2(x+6)(x-6)}$ Subtract numerators.

$\quad = \dfrac{12 - (x^2 - x - 30)}{2(x+6)(x-6)}$ Use FOIL.

$\quad = \dfrac{-x^2 + x + 42}{2(x+6)(x-6)}$ Simplify the numerator.

$\quad = \dfrac{-(x^2 - x - 42)}{2(x+6)(x-6)} = -\dfrac{(x+6)(x-7)}{2(x+6)(x-6)}$ Factor the numerator.

$\quad = -\dfrac{\cancel{(x+6)}(x-7)}{2\cancel{(x+6)}(x-6)} = -\dfrac{x-7}{2(x-6)}$ Eliminate common factors.

⬤ **EXAMPLE 5   Combining Operations with Rational Expressions**

Let

$$P = \frac{2}{2x - 1} \qquad Q = \frac{6}{x^2} \qquad R = \frac{3}{2x^2 - x}$$

Perform the indicated operations and simplify.

(a) $P - Q + R$ **(b)** $P + Q \div R$

SOLUTION

(a) $P - Q + R = \dfrac{2}{2x - 1} - \dfrac{6}{x^2} + \dfrac{3}{2x^2 - x}$

$$= \frac{2}{2x - 1} - \frac{6}{x^2} + \frac{3}{x(2x - 1)} \qquad \text{Factor denominators.}$$

$$= \frac{2x^2}{x^2(2x - 1)} - \frac{6(2x - 1)}{x^2(2x - 1)} + \frac{3x}{x^2(2x - 1)} \qquad \begin{array}{l}\text{The LCD is}\\ x^2(2x - 1).\end{array}$$

$$= \frac{2x^2 - 6(2x - 1) + 3x}{x^2(2x - 1)} \qquad \text{Combine numerators.}$$

$$= \frac{2x^2 - 9x + 6}{x^2(2x - 1)} \qquad \begin{array}{l}\text{Simplify the}\\ \text{numerator.}\end{array}$$

(b) Since multiplication and division are performed before addition and subtraction, we first determine $Q \div R$ and then add P:

$$P + Q \div R = \frac{2}{2x - 1} + \frac{6}{x^2} \div \frac{3}{2x^2 - x}$$

$$= \frac{2}{2x - 1} + \frac{6}{x^2} \cdot \frac{2x^2 - x}{3} \qquad \begin{array}{l}\text{Multiply by the reciprocal}\\ \text{of the divisor.}\end{array}$$

$$= \frac{2}{2x - 1} + \frac{2 \cdot 3}{x^2} \cdot \frac{x(2x - 1)}{3} \qquad \text{Factor.}$$

$$= \frac{2}{2x - 1} + \frac{2 \cdot \cancel{3}\cancel{x}(2x - 1)}{\cancel{3}\cancel{x} \cdot x} \qquad \begin{array}{l}\text{Multiply and eliminate}\\ \text{common factors.}\end{array}$$

$$= \frac{2x}{x(2x - 1)} + \frac{2(2x - 1)^2}{x(2x - 1)} \qquad \text{The LCD is } x(2x - 1).$$

$$= \frac{2x + 2(4x^2 - 4x + 1)}{x(2x - 1)} \qquad \text{Add numerators.}$$

$$= \frac{8x^2 - 6x + 2}{x(2x - 1)} \qquad \text{Simplify the numerator.}$$

$$= \frac{2(4x^2 - 3x + 1)}{x(2x - 1)} \qquad \text{Factor.} \qquad ⬤$$

Polynomial Division

Dividing a polynomial by a monomial is fairly simple. Dividing a polynomial by a polynomial with two or more terms is more tedious. In the next three examples we consider both types of division.

■ EXAMPLE 6 Division by a Monomial

Perform each indicated division.

(a) $\dfrac{5x - 2}{x}$ (b) $\dfrac{4x^3 - 8x^2 + x + 12}{4x^2}$

SOLUTION

(a) $\dfrac{5x - 2}{x} = \dfrac{5x}{x} - \dfrac{2}{x} = 5 - \dfrac{2}{x}$

(b) $\dfrac{4x^3 - 8x^2 + x + 12}{4x^2} = \dfrac{4x^3}{4x^2} - \dfrac{8x^2}{4x^2} + \dfrac{x}{4x^2} + \dfrac{12}{4x^2}$

$$= x - 2 + \dfrac{1}{4x} + \dfrac{3}{x^2}$$

The method in Example 6 only applies to division by a monomial. To divide by a polynomial with more than one term, we use long division. Long division with polynomials is similar to long division with positive integers.

■ EXAMPLE 7 Long Division with Polynomials

Use long division to divide: $\dfrac{9x + 8}{3x - 4}$

SOLUTION

We set up the division similar to long division of positive integers. The polynomials should be written in decreasing powers of x.

$$3x - 4 \overline{)9x + 8}$$

The leading term of the dividend, $9x + 8$, is $9x$, and the leading term of the divisor, $3x - 4$, is $3x$. Dividing $9x$ by $3x$ provides the first term of the quotient:

$$\begin{array}{r} 3 \\ 3x - 4 \overline{)9x + 8} \end{array} \quad \text{Since } 9x \div 3x = 3$$

Next, we multiply the divisor by the first term of the quotient, and place the product under the dividend so that like terms line up:

$$\begin{array}{r} 3 \\ 3x - 4 \overline{)9x + 8} \\ \underline{9x - 12} \end{array} \quad \text{Since } 3(3x - 4) = 9x - 12$$

Now we subtract:

$$
\begin{array}{r}
3 \\
3x - 4\overline{\smash{)}9x + 8} \\
\underline{9x - 12} \\
20
\end{array}
\quad \text{Subtract } (9x + 8) - (9x - 12).
$$

Since the degree of 20 is less than the degree of $3x - 4$, the remainder is 20. Thus,

$$
\frac{9x + 8}{3x - 4} = 3 + \frac{20}{3x - 4}
$$

After arranging the dividend in decreasing powers, it is a good idea to insert any missing terms with coefficients of 0.

EXAMPLE 8   Long Division with Terms Missing

Use long division to divide:

(a) $\dfrac{2x}{5x + 1}$ (b) $\dfrac{4x^3 - 6x - 13}{2x^2 - x}$

SOLUTION

(a) The constant term is missing from the dividend, $2x$, so we write $2x$ as $2x + 0$:

$$5x + 1 \overline{\smash{)}2x + 0}$$

Dividing the leading term of the dividend by the leading term of the quotient, we get

$$
\begin{array}{r}
\frac{2}{5} \\
5x + 1\overline{\smash{)}2x + 0}
\end{array}
\quad \text{Since } 2x \div 5x = \tfrac{2}{5}
$$

Multiplying this quotient by the divisor and subtracting, we have

$$
\begin{array}{r}
\frac{2}{5} \\
5x + 1\overline{\smash{)}2x + 0} \\
\underline{2x + \tfrac{2}{5}} \\
-\tfrac{2}{5}
\end{array}
\quad \text{Since } \tfrac{2}{5}(5x + 1) = 2x + \tfrac{2}{5}
$$

This gives

$$
\frac{2x}{5x + 1} = \frac{2}{5} + \frac{-\tfrac{2}{5}}{5x + 1}
$$

which can also be written as

$$\frac{2x}{5x + 1} = \frac{2}{5} - \frac{2}{5(5x + 1)}$$

(b) We place the term $0x^2$ in the dividend to keep the terms aligned:

$$2x^2 - x\overline{)4x^3 + 0x^2 - 6x - 13}$$

Dividing the leading term of the dividend by the leading term of the quotient, we get

$$\begin{array}{r} 2x \\ 2x^2 - x\overline{)4x^3 + 0x^2 - 6x - 13} \end{array}$$

We multiply the divisor by $2x$, and align the product under the dividend so that like terms line up. After subtracting and bringing down the next term, we get

$$\begin{array}{r} 2x \\ 2x^2 - x\overline{)4x^3 + 0x^2 - 6x - 13} \\ \underline{4x^3 - 2x^2} \\ 2x^2 - 6x \end{array}$$

Repeating the process leads to

$$\begin{array}{r} 2x + 1 \\ 2x^2 - x\overline{)4x^3 + 0x^2 - 6x - 13} \\ \underline{4x^3 - 2x^2} \\ 2x^2 - 6x \\ \underline{2x^2 - x} \\ -5x - 13 \end{array}$$

Therefore,

$$\frac{4x^3 - 6x - 13}{2x^2 - x} = 2x + 1 + \frac{-5x - 13}{2x^2 - x}$$

EXERCISE SET 4

A

In Exercises 1–12, reduce each rational expression to lowest terms.

1. $\dfrac{10x^2 + 5x}{10x}$

2. $\dfrac{14x}{6x^2 - 2x}$

3. $\dfrac{4x^2 - 12x}{x^2 - 9}$

4. $\dfrac{x^2 - 2x}{2x^2 - 8}$

5. $\dfrac{-5x - 15}{x^2 + 5x + 6}$

6. $\dfrac{-3x - 18}{x^2 + 7x + 6}$

7. $\dfrac{3x^2 - 3x - 18}{6x^3 - 24x}$

8. $\dfrac{4x^2 - 20x + 16}{2x^3 - 32x}$

9. $\dfrac{3x - 6}{24x - 12x^2}$

10. $\dfrac{20x - 4x^2}{8x - 40}$

11. $\dfrac{2x^2 + 7x + 6}{9 - 4x^2}$

12. $\dfrac{2x^2 - x - 10}{25 - 4x^2}$

B

In Exercises 13–34, perform the indicated operations and simplify.

13. $\dfrac{12x^2}{x^2 + x - 6} \cdot \dfrac{x - 2}{10x}$

14. $\dfrac{9x^3}{x^2 - 3x - 4} \cdot \dfrac{x + 1}{6x}$

15. $\dfrac{4x^2 - 25}{3x^2 - 5x - 2} \cdot \dfrac{6x + 2}{2x^2 + 5x}$

16. $\dfrac{16x^2 - 9}{2x^2 + 11x - 6} \cdot \dfrac{6x - 3}{8x^2 + 6x}$

17. $\dfrac{4x^2 + 4x + 1}{2x^2 - x - 1} \cdot \dfrac{2x^2 + x - 1}{4x^2 - 1}$

18. $\dfrac{2x^2 + 11x + 15}{2x^2 + 5x} \cdot \dfrac{4x^2 + 3x - 1}{x^2 + 4x + 3}$

19. $\dfrac{3x}{x + 2} \div \dfrac{x^2}{x^2 - 4}$

20. $\dfrac{y}{2y - 2} \div \dfrac{5y^3}{y^2 - 1}$

21. $\dfrac{2a^2 - 5a - 3}{a^2 - 6a + 9} \div \dfrac{4a^2 - 1}{a^2 - 5a + 6}$

22. $\dfrac{x^2 - 16}{x^2 + x - 12} \div \dfrac{x^2 - 2x - 8}{x^2 + 5x + 6}$

23. $\dfrac{2}{x - 1} + \dfrac{3}{x + 4}$

24. $\dfrac{1}{x - 7} + \dfrac{4}{x + 2}$

25. $\dfrac{7}{2x} - \dfrac{2}{x + 1}$

26. $\dfrac{3}{4x} - \dfrac{1}{x + 1}$

27. $\dfrac{2x}{2x^2 + 3x + 1} + \dfrac{3}{2x + 1}$

28. $\dfrac{5x}{3x^2 - 5x - 2} + \dfrac{1}{x - 2}$

29. $\dfrac{x + 4}{2x + 10} - \dfrac{7x}{x^2 - 2x - 35}$

30. $\dfrac{5x}{2x + 8} - \dfrac{x}{x^2 - 4x - 32}$

31. $\dfrac{5x}{2x + 14} - \dfrac{3}{x + 7} + \dfrac{x^2}{x^2 - 49}$

32. $\dfrac{4}{3x + 18} + \dfrac{2x}{x - 6} - \dfrac{x^2 + 1}{x^2 - 36}$

33. $(x + 1) + 3x^{-1} - 5(x - 1)^{-1}$

34. $(x - 2) + 6x^{-1} - 7(x + 2)^{-1}$

In Exercises 35–40, let

$$P = \dfrac{5}{2x} \qquad Q = \dfrac{4x + 4}{x - 1} \qquad R = \dfrac{x}{x^2 - 1}$$

Perform the indicated operations and simplify.

35. $P + QR$

36. $Q + PR$

37. $P + Q \div R$

38. $Q + P \div R$

39. $(6P - Q)R$

40. $(4P + R)Q$

In Exercises 41–46, simplify each compound fraction.

41. $\dfrac{2 + \dfrac{1}{x}}{3}$

42. $\dfrac{5 - \dfrac{1}{x}}{2}$

43. $\dfrac{\dfrac{1}{x + 3} - \dfrac{1}{x}}{3}$

44. $\dfrac{\dfrac{1}{x - 4} - \dfrac{1}{x}}{4}$

45. $\dfrac{2 + \dfrac{1}{x}}{1 - \dfrac{4}{x + 1}}$

46. $\dfrac{1 + \dfrac{x}{3}}{1 + \dfrac{5}{x + 2}}$

In Exercises 47–68, perform the indicated division.

47. $\dfrac{18x^2 + 27x - 15}{3x}$

48. $\dfrac{8x^2 - 22x + 10}{2x}$

49. $\dfrac{15x^4 - 7x^2 + 10x + 4}{5x}$

50. $\dfrac{12x^4 - 8x^2 + 6x + 1}{4x}$

51. $\dfrac{3x + 1}{x - 4}$

52. $\dfrac{6x + 5}{x - 3}$

53. $\dfrac{8x}{2x + 5}$

54. $\dfrac{14x}{7x + 1}$

55. $\dfrac{3x + 10}{2x + 6}$

56. $\dfrac{4x - 5}{3x - 6}$

57. $\dfrac{3x^2 + 2x - 5}{x + 2}$

58. $\dfrac{2x^2 + 12x + 13}{x + 5}$

59. $\dfrac{4x^2 + 2x + 7}{2x - 1}$

60. $\dfrac{6x^2 + 5x + 1}{3x - 2}$

61. $\dfrac{3x^3 + 10x^2 + 7x - 5}{x^2 + 2x}$

62. $\dfrac{4x^3 - 9x^2 + x - 7}{x^2 - 3x}$

63. $\dfrac{4x^3 - 2x^2 + 3}{2x^2 + 1}$

64. $\dfrac{6x^3 + 3x^2 - 11}{3x^2 - 1}$

65. $\dfrac{4x^5 + 15x^4 - 33x^2 + x + 12}{x^2 + 3x + 1}$

66. $\dfrac{5x^5 + 8x^4 - 3x^3 - 7x + 6}{x^2 + 2x - 1}$

67. $\dfrac{2x^2 - 7}{2x - 3}$

68. $\dfrac{4x^2 + 19x + 6}{2x + 9}$

C

69. Determine the value of k such that $2x + 3$ is a factor of $2x^3 - 15x^2 + kx + 3$.

70. Let n be an integer greater than 1. The rational expression

$$\frac{x^n - nx + (n - 1)}{x^2 - 2x + 1}$$

reduces to a polynomial of degree $n - 2$.

(a) Determine the resulting polynomial for $n = 2, 3, 4,$ and 5.

(b) Look for a pattern in the results of part (a) and conjecture the resulting polynomial for any integer n greater than 1.

(c) Prove your conjecture in part (b).

| SECTION 5 | RADICALS AND RATIONAL EXPONENTS |

- Radicals
- Rational Exponents
- Algebraic Fractions

In this section, we extend the definition of exponent to include rational numbers. We also discuss how to simplify the quotient of two algebraic expressions.

Radicals

For any real number a, there is only one real number b such that

$$b^3 = a$$

The number whose cube is a is called the *third root* of a, denoted as $\sqrt[3]{a}$. For example, since $2^3 = 8$, the third root of 8 is 2:

$$\sqrt[3]{8} = 2$$

For any positive real number a, there are two real numbers b such that

$$b^4 = a$$

The two numbers whose fourth powers are a are called the *fourth roots* of a. For example, the fourth roots of 81 are 3 and -3, since $3^4 = 81$ and $(-3)^4 = 81$.

The positive fourth root of a is denoted as $\sqrt[4]{a}$. Thus, the two fourth roots are written as follows:

$$\sqrt[4]{81} = 3 \quad \text{and} \quad -\sqrt[4]{81} = -3$$

We generalize to nth roots in the following definition:

Definition of $\sqrt[n]{a}$

Suppose that a is a real number and n is an integer greater than 1.

1. If n is even and $a > 0$, then $\sqrt[n]{a}$ is the positive real number b such that $b^n = a$. We call $\sqrt[n]{a}$ the **positive nth root of a.**
2. If n is even and $a < 0$, then $\sqrt[n]{a}$ is not a real number.
3. If n is odd, then $\sqrt[n]{a}$ is the real number b such that $b^n = a$. We call $\sqrt[n]{a}$ the **nth root of a.**

■ **NOTE:** *The positive nth root of a is also called the* principal *nth root of a.*

For the expression $\sqrt[n]{a}$, the symbol $\sqrt{}$ is called a *radical sign*, n is the *index* of the radical, and a is the *radicand*. If $n = 2$, we write $\sqrt{a}$ instead of $\sqrt[2]{a}$. We call $\sqrt{a}$ the positive *square root* of a. The third root of a is also called the *cube root* of a.

■ **EXAMPLE 1** **Evaluating nth Roots**

Evaluate each expression.

(a) $\sqrt{\frac{1}{9}}$ (b) $\sqrt[3]{125}$ (c) $\sqrt[3]{-27}$ (d) $-\sqrt[4]{16}$ (e) $\sqrt[4]{-16}$

SOLUTION

(a) $\sqrt{\frac{1}{9}}$ is the positive number whose square is $\frac{1}{9}$. Since $\left(\frac{1}{3}\right)^2 = \frac{1}{9}$, $\sqrt{\frac{1}{9}} = \frac{1}{3}$.

(b) $\sqrt[3]{125}$ is the number whose cube is 125. Since $5^3 = 125$, $\sqrt[3]{125} = 5$.

(c) $\sqrt[3]{-27} = -3$, since $(-3)^3 = -27$.

(d) First note that $\sqrt[4]{16} = 2$, since $2^4 = 16$. Therefore, $-\sqrt[4]{16} = -2$.

(e) When a real number is raised to the fourth power, the result is never negative. Therefore, $\sqrt[4]{-16}$ is not a real number. ●

The manipulation and simplification of expressions involving radicals is aided by the following properties:

Properties of Radicals

Suppose that a and b are nonnegative real numbers, and that n and m are integers greater than 1.

1. $\sqrt[n]{ab} = \sqrt[n]{a}\,\sqrt[n]{b}$

2. $\sqrt[n]{\dfrac{a}{b}} = \dfrac{\sqrt[n]{a}}{\sqrt[n]{b}}, \quad b \neq 0$

3. $\sqrt[n]{a^m} = (\sqrt[n]{a})^m$

4. $(\sqrt[n]{a})^n = \sqrt[n]{a^n} = a$

5. $\sqrt[m]{\sqrt[n]{a}} = \sqrt[mn]{a}$

6. $\sqrt[mn]{a^m} = \sqrt[n]{a}$

7. If n is odd, then $\sqrt[n]{-a} = -\sqrt[n]{a}$.

■ EXAMPLE 2 Simplifying Expressions with Radicals

Use the properties of radicals to simplify each expression. Assume that all variables are positive.

(a) $\sqrt{50x^3y}$ **(b)** $\sqrt[3]{-16a^6b^{10}}$ **(c)** $\sqrt[4]{\dfrac{16x^2}{81y^8}}$ **(d)** $\sqrt[3]{4x\sqrt{y}}(\sqrt[3]{x})^2$

SOLUTION

(a) $\sqrt{50x^3y} = \sqrt{25x^2 \cdot 2xy} = \sqrt{(5x)^2 \cdot 2xy}$

$\qquad\qquad\qquad = \sqrt{(5x)^2}\,\sqrt{2xy}$ Use property 1.

$\qquad\qquad\qquad = 5x\sqrt{2xy}$ Use property 4.

(b) $\sqrt[3]{-16a^6b^{10}} = \sqrt[3]{-8a^6b^9 \cdot 2b}$

$\qquad\qquad\qquad = \sqrt[3]{-(2a^2b^3)^3 \cdot 2b}$

$\qquad\qquad\qquad = -\sqrt[3]{(2a^2b^3)^3}\,\sqrt[3]{2b}$ Use properties 1 and 7.

$\qquad\qquad\qquad = -2a^2b^3\sqrt[3]{2b}$ Use property 4.

(c) $\sqrt[4]{\dfrac{16x^2}{81y^8}} = \dfrac{\sqrt[4]{16x^2}}{\sqrt[4]{81y^8}}$ Use property 2.

$\qquad\qquad = \dfrac{\sqrt[4]{(2)^4x^2}}{\sqrt[4]{(3y^2)^4}}$

$\qquad\qquad = \dfrac{\sqrt[4]{(2)^4}\,\sqrt[4]{x^2}}{\sqrt[4]{(3y^2)^4}}$ Use property 1.

$\qquad\qquad = \dfrac{2\sqrt{x}}{3y^2}$ Use properties 4 and 6.

(d) $\sqrt[3]{4x\sqrt{y}}\,(\sqrt[3]{x})^2 = \sqrt[3]{4x}\,\sqrt[3]{\sqrt{y}}\,\sqrt[3]{x^2}$ Use properties 1 and 3.

$\qquad\qquad\qquad = \sqrt[3]{4x}\,\sqrt[3]{x^2}\,\sqrt[3]{\sqrt{y}}$

$\qquad\qquad\qquad = \sqrt[3]{4x^3}\,\sqrt[3]{\sqrt{y}}$ Use property 1.

$\qquad\qquad\qquad = x\sqrt[3]{4}\,\sqrt[6]{y}$ Use properties 1, 4, and 5.

Expressions with radicals that are added or subtracted can be combined as like terms provided they have the same index and radicand.

■ EXAMPLE 3   Combining Like Radical Expressions

Perform the indicated operations.

(a) $4\sqrt[3]{2x} + 5\sqrt[3]{6x} - \sqrt[3]{16x}$ (b) $(x - \sqrt{x})(1 + 2\sqrt{x})$

SOLUTION

(a) $4\sqrt[3]{2x} + 5\sqrt[3]{6x} - \sqrt[3]{16x} = 4\sqrt[3]{2x} + 5\sqrt[3]{6x} - \sqrt[3]{8}\sqrt[3]{2x}$ Use property 1.

$\qquad\qquad\qquad\qquad\qquad = 4\sqrt[3]{2x} + 5\sqrt[3]{6x} - 2\sqrt[3]{2x}$

$\qquad\qquad\qquad\qquad\qquad = (4\sqrt[3]{2x} - 2\sqrt[3]{2x}) + 5\sqrt[3]{6x}$ Group like terms.

$\qquad\qquad\qquad\qquad\qquad = (4 - 2)\sqrt[3]{2x} + 5\sqrt[3]{6x}$ Factor out $\sqrt[3]{2x}$ to combine like terms.

$\qquad\qquad\qquad\qquad\qquad = 2\sqrt[3]{2x} + 5\sqrt[3]{6x}$

(b) $(x - \sqrt{x})(1 + 2\sqrt{x}) = x + 2x\sqrt{x} - \sqrt{x} - 2(\sqrt{x})^2$ Use the FOIL method.

$\qquad\qquad\qquad\qquad\quad = x + 2x\sqrt{x} - \sqrt{x} - 2x$ Use property 4.

$\qquad\qquad\qquad\qquad\quad = -x + 2x\sqrt{x} - \sqrt{x}$ Combine like terms.

Rational Exponents

The concept of the nth root of a number leads to a natural way to define exponents that are rational numbers. For example, consider the expression $2^{1/3}$. If we let

$$a = 2^{1/3}$$

and assume that the properties of exponents are valid for rational exponents, then

$$a^3 = (2^{1/3})^3 = 2^1 = 2$$

This equation implies that

$$a = \sqrt[3]{2}$$

Since we originally let $a = 2^{1/3}$, it follows that

$$2^{1/3} = \sqrt[3]{2}$$

We can extend this idea to give meaning to an expression such as $2^{5/3}$ as follows:

$$2^{5/3} = (2^{1/3})^5 = (\sqrt[3]{2})^5$$

This result motivates the following definition:

Definition of $a^{1/n}$ and $a^{m/n}$

Suppose that a is a real number, and n is an integer greater than 1. If $\sqrt[n]{a}$ is a real number, then

$$a^{1/n} = \sqrt[n]{a}$$

Furthermore, if m is an integer, then

$$a^{m/n} = (\sqrt[n]{a})^m = \sqrt[n]{a^m}$$

EXAMPLE 4   Evaluating Expressions with Rational Exponents

Evaluate each expression, if possible.

(a) $36^{1/2}$ **(b)** $(-16)^{1/2}$ **(c)** $-81^{1/4}$ **(d)** $(-27)^{2/3}$ **(e)** $64^{-4/3}$

SOLUTION

(a) $36^{1/2} = \sqrt{36} = 6$

(b) Since $(-16)^{1/2} = \sqrt{-16}$ and $\sqrt{-16}$ is not a real number, $(-16)^{1/2}$ is not a real number.

(c) $-81^{1/4} = -\sqrt[4]{81} = -3$

(d) $(-27)^{2/3} = (\sqrt[3]{-27})^2 = (-3)^2 = 9$

(e) $64^{-4/3} = \dfrac{1}{64^{4/3}} = \dfrac{1}{(\sqrt[3]{64})^4} = \dfrac{1}{4^4} = \dfrac{1}{256}$

An expression with rational exponents is simplified in the same way as an expression with integer exponents.

EXAMPLE 5   Simplifying Expressions with Rational Exponents

Use the properties of exponents to simplify each expression.

(a) $6x^{1/5}(2x^{1/3})^2$ **(b)** $2x(x^2 + 4)^{3/2}(x^2 + 4)^{-1/2} - x^3$

(c) $(x^{1/2} + y^{5/2})^2$

SOLUTION

(a) $6x^{1/5}(2x^{1/3})^2 = 6x^{1/5}[2^2(x^{1/3})^2]$

$= 6x^{1/5}(4x^{2/3})$

$= 24x^{(1/5)+(2/3)}$

$= 24x^{13/15}$

(b) $2x(x^2 + 4)^{3/2}(x^2 + 4)^{-1/2} - x^3 = 2x(x^2 + 4)^{(3/2)+(-1/2)} - x^3$

$= 2x(x^2 + 4) - x^3$

$= x^3 + 8x$

(c) $(x^{1/2} + y^{5/2})^2 = (x^{1/2} + y^{5/2})(x^{1/2} + y^{5/2})$

$= x^{1/2}x^{1/2} + 2x^{1/2}y^{5/2} + y^{5/2}y^{5/2}$ Use the FOIL method.

$= x + 2x^{1/2}y^{5/2} + y^5$

An expression with radicals can be simplified by rewriting it in terms of rational exponents and using the properties of exponents.

◼ EXAMPLE 6 Using Rational Exponents to Simplify Radicals

Rewrite each expression using rational exponents, and use the properties of exponents to simplify. Leave answers in rational exponent form.

(a) $\dfrac{\sqrt[3]{x^2}}{\sqrt{\sqrt[3]{x}}}$ **(b)** $\sqrt[3]{\dfrac{8x^2}{x^{1/2}}}$

SOLUTION

(a) $\dfrac{\sqrt[3]{x^2}}{\sqrt{\sqrt[3]{x}}} = \dfrac{x^{2/3}}{(x^{1/3})^{1/2}} = \dfrac{x^{2/3}}{x^{1/6}} = x^{(2/3)-(1/6)} = x^{1/2}$

(b) $\sqrt[3]{\dfrac{8x^2}{x^{1/2}}} = \left(\dfrac{8x^2}{x^{1/2}}\right)^{1/3} = (8x^{2-(1/2)})^{1/3} = (8x^{3/2})^{1/3} = 8^{1/3}(x^{3/2})^{1/3} = 2x^{1/2}$ ◼

Algebraic Fractions

An *algebraic fraction* is the quotient of two algebraic expressions. Examples of algebraic fractions are

$$\frac{2x}{\sqrt{x}+3x} \qquad \frac{x^{3/2}(x+1)^2 - 4x^{-1/2}(x+1)}{x^3+1} \qquad \frac{5-\sqrt{2x}}{\sqrt[3]{x}}$$

Algebraic fractions are reduced in the same way that rational expressions are reduced — by factoring the numerator and denominator and eliminating common factors.

◼ EXAMPLE 7 Simplifying a Fraction with Radicals

Simplify the algebraic fraction: $\dfrac{-10+\sqrt{48}}{8}$

SOLUTION

$$\frac{-10+\sqrt{48}}{8} = \frac{-10+\sqrt{16\cdot 3}}{8}$$

$$= \frac{-10+4\sqrt{3}}{8} \qquad \text{Use radical properties 1 and 4.}$$

$$= \frac{\cancel{2}(-5 + 2\sqrt{3})}{\cancel{2} \cdot 4} \qquad \text{Factor and eliminate common factors.}$$

$$= \frac{-5 + 2\sqrt{3}}{4}$$

It is sometimes convenient or necessary to rewrite an algebraic fraction so that the denominator contains no radicals. The process of eliminating radicals from the denominator is called *rationalizing the denominator*.

If the denominator has only one square root term, the denominator can often be rationalized by multiplying both the numerator and denominator by the denominator. A denominator with two terms often can be rationalized by multiplying the numerator and denominator by the conjugate of the denominator. The *conjugate* of a sum $a + b$ is $a - b$. For example, the conjugate of $2 + \sqrt{3}$ is $2 - \sqrt{3}$, and the conjugate of $\sqrt{x} - 5x$ is $\sqrt{x} + 5x$.

● EXAMPLE 8   Rationalizing a Denominator

Rationalize the denominator of the algebraic fraction.

(a) $\dfrac{5}{\sqrt{2x}}$ **(b)** $\dfrac{1}{x + \sqrt{x}}$

SOLUTION

(a) $\dfrac{5}{\sqrt{2x}} = \dfrac{5}{\sqrt{2x}} \cdot \dfrac{\sqrt{2x}}{\sqrt{2x}} = \dfrac{5\sqrt{2x}}{2x}$

(b) $\dfrac{1}{x + \sqrt{x}} = \dfrac{1}{x + \sqrt{x}} \cdot \dfrac{x - \sqrt{x}}{x - \sqrt{x}}$ Multiply the numerator and denominator by the conjugate of $x + \sqrt{x}$.

$$= \frac{x - \sqrt{x}}{x^2 - (\sqrt{x})^2} \qquad \text{Multiply.}$$

$$= \frac{x - \sqrt{x}}{x^2 - x} \qquad \text{Simplify.}$$

Our last example deals with a type of algebraic fraction that is typically encountered in calculus.

● EXAMPLE 9   Simplifying an Algebraic Fraction

Simplify: $\dfrac{3x^{1/2}(x + 1)^2 - 4x^{3/2}(x + 1)}{[(x + 1)^2]^2}$

SOLUTION

$$\frac{3x^{1/2}(x+1)^2 - 4x^{3/2}(x+1)}{[(x+1)^2]^2}$$

$$= \frac{3x^{1/2}(x+1)^2 - 4x^{3/2}(x+1)}{(x+1)^4} \qquad \text{Use properties of exponents.}$$

$$= \frac{x^{1/2}(x+1)[3(x+1) - 4x]}{(x+1)^4} \qquad \text{Factor } x^{1/2}(x+1) \text{ out of the numerator.}$$

$$= \frac{x^{1/2}\cancel{(x+1)}[3(x+1) - 4x]}{\cancel{(x+1)}(x+1)^3} \qquad \text{Eliminate common factors.}$$

$$= \frac{x^{1/2}(3-x)}{(x+1)^3} \qquad \text{Simplify.}$$

EXERCISE SET 5

A

In Exercises 1–12, evaluate each expression, if possible.

1. $\sqrt{\frac{4}{49}}$ 2. $\sqrt{\frac{36}{25}}$ 3. $\sqrt[3]{64}$

4. $\sqrt[3]{216}$ 5. $\sqrt[3]{-\frac{1}{8}}$ 6. $\sqrt[3]{-\frac{1}{27}}$

7. $\sqrt{-9}$ 8. $\sqrt{-16}$ 9. $-\sqrt[4]{81}$

10. $-\sqrt[4]{256}$ 11. $\sqrt{9+16}$ 12. $\sqrt{25+144}$

In Exercises 13–22, evaluate each expression, if possible.

13. $-9^{1/2}$ 14. $-25^{1/2}$ 15. $(-36)^{1/2}$

16. $(-49)^{1/2}$ 17. $8^{1/3}$ 18. $27^{1/3}$

19. $125^{2/3}$ 20. $8^{4/3}$ 21. $27^{-4/3}$

22. $216^{-2/3}$

B

In Exercises 23–40, use the properties of radicals to simplify each expression. Assume that all variables are positive.

23. $\sqrt{8x^3y^2}$ 24. $\sqrt{18x^4y^5}$

25. $\sqrt[3]{54x^4}$ 26. $\sqrt[3]{40x^5}$

27. $\sqrt[4]{16x^{10}}$ 28. $\sqrt[4]{81x^{14}}$

29. $\sqrt[3]{\frac{-3a^{11}}{8b^9}}$ 30. $\sqrt[3]{\frac{-2a^{10}}{27b^6}}$

31. $\sqrt{\sqrt[3]{x^4}}$ 32. $\sqrt[4]{\sqrt[3]{x^9}}$

33. $\sqrt{75x} \cdot \sqrt{3x^3}$ 34. $\sqrt{18x} \cdot \sqrt{8x}$

35. $\frac{\sqrt{2x^5y}}{\sqrt{18x^4y^3}}$ 36. $\frac{\sqrt{3x^9y^3}}{\sqrt{48x^6y^7}}$

37. $\sqrt[3]{y^2\sqrt{y}} \cdot \sqrt[3]{y}$ 38. $\sqrt[3]{a\sqrt[4]{a}} \cdot \sqrt[3]{a^5}$

39. $\sqrt{\frac{x^2+y^2}{(a+b)^2}}$ 40. $\sqrt{\frac{x^2-y^2}{(a-b)^2}}$

In Exercises 41–52, perform the indicated operations.

41. $6\sqrt{3x} + 7\sqrt{2x} + \sqrt{12x}$ 42. $9\sqrt{2x} - \sqrt{18x} + \sqrt{12x}$

43. $\sqrt[3]{54x^4} + \sqrt[3]{16x^4}$ 44. $\sqrt[3]{24x^5} + \sqrt[3]{81x^5}$

45. $10x + \sqrt{x}(2x - 3\sqrt{x})$ 46. $9x + 2\sqrt{x}(\sqrt{x} + x)$

47. $(x + \sqrt{3})(x - \sqrt{3})$ 48. $(\sqrt{x} - 4)(\sqrt{x} + 4)$

49. $(\sqrt{x+1} + 2)^2$ 50. $(\sqrt{x-2} + 3)^2$

51. $(3\sqrt{x+2})^2$ 52. $(2\sqrt{x+1})^2$

In Exercises 53–62, use the properties of exponents to simplify each expression.

53. $(9x^2y^4)^{1/2}$ 54. $(16x^6y^2)^{1/2}$

55. $3x(2x^{3/4}y^{1/2})^2$ 56. $4y(3x^{2/3}y^{1/2})^3$

57. $(x^2 + 3)^{2/3}(x^2 + 3)^{4/3} - x^4$

58. $(2x + 1)^{4/5}(2x + 1)^{6/5} - 4x^2$

59. $(x^{3/2} - y^{1/2})^2$

60. $(x^{3/4} + y^{1/2})^2$

61. $3x^{4/3} - x^{2/3}(x^{2/3} - 2)$

62. $x^{1/2}(x^2 - 2x^{3/2}) + x^{5/2}$

In Exercises 63–66, rewrite each expression using rational exponents, and use the properties of exponents to simplify. Leave answers in rational exponent form.

63. $\sqrt[6]{x^5} \cdot \sqrt[3]{\sqrt{x}}$

64. $\sqrt[4]{x^3} \cdot \sqrt{\sqrt{x}}$

65. $\dfrac{\sqrt[3]{x^2}}{\sqrt[6]{\sqrt{x}}}$

66. $\dfrac{\sqrt[3]{x^2}}{\sqrt{\sqrt[3]{x^2}}}$

In Exercises 67–74, simplify each algebraic fraction.

67. $\dfrac{-4 + \sqrt{24}}{2}$

68. $\dfrac{-3 + \sqrt{72}}{3}$

69. $\dfrac{9 - \sqrt{18}}{12}$

70. $\dfrac{14 - \sqrt{8}}{6}$

71. $\dfrac{1}{2}(-8 \pm \sqrt{56 - 4x})$

72. $\dfrac{1}{2}(12 \pm \sqrt{20 + 8x^2})$

73. $\sqrt{1 + \left[\dfrac{1}{2}\left(x - \dfrac{1}{x}\right)\right]^2}$

74. $\sqrt{2 + \left[\dfrac{1}{2}\left(2x - \dfrac{1}{x}\right)\right]^2}$

In Exercises 75–84, rationalize the denominator of each algebraic fraction.

75. $\dfrac{5}{\sqrt{6}}$

76. $\dfrac{9}{\sqrt{5}}$

77. $\dfrac{4}{3\sqrt{2x}}$

78. $\dfrac{6}{5\sqrt{3y}}$

79. $\dfrac{7}{3 + \sqrt{5}}$

80. $\dfrac{8}{2 + \sqrt{3}}$

81. $\dfrac{\sqrt{2}}{\sqrt{6} - \sqrt{2}}$

82. $\dfrac{\sqrt{6}}{\sqrt{10} - \sqrt{6}}$

83. $\dfrac{x}{x + \sqrt{x}}$

84. $\dfrac{x^2}{\sqrt{x} - x}$

C

In Exercises 85 and 86, simplify each algebraic fraction.

85. $\dfrac{8x^{1/3}(x - 2)^2 - 12x^{4/3}(x - 2)}{[(x - 2)^2]^2}$

86. $\dfrac{7x^{3/4}(x + 3)^3 - 12x^{7/4}(x + 3)^2}{[(x + 3)^3]^2}$

87. (a) Give an example of a rational number raised to a rational power in which the result is rational.
(b) Give an example of a rational number raised to a rational power in which the result is irrational.
(c) Give an example of an irrational number raised to a rational power in which the result is rational.
(d) Give an example of an irrational number raised to a rational power in which the result is irrational.

88. Without using a calculator, show that $\sqrt{6} + \sqrt{2} = 2\sqrt{2 + \sqrt{3}}$.

89. Without using a calculator, show that $\sqrt[4]{3\sqrt{5} + 6\sqrt[3]{25} - 19} = \sqrt[3]{5} - 1$.

90. Property 1 of radicals, $\sqrt[n]{ab} = \sqrt[n]{a}\,\sqrt[n]{b}$, can be expressed in terms of rational exponents as $(ab)^{1/n} = a^{1/n}b^{1/n}$, which corresponds to the distributive law of exponents over multiplication (see Section 2). Rewrite each of the following properties of radicals in terms of rational exponents and state the corresponding law of exponents from Section 2.
(a) $\sqrt[n]{\dfrac{a}{b}} = \dfrac{\sqrt[n]{a}}{\sqrt[n]{b}}$
(b) $\sqrt[m]{\sqrt[n]{a}} = \sqrt[mn]{a}$

91. One solution to the third-degree equation $x^3 + ax + b = 0$ is

$$x = \sqrt[3]{-\dfrac{b}{2} + \sqrt{\dfrac{b^2}{4} + \dfrac{a^3}{27}}} + \sqrt[3]{-\dfrac{b}{2} - \sqrt{\dfrac{b^2}{4} + \dfrac{a^3}{27}}}$$

Use this formula to determine a solution of:
(a) $x^3 - 3x + 1 = 0$
(b) $2x^3 + 10x - 6 = 0$

SECTION 6

- Linear Equations
- Literal Equations
- Linear Inequalities
- Applications

LINEAR EQUATIONS AND INEQUALITIES

A *solution* of an equation or inequality in one variable is a value for the variable that makes the equation or inequality a true statement. For example, -2 is a solution to the equation

$$3x + 5 = 1 + x$$

NOTE: *A solution to an equation is also called a* root *of the equation.*

because when we substitute -2 for x in the equation, we have

$$3(-2) + 5 \stackrel{?}{=} 1 + (-2)$$
$$-1 = -1 \qquad \text{True}$$

As another example, 0 and 3 are solutions to the inequality

$$y \leq 9 - 2y$$

since when y is replaced by 0, we get

$$0 \stackrel{?}{\leq} 9 - 2(0)$$
$$0 \leq 9 \qquad \text{True}$$

and when we replace y with 3, we get

$$3 \stackrel{?}{\leq} 9 - 2(3)$$
$$3 \leq 3 \qquad \text{True}$$

To *solve* an equation or inequality in one variable means to determine all of its solutions. In this section, we review the methods for solving linear equations and inequalities.

Two equations are *equivalent* if they have the same set of solutions. Solving an equation typically involves replacing it with an equivalent equation with roots that are apparent. The following two fundamental operations produce an equivalent equation:

Operations That Produce
Equivalent Equations

> Adding (subtracting) the same quantity to (from) each side of an equation produces an equivalent equation.
>
> Multiplying or dividing each side of an equation by the same nonzero quantity produces an equivalent equation.

Linear Equations

A *linear equation in one variable* is an equation that can be written in the form

$$ax + b = 0$$

where a and b are real numbers and $a \neq 0$.

A linear equation has exactly one solution. The strategy for solving a linear equation is to simplify both sides, isolate the terms containing the variable on the same side, and then isolate the variable. Of course, you also should check your solution.

⬤ EXAMPLE 1 Solving Linear Equations

Solve each equation.

(a) $3(2x + 9) = 10x + 19$ **(b)** $5x + 2x(x + 3) = (2x - 7)(x + 2)$

SOLUTION

(a) $3(2x + 9) = 10x + 19$

$\quad\quad 6x + 27 = 10x + 19$ Use the distributive property.

$\quad\quad\quad\quad 27 = 4x + 19$ Subtract $6x$ from each side.

$\quad\quad\quad\quad\ \ 8 = 4x$ Subtract 19 from each side.

$\quad\quad\quad\quad\ \ 2 = x$ Divide each side by 4.

CHECK: To verify that 2 is a solution, we substitute 2 for x in the original equation:

$$3[2(2) + 9] \overset{?}{=} 10(2) + 19$$
$$3(13) \overset{?}{=} 20 + 19$$
$$39 = 39 \quad\quad\quad \text{True}$$

The solution to the equation is $x = 2$.

(b) $5x + 2x(x + 3) = (2x - 7)(x + 2)$

$\quad 5x + 2x^2 + 6x = 2x^2 - 3x - 14$ Use the distributive property and the FOIL method.

$\quad\quad\ 2x^2 + 11x = 2x^2 - 3x - 14$ Add like terms.

$\quad\quad\quad\quad\ 11x = -3x - 14$ Subtract $2x^2$ from each side.

$\quad\quad\quad\quad\ 14x = -14$ Add $3x$ to each side.

$\quad\quad\quad\quad\quad\ x = -1$ Divide each side by 14.

CHECK: We check this by replacing x with -1 in the original equation:

$$5(-1) + 2(-1)[(-1) + 3] \overset{?}{=} [2(-1) - 7][(-1) + 2]$$
$$-5 + (-2)(2) \overset{?}{=} (-2 - 7)(1)$$
$$-9 = -9 \quad\quad\quad \text{True}$$

The solution to the equation is $x = -1$.   ⬤

Equations with fractions or decimals are often difficult to manipulate. Multiplying each side of an equation by the least common denominator produces an equivalent equation without fractions or decimals.

◯ EXAMPLE 2   Solving Linear Equations with Fractions

Solve each equation.

(a) $\dfrac{3x}{4} + 1 = \dfrac{x}{2}$ **(b)** $4.2x - 5 = 1.6(2x + 3)$

SOLUTION

(a) We clear out fractions in the equation by multiplying each side of the equation by 4:

$$\frac{3x}{4} + 1 = \frac{x}{2}$$

$$4 \cdot \left(\frac{3x}{4} + 1\right) = 4 \cdot \left(\frac{x}{2}\right) \qquad \text{Multiply each side by 4.}$$

$$3x + 4 = 2x \qquad \text{Use the distributive property and multiply.}$$

$$x + 4 = 0 \qquad \text{Subtract } 2x \text{ from each side.}$$

$$x = -4 \qquad \text{Subtract 4 from each side.}$$

CHECK: A check of our result follows:

$$\frac{3(-4)}{4} + 1 \stackrel{?}{=} \frac{(-4)}{2}$$

$$-2 = -2 \qquad \text{True}$$

Thus, $x = -4$ is the solution.

(b) We can eliminate the decimals in the equation by multiplying each side of the equation by 10:

$$4.2x - 5 = 1.6(2x + 3)$$

$$10(4.2x - 5) = 10[1.6(2x + 3)] \qquad \text{Multiply each side by 10.}$$

$$42x - 50 = (10 \cdot 1.6)(2x + 3) \qquad \text{Use the distributive property on the left side, and use the associative property on the right side.}$$

$$42x - 50 = 16(2x + 3)$$

$$42x - 50 = 32x + 48 \qquad \text{Use the distributive property.}$$

$$10x - 50 = 48 \qquad \text{Subtract } 32x \text{ from each side.}$$

$$10x = 98 \qquad \text{Add 50 to each side.}$$

$$x = 9.8 \qquad \text{Divide each side by 10.}$$

CHECK: We verify that $x = 9.8$ is the solution of the equation by substituting in the original equation:

$$4.2(9.8) - 5 \stackrel{?}{=} 1.6[2(9.8) + 3]$$

$$36.16 = 36.16 \qquad \text{True} \qquad ◯$$

Literal Equations

Many problems involve equations or formulas with more than one variable. For example, the gravitational attraction F between two objects is given by

$$F = \frac{GmM}{r^2}$$

where m and M are the masses of the objects, r is the distance between them, and G is the universal gravitational constant.

There are situations when it is necessary or convenient to solve the equation for a quantity such as m. Multiplying each side by r^2 gives

$$Fr^2 = GmM$$

Dividing each side by GM, we get

$$\frac{Fr^2}{GM} = m$$

In solving for m, we treat F, r, G and M as if they are constants. An equation in which constants are represented by letters is called a *literal equation*.

● EXAMPLE 3   Solving a Literal Equation

Solve each equation for the specified variable.

(a) $S = 2\pi rh + \pi r^2$, for h **(b)** $C = wh + wl + hl$, for w

SOLUTION

(a) We first isolate the term containing h, and then we isolate h:

$$S = 2\pi rh + \pi r^2$$

$$S - \pi r^2 = 2\pi rh \qquad \text{Subtract } \pi r^2 \text{ from each side.}$$

$$\frac{S - \pi r^2}{2\pi r} = \frac{2\pi rh}{2\pi r} \qquad \text{Divide each side by } 2\pi r.$$

or

$$h = \frac{S - \pi r^2}{2\pi r}$$

(b) Our first step is to isolate the two terms that contain w:

$$C = wh + wl + hl$$

$$C - hl = wh + wl \qquad \text{Subtract } hl \text{ from each side.}$$

$$C - hl = w(h + l) \qquad \text{Factor out } w \text{ on the right side.}$$

$$\frac{C - hl}{h + l} = w \qquad \text{Divide each side by } h + l.$$

or

$$w = \frac{C - hl}{h + l}$$

Linear Inequalities

A *linear inequality in one variable* is an inequality that can be written in one of the following forms:

$$ax + b > 0 \qquad ax + b < 0 \qquad ax + b \geq 0 \qquad ax + b \leq 0$$

where a and b are real numbers and $a \neq 0$.

Our approach to solving linear inequalities is similar to our method of solving linear equations—we perform operations on each side of the inequality to produce equivalent inequalities. The operations that produce equivalent inequalities are the same as for equations, with two important exceptions.

Operations That Produce
Equivalent Inequalities

> Adding (subtracting) the same quantity to (from) each side of an inequality produces an equivalent inequality.
>
> Multiplying or dividing each side of an inequality by the same *positive* quantity produces an equivalent inequality.
>
> Multiplying or dividing each side of an inequality by the same *negative* quantity produces an equivalent inequality if the sense (direction) of the inequality is reversed.

The solution of a linear inequality is an interval of real numbers.

EXAMPLE 4   Solving a Linear Inequality

Solve each inequality. Express the solution in interval notation.

(a) $7x + 5 \leq 3(x + 11)$ **(b)** $\dfrac{x}{3} < 7 + \dfrac{3x}{2}$

SOLUTION

(a) $7x + 5 \leq 3(x + 11)$

$\qquad 7x + 5 \leq 3x + 33$ Use the distributive property.

$\qquad 4x + 5 \leq 33$ Subtract $3x$ from each side.

$\qquad\quad 4x \leq 28$ Subtract 5 from each side.

$\qquad\quad\ x \leq 7$ Divide each side by 4.

The solution in interval notation is $(-\infty, 7]$.

(b) $\dfrac{x}{3} < 7 + \dfrac{3x}{2}$

$\quad\quad 2x < 42 + 9x$ Multiply each side by 6.

$\quad\quad -7x < 42$ Subtract $9x$ from each side.

$\quad\quad\quad x > -6$ Divide each side by -7 and reverse the inequality symbol.

The solution is $(-6, +\infty)$.

The compound inequality

$$-3 \le 2x - 1 \le 5$$

is equivalent to the two inequalities

$$-3 \le 2x - 1 \quad\quad \text{and} \quad\quad 2x - 1 \le 5$$

Solving the first inequality, we have

$\quad\quad -3 \le 2x - 1$

$\quad\quad -2 \le 2x$ Add 1 to each side.

$\quad\quad -1 \le x$ Divide each side by 2.

The second inequality is solved as follows:

$\quad\quad 2x - 1 \le 5$

$\quad\quad\quad 2x \le 6$ Add 1 to each side.

$\quad\quad\quad x \le 3$ Divide each side by 2.

The solution is

$$-1 \le x \quad\quad \text{and} \quad\quad x \le 3$$

which is equivalent to

$$-1 \le x \le 3$$

Notice that the steps in solving each inequality are identical. A more efficient way to solve a compound inequality is to deal with both inequalities at the same time, as in the next example.

● EXAMPLE 5   Solving a Compound Inequality

Solve each compound inequality. Express the solution in interval notation.

(a) $1 \le 3x + 4 \le 10$ **(b)** $-7 < \dfrac{1 - 3x}{2} < 5$

SOLUTION

Our goal is to isolate x in the middle member of each compound inequality.

(a) $1 \le 3x + 4 \le 10$

$-3 \le 3x \le 6$	Subtract 4 from the left, middle, and right members of the compound inequality.
$-1 \le x \le 2$	Divide each member of the compound inequality by 3.

The solution is $[-1, 2]$.

(b) $-7 < \dfrac{1 - 3x}{2} < 5$

$-14 < 1 - 3x < 10$	Multiply each member of the compound inequality by 2.
$-15 < -3x < 9$	Subtract 1 from each member of the compound inequality.
$5 > x > -3$	Divide each member of the compound inequality by -3 and reverse the inequality symbols.

or

$$-3 < x < 5$$

The solution is $(-3, 5)$.

There are two special cases associated with solving equations and inequalities:

1. Some equations and inequalities have no solution.
2. Some equations and inequalities have all real numbers as the set of solutions.

Examples of these special cases follow.

Some equations and inequalities have no solution. For example, when we solve the equation

$$2(3x - 1) = 6x + 5$$

for x, we get

$6x - 2 = 6x + 5$	
$-2 = 5$	Subtract $6x$ from each side.

Since this is a false statement, any number we substitute for x in the original equation will also give a false statement. This means the original equation has no solution.

In the same way, when we solve the inequality

$$2(3x - 1) > 6x + 5$$

we get

$6x - 2 > 6x + 5$	
$-2 > 5$	False

Since this statement is false, the original inequality has no solutions.

Some equations and inequalities have all real numbers as the set of solutions. For example, when we solve the inequality

$$2(3x - 1) < 6x + 5$$

we get

$$6x - 2 < 6x + 5$$
$$-2 < 5 \qquad \text{True}$$

Since this is a true statement, the original inequality is true for all real numbers.

An equation that is true for all real numbers for which each member is defined is an *identity*. For example, when solving the equation

$$2(3x - 1) = 6x - 2$$

we get

$$6x - 2 = 6x - 2 \quad \text{Use the distributive property.}$$

Since this equation is true for any real number x, the set of solutions to the original equation is the set of all real numbers.

We frequently write identities when we use properties. For instance, using the distributive property to simplify $2(3x - 1)$, we write the identity

$$2(3x - 1) = 6x - 2$$

When we use the commutative and associative properties to simplify $3 + (x + 2)$, we write the identity

$$3 + (x + 2) = x + 5$$

Applications

We conclude this section with two applications—one that involves a linear equation and one that involves a linear inequality.

◑ EXAMPLE 6   Application: Finding the Right Amount for a Mixture

A radiator contains 5 quarts of 50% antifreeze mixture. How much of this mixture must be drained and replaced with pure antifreeze to obtain a 90% mixture?

SOLUTION

A fundamental concept for this problem is that

$$\begin{pmatrix} \text{Amount of} \\ \text{antifreeze} \end{pmatrix} = \begin{pmatrix} \text{Percentage of} \\ \text{antifreeze} \end{pmatrix} \begin{pmatrix} \text{Volume of} \\ \text{mixture} \end{pmatrix}$$

We start by considering the amounts of (pure) antifreeze that get drained and replaced.

$$\begin{pmatrix} \text{Final} \\ \text{amount of} \\ \text{antifreeze} \end{pmatrix} = \begin{pmatrix} \text{Amount} \\ \text{of original} \\ \text{antifreeze} \end{pmatrix} - \begin{pmatrix} \text{Amount} \\ \text{of drained} \\ \text{antifreeze} \end{pmatrix} + \begin{pmatrix} \text{Amount} \\ \text{of added} \\ \text{antifreeze} \end{pmatrix}$$

$$\begin{pmatrix} \text{Final} \\ \text{percentage of} \\ \text{antifreeze} \end{pmatrix}\begin{pmatrix} \text{Volume} \\ \text{of final} \\ \text{mixture} \end{pmatrix} = \begin{pmatrix} \text{Percentage} \\ \text{of original} \\ \text{antifreeze} \end{pmatrix}\begin{pmatrix} \text{Volume} \\ \text{of original} \\ \text{mixture} \end{pmatrix} - \begin{pmatrix} \text{Percentage} \\ \text{of drained} \\ \text{antifreeze} \end{pmatrix}\begin{pmatrix} \text{Volume} \\ \text{of drained} \\ \text{mixture} \end{pmatrix} + \begin{pmatrix} \text{Percentage} \\ \text{of added} \\ \text{antifreeze} \end{pmatrix}\begin{pmatrix} \text{Volume} \\ \text{of added} \\ \text{mixture} \end{pmatrix}$$

$$(0.90)(5) = (0.50)(5) \qquad - (0.50)\begin{pmatrix} \text{Volume} \\ \text{of drained} \\ \text{mixture} \end{pmatrix} + (1.00)\begin{pmatrix} \text{Volume} \\ \text{of added} \\ \text{mixture} \end{pmatrix}$$

If we let x represent the volume (in quarts) of drained mixture, then the volume of added mixture is also x. This gives us

$$(0.90)(5) = (0.50)(5) - (0.50)x + (1.00)x$$

Solving for x, we have

$$4.5 = 2.5 + 0.50x \quad \text{Simplify each side.}$$
$$45 = 25 + 5x \quad \text{Multiply each side by 10.}$$
$$20 = 5x \quad \text{Subtract 25 from each side.}$$
$$4 = x \quad \text{Divide each side by 5.}$$

Four quarts must be drained and replaced with pure antifreeze.

EXAMPLE 7   Application: Salary and Commission

Each sales representative for a health club decides on one of two salary options: Option A is a flat salary of $750 per week. Option B is $580 per week plus a commission of $15 for every new membership sold. During a particular week, how many memberships must be sold to make the salary for option B greater than that for option A? Over a period of many weeks, how many memberships must be sold per week to make the salary for option B greater than that for option A?

SOLUTION

Comparing the two options, we have

$$\begin{pmatrix} \text{Weekly salary earned} \\ \text{under option B} \end{pmatrix} > \begin{pmatrix} \text{Weekly salary earned} \\ \text{under option A} \end{pmatrix}$$

$$15\begin{pmatrix} \text{Number of memberships} \\ \text{sold per week} \end{pmatrix} + 580 > 750$$

Letting x represent the number of memberships sold per week, we get

$$15x + 580 > 750$$
$$15x > 170 \quad \text{Subtract 580 from each side.}$$
$$x > 11\tfrac{1}{3} \quad \text{Divide each side by 15.}$$

During a particular week, x must be a positive integer, so at least 12 memberships must be sold to make the salary for option B greater than that for option A. Over a period of many weeks, an average of more than $11\frac{1}{3}$ memberships must be sold per week to make the salary for option B greater than that for option A.

EXERCISE SET 6

A

In Exercises 1–22, solve each equation.

1. $12 - x = 5$ **2.** $14 - x = 4$

3. $6x + 11 = 3$ **4.** $4x + 15 = 9$

5. $2 - 3x = 2$ **6.** $1 - 5x = 1$

7. $2(5x + 4) = 7x + 17$ **8.** $3(x + 3) = 10x - 9$

9. $12 + 6x(x + 1) = (3x + 7)(2x - 2)$

10. $(4x + 1)(x + 2) = (2x - 5)(2x + 2) + 3x$

11. $x^2 - 3(x + 10) = x(x - 1) + 8$

12. $4x(x - 3) + 9 = x^2 - 3x(2 - x)$

13. $\frac{1}{2}x + 3 = 7$ **14.** $\frac{1}{4}x - 5 = 11$

15. $\frac{3x}{4} + 1 = \frac{x}{2}$ **16.** $\frac{5x}{9} + 2 = \frac{x}{3}$

17. $1 - \frac{5 - x}{6} = \frac{x + 2}{4}$ **18.** $2 - \frac{4 - x}{9} = \frac{x + 7}{6}$

19. $2.6x - 3 = 0.8(2x - 1)$

20. $4.9x - 8 = 3.7(x + 4)$

21. $0.2x + 0.7(6 - x) = 3.2$

22. $0.1x + 0.5(12 - x) = 4.8$

In Exercises 23–34, solve each inequality. Express the solution in interval notation.

23. $4x - 3 \le 9$ **24.** $6x - 7 \le 11$

25. $11 - 2x \ge 7$ **26.** $16 - 5x \ge 1$

27. $8x - 2 \le 2(x + 5)$ **28.** $5x + 9 < 3(x - 1)$

29. $3x + 6x(x + 2) > (2x + 1)(3x + 2)$

30. $x + 5x(x - 1) > (x - 1)(5x - 3)$

31. $\frac{2x}{3} > 5 - \frac{x}{6}$ **32.** $\frac{3x}{4} > 10 - \frac{x}{2}$

33. $9.2x - 6 \le 2.5(3x + 1)$

34. $0.8x + 2 \le 1.4(5x - 3)$

In Exercises 35–40, solve each compound inequality. Express the solution in interval notation.

35. $3 \le 2x - 5 \le 11$ **36.** $7 \le 3x + 4 \le 16$

37. $-1 < \frac{5x + 4}{3} < 8$ **38.** $-7 \le \frac{3x - 1}{4} \le 5$

39. $-7 < 1 - 2x \le 0$ **40.** $-10 \le 2 - 3x < 0$

B

In Exercises 41–54, solve each equation for the specified variable.

41. $PV = nRT$, for R **42.** $V = wlh$, for w

43. $P = 2w + 2l$, for w **44.** $y = mx + b$, for x

45. $x = 5400 - 800p$, for p

46. $x = 6500 - 260p$, for p

47. $\frac{1}{R} = \frac{1}{R_1} + \frac{1}{R_2}$, for R **48.** $\frac{1}{f} = \frac{1}{d_1} + \frac{1}{d_2}$, for f

49. $A = (w - 2)(l - 4)$, for l

50. $A = (w - 3)(l - 4)$, for w

51. $A = P + Prt$, for P

52. $S = 2(wh + wl + hl)$, for h

53. $z = \frac{x - \mu}{s}$, for x **54.** $z = \frac{x - \mu}{s}$, for μ

In Exercises 55–60, determine the values of x for which each expression is a real number.

55. $\sqrt{x - 3}$ **56.** $\sqrt{x + 2}$ **57.** $\sqrt{9 - 2x}$

58. $\sqrt{8 - 3x}$ **59.** $\frac{3}{\sqrt{10 - 4x}}$ **60.** $\frac{8}{\sqrt{6 - 9x}}$

61. A radiator contains 5 quarts of 20% antifreeze mixture. How much of this mixture must be drained and replaced with pure antifreeze to obtain a 60% mixture?

62. A tank contains 45 gallons of 20% acid mixture. How much of this mixture must be drained and replaced with pure acid to obtain a 70% mixture?

63. Two car rental agencies have the following weekly rates: Agency A charges $100 per week plus $0.20 per mile. Agency B charges $135 per week with no charge for mileage. How many miles must be driven for the weekly cost to be greater for a car rented from agency A?

64. An investor wants to place part of $30,000 in a savings account that pays 5% per year and the rest in high-risk stocks that pay 12% per year. If the total annual return on the investment must be at least $2000, how much can be invested in the stocks?

65. A candy store sells boxes of candy containing chocolates and caramels. The caramels cost $0.15 each and the chocolates cost $0.25 each. How many of the chocolates can be put in each box if it is to consist of 30 pieces for a total cost of $5.00 or less?

66. A carpet cleaning specialist can clean the carpet of a typical house in 6 hours. The specialist hires a helper, and together they can do the same job in 4 hours. If the helper works alone, how long will it take the helper to clean the carpet of a typical house?

67. An old printer can print 19 mailing labels per minute, and a newer one can print 45 labels per minute. How long will it take both printers together to print a list of 2000 labels?

68. Ohm's law states that the voltage E (in volts) in an electric circuit is the product of the current I (in amperes) and the resistance R (in ohms):

$$E = IR$$

If an electric appliance lists a resistance of 15 ohms, and the voltage varies from 110 to 130 volts, what is the range of current that the appliance will draw?

C

69. Explain what is wrong with the following solution of the inequality:

$$\frac{5}{x} \geq 2$$

$$\frac{5}{x} \cdot x \geq 2 \cdot x$$

$$5 \geq 2x$$

$$\frac{5}{2} \geq x$$

70. Solve the inequality in Problem 69.

71. Suppose you leave your house the same time each day to drive to work. One day you drive to work averaging 54 miles per hour and arrive 2 minutes early. Another day you average 50 miles per hour and arrive 2 minutes late. How far do you live from work?

QUADRATIC EQUATIONS

A *quadratic equation* is an equation that can be written in the form

$$ax^2 + bx + c = 0$$

where a, b, and c are real numbers and $a \neq 0$. This form is the *general form* of a quadratic equation. (The condition that $a \neq 0$ ensures that the expression on the left side of the equation is a second-degree polynomial.)

In this section we discuss several methods of solving quadratic equations.

Solving Quadratic Equations by Factoring

The method of factoring uses the fact that if the product of two real numbers is 0, then at least one of the factors is 0.

Zero Factor Property

For real numbers A and B,
$AB = 0$ if and only if $A = 0$ or $B = 0$

■ **NOTE:** *The zero factor property is valid only for products that equal* 0.

We can use the zero factor property to solve a quadratic equation in the general form $ax^2 + bx + c = 0$ if the left side can be factored.

EXAMPLE 1   Solving by Factoring

Solve each equation by factoring.

(a) $2x^2 + 9x - 5 = 0$ **(b)** $5x^2 - 7x = 3x$ **(c)** $9 + 4x^2 = 12x$

SOLUTION

(a)

$\quad 2x^2 + 9x - 5 = 0$ Equation is in general form.

$\quad (2x - 1)(x + 5) = 0$ Factor the left side of the equation.

$2x - 1 = 0 \mid x + 5 = 0$ Use the zero factor property.

$\quad 2x = 1 \mid \quad x = -5$ Solve the two resulting equations.

$\quad x = \frac{1}{2} \mid$

The solutions are $\frac{1}{2}$ and -5.

(b) $5x^2 - 7x = 3x$

$\quad 5x^2 - 10x = 0$ Subtract $3x$ from each side to write the equation in general form.

$\quad 5x(x - 2) = 0$ Factor $5x$ out of the left side.

$5x = 0 \mid x - 2 = 0$ Use the zero factor property.

$x = 0 \mid \quad x = 2$ Solve the two resulting equations.

The solutions are 0 and 2.

(c) $9 + 4x^2 = 12x$

$\quad 4x^2 - 12x + 9 = 0$ Subtract $12x$ from each side and write the equation in general form.

$\quad (2x - 3)(2x - 3) = 0$ Factor the left side of the equation.

$\quad 2x - 3 = 0$ Since the factors are identical, set the repeated factor equal to 0.

$\quad x = \frac{3}{2}$ Solve the resulting equation.

The solution is $\frac{3}{2}$.

Solving Quadratic Equations by Taking Square Roots

Consider the equation

$$x^2 = \tfrac{9}{4}$$

It takes several steps to solve this equation by factoring. Instead, we observe that the solutions are the square roots of $\tfrac{9}{4}$:

$$x = \sqrt{\tfrac{9}{4}} = \tfrac{3}{2} \quad \text{or} \quad x = -\sqrt{\tfrac{9}{4}} = -\tfrac{3}{2}$$

which typically is written as

$$x = \pm\sqrt{\tfrac{9}{4}} = \pm\tfrac{3}{2}$$

Many quadratic equations can be conveniently solved this way, using the following property:

Square Root Property

> For any nonnegative real number k, the equation
>
> $$u^2 = k$$
>
> is equivalent to
>
> $$u = \pm\sqrt{k}$$

The technique of solving quadratic equations by taking square roots amounts to writing the equation in the form $u^2 = k$, and then applying the square root property.

● EXAMPLE 2 Solving by Taking Square Roots

Solve each equation by taking square roots.

(a) $3x^2 = 24$ **(b)** $9(x - 2)^2 - 5 = 0$ **(c)** $4(x - 3)^2 + 8 = 0$

SOLUTION

(a) $3x^2 = 24$

$$\begin{aligned}
x^2 &= 8 && \text{Divide each side by 3.} \\
x &= \pm\sqrt{8} && \text{Take square roots of each side.} \\
x &= \pm 2\sqrt{2} && \text{Simplify the radical.}
\end{aligned}$$

The solutions are $2\sqrt{2}$ and $-2\sqrt{2}$.

(b) $9(x - 2)^2 - 5 = 0$

$$9(x - 2)^2 = 5 \qquad \text{Subtract 5 from each side.}$$

$$(x - 2)^2 = \tfrac{5}{9} \qquad \text{Divide each side by 9.}$$

$$x - 2 = \pm\sqrt{\tfrac{5}{9}} \qquad \text{Take square roots of each side.}$$

$$x = 2 \pm \frac{\sqrt{5}}{3} \qquad \text{Add 2 to each side and simplify the radical.}$$

$$x = \frac{6}{3} \pm \frac{\sqrt{5}}{3} = \frac{6 \pm \sqrt{5}}{3} \qquad \text{Combine terms on the right side.}$$

The solutions are $\dfrac{6 + \sqrt{5}}{3}$ and $\dfrac{6 - \sqrt{5}}{3}$.

(c) $4(x - 3)^2 + 8 = 0$

$$4(x - 3)^2 = -8 \qquad \text{Subtract 8 from each side.}$$

$$(x - 3)^2 = -2 \qquad \text{Divide each side by 4.}$$

$$x - 3 = \pm\sqrt{-2} \qquad \text{Take square roots of each side.}$$

Since $\sqrt{-2}$ is not a real number, the equation has no real solutions.

Solving Quadratic Equations by the Quadratic Formula

The methods of factoring or taking square roots only apply directly in certain instances. A more general method of solving quadratic equations involves manipulating an equation in the form

$$ax^2 + bx + c = 0$$

into the form

$$(x - h)^2 = k$$

This process is called *completing the square*.

For example, consider the equation

$$x^2 - 8x = 1$$

If we add 16 to each side,

$$x^2 - 8x + 16 = 1 + 16$$

the left side factors as a perfect square:

$$(x - 4)^2 = 17$$

The key is knowing what to add to each side (in this case 16) to make the left side a trinomial that factors as a perfect square. In general, if the left side of the equation is

$$x^2 + nx$$

■ **NOTE:** *Here's a simple rule for determining what to add to $x^2 + nx$: Take half of the coefficient of the x term, and then square it.*

we add $(n/2)^2$ to form a perfect square trinomial:

$$x^2 + nx + \left(\frac{n}{2}\right)^2 = \left(x + \frac{n}{2}\right)^2$$

Once we complete the square, we can solve the quadratic equation by taking square roots.

■ **EXAMPLE 3   Solving by Completing the Square**

Solve each equation by completing the square.

(a) $x^2 + 12x + 33 = 0$ **(b)** $4x^2 = 8x + 41$

SOLUTION

(a) $x^2 + 12x + 33 = 0$

$\qquad\quad x^2 + 12x = -33$ Subtract 33 from each side.

$\qquad x^2 + 12x + 36 = -33 + 36$ Half of 12 is 6, and the square of 6 is 36. Add 36 to each side.

$\qquad\qquad\quad (x + 6)^2 = 3$ The left side factors as a perfect square.

$\qquad\qquad\quad x + 6 = \pm\sqrt{3}$ Take square roots of each side.

$\qquad\qquad\qquad x = -6 \pm \sqrt{3}$ Subtract 6 from each side.

The solutions are $-6 + \sqrt{3}$ and $-6 - \sqrt{3}$.

(b) $4x^2 = 8x + 41$

$\qquad\qquad 4x^2 - 8x = 41$ Subtract 8x from each side.

$\qquad\qquad x^2 - 2x = \frac{41}{4}$ Divide each side by 4.

$\qquad x^2 - 2x + 1 = \frac{41}{4} + 1$ Half of -2 is -1, and the square of -1 is 1. Add 1 to each side.

$\qquad\qquad (x - 1)^2 = \frac{45}{4}$ The left side is a perfect square.

$\qquad\qquad\quad x - 1 = \pm\sqrt{\frac{45}{4}}$ Take square roots of each side.

$\qquad\qquad\qquad x = 1 \pm \sqrt{\frac{45}{4}}$ Add 1 to each side.

$\qquad\qquad\qquad x = 1 \pm \dfrac{3\sqrt{5}}{2} = \dfrac{2 \pm 3\sqrt{5}}{2}$ Simplify.

The solutions are $\dfrac{2 + 3\sqrt{5}}{2}$ and $\dfrac{2 - 3\sqrt{5}}{2}$.

The method of completing the square can be applied to solve the quadratic equation

$$ax^2 + bx + c = 0$$

in general for x. The result is the following formula, called the *quadratic formula*, for the solution of any quadratic equation:

Quadratic Formula

The solutions of the equation $ax^2 + bx + c = 0$, with $a \neq 0$, are

$$x = \frac{-b \pm \sqrt{b^2 - 4ac}}{2a}$$

NOTE: *The quadratic formula is used so widely that you should commit it to memory.*

The derivation of the quadratic formula is left as Exercise 66.

EXAMPLE 4   Solving by Using the Quadratic Formula

Solve each equation using the quadratic formula.

(a) $\frac{2}{3}x^2 + 1 = 2x$ **(b)** $x(4x + 1) + 3 = 0$ **(c)** $2x^2 - 9x - 5 = 0$

SOLUTION

(a) $\frac{2}{3}x^2 + 1 = 2x$

$\frac{2}{3}x^2 - 2x + 1 = 0$ Subtract $2x$ from each side to write the equation in general form.

$2x^2 - 6x + 3 = 0$ Clear fractions by multiplying each side by 3.

$x = \frac{-(-6) \pm \sqrt{(-6)^2 - 4(2)(3)}}{2(2)}$ Use the quadratic formula with $a = 2$, $b = -6$, and $c = 3$.

$x = \frac{6 \pm \sqrt{12}}{4} = \frac{6 \pm 2\sqrt{3}}{4} = \frac{2(3 \pm \sqrt{3})}{2 \cdot 2}$ Simplify.

The solutions are $\frac{3 + \sqrt{3}}{2}$ and $\frac{3 - \sqrt{3}}{2}$.

(b) $x(4x + 1) + 3 = 0$

$4x^2 + x + 3 = 0$ Use the distributive property to write the equation in general form.

$x = \frac{-1 \pm \sqrt{1^2 - 4(4)(3)}}{2(4)}$ Use the quadratic formula with $a = 4$, $b = 1$, and $c = 3$.

$x = \frac{-1 \pm \sqrt{-47}}{8}$

Since the square root of a negative number is not a real number,

$$\frac{-1 + \sqrt{-47}}{8} \quad \text{and} \quad \frac{-1 - \sqrt{-47}}{8}$$

are not real numbers. The equation has no real solutions.

(c) $2x^2 + 9x - 5 = 0$

$$x = \frac{-9 \pm \sqrt{9^2 - 4(2)(-5)}}{2(2)}$$ Use the quadratic formula with $a = 2$, $b = 9$, and $c = -5$.

$$x = \frac{-9 \pm \sqrt{121}}{4} = \frac{-9 \pm 11}{4}$$

Writing the last expression as two separate expressions, the solutions are

$$x = \frac{-9 + 11}{4} = \frac{1}{2} \quad \text{and} \quad x = \frac{-9 - 11}{4} = -5$$

■ **NOTE:** *Since the solutions of this quadratic equation are rational numbers, the equation can be solved by factoring. In fact, we solved this equation by factoring in part (a) of Example 1.*

In the quadratic formula,

$$x = \frac{-b \pm \sqrt{b^2 - 4ac}}{2a}$$

the radicand

$$b^2 - 4ac$$

can be used to determine the number of real solutions of a quadratic equation. For instance, the solutions of $3x^2 + 6x + 5 = 0$ are given by

$$x = \frac{-6 \pm \sqrt{6^2 - 4(3)(5)}}{2(3)}$$

The radicand is

$$b^2 - 4ac = 6^2 - 4(3)(5) = -24$$

Since this expression is negative, the quadratic equation has no real solutions.

As another example, if we use the quadratic formula to solve the equation $4x^2 - 12x + 9 = 0$, we find that the radicand is

$$b^2 - 4ac = (-12)^2 - 4(4)(9) = 0$$

This implies that the equation has exactly one real solution, namely

$$x = \frac{12 \pm \sqrt{0}}{2(4)} = \frac{3}{2}$$

You may have recognized this quadratic equation from part (c) of Example 1, where we found by factoring that the only solution was $\frac{3}{2}$.

The following box summarizes how to determine the number of real solutions of a quadratic equation:

The Discriminant

> The expression $b^2 - 4ac$ is the **discriminant** of the quadratic equation $ax^2 + bx + c = 0$.
>
> **1.** If $b^2 - 4ac > 0$, then the quadratic equation has two distinct real roots.
> **2.** If $b^2 - 4ac = 0$, then the quadratic equation has exactly one real root.
> **3.** If $b^2 - 4ac < 0$, then the quadratic equation has no real roots.

Applications

There are many applications that require solving quadratic equations. The first application we consider makes use of the method of taking square roots; the second application uses the quadratic formula.

EXAMPLE 5 Application: Speed of a Falling Object

For a sales promotion, a sports equipment manufacturer hires a well-known major league baseball player to catch a baseball dropped from the top of the Arco Tower in Denver, Colorado. The catch is to be made at ground level, and the Arco Tower is 527 feet high. The speed s (in feet per second) of an object that falls d feet from rest is given by

$$s^2 = 64d$$

Determine the speed of the baseball immediately before the moment of impact.

SOLUTION
Since the ball falls 527 feet, we have

$$s^2 = 64(527) = 33{,}728$$
$$s = \pm\sqrt{33{,}728}$$

Since speed is nonnegative, the speed is $\sqrt{33{,}728}$ feet per second, or approximately 184 feet per second.

NOTE: *A speed of 184 feet per second is over 125 miles per hour. Even a major league player should question the wisdom of attempting such a catch.*

EXAMPLE 6 Application: Finding the Width of a Border

The owner of a corner lot plans to sell a strip of uniform width along two of the sides for road and sidewalk development. The lot is 28 yards by 40 yards (Figure 13). If the owner wants the remaining piece to have an area of 900 square yards, how wide should the strip be?

SOLUTION
We want to determine the width of the strip so that the area of the remaining (inner) rectangle is 900 square yards.

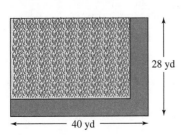

28 yd

40 yd

Figure 13

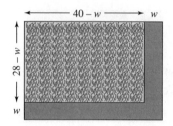

Figure 14

$$\left(\begin{array}{c}\text{Length of remaining}\\ \text{rectangle}\end{array}\right)\left(\begin{array}{c}\text{Width of remaining}\\ \text{rectangle}\end{array}\right) = 900$$

If we let w represent the width of the border (Figure 14), we have

$$(40 - w)(28 - w) = 900$$

Simplifying and then solving for w, we get

$w^2 - 68w + 1120 = 900$	Use FOIL.
$w^2 - 68w + 220 = 0$	Write the equation in general form.
$w = \dfrac{68 \pm \sqrt{(-68)^2 - 4(1)(220)}}{2(1)}$	Use the quadratic formula with $a = 1$, $b = -68$, $c = 220$.
$= \dfrac{68 \pm 12\sqrt{26}}{2}$	Simplify the radical.
$= 34 \pm 6\sqrt{26}$	Factor 2 out of the numerator and reduce.

Since $34 + 6\sqrt{26}$ is more than the width of the original lot, $34 + 6\sqrt{26}$ is not a feasible solution. The width of the border is $34 - 6\sqrt{26}$ yards. This is approximately 3.4 yards.

EXERCISE SET 7

A

In Exercises 1–10, solve each equation by factoring.

1. $x^2 + 5x - 6 = 0$ **2.** $x^2 - 5x + 6 = 0$

3. $2(x^2 + 3) = 7x$ **4.** $2(x^2 - 1) = 3x$

5. $6x^2 = 20x$ **6.** $12x^2 = 15x$

7. $4x^2 = 9$ **8.** $9x^2 = 25$

9. $x^2 - 14x + 49 = 0$ **10.** $x^2 + 8x + 16 = 0$

In Exercises 11–22, solve each equation by taking square roots.

11. $4x^2 = 25$ **12.** $16x^2 = 9$

13. $2x^2 = 6$ **14.** $4x^2 = 20$

15. $(x - 5)^2 = 12$ **16.** $(x - 6)^2 = 18$

17. $4(x + 1)^2 = 13$ **18.** $9(x + 2)^2 = 10$

19. $(2x + 1)^2 + 4 = 10$ **20.** $(2x - 3)^2 + 1 = 16$

21. $(x - 3)^2 + 7 = 5$ **22.** $(x + 2)^2 + 6 = 3$

In Exercises 23–30, solve each equation by completing the square.

23. $x^2 + 6x = 1$ **24.** $x^2 + 8x = 3$

25. $x^2 - 4x + 2 = 0$ **26.** $x^2 - 10x + 22 = 0$

27. $4x^2 + 8x = 1$ **28.** $3x^2 + 12x = 3$

29. $2x^2 - 6x - 7 = 0$ **30.** $2x^2 + 2x - 5 = 0$

In Exercises 31–38, solve each equation using the quadratic formula.

31. $x^2 + 3x + 1 = 0$ **32.** $x^2 + 7x + 2 = 0$

33. $x^2 - 5x = 5$ **34.** $x^2 = 7x - 1$

35. $2x^2 - 5 = 6x$ **36.** $2x^2 - 3 = 4x$

37. $5x^2 = 2x - 6$ **38.** $3x^2 = 4x - 9$

B

In Exercises 39–58, solve each equation using any method.

39. $x^2 + x - 1 = 0$

40. $x^2 + x - 3 = 0$

41. $x^2 + 3x - 4 = 0$

42. $x^2 + 2x - 8 = 0$

43. $9x^2 - 5 = 0$

44. $12x^2 - 21 = 0$

45. $4x^2 + 12x + 9 = 0$

46. $9x^2 - 6x + 1 = 0$

47. $x^2 = 5x + 1$

48. $x^2 = 3x + 3$

49. $2x^2 + 3x + 4 = 0$

50. $3x^2 + 2x + 6 = 0$

51. $(2x - 1)(x - 4) = 22$

52. $(2x + 1)(x - 2) = 18$

53. $\frac{1}{3}x^2 + x - \frac{3}{2} = 0$

54. $\frac{2}{3}x^2 + 2x - \frac{5}{2} = 0$

55. $2x^2 = 0.5x + 1.5$

56. $0.1x^2 + 2.4 = 1.1x$

57. $x^2 - 2\sqrt{3}x = 1$

58. $x^2 + \sqrt{17}x = 4$

59. A printed poster is to have 3 inch margins at the top and bottom and 2 inch margins on the sides. The area of the printed portion is 48 square inches, and the total area of the poster is 160 square inches. Find the dimensions of the poster (Figure 15).

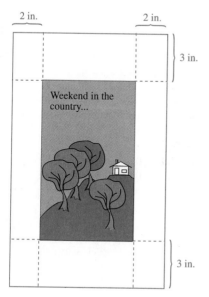

Figure 15

60. If an object is thrown upward with an initial velocity of 60 feet per second, the height h (in feet) t seconds after it is released is given by

$$h = 60t - 16t^2$$

(a) When is the object 80 feet high?

(b) When will it hit the ground?

61. An open-top box is to be constructed from a square sheet of cardboard by cutting squares with sides of 4 inches from each corner and turning up the edges (Figure 16). If the box is to hold a volume of 1536 cubic inches, determine the dimensions of the cardboard sheet.

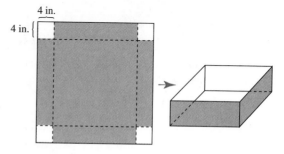

Figure 16

62. A boat that can travel 18 miles per hour in still water travels 32 miles upstream and returns downstream in a total of 4 hours. What is the speed of the current?

C

63. A polygon with n sides has a total of $\frac{1}{2}n(n - 3)$ diagonals. Determine the number of sides for a polygon with a total of 135 diagonals. Explain why it is impossible for a polygon to have a total of 200 diagonals.

64. For the equation $x^2 + 2y^2 + xy - y = 0$:

(a) Solve for x in terms of y.

(b) Solve for y in terms of x.

65. Solve: $x^4 - 12x^2 + 33 = 0$

66. **Derivation of the Quadratic Formula**

(a) Show that the quadratic equation $ax^2 + bx + c = 0$ can be written as

$$x^2 + \frac{b}{a}x = -\frac{c}{a}$$

(b) Complete the square to get

$$\left(x + \frac{b}{2a}\right)^2 = \frac{b^2 - 4ac}{4a^2}$$

(c) Solve for x by taking square roots.

67. Find all values of k such that

$$kx^2 - x + k = 1$$

has exactly one real root.

68. Suppose r_1 and r_2 are two distinct real roots of the equation $ax^2 + bx + c = 0$. Show that

$$\frac{1}{r_1} + \frac{1}{r_2} = -\frac{b}{c}$$

QUICK REFERENCE

Topic	Page	Remarks
Integers, rational numbers, irrational numbers, real numbers	2–3	The set of integers is the set of numbers $$\dots, -4, -3, -2, -1, 0, 1, 2, 3, 4, \dots$$ The set of rational numbers is the set of all numbers that can be expressed as a ratio of two integers. The decimal representation of a rational number either terminates or repeats. The set of irrational numbers is the set of all real numbers that are not rational.
Commutative, associative, identity, inverse, and distributive properties	4	See the box on page 4.
Inequality notation and interval notation	5–6	See the table on page 7.
Absolute value	7–8	The absolute value of real number a, denoted $\lvert a \rvert$, is the distance between 0 and a on the number line. Equivalently, $$\lvert a \rvert = \begin{cases} a & \text{if } a > 0 \\ 0 & \text{if } a = 0 \\ -a & \text{if } a < 0 \end{cases}$$
Equations and inequalities with absolute value	8–9	See the table on page 9. $\lvert u \rvert < a$ is equivalent to $-a < u < a$ $\lvert u \rvert \le a$ is equivalent to $-a \le u \le a$ $\lvert u \rvert = a$ is equivalent to $u = a$ or $u = -a$ $\lvert u \rvert > a$ is equivalent to $u < -a$ or $u > a$ $\lvert u \rvert \ge a$ is equivalent to $u \le -a$ or $u \ge a$
Distance between two numbers	9	The distance between two real numbers a and b, denoted $d(a, b)$, is given by $$d(a, b) = \lvert a - b \rvert$$

Topic	Page	Remarks
Zero exponent	13	If a is a nonzero real number, then $a^0 = 1$.
Negative exponents	13	If n is a positive integer and $a \neq 0$, then $$a^{-n} = \frac{1}{a^n}$$
Laws of integer exponents	14	Product law: $a^m a^n = a^{m+n}$ Quotient law: $\dfrac{a^m}{a^n} = a^{m-n}, a \neq 0$ Power law: $(a^m)^n = a^{mn}$ Distributive laws: $(ab)^n = a^n b^n$ $\left(\dfrac{a}{b}\right)^n = \dfrac{a^n}{b^n}, b \neq 0$
Scientific notation	17	Scientific notation is a practical way to express very large numbers and very small numbers. Scientific notation is used with the laws of exponents to perform calculations with large and small numbers; see Example 3 on page 17.
Polynomial	18	A polynomial in x is an expression that can be written in the form $a_n x^n + a_{n-1} x^{n-1} + \cdots + a_2 x^2 + a_1 x + a_0$, where n is a nonnegative integer and $a_n \neq 0$.
Special factoring formulas	26	Perfect square: $u^2 + 2uv + v^2 = (u + v)^2$ $u^2 - 2uv + v^2 = (u - v)^2$ Difference of squares: $u^2 - v^2 = (u - v)(u + v)$ Difference of cubes: $u^3 - v^3 = (u - v)(u^2 + uv + v^2)$ Sum of cubes: $u^3 + v^3 = (u + v)(u^2 - uv + v^2)$
Rational expression	29	A rational expression is the ratio of two polynomials. Simplifying, adding, subtracting, multiplying, and dividing rational expressions is similar to performing these operations with rational numbers.
Fundamental principle of rational expressions	30	If P, Q, and R are polynomials with $Q \neq 0$ and $R \neq 0$, then $$\frac{PR}{QR} = \frac{P}{Q}$$
Properties of radicals	41	The properties listed in the box on page 41 are fundamental for simplifying radical expressions.
Rational exponents	42–43	For the real number a and integers m and n, $n > 1$: $$a^{1/n} = \sqrt[n]{a}$$ $$a^{m/n} = (\sqrt[n]{a})^m = \sqrt[n]{a^m}$$
Operations that produce equivalent equations	48	Adding (subtracting) the same quantity to (from) each side of an equation produces an equivalent equation. Multiplying or dividing each side of an equation by the same nonzero quantity produces an equivalent equation.

Continued

Topic	Page	Remarks
Linear equation in one variable	48	A linear equation in the variable x is an equation that can be expressed in the form $ax + b = 0$, where a and b are real numbers and $a \neq 0$.
Linear inequality in one variable	52	A linear inequality in the variable x is an inequality that can be expressed in one of the following forms: $ax + b < 0$, $ax + b > 0$, $ax + b \leq 0$, or $ax + b \geq 0$, where a and b are real numbers and $a \neq 0$.
Operations that produce equivalent inequalities	52	Adding (subtracting) the same quantity to (from) each side of an inequality produces an equivalent inequality. Multiplying or dividing each side of an inequality by the same positive quantity produces an equivalent inequality. Multiplying or dividing each side of an inequality by the same negative quantity produces an equivalent inequality if the sense of the inequality is reversed.
Quadratic equation in one variable	58	A quadratic equation in the variable x is an equation that can be written in the form $ax^2 + bx + c = 0$, where a, b, and c are real numbers and $a \neq 0$.
Zero factor property	59	For real numbers A and B, $$AB = 0 \quad \text{if and only if} \quad A = 0 \quad \text{or} \quad B = 0$$
Square root property	60	For any nonnegative real number k, the equation $$u^2 = k \quad \text{is equivalent to} \quad u = \pm\sqrt{k}$$
Quadratic formula	63	The solutions to $ax^2 + bx + c = 0$ are $$x = \frac{-b \pm \sqrt{b^2 - 4ac}}{2a}$$

◉ MISCELLANEOUS EXERCISES

In Exercises 1 and 2, for each set of real numbers, determine which are integers, which are rational numbers, and which are irrational numbers. Plot the numbers on a coordinate line.

1. $-3, 5.1, -\sqrt{30}, 2\pi, 9, \dfrac{7}{13}$

2. $-4, \dfrac{\pi}{2}, -1.2, 0, \dfrac{11}{7}, 7$

In Exercises 3 and 4, identify the property of real numbers that is exhibited in each statement.

3. (a) $2x + 5 = 5 + 2x$ (b) $5(x - y) = 5x - 5y$

     (c) $0 + \dfrac{3}{4}b = \dfrac{3}{4}b$

4. (a) $x(x + 2) = (x + 2)x$

     (b) $(x + 2) \cdot \dfrac{1}{x + 2} = 1$

     (c) $3(2x) = (3 \cdot 2)x$

In Exercises 5–8, sketch the graph of the set described by each inequality and describe the set with interval notation.

5. (a) $x \geq 2$ (b) $1 \leq x < 5$

6. (a) $x < -3$ (b) $-3 < x \leq 4$

7. (a) $|x| < 2$ (b) $|x - 2| > 0.5$

8. (a) $|x| \geq 5$ (b) $|x + 1| < 0.2$

In Exercises 9–16, simplify each expression. (Assume all variables are positive.)

9. (a) $-3x^{-3}(2x)^2$ (b) $\dfrac{y^0}{2x^{-1}}$

10. (a) $5x^{-3}(-x^2)^3$ (b) $\dfrac{1}{5a^0b^{-1}}$

11. (a) $(36t^4)^{-1/2}$ (b) $\dfrac{12r^3t^{-2}}{3r^{-6}t^5}$

12. (a) $(8t^{-6})^{1/3}$ (b) $\dfrac{-m^4n^{-1}}{2m^{-1}n^4}$

13. (a) $\sqrt[3]{24a^8}$ (b) $\sqrt[3]{\dfrac{-2x^{13}}{y^{12}}}$

14. (a) $\sqrt[4]{16t^6}$ (b) $\sqrt[3]{\dfrac{5b^{10}}{(-c)^6}}$

15. (a) $\sqrt[4]{\sqrt[3]{x^{24}}}$ (b) $(x^5\sqrt{x^{-4}})^{-1/3}$

16. (a) $\sqrt{\sqrt[5]{32x^0}}$ (b) $(z^5\sqrt{z^{-2}})^{1/4}$

In Exercises 17–24, write each expression in simplified form.

17. $(2x^3 + 3x^2 - 10x + 3) - 2(x^2 - 3x + 4)$

18. $3(x^3 + 2x - 6) - (x^2 - 12x + 1)$

19. $(a^2 + 5a)(4a - 1)$

20. $x(x^2 - 1)(x + 4)$

21. $(x + 5)(2x - 3) - x^2$

22. $6x^2 + (-3x + 4)(2x + 3)$

23. $(2x^{1/2} - x^{5/2})^2$

24. $(a^{3/2} - a^{1/2})(a^{1/2} + a^{3/2})$

In Exercises 25–34, factor each polynomial completely.

25. $a^3b^4 + 7a^4b^2 - a^2b^5$

26. $(x^2 + 6)x^3 - (x^2 + 6)x^2$

27. $16s^4 - t^2$

28. $m^4 - 9n^2$

29. $4t^5 - t^4 - 3t^3$

30. $y^3 + y^2 - 9y - 9$

31. $(x + 2)^3 - (x + 2)$

32. $(x + 1)^4 - (x - 1)^4$

33. $2x^3(x^2 + 10)(2x) + 3x^2(x^2 + 10)^2$

34. $4x^2(x^2 + 3x + 1)^3(2x + 3) + 2x(x^2 + 3x + 1)^4$

In Exercises 35–46, simplify and reduce each rational expression.

35. $\dfrac{4x^2 - 20x}{x - 5}$ **36.** $\dfrac{x^2 - 9}{2x + 6}$

37. $\dfrac{24}{(x - a)^2} \cdot \dfrac{x^2 - a^2}{6}$ **38.** $\dfrac{x^2 - 4}{x^2 - 1} \cdot \dfrac{x^2 - x - 2}{(x - 2)^2}$

39. $(t^2 + t - 6) \div \dfrac{t + 3}{t - 2}$

40. $\dfrac{x^2 - 2xy}{xy^2} \div (x^2y - 2x^2y)$

41. $\dfrac{1}{2x + 1} + \dfrac{3}{x - 2}$ **42.** $\dfrac{4x}{x - 1} - \dfrac{5x + 2}{2x}$

43. $\dfrac{\dfrac{2}{x + 1} + \dfrac{1}{x}}{\dfrac{3}{x}}$ **44.** $\dfrac{x^2 - \dfrac{1}{9}}{2x + \dfrac{2}{3}}$

45. $\dfrac{x^{1/2} + x^{-3/2}}{x^{-1/2}}$ **46.** $\dfrac{x^{3/2}}{x^{5/2} - x^{-1/2}}$

In Exercises 47–52, perform each algebraic division.

47. $\dfrac{8x^3 - 4x^2 + 7}{2x^2}$ **48.** $\dfrac{9x^4 - 2x + 6}{3x}$

49. $\dfrac{4x - 5}{2x + 1}$ **50.** $\dfrac{2x^2 - 5}{x^2 - 2}$

51. $\dfrac{2x^2 - 5x + 12}{x - 4}$ **52.** $\dfrac{4x^2 - 6x + 5}{2x + 3}$

In Exercises 53–60, simplify each radical expression.

53. $(\sqrt{x^3} - 6)(\sqrt{x^3} + 6)$ **54.** $(\sqrt{x} - \sqrt{2})^2$

55. $5\sqrt{3t} - \sqrt{12t} + 6\sqrt{\dfrac{t}{3}}$

56. $2\sqrt{\dfrac{t}{2}} - 4t\sqrt{\dfrac{2}{t}} + 6\sqrt{2t}$

57. $\dfrac{-6 - \sqrt{108}}{2}$ **58.** $\dfrac{21 - \sqrt{98}}{14}$

59. $\dfrac{1}{2}(-2x + \sqrt{4x^2 - 8})$

60. $\dfrac{1}{2}(-6x + \sqrt{36x^2 + 16})$

In Exercises 61–72, solve each equation.

61. $\dfrac{2x}{3} - \dfrac{1}{2} = 3x$

62. $4 + \dfrac{2x}{5} = \dfrac{x}{2}$

63. $3(x - 2)(2x + 1) = 6x^2 - 1$

64. $(x - 5)^2 - 20 = 2x(x + 2) - x^2$

65. $2x^2 - x = 15$

66. $(x + 3)(x - 1) = 12$

67. $9(x + 1)^2 = 4$

68. $(2x - 3)^2 = 49$

69. $x^2 - 3x = 5$

70. $2x = 6 - x^2$

71. $\dfrac{x^2}{2} - \dfrac{x}{4} - \dfrac{1}{4} = 0$

72. $\dfrac{x^2}{2} - x - 4 = 0$

In Exercises 73–78, solve each inequality.

73. $2x + (x - 2) \le 5$

74. $\dfrac{3x}{4} - x > 4$

75. $1 \le 1 - 5x < 16$

76. $-13 \le 5(4 - x) < 6$

77. $|2x - 1| < 3$

78. $|3 - x| < 4$

In Exercises 79–82, solve each formula for the given variable.

79. $PV = nRT$, for n

80. $A = \frac{1}{2}(b + B)h$, for B

81. $A = \pi r^2$, for r $(r > 0)$

82. $A = \frac{4}{3}\pi r^3$, for r $(r > 0)$

CHAPTER TEST

1. For the set of real numbers

$$\dfrac{\sqrt{3}}{2}, 0, \dfrac{4}{7}, -0.3, -\sqrt{9}, \dfrac{5}{\pi}$$

determine which are integers, which are rational numbers, and which are irrational numbers.

2. Write $|2\pi - 8|$ without absolute value symbols.

3. Sketch the graph of $|x + 1| \le 3$.

4. Describe the graph with interval notation:

In Exercises 5–8, simplify each expression. Express your answer without negative exponents.

5. $(-2x^2y)(3xy^{-4})$

6. $\left(\dfrac{25a^6}{b^2}\right)^{-1/2}$

7. $\dfrac{p^3q^{-4}}{-5p^{-2}q}$

8. $\dfrac{14x^{3n+1}}{(5x^n)(4x^{n-2})}$

9. Evaluate the expression $(3.8 \times 10^{17})(1.5 \times 10^{-12})$, and express the answer in scientific notation.

In Exercises 10 and 11, write each expression in simplified form.

10. $x^2 - (x + 4)(2x^2 + x - 1)$

11. $(a^{3/2} + 2a^{5/2})(a^{3/2} - 2a^{5/2}) + 3a^5$

In Exercises 12–15, factor each polynomial completely.

12. $2a^3b - 18ab^3$

13. $2x^2 + 7x - 4$

14. $x^2(x + 2) + 5(x + 2)$

15. $3y^3 + 5y^2 - 2y$

In Exercises 16–20, simplify and reduce each rational expression.

16. $\dfrac{2 - x}{3x^2 - 2x - 8}$

17. $\dfrac{6x^2 + 3x}{x^2 - \frac{1}{4}}$

18. $\dfrac{7x - 1}{2x} - \dfrac{6x}{x + 1}$

19. $\dfrac{x^2 - 4x + 4}{x^2 - 1} \div \dfrac{x^2 - 4}{x^2 - 2x + 1}$

20. $\dfrac{x}{2x + 3} + \dfrac{5x + 1}{4x^2 - 9}$

In Exercises 21–23, simplify each radical expression.

21. $\dfrac{6 - \sqrt{108}}{12}$

22. $\dfrac{1}{3}\left(-12x + \sqrt{18x^2 - 45}\right)$

23. $(\sqrt{x} + \sqrt{3})^2 + \sqrt{2}(\sqrt{6x} - \sqrt{2})$

In Exercises 24–27, solve each equation.

28. Solve for r: $s = \dfrac{a}{1-r}$

24. $\dfrac{3x}{2} + 5 = \dfrac{x-2}{6}$

25. $(2x + 1)(x - 1) = 2$

In Exercises 29 and 30, solve each inequality.

26. $(3x + 4)^2 = 81$

27. $\dfrac{x^2}{3} + x = \dfrac{1}{2}$

29. $2(7 - 3x) \le 17$

30. $0 < \dfrac{2x + 11}{3} \le 5$

**FUNCTIONS
AND
GRAPHS**

In this chapter we introduce the idea of a function and its graph, and show how this concept can be applied in situations such as predicting the effects of pollution on wildlife or investigating wind-chill at various temperatures. After we discuss how to use a graphing utility to investigate the graph of a function, we will use functions to set up mathematical models for applied problems. For example, minimizing the installation cost of underground cable and maximizing revenue in sales are two problems that can be described with functions and solved using a graphing utility. We will also discuss how to combine functions to create another function. We encounter such functions in many applications, such as determining a mathematical model for the spread of an oil spill or predicting the profits from selling a product at various prices.

The function concept is very important in using mathematics; this chapter is the foundation for the rest of the text and for your ensuing studies in mathematics. Taking the time to master the ideas in this chapter will be well worth the effort in the long run.

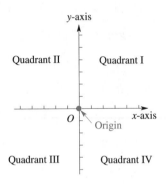

Figure 1.1

> ● **NOTE:** *The coordinate plane is also called the* rectangular plane *or the* Cartesian plane *[after René Descartes (1596–1659), the French mathematician who is credited with the introduction of this idea].*

THE COORDINATE PLANE

The power of a mathematical idea is judged by two qualities—its simplicity and its utility. When measured by these two standards, the idea of the *coordinate plane* is indeed a powerful tool of mathematics and science. Its basic concept is simple enough to be part of a beginning algebra course, yet it is used throughout the study of mathematics and science. The coordinate plane gives us a means to look at algebraic concepts in a geometric setting and also a way to solve tough geometric problems using algebraic tools.

The basic ideas of the coordinate plane are probably already familiar to you. A horizontal axis and a vertical axis intersect at a point O, the *origin*. In most contexts, the horizontal axis is the *x-axis*, and the vertical axis is the *y-axis*. The axes divide the plane into four *quadrants*. Each axis is marked off as a number line, with 0 for each at the origin. The positive direction is to the right for the horizontal axis and up for the vertical axis (Figure 1.1).

Just as each real number corresponds to a point on the coordinate line, each *ordered pair* of real numbers (h, k) corresponds to a point P on the coordinate plane (Figure 1.2). The numbers h and k are the *coordinates* of the point. The point P is h units horizontally and k units vertically from the origin. Figure 1.3 shows examples of ordered pairs and the points on the plane they represent.

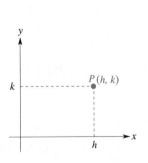

Figure 1.2

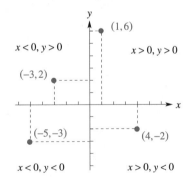

Figure 1.3

Graph of an Equation

An ordered pair (h, k) is a *solution* of an equation in two variables x and y if substituting $x = h$ and $y = k$ makes the equation true. For example, the ordered pair $(-5, 1)$ is a solution of $y^2 - x = 6$, because substituting $x = -5$ and $y = 1$ into the equation makes the equation true:

$$1^2 - (-5) \stackrel{?}{=} 6$$
$$1 + 5 = 6$$

On the other hand, the ordered pair $(3, 1)$ is not a solution of $y^2 - x = 6$, because

$$1^2 - 3 \overset{?}{=} 6$$
$$1 - 3 \neq 6$$

The *graph* of an equation in two variables is the set of all points that correspond to solutions of the equation. A graph is a geometric picture that can tell us many things about its equation.

For example, the graph of the equation $y^2 - x = 6$ is shown in Figure 1.4. The point $(-5, 1)$ is on the graph, because, as we just saw, it is a solution of the equation. Three more points on the graph are marked with their coordinates. Take a few minutes to verify that $(-2, -2)$, $(3, 3)$, and $(10, -4)$ are points on the graph by showing that these ordered pairs are also solutions of the equation.

In Figure 1.4, you can see that the graph of $y^2 - x = 6$ crosses the x-axis once and the y-axis twice. The points where a graph intersects the x-axis are the *x-intercepts* of the graph; an x-intercept has a y-coordinate of 0. Likewise, the *y-intercepts* of a graph are the points where the graph intersects the y-axis; a y-intercept has an x-coordinate of 0.

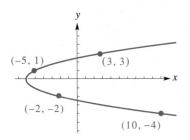

Figure 1.4
$y^2 - x = 6$

EXAMPLE 1   Finding the Intercepts of a Graph

Determine the intercepts of the graph of $y^2 - x = 6$.

SOLUTION

The x-intercept is determined by substituting $y = 0$ into the equation and solving for x:

$$y^2 - x = 6$$
$$0^2 - x = 6$$
$$-x = 6$$
$$x = -6$$

The point $(-6, 0)$ is the x-intercept.

To find the y-intercepts, we substitute $x = 0$ into the equation:

$$y^2 - x = 6$$
$$y^2 - 0 = 6$$
$$y^2 = 6$$
$$y = \pm\sqrt{6}$$

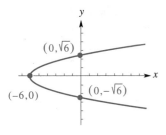

Figure 1.5
$y^2 - x = 6$

$(0, \sqrt{6})$ and $(0, -\sqrt{6})$ are the y-intercepts (Figure 1.5).

The Distance Formula

The next example shows how the Pythagorean theorem can be used to solve a geometric problem algebraically.

⬤ EXAMPLE 2   Computing the Distance Between Two Points

Determine the distance $d(A, B)$ between the points $A(-1, 2)$ and $B(3, 8)$.

SOLUTION

The first step is to plot the points and draw the line segment AB (Figure 1.6). The point C in the figure is the point on the same horizontal line as A and the same vertical line as B. The triangle formed by A, B, and C is a right triangle with its right angle at C. The horizontal leg of the triangle is of length 4; this is the change in x from A to B. The change in y from A to B is 6, the length of the vertical leg of the triangle. The length of the hypotenuse is $d(A, B)$. Thus,

$$d(A, B) = \sqrt{4^2 + 6^2} = \sqrt{52} = 2\sqrt{13}$$

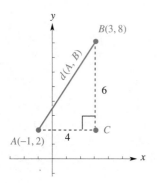

Figure 1.6

Look again at the process used in Example 2. Suppose that the points had been $A(x_1, y_1)$ and $B(x_2, y_2)$ instead of $A(-1, 2)$ and $B(3, 8)$. Refer to Figure 1.7. The change in x from A to B is $x_2 - x_1$, so the length of the horizontal leg of the triangle is $|x_2 - x_1|$. Likewise, the change in y from A to B is $y_2 - y_1$, so the length of the vertical leg is $|y_2 - y_1|$. Thus,

$$d(A, B) = \sqrt{|x_2 - x_1|^2 + |y_2 - y_1|^2}$$
$$= \sqrt{(x_2 - x_1)^2 + (y_2 - y_1)^2} \quad \text{For any quantity } u, |u|^2 = u^2.$$

⬛ **NOTE:** *The absolute value is used because a distance must be nonnegative.*

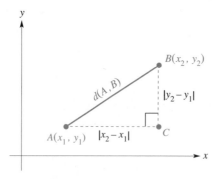

Figure 1.7

⬛ **NOTE:** *Read Δx as "delta x" and Δy as "delta y."*

The change in x is often denoted by Δx and the change in y by Δy. This result is summarized here:

The Distance Formula

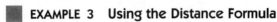

The distance between the points $A(x_1, y_1)$ and $B(x_2, y_2)$ is

$$d(A, B) = \sqrt{(x_2 - x_1)^2 + (y_2 - y_1)^2}$$

Alternatively, this distance is written as

$$d(A, B) = \sqrt{(\Delta x)^2 + (\Delta y)^2}$$

where $\Delta x = x_2 - x_1$ and $\Delta y = y_2 - y_1$. (See Figure 1.7.)

EXAMPLE 3　Using the Distance Formula

Determine whether the triangle with vertices $P(6, -3)$, $Q(0, 5)$, and $R(-2, 1)$ is a right triangle.

SOLUTION

The first step is to plot the points and draw the triangle (Figure 1.8). To determine if the triangle is a right triangle, we can use the converse of the Pythagorean theorem: If the two shortest sides, a and b, and the longest side, c, are such that $a^2 + b^2 = c^2$, then the triangle is a right triangle.

The lengths of the sides are computed using the distance formula:

$$d(P, Q) = \sqrt{(0 - 6)^2 + [5 - (-3)]^2} = \sqrt{(-6)^2 + 8^2} = \sqrt{100} = 10$$
$$d(Q, R) = \sqrt{(-2 - 0)^2 + (1 - 5)^2} = \sqrt{(-2)^2 + (-4)^2} = \sqrt{20} = 2\sqrt{5}$$
$$d(R, P) = \sqrt{[6 - (-2)]^2 + [(-3) - 1]^2} = \sqrt{8^2 + (-4)^2} = \sqrt{80} = 4\sqrt{5}$$

The side PQ is the longest side, so the triangle is a right triangle if

$$[d(Q, R)]^2 + [d(R, P)]^2 = [d(P, Q)]^2$$
$$[2\sqrt{5}]^2 + [4\sqrt{5}]^2 \overset{?}{=} [10]^2$$
$$20 + 80 = 100$$

It follows that triangle PQR is a right triangle.

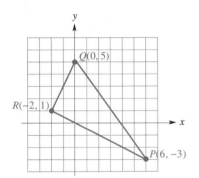

Figure 1.8

■ **NOTE:** *You might have guessed that the triangle was a right triangle from the picture, but to be absolutely sure, we need to perform the algebraic computations.*

Equation of a Circle

In elementary geometry, a circle is defined to be the set of all points that are the same given distance r (the *radius*) from a given fixed point C (the *center*). If we know the radius and the coordinates of the center, can we find an equation for the graph of the circle?

EXAMPLE 4　Finding the Equation of a Circle

Determine an equation for the circle in the coordinate plane with center $C(2, 4)$ and radius 6.

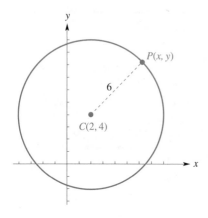

Figure 1.9
$(x - 2)^2 + (y - 4)^2 = 36$

Standard Equation of a Circle

SOLUTION

Suppose that $P(x, y)$ is a point on the circle. The distance between the center C and P is equal to the radius of the circle:

$$d(C, P) = 6$$
$$\sqrt{(x - 2)^2 + (y - 4)^2} = 6$$

Squaring both sides of this equation gives an equivalent equation without a radical:

$$(x - 2)^2 + (y - 4)^2 = 36$$

This is an equation whose graph is the circle in Figure 1.9.

There is nothing special about the circle in Example 4. If the circle has a center with coordinates (h, k) instead of $(2, 4)$ and radius r instead of 6, we get the following general result:

The **standard form** of a circle with center $C(h, k)$ and radius r is

$$(x - h)^2 + (y - k)^2 = r^2$$

In other words, a point $P(x, y)$ is on the circle if and only if (x, y) is a solution of this equation.

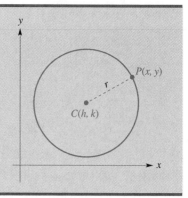

An equation whose graph is a circle also can be written in the *general form* $x^2 + y^2 + Dx + Ey + F = 0$, where D, E, and F are real numbers. For example, the equation in Example 4,

$$(x - 2)^2 + (y - 4)^2 = 36$$

can be written as

$$x^2 + y^2 - 4x - 8y - 16 = 0$$

(Pause here to verify this result.)

Likewise, an equation of a circle in the general form can be rewritten in standard form. The next example shows how completing the square (see Section 7, page 58) is used to accomplish this.

EXAMPLE 5   Writing the Equation of a Circle in Standard Form
Consider the equation $x^2 + y^2 - 6x + 12y + 32 = 0$.

(a) Show that this equation can be rewritten as a standard equation of a circle, and determine the center and radius of the circle.

(b) Compute the coordinates of the intercepts and sketch the circle.

SOLUTION

(a) To get the equation in the form $(x - h)^2 + (y - k)^2 = r^2$, we complete the square on the x terms and the y terms:

$$x^2 + y^2 - 6x + 12y + 32 = 0$$

$(x^2 - 6x \quad) + (y^2 + 12y \quad) = -32$ Group the x terms and the y terms together.

$(x^2 - 6x + 9) + (y^2 + 12y + 36) = -32 + 9 + 36$ Complete the squares on the x terms and the y terms.

$$(x - 3)^2 + (y + 6)^2 = 13$$

The circle has center $(h, k) = (3, -6)$ and radius $r = \sqrt{13}$.

(b) To find the y-intercepts, we let $x = 0$ and solve the equation for y:

$$(x - 3)^2 + (y + 6)^2 = 13$$
$$(0 - 3)^2 + (y + 6)^2 = 13$$
$$9 + (y + 6)^2 = 13$$
$$(y + 6)^2 = 4$$
$$y + 6 = \pm 2$$
$$y = -6 \pm 2$$
$$y = -4, \, -8$$

The y-intercepts are $(0, -4)$ and $(0, -8)$. To find the x-intercepts, we let $y = 0$ and solve the equation for x:

$$(x - 3)^2 + (y + 6)^2 = 13$$
$$(x - 3)^2 + (0 + 6)^2 = 13$$
$$(x - 3)^2 + 36 = 13$$
$$(x - 3)^2 = -23$$

This last equation has no real solutions, so the graph has no x-intercepts, as can be seen in Figure 1.10.

Quite often, the graph of a given equation can be determined from a graph that is already known. The next example shows how the graph of a circle is used to determine the graph of another equation.

EXAMPLE 6 The Graph of a Semicircle

Use the graph of $x^2 + y^2 = 16$ to determine the graph of $y = \sqrt{16 - x^2}$.

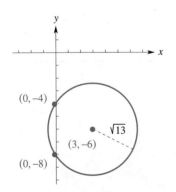

Figure 1.10
$(x - 3)^2 + (y + 6)^2 = 13$

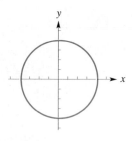

Figure 1.11
$x^2 + y^2 = 16$

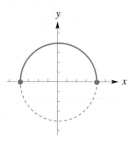

Figure 1.12
$y = \sqrt{16 - x^2}$

SOLUTION

Because $x^2 + y^2 = 16$ is equivalent to $(x - 0)^2 + (y - 0)^2 = 4^2$, its graph is a circle centered at $(0, 0)$ with radius 4 (Figure 1.11). Solving the equation for y yields

$$x^2 + y^2 = 16$$
$$y^2 = 16 - x^2$$
$$y = \pm\sqrt{16 - x^2}$$

Any solution (x, y) to $x^2 + y^2 = 16$ is a solution to either $y = \sqrt{16 - x^2}$ or $y = -\sqrt{16 - x^2}$. Specifically, the solutions to $y = \sqrt{16 - x^2}$ are those solutions of $x^2 + y^2 = 16$ such that $y \geq 0$. (This is true because the principal square root of a quantity is nonnegative.) Thus, the graph of $y = \sqrt{16 - x^2}$ is simply the half of the graph from Figure 1.11 that is on or above the x-axis. Figure 1.12 shows this upper semicircle. (The lower semicircle is the graph of $y = -\sqrt{16 - x^2}$.)

Midpoint of a Line Segment

Many problems that arise in physics and engineering require finding the midpoint of a line segment. (The *midpoint* is the point on the line segment that is the same distance from both endpoints.) The coordinates of the midpoint of a line segment with endpoints $A(x_1, y_1)$ and $B(x_2, y_2)$ are given by the formula below. The proof of this formula is left to you (Exercises 68 and 69).

Midpoint Formula

The midpoint M of the line segment with endpoints $A(x_1, y_1)$ and $B(x_2, y_2)$ has the coordinates

$$\left(\frac{x_1 + x_2}{2}, \frac{y_1 + y_2}{2}\right)$$

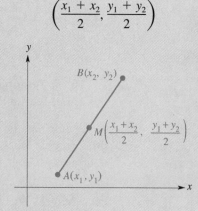

EXAMPLE 7 Using the Midpoint Formula

Find the coordinates of the midpoint M of the line segment with endpoints $A(-2, 4)$ and $B(8, -6)$, and show that $d(A, M) = d(M, B)$.

SOLUTION

The coordinates of the midpoint M of the line segment are

$$\left(\frac{(-2) + 8}{2}, \frac{4 + (-6)}{2}\right) = (3, -1) \quad \text{Figure 1.13}$$

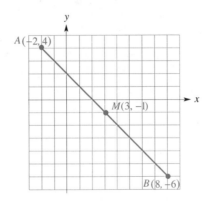

Figure 1.13

The distances from M to the endpoints A and B are

$$d(A, M) = \sqrt{[3 - (-2)]^2 + [(-1) - 4]^2} = \sqrt{5^2 + (-5)^2} = \sqrt{50} = 5\sqrt{2}$$
$$d(M, B) = \sqrt{(8 - 3)^2 + [(-6) - (-1)]^2} = \sqrt{5^2 + (-5)^2} = \sqrt{50} = 5\sqrt{2}$$

The midpoint M is the same distance from each endpoint of the line segment.

Symmetry of a Graph

A graph is more than a collection of plotted points. When considered as a whole, it has certain qualities and characteristics that provide information about its equation. For example, look at the graph of $y = x^4 - 6x^2$ in Figure 1.14. Notice that the y-axis divides the graph into two parts. The part of the graph on the right is the same as the part on the left, except that it is reversed. The part on the left is the *reflection* through the y-axis of the part on the right. We say that this graph is *symmetric with respect to the y-axis*.

A graph also may be *symmetric with respect to the x-axis*, or *symmetric with respect to the origin*, or it may have none of these symmetries. The following box and examples define these ideas more precisely and show how they are related to the graphs of equations.

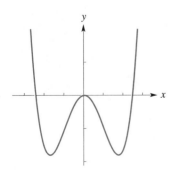

Figure 1.14
$y = x^4 - 6x^2$

Symmetry of a Graph

Symmetry about the x-axis

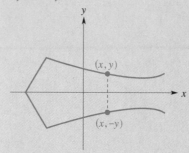

Definition

A graph is symmetric with respect to the *x*-axis if for any point (x, y) on the graph, the point $(x, -y)$ is also on the graph.

To test the graph of an equation for symmetry about the x-axis:
If replacing (x, y) with $(x, -y)$ in the equation results in an equivalent equation, then the graph of the equation is symmetric with respect to the *x*-axis.

Symmetry about the y-axis

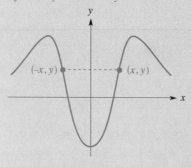

Definition

A graph is symmetric with respect to the *y*-axis if for any point (x, y) on the graph, the point $(-x, y)$ is also on the graph.

To test the graph of an equation for symmetry about the y-axis:
If replacing (x, y) with $(-x, y)$ in the equation results in an equivalent equation, then the graph of the equation is symmetric with respect to the *y*-axis.

Symmetry about the origin

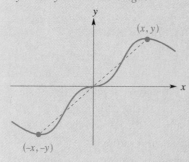

Definition

A graph is symmetric with respect to the origin if for any point (x, y) on the graph, the point $(-x, -y)$ is also on the graph.

To test the graph of an equation for symmetry about the origin:
If replacing (x, y) with $(-x, -y)$ in the equation results in an equivalent equation, then the graph of the equation is symmetric with respect to the origin.

For example, to verify that the graph of the equation $y = x^4 - 6x^2$ (Figure 1.14) is symmetric with respect to the y-axis, we substitute $(-x, y)$ for (x, y) in $y = x^4 - 6x^2$:

$$y = (-x)^4 - 6(-x)^2$$
$$y = x^4 - 6x^2$$

Because this last equation is the same as the original equation, the graph is indeed symmetric with respect to the y-axis.

EXAMPLE 8   Completing a Graph Using Symmetry

Figure 1.15 shows the portion of the graph of

$$y = \frac{5x}{x^2 + 1}$$

that is in quadrant I. Use the symmetry tests to complete the graph.

SOLUTION

The plan of attack here is to test the equation for symmetry with respect to the x-axis, the y-axis, and the origin. If the equation passes one of these tests, then the graph can be completed by sketching the corresponding reflection of the part given.

To test whether the graph is symmetric with respect to the x-axis, we replace (x, y) with $(x, -y)$ in the equation and see if an equivalent equation results:

$$-y = \frac{5x}{x^2 + 1}$$

This equation is not equivalent to the original, so the complete graph is not symmetric with respect to the x-axis.

We replace (x, y) with $(-x, y)$ in the equation to test whether the graph is symmetric with respect to the y-axis:

$$y = \frac{5(-x)}{(-x)^2 + 1}$$
$$y = \frac{-5x}{x^2 + 1}$$

This last equation is not equivalent to the original equation, so the complete graph is not symmetric with respect to the y-axis.

To test whether the complete graph is symmetric with respect to the origin,

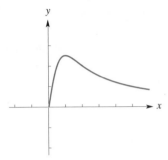

Figure 1.15

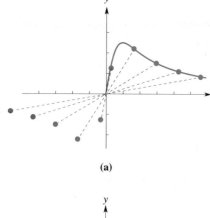

(a)

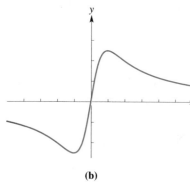

(b)

Figure 1.16

we replace (x, y) with $(-x, -y)$ in the equation and see if an equivalent equation results:

$$-y = \frac{5(-x)}{(-x)^2 + 1}$$

$$-y = \frac{-5x}{x^2 + 1}$$

$$y = \frac{5x}{x^2 + 1} \qquad \text{Multiply each side by } -1.$$

This last equation is equivalent to the original, so the graph is symmetric with respect to the origin.

To complete the graph, we choose a few points on the given graph and plot their reflections with respect to the origin. These points are in quadrant III (Figure 1.16a). Extending the graph through these points gives the complete graph (Figure 1.16b). ■

Applications

In the next example, the coordinate plane is used to apply a result from geometry to the triangle from Example 3.

EXAMPLE 9   Application: Finding the Circumscribing Circle of a Right Triangle

The circle that passes through the vertices of a triangle is the *circumscribing circle* of the triangle. A theorem from geometry states that the center of the circumscribing circle of a right triangle is the midpoint of the hypotenuse of the triangle.

(a) Show that the midpoint of the hypotenuse of the triangle in Example 3 (Figure 1.17) is the same distance from each vertex of the triangle.

(b) Find the standard equation of the circumscribing circle for this triangle.

SOLUTION

(a) From Example 3, the side PQ is the hypotenuse of the triangle. Using the midpoint formula to compute the coordinates of the midpoint M of the hypotenuse gives us

$$M = \left(\frac{6 + 0}{2}, \frac{-3 + 5}{2}\right) = (3, 1)$$

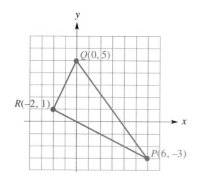

Figure 1.17

The distances from $M(3, 1)$ to the vertices $P(6, -3)$, $Q(0, 5)$, and $R(-2, 1)$ are

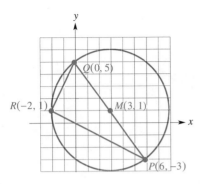

Figure 1.18
$(x - 3)^2 + (y - 1)^2 = 25$

$$d(M, P) = \sqrt{(6 - 3)^2 + [(-3) - 1]^2} = \sqrt{3^2 + (-4)^2} = \sqrt{25} = 5$$
$$d(M, Q) = \sqrt{(0 - 3)^2 + (5 - 1)^2} = \sqrt{(-3)^2 + 4^2} = \sqrt{25} = 5$$
$$d(M, R) = \sqrt{[(-2) - 3]^2 + (1 - 1)^2} = \sqrt{(-5)^2 + 0^2} = \sqrt{25} = 5$$

This shows that the midpoint of the hypotenuse is the same distance from each vertex of the triangle.

(b) The midpoint of the hypotenuse is the center of the circle, and the common distance, 5, found in part (a) is the radius. Letting $r = 5$ and $(h, k) = (3, 1)$ in the standard equation of a circle yields

$$(x - 3)^2 + (y - 1)^2 = 5^2 \quad \text{or} \quad (x - 3)^2 + (y - 1)^2 = 25$$

The graph of this circle is shown in Figure 1.18.

EXERCISE SET 1.1

A

In Exercises 1–6, the graph of the given equation is shown. Determine the coordinates of the intercepts of each graph.

1. $2x + 3y = 12$

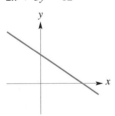

2. $2x - 5y = 20$

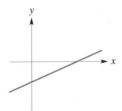

3. $x^2 - y = 9$

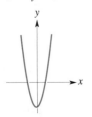

4. $x + 4 = y^2$

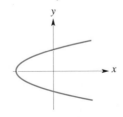

5. $y = (x - 2)(x^2 - 4x - 5)$

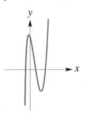

6. $y = (x^2 - 12)(x^2 + 2)$

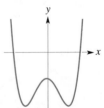

In Exercises 7–12:

(a) Determine the distance $d(A, B)$ between the points.
(b) Determine the midpoint M of the line segment with endpoints A and B.

7. $A(2, 3); B(5, 7)$

8. $A(12, -3); B(6, 0)$

9. $A\left(-\frac{3}{2}, \frac{4}{5}\right); B\left(\frac{7}{2}, -\frac{6}{5}\right)$

10. $A\left(-\frac{5}{12}, 1\right)$; $B(0, 0)$

11. $A(\sqrt{3}, 0)$; $B(0, \sqrt{6})$

12. $A(\sqrt{2}, 1)$; $B(-\sqrt{2}, 5)$

In Exercises 13–18, complete the square to determine the standard equation of each circle. Determine the center and radius of the circle.

13. $x^2 + 8x + y^2 - 4y + 4 = 0$

14. $x^2 - 4x + y^2 + 10y + 20 = 0$

15. $x^2 - 10x + y^2 + 5 = 0$

16. $x^2 + y^2 - 4y - 14 = 0$

17. $x^2 - 6x + y^2 + 8y = 0$

18. $x^2 - 3x + y^2 + y = 0$

In Exercises 19–26, the graph of the given equation is shown. Use the symmetry tests to verify any of the three symmetries (x-axis, y-axis, or origin) the graph suggests.

19. $y = 2x$

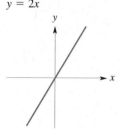

20. $x = 2y^2$

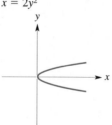

21. $x^2 + 3y^2 = 12$

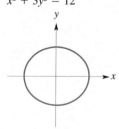

22. $2x + 3y = 12$

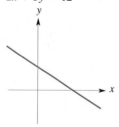

23. $y = \frac{1}{2}x^2 - 8$

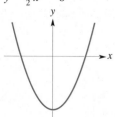

24. $y = x|x|$

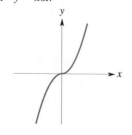

25. $y = x^3 - 4x - 6$

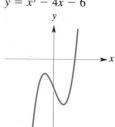

26. $y^{2/3} - x = 8$

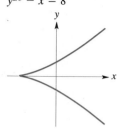

B

In Exercises 27–30, a portion of the graph of the given equation is shown. Use the symmetry tests to complete each graph, and find the coordinates of the intercepts.

27. $y = x^3 - 3x$

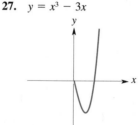

28. $y = 4x^2 - x^4$

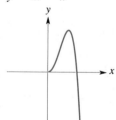

29. $5x^2 - 8xy + 5y^2 = 45$

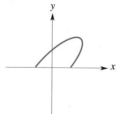

30. $x^2 - y^2 = 4$

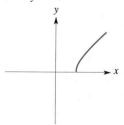

In Exercises 31–40, sketch the circle described and determine its standard equation.

31. The circle is in quadrant IV, is tangent to both the x-axis and the y-axis, and has radius 3.

32. The circle is in quadrant II, is tangent to both the x-axis and the y-axis, and has radius 5.

33. The circle is tangent to the x-axis at $(6, 0)$ and tangent to the y-axis at $(0, 6)$.

34. The circle is tangent to the x-axis at $(-4, 0)$ and tangent to the y-axis at $(0, 4)$.

35. The center of the circle is $(4, -3)$, and the circle passes through the origin.

36. The center of the circle is the origin, and the circle passes through $(5, 12)$.

37. The center of the circle is the origin, and the circle is tangent to the vertical line passing through $(-3, 6)$.

38. The center of the circle is at $(4, 1)$, and the circle is tangent to the horizontal line passing through $(6, -4)$.

39. The line segment with endpoints $(5, -4)$ and $(1, 8)$ is a diameter of the circle.

40. The line segment with endpoints $(0, 4)$ and $(8, 0)$ is a diameter of the circle.

In Exercises 41–48, draw a picture as part of your solution.

41. Points that lie on the same line are *collinear*. Three points A, B, and C are collinear, with B in the middle, if and only if $d(A, B) + d(B, C) = d(A, C)$. Determine whether the points $A(1, 3)$, $B(4, 7)$, and $C(10, 15)$ are collinear.

42. Determine whether the points $S(-8, 10)$, $T(-5, -1)$, and $U(-4, -5)$ are collinear. (See Exercise 41.)

43. A triangle is *equilateral* if all three sides are of equal length, *isosceles* if exactly two sides are of equal length, and *scalene* if all sides are of different lengths. Determine whether the triangle with vertices $A(7, 3)$, $B(3, 5)$, and $C(5, -1)$ is equilateral, isosceles, or scalene.

44. Determine whether the triangle with vertices $P(-4, 7)$, $Q(23, 6)$, and $R(8, -17)$ is equilateral, isosceles, or scalene. (See Exercise 43.)

45. According to the Pythagorean theorem, a triangle with sides a, b, and c (c the longest side) is a right triangle if $a^2 + b^2 = c^2$; the triangle is *obtuse* if $a^2 + b^2 < c^2$; and it is *acute* if $a^2 + b^2 > c^2$. Determine whether the triangle with vertices $A(1, -6)$, $B(6, -8)$, and $C(3, 0)$ is a right triangle, an obtuse triangle, or an acute triangle.

46. Determine whether the triangle with vertices $P(1, 1)$,

$Q(7, 0)$, and $R(8, 18)$ is a right triangle, an obtuse triangle, or an acute triangle. (See Exercise 45.)

47. A quadrilateral is a parallelogram if and only if its diagonals have the same midpoint. Use this fact to determine whether the quadrilateral with vertices $A(3, -2)$, $B(6, 9)$, $C(17, 8)$, and $D(14, -3)$ is a parallelogram.

48. A quadrilateral is a parallelogram if and only if each pair of opposite sides are of equal length. Use this fact to determine whether the quadrilateral with vertices $P(-8, 7)$, $Q(2, 8)$, $R(5, 1)$, and $S(-5, 2)$ is a parallelogram.

49. Write $(x - y)^2 = 2x(1 - y)$ as the standard equation of a circle and sketch its graph.

50. Write $(x + y)^2 = 2xy + 16$ as the standard equation of a circle, and sketch its graph.

51. Explain why the graph of $x^2 + y^2 - 14x - 6y + 58 = 0$ is not a circle. What is it?

52. Explain why the graph of $x^2 + y^2 + 2x + 4y + 14 = 0$ is not a circle. What is it?

53. What point is symmetric to the point $A(3, 7)$ with respect to the point $P(2, 3)$?

54. What point is symmetric to the point $B(-6, 0)$ with respect to the point $Q(3, 5)$?

55. What point is symmetric to the point $A(3, 7)$ with respect to the vertical line through the point $P(2, 3)$?

56. What point is symmetric to the point $B(-6, 0)$ with respect to the horizontal line through the point $Q(3, 5)$?

57. Use the graph of $y^2 - x = 6$ in Example 1 to sketch the graph of $y = \sqrt{x + 6}$.

58. Use the graph of $y^2 - x = 6$ in Example 1 to sketch the graph of $y = -\sqrt{x + 6}$.

59. Use the graph of a circle to sketch the graph of $y = \sqrt{16 - (x - 2)^2}$.

60. Use the graph of a circle to sketch the graph of $x = 3 + \sqrt{16 - y^2}$.

In Exercises 61–64, each graph shown is a semicircle. Determine an equation for each graph.

61.

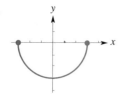

62.

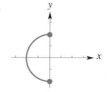

63.

64.

C

65. Suppose that points A and B are symmetric with respect to the line that is the graph of $y = x$. What can you say about the coordinates of A and B?

66. The coordinates of the midpoints of the sides of a triangle are $(5, 3)$, $(0, -3)$, and $(-3, 1)$. What are the vertices of the triangle?

67. **(a)** Explain why a graph that is symmetric with respect to the x-axis and y-axis also is symmetric with respect to the origin.
(b) Explain why a graph that is symmetric with respect to the x-axis and the origin also is symmetric with respect to the y-axis.

(c) Explain why a graph that is symmetric with respect to the y-axis and the origin also is symmetric with respect to the x-axis.
(d) Generalize from parts (a), (b), and (c).

68. **Proof of the Midpoint Formula**   Use the figure given to show that the midpoint M of the line segment with endpoints $A(x_1, y_1)$ and $B(x_2, y_2)$ has the coordinates

$$\left(\frac{x_1 + x_2}{2}, \frac{y_1 + y_2}{2} \right)$$

HINT: First show that triangles AMP and MBQ are congruent.

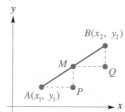

69. **Proof of the Midpoint Formula**   Consider the points $A(x_1, y_1)$, $B(x_2, y_2)$, and $M\left(\dfrac{x_1 + x_2}{2}, \dfrac{y_1 + y_2}{2} \right)$.

(a) Show that A, M, and B are collinear. (See Exercise 41.)
(b) Use the distance formula to show that M is the same distance from A and B.
(c) Explain why this proves the midpoint formula.

SECTION 1.2   FUNCTIONS

- Functions and Functional Notation
- Domain and Zeros of a Function
- Expressions with Functional Notation
- Piecewise-Defined Functions
- Applications

Successful skydivers know that the distance a body falls depends upon how long the body falls. Table 1.1 shows the connection between t, the time falling (in seconds), and the distance traveled (in meters), for a skydiver on a particular jump. The table tells us, for example, that after 4 seconds the distance traveled is 73 meters.

TABLE 1.1

Time, t	0	2	4	6	8	10	12	14	16	18	20
Distance, $d(t)$	0	19	73	153	251	359	473	589	708	827	946

■ **NOTE:** *Read d(t) as "d of t."*

If we use $d(t)$ to represent the distance (in meters) the skydiver falls in t seconds, then it follows that $d(4) = 73$. Table 1.1 represents a process that accepts a value t and assigns to it the value $d(t)$. This correspondence between t and $d(t)$ is an example of a *function*.

Functions and Functional Notation

Roughly speaking, a function is a rule that describes how one quantity depends upon another. For instance, the area of a circle is dependent on its radius, the number of sales of a particular model of an automobile depends upon its selling price, and the number of seconds that an earthworm takes to react to an electrical stimulus depends upon its body temperature. These correspondences between quantities are the basis for applications of mathematics to many situations.

Function

> A **function** is a rule that assigns to each element of a set of inputs exactly one element of a set of outputs. The set of inputs is the **domain** of the function, and the set of outputs is the **range** of the function.

The key word in this definition—*exactly*—indicates that a particular input is assigned to one and only one output.

■ **NOTE:** *The output f(x) is also called the* image *of x under f.*

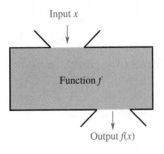

Figure 1.19

You can think of a function as a machine in which there is an input x, a process f, and an output $f(x)$ (Figure 1.19). The output $f(x)$ is called the *value* of f at x.

A function can be represented by a table (such as Table 1.1), a mathematical expression (as we see next), or a graph (the subject of the next section). Our concern in this text is with functions that have real numbers as inputs and outputs.

Functions defined by mathematical expressions can be described by *functional notation*. For example, consider a function g such that its output $g(x)$ is the square of the input x. The function g is described precisely by the functional notation $g(x) = x^2$. The choice of x as the variable in this formula is incidental; the function also can be defined equally well by $g(t) = t^2$, by $g(n) = n^2$, or even by $g(☆) = ☆^2$.

■ EXAMPLE 1   Using Functional Notation

(a) Given the function $f(x) = x^2 + 5x$, evaluate $f(-6)$, $f(2)$, and $f(1)$.

(b) Given the function $G(x) = (x - 1)\sqrt{x + 4}$, evaluate $G(3)$, $G(0)$, and $G(-6)$.

SOLUTION

(a)
$$f(-6) = (-6)^2 + 5(-6) = 36 - 30 = 6$$
$$f(2) = (2)^2 + 5(2) = 4 + 10 = 14$$
$$f(1) = (1)^2 + 5(1) = 1 + 5 = 6$$

(b)
$$G(3) = (3 - 1)\sqrt{3 + 4} = 2\sqrt{7}$$
$$G(0) = (0 - 1)\sqrt{0 + 4} = (-1)2 = -2$$

Attempting to evaluate $G(-6)$ gives

$$G(-6) = (-6 - 1)\sqrt{-6 + 4} = (-7)\sqrt{-2}$$

Because $\sqrt{-2}$ is not a real number, $G(-6)$ is undefined.

It may appear to you that function f in part (a) in Example 1 violates the definition of a function because f assigns the value 6 to both -6 and 1. However, careful reading of the definition should convince you that this is okay—exactly one value is assigned to each of these elements of the domain. There is nothing in the definition that says the range values for different domain values can't be the same number.

Domain and Zeros of a Function

The domain of a function is the set of those real numbers for which the function is defined. Determining the domain of a function f often reduces to finding any values of x such that $f(x)$ is undefined—for example, any values of x that make a denominator equal to 0 or the radicand of a square root negative. A *zero* of a function f is a number a in the domain of f such that $f(a) = 0$. The next example illustrates these two ideas.

EXAMPLE 2   Finding the Domain and Zeros of a Function
Determine the domain and zeros of each function.

(a) $f(x) = x^2 + 5x$ **(b)** $G(x) = (x - 1)\sqrt{x + 4}$

SOLUTION

(a) Because the expression $x^2 + 5x$ is defined for all real numbers, the domain of f is the set of all real numbers. In interval notation, this is written as $(-\infty, +\infty)$. The zeros of f are those values a in the domain of f such that $f(a) = 0$. Solving this equation gives

$$f(a) = 0$$
$$a^2 + 5a = 0$$

We solve this quadratic equation by factoring:

$$a(a + 5) = 0 \qquad \text{Apply the zero factor property (Section 7, page 58).}$$

$$a = 0 \quad \bigg| \quad a + 5 = 0$$
$$a = -5$$

The zeros of f are 0 and -5.

(b) The domain of G is the set of all real numbers x such that the expression $(x - 1)\sqrt{x + 4}$ is defined. Because the radicand of the square root, $x + 4$, must be nonnegative, the domain is the set of all x such that $x + 4 \geq 0$, or $x \geq -4$. In interval notation, the domain of G is $[-4, +\infty)$. The zeros of G are those values a in the domain of G such that $G(a) = 0$:

$$G(a) = 0$$
$$(a - 1)\sqrt{a + 4} = 0$$

By the zero factor property (Section 7, page 58), either

$$a - 1 = 0 \quad \text{or} \quad \sqrt{a + 4} = 0$$

If $a - 1 = 0$, then $a = 1$. The second equation, $\sqrt{a + 4} = 0$, is solved by squaring both sides:

$$\sqrt{a + 4} = 0$$
$$(\sqrt{a + 4})^2 = (0)^2$$
$$a + 4 = 0$$
$$a = -4$$

The zeros of G are 1 and -4.

Expressions with Functional Notation

In addition to evaluating functions at particular real numbers, it is also important to be able to interpret and simplify functional notation involving algebraic expressions. The next three examples will give you an idea of the techniques you will need in later chapters of this text and in courses such as trigonometry and calculus.

EXAMPLE 3   Functional Notation with Algebraic Expressions

Let

$$h(x) = \frac{4}{x^2 - 3}$$

Find $h(2t)$, $2h(t)$, $h(t + 3)$, and $h(t) + h(3)$.

SOLUTION

$$h(2t) = \frac{4}{(2t)^2 - 3} = \frac{4}{4t^2 - 3}$$

$$2h(t) = 2\left(\frac{4}{t^2 - 3}\right) = \frac{8}{t^2 - 3}$$

$$h(t + 3) = \frac{4}{(t + 3)^2 - 3} = \frac{4}{t^2 + 6t + 6}$$

$$h(t) + h(3) = \frac{4}{t^2 - 3} + \frac{4}{3^2 - 3} = \frac{4}{t^2 - 3} + \frac{2}{3}$$

In the last example, you can see that notation that looks similar may give very different results. Note that $h(2t)$ and $2h(t)$ are not the same, and that $h(t + 3)$ and $h(t) + h(3)$ are not the same.

For a function f, expressions of the form

$$\frac{f(b) - f(a)}{b - a} \quad \text{or} \quad \frac{f(x + h) - f(x)}{h}$$

■ **NOTE:** *Example 5 may give you a hint of the significance of difference quotients.*

are called *difference quotients* of f. These expressions play an important role in the development of calculus. The next two examples demonstrate the algebraic techniques used to simplify them.

EXAMPLE 4 Simplifying Difference Quotients

(a) Given $f(t) = 2t + 7$, simplify the difference quotient

$$\frac{f(b) - f(a)}{b - a}$$

(b) Given $g(x) = x^2 - 4x$, simplify the difference quotient

$$\frac{g(x + h) - g(x)}{h}$$

SOLUTION

(a) Because $f(b) = 2b + 7$ and $f(a) = 2a + 7$, it follows that

$$\frac{f(b) - f(a)}{b - a} = \frac{(2b + 7) - (2a + 7)}{b - a}$$

$$= \frac{2b + 7 - 2a - 7}{b - a}$$

$$= \frac{2b - 2a}{b - a}$$

$$= \frac{2(b - a)}{b - a}$$

$$= 2$$

(b) Evaluating g at $x + h$ gives $g(x + h) = (x + h)^2 - 4(x + h)$, so

$$\frac{g(x + h) - g(x)}{h} = \frac{[(x + h)^2 - 4(x + h)] - [x^2 - 4x]}{h}$$

$$= \frac{[x^2 + 2xh + h^2 - 4x - 4h] - [x^2 - 4x]}{h}$$

$$= \frac{x^2 + 2xh + h^2 - 4x - 4h - x^2 + 4x}{h}$$

$$= \frac{2xh + h^2 - 4h}{h}$$

$$= \frac{h(2x + h - 4)}{h}$$

$$= 2x + h - 4$$

EXAMPLE 5 Application: Using Difference Quotients

Table 1.1 at the beginning of this section shows the distance $d(t)$ that a skydiver falls in t seconds.

(a) Compute the difference quotient

$$\frac{d(12) - d(4)}{12 - 4}$$

(b) Explain the significance of this value in terms of the skydiver's fall.

SOLUTION

(a) From the table, $d(12) = 473$ and $d(4) = 73$. Thus,

$$\frac{d(12) - d(4)}{12 - 4} = \frac{473 - 73}{12 - 4} = \frac{400}{8} = 50$$

(b) The value of $d(12) - d(4)$ is the distance that the skydiver falls over the time interval from $t = 4$ to $t = 12$. The value of $12 - 4$ is the number of seconds traveled over this same interval. The difference quotient represents the ratio of distance traveled to time traveled, which is the *average velocity* over this interval of time.

Piecewise-Defined Functions

Many functions are defined by two or more mathematical expressions. For example, the amount of income tax paid by an individual to the federal government is a function of the individual's taxable income x (in dollars). In 1992, the tax (in dollars) was

$$T(x) = \begin{cases} 0.15x & 0 \le x \le 21{,}450 \\ 0.28x - 2788.50 & 21{,}450 < x \le 51{,}900 \\ 0.31x - 4345.50 & x > 51{,}900 \end{cases}$$

Notice that three different expressions are used to define this function. The choice of which expression to use for a particular individual is determined by that individual's income. For example, if an individual's taxable income is $10,000, the tax is

$$T(10{,}000) = 0.15(10{,}000) = 1500 \qquad \text{because} \qquad 0 \le 10{,}000 \le 21{,}450$$

Likewise, for an income of $25,000, the tax is

$$T(25{,}000) = 0.28(25{,}000) - 2788.50 = 4211.50$$
$$\text{because} \qquad 21{,}450 < 25{,}000 < 51{,}900$$

Also,

$$T(80{,}000) = 0.31(80{,}000) - 4345.50 = 20{,}454.50$$
$$\text{because} \qquad 80{,}000 > 51{,}900$$

This function T is an example of a piecewise-defined function. In general, a *piecewise-defined function f* is defined by two or more expressions. Each expression has its own interval in the domain of f. Depending upon the interval that contains an input x, we select the corresponding expression to compute the output $f(x)$.

EXAMPLE 6   Using Piecewise-Defined Functions
Given

$$f(x) = \begin{cases} 2x & x \le 1 \\ \dfrac{6x}{x + 4} & x > 1 \end{cases}$$

evaluate:

(a) $f(3)$ **(b)** $f(-4)$ **(c)** $f(1)$

SOLUTION
The function f is defined by two expressions,

$$2x \qquad \text{and} \qquad \frac{6x}{x + 4}$$

We will determine which expression to use to compute $f(x)$ based upon whether $x \le 1$ or $x > 1$.

(a) Because $3 > 1$,

$$f(3) = \frac{6(3)}{(3) + 4} = \frac{18}{7}$$

(b) Since $-4 \leq 1$, we get

$$f(-4) = 2(-4) = -8$$

(c) Since $1 \leq 1$,

$$f(1) = 2(1) = 2$$

Applications

Functions are used to model the situations of real life. In physics, the distance that an elastic spring stretches is a function of the force pulling on it. In chemistry, the time required to dissolve a solid in a liquid is a function of the temperature of the liquid. In marketing, the number of products sold is a function of how much money is spent on advertising. Using functions such as these to explain and understand connections between quantities is fundamental to many disciplines.

● EXAMPLE 7 Application: Pollution and Wildlife

Environmental science students at a local college find that sewage dumped in a nearby lake is causing a decline in the frog population of the lake. By collecting data over a period of time, the students determine that the number of frogs in the lake (in thousands) after d tons of sewage are dumped is given approximately by

$$F(d) = 10.79 - 1.30d$$

(a) How many frogs remain after 6 tons of sewage have been dumped in the lake?

(b) Determine the zero of F. What is the significance of this number?

SOLUTION

(a) The number of frogs remaining after 6 tons have been dumped in the lake is given by

$$F(6) = 10.79 - 1.30(6) = 2.99$$

The frog population will decline to approximately 2990 if 6 tons of sewage are dumped in the lake.

(b) The zero of F is the solution to the equation $F(d) = 0$.

$$F(d) = 0$$
$$10.79 - 1.30d = 0$$
$$-1.30d = -10.79$$
$$d = \frac{-10.79}{-1.30} = 8.3$$

Because $F(d) = 0$, dumping 8.3 tons of sewage into the lake will cause the frog population to completely disappear.

EXAMPLE 8   Application: The Cost of Manufacturing Radios

From past experience, a manufacturer of portable radios has found that in a particular month, the cost C (in dollars) of manufacturing n units is given by

$$C(n) = 0.06n^2 + 28n + 3200$$

(a) What is the cost of manufacturing 200 units in the month?
(b) How many units can be manufactured during the month if the cost cannot exceed $15,000?
(c) During that month, what is the cost of manufacturing the 200th unit?

SOLUTION
(a) Evaluating the function C at 200 gives

$$C(200) = 0.06(200)^2 + 28(200) + 3200 = 11{,}200$$

Thus, the cost of manufacturing 200 units during the month is $11,200.
(b) If n units can be manufactured for $15,000, then $C(n) = 15{,}000$, or $0.06n^2 + 28n + 3200 = 15{,}000$. We solve this quadratic equation using the quadratic formula:

$$0.06n^2 + 28n + 3200 = 15{,}000$$
$$0.06n^2 + 28n - 11{,}800 = 0$$

So,

$$n = \frac{-28 \pm \sqrt{28^2 - 4(0.06)(-11{,}800)}}{2(0.06)} = 267.7766\ldots \quad \text{or} \quad -734.4432\ldots$$

Our answer must be positive, so we discard $-734.4432\ldots$, the negative solution to the equation. Because it is reasonable to manufacture only complete radios, 267.7766... should be rounded to an integer. Normal rounding rules dictate that this value should be rounded to 268 units, but think about what makes sense in this situation. The cost of manufacturing 268 units during the month would exceed $15,000. The greatest number of radios that can be manufactured during the month so that the cost does not exceed $15,000 is actually 267 units.

(c) The cost of manufacturing the 200th unit is the difference between the costs of manufacturing 200 units and 199 units, or $C(200) - C(199)$. From part (a), we know that $C(200) = 11,200$. We compute $C(199)$ in a similar way:

$$C(199) = 0.06(199)^2 + 28(199) + 3200 = 11,148.06$$

Then

$$C(200) - C(199) = 11,200.00 - 11,148.06 = 51.94$$

The cost of manufacturing the 200th unit is $51.94.

EXERCISE SET 1.2

A

In Exercises 1–4, use the given function to evaluate each expression.

1. $f(t) = 3t - 6$
    (a) $f(4)$ (b) $f(2)$ (c) $f(-2)$
    (d) $f(-3)$

2. $g(x) = -\frac{1}{2}x^2 + 6$
    (a) $g(2)$ (b) $g(0)$ (c) $g(\sqrt{3})$
    (d) $g(-3)$

3. $F(x) = \dfrac{2x}{x^2 - 25}$
    (a) $F(3)$ (b) $F(-5)$ (c) $F(0)$
    (d) $F(-7)$

4. $b(x) = \sqrt{64 - x^2}$
    (a) $b(0)$ (b) $b(3)$ (c) $b(10)$
    (d) $b(\sqrt{15})$

In Exercises 5–12:

(a) Determine the domain of the function and express it in interval notation.

(b) Find the zeros of the function.

5. $f(x) = 2x + 4$ **6.** $g(x) = x^2 - 3x - 10$

7. $p(t) = \sqrt{t + 8}$ **8.** $q(n) = -\sqrt{2n - 4}$

9. $F(y) = \dfrac{3y - 5}{y^2 - y - 6}$ **10.** $h(x) = \dfrac{4x + 1}{x^2 - 7}$

11. $D(t) = \dfrac{\sqrt{5 + t^2}}{4t^2 + 25}$ **12.** $C(x) = \dfrac{\sqrt{2x^2 + 7}}{x^2 + 3}$

In Exercises 13–16, use the given function to simplify each expression.

13. $P(x) = x^2 + 3$
    (a) $P(t + 4)$ (b) $P(t) + P(4)$ (c) $P(t) + 4$
    (d) $t + P(4)$

14. $Q(x) = \dfrac{3}{2x}$
    (a) $Q(2n)$ (b) $2Q(n)$ (c) $Q(2) \cdot Q(n)$
    (d) $Q(2) \cdot n$

15. $R(p) = 5p^3 - 4$
    (a) $R\left(\dfrac{x}{2}\right)$ (b) $\dfrac{R(x)}{R(2)}$ (c) $\dfrac{x}{R(2)}$
    (d) $\dfrac{R(x)}{2}$

16. $T(x) = \dfrac{x + 1}{2x}$
    (a) $T(y - 2)$ (b) $T(2 - y)$ (c) $y - T(2)$
    (d) $T(y) - T(2)$

In Exercises 17–24, use the functions $f(x) = 2x - 6$, $g(x) = 8 - x$, and $P(x) = x^2$ to simplify the difference quotient given.

17. $\dfrac{f(a) - f(5)}{a - 5}$

18. $\dfrac{g(a + 2) - g(a)}{2}$

19. $\dfrac{P(a) - P(5)}{a - 5}$

20. $\dfrac{g(2 + h) - g(2)}{h}$

21. $\dfrac{f(x + 2) - f(x)}{2}$

22. $\dfrac{g(x) - g(x - 4)}{4}$

23. $\dfrac{P(x + 3) - P(x)}{3}$

24. $\dfrac{P(x + 1) - P(x - 1)}{2}$

B

In Exercises 25–28, simplify

$$\frac{f(x + h) - f(x)}{h}$$

for each function f.

25. $f(x) = 4x + 13$

26. $f(x) = 5 - 3x$

27. $f(x) = x^2 - 2x + 7$

28. $f(x) = x - 3x^2$

29. Given the function

$$f(x) = \begin{cases} \frac{1}{2}x & x < 2 \\ 4 - x^2 & x \geq 2 \end{cases}$$

evaluate:

(a) $f(-3)$ (b) $f(0)$ (c) $f(2)$ (d) $f(3)$

30. Given the function

$$g(x) = \begin{cases} \sqrt{-x} & x \leq 0 \\ 2x - 7 & x > 0 \end{cases}$$

evaluate:

(a) $g(-4)$ (b) $g(0)$ (c) $g(4)$ (d) $g(6)$

31. Given the function

$$M(x) = \begin{cases} -\frac{2}{3}x & x < -4 \\ 5 & -4 \leq x < 2 \\ \sqrt{2x} & x \geq 2 \end{cases}$$

evaluate:

(a) $M(-7)$ (b) $M(-1)$ (c) $M(2)$ (d) $M(18)$

32. Given the function

$$n(x) = \begin{cases} -\sqrt{2} & x < -3 \\ x + 5 & -3 \leq x \leq 3 \\ x^2 - 4 & x > 3 \end{cases}$$

evaluate:

(a) $n(-4)$ (b) $n(-3)$ (c) $n(3)$ (d) $n(6)$

In Exercises 33–40, determine the zeros of each function.

33. $W(t) = \frac{1}{2}(t - 6) - 3$

34. $V(n) = 3(n - 4) + 8$

35. $D(x) = x^2 - 8x + 5$

36. $B(y) = y^2 + 2y - 5$

37. $H(x) = \dfrac{x^2 - 7}{x + 3}$

38. $D(p) = \dfrac{p^2 - 6}{p^2 + 3p - 4}$

39. $R(t) = \sqrt{16 - t^2}$

40. $R(x) = \frac{3}{4}\sqrt{x^2 - 12}$

41. The cost (in dollars) of manufacturing x units per day (for $0 \leq x \leq 90$) is given by $C(x) = -0.25x^2 + 50x + 1200$. The revenue (in dollars) generated by x units is given by $R(x) = 55x$. The profit (or loss) generated by x units is the difference between the revenue generated by x units and the cost of manufacturing x units.

(a) How many units can be manufactured for a cost of 3075? What is the revenue generated at this level of production?

(b) Determine a function P that represents the profit (or loss) generated by x units.

(c) Determine the zero of P in part (b). Explain the significance of this value.

42. Consider the function d defined by the distance traveled by the skydiver in the discussion at the beginning of this section and in Example 5. Compute the difference quotient

$$\frac{d(t + 2) - d(t)}{2}$$

for $t = 0, 2, 4, \ldots, 14, 16, 18$. Explain the significance of these values.

43. The per capita federal debt (pcfd) is each U.S. citizen's share of the total federal debt. The table at the top of the next page shows the pcfd for each presidential election year 1968–1992.

Year	1968	1972	1976	1980
pcfd	$1,739	$2,057	$2,852	$3,985

Year	1984	1988	1992
pcfd	$6,640	$10,534	$14,993

Source: U.S. Department of Treasury

Let $d(t)$ be the pcfd in year t.
(a) What is $d(1984)$?
(b) Write an expression using d to represent the increase in pcfd during the years of Presidential Carter's administration (1977–1980).
(c) Compute the difference quotient

$$\frac{d(t+4) - d(t)}{4}$$

for $t = 1968, 1972, 1976, 1980, 1984, 1988$. Explain the significance of these values.

44. A ball is thrown upward from an initial height of h_0 feet with an initial velocity of v_0 feet per second. Its height (in feet) above ground level t seconds after it is thrown is given by the function

$$h(t) = -16t^2 + v_0 t + h_0$$

Suppose that a ball is thrown upward from a tower such that $v_0 = 64$ and $h_0 = 24$.
(a) How high was the ball after 3 seconds?
(b) When did the ball hit the ground (nearest 0.01 second)?
(c) Compute

$$\frac{h(2) - h(0)}{2}$$

What is the significance of this number?

45. A ball is dropped from the top of the Empire State Building (1250 feet tall) in New York City. Its height t seconds after it is dropped is given by $h(t) = 1250 - 16t^2$. After t_0 seconds, it hits the ground.
(a) What is the value of t_0 (nearest 0.01 second)?

(b) Compute

$$\frac{h(t_0) - h(0)}{t_0} \quad \text{and} \quad \frac{h(t_0) - h(t_0 - 1)}{1}$$

Suppose that you were standing on the ground, waiting to catch the dropped ball. Which of these two numbers would be most important to you?

In Exercises 46–49, a table of values has been generated with function f of the form $f(x) = nx$, $f(x) = n/x$, $f(x) = x^n$, or $f(x) = n^x$ (n is constant). Determine f for each table.

46.

x	2	3	6	8	12
$f(x)$	6	4	2	1.5	1

47.

x	0	1	3	5	10
$f(x)$	0	1	27	125	1000

48.

x	0	1	2	3	5
$f(x)$	1	3	9	27	243

49.

x	4	6	10	12	16
$f(x)$	6	9	15	18	24

In Exercises 50–53, a table of values has been generated with function f of the form $f(x) = mx + b$ (m and b are constants) or $f(x) = ax^n$ (a and n are constants). Determine f for each table.

50.

x	0	1	3	5	10
$f(x)$	5	8	14	20	35

51.

x	0	1	4	10	12
$f(x)$	0	$\frac{1}{2}$	8	50	72

52.

x	-3	0	3	4	5
$f(x)$	54	0	-54	-128	-250

53.

x	2	6	9	12	25
$f(x)$	7	9	10.5	12	18.5

C

54. If $f(x + 2) = x^2 + 3x - 2$, what is $f(x)$?

55. If $f(x + 4) = \dfrac{1}{x - 5}$, what is $f(x)$?

56. Given the function $p(x) = -2/x$, evaluate and simplify

$$\frac{p(x + h) - p(x)}{h}$$

57. Given the function $q(x) = x/(x + 1)$, evaluate and simplify

$$\frac{q(x + h) - q(x)}{h}$$

58. Given $f(x) = \sqrt{x}$, evaluate the difference quotient

$$\frac{f(x + h) - f(x)}{h}$$

and simplify the result so that the numerator has no radicals.

59. Given $f(x) = 2/\sqrt{x}$, evaluate the difference quotient

$$\frac{f(x + h) - f(x)}{h}$$

and simplify the result so that the numerator has no radicals.

GRAPHS OF FUNCTIONS

In the last section, functions were represented by tables and also by mathematical expressions. The focus of this section is the third and most revealing representation of a function—its graph.

Graph of a Function

For a function f, the graph of the equation $y = f(x)$ gives a picture that allows us to investigate properties of the function. The graph of a function is more than a collection of plotted points. For example, we will use a graph to find the zeros of a function and to determine its domain and range. Whenever you investigate a function and its characteristics, you should use its graph as a primary part of your investigation.

Graph of a Function

> The **graph of a function** f is the graph of the equation $y = f(x)$.

The points on the graph of the equation $y = f(x)$ are given by the coordinates $(x, f(x))$, as indicated in Figure 1.20. These coordinates are determined by an input x in the domain of f and its corresponding output $f(x)$. Because the y-coordinate, $f(x)$, is determined by the x-coordinate, x is called the *independent*

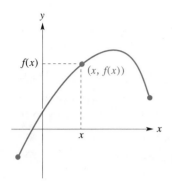

Figure 1.20

variable and *y* is called the *dependent variable*. We also say that *y is a function of x*.

Because each input *x* determines a point on the graph of a function *f*, the domain of *f* is represented by those values on the *x*-axis that correspond to points on the graph (Figure 1.21). Similarly, the range of the function is represented by those values on the *y*-axis that correspond to points on the graph (Figure 1.22).

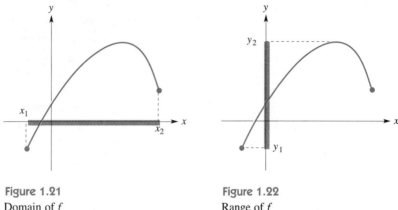

Figure 1.21
Domain of *f*

Figure 1.22
Range of *f*

EXAMPLE 1   Using the Graph of a Function

The graph of a function *f* is shown in Figure 1.23.

(a) Determine the value of $f(-2), f(1)$, and $f(6)$ from the graph.
(b) Determine the domain and range of *f* from the graph.

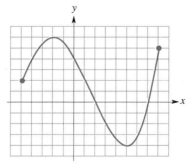

Figure 1.23
$y = f(x)$

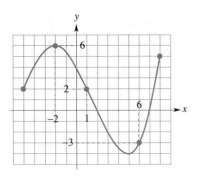

Figure 1.24

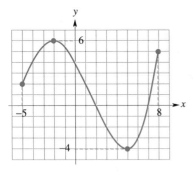

Figure 1.25

SOLUTION

(a) The value of $f(-2)$ is the y-coordinate of the point on the graph with -2 as its x-coordinate. From the graph, it appears that $f(-2) = 6$. Likewise, $f(1) = 2$ and $f(6) = -3$ (Figure 1.24).

(b) The domain is the set of x-coordinates for the points on the graph. From the graph, the domain appears to be $-5 \leq x \leq 8$. The range is the set of y-coordinates for the points on the graph. The range appears to be $-4 \leq y \leq 6$ (Figure 1.25). ■

Not every graph is the graph of a function. Suppose that two points on the graph of a function lie on a vertical line; it follows that one value of x corresponds to two outputs y_1 and y_2 (Figure 1.26). But for a function f, each input x has only one output $f(x)$. Because of this contradiction, this graph cannot be the graph of a function.

On the other hand, if every vertical line intersects the graph at only one point, then the graph does represent a function. If a value x corresponds to a value y, then $y = f(x)$ (Figure 1.27).

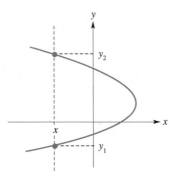

Figure 1.26
Not the graph of a function

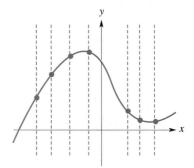

Figure 1.27
Graph of a function

Vertical-Line Test

> A set of points in a coordinate plane is the graph of a function if and only if no two points on the graph lie on the same vertical line. That is, no vertical line intersects the graph at more than one point.

■ EXAMPLE 2   The Vertical-Line Test

Determine whether each graph given is the graph of a function.

(a)

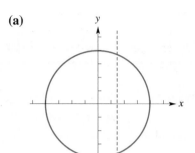

(b)

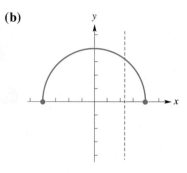

(c)

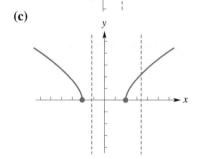

SOLUTION

(a) This graph does not represent a function because the vertical line shown intersects the graph at two points.

(b) This graph represents a function because no vertical line intersects the graph at more than one point (one line is shown).

(c) This graph represents a function. One of the vertical lines shown does not intersect the graph, but this only means that the function is undefined for that value of x.

Intercepts and Turning Points

Because of the vertical-line test, the graph of a function f can have at most one y-intercept, namely $(0, f(0))$ (Figure 1.28). The y-intercept of the graph of f is determined by evaluating $f(0)$. If $f(0)$ is undefined, then 0 is not in the domain of f, and the graph of f has no y-intercept.

An x-intercept on the graph of f is a point $(a, f(a))$ such that $f(a) = 0$ (Figure 1.28). Recall from the last section that the solutions to this equation are the zeros of f.

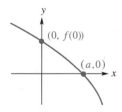

Figure 1.28

Zeros and x-Intercepts of a Function

Given that a is a value in the domain of function f, $(a, f(a))$ is an x-intercept of the graph of f if and only if a is a zero of f.

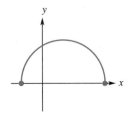

Figure 1.29
$y = \sqrt{16 - (x - 2)^2}$

EXAMPLE 3 Finding the Intercepts and the Domain of a Function

The graph of $h(x) = \sqrt{16 - (x - 2)^2}$ is shown in Figure 1.29. Determine:

(a) The y-intercept   **(b)** The x-intercepts

SOLUTION

(a) The y-intercept of the graph of h is $(0, h(0))$. Because

$$h(0) = \sqrt{16 - (0 - 2)^2} = \sqrt{16 - 4} = \sqrt{12} = 2\sqrt{3}$$

the y-intercept is $(0, 2\sqrt{3})$.

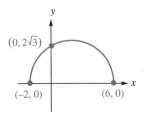

Figure 1.30

(b) Figure 1.29 shows two x-intercepts on the graph of h. Each of these intercepts is of the form $(a, 0)$, in which a is a zero of h. Finding the zeros of h means solving the equation $h(x) = 0$:

$$h(x) = 0$$
$$\sqrt{16 - (x - 2)^2} = 0$$
$$16 - (x - 2)^2 = 0 \qquad \text{Square each side.}$$
$$(x - 2)^2 = 16$$
$$x - 2 = \pm 4 \qquad \text{Take square roots.}$$
$$x = 2 \pm 4$$
$$x = -2 \quad \text{or} \quad 6$$

Because the zeros of h are -2 and 6, the x-intercepts are $(-2, 0)$ and $(6, 0)$. Figure 1.30 shows the graph of h, including its intercepts.

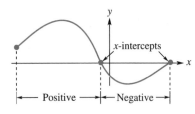

Figure 1.31

Suppose that a number c is in the domain of a function f. If $f(c) > 0$, then f is *positive* at c, and the point $(c, f(c))$ on the graph of f is above the x-axis. Likewise, if $f(c) < 0$, then f is *negative* at c, and the point $(c, f(c))$ on the graph is below the x-axis. If a function is positive (or negative) at each number in a given interval, then the function is *positive* (or *negative*) *on the interval* (Figure 1.31).

A function f is *increasing* over an open interval of its domain if $f(x)$ increases as x increases (Figure 1.32). The graph of f rises from left to right over an interval for which f is increasing. Likewise, f is *decreasing* over an open interval if $f(x)$ decreases as x increases. The graph of f falls from left to right if f is decreasing. The remaining possibility is that $f(x)$ remains *constant* over a given open interval; the graph of f in this case is a horizontal line (Figure 1.32).

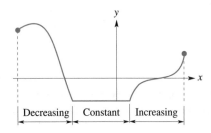

Figure 1.32

These terms are defined more precisely below. As you read these definitions, confirm that they are consistent with the descriptions given above.

Increasing, Decreasing, and
Constant Functions

> A function f is **increasing** over an open interval in the domain of f if, for any two values x_1 and x_2 in the interval, $x_1 < x_2$ implies that $f(x_1) < f(x_2)$.
>
> A function f is **decreasing** over an open interval if, for any x_1 and x_2 in the interval, $x_1 < x_2$ implies that $f(x_1) > f(x_2)$.
>
> A function f is **constant** over an open interval if, for any x_1 and x_2 in the interval, $f(x_1) = f(x_2)$.

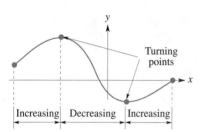

Figure 1.33

A point $(c, f(c))$ is a *turning point* of the graph of f if it separates a portion of the graph for which f is increasing from a portion for which f is decreasing, or vice versa (Figure 1.33). As the name suggests, a turning point is where the graph changes direction. In general, discovering the exact coordinates of the turning points of a graph is difficult or impossible without using calculus. However, we can make a relatively accurate guess from the graph.

⬤ EXAMPLE 4   Finding x-Intercepts and Turning Points

Consider the function f described by the graph in Figure 1.34. Assume that the coordinates of the x-intercepts and turning points are integers. (This is the same graph we discussed in Example 1.)

(a) Determine the x-intercepts.
(b) Determine the intervals over which f is positive and the intervals over which f is negative. (Use inequality notation.)
(c) Determine the turning points of f.
(d) Determine the open intervals over which f is increasing and the open intervals over which f is decreasing. (Use inequality notation.)

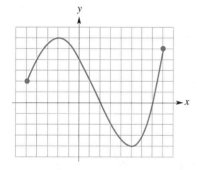

Figure 1.34

SOLUTION

(a) The graph crosses the x-axis at 2 and at 7. The x-intercepts are the points $(2, 0)$ and $(7, 0)$.

(b) The graph of f is above the x-axis over the interval $-5 \le x < 2$ and the interval $7 < x \le 8$. Therefore, f is positive over these intervals. Likewise, f is negative over the interval $2 < x < 7$ because the graph of f lies below the x-axis over this interval.

(c) The graph changes direction at $(-2, 6)$ and $(5, -4)$. These points are the turning points.

(d) The graph of f rises to the turning point $(-2, 6)$, falls to the turning point $(5, -4)$, and rises again. Thus, f is increasing over the open intervals $-5 < x < -2$ and $5 < x < 8$, and it is decreasing over the open interval $-2 < x < 5$.

Directory of Graphs

The graphs of certain functions that arise frequently in mathematics and many applications are shown in the following directory of graphs. Being familiar with these basic graphs will help you analyze more complicated graphs later in this course. Practice sketching each of these eight graphs until you can do it quickly and without difficulty.

Directory of Graphs

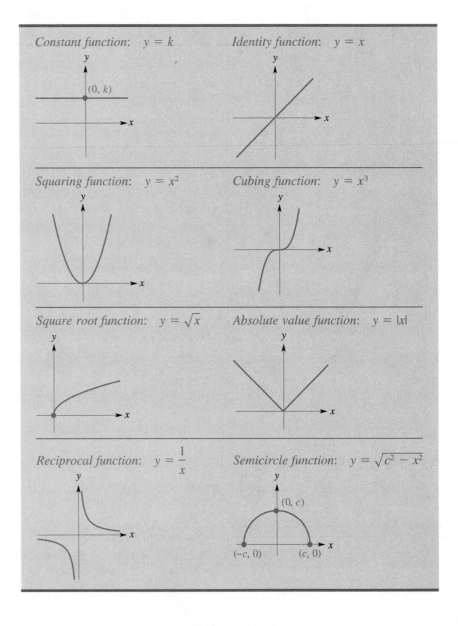

In the next example, graphs from the directory are used to construct the graph of a piecewise-defined function.

● EXAMPLE 5   Sketching a Piecewise-Defined Function

Sketch the graph of $f(x) = \begin{cases} 2 & x < -3 \\ x & -3 \leq x \leq 1 \\ \dfrac{1}{x} & x > 1 \end{cases}$

SOLUTION

The first step in sketching the graph of this piecewise-defined function is to sketch the graphs of the constant function $y = 2$, the identity function $y = x$, and the reciprocal function $y = 1/x$ in the same coordinate plane (Figure 1.35). The graph of f is made up of portions of these three graphs. It is the same as $y = 2$ for $x < -3$, because $f(x) = 2$ over this interval. Likewise, the graph is the same as $y = x$ over the interval $-3 \leq x \leq 1$, and the same as $y = 1/x$ over the interval $x > 1$ (Figure 1.36).

● **NOTE:** *Because $f(-3) = -3$, the point $(-3, -3)$ is on the graph of f. The open circle at $(-3, 2)$ indicates that this point is not on the graph.*

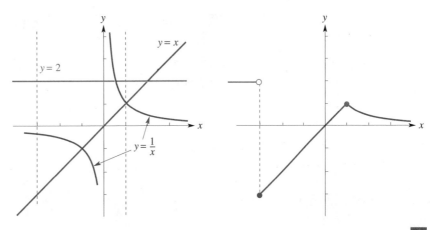

Figure 1.35   Figure 1.36

Odd and Even Functions

In Section 1.1, we discussed symmetry and how it reduced the labor in sketching the graph of an equation. We saw that the graph of an equation is symmetric with respect to the origin if replacing (x, y) with $(-x, -y)$ in the equation results in an equivalent equation. Thus, the graph of a function f is symmetric with respect

Can the graph of a function be symmetric with respect to the x-axis? See Exercise 59.

to the origin if $y = f(x)$ is equivalent to $-y = f(-x)$, or $y = -f(-x)$. A function with a graph that is symmetric with respect to the origin is called an *odd function*.

We also saw that the graph of an equation is symmetric with respect to the y-axis if replacing (x, y) with $(-x, y)$ in the equation results in an equivalent equation. For a function f, if $y = f(x)$ is equivalent to $y = f(-x)$, then its graph is symmetric with respect to the y-axis. A function with a graph that is symmetric with respect to the y-axis is called an *even function*.

Determining that a given function is odd or even will help you in your investigation of a function. The test for odd and even functions is summarized here:

Test for Odd and Even Functions

To determine whether a function f is an odd function or an even function, evaluate and simplify $f(-x)$. Then:

f is an **odd function** if $f(-x) = -f(x)$

f is an **even function** if $f(-x) = f(x)$

f is neither even nor odd if $f(-x) \neq -f(x)$ and $f(-x) \neq f(x)$

Example 6 shows how the graph of a function will suggest whether it is odd or even, and how the test stated above can verify this conjecture.

EXAMPLE 6 Testing for Odd and Even Functions

Given each function and its graph, determine whether the function is even, odd, or neither.

(a) $f(x) = x^2 - 4$ **(b)** $g(x) = x^3 - 4x$

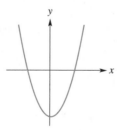   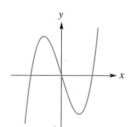

(c) $h(x) = x^2 - 4x$

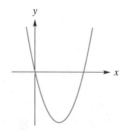

SOLUTION

(a) The graph appears to be symmetric with respect to the y-axis, so we suspect that f is an even function, and we check to see if $f(-x) = f(x)$:

$$f(-x) = (-x)^2 - 4 = x^2 - 4 = f(x)$$

Because $f(-x) = f(x)$, the function f is an even function.

(b) The graph of g suggests symmetry with respect to the origin, so we check to see if $g(-x) = -g(x)$:

$$g(-x) = (-x)^3 - 4(-x) = -x^3 + 4x = -(x^3 - 4x) = -g(x)$$

The function g is an odd function, since $g(-x) = -g(x)$.

(c) Neither symmetry is apparent in the graph of h.

$$h(-x) = (-x)^2 - 4(-x)$$
$$= x^2 + 4x$$

Because $x^2 + 4x$ is neither $h(x)$ nor $-h(x)$, the function h is neither even nor odd.

Qualitative Graphs

The graph of a function is more than just a set of plotted points. It shows the qualities of the function, such as where it increases and decreases, where it is positive and negative, and how many zeros it has.

A rough sketch of a graph can reveal these qualities, even without accurately plotting points on the graph. Such a sketch is called a *qualitative graph of the function*, because it emphasizes the broader qualities of the graph rather than the precise numerical values of the function. The next example shows how a qualitative graph can describe the temperature of a cabin as a function of time.

EXAMPLE 7 Sketching a Qualitative Graph

From the following scenario, sketch a qualitative graph of the room temperature of the cabin as a function of time.

Dawn breaks. Robert wakes up, and gets out of bed. His cabin is cold from the night's chill, so he lights his wood stove. The cabin warms slowly at first, but then rapidly reaches a comfortable temperature. Robert adds fuel to the wood stove to maintain this temperature throughout the day. Later in the day, he leaves the cabin for a long hike. While he is gone, the fire burns out and the cabin cools. When he returns, he relights the fire and the cabin warms again to a comfortable temperature.

SOLUTION

Of all that we are told that happens, only certain events affect the temperature of Robert's cabin. We do not need to consider, for example, that dawn breaks or that Robert gets up. The events that affect the temperature of the cabin are these:

1. Robert lights the wood stove.
2. The room reaches a comfortable temperature.
3. The temperature begins to drop after the fire burns out.
4. Robert returns and relights the fire.
5. The room again reaches a comfortable temperature.

How do each of these events affect the temperature of the cabin? After event 1, the temperature increases slowly at first and then rapidly. After event 2, the temperature levels off. The temperature decreases after event 3 until event 4, and then increases until event 5. These events are shown in Figure 1.37.

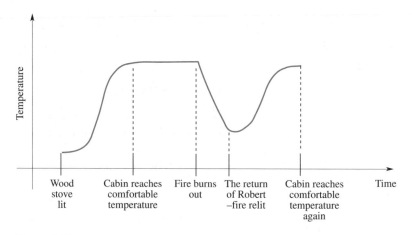

Figure 1.37

Applications

The next example shows that the graph of a function can reveal a great deal about the situation that the function represents.

◼ EXAMPLE 8   Application: Wind-Chill

When the wind blows strongly on a cold day, a person feels even colder than the actual air temperature because the combination of the wind and cold causes more heat loss than just the cold alone. This combination is described by the temperature that would cause the same heat loss with no wind; this number is called the

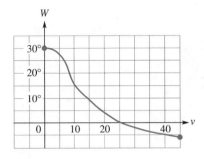

Figure 1.38

Source: National Weather Service

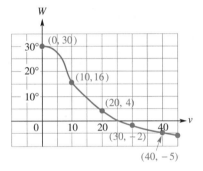

Figure 1.39

wind-chill temperature. The graph in Figure 1.38 shows the wind-chill temperature W when the actual temperature is 30°F for winds blowing v miles per hour for $0 \leq v \leq 45$. (Winds greater than 45 mph have little extra chilling effect.)

(a) Use the graph to construct a table for the function using the values $v = 0, 10, 20, 30, 40$. (Approximate the values of W to the nearest degree.)

(b) Determine the domain of the function.

(c) Determine the intervals over which the function is positive and the intervals over which it is negative.

(d) Determine the open intervals over which the function is increasing and the intervals over which it is decreasing.

SOLUTION

(a) The points $(0, 30)$, $(10, 16)$, $(20, 4)$, $(30, -2)$, and $(40, -5)$ are on the graph (Figure 1.39). These ordered pairs are used to create a table:

v	0	10	20	30	40
W	30	16	4	-2	-5

(b) The domain is the interval $0 \leq v \leq 45$, the set of all values v for which the function is defined.

(c) The function is positive for $0 \leq v < 25$, because the graph is above the v-axis over this interval. It is negative for $25 < v \leq 45$, because the graph is below the v-axis on this interval.

(d) Because the graph falls from left to right over its complete domain, the function decreases over the open interval $0 < v < 45$. It does not increase anywhere.

EXERCISE SET 1.3

A

In Exercises 1–8, use the graphs shown for $y = f(x)$ and $y = g(x)$.

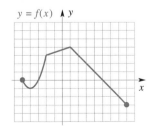

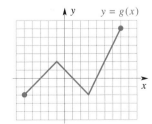

1. Determine the value of $f(-4)$, $f(-2)$, $f(2)$, $f(4)$, and $f(7)$ from the graph of f.

2. Determine the value of $g(-5)$, $g(-1)$, $g(2)$, $g(5)$, and $g(7)$ from the graph of g.

3. Determine the domain and range of f from the graph of f.

4. Determine the domain and range of g from the graph of g.

5. **(a)** Determine the x-intercepts of the graph of f. (Assume that the coordinates are integers.)

(b) Determine the intervals over which f is positive and the intervals over which f is negative.

6. (a) Determine the x-intercepts of the graph of g. (Assume that the coordinates are integers.)
   (b) Determine the intervals over which g is positive and the intervals over which g is negative.

7. (a) Determine the turning points of the graph of f. (Assume that the coordinates are integers.)
   (b) Determine the open intervals over which f is increasing and the open intervals over which f is decreasing.

8. (a) Determine the turning points of the graph of g. (Assume that the coordinates are integers.)
   (b) Determine the open intervals over which g is increasing and the open intervals over which g is decreasing.

In Exercises 9–16, determine whether each graph represents a function.

9.
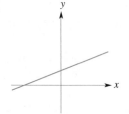

10.
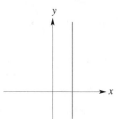

11.
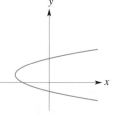

12.
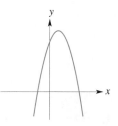

13.
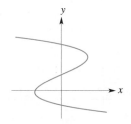

14.
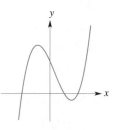

15.
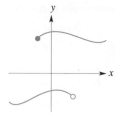

16.
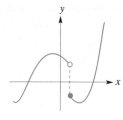

In Exercises 17–24, simplify $f(-x)$, and determine whether f is an even function, an odd function, or neither.

17. $f(x) = x^3 + 2x$
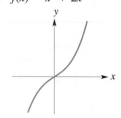

18. $f(x) = 7 - x^3$
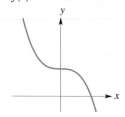

19. $f(x) = |x + 8|$
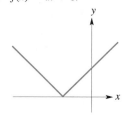

20. $f(x) = |x| + 8$
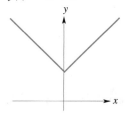

21. $f(x) = \dfrac{12}{x^2 + 4}$
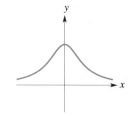

22. $f(x) = \dfrac{3x}{x^2 + 4}$
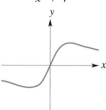

23. $f(x) = x^{3/5} + 5$

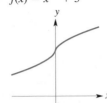

24. $f(x) = x^{2/3} - 4$

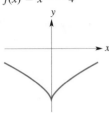

33. $f(x) = \begin{cases} -3 & x < 0 \\ x & 0 \le x < 4 \\ 1 & x \ge 4 \end{cases}$

34. $f(x) = \begin{cases} -5 & x \le -3 \\ 4 & -3 < x < 0 \\ \sqrt{x} & x \ge 0 \end{cases}$

In Exercises 25–28, the graph of the given function is shown. Determine:

(a) The y-intercept   **(b)** The x-intercepts

25. $D(x) = 2x - 8$

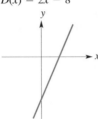

26. $p(x) = -\frac{1}{2}x + 3$

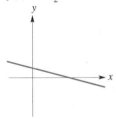

27. $f(x) = x^2 - 6$

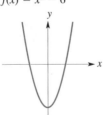

28. $Q(x) = 4x - x^2$

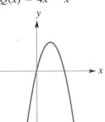

B

In Exercises 35–38, the graph of the given function is shown with its turning points labeled.

(a) Determine the x-intercepts.
(b) Determine the intervals over which the function is positive and the intervals over which it is negative.
(c) Determine the open intervals over which the function is increasing and the open intervals over which it is decreasing.

35. $g(x) = 12x - x^3$

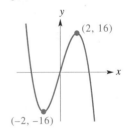

36. $L(x) = (4 - x^2)(2 + x^2)$

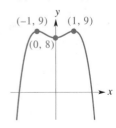

In Exercises 29–34, use the directory of graphs to sketch the graph of each piecewise-defined function.

29. $f(x) = \begin{cases} x & x < 4 \\ 4 & x \ge 4 \end{cases}$ **30.** $f(x) = \begin{cases} -3 & x < -3 \\ x & x \ge -3 \end{cases}$

31. $f(x) = \begin{cases} 4 & x \le -4 \\ |x| & -4 < x < 2 \\ 2 & x \ge 2 \end{cases}$

32. $f(x) = \begin{cases} 0 & x < -5 \\ \sqrt{25 - x^2} & -5 \le x < 3 \\ 4 & x \ge 3 \end{cases}$

37. $h(x) = \dfrac{x^2 + 4}{x}$

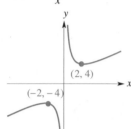

38. $C(x) = -x^2 + 2|x| + 5$

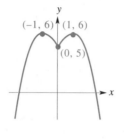

45. $y = x^{2/3}$

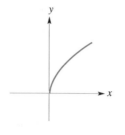

46. $y = -2x^{3/5}$

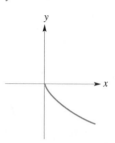

In Exercises 39–46, a portion of the graph of the given function is shown. Determine whether the function is even or odd and use that fact to complete the graph.

39. $y = 9x - x^3$

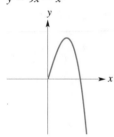

40. $y = 9x^2 - x^4$

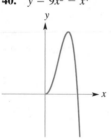

41. $y = |x| - 6$

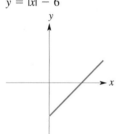

42. $y = x|x| - 2x$

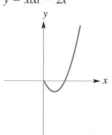

43. $y = \dfrac{12x}{x^2 + 1}$

44. $y = \dfrac{8x^2}{x^2 + 4}$

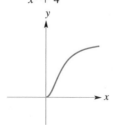

47. From the following scenario, sketch a qualitative graph for productivity as a function of the number of employees.

> *The productivity at an understaffed factory increases when additional employees are hired. At a certain level of employment, the factory achieves its maximum productivity. Adding more employees after this point actually decreases productivity.*

48. From the following scenario, sketch a qualitative graph for fuel efficiency as a function of speed.

> *The fuel efficiency of a car (at low speeds) increases as speed increases. Efficiency begins to level off, however, at moderate speeds. It then begins to decrease at high speeds. Efficiency is lowest at the maximum speed of the car.*

49. From the following scenario, sketch a qualitative graph for Neil's distance from home as a function of time.

> *Neil leaves home in his car, bound for class. Since he is early, he starts off slowly. Suddenly, he remembers that his homework is at home on the table, so he turns around and returns home to get his homework. Running short of time, he drives faster now, arriving at school just in time for class.*

50. From the following scenario, sketch a qualitative graph for water temperature as a function of time.

> *Ice is removed from a freezer, and heat is applied. The ice warms to melting temperature at a constant rate. It stays at this melting temperature until it is completely melted. It warms again at a constant rate until it reaches boiling temperature. The temperature remains constant until the water completely boils away.*

51. The flow F of a river (measured in hundred cubic feet per second) on day d is represented by the graph:

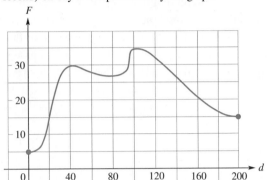

   (a) Use the graph to construct a table for the function using the values $d = 0, 40, 80, 120, 160, 200$. (Approximate the values of F to the nearest hundred cubic feet per second.)

   (b) Determine the open intervals over which the flow of the river is increasing and over which it is decreasing.

   (c) What is the maximum flow according to this graph? What is the minimum? For what values of d do these occur?

52. A small software company finds that their profit P (in thousands of dollars) is a function of how many hours h (in hundreds of hours) their employees work each week. The figure shows the graph of this function:

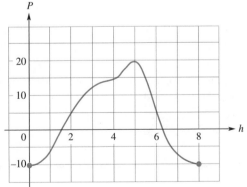

   (a) Use the graph to construct a table for the function using the values $h = 0, 2, 4, 6, 8$. (Approximate the values of P to the nearest thousand dollars.)

   (b) Determine the intervals over which the profit is positive and the intervals over which it is negative.

   (c) Determine the open intervals over which the profit is increasing and the open intervals over which it is decreasing.

   (d) What is the maximum profit possible according to this graph? What is the minimum? For what values of h do these occur?

C

53. The cost of first class postage in the United States is a function of the weight of the letter. Letters that are 1 ounce or less cost 32¢. Each additional ounce, or any fraction of an ounce, costs another 23¢. Sketch this function for 0–5 ounces.

54. The fee schedule for a certain taxi cab is $2.50 plus $0.75 for each mile, or any fraction of a mile, traveled over 2 miles. Sketch the cost of a ride as a function of distance for 0 to 6 miles.

55. Consider the function $f(x) = x^n$, where n is a positive integer. For what n is f an odd function? An even function? Explain why this suggests the names "odd" and "even."

56. Which of the functions shown in the directory of graphs in this section are odd functions? Even functions? Neither?

57. **(a)** If an odd function f has a zero a ($a \neq 0$), can you determine another zero of f? Explain your reasoning.

   (b) If an even function f has a zero a ($a \neq 0$), can you determine another zero of f? Explain your reasoning.

58. **(a)** If the graph of an odd function f has a turning point P (P is not the origin), can you determine another turning point of f? Explain your reasoning.

   (b) If the graph of an even function f has a turning point P (P is not the origin), can you determine another turning point of f? Explain your reasoning.

59. Consider a function with the set of all real numbers as its domain. Explain why, with one exception, the graph of this function is not symmetric with respect to the x-axis. What is the exception?

60. The *greatest-integer function*, $\lfloor x \rfloor$, is a real function that assigns to each real number x the greatest integer that is less than or equal to x. For example, $\lfloor \pi \rfloor = 3$, $\lfloor 7 \rfloor = 7$, and $\lfloor -4.2 \rfloor = -5$. Sketch the graph of $y = \lfloor x \rfloor$.

SECTION 1.4

- The Viewing Window
- Complete Graphs
- Zoom and Trace Features
- Approximating Zeros of a Function Graphically
- Applications

GRAPHS AND GRAPHING UTILITIES

In the last section, we saw that the graph of a function reveals much information about the function. Traditionally, the method for sketching the graph of a function was to first evaluate the function at some values in its domain, plot the corresponding ordered pairs on the coordinate plane, and then sketch a curve passing through the points. The problem with this method is that plotting too few points means that this sketched curve could be inaccurate or even wrong. On the other hand, plotting a great number of points to get an accurate curve can be a grueling, boring task.

A *graphing utility* solves this dilemma. Simply stated, a graphing utility is a graphics calculator or a computer software program that can accept a function in the form of a mathematical expression, plot many points on the graph, and connect them with a curve. A graphing utility usually will give a highly accurate graph quickly and with little effort on your part.

No technological solution is without its own new problems, however. Because a graphing utility plots only a certain number of points, an important part of the graph (such as an intercept or a turning point) may not be represented exactly by a graphing utility. Also, there is no guarantee that two different graphing utilities (or two identical utilities with different settings) will represent the graph of a function in exactly the same way. It is important to be aware of the features of the graph ahead of time and add them to your own sketch of the function even though your graphing utility may not show them.

● **GRAPHING NOTE:** *The screen of a graphing utility is a grid of small squares called pixels. The plotted points and the connecting curve are represented by these squares on the screen. The size and number of the pixels depends upon the graphing utility. In general, screens with more pixels give more precise graphs.*

The Viewing Window

The screen of a graphing utility shows only a portion of the coordinate plane. This portion, called the *viewing window*, is determined by four values. The left side of the viewing window is defined by *Xmin*, the minimum value of x on the screen, and the right side is defined by *Xmax*, the maximum value of x on the screen. Similarly, the top and the bottom are defined by *Ymin* and *Ymax*. Also, the distance between the marks on the x-axis is the *Xscale* of the window, and the distance between the marks on the y-axis is the *Yscale* of the window.

Describing a Viewing Window

The viewing window of the coordinate plane that is described by

$$[Xmin, Xmax]_{Xscale} \times [Ymin, Ymax]_{Yscale}$$

is the set of all points (x, y) on the plane such that $Xmin \leq x \leq Xmax$ and $Ymin \leq y \leq Ymax$. The marks on the x-axis and y-axis are defined by

Xscale and Yscale. If either Xscale or Yscale is omitted in the description of a viewing window, then we assume that this value is 1.

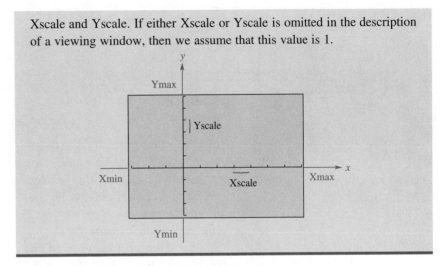

For example, $[-4, 6] \times [-20, 30]_5$ defines the viewing window in Figure 1.40. The *standard viewing window* is described by $[-10, 10] \times [-10, 10]$ (Figure 1.41).

One advantage of using a graphing utility is that you can easily change the viewing window to examine the different aspects of the graph of a function. The next example shows how different viewing windows give vastly different impressions of the graph of a function.

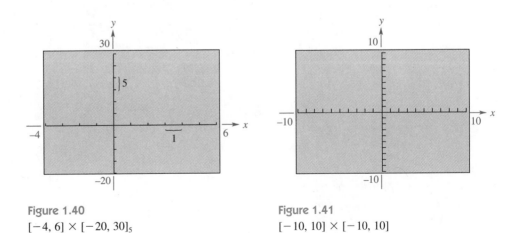

Figure 1.40
$[-4, 6] \times [-20, 30]_5$

Figure 1.41
$[-10, 10] \times [-10, 10]$

⬤ EXAMPLE 1   Changing the Viewing Window

Use a graphing utility to determine the graph of $f(x) = x^3 - 8x - 2^x$ in each viewing window.

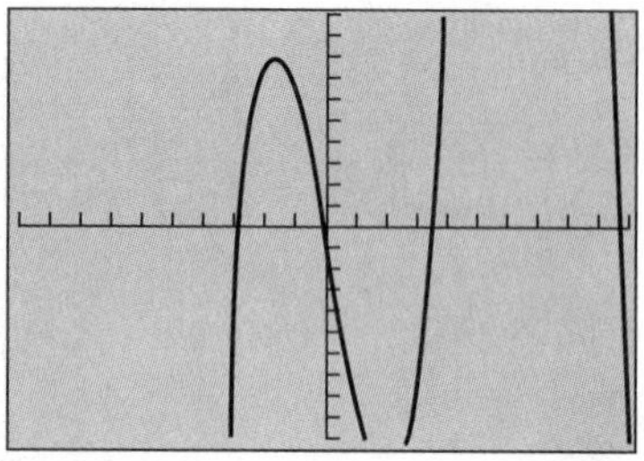

Figure 1.42
$y = x^3 - 8x - 2^x$
$[-10, 10] \times [-10, 10]$

(a) $[-10, 10] \times [-10, 10]$ **(b)** $[3, 4]_{0.1} \times [-5, 5]$
(c) $[-20, 20]_{10} \times [-3000, 1000]_{500}$ **(d)** $[-6, 12]_3 \times [-100, 200]_{50}$

SOLUTION

Enter $y = x^3 - 8x - 2^x$ into the graphing utility.

(a) The viewing window $[-10, 10] \times [-10, 10]$ is the standard viewing window. (Your graphing utility may have a feature that automatically sets this window.) This viewing window shows four x-intercepts and one turning point, but it appears that the graph has at least two other turning points not shown in the viewing window (Figure 1.42).

(b) The viewing window $[3, 4]_{0.1} \times [-5, 5]$ in Figure 1.43 shows the intercept that appears between 3 and 4 on the x-axis. Because Xscale is 0.1, the marks on the x-axis indicate 3.0, 3.1, 3.2, and so on to 4.0. This window shows that the intercept is between 3.3 and 3.4.

(c) The viewing window $[-20, 20]_{10} \times [-3000, 1000]_{500}$ in Figure 1.44 shows the turning points and x-intercepts of the graph, but the difference between Ymin and Ymax makes these features of the graph difficult to distinguish.

GRAPHING NOTE: *If the y-axis does not intersect the window, most graphing utilities will mark the left side of the screen (as in Figure 1.43) according to the value of Yscale.*

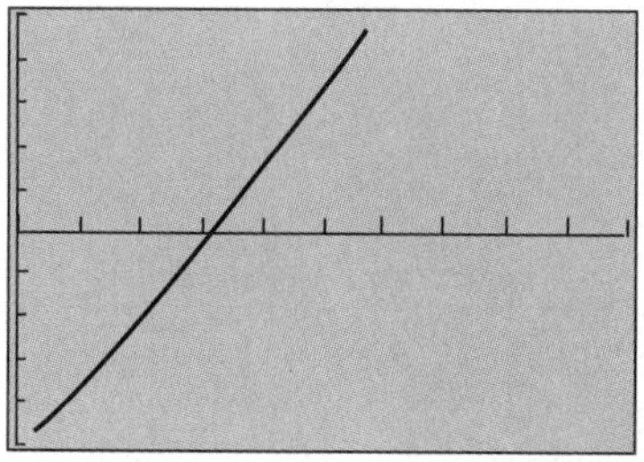

Figure 1.43
$y = x^3 - 8x - 2^x$
$[3, 4]_{0.1} \times [-5, 5]$

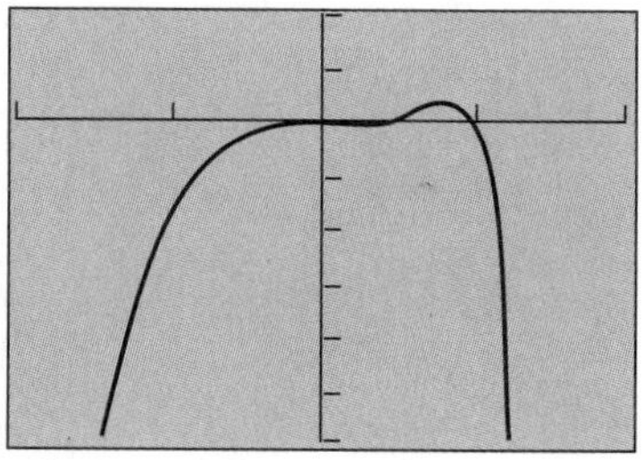

Figure 1.44
$y = x^3 - 8x - 2^x$
$[-20, 20]_{10} \times [-3000, 1000]_{500}$

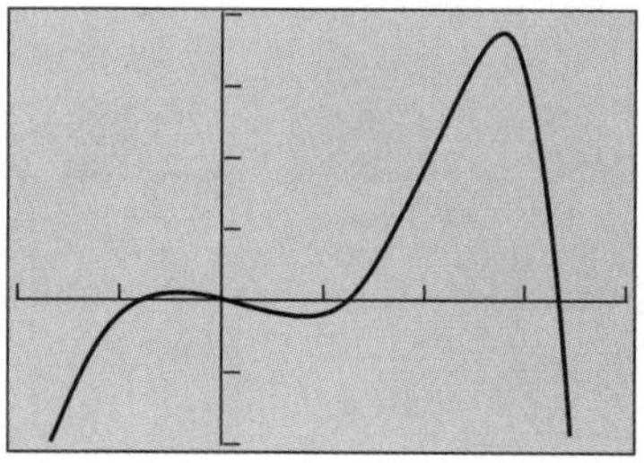

Figure 1.45
$y = x^3 - 8x - 2^x$
$[-6, 12]_3 \times [-100, 200]_{50}$

(d) The viewing window $[-6, 12]_3 \times [-100, 200]_{50}$ in Figure 1.45 shows the same features of the graph as Figure 1.44, but because the difference between Ymin and Ymax is smaller, these features appear more prominently.

■ **GRAPHING NOTE:** *Parts (c) and (d) of Example 1 make a very important point. The shape of the graph shown by the graphing utility depends greatly on the viewing window.*

Complete Graphs

The graph of a function has intercepts, turning points, and other interesting features. A viewing window that best shows all these features of the graph is called a *complete view of the graph*, or more simply, a *complete graph*. For example, Figure 1.45 shows a complete graph of $y = x^3 - 8x - 2^x$. (Figure 1.44 also shows these features, but not as clearly.)

This definition is not precise, but for our purposes it will be precise enough. As you learn more about the graphs of functions, you will be able to determine a complete graph more easily. Example 2 shows how experimenting with the viewing window will lead to a complete graph of a function.

■ **EXAMPLE 2 Finding a Complete Graph of a Function**

Determine the complete graph of $f(x) = \frac{1}{3}x^3 - 12x$.

SOLUTION

After entering $y = \frac{1}{3}x^3 - 12x$ into the graphing utility, we set the standard viewing window $[-10, 10] \times [-10, 10]$ shown in Figure 1.46. This window does not capture the complete graph of the equation; it appears that there are turning points, and there may also be other x-intercepts that don't show in this viewing window. So we enlarge the viewing window to $[-40, 40]_{10} \times [-40, 40]_{10}$ to get more of the graph on the screen (Figure 1.47).

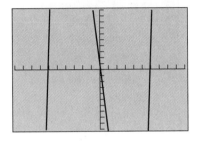

Figure 1.46
$y = \frac{1}{3}x^3 - 12x$
$[-10, 10] \times [-10, 10]$

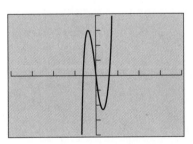

Figure 1.47
$y = \frac{1}{3}x^3 - 12x$
$[-40, 40]_{10} \times [-40, 40]_{10}$

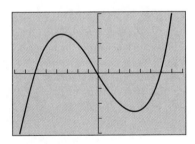

Figure 1.48
$y = \frac{1}{3}x^3 - 12x$
$[-8, 8] \times [-40, 40]_{10}$

■ **GRAPHING NOTE:**

Certainly, other viewing windows would also show the complete graph of this equation. Try $[-9, 10] \times [-30, 30]_{10}$, $[-12, 12] \times [-45, 50]_{10}$, and $[-15, 10] \times [-40, 30]_{10}$. These viewing windows all convey the same information as Figure 1.48.

Enlarging the viewing window even more should persuade you that the intercepts and the turning points shown in Figure 1.47 are the only intercepts and turning points of the graph. The viewing window $[-8, 8] \times [-40, 40]_{10}$ allows a closer look at the graph while still showing the important features (Figure 1.48). This viewing window is a complete graph of $y = \frac{1}{3}x^3 - 12x$. ●

Zoom and Trace Features

Most graphing utilities have a feature that allows you to change the viewing window rapidly. With the *zoom* features, you can see a larger portion of the coordinate plane, or focus more closely on a smaller portion of the current screen.

The *zoom-out* feature expands the viewing window about a fixed point in the current viewing window (Figure 1.49).

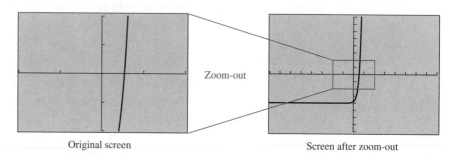

Original screen   Zoom-out   Screen after zoom-out

Figure 1.49

● **GRAPHING NOTE:** *The factors by which these expansions and reductions occur are determined by the graphing utility. Your utility may have other zoom features in addition to zoom-in and zoom-out.*

Similarly, the *zoom-in* feature reduces the viewing window about a chosen point in the current viewing window (Figure 1.50).

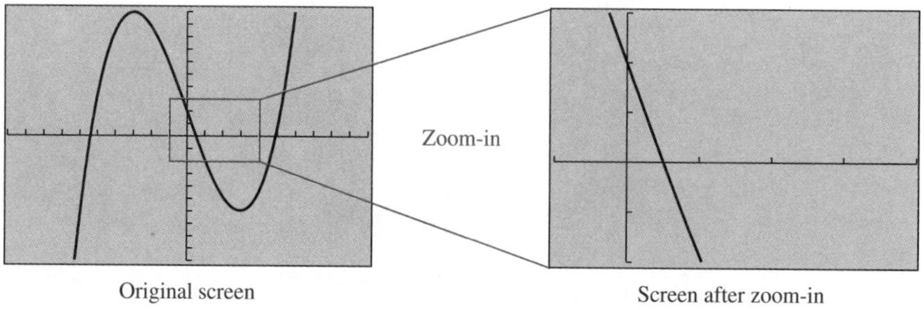

Original screen   Zoom-in   Screen after zoom-in

Figure 1.50

■ **EXAMPLE 3 Using Zoom to Find a Complete Graph**

Use the zoom feature to find a complete graph of $f(x) = \frac{1}{4} x^4 - 2x^2 - 5x - 12$.

SOLUTION

We enter $y = \frac{1}{4} x^4 - 2x^2 - 5x - 12$ in the graphing utility and use the standard viewing window (Figure 1.51a). The graph has *x*-intercepts near $x = -3$ and $x = 4$. There appears to be a turning point below this viewing window. Using the zoom-out feature (Figure 1.51b) allows more of the graph to be shown. (We also changed the Xscale and Yscale to 10.)

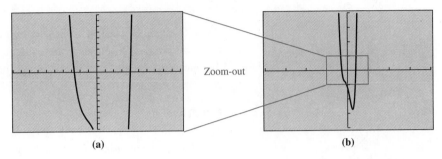

(a)   (b)

Figure 1.51
$y = \frac{1}{4}x^4 - 2x^2 - 5x - 12$

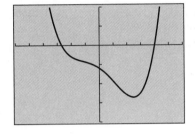

Figure 1.52
$y = \frac{1}{4}x^4 - 2x^2 - 5x - 12$
$[-6, 6] \times [-40, 20]_{10}$

We change the viewing window to $[-6, 6] \times [-40, 20]_{10}$ to get a better view of this portion of the graph (Figure 1.52). This is a complete graph.

Another important feature, the *trace* feature, enables you to determine the coordinates of the plotted points used by the graphing utility to draw the graph. The trace feature can help you approximate the coordinates of intercepts and turning points.

EXAMPLE 4   Approximating the Turning Point of a Graph

Use the trace feature to approximate the turning point in quadrant IV of the graph of $f(x) = \frac{1}{3}x^3 - 12x$ in the viewing window $[-8, 8] \times [-40, 40]_{10}$. (This graph was discussed in Example 2.)

SOLUTION

The graph of $f(x) = \frac{1}{3}x^3 - 12x$ in this viewing window is shown in Figure 1.53. Activating the trace feature displays the trace cursor on the graph of the function and the coordinates of the cursor. Moving the cursor along the curve gives the coordinates of the plotted points on the graph. The point with the smallest y-coordinate is the best approximation of the turning point in this viewing window. To the nearest 0.1, the coordinates of the turning point are $(3.5, -27.7)$.

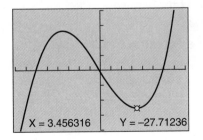

X = 3.456316   Y = −27.71236

Figure 1.53
$y = \frac{1}{3}x^3 - 12x$

Approximating Zeros of a Function Graphically

In the last section, we saw that finding the intercepts of the graph of a function f involved algebraic techniques. Finding the y-intercept required evaluating $f(0)$.

Finding the x-intercepts called for determining the zeros of f by solving the equation $f(x) = 0$. These algebraic solutions may be aided and verified by using the graph of the function furnished by a graphing utility.

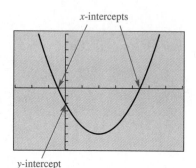

x-intercepts

y-intercept

Figure 1.54
$y = x^2 - 4x - 2$
$[-3, 7] \times [-8, 7]$

⬤ EXAMPLE 5   Using a Graph to Find the Zeros of a Function

Use a graphing utility to determine the graph of $f(x) = x^2 - 4x - 2$. Determine the intercepts exactly, and then use a graphing utility to approximate the x-intercepts to the nearest 0.1.

SOLUTION

The graph of $y = x^2 - 4x - 2$ is shown in Figure 1.54. The graph appears to have a y-intercept close to -2 on the y-axis, and it appears to have two x-intercepts, one between -1 and 0 and one between 4 and 5 on the x-axis. Since $f(0) = (0)^2 - 4(0) - 2 = -2$, the y-intercept is $(0, -2)$.

Solving $x^2 - 4x - 2 = 0$ with the quadratic formula yields

$$x = \frac{-(-4) \pm \sqrt{(-4)^2 - 4(1)(-2)}}{2(1)}$$

$$= \frac{4 \pm \sqrt{24}}{2} = \frac{4 \pm 2\sqrt{6}}{2} = \frac{2(2 \pm \sqrt{6})}{2} = 2 \pm \sqrt{6}$$

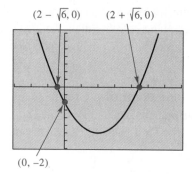

$(2 - \sqrt{6}, 0)$ $(2 + \sqrt{6}, 0)$

$(0, -2)$

Figure 1.55

The x-intercepts are $(2 + \sqrt{6}, 0)$ and $(2 - \sqrt{6}, 0)$, as shown in Figure 1.55.

To the nearest 0.1, the x-intercepts are $(4.4, 0)$ and $(-0.4, 0)$. These computations support our earlier conjectures about the intercepts based on the graph.   ⬤

The algebraic methods we used in Example 5 to find the x-intercepts of the graph of a function f fail if we cannot solve the equation $f(x) = 0$ to find the zeros of f. The next example shows how x-intercepts can be determined graphically and how this helps determine the zeros of the function.

⬤ EXAMPLE 6   Finding the Zeros of a Function Graphically

The zero of the function $b(x) = 2^x - 8$ is an integer. Use the graph of the function in the standard viewing window to guess the zero, and then verify your guess.

SOLUTION

Figure 1.56 shows the graph of $y = 2^x - 8$ in the standard viewing window. From the graph, it appears that the x-intercept is $(3, 0)$. This implies that the zero of the graph is 3. Evaluating $b(3)$, we get

$$b(3) = 2^3 - 8 = 8 - 8 = 0$$

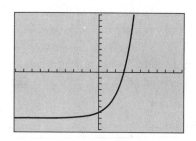

Figure 1.56
$y = 2^x - 8$
$[-10, 10] \times [-10, 10]$

This calculation verifies that 3 is a zero of b.   ⬤

Of course, a zero of a function is not always an integer. We can, however, approximate the x-coordinate of an x-intercept, which will give us an approximation to the corresponding zero of the function.

Many graphing utilities have a *root* feature, which gives a very precise approximation to the x-coordinate of an x-intercept. If your graphing utility has this feature, it is certainly worth your effort to learn how to use it. If your graphing utility does not have a root feature, you can still approximate a zero by using the zoom and trace features to any degree of precision, such as to the nearest 0.01 or to the nearest 0.0001.

First, though, we need to define what we mean by *precision*. If two numbers a and b when rounded to a given precision yield equal results, we will say that a and b are *equal to the given precision*. For example, 2.3157895 and 2.3243158 both yield 2.32 when rounded to the nearest 0.01. Thus, 2.3157895 and 2.3243158 are equal to the nearest 0.01.

Example 7 demonstrates a procedure that you can use to approximate a zero of a function if your graphing utility does not have a root feature.

EXAMPLE 7   Approximating the Zeros of a Function

Approximate the zero of $f(x) = x^2 - 6x + 7$ in the interval $[4, 5]$ to the nearest 0.01.

SOLUTION

Our plan of attack is to start with the graph of $y = x^2 - 6x + 7$ in the standard viewing window, move in on the intercept in question by using the zoom-in feature, and then use the trace feature to investigate the graph around this intercept. If we can determine two points of the graph, one on each side of the intercept, such that their x-coordinates are equal to the nearest 0.01, we have accomplished our goal. If not, we use the zoom-in feature once more and try again.

Figure 1.57 shows the screens of a graphing utility that took three iterations of this process.

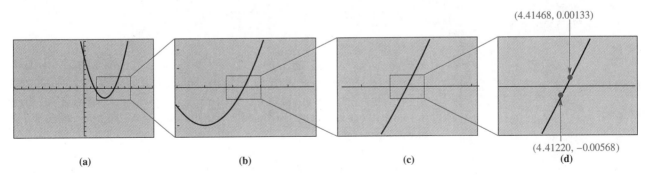

(a)   (b)   (c)   (4.41468, 0.00133)   (4.41220, −0.00568)   (d)

Figure 1.57
$y = x^2 - 6x + 7$

In Figure 1.57d, a point to the left and below the x-intercept has an x-coordinate of 4.41220, and a point to the right and above the x-intercept has an x-coordinate of 4.41468. To the nearest 0.01, they are both 4.41; that is, they are equal to the nearest 0.01. Because the x-intercept is between these two points, it follows that it is also 4.41 to the nearest 0.01.

The procedure suggested by Example 7 is outlined below.

Approximating a Zero of a
Function Graphically

If your graphing utility does not have a root feature, then you can approximate a zero of a function by following these steps:

Step 1: Graph the function in a viewing window that includes the x-intercept corresponding to the zero you seek to approximate.

Step 2: Use the zoom-in feature to move in on the x-intercept.

Step 3: Use the trace feature to examine the region about the x-intercept.

Step 4: If you can locate two points, one on each side of the intercept, that are equal to the desired precision, then the zero that corresponds to this x-intercept is equal to both of these points to the desired precision. If you can't locate such points, repeat steps 2 and 3.

The next example shows how the solutions to an equation can be approximated with this technique.

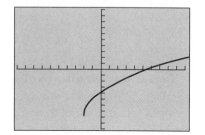

Figure 1.58
$y = 3\sqrt{x + 2} - 8$
$[-10, 10] \times [-10, 10]$

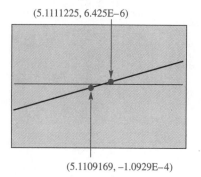

(5.1111225, 6.425E–6)

(5.1109169, –1.0929E–4)

Figure 1.59

EXAMPLE 8   Solving an Equation Graphically

Approximate, to the nearest 0.001, the solutions of the equation

$$3\sqrt{x + 2} - 8 = 0$$

by using the graph of $f(x) = 3\sqrt{x + 2} - 8$.

SOLUTION

The solutions of the equation $3\sqrt{x + 2} - 8 = 0$ are the zeros of $f(x) = 3\sqrt{x + 2} - 8$, and these correspond to the x-intercepts of the graph. Figure 1.58 shows a complete graph of f in the standard viewing window. Because the graph has only one x-intercept, the function has only one zero, and the equation has only one solution.

We move in on the x-intercept in the interval $[5, 6]$ and use the trace feature to determine two points such as those shown in Figure 1.59. The intercept, to the nearest 0.001, is (5.111, 0). The only solution of the equation is 5.111 (to the nearest 0.001).

Applications

In the last section, we saw that the graph of a function that represents a particular situation tells us a great deal about the situation. The next example shows how a graphing utility can be used to help answer questions about a function and the situation it represents.

■ EXAMPLE 9 Application: Profit and Manufacturing

From past experience, a manufacturer of large trucks has found that the profit (or loss) P, in thousands of dollars, realized from manufacturing n units annually is given by

$$P = -1.05^n + 42.5n - 2420$$

(a) Determine a complete graph of P.

(b) What is the minimum number of units that should be manufactured to produce a profit? What is the maximum number of units?

(c) What is the number of units that should be manufactured to produce the maximum profit? What is this profit (to the nearest hundred dollars)?

SOLUTION

(a) In the function given, the independent variable is n and the dependent variable is P. Most graphing utilities require x and y as the independent and dependent variables, so we enter $y = -1.05^x + 42.5x - 2420$. Because the function is defined only for $n \geq 0$ (n is a number of trucks), we consider only graphs such that Xmin ≥ 0. Experimenting with different viewing windows leads to a complete graph such as the one shown in Figure 1.60.

(b) A profit is produced only if $P > 0$, which is true only if the graph is above the x-axis. The graph in Figure 1.61 shows two intercepts, one at $x = 57.3$ and one at $x = 174.6$ (to the nearest 0.1). The graph is above the x-axis between these intercepts. Because the manufacturer makes only complete trucks, the minimum number of units n required for P to be positive is 58.

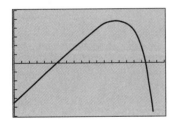

Figure 1.60
$y = -1.05^x + 42.5x - 2420$
$[0, 200]_{10} \times [-3000, 3000]_{500}$

■ **GRAPHING NOTE:** *Use the root feature of your graphing utility or the procedure demonstrated in Example 7 to verify the x-intercepts in Figure 1.61.*

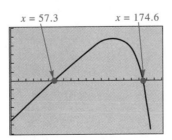

$x = 57.3$ $x = 174.6$

Figure 1.61

Likewise, the maximum number of trucks for the profit P to be positive is 174.

(c) The maximum value of the function occurs at the turning point, because the y-coordinate of this point is greater than any other point on the graph. By using the trace and zoom features, we see that the x-coordinate of the turning point is between 138 and 139 (Figure 1.62). Because the manufacturer makes only complete trucks, the maximum profit is either at $n = 138$ or $n = 139$. If $n = 138$, then

$$P = -1.05^{138} + 42.5(138) - 2420 \approx 2605.3$$

If $n = 139$, then

$$P = -1.05^{139} + 42.5(139) - 2420 \approx 2605.8$$

Because $2605.8 > 2605.3$, the maximum profit is produced by manufacturing 139 trucks. The profit (to the nearest hundred) is \$2,605,800. ■

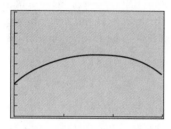

Figure 1.62
$[137, 140] \times [2600, 2610]$

EXERCISE SET 1.4

A

In Exercises 1–6, use a graphing utility to determine the graph of each function in the viewing windows given.

1. $f(x) = x^2 - 12x$
   (a) $[-10, 10] \times [-10, 10]$
   (b) $[-3, 17]_2 \times [-50, 50]_{10}$
   (c) $[10, 14] \times [-4, 4]$

2. $f(x) = x^3 - 24x$
   (a) $[-10, 10] \times [-10, 10]$
   (b) $[-6, 6] \times [-50, 50]_{10}$
   (c) $[-6, -4] \times [-8, 8]$

3. $f(x) = 3^x - 20$
   (a) $[-10, 10] \times [-10, 10]$
   (b) $[-5, 5] \times [-25, 20]_5$
   (c) $[1, 5] \times [-5, 5]$

4. $f(x) = 30 - 2^x$
   (a) $[-10, 10] \times [-10, 10]$
   (b) $[-10, 10] \times [-20, 40]_{10}$
   (c) $[4, 6]_{0.5} \times [-8, 8]$

5. $f(x) = 8 - 2\sqrt{4 - x}$
   (a) $[-10, 10] \times [-10, 10]$
   (b) $[-5, 5] \times [0, 10]$
   (c) $[-1, 1] \times [3, 5]_{0.5}$

6. $f(x) = -2 + \sqrt{x + 6}$
   (a) $[-10, 10] \times [-10, 10]$
   (b) $[-7, 6] \times [-3, 2]_{0.5}$
   (c) $[-3, -1] \times [-0.5, 0.5]_{0.1}$

In Exercises 7–18, use a graphing utility to determine a complete graph of each function and describe your viewing window.

7. $f(x) = x^3 - 4x - 16$ 8. $f(x) = -x^3 + 6x - 10$

9. $f(x) = (x - 12)(x^2 - 20)$

10. $f(x) = (x + 15)(32 - x^2)$

11. $f(x) = (9 - x^2)(x^2 - 20)$

12. $f(x) = (x^2 - 5x - 24)(x^2 - 9)$

13. $f(x) = x - 4^x + 24$ 14. $f(x) = 3^x - x - 60$

15. $f(x) = |x - 15| - |x + 15| + 20$

16. $f(x) = x|x - 21|$

17. $f(x) = \dfrac{500}{x^2 + 2}$

18. $f(x) = \dfrac{30 + 3x^2}{x^2 + 5}$

In Exercises 19–24, use a graphing utility to determine the complete graph of the given function and describe your viewing window. Determine the x-intercepts exactly using algebraic methods, and approximate them graphically (to the nearest 0.01).

19. $f(x) = -3x + 8$

20. $f(x) = \frac{5}{2}x - 7$

21. $f(x) = x^2 - 2x - 5$

22. $f(x) = 7 - \frac{1}{2}x^2$

23. $f(x = x^3 - 2x^2 - 4x$

24. $f(x) = 2x^3 - 4x^2 - 12x$

In Exercises 25–30, the zeros of the graph of the given function are integers. Use the graph of the function in the standard viewing window to guess each zero, and then verify your guess algebraically.

25. $f(x) = -\frac{1}{2}x - 4$

26. $f(x) = \frac{3}{2}x - 9$

27. $f(x) = \frac{1}{2}x^2 - 3x + 4$

28. $f(x) = -x^2 + 5x - 6$

29. $f(x) = x^3 - 3x^2 - 9x - 5$

30. $f(x) = -x^3 + 9x$

B

In Exercises 31–36, determine the graph of each function with the viewing window given, and use the trace feature to approximate (to the nearest 0.1) the coordinates of the turning point of the graph that appears in the viewing window. You may need to use the zoom feature.

31. $f(x) = 2x + \dfrac{3}{x}$; $[0, 4] \times [0, 10]$

32. $f(x) = 8x - 2^x$; $[0, 7] \times [0, 20]$

33. $f(x) = \dfrac{3x}{x^2 + 1}$; $[-1, 3] \times [-1, 2]$

34. $f(x) = \dfrac{x}{x^3 - 1}$; $[-3, 1] \times [-1, 1]_{0.2}$

35. $f(x) = |(x - 4)^3 + 7|$; $[-3, 7] \times [-2, 8]$

36. $f(x) = \sqrt{|2x - 7|}$; $[-1, 6] \times [-2, 4]$

37. Approximate graphically (to the nearest 0.0001) the zero of $f(x) = x^3 - 14$ in the interval $[2, 3]$.

38. Approximate graphically (to the nearest 0.0001) the zero of $g(x) = x^3 + 30$ in the interval $[-4, -3]$.

39. Approximate (to the nearest 0.0001) the solution of the equation $x^3 - 2x + 3 = 0$ by using the graph of the function $f(x) = x^3 - 2x + 3$.

40. Approximate (to the nearest 0.0001) the solution of the equation $x^3 + 5x - 4 = 0$ by using the graph of the function $f(x) = x^3 + 5x - 4$.

41. Approximate graphically (to the nearest 0.0001) the zero a of $f(x) = 22 - 3^x$, and check your approximation by computing $f(a)$.

42. Repeat Exercise 41 with $g(x) = 2^x - 13$.

43. **(a)** Approximate graphically (to the nearest 0.001) the zeros of

$$h(x) = \dfrac{x^2 - 6x + 2}{x^2 + 2}$$

(b) Find the zeros of h exactly using algebraic methods.

44. **(a)** Approximate graphically (to the nearest 0.001) the zeros of $G(x) = 2^x(x^2 - 4x + 3)$.

(b) Find the zeros exactly using algebraic methods.

45. The theory of relativity predicts that the mass of an object changes with velocity according to the formula

$$m = \dfrac{m_0}{\sqrt{1 - (v^2/c^2)}}$$

In this formula, m_0 is the mass of the object at rest, v is the velocity of the object, m is the mass of the object at that velocity, and c is the speed of light (1.1×10^9 kilometers per hour). Suppose that the mass of the object at rest is 80 kilograms.

(a) Determine the graph of this equation with viewing window $[0, 1000]_{100} \times [0, 300]_{50}$. Use the trace feature to investigate the graph.

(b) Determine the graph of this equation with viewing window $[0, 1.2 \times 10^9]_{10^8} \times [0, 300]_{50}$.

(c) Comment on your observations in parts (a) and (b). When is this change in mass significant?

46. The distance d (in feet) that an automobile driver needs to come to a complete stop from a velocity of v miles per hour under ideal conditions is given by the formula

$$d = v + \frac{v^2}{25}$$

Determine the graph of this equation with viewing window $[0, 40]_5 \times [0, 100]_{10}$. Comment on the shape of the graph and its interpretation in terms of braking distances at high speeds.

47. A *catenary arch* is an architectural structure whose shape is modeled by the equation $y = L - k(a^x + a^{-x})$. The most famous catenary arch is the Gateway Arch in St. Louis. For this arch, $L = 757.7$, $k = 63.85$, and $a = 1.008$. Determine the graph of the arch for $y \geq 0$. How high is the arch?

48. Two lamps are 10 meters apart. The combined brightness L (in lumens) of a point between them that is d meters from the brighter lamp is

$$L = \frac{5}{(10 - d)^2} + \frac{30}{d^2}$$

(a) Determine the complete graph of L for $0 \leq d \leq 10$.
(b) At what point (to the nearest 0.1 meter) between the lamps is the brightness the least? What is the brightness at that point?
(c) Over what interval between the lamps is the brightness less than 6 lumens?

C

49. The equation $x = y^2 - 4$ cannot be graphed directly on a graphing utility. However, solving this equation for y yields $y = \pm\sqrt{x + 4}$, and graphing both $y = \sqrt{x + 4}$ and $y = -\sqrt{x + 4}$ simultaneously gives us the graph of $x = y^2 - 4$. Use a graphing utility to determine the graph of $x = y^2 - 4$ using this technique.

50. Use the technique outlined in Exercise 49 to determine the graph of $4x^2 + 9y^2 = 36$.

51. Use the technique outlined in Exercise 49 to determine the graph of $x^2 + (y - 2)^2 = 16$.

52. When a disease is introduced into an isolated community, the number of persons N with the disease at day t of the epidemic is given by

$$N = \frac{P}{1 + (P - 1)2^{-kt}}$$

where P is the population of the community and k is a constant that depends upon the contagious nature of the disease. Suppose that $P = 2000$. Determine the complete graph of this equation for $k = 1, 2, 3, 4,$ and 5. Comment on the significance of k. What might determine the value of k in the application of this formula?

53. Determine the graphs of

$$y = \frac{x^2 - 6x + 2}{x^2 + 2} \quad \text{and} \quad y = 1 - \frac{6x}{x^2 + 2}$$

Compare the graphs and comment on your observations.

54. Determine the graphs of $y = x^3 - 3x^2 - 9x - 5$ and $y = (x + 1)^2(x - 5)$. Compare the graphs and comment on your observations.

55. The graph of $y = \sqrt{25 - x^2}$ in the standard viewing window on a particular graphing utility and the actual graph from the directory of graphs in Section 1.3 are shown below. Explain why the graphing utility does not give the actual graph exactly.

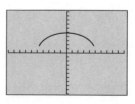

Graphing utility screen

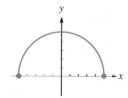

Actual graph

APPLICATIONS OF FUNCTIONS

In the four previous sections of this chapter, the focus has been on functions and their graphs. In this section, we use these ideas to solve problems. Specifically, we examine real situations in which one quantity depends on another and express those relationships as mathematical functions. In addition, we will see how the attributes of the graph, such as intercepts and turning points, can be interpreted and applied. A graphing utility will be needed throughout this discussion, so you should have a graphing utility readily available as you read this section.

Mathematical Modeling

The process of describing a physical situation in mathematical terms is called *mathematical modeling*. In general, this process may involve constructing one or more functions, equations, or inequalities. It may also require collecting data. Once a model is established, it can be used to analyze the situation and make predictions. Sometimes the results derived from a model do not predict what actually happens, and we must modify the model. There is often an ongoing process of revision.

Some mathematical models are very accurate. For example, the focal length of a lens, the distance that an object falls in a vacuum, and the volume of a sphere all have precise mathematical models. However, many real-life situations, such as inventory costs in business, or the time required for mice to learn a maze, are too complicated to be precisely modeled. In these cases, it is often necessary to make certain assumptions or rely on data from experimentation.

Precise Models

Some models predict the exact outcome of a situation. Such *precise models* depend only on laws that are known to be certain, such as laws of physical science or facts of mathematics.

EXAMPLE 1 Application: The Area of a Square

(a) Determine a model for the area $A(x)$ of a square as a function of the length of its diagonal x.

(b) Use the model in part (a) to determine the area of a square with diagonal 6 centimeters in length.

SOLUTION

(a) Our first step is to draw a picture and label the diagonal length x, as shown in Figure 1.63. We know that

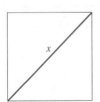

Figure 1.63

$$\left(\begin{array}{c}\text{Area of}\\\text{square}\end{array}\right) = \left(\begin{array}{c}\text{Length of}\\\text{side}\end{array}\right)^2$$

This model expresses the area as a function of the side. But since the input of the function is to be the length of the diagonal, we need a relationship between the length of the side and the diagonal. Since the diagonal divides the square into two right triangles, we use the Pythagorean theorem:

$$\left(\begin{array}{c}\text{Length of}\\\text{side}\end{array}\right)^2 + \left(\begin{array}{c}\text{Length of}\\\text{side}\end{array}\right)^2 = \left(\begin{array}{c}\text{Length of}\\\text{diagonal}\end{array}\right)^2$$

or

$$\left(\begin{array}{c}\text{Length of}\\\text{side}\end{array}\right)^2 = \frac{\left(\begin{array}{c}\text{Length of}\\\text{diagonal}\end{array}\right)^2}{2}$$

Substituting this into our original equation, we have

$$\left(\begin{array}{c}\text{Area of}\\\text{square}\end{array}\right) = \frac{\left(\begin{array}{c}\text{Length of}\\\text{diagonal}\end{array}\right)^2}{2}$$

So

$$A(x) = \frac{x^2}{2}$$

(b) Letting $x = 6$, we have

$$A(6) = \frac{6^2}{2} = 18$$

The area is 18 square centimeters.

There is no specific set of steps for finding functions to model applied situations. However, here are some general strategies and guidelines that will help.

Strategies for Developing Models
with Functions

1. Draw a picture whenever appropriate. Label the relevant quantities, known and unknown.
2. Determine which quantity is the input of the function and which quantity is the output of the function. Write down relationships that involve these quantities.
3. Find a relationship involving the input and the output of the function.
4. Rewrite this relationship as a function.

EXAMPLE 2 Application: The Construction Cost of a Package

An open-top box with a square base is to be constructed from two materials, one for the bottom and one for the sides. The volume of the box is to be 9 cubic feet. The cost of the material for the bottom is $4 per square foot, and the cost of the material for the sides is $3 per square foot.

(a) Determine a model for the cost of the box as a function of its height h. What is the domain of the function?

(b) Which will be most expensive to construct, a box with a height of 1 foot, 2 feet, or 3 feet?

(c) Use a graphing utility to determine a graph of the model. Label the points on the graph that correspond to the three heights in part (b). Determine (to the nearest 0.1 foot) the dimensions of the least expensive box. What is the minimum cost?

SOLUTION

(a) The first step is to draw a picture (Figure 1.64). Since the cost per square foot for the bottom is different from the cost per square foot for the sides, our strategy is to consider the total cost in terms of the cost of the bottom and the cost of the sides:

$$\binom{\text{Total}}{\text{cost}} = \binom{\text{Cost of}}{\text{bottom}} + \binom{\text{Cost of}}{\text{sides}}$$

$$= \binom{\$4 \text{ per}}{\text{square foot}}\binom{\text{Area of}}{\text{bottom}} + \binom{\$3 \text{ per}}{\text{square foot}}\binom{\text{Area of}}{\text{sides}}$$

$$= 4\binom{\text{Area of}}{\text{bottom}} + 3\binom{\text{Area of}}{\text{sides}}$$

$$= 4\binom{\text{Area of}}{\text{bottom}} + 3\left[4\binom{\text{Area of}}{\text{one side}}\right]$$

$$= 4(\text{Width})^2 + 12(\text{Width})(\text{Height})$$

Now, since the base is square and the volume is 9 cubic feet,

$$(\text{Width})^2(\text{Height}) = 9$$

Solving for the width, and replacing the height with h, we get

$$(\text{Width}) = \frac{3}{\sqrt{h}}$$

If we name the cost function C, then replacing width by $3/\sqrt{h}$ and height by h in the equation we found above, we obtain

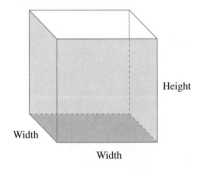

Width

Width

Height

Figure 1.64

or

$$C(h) = 4\left[\left(\frac{3}{\sqrt{h}}\right)^2\right] + 12\left[\left(\frac{3}{\sqrt{h}}\right)h\right]$$

$$C(h) = \frac{36}{h} + 36\sqrt{h}$$

Since the height h must be positive, the domain is $(0, +\infty)$.

(b) The cost of constructing a box with a height of 1 foot, 2 feet, and 3 feet is $C(1)$, $C(2)$, and $C(3)$, respectively:

$$C(1) = \tfrac{36}{1} + 36\sqrt{1} = 72$$
$$C(2) = \tfrac{36}{2} + 36\sqrt{2} \approx 68.91$$
$$C(3) = \tfrac{36}{3} + 36\sqrt{3} \approx 74.35$$

Of the three, the most expensive to construct is a box with a height of 3 feet.

(c) The graph of C is shown in Figure 1.65. The least expensive box corresponds to the lowest point on the graph; in this case it is a turning point. From the graph in Figure 1.65, the turning point apparently occurs between $h = 1$ and $h = 2$. Using the zoom and trace features on a graphing utility (Figure 1.66), we can approximate the coordinates of the turning point:

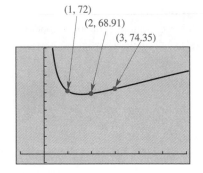

Figure 1.65
$$y = \frac{36}{x} + 36\sqrt{x}$$
$[-1, 6] \times [-10, 120]_{10}$

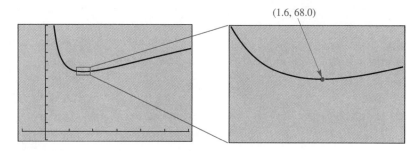

Figure 1.66
$$y = \frac{36}{x} + 36\sqrt{x}$$

The turning point occurs when h is about 1.6 feet, and the minimum cost is approximately $C(1.6) = 68$. The corresponding width of the base is determined by

$$\frac{3}{\sqrt{h}} \approx \frac{3}{\sqrt{1.6}} \approx 2.4$$

The height and width of the least expensive box is 1.6 feet and 2.4 feet, respectively. The minimum cost is $68.

● EXAMPLE 3 Application: The Dimensions of a Pen

A farmer wishes to build a pen in the shape of a rectangle, using the bank of a river as one of the sides. The farmer has 210 meters of fencing.

(a) Determine a model for the area of the pen as a function of its width. What is the domain of the function?

(b) Use a graphing utility to determine a graph of the model. What are the intercepts? Over what (approximate) intervals is the function increasing? Decreasing? Interpret your results.

(c) Using a graphing utility, determine (to the nearest 0.1 meter) the dimensions of the pen with greatest area. What is the maximum area?

SOLUTION

(a) Our first step is to draw a picture of the situation (Figure 1.67). Since there is a fixed amount of fencing, once the farmer decides on the width, the length is determined as well. We want a function for the area of this rectangle, so we start with

$$\left(\begin{matrix}\text{Area of} \\ \text{rectangle}\end{matrix}\right) = \left(\begin{matrix}\text{Width of} \\ \text{rectangle}\end{matrix}\right)\left(\begin{matrix}\text{Length of} \\ \text{rectangle}\end{matrix}\right)$$

So far, the area is expressed in terms of width and length. We want to express the area in terms of width only, so we need a relationship between width and length. Since there are 210 meters of fencing,

$$\left(\begin{matrix}\text{Length of} \\ \text{rectangle}\end{matrix}\right) = 210 - 2\left(\begin{matrix}\text{Width of} \\ \text{rectangle}\end{matrix}\right)$$

Replacing

$$\left(\begin{matrix}\text{Length of} \\ \text{rectangle}\end{matrix}\right) \qquad \text{by} \qquad 210 - 2\left(\begin{matrix}\text{Width of} \\ \text{rectangle}\end{matrix}\right)$$

in our original equation, we have

$$\left(\begin{matrix}\text{Area of} \\ \text{rectangle}\end{matrix}\right) = \left(\begin{matrix}\text{Width of} \\ \text{rectangle}\end{matrix}\right)\left[210 - 2\left(\begin{matrix}\text{Width of} \\ \text{rectangle}\end{matrix}\right)\right]$$

If we name this function A and let x represent the width, then

$$A(x) = x(210 - 2x)$$

The domain for this function is determined by the situation. Certainly, the width must be positive. In addition, since there are only 210 meters of fencing

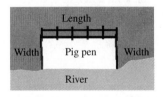

Figure 1.67

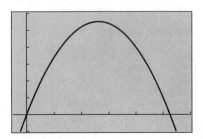

Figure 1.68
$y = x(210 - 2x)$
$[-10, 120]_{20} \times [-1000, 6000]_{1000}$

available, the width must be less than 105 meters. Thus, the domain is the interval $0 < x < 105$.

(b) The graph of A is shown in Figure 1.68. Over the domain $0 < x < 105$ there are no intercepts. Therefore, the area cannot be 0. The function is increasing on the approximate interval $0 < x < 52$. Therefore, as we increase the width from $x = 0$ to $x = 52$, the area of the pen increases as well. The function is decreasing over the approximate interval $53 < x < 105$, meaning that as we increase the width from 53 to 105, the area of the pen diminishes.

(c) The pen with greatest area corresponds to the highest point on the graph. In Figure 1.68, the turning point apparently occurs between $x = 52$ and $x = 53$. Using the zoom and trace features on a graphing utility (Figure 1.69), we can approximate the coordinates of the turning point:

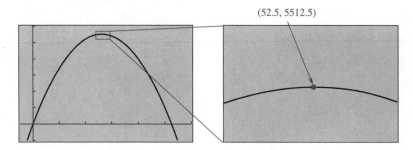

Figure 1.69
$y = x(210 - 2x)$

The turning point occurs when x is 52.5, and the maximum area is 5512.5 (to the nearest 0.1). The corresponding length of the pen is

$$210 - 2(52.5) = 105$$

The pen with maximum area has a width of 52.5 meters and length of 105 meters. The maximum area is 5512.5 square meters.

In Examples 2 and 3, we used the graphs of the functions to determine the minimum cost and the maximum area, respectively. Many real-life problems call for finding the maximum or minimum values of a function; such problems are called *max–min*, or *optimization*, problems. Optimizing a function amounts to locating the highest or lowest points on the graph within its domain. We will approximate such points with a graphing utility. The determination of the *exact* maximum or minimum values of a function usually requires calculus. However, in Section 2.2 we will see how to find the exact maximum or minimum values of functions that are second-degree polynomials.

Approximate Models

As we mentioned earlier, many real-life situations are too complicated to be precisely modeled. When developing a model to represent actual data, we must often compromise accuracy in order to keep the model simple enough to be workable.

■ EXAMPLE 4   Application: Blood Alcohol Level

In the United States, each state sets the level for the amount of alcohol in the bloodstream that designates an individual as intoxicated. The blood alcohol level measures the percentage of alcohol in the bloodstream. In most states, a person with a blood alcohol level of 0.05% or greater is considered intoxicated.

Table 1.2 indicates a relationship between a person's weight and the number of drinks (in 1 hour) that will lead to a blood alcohol level equal to or greater than 0.05%. (A drink is defined here as one containing 0.5 ounce of alcohol.) The number of drinks $N(x)$ can be related to the person's weight x (in pounds) approximately by the model

$$N(x) = 0.008x$$

TABLE 1.2

Weight	Number of drinks
100	0.75
120	1.00
140	1.25
160	1.30
180	1.50
200	1.60
220	1.80

(a) Sketch the graph of N, and plot the data in the table on the same axes. Is the model a good fit?

(b) Use the model to predict the number of drinks that would put a 146 pound person over this legal limit.

(c) Other factors affect the accuracy of the data. For example, metabolism, amount or type of food eaten, or medications taken also determine the blood alcohol level. Suppose the number of drinks can vary by 10%; for example, a 120 pound person may be intoxicated by as little as 0.90 of a drink or as much as 1.10 drinks. Under this assumption, if a 200 pound person consumed 1.50 drinks instead of 1.60, would this person be assured of not being intoxicated?

(d) Assuming the number of drinks listed in the table can vary by 10%, modify the model to determine a model for how little a person might drink and a model for how much a person might drink before being intoxicated.

SOLUTION

(a) Figure 1.70 shows the graph of $N(x) = 0.008x$ along with the ordered pairs (100, 0.75), (120, 1.00), and so on, from Table 1.2. The model appears to fit the actual data very well.

(b) According to the model, a 146 pound person needs only about

$$N(146) = 0.008(146) = 1.168$$

drinks to reach the 0.05% blood alcohol level.

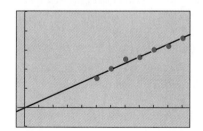

Figure 1.70
$y = 0.008x$
$[-10, 230]_{20} \times [-0.5, 2.5]_{0.5}$

(c) If the actual number of drinks can vary by 10%, a 200 pound person may be intoxicated after only 90% of 1.60, or

$$(0.90)1.60 = 1.44$$

drinks. Therefore, even though the set of data indicates that it would take 1.60 drinks, a 200 pound person *might* be intoxicated after 1.50 drinks.

(d) Starting with the model $N(x) = 0.008x$, a model for how little a person might drink before being intoxicated is 90% of $N(x)$, or

$$(0.90)N(x) = (0.90)0.008x = 0.0072x$$

A model for how much a person might be able to drink before being intoxicated is 110% of $N(x)$:

$$(1.10)N(x) = (1.10)0.008x = 0.0088x$$

⬤ EXAMPLE 5 Application: Rescue Time

A swimmer in distress is 150 feet from the nearest point P on a straight shoreline. A lifeguard standing on the shore 200 feet from point P must run some distance along the shoreline and then swim the remaining distance. The lifeguard can run 21 feet per second and swim 5 feet per second.

(a) Determine the time it takes the lifeguard to reach the swimmer as a function of the distance x from point P. What is the domain of this function?

(b) Use a graphing utility to graph the model obtained in part (a). Determine (to the nearest second) the longest time it takes the lifeguard to reach the swimmer. Determine the shortest time.

SOLUTION

(a) Draw and label a picture (Figure 1.71). The time (in seconds) required is

$$\begin{pmatrix} \text{Total} \\ \text{time} \end{pmatrix} = \begin{pmatrix} \text{Time} \\ \text{running} \end{pmatrix} + \begin{pmatrix} \text{Time} \\ \text{swimming} \end{pmatrix}$$

$$= \frac{(\text{Distance running})}{(\text{Rate running})} + \frac{(\text{Distance swimming})}{(\text{Rate swimming})}$$

We are given that the rate running is 21 feet per second and the rate swimming is 5 feet per second. From our picture we know that

$$\begin{pmatrix} \text{Distance} \\ \text{running} \end{pmatrix} = 200 - x$$

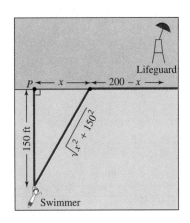

Figure 1.71

and, by the Pythagorean theorem,

$$\left(\begin{matrix} \text{Distance} \\ \text{swimming} \end{matrix}\right) = \sqrt{x^2 + 150^2}$$

Therefore,

$$\left(\begin{matrix} \text{Total} \\ \text{time} \end{matrix}\right) = \frac{200 - x}{21} + \frac{\sqrt{x^2 + 150^2}}{5}$$

If we name this function T, then

$$T(x) = \frac{200 - x}{21} + \frac{\sqrt{x^2 + 22{,}500}}{5}$$

The domain is determined by the situation. From our picture the domain of T is $0 \le x \le 200$.

(b) The graph of T is shown in Figure 1.72. The longest time corresponds to the highest point on the graph over the domain $0 \le x \le 200$; in this case when $x = 200$. The longest time is

$$T(200) = \frac{200 - 200}{21} + \frac{\sqrt{200^2 + 22{,}500}}{5} = 50$$

Swimming directly, the lifeguard will take 50 seconds to reach the swimmer. The shortest time corresponds to the lowest point on the graph in Figure 1.72. This apparently occurs between $x = 35$ and $x = 40$. Using the zoom and trace features on a graphing utility (Figure 1.73), we can approximate the coordinates of the turning point:

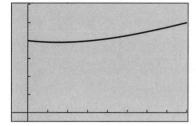

Figure 1.72
$$y = \frac{200 - x}{21} + \frac{\sqrt{x^2 + 22{,}500}}{5}$$
$[-20, 200]_{25} \times [-5, 60]_{10}$

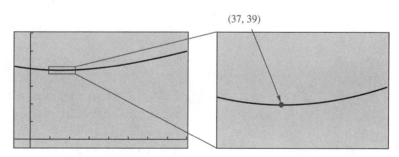

(37, 39)

Figure 1.73
$$y = \frac{200 - x}{21} + \frac{\sqrt{x^2 + 22{,}500}}{5}$$

The turning point occurs when x is about 37 feet. The corresponding time, about 39 seconds, is the shortest time required to reach the swimmer.

The model in Example 5 is not a precise model because it does not include the possibilities of wind or current affecting the running or swimming speeds. For example, assume there is no wind, but that the current can increase or decrease the swimming speed by as much as 20%. The lifeguard's swimming rate may be

$$(1.20)5 = 6 \text{ feet per second} \qquad \text{or} \qquad (0.80)5 = 4 \text{ feet per second}$$

The models for the time it takes the lifeguard to reach the swimmer with each of these swimming rates are

$$T_1(x) = \frac{200 - x}{21} + \frac{\sqrt{x^2 + 22{,}500}}{6} \quad \text{and} \quad T_2(x) = \frac{200 - x}{21} + \frac{\sqrt{x^2 + 22{,}500}}{4}$$

Business Models

Two functions are of critical importance to a manufacturer: the *revenue function* R, and the *cost function* C. Suppose a business manufactures x units of a product that sells for p dollars per unit. The revenue and cost are defined as follows:

Revenue function: $R(x) = xp$ = Amount collected from selling x units
Cost function: $C(x)$ = Cost of producing and selling x units

The cost $C(x)$ typically consists of two parts. One part is the *fixed cost*, which includes rent, utilities, design of the product, and purchase of equipment. We assume that the fixed cost does not change as more items are produced. The other part of the cost is the *variable cost*, including labor, materials, packaging, and distribution. The variable cost increases as the number of units produced increases. The cost function can be modeled by

$$C(x) = mx + b$$

where m represents the variable cost per unit, and b represents the fixed cost.

◉ EXAMPLE 6   Application: Cost Analysis

A company that sells video cassettes is going to select a manufacturer. One manufacturer quotes $2.30 per cassette and a start-up cost of $12,000. Another manufacturer quotes $2.90 per cassette and a start-up cost of $9000.

(a) Determine models $C_1(x)$ and $C_2(x)$ for the total cost of producing x cassettes from each manufacturer. What is the cost of producing 4000 cassettes from each manufacturer?

(b) Use a graphing utility to determine a graph of each model in the same coordinate plane.

(c) At what level of production, x, are the costs from the two manufacturers the same? Which manufacturer should the company select if the level of production will be less than this amount?

SOLUTION

(a) The total cost of producing x cassettes is

$$\begin{pmatrix} \text{Total cost of} \\ \text{producing } x \text{ cassettes} \end{pmatrix} = \begin{pmatrix} \text{Variable cost of} \\ \text{producing } x \text{ cassettes} \end{pmatrix} + \begin{pmatrix} \text{Fixed} \\ \text{cost} \end{pmatrix}$$

$$= \begin{pmatrix} \text{Variable cost} \\ \text{per cassette} \end{pmatrix} \begin{pmatrix} \text{Number of} \\ \text{cassettes} \end{pmatrix} + \begin{pmatrix} \text{Start-up} \\ \text{cost} \end{pmatrix}$$

For the first manufacturer, the variable cost per unit is \$2.30 and the fixed cost is \$12,000. Therefore, our cost model is

$$C_1(x) = 2.30x + 12,000$$

For the other manufacturer, the variable cost per unit is \$2.90 and the fixed cost is \$9000. Thus,

$$C_2(x) = 2.90x + 9000$$

The costs of producing 4000 cassettes from each manufacturer are given by

$$C_1(4000) = 2.30(4000) + 12,000 = 21,200$$
$$C_2(4000) = 2.90(4000) + 9000 = 20,600$$

Therefore, the costs are \$21,200 and \$20,600.

(b) The graphs of both models are shown in Figure 1.74.

(c) We want to determine x such that $C_1(x) = C_2(x)$, so we set

$$2.30x + 12,000 = 2.90x + 9000$$

and solve for x:

$$12,000 = 0.6x + 9000$$
$$3000 = 0.6x$$
$$5000 = x$$

Therefore, the cost of production will be the same when each manufacturer produces 5000 cassettes. Notice that this result checks with the graph in Figure 1.74; the graphs appear to cross at about $x = 5000$. Now, consider the graphs of the models in Figure 1.74. For $x < 5000$, the graph of C_2 is below the graph of C_1. Thus, at a level of production under 5000 cassettes, the models predict that the second manufacturer will supply them at a lower cost. The company should select the second manufacturer.

The next example develops a business cost model that does not have the form $C(x) = mx + b$.

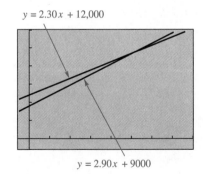

$y = 2.30x + 12,000$

$y = 2.90x + 9000$

Figure 1.74

$[-500, 8000]_{1000} \times$
$[-1000, 30,000]_{5000}$

⬤ EXAMPLE 7   Application: Delivery Costs

A company makes frequent deliveries to a customer 25 miles away. The cost of operating a delivery van (gas, oil, tire wear, etc.) is $0.34 per mile. The driver is paid $10 per hour.

(a) Determine a model for the total cost of making the delivery as a function of the average speed the van is driven.

(b) Assume the speed limit is 55 miles per hour. Use a graphing utility to graph the model over the interval (0, 55]. What are the intercepts? Over what intervals is the function increasing? Decreasing? Interpret your results.

(c) Determine (to the nearest mile per hour) the speed the driver should average to minimize the cost of delivery. What is the minimum cost?

SOLUTION

(a) Our strategy is to consider the total cost in terms of the operating cost and the cost of the driver:

$$\begin{pmatrix} \text{Total} \\ \text{cost} \end{pmatrix} = \begin{pmatrix} \text{Operating} \\ \text{cost} \end{pmatrix} + \begin{pmatrix} \text{Driver} \\ \text{cost} \end{pmatrix}$$

$$= \begin{pmatrix} \text{Cost per} \\ \text{mile} \end{pmatrix} \begin{pmatrix} \text{Number} \\ \text{of miles} \end{pmatrix} + \begin{pmatrix} \text{Cost per} \\ \text{hour} \end{pmatrix} \begin{pmatrix} \text{Number of} \\ \text{hours driving} \end{pmatrix}$$

$$= \begin{pmatrix} \$0.34 \\ \text{per mile} \end{pmatrix} \begin{pmatrix} \text{Number} \\ \text{of miles} \end{pmatrix} + \begin{pmatrix} \$10 \\ \text{per hour} \end{pmatrix} \begin{pmatrix} \text{Number of} \\ \text{hours driving} \end{pmatrix}$$

The destination is 25 miles away, so the (round trip) distance traveled is 50 miles. Thus,

$$\begin{pmatrix} \text{Total} \\ \text{cost} \end{pmatrix} = (0.34)(50) + (10) \left(\frac{\text{Number of miles driving}}{\text{Speed}} \right)$$

$$= 17 + 10 \left(\frac{50}{\text{Speed}} \right)$$

If we name this function C and let s represent the speed, we have

$$C(s) = 17 + 10 \left(\frac{50}{s} \right) = 17 + \frac{500}{s}$$

(b) The graph of C is shown in Figure 1.75. There are no intercepts, meaning that neither the cost nor the average speed can be 0. The function is decreasing on the interval $0 < s < 55$. The function is never increasing, so the faster the van is driven, the lower the delivery cost.

(c) The minimum cost corresponds to the lowest point on the graph over the specified domain. Since the function is decreasing throughout the domain, the lowest point occurs at $s = 55$, and the corresponding cost is 26.09. Hence,

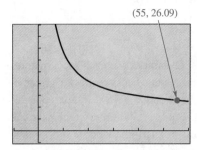

(55, 26.09)

Figure 1.75

$y = 17 + \dfrac{500}{x}$

$[-10, 60]_{10} \times [-10, 90]_{10}$

■ **NOTE:** *Under these conditions, common sense suggests that the delivery costs will be lowest when the driver goes as fast as allowable. Under different conditions, this may not be true. (See Exercise 23.)*

the driver should average 55 miles per hour for a minimum cost of approximately $26.09. ■

Each of the last two examples modeled the costs $C(x)$ involved in a business venture. The revenue $R(x)$ is just as important to a business.

■ **EXAMPLE 8   Application: Newspaper Sales**

A newspaper currently charges $0.35 per copy and sells 40,000 copies daily. The newspaper plans to increase the price per copy. Past experience indicates that each $0.05 price increase lowers daily sales by 2200 copies.

(a) Determine an equation relating the selling price p and the number x of copies sold daily.

(b) Determine a model for the daily revenue $R(x)$ as a function of the number of copies sold.

(c) Using a graphing utility, determine a graph of R.

(d) Using a graphing utility, determine the value of x that maximizes R.

(e) What price p (to the nearest 0.05) should the newspaper charge to maximize revenue?

SOLUTION

(a) We are given that

$$\begin{pmatrix} \text{Future number} \\ \text{of copies sold} \end{pmatrix} = 40{,}000 - 2200 \begin{pmatrix} \text{Number of \$0.05} \\ \text{increases} \end{pmatrix}$$

Think about how the number of $0.05 increases is related to the increase in price per paper:

$$\begin{pmatrix} \text{Number of \$0.05} \\ \text{increases} \end{pmatrix} (\$0.05) = \begin{pmatrix} \text{Increase in price} \\ \text{per paper} \end{pmatrix}$$

Since the increase in the price per paper is $p - 0.35$, we have

$$\begin{pmatrix} \text{Number of \$0.05} \\ \text{increases} \end{pmatrix} (\$0.05) = p - 0.35$$

Dividing each side by 0.05 gives

$$\begin{pmatrix} \text{Number of \$0.05} \\ \text{increases} \end{pmatrix} = \frac{p - 0.35}{0.05}$$

Returning to our original equation, we have

$$\left(\begin{array}{c}\text{Future number}\\ \text{of copies sold}\end{array}\right) = 40{,}000 - 2200\left(\frac{p - 0.35}{0.05}\right)$$

$$x = 40{,}000 - 44{,}000(p - 0.35)$$

$$x = 55{,}400 - 44{,}000p$$

(b) The current revenue is $(0.35)(40{,}000) = \$14{,}000$. After a price increase, the revenue will be

$$\left(\begin{array}{c}\text{Future}\\ \text{revenue}\end{array}\right) = \left(\begin{array}{c}\text{Future number}\\ \text{of copies sold}\end{array}\right)\left(\begin{array}{c}\text{Future price}\\ \text{per copy}\end{array}\right)$$

$$= xp$$

Since we want to express revenue as a function of x, we need to express p in terms of x. Using the result from part (a),

$$x = 55{,}400 - 44{,}000p$$

we can solve for p to get

$$44{,}000p + x = 55{,}400$$

$$p = \frac{(55{,}400 - x)}{44{,}000}$$

If we substitute the expression on the right of this equation for p into

$$\left(\begin{array}{c}\text{Future}\\ \text{revenue}\end{array}\right) = xp$$

we have

$$\left(\begin{array}{c}\text{Future}\\ \text{revenue}\end{array}\right) = x\,\frac{(55{,}400 - x)}{44{,}000}$$

Thus,

$$R(x) = x\,\frac{(55{,}400 - x)}{44{,}000}$$

(c) A graph of R is shown in Figure 1.76.
(d) The maximum corresponds to the highest point on the graph. Looking at Figure 1.76, R is maximized when x is approximately 28,000. Using the zoom and trace features on a graphing utility, we can improve this estimate (Figure 1.77). The maximum value of R occurs when x is 27,700.

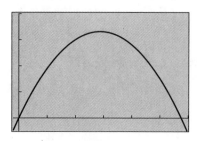

Figure 1.76

$$y = x\,\frac{(55{,}400 - x)}{44{,}000}$$

$[-1000, 60{,}000]_{10{,}000} \times$
$[-1000, 20{,}000]_{5000}$

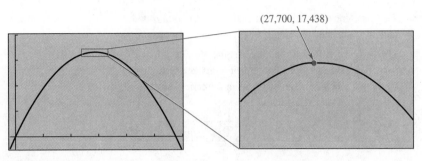

(27,700, 17,438)

Figure 1.77
$$y = x\frac{(55,400 - x)}{44,000}$$

(e) Using the relationship

$$p = \frac{(55,400 - x)}{44,000}$$

the price that corresponds to $x = 27,700$ is

$$p = \frac{(55,400 - 27,700)}{44,000} = 0.6295... \approx 0.65$$

The newspaper should charge $0.65 to maximize revenue.

EXERCISE SET 1.5

A

1. Express the area of a square as a function of its perimeter p.

2. Express the area of a circle as a function of its circumference c.

3. A rectangle has an area of 72 square centimeters. Determine the length as a function of the width w.

4. A rectangle has an area of 38 square feet. Determine the perimeter as a function of the length l.

5. Express the volume of a cube as a function of its surface area s.

6. The volume of a right circular cylinder is 540 cubic centimeters. Express the height of the cylinder as a function of its radius r.

7. A ladder of length 8 meters leans against a wall (Figure 1.78). Express the height of the top of the ladder as a function of the distance x from the foot of the wall. What is the domain?

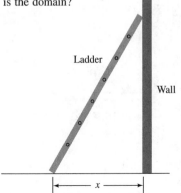

Ladder

Wall

x

Figure 1.78

8. The hypotenuse of a right triangle is twice as long as the height. Determine the height as a function of the base b.

9. The percentage $S(x)$ of n-butane gas that dissolves in a particular acid at various Fahrenheit temperatures x must be known when designing petroleum refineries. The following set of data is available:

Temperature (°F), x	Solubility (% weight), S
77	2.4
100	3.4
185	7.0 .
239	11.1

The data can be related (approximately) by the function

$S(x) = 0.003\sqrt{x^3}$

(a) Use a graphing utility to graph S, and plot the points in the table on the same set of axes. Is the model a good fit?

(b) Use the model to predict the solubility at 60°F and at 120°F.

10. To study learning retention, German philosopher Hermann Ebbinghaus (1850–1909) practiced a list of nonsense syllables x times each day and recorded the time $t(x)$ (in minutes) it took to relearn them the next day. The table indicates the results:

Repetitions one day, x	Time to relearn next day, t(x)
8	19.5
16	18
24	17
32	14
42	12
53	10

The data can be related (approximately) by the model

$t(x) = -0.22x + 21.5$

(a) Sketch the graph of t, and plot the points in the table on the same set of axes. Is the model a good fit?

(b) Suppose the list is practiced 30 times one day. Use the model to predict how many minutes it will take to relearn the list the following day.

(c) Assume that the number of minutes $t(x)$ can vary by as much as 10%. For example, 42 repetitions of the list one day might require as much as 13.2 or as little as 10.8 minutes to relearn the list the next day. Under this assumption, is 18 minutes enough time to relearn the list the next day if the list is practiced 24 times one day?

(d) Assuming that the number of minutes $t(x)$ can vary by as much as 10%, modify the model to determine a model for the least amount of time and a model for the most amount of time expected to relearn the list the next day.

11. In a learning study, several groups of rats were trained to press a bar for a food reward. For some, a bar press was followed immediately by the food. Others were trained with longer delay times. The learning score $S(x)$ and the delay times x (in seconds) are shown in the table:

Reward delay, x	Learning score, S(x)
2.5	52
6.7	31
20	22
40	15
75	7

The data can be related (approximately) by the function

$S(x) = \dfrac{85}{\sqrt{x}}$

(a) Use a graphing utility to graph S, and plot the points in the table on the same set of axes.

(b) Suppose the food reward is delayed for 30 seconds. Use the model to predict the learning score.

(c) Assume that the score $S(x)$ can vary by as much as 5%. For example, the score after a 20 second delay might be as much as 23.1 or as little as 20.9. Under this assumption, if the delay time is 40 seconds, will the score be at least 14?

(d) Assuming that the score $S(x)$ can vary by as much as 5%, modify the model to determine a model for the least score and a model for the highest score expected after a delay time of x seconds.

12. A company that sells sunglasses is deciding on a manufacturer for the frames. One manufacturer quotes $1.60 per frame with a start-up cost of $6000. Another manufacturer quotes $1.10 per frame and a start-up cost of $7500.
    (a) Determine models $C_1(x)$ and $C_2(x)$ for the total cost of producing x frames from each manufacturer. What is the cost of producing 2000 frames from each manufacturer?
    (b) Use a graphing utility to determine a graph of each model in the same viewing window.
    (c) At what level of production x is the cost the same? Which manufacturer should the company select if the level of production will be less than this amount?

13. A company that sells backpacks must decide on a supplier. One supplier bids $13 per backpack and a fixed cost of $15,000. Another supplier bids $15.50 per backpack and a fixed cost of $12,000.
    (a) Determine models $C_1(x)$ and $C_2(x)$ for the total cost of producing x backpacks. What is the cost of producing 2500 backpacks from each manufacturer?
    (b) Use a graphing utility to determine a graph of each model in the same viewing window.
    (c) At what level of production x are the costs from the two manufacturers the same? Which manufacturer should the company select if the level of production will be less than this amount?

14. A garden in the shape of a rectangle must have a perimeter of 100 feet.
    (a) Determine a model for the area of the garden as a function of its width w. What is the domain of the function?
    (b) Use a graphing utility to determine a graph of the model. What are the intercepts? Over what (approximate) intervals is the function increasing? Decreasing? Interpret your results.
    (c) Using a graphing utility, determine (to the nearest foot) the dimensions of the garden with greatest area. What is the maximum area?

15. A rancher wants to fence off a feedlot in the shape of a rectangle against an existing wall, so that only three sides need to be fenced. The rancher has 280 yards of fencing.

    (a) Determine a model for the area of the feedlot as a function of its width w. What is the domain of the function?
    (b) Use a graphing utility to determine a graph of the model. What are the intercepts? Over what (approximate) intervals is the function increasing? Decreasing? Interpret your results.
    (c) Using a graphing utility, determine (to the nearest yard) the dimensions of the feedlot with greatest area. What is the maximum area?

B

16. A rectangular exhibit area is to be roped off with 120 feet of rope. An existing 14 foot wall is to be used as part of the boundary (Figure 1.79).

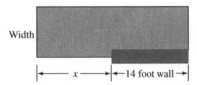

Figure 1.79

    (a) Determine a model for the area of the exhibit area as a function x. What is the domain of the function?
    (b) Use a graphing utility to graph the model. What are the intercepts? Over what (approximate) intervals is the function increasing? Decreasing? Interpret your results.
    (c) Using a graphing utility, determine (to the nearest 0.5 foot) the value of x that will give the greatest exhibit area. What is the maximum area?

17. A tank with a volume of 30 cubic feet has the shape of a closed-top box with a square base.
    (a) Determine the surface area (total area of the sides, top, and bottom) as a function of the length x of a side of the square base.
    (b) What is the domain of the function? Which has less surface area, a tank with a base that is 2 feet or 3 feet on a side?
    (c) Use a graphing utility to graph the model. Label the points on the graph that correspond to the two lengths

in part (b). Determine (to the nearest 0.1 foot) the length x that corresponds to the minimum surface area. What is the minimum area?

18. An open-top box with square base is to be constructed from two materials: one for the bottom and one for the sides. The volume of the box is to be 6 cubic feet. The cost of the material for the bottom is $3 per square foot, and the cost of the material for the sides is $2 per square foot.

   (a) Determine a model for the cost of the box as a function of its height h.

   (b) What is the domain of the function? Which will be more expensive to construct, a box with a height of 1 foot or 2 feet?

   (c) Use a graphing utility to graph the model. Label the points on the graph that correspond to the two heights in part (b). Determine (to the nearest 0.1 foot) the dimensions of the least expensive box. What is the minimum cost?

19. A wall is to be constructed around a rectangle of 240 square yards with a fence running the length, as shown in Figure 1.80. The construction cost for the fence is $5 per yard and for the wall is $8 per yard.

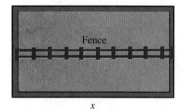

Figure 1.80

   (a) Determine a model for the cost of the total project as a function of the length x of the fence.

   (b) What is the domain of the function? Which leads to a less expensive construction cost, a length of 10 yards or 12 yards?

   (c) Use a graphing utility to graph the model. Label the points on the graph that correspond to the two lengths in part (b). Determine (to the nearest 0.1 yard) the dimensions that minimize the cost of the project. What is the minimum cost?

20. An underground telephone line is to be installed from a new house to the nearest junction box. The house is 50 meters from the main road, and the junction box is 180 meters down the road (Figure 1.81).

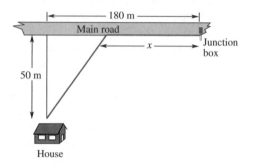

Figure 1.81

   (a) Determine a model for the length of the line as a function of the distance x the line is laid along the road.

   (b) What is the domain of the function? Is less line used when $x = 50$ or when $x = 60$? What is the domain of the function?

   (c) Use a graphing utility to graph the model. Label the points on the graph that correspond to the two lengths in part (b). Determine (to the nearest 0.1 meter) the value of x that will minimize the amount of line used. What is the minimum amount of line?

21. Suppose that in Exercise 20 the installation cost of the line is $2 per meter along the road and $3 per meter off the road.

   (a) Determine a model for the total cost of installing the line as a function of the distance x the line is laid along the road.

   (b) Use a graphing utility to graph the model. Determine (to the nearest 0.1 meter) the value of x that will minimize the installation cost. What is the minimum cost? What value of x corresponds to the greatest cost?

22. A bus company provides transportation from a train depot to an amusement park 22 miles away. The cost of operating a bus is $0.56 per mile. The bus driver is paid $15 per hour.

(a) Determine a model for the total cost of transporting passengers to the park and returning to the depot as a function of the average speed s the bus is driven.

(b) Suppose the speed limit is 45 miles per hour. Use a graphing utility to graph the model over the interval $(0, 45]$. What are the intercepts? Over what intervals is the function increasing? Decreasing? Interpret your results.

(c) Using a graphing utility, determine (to the nearest mile per hour) the speed that the bus driver should average to minimize the total cost. What is the minimum cost?

23. The operating cost of driving a vehicle may increase slightly as the speed increases (gas mileage suffers). Suppose the operating cost of the delivery van in Example 7 is $\$0.30 + 0.004s$ per mile, where s is the average speed.

(a) Determine a model for the total cost of delivery as a function of the average speed if everything else is the same as described in Example 7.

(b) Use a graphing utility to graph the model over the interval $(10, 55)$. What are the intercepts? Over what intervals is the function increasing? Decreasing? Interpret your results.

(c) Using a graphing utility, determine (to the nearest mile per hour) the speed the driver should average to minimize the total cost. What is the minimum cost?

24. Suppose the operating cost of the van in Exercise 23 is $\$0.45 + 0.01s$ per mile, where s is the average speed.

(a) Determine a model for the total cost as a function of the average speed if everything else is the same.

(b) Use a graphing utility to graph the model over the interval $(10, 45)$. What are the intercepts? Over what intervals is the function increasing? Decreasing? Interpret your results.

(c) Using a graphing utility, determine (to the nearest mile per hour) the speed the driver should average to minimize the total cost. What is the minimum cost?

25. A tire company currently sells 3200 tires a month at a selling price of $\$32$. Past experience indicates that for each $\$3$ increase in selling price, the monthly sales will decrease by 500. Determine an equation relating the selling price p and the number x of tires sold.

26. A cosmetics company sells 12,800 bottles of lotion a month at a selling price of $\$6.50$. A marketing survey indicates that for each $\$0.25$ decrease in selling price, the monthly sales will increase by 900. Determine an equation relating the selling price p and the number x of bottles sold.

27. A coffee company currently charges $\$2.40$ per bag and sells 4000 bags per week. The company plans to decrease the price per bag. Past experience indicates that each $\$0.10$ price decrease raises weekly sales by 400 bags.

(a) Determine an equation relating the selling price p and the number x of bags sold.

(b) Determine a model for the weekly revenue $R(x)$ as a function of the number of bags sold.

(c) Use a graphing utility to determine a graph for the revenue model.

(d) Using a graphing utility, determine (to the nearest 100 bags) the number x of bags that will maximize weekly revenue.

(e) To the nearest $\$0.10$, what price p should the coffee company charge to maximize revenue?

28. A video store rents 280 movies a day for a charge of $\$2.75$ per movie, but plans to decrease the charge. A survey of other stores indicates that each $\$0.25$ decrease in the rental fee raises daily rentals by 30.

(a) Determine an equation relating the rental price p and the number x of videos rented.

(b) Determine a model for the daily revenue $R(x)$ as a function of the number x of videos rented.

(c) Use a graphing utility to graph the model for revenue.

(d) Using a graphing utility, determine (to the nearest 10) the number of videos rented that will maximize revenue.

(e) To the nearest $\$0.25$, what rental fee p will maximize revenue?

C

29. An oil can in the shape of a right circular cylinder is to be constructed from two materials: one for the top and bottom and one for the sides. The volume of the can is to be

60 cubic inches. The cost for the top and bottom is $0.15 per square inch, and the cost for the sides is $0.09 per square inch.

(a) Determine a model for the cost of the can as a function of the radius x of the can.

(b) Using a graphing utility, determine (to the nearest 0.1 inch) the radius of the least expensive can. What is the minimum cost?

30. Laurie and Sheri are approaching the same intersection. Laurie is 2 miles north of the intersection traveling at 45 miles per hour. Sheri is 1 mile east of the intersection traveling at 30 miles per hour. Express the distance between them as a function of the time traveled, t.

31. A poster is to have 1200 square centimeters of text and graphics with 8 centimeter margins on top and bottom and 4 centimeter margins on each side (Figure 1.82). Find the area of the entire poster as a function of the width w of the poster.

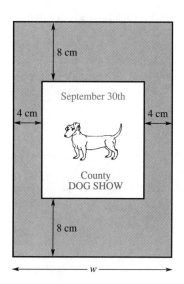

Figure 1.82

32. Determine the distance D from the point $P(1, 0)$ to a point $Q(x, y)$ on the graph of $x^2 + y^2 = 4$ as a function of the x-coordinate of Q (Figure 1.83). What is the domain of D? What is $D(2)$?

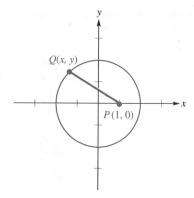

Figure 1.83

33. A water trough 8 feet long is constructed in such a way that the ends are isosceles triangles and the sides are rectangles that are 8 feet by 4 feet (Figure 1.84). Express the volume as a function of the width across the top.

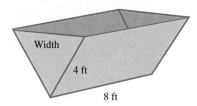

Figure 1.84

34. Water is poured into a conical tank at a rate of 24 cubic feet per hour. The tank is 12 feet high, and the top radius is 6 feet (Figure 1.85). Express the depth of the water as a function of the number of hours the water has been pouring into the tank.

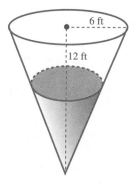

Figure 1.85

MORE ON GRAPHING

In this section we will examine how the graph of an equation is affected when we modify the equation. In addition, we will see how to determine an equation for a graph that is similar but not identical to a graph in the directory of graphs on page 108. Our discussion will rely on your familiarity with those eight graphs.

Horizontal and Vertical Translations

The graph of a function may have the same size and shape as another, but be moved to a different position in the coordinate plane. Such a shift in position is called a *translation*.

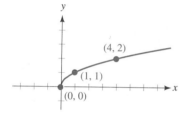

Figure 1.86
$f(x) = \sqrt{x}$

EXAMPLE 1 Translating a Known Graph Horizontally

Use a graphing utility to graph each function. Compare each with the graph of $f(x) = \sqrt{x}$ from the directory of graphs (repeated here as Figure 1.86).

(a) $y = \sqrt{x - 3}$ **(b)** $y = \sqrt{x + 2}$

SOLUTION

(a) Figure 1.87 shows the graph of $y = \sqrt{x - 3}$. The graph is identical to the graph of f except that it is translated (shifted) 3 units to the right. Three points that correspond to those labeled in Figure 1.86 are shown.

(b) Figure 1.88 shows the graph of $y = \sqrt{x + 2}$. The graph is the same as the graph of f except that it is translated (shifted) 2 units to the left. Three points that correspond to those labeled in Figure 1.86 are shown.

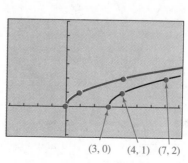

Figure 1.87
$y = \sqrt{x - 3}$
$[-4, 8] \times [-2, 6]$

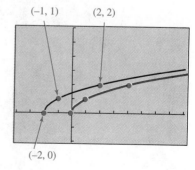

Figure 1.88
$y = \sqrt{x + 2}$
$[-4, 8] \times [-2, 6]$

Notice that each equation in Example 1 can be written in the form $y = f(x - h)$:

$$y = \sqrt{x - 3} \quad \text{can be written as} \quad y = f(x - 3)$$
$$y = \sqrt{x + 2} \quad \text{can be written as} \quad y = f(x - (-2))$$

In general, the graph of $y = f(x - h)$ is identical to the graph of f translated the *directed distance* of h units horizontally. By "directed distance," we mean that if h is positive, then the horizontal translation is to the right, and if h is negative, then the horizontal translation is to the left.

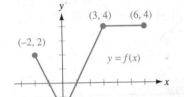

Figure 1.89

◼ EXAMPLE 2 Translating a Given Graph Horizontally

Given the graph of $y = f(x)$ in Figure 1.89, sketch the graph of each function.

(a) $y = f(x - 2)$ **(b)** $y = f(x + 1)$

SOLUTION

(a) The equation $y = f(x - 2)$ has the form $y = f(x - h)$, where $h = 2$. Therefore, the graph of $y = f(x - 2)$ is that of $y = f(x)$ translated 2 units to the right (Figure 1.90).

(b) Thinking of the form $y = f(x - h)$, we first rewrite the given equation $y = f(x + 1)$ as

$$y = f(x - (-1))$$

Thus, the value of h is -1. The graph is that of $y = f(x)$ translated 1 unit to the left (Figure 1.91).

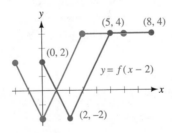

Figure 1.90

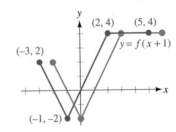

Figure 1.91

◼ EXAMPLE 3 Translating a Known Graph Vertically

Use a graphing utility to graph each function. Compare each with the graph of $f(x) = \sqrt{x}$ (Figure 1.86).

(a) $y = \sqrt{x} + 1$ **(b)** $y + 3 = \sqrt{x}$

SOLUTION

(a) Figure 1.92 shows the graph of $y = \sqrt{x} + 1$. The graph is identical to the graph of f except that it is translated 1 unit up.

(b) Our first step is to rewrite the equation with y isolated on one side:

$$y + 3 = \sqrt{x}$$
$$y = \sqrt{x} - 3$$

Figure 1.93 shows the graph of $y = \sqrt{x} - 3$. This graph is the same as the graph of f except that it is translated 3 units down.

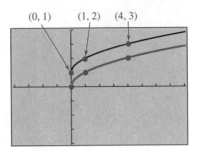

Figure 1.92
$y = \sqrt{x} + 1$
$[-4, 8] \times [-4, 4]$

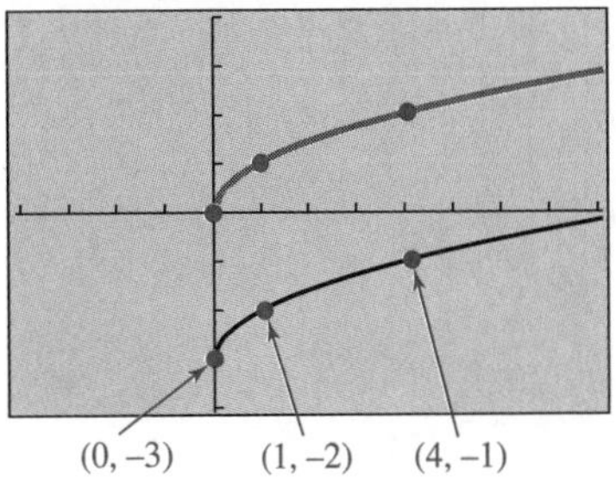

Figure 1.93
$y = \sqrt{x} - 3$
$[-4, 8] \times [-4, 4]$

Notice that each equation in Example 3 can be written in the form $y = f(x) + k$:

$$y = \sqrt{x} + 1 \quad \text{can be written as} \quad y = f(x) + 1$$
$$y = \sqrt{x} - 3 \quad \text{can be written as} \quad y = f(x) + (-3)$$

In general, the graph of $y = f(x) + k$ is the same as the graph of $y = f(x)$ translated the directed distance of k units vertically. If k is positive, the vertical translation is up, and if k is negative, the vertical translation is down.

EXAMPLE 4   Translating a Given Graph Vertically

Given the graph of $y = f(x)$ in Figure 1.94, sketch the graph of each equation.

(a) $y = f(x) + 2$ **(b)** $y = f(x) - 1$

SOLUTION

(a) The equation $y = f(x) + 2$ has the form $y = f(x) + k$, where $k = 2$. Therefore, the graph is that of $y = f(x)$ translated 2 units up, as shown in Figure 1.95 (page 154).

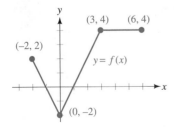

Figure 1.94

(b) The equation $y = f(x) - 1$ can be written as $y = f(x) + (-1)$, which has the form $y = f(x) + k$, where $k = -1$. The graph is that of $y = f(x)$ translated 1 unit down (Figure 1.96).

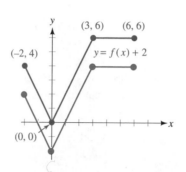

Figure 1.95

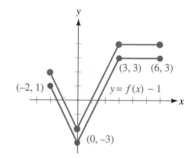

Figure 1.96

The next example involves both vertical and horizontal translation. (Try to anticipate the graph before seeing the result on your graphing utility.)

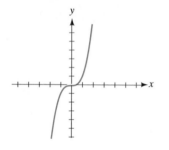

Figure 1.97
$f(x) = x^3$

EXAMPLE 5   Translating a Known Graph Horizontally and Vertically

Use a graphing utility to graph

$$y = (x - 3)^3 + 2$$

Compare the result with the graph of $f(x) = x^3$ from the directory of graphs in Section 1.3 (repeated here as Figure 1.97).

SOLUTION

The graph of $y = (x - 3)^3 + 2$ is shown in Figure 1.98; it is the same as the graph of f translated 3 units to the right and 2 units up.

The equation in Example 5 can be written as $y = f(x - h) + k$, where $h = 3$ and $k = 2$:

$$y = (x - 3)^3 + 2 \quad \text{can be written as} \quad y = f(x - 3) + 2$$

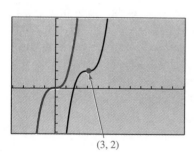

(3, 2)

Figure 1.98
$y = (x - 3)^3 + 2$
$[-4, 12] \times [-5, 8]$

In general, the graph of $y = f(x - h) + k$ is identical to the graph of f translated the directed distance of h units horizontally and k units vertically.

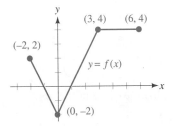

Figure 1.99

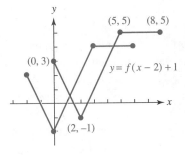

Figure 1.100

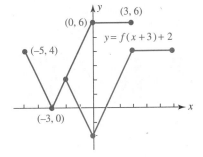

Figure 1.101

EXAMPLE 6   Translating a Given Graph Horizontally and Vertically

Given the graph of $y = f(x)$ in Figure 1.99, sketch the graph of each function.

(a) $y = f(x - 2) + 1$ **(b)** $y - 2 = f(x + 3)$

SOLUTION

(a) The equation $y = f(x - 2) + 1$ has the form $y = f(x - h) + k$, where $h = 2$ and $k = 1$. Therefore, the graph is the same as the graph of $y = f(x)$ translated 2 units right and 1 unit up, as shown in Figure 1.100.

(b) We want to write the equation in the form $y = f(x - h) + k$. Our first step is to rewrite $y - 2 = f(x + 3)$ with y isolated on the left side:

$$y = f(x + 3) + 2$$

Now we write this equation in the form $y = f(x - h) + k$:

$$y = f(x - (-3)) + 2$$

This equation has the form $y = f(x - h) + k$, where $h = -3$ and $k = 2$. Therefore, the graph is that of $y = f(x)$ translated 3 units to the left and 2 units up (Figure 1.101).

A summary of our discussion about translations follows:

Horizontal and Vertical Translations

Given the graph of $y = f(x)$:

1. To graph $y = f(x - h)$, translate the graph of $y = f(x)$ a directed distance of h units horizontally. If $h > 0$, the translation is to the right. If $h < 0$, the translation is to the left.

2. To graph $y = f(x) + k$, translate the graph of $y = f(x)$ a directed distance of k units vertically. If $k > 0$, the translation is upward. If $k < 0$, the translation is downward.

3. To graph $y = f(x - h) + k$, translate the graph of $y = f(x)$ a directed distance of h units horizontally (right if $h > 0$, left if $h < 0$) and k units vertically (up if $k > 0$, down if $k < 0$).

Translating a graph only changes the position of a graph—it does not change the shape of the graph. Now we investigate distortions of the shape of a graph.

Horizontal and Vertical Compressions or Expansions

Sometimes the graph of a function is similar to a graph we already know, except that it appears to be stretched or compressed.

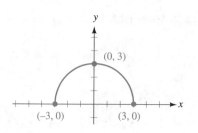

Figure 1.102
$f(x) = \sqrt{9 - x^2}$

EXAMPLE 7 Expanding or Compressing a Known Graph Vertically

Use a graphing utility to graph each function. Compare each with the graph of $f(x) = \sqrt{9 - x^2}$ shown in Figure 1.102.

(a) $y = 3\sqrt{9 - x^2}$ **(b)** $y = \frac{1}{3}\sqrt{9 - x^2}$

SOLUTION

(a) Figure 1.103 shows the graph of $y = 3\sqrt{9 - x^2}$. Notice that this graph resembles the graph of f, but it is 3 times as "tall." For each value of x, the y-coordinate of the graph of $y = 3\sqrt{9 - x^2}$ is 3 times the y-coordinate of the graph of f. In other words, the graph of $y = 3\sqrt{9 - x^2}$ is a *vertical expansion* of the graph of f from the x-axis.

(b) Figure 1.104 shows the graph of $y = \frac{1}{3}\sqrt{9 - x^2}$. This graph resembles the graph of f, except that it is $\frac{1}{3}$ as "tall." For each value of x, the y-coordinate of the graph of $y = \frac{1}{3}\sqrt{9 - x^2}$ is $\frac{1}{3}$ times the y-coordinate of the graph of f. In other words, the graph of $y = \frac{1}{3}\sqrt{9 - x^2}$ is a *vertical compression* of the graph of f toward the x-axis.

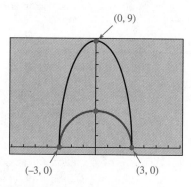

Figure 1.103
$y = 3\sqrt{9 - x^2}$
$[-7, 7] \times [0, 9]$

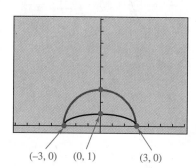

Figure 1.104
$y = \frac{1}{3}\sqrt{9 - x^2}$
$[-7, 7] \times [0, 9]$

Notice that if $f(x) = \sqrt{9 - x^2}$, each equation in Example 7 can be written in the form $y = Cf(x)$:

$$y = 3\sqrt{9 - x^2} \quad \text{can be written as} \quad y = 3f(x)$$
$$y = \frac{1}{3}\sqrt{9 - x^2} \quad \text{can be written as} \quad y = \frac{1}{3}f(x)$$

In general, consider how the graph of $y = Cf(x)$ is related to the graph of $y = f(x)$. If $C > 1$, the graph of $y = Cf(x)$ will look like that of $y = f(x)$ expanded vertically from the x-axis. If $0 < C < 1$, the graph of $y = Cf(x)$ will look like that of $y = f(x)$ compressed vertically toward the x-axis.

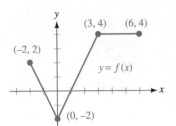

Figure 1.105

◼ EXAMPLE 8   Expanding or Compressing a Given Graph Vertically

Given the graph of $y = f(x)$ in Figure 1.105, sketch the graph of each of the following equations:

(a) $y = 2f(x)$ **(b)** $y = \frac{1}{2}f(x)$

SOLUTION

(a) The equation $y = 2f(x)$ has the form $y = Cf(x)$, where $C = 2$. Since $C > 1$, the graph of $y = 2f(x)$ is that of $y = f(x)$ expanded away from the x-axis. For each value of x, the y-coordinate of the graph of $y = 2f(x)$ is twice the y-coordinate of the graph of $y = f(x)$. To sketch the graph, we replace each point (a, b) on the graph of $y = f(x)$ with $(a, 2b)$. For example, $(-2, 2)$ corresponds to $(-2, 4)$, and $(0, -2)$ corresponds to $(0, -4)$. The graph of $y = 2f(x)$ is shown in Figure 1.106. Notice that the x-intercepts do not change.

(b) The equation $y = \frac{1}{2}f(x)$ has the form $y = Cf(x)$, where $C = \frac{1}{2}$. Since $C < 1$, the graph is a compression toward the x-axis. For each value of x, the y-coordinate of the graph of $y = \frac{1}{2}f(x)$ is $\frac{1}{2}$ the y-coordinate of the graph of $y = f(x)$. To sketch the graph, we replace each point (a, b) on the graph of $y = f(x)$ with $(a, \frac{1}{2}b)$. For example, $(-2, 2)$ corresponds to $(-2, 1)$, and $(0, -2)$ corresponds to $(0, -1)$. The graph of $y = \frac{1}{2}f(x)$ is shown in Figure 1.107. Notice that the x-intercepts do not change.

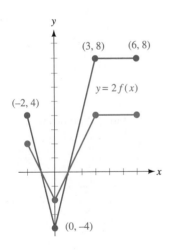

Figure 1.106

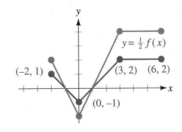

Figure 1.107

A summary of the relationship between the graph of $y = f(x)$ and the graph of $y = Cf(x)$, where $C > 0$, follows:

Vertical Expansions
and Compressions

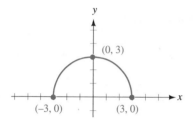

Figure 1.108
$f(x) = \sqrt{9 - x^2}$

Given the graph of $y = f(x)$:

To graph $y = Cf(x)$, replace each point (a, b) on the graph of $y = f(x)$ with (a, Cb).

1. If $C > 1$, the resulting graph is a vertical expansion away from the x-axis.

2. If $0 < C < 1$, the resulting graph is a vertical compression toward the x-axis.

Now we consider how the graph of $y = f(Cx)$ compares to that of $y = f(x)$. (Try to anticipate the graph before seeing it on your graphing utility.)

● EXAMPLE 9   Expanding or Compressing a Known Graph Horizontally

Use a graphing utility to graph each function. Compare each with the graph of $f(x) = \sqrt{9 - x^2}$ shown in Figure 1.108.

(a) $y = \sqrt{9 - (2x)^2}$ **(b)** $y = \sqrt{9 - \left(\frac{1}{2}x\right)^2}$

SOLUTION

(a) Figure 1.109 shows the graph of $y = \sqrt{9 - (2x)^2}$. This graph is similar to the graph of f, but it is $\frac{1}{2}$ as "wide." In other words, the graph of $y = \sqrt{9 - (2x)^2}$ is a *horizontal compression* of the graph of f toward the y-axis.

(b) Figure 1.110 shows the graph of $y = \sqrt{9 - \left(\frac{1}{2}x\right)^2}$. Notice that this graph is similar to the graph of f, except that it is 2 times as "wide." In other words, the graph of $y = \sqrt{9 - \left(\frac{1}{2}x\right)^2}$ is a *horizontal expansion* of the graph of f from the y-axis.   ●

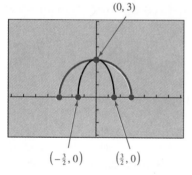

Figure 1.109
$y = \sqrt{9 - (2x)^2};\ [-7, 7] \times [-3, 6]$

If $f(x) = \sqrt{9 - x^2}$, each equation in Example 9 can be written in the form $y = f(Cx)$:

$$y = \sqrt{9 - (2x)^2} \quad \text{can be written as} \quad y = f(2x)$$
$$y = \sqrt{9 - \left(\tfrac{1}{2}x\right)^2} \quad \text{can be written as} \quad y = f\left(\tfrac{1}{2}x\right)$$

In general, how is the graph of $y = f(Cx)$ related to the graph of $y = f(x)$? If $C > 1$, the graph of $y = f(Cx)$ looks like that of $y = f(x)$ compressed horizontally toward the y-axis. If $0 < C < 1$, the graph of $y = f(Cx)$ looks like the graph of $y = f(x)$ expanded horizontally from the y-axis.

If (a, b) is any point on the graph of $y = f(x)$, then $(a/C, b)$ is on the graph

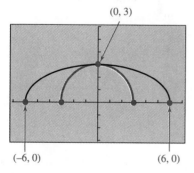

Figure 1.110
$y = \sqrt{9 - \left(\frac{1}{2}x\right)^2};\ [-7, 7] \times [-3, 6]$

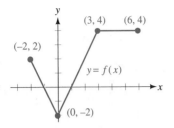

Figure 1.111

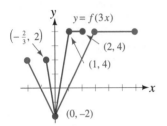

Figure 1.112

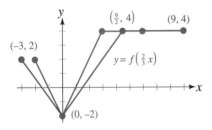

Figure 1.113

of $y = f(Cx)$. Therefore, if we know the graph of $y = f(x)$, we can sketch the graph of $y = f(Cx)$ by replacing each point (a, b) on the graph of $y = f(x)$ with the point $(a/C, b)$.

⬤ **EXAMPLE 10   Expanding or Compressing a Given Graph Horizontally**

Given the graph of $y = f(x)$ in Figure 1.111, sketch the graph of each equation.

(a) $y = f(3x)$ **(b)** $y = f(\frac{2}{3}x)$

SOLUTION

(a) The equation $y = f(3x)$ has the form $y = f(Cx)$, where $C = 3$. Since $C > 1$, the graph is that of $y = f(x)$ compressed toward the y-axis. We can sketch the graph (Figure 1.112) by replacing each point (a, b) on the graph of $y = f(x)$ with $(a/3, b)$. For example, $(-2, 2)$ corresponds to $\left(-\frac{2}{3}, 2\right)$, and $(3, 4)$ corresponds to $(1, 4)$. Notice that the y-intercept does not change.

(b) The equation $y = f(\frac{2}{3}x)$ has the form $y = f(Cx)$, where $C = \frac{2}{3}$. Since $C < 1$, the graph is an expansion from the y-axis. We replace each point (a, b) on the graph of $y = f(x)$ with

$$\left(\frac{a}{\frac{2}{3}}, b\right) \qquad \text{or} \qquad \left(\frac{3a}{2}, b\right)$$

For example, $(-2, 2)$ corresponds to $(-3, 2)$, and $(3, 4)$ corresponds to $\left(\frac{9}{2}, 4\right)$. The graph is shown in Figure 1.113. ⬤

A summary of the relationship between the graph of $y = f(x)$ and the graph of $y = f(Cx)$, where $C > 0$, follows:

Horizontal Expansions
and Compressions

Given the graph of $y = f(x)$:
To graph $y = f(Cx)$, replace each point (a, b) on the graph of $y = f(x)$ with $(a/C, b)$.

1. If $C > 1$, the resulting graph is a horizontal compression toward the y-axis.
2. If $0 < C < 1$, the resulting graph is a horizontal expansion from the y-axis.

We conclude this section with an investigation of the graphs of $y = Cf(x)$ and $y = f(Cx)$ where $C = -1$.

Reflections About the x-Axis or y-Axis

The graph of a function may be a mirror image, or *reflection*, of another graph.

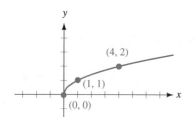

Figure 1.114
$f(x) = \sqrt{x}$

EXAMPLE 11   Reflecting a Known Graph

Use a graphing utility to graph each function. Compare each with the graph of $f(x) = \sqrt{x}$ (Figure 1.114).

(a) $y = -\sqrt{x}$ **(b)** $y = \sqrt{-x}$

SOLUTION

(a) Figure 1.115 shows the graph of $y = -\sqrt{x}$. The graph is a reflection of the graph of $f(x) = \sqrt{x}$ about the x-axis.

(b) The graph of $y = \sqrt{-x}$ is shown in Figure 1.116. The graph is a reflection of the graph of f about the y-axis.

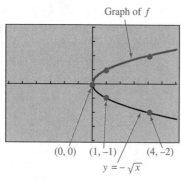

Figure 1.115
$[-6, 6] \times [-4, 4]$

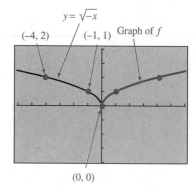

Figure 1.116
$[-6, 6] \times [-4, 4]$

The equations in Example 11 can be expressed in terms of $f(x) = \sqrt{x}$:

$$y = -\sqrt{x} \quad \text{can be written as} \quad y = -f(x)$$
$$y = \sqrt{-x} \quad \text{can be written as} \quad y = f(-x)$$

In general, the graph of $y = -f(x)$ is the reflection of the graph of $y = f(x)$ through the x-axis; the graph of $y = f(-x)$ is the reflection of the graph of $y = f(x)$ through the y-axis.

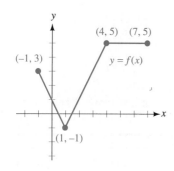

Figure 1.117

EXAMPLE 12   Reflecting a Given Graph

The graph of $y = f(x)$ is shown in Figure 1.117. Sketch the graph of each of the following functions:

(a) $y = -f(x)$ **(b)** $y = f(-x)$

SOLUTION

(a) The graph of $y = -f(x)$ is the reflection of the graph of $y = f(x)$ through the x-axis. We can graph $y = -f(x)$ by replacing each point (a, b) on the graph of $y = f(x)$ with $(a, -b)$. For example, $(4, 5)$ corresponds to $(4, -5)$, and $(-1, 3)$ corresponds to $(-1, -3)$. The graph of $y = -f(x)$ is shown in Figure 1.118.

(b) The graph of $y = f(-x)$ is the reflection of the graph of $y = f(x)$ through the y-axis. To graph $y = f(-x)$, we replace each point (a, b) on the graph of $y = f(x)$ with $(-a, b)$. For example, $(4, 5)$ corresponds to $(-4, 5)$, and $(-1, 3)$ corresponds to $(1, 3)$, as shown in Figure 1.119.

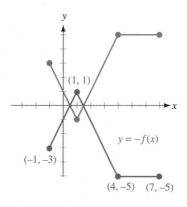

Figure 1.118   **Figure 1.119**

A summary of our discussion of the graphs of $y = -f(x)$ and $y = f(-x)$ follows:

Reflecting About the x-Axis
and y-Axis

Given the graph of $y = f(x)$:

1. To graph $y = -f(x)$, replace each point (a, b) on the graph of $y = f(x)$ with $(a, -b)$. The resulting graph is a reflection about the x-axis.

2. To sketch the graph of $y = f(-x)$, replace each point (a, b) on the graph of $y = f(x)$ with $(-a, b)$. The resulting graph is a reflection about the y-axis.

We can apply the techniques of this section to determine the equation for a given graph.

EXAMPLE 13 Finding an Equation for a Graph

Each graph is a translation, a reflection, or both a translation and a reflection of one of the graphs in the directory of graphs on page 108. Determine an equation for each graph.

■ **NOTE:** *The two dashed lines in the graph in part (a) are guidelines to help see the translation.*

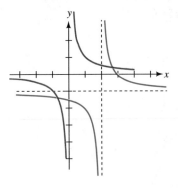

Figure 1.120

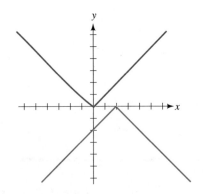

Figure 1.121

(a)

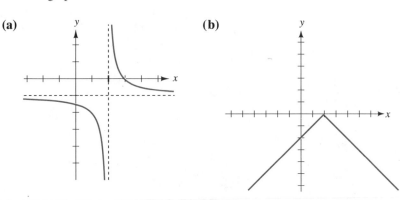

(b)

SOLUTION

(a) This graph resembles the graph of $f(x) = 1/x$. However, it has been translated 2 units to the right and 1 unit down (Figure 1.120). Thinking in terms of

$$y = f(x - h) + k$$

we see that $h = 2$ and $k = -1$. Therefore, an equation for this graph is

$$y = \frac{1}{x - 2} + (-1)$$

$$y = \frac{1}{x - 2} - 1$$

(b) This is the graph of $f(x) = |x|$ translated 2 units to the right, followed by a reflection about the x-axis (Figure 1.121). Thus, an equation for the graph is

$$y = -f(x - h)$$
$$y = -|x - 2|$$

● EXERCISE SET 1.6

A

In Exercises 1–10, use a graphing utility to graph each function. Compare your graph with one of the graphs in the directory of graphs on page 108.

1. $y = |x - 2|$

2. $y = |x + 3|$

3. $y = \sqrt{4 - (x + 5)^2}$

4. $y = \sqrt{9 - (x - 2)^2}$

5. $y = x^2 - 4$

6. $y = x^2 + 2$

7. $y = \dfrac{1}{x - 2} + 3$

8. $y = \dfrac{1}{x - 1} - 2$

9. $y = (x - 2)^3 - 1$

10. $y = (x - 1)^3 + 2$

In Exercises 11–18, given the graph of $y = f(x)$ in Figure 1.122, sketch the graph of each equation.

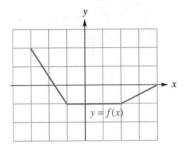

Figure 1.122

11. $y = f(x + 2)$

12. $y = f(x - 4)$

13. $y = f(x) - 1$

14. $y = f(x) + 2$

15. $y = f(x - 1) - 2$

16. $y = f(x - 3) + 2$

17. $y - 3 = f(x - 2)$

18. $y - 5 = f(x + 4)$

In Exercises 19–30, use a graphing utility to graph each function. Compare your graph with one of the graphs in the directory of graphs on page 108.

19. $y = 2\sqrt{x}$

20. $y = 3\sqrt{x}$

21. $y = \frac{1}{2}\sqrt{16 - x^2}$

22. $y = \frac{3}{5}\sqrt{25 - x^2}$

23. $y = \frac{1}{2}x^3$

24. $y = \frac{1}{3}x^3$

25. $y = \frac{4}{3}|x|$

26. $y = \frac{5}{2}|x|$

27. $y = \sqrt{9 - (3x)^2}$

28. $y = \sqrt{36 - (2x)^2}$

29. $y = \sqrt{36 - \left(\dfrac{x}{2}\right)^2}$

30. $y = \sqrt{4 - \left(\dfrac{x}{3}\right)^2}$

In Exercises 31–38, given the graph of $y = g(x)$ in Figure 1.123, sketch the graph of each equation.

31. $y = 3g(x)$

32. $y = 2g(x)$

33. $y = \frac{2}{3}g(x)$

34. $y = \frac{3}{4}g(x)$

35. $y = g(3x)$

36. $y = g(2x)$

37. $y = g(\frac{1}{2}x)$

38. $y = g(\frac{1}{3}x)$

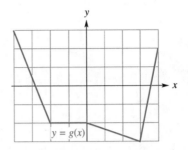

Figure 1.123

In Exercises 39–44 given the graph of $y = f(x)$ in Figure 1.122 and $y = g(x)$ in Figure 1.123, sketch the graph of each equation.

39. $y = -f(x)$

40. $y = -g(x)$

41. $y = f(-x)$

42. $y = g(-x)$

43. $y = -f(-x)$

44. $y = -g(-x)$

In Exercises 45–52, use a graphing utility to graph each function. Compare your result with one of the graphs in the directory of graphs on page 108.

45. $y = -\sqrt{9 - x^2}$

46. $y = -\sqrt{-x}$

47. $y = -\frac{1}{2}|x|$

48. $y = -2|x|$

49. $y = -(x + 1)^2 - 2$

50. $y = -(x - 1)^2 - 3$

51. $y = \sqrt{-(x + 4)}$

52. $y = \sqrt{-(x - 2)}$

B

In Exercises 53–58, each graph is a translation, a reflection, or both a translation and a reflection of one of the graphs in the directory of graphs on page 108. Determine an equation for each graph.

53.

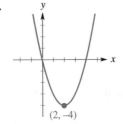

54.

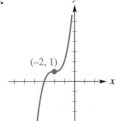

55.

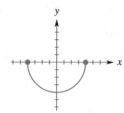

56.

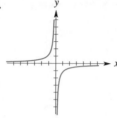

57.

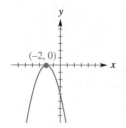

58.

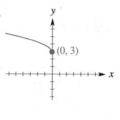

In Exercises 59–62, each graph is a compression or expansion of one of the graphs in the directory of graphs on page 108. Determine an equation for each graph.

59.

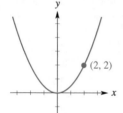

60.

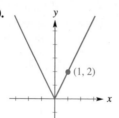

61.

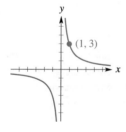

62.

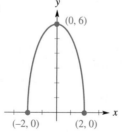

63. Describe the graph of

$$y + 3 = \frac{1}{2 - x}$$

relative to one of the graphs in the directory of graphs on page 108.

HINT: $\dfrac{1}{2 - x} = \dfrac{1}{-(x - 2)}$

64. Describe the graph of

$$y = \frac{x - 1}{x - 2}$$

relative to one of the graphs in the directory of graphs on page 108.

HINT: $\dfrac{x - 1}{x - 2} = \dfrac{x - 2}{x - 2} + \dfrac{1}{x - 2}$

65. Describe the graph of $y = \sqrt{7 - x^2 - 6x}$ relative to one of the graphs in the directory of graphs on page 108.
HINT: $\sqrt{7 - x^2 - 6x} = \sqrt{7 - (x^2 + 6x)}$; complete the square.

66. Consider the graph of $y = E(x)$ in Figure 1.124. Reflect the graph about the y-axis, and then expand away from the y-axis by a factor of 2. Sketch the resulting graph and give an equation for this graph.

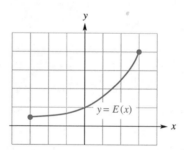

Figure 1.124

67. Consider the graph of $y = S(x)$ in Figure 1.125. Reflect the graph about the y-axis, and then expand away from the x-axis by a factor of 3. Sketch the resulting graph and give an equation for this graph.

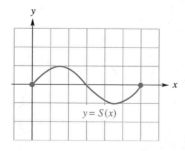

Figure 1.125

68. A skydiver jumps from an airplane that is 15,000 feet above ground level. Her height (in feet) above ground level t seconds after she jumps is given by

$$h(t) = -16t^2 + 15,000$$

(a) In terms of translations, reflections, compressions, or expansions, describe how the graph of h is related to that of $f(t) = t^2$ in the directory of graphs (page 108).

(b) Suppose there is another skydiver in the airplane, and he jumps 3 seconds after the first skydiver. Determine a model for his height $H(t)$ (in feet) above ground level as a function of t.

(c) In terms of translations, reflections, compressions, or expansions, describe how the graph of H is related to the graph of h.

69. A speedboat travels on a river at a rate of r miles per hour to a destination 10 miles away. The time (in hours) that it will take to reach the destination can be modeled by

$$T(r) = \frac{10}{r}$$

(a) In terms of translations, reflections, compressions, or expansions, describe how the graph of T is related to that of $f(r) = 1/r$ in the directory of graphs (page 108).

(b) The model given assumes that there is no current. Suppose the speedboat travels against a current of 5 miles per hour. Determine a model for the time it will take to reach the destination as a function of r.

(c) In terms of translations, reflections, compressions, or expansions, describe how the graph of the model in part (b) is related to the graph of $T(r) = 10/r$.

70. Two sides of a triangle are 3 units long and the base is x units long, as shown in Figure 1.126:

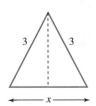

Figure 1.126

(a) Determine a model for the height of the triangle as a function of x.

(b) In terms of translations, reflections, compressions, or expansions, describe how the graph of the model in part (a) is related to one of the graphs in the directory of graphs (page 108).

71. Consider the cost model

$$C(s) = 17 + \frac{500}{s}$$

developed in Example 7 of Section 1.5.

(a) In terms of translations, compressions, reflections, or expansions, describe how the graph of C is related to one of the graphs in the directory of graphs (page 108).

(b) Use a graphing utility to graph C.

C

72. Sketch the graph of

$$f(x) = \begin{cases} x + 1 & x < 0 \\ x^2 & x \geq 0 \end{cases}$$

Reflect the graph about the y-axis; then translate 3 units to the left. Sketch the resulting graph and give an equation for this graph.

73. Refer to the graph of $y = g(x)$ given below:

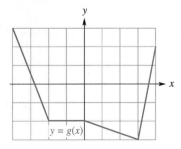

(a) Sketch the graph of $y = |g(x)|$.

(b) Sketch the graph of $y = g(|x|)$.

OPERATIONS WITH FUNCTIONS

Many real-life situations call for combining functions. For example, in Section 1.5 we developed a model for business costs $C(x)$ and revenue $R(x)$; the model for profit is the difference of these two functions. Radio transmissions can be modeled as functions of time, and when two transmissions occupy the same space, the resulting transmission is the sum of the two original functions. In this section we will see how to combine functions to create new functions.

Combining Functions with the Four Arithmetic Operations

NOTE: *When we combine real numbers or functions by division, we must exclude division by 0.*

When two real numbers are combined by addition, subtraction, multiplication, or division, the result is always a real number. Similarly, when two functions are combined by using these four operations, the result is always a function.

For example, suppose that

$$f(x) = 2x^2 - x + 3 \quad \text{and} \quad g(x) = 4 - 7x$$

The difference of these two functions is another function, $f - g$, defined as

$$(f - g)(x) = f(x) - g(x)$$
$$= (2x^2 - x + 3) - (4 - 7x)$$
$$= 2x^2 + 6x - 1$$

The Four Arithmetic Operations on Functions

> Given the functions f and g, the **sum**, **difference**, **product**, and **quotient** are defined as follows:
>
> **1.** *Sum:* $\qquad (f + g)(x) = f(x) + g(x)$
> **2.** *Difference:* $\quad (f - g)(x) = f(x) - g(x)$
> **3.** *Product:* $\qquad (fg)(x) = f(x)g(x)$
> **4.** *Quotient:* $\qquad \left(\dfrac{f}{g}\right)(x) = \dfrac{f(x)}{g(x)}, \quad g(x) \neq 0$
>
> The domain of the sum, difference, or product of f and g is the set of all real numbers common to the domain of f and the domain of g. The domain of the quotient is also the set of all real numbers common to the domain of f and the domain of g, with the additional restriction that $g(x) \neq 0$.

EXAMPLE 1 The Sum and Difference of Two Functions

Let $f(x) = 3x^2 + 2x - 1$ and $g(x) = x^2 - 4$.

(a) Find $(f + g)(x)$ and $(f - g)(x)$.
(b) Evaluate $(f + g)(3)$ and $(f - g)(3)$.

SOLUTION

(a)
$$(f + g)(x) = f(x) + g(x)$$
$$= (3x^2 + 2x - 1) + (x^2 - 4)$$
$$= 4x^2 + 2x - 5$$

■ **NOTE:** *When finding* $f(x) - g(x)$, *it is important to enclose* $g(x)$ *in parentheses.*

$$(f - g)(x) = f(x) - g(x)$$
$$= (3x^2 + 2x - 1) - (x^2 - 4)$$
$$= 2x^2 + 2x + 3$$

(b) Substituting $x = 3$ into our result for $(f + g)(x)$ and $(f - g)(x)$ in part (a), we get
$$(f + g)(3) = 4(3)^2 + 2(3) - 5 = 37$$
$$(f - g)(3) = 2(3)^2 + 2(3) + 3 = 27$$

● **EXAMPLE 2   The Product and Quotient of Two Functions**
Let $f(x) = 3x^2 + 2x - 1$ and $g(x) = x^2 - 4$.

(a) Find $(fg)(x)$ and $(f/g)(x)$.
(b) Evaluate $(fg)(-4)$ and $(f/g)(-4)$.

SOLUTION

(a)
$$(fg)(x) = f(x)g(x)$$
$$= (3x^2 + 2x - 1)(x^2 - 4)$$
$$= 3x^4 + 2x^3 - 13x^2 - 8x + 4$$

$$\left(\frac{f}{g}\right)(x) = \frac{f(x)}{g(x)}$$
$$= \frac{3x^2 + 2x - 1}{x^2 - 4}$$

(b) We substitute $x = -4$ into our results for $(fg)(x)$ and $(f/g)(x)$ in part (a):
$$(fg)(-4) = 3(-4)^4 + 2(-4)^3 - 13(-4)^2 - 8(-4) + 4 = 468$$
$$\left(\frac{f}{g}\right)(-4) = \frac{3(-4)^2 + 2(-4) - 1}{(-4)^2 - 4} = \frac{39}{12} = \frac{13}{4}$$

In Examples 1 and 2, notice that the domain of $f(x) = 3x^2 + 2x - 1$ is $(-\infty, +\infty)$ and the domain of $g(x) = x^2 - 4$ is $(-\infty, +\infty)$. Therefore, the domain of $f + g, f - g$, and fg is $(-\infty, +\infty)$. To determine the domain of f/g, we must exclude all real values of x such that $g(x) = 0$:

$$x^2 - 4 = 0$$
$$(x + 2)(x - 2) = 0$$

$$x + 2 = 0 \qquad | \qquad x - 2 = 0$$
$$x = -2 \qquad | \qquad x = 2$$

Thus, the domain of f/g is the set of all real numbers except for -2 and 2. In interval notation the domain is

$$(-\infty, -2) \quad \text{or} \quad (-2, 2) \quad \text{or} \quad (2, +\infty)$$

The next example discusses functions with more restricted domains.

■ EXAMPLE 3   The Domain of Sum, Difference, Product, and Quotient

Let $f(x) = \sqrt{9 - x^2}$ and $g(x) = \sqrt{x + 2}$.

(a) Determine the domain of $f + g$, $f - g$, and fg. Write each answer in interval notation.

(b) Determine the domain of f/g. Write the answer in interval notation.

SOLUTION

(a) We start by determining the domain of f and the domain of g. Because f and g are easily graphed, we will determine their domains from their graphs. The graph of f is shown in Figure 1.127. From the graph, we see that $f(x)$ is defined for $-3 \le x \le 3$, which means the domain of f is $[-3, 3]$. The graph of g is the graph of $y = \sqrt{x}$ translated 2 units to the left (Figure 1.128). From this graph we see that the domain of g is $[-2, +\infty)$. The domain of $f + g$, $f - g$, and fg is the set of all real numbers common to both the domain of f and the domain of g. We find that the intersection of $[-3, 3]$ and $[-2, +\infty)$ is $[-2, 3]$. Therefore, the domain of $f + g$, $f - g$, and fg is $[-2, 3]$.

■ **GRAPHING NOTE:** *To check the result of part (a), use a graphing utility to graph $f + g$, $f - g$, or fg. For example, the graph of $f - g$ does not extend to the left of $x = -2$ or to the right of $x = 3$ (see the figure below).*

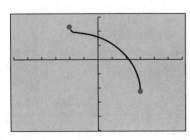

$y = \sqrt{9 - x^2} - \sqrt{x + 2}$
$[-6, 6] \times [-5, 3]$

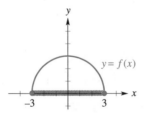

Figure 1.127

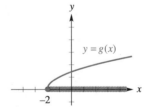

Figure 1.128

(b) From part (a), the intersection of $[-3, 3]$ and $[-2, +\infty)$ is $[-2, 3]$. In addition, we must exclude any real values of x that will cause the denominator to be 0. We can locate these values by setting $g(x) = 0$ and solving for x:

$$g(x) = 0$$
$$\sqrt{x + 2} = 0$$
$$x + 2 = 0 \qquad \text{Square each side.}$$
$$x = -2$$

We must exclude $x = -2$ from $[-2, 3]$, so the domain for f/g is $(-2, 3]$.

A common error is to simplify $f + g, f - g, fg$, or f/g and then determine the domain based only on the final result. For example, if

$$f(x) = x - \sqrt{x} + 2 \qquad \text{and} \qquad g(x) = \sqrt{x}$$

then

$$(f + g)(x) = (x - \sqrt{x} + 2) + \sqrt{x} = x + 2$$

If we only consider the final result, the domain appears to be $(-\infty, +\infty)$. However, taking the restrictions on the domains of f and g into account, the domain for $f + g$ is actually $[0, +\infty)$. Remember: First find the domains of the original functions f and g.

Graphing by Adding y-Coordinates

We can describe the sum of two functions geometrically. For example, consider the function $h(x) = |x - 3| + \sqrt{x + 2}$, which is the sum of the two functions $f(x) = |x - 3|$ and $g(x) = \sqrt{x + 2}$. If we sketch the graph of $y = f(x)$ and $y = g(x)$ on the same axes (Figure 1.129), we can visualize the graph of their sum as follows: Since $h(x) = f(x) + g(x)$, we add the y values at each value of x to obtain the graph of $y = f(x) + g(x)$. For example, at $x = 2, f(2) = |2 - 3| = 1$ and $g(2) = \sqrt{2 + 2} = 2$. In Figure 1.130, these two values can be seen as the "height" of each of the graphs at $x = 2$. To determine the point on the graph of $y = |x - 3| + \sqrt{x + 2}$ at $x = 2$, we add these two "heights" and find the point $(2, 3)$. Doing this for other values of x, we complete the graph (Figure 1.131).

Graphing the sum of two functions this way is a technique of graphing called *adding y-coordinates*.

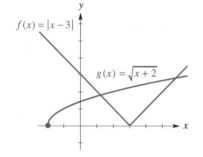

Figure 1.129

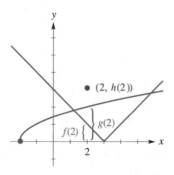

Figure 1.130

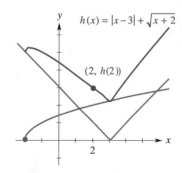

Figure 1.131

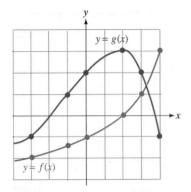

Figure 1.132

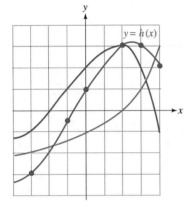

Figure 1.133
$h(x) = f(x) + g(x)$

EXAMPLE 4 Graphing the Sum of Two Functions

The graphs of $y = f(x)$ and $y = g(x)$ are shown in Figure 1.132. Let $h(x) = f(x) + g(x)$.

(a) Approximate the values of $h(-3)$, $h(-1)$, and $h(2)$.

(b) Sketch the graph of $y = h(x)$ by adding y-coordinates.

SOLUTION

(a) From the graphs we get

$$h(-3) = f(-3) + g(-3) = -2 + (-1) = -3$$
$$h(-1) = f(-1) + g(-1) \approx -1.5 + 1 = -0.5$$
$$h(2) = f(2) + g(2) = 0 + 3 = 3$$

(b) Using the results from part (a), we plot the points $(-3, -3)$, $(-1, -0.5)$, and $(2, 3)$. In addition,

$$h(0) = f(0) + g(0) = -1 + 2 = 1$$
$$h(3) = f(3) + g(3) = 1 + 2 = 3$$
$$h(4) = f(4) + g(4) = 3 + (-1) = 2$$

We plot the points $(0, 1)$, $(3, 3)$, and $(4, 2)$; then we connect the points with a curve, as shown in Figure 1.133.

Composition and Decomposition

In Section 1.2, we mentioned that a function g is a process that assigns to each input x exactly one output $g(x)$. Suppose we use the value $g(x)$ as an input for another function f. For example, let

$$g(x) = |x - 1| \quad \text{and} \quad f(x) = \sqrt{x}$$

Start with an input, say $x = 5$, for g:

	Input	*Output*
	$x = 5$	$g(5) = |5 - 1| = 4$

Now use $g(5) = 4$ as the input for f:

	Input	*Output*
	4	$f(4) = \sqrt{4} = 2$

Notice that for our original input, 5, there is exactly one final output, 2. In fact, for any original input, there will be exactly one final output. This last statement qualifies the entire process as a function, called the *composition* of the functions f and g.

Composition of Functions

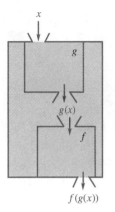

$f(g(x))$

Figure 1.134

> Given the functions f and g, the **composition of f with g** is a function defined as
> $$(f \circ g)(x) = f(g(x))$$
> The domain of $f \circ g$ is the set of all real numbers x in the domain of g such that $g(x)$ is in the domain of f.

The composition of f with g can be thought of as a machine whose inner components are the functions f and g (Figure 1.134). If x is the input to the machine, it is processed first by g. The output, $g(x)$, becomes the input for the function f. The final output is $f(g(x))$. Figure 1.135 may help you understand the domain of $f \circ g$.

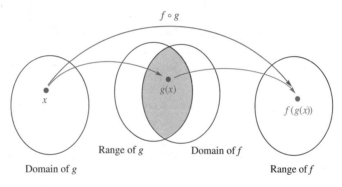

Figure 1.135

EXAMPLE 5   Composition of Functions

Let $f(x) = x^2$ and $g(x) = \sqrt{1 - x}$. Determine:

(a) $(f \circ g)(x)$ **(b)** $(g \circ f)(x)$

SOLUTION

NOTE: *When we write $f(x) = x^2$, the choice of the variable x is not essential. The function f can just as well be written as $f(u) = u^2$.*

(a) $(f \circ g)(x) = f(g(x))$

$\qquad\qquad\quad = f(\sqrt{1 - x})$ \qquad Substitute $\sqrt{1 - x}$ for $g(x)$.

$\qquad\qquad\quad = (\sqrt{1 - x})^2$ \qquad Definition of f is $f(u) = u^2$.

$\qquad\qquad\quad = 1 - x, \quad x \le 1$ \quad Since $(\sqrt{u})^2 = u$, $u \ge 0$.

Therefore, $(f \circ g)(x) = 1 - x$, $x \le 1$.

(b) $(g \circ f)(x) = g(f(x))$

$\qquad\qquad = g(x^2)$ Substitute x^2 for $f(x)$.

$\qquad\qquad = \sqrt{1 - x^2}$ Definition of g is $g(u) = \sqrt{1 - u}$.

Therefore, $(g \circ f)(x) = \sqrt{1 - x^2}$.

Notice that in Example 5,

$$(f \circ g)(x) \neq (g \circ f)(x)$$

This is usually the case. However, there are functions f and g such that $(f \circ g)(x)$ and $(g \circ f)(x)$ are equal (see Exercises 52–55).

Thus far, we have started with two functions and found their composition. In calculus it is important to be able to *decompose* a function—that is, to recognize a function as the composition of two functions.

EXAMPLE 6   The Decomposition of Functions

Express

$$h(x) = \frac{2}{\sqrt{x + 1}}$$

as the composition of two functions $f(g(x))$. Do not use $f(x) = x$ or $g(x) = x$.

SOLUTION

The expression $x + 1$ under the radical can be viewed as the function $g(x) = x + 1$. If we let $f(x) = 2/\sqrt{x}$, then

$$f(g(x)) = f(x + 1) = \frac{2}{\sqrt{x + 1}}$$

NOTE: *Often, you will see more than one way to decompose a function.*

ALTERNATE SOLUTION: Consider the denominator of $h(x)$ as the function $g(x) = \sqrt{x + 1}$. If we let $f(x) = 2/x$, then

$$f(g(x)) = f(\sqrt{x + 1}) = \frac{2}{\sqrt{x + 1}}$$

Applications

Operations with functions are often encountered in mathematical models. In Section 1.5, the models developed in Examples 2 and 5 were sums of two functions.

We close this section with applications of the composition of functions, the difference of two functions, and adding y-coordinates.

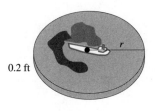

0.2 ft

Figure 1.136

● EXAMPLE 7 Application: The Spread of an Oil Spill

Suppose a wrecked oil tanker spills oil at a rate of 70 cubic feet per minute in the shape of a right circular cylinder that is 0.2 foot thick (Figure 1.136).

(a) Use the formula for the volume of a circular cylinder to express the radius $g(v)$ as a function of the volume v.

(b) Express the volume of the spill $f(t)$ as a function of the number of minutes t of oil flow.

(c) Determine a model for the radius $h(t)$ as a function of t.

(d) Suppose the center of the spill is $\frac{1}{8}$ mile away from the shore. According to our model, how long (to the nearest minute) will it take for the spill to reach the shore?

SOLUTION

(a) The volume of a circular cylinder is given by the formula

$$\pi(\text{Radius})^2(\text{Height}) = (\text{Volume})$$

Since the thickness is given as 0.2 foot, we substitute 0.2 for the height. Letting r and v represent the radius and volume, respectively, we have

$$\pi r^2(0.2) = v$$

We want to determine the radius as a function of the volume. Solving for r gives

$$r^2 = \frac{v}{0.2\pi}$$

$$r^2 = \frac{5v}{\pi}$$

$$r = \sqrt{\frac{5v}{\pi}} \quad \text{Take the positive square root of each side, since the radius is positive.}$$

Thus, the radius $g(v)$ is a function of volume:

$$g(v) = \sqrt{\frac{5v}{\pi}}$$

(b) The volume of the spill is given by

$$\left(\begin{array}{c}\text{Volume} \\ \text{of spill}\end{array}\right) = \left(\begin{array}{c}\text{Rate} \\ \text{of spill}\end{array}\right)\left(\begin{array}{c}\text{Time} \\ \text{of spill}\end{array}\right)$$

$$v = 70t$$

So the volume $f(t)$ is a function of time:

$$f(t) = 70t$$

(c) The radius is a function of volume, and the volume is a function of time:

$$\begin{aligned} g(v) &= g(f(t)) \\ &= g(70t) \\ &= \sqrt{\frac{5(70t)}{\pi}} \\ &= \frac{5}{\pi}\sqrt{14\pi t} \end{aligned}$$

Thus,

$$h(t) = \frac{5}{\pi}\sqrt{14\pi t}$$

(d) Since our model for the radius is in terms of feet, we convert $\frac{1}{8}$ mile into feet:

$$\tfrac{1}{8}\text{ mile} = (\tfrac{1}{8}\text{ mile})(5280\text{ feet per mile}) = 660\text{ feet}$$

The oil will reach the shore when the radius is 660 feet. Substituting 660 for $h(t)$ into our model, we have

$$h(t) = \frac{5}{\pi}\sqrt{14\pi t}$$

$$660 = \frac{5}{\pi}\sqrt{14\pi t}$$

Solving for t gives

$$660\,\frac{\pi}{5} = \sqrt{14\pi t}$$

$$132\pi = \sqrt{14\pi t}$$

$$(132\pi)^2 = 14\pi t$$

$$\frac{(132\pi)^2}{14\pi} = t$$

$$3910 \approx t$$

The oil will reach the shore in about 3910 minutes, or 65 hours 10 minutes.

In Section 1.5, we discussed two functions that are important to a business manufacturing x units of a product selling for p dollars per unit:

Revenue function: $R(x) = xp = $ Amount collected from selling x units

Cost function: $C(x) = $ Cost of producing and selling x units

The *profit function P* is defined as:

Profit function: $P(x) = R(x) - C(x)$

$\qquad\qquad\qquad\quad = $ Net income from producing and selling x units

EXAMPLE 8   Application: Profit Analysis

A clothing company that currently sells 8400 blouses per month at a price of $18.50 per blouse plans to increase the price. A marketing survey indicates that the monthly sales will decrease by 600 for each $0.50 increase in price. It follows that the selling price p and the number x of blouses sold is related by

$$x = 8400 - 600\left(\frac{p - 18.50}{0.50}\right)$$

(a) Determine a model for the monthly revenue $R(x)$ as a function of the number of blouses sold.

(b) Suppose the cost of producing x blouses per month is

$$C(x) = 10{,}000 + 7x$$

Determine a model for the monthly profit $P(x)$.

(c) Use a graphing utility to graph the model for profit. Determine (to the nearest 100) the number of blouses that will maximize profit.

SOLUTION

(a) After a price increase, the revenue will be

$$R(x) = \left(\begin{matrix}\text{Future number} \\ \text{of blouses sold}\end{matrix}\right)\left(\begin{matrix}\text{Future price} \\ \text{per blouse}\end{matrix}\right)$$

$$= xp$$

Since we want to express revenue as a function of x, we need to express p in terms of x. We are given that

$$x = 8400 - 600\left(\frac{p - 18.50}{0.50}\right)$$

Simplifying the right side gives

$$x = 30,600 - 1200p$$

We can solve this equation for p:

$$1200p + x = 30,600$$

$$p = \frac{(30,600 - x)}{1200}$$

If we substitute this expression for p into

$$R(x) = xp$$

we have

$$R(x) = x\frac{(30,600 - x)}{1200}$$

(b) The monthly profit is

$$P(x) = R(x) - C(x)$$

$$= x\left(\frac{30,600 - x}{1200}\right) - (10,000 + 7x)$$

(c) A graph for P is shown in Figure 1.137. The maximum profit corresponds to the highest point on the graph. Looking at Figure 1.137, profit is maximized when $x \approx 11,000$ blouses are sold. Using the zoom and trace features on a graphing utility, we can improve this estimate. As indicated in Figure 1.138, the number of blouses that will maximize profit is 11,100 (to the nearest 100).

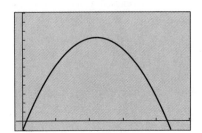

Figure 1.137

$$y = x\left(\frac{30,600 - x}{1200}\right) - (10,000 + 7x)$$

$$[-1000, 25,000]_{5000} \times$$
$$[-10,000, 120,000]_{10,000}$$

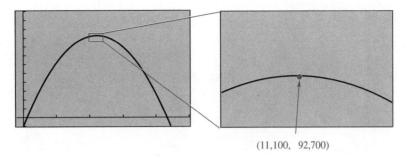

Figure 1.138

APPLICATION OF ADDING *y*-COORDINATES—DIRECTIONAL ANTENNAS:
Radio signals travel in the form of a wave called a *carrier wave*. At a particular
instant in time, the graph of a carrier wave looks something like the graph in
Figure 1.139. The graph is symmetric with respect to the vertical axis through
the transmitter, and continues the oscillating pattern.

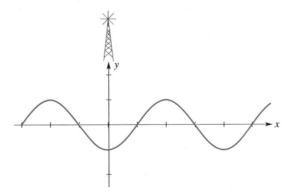

Figure 1.139

When two carrier waves *f* and *g* are transmitted, the resulting wave is the sum
of *f* and *g*. Radio stations that want a strong signal in one direction and no signal
in the opposite direction place two antennas *h* units apart (Figure 1.140, page
178). The resulting carrier wave is obtained by adding the *y*-coordinates of *f* and
g. To the right the result is a strong signal, and to the left the resulting carrier
wave is 0 (no signal).

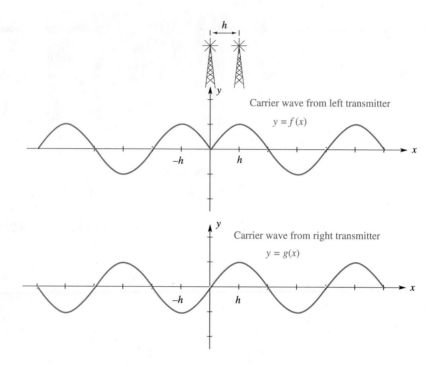

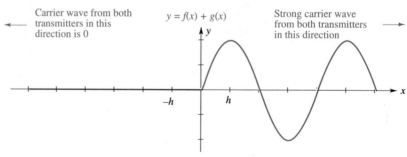

Figure 1.140

EXERCISE SET 1.7

A

In Exercises 1–6:

(a) Determine the functions $(f + g)(x)$ and $(f - g)(x)$.

(b) Evaluate $(f + g)(2)$ and $(f - g)(2)$.

  1. $f(x) = x^2 - 2x - 4$; $g(x) = 4x - 1$

  2. $f(x) = 2x^2 + x - 2$; $g(x) = 3x^2 - 4x$

  3. $f(x) = x^2 - 9$; $g(x) = \sqrt{2x}$

  4. $f(x) = \sqrt{x + 7}$; $g(x) = x^2 - 5x + 6$

  5. $f(x) = \dfrac{1}{x + 1}$; $g(x) = \dfrac{2}{x}$

6. $f(x) = \dfrac{5}{x-3}$; $g(x) = \dfrac{1}{x}$

In Exercises 7–12:

(a) Determine the functions $(fg)(x)$ and $(f/g)(x)$.
(b) Evaluate $(fg)(-3)$ and $(f/g)(-3)$.

7. $f(x) = 2 - x^2$; $g(x) = 2x + 3$
8. $f(x) = 2x^2 + x$; $g(x) = 5 - x$
9. $f(x) = x^2 - 9$; $g(x) = \sqrt{x+4}$
10. $f(x) = x^2 - 5x + 6$; $g(x) = \sqrt{1-x}$
11. $f(x) = 2x - 7$; $g(x) = x^2 + 3x$
12. $f(x) = x^2 - 5$; $g(x) = \sqrt{2x+6}$

In Exercises 13–16:

(a) Determine the domain of $f + g$, $f - g$, and fg.
(b) Determine the domain of f/g.

Write your answers in interval notation.

13. $f(x) = x^2 + 1$; $g(x) = \sqrt{x+3}$
14. $f(x) = |x|$; $g(x) = \sqrt{x-1}$
15. $f(x) = \dfrac{1}{x}$; $g(x) = \sqrt{4 - x^2}$
16. $f(x) = \dfrac{1}{x-2}$; $g(x) = \sqrt{9 - x^2}$

17. Let

$$f(x) = \frac{x-1}{x+1} \quad \text{and} \quad g(x) = 2(x+1)$$

Graph $y = (fg)(x)$.

18. Let

$$f(x) = \sqrt{x+4} \quad \text{and} \quad g(x) = x - \sqrt{x}$$

Graph $y = (f + g)(x)$.

In Exercises 19–26:

(a) Find $(f \circ g)(x)$.　　　　**(b)** Find $(g \circ f)(x)$.

19. $f(x) = 2x + 3$; $g(x) = x^2 - x$
20. $f(x) = 3x + 1$; $g(x) = x^2 + 2x$
21. $f(x) = \dfrac{9}{x}$; $g(x) = \sqrt{x}$

22. $f(x) = \dfrac{1}{4x}$; $g(x) = \sqrt{x}$
23. $f(x) = x^2 + 3$; $g(x) = 3x - x^2$
24. $f(x) = 2x^2 + 1$; $g(x) = x - 2x^2$
25. $f(x) = 3$; $g(x) = 2x^2 + 5$
26. $f(x) = 3|x - 2|$; $g(x) = 6$

In Exercises 27–32, use a graphing utility to graph the functions f, g, and $f + g$ in the same coordinate plane.

27. $f(x) = |x + 1|$; $g(x) = -x$
28. $f(x) = |x - 3|$; $g(x) = x - 1$
29. $f(x) = \frac{1}{2}x^3$; $g(x) = -x$
30. $f(x) = \frac{1}{2}x^2$; $g(x) = x - 1$
31. $f(x) = \sqrt{9 - x^2}$; $g(x) = x - 1$
32. $f(x) = \sqrt{16 - x^2}$; $g(x) = -x$

B

In Exercises 33–36, the graphs of f, g, and $f + g$ are given. Determine which is the graph of $f + g$.

33.

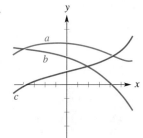

34.

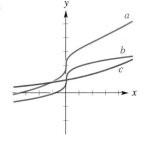

35.

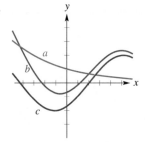

36.

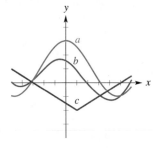

In Exercises 37–44, use the graphs of f, g, and h in Figure 1.141 to sketch the graph of each equation.

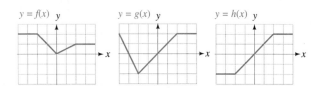

Figure 1.141

37. $y = f(x) + g(x)$ **38.** $y = f(x) + h(x)$

39. $y = g(x) + h(x)$ **40.** $y = 2f(x) + g(x)$

41. $y = f(x) + 2g(x)$ **42.** $y = f(x) - g(x)$

43. $y = f(x) + x$ **44.** $y = h(x) + |x|$

In Exercises 45–51, express $h(x)$ as the composition of two functions $(f \circ g)(x)$. Do not use $g(x) = x$ or $f(x) = x$.

45. $h(x) = \sqrt{2x + 1}$ **46.** $h(x) = |5 - x^2|$

47. $h(x) = \dfrac{3}{\sqrt{4 - x}}$ **48.** $h(x) = \dfrac{|x|}{|x| + 5}$

49. $h(x) = \dfrac{x^4}{x^4 - 2}$ **50.** $h(x) = (3x^2 + x - 1)^3$

51. $h(x) = (x^3 - 4x^2 + x)^5$

In Exercises 52–55:

(a) Show that $(f \circ g)(x) = x$.
(b) Show that $(g \circ f)(x) = x$.

52. $f(x) = \frac{1}{3}x - 2$; $g(x) = 3x + 6$

53. $f(x) = \frac{2}{5}x + 3$; $g(x) = \frac{5}{2}(x - 3)$

54. $f(x) = \dfrac{1}{x + 2}$; $g(x) = \dfrac{1}{x} - 2$

55. $f(x) = \dfrac{x + 3}{x}$; $g(x) = \dfrac{3}{x - 1}$

In Exercises 56 and 57, find $(f \circ g)(x)$ and sketch the graph of $y = (f \circ g)(x)$.

56. $f(x) = \frac{1}{2}x^2$; $g(x) = \sqrt{x - 4}$

57. $f(x) = \dfrac{1}{x}$; $g(x) = \dfrac{1}{x + 1}$

58. A coffee company currently charges \$2.40 per bag and sells 4000 bags per week. The company plans to decrease the price per bag. Past experience indicates that each \$0.10 price decrease raises weekly sales by 400 bags. It follows that the selling price p and the number x of bags sold is related by

$$x = 4000 + 400\left(\frac{2.40 - p}{0.10}\right)$$

(a) Determine a model for the weekly revenue $R(x)$ as a function of the number x of bags sold.
(b) Suppose the cost of producing x bags per week is $C(x) = 2000 + 0.80x$. Determine a model for the weekly profit $P(x) = R(x) - C(x)$.
(c) Use a graphing utility to determine (to the nearest 50) the number of bags that will maximize profit.

59. A video store rents 280 movies a day for a charge of \$2.75 per movie, but plans to increase the charge. A survey of other stores indicates that each \$0.25 increase lowers daily rentals by 30.

(a) Determine an equation relating the rental price p and the number x of videos rented.

(b) Determine a model for the weekly revenue $R(x)$ as a function of the number x of videos rented.

(c) Suppose the cost of maintaining x videos per day is $C(x) = 300 + 0.5x$. Determine a model for the weekly profit $P(x) = R(x) - C(x)$.

(d) Use a graphing utility to determine (to the nearest 5) the number of videos rented that will maximize profit.

60. The cost model

$$C(h) = \frac{36}{h} + 36\sqrt{h}$$

developed in Example 2 of Section 1.5 can be sketched by adding y-coordinates and then expanding from the x-axis.

(a) Sketch $f(x) = 1/x$ and $g(x) = \sqrt{x}$ on the same axes for $0 \leq x \leq 9$.

(b) Sketch $h(x) = f(x) + g(x)$ by the method of adding y-coordinates.

(c) Sketch $C(x) = 36h(x)$ by expanding the graph of h from the x-axis. You will need to adjust the scale for the y-axis.

61. Refer to the application of directional antennas discussed at the end of this section. Assume everything remains the same, except that instead of placing the antennas h units apart, it is more desirable to place them farther apart. What are the possible distances?

62. The radius of a balloon is initially 4 inches and grows at a rate of 2 inches per minute.

(a) Determine a function $r(t)$ that gives the radius at t seconds.

(b) Use $r(t)$ from part (a) to find a function $V(t)$ that gives the volume at t minutes.

63. A car travels north from an intersection at a rate of 40 miles per hour. A town is located 80 miles due west of the intersection, as shown in Figure 1.142.

(a) Determine the distance $g(t)$ from the car to the intersection t hours after the car passes the intersection.

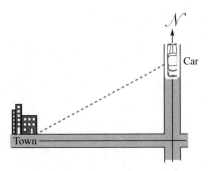

Figure 1.142

(b) Determine a function $f(t)$ that gives the distance from the car to the town t hours after the car passes the intersection.

64. Figure 1.143 shows the graphs of $y = f(x)$ and $y = g(x)$:

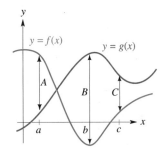

Figure 1.143

Let A = the vertical distance between the curves at $x = a$, B = the vertical distance between the curves at $x = b$, and C = the vertical distance between the curves at $x = c$. Express A, B, and C in terms of f, g, a, b, and c.

C

65. Graph the function obtained in part (c) of Example 7:

$$h(t) = \frac{5}{\pi}\sqrt{14\pi t}$$

Discuss the rate at which the radius grows as time passes. Does the rate of growth increase, decrease, or remain constant? When does the rate of growth appear to be greatest?

66. **(a)** Determine two functions f and g such that $f(g(x)) = x$. Graph f and g on the same coordinate axes.
  (b) Find another pair of functions p and q such that $p(q(x)) = x$. Graph p and q on the same coordinate axes.
  (c) Discuss how the graphs of the pairs of functions in parts (a) and (b) are related.

67. Bill is attached to an elastic chord and jumps from a 250 foot bridge. Figure 1.144 shows Bill's height during the initial fall as a function of time. Figure 1.145 shows the graph of Bill's velocity during the initial fall as a function of height. Sketch the graph of Bill's velocity as a function of time.

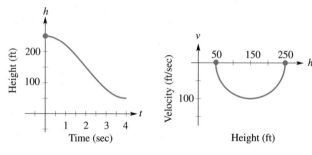

Figure 1.144   Figure 1.145

QUICK REFERENCE

Topic	Page	Remarks
Graph of an equation in two variables	77	The graph of an equation in two variables x and y is the set of all points with coordinates (x, y) that are solutions of the equation.
x-Intercept of a graph	77	An x-intercept is a point where the graph intersects the x-axis; the x-intercepts can be determined by substituting $y = 0$ into the equation and solving for x.
y-Intercept of a graph	77	A y-intercept is a point where the graph intersects the y-axis; the y-intercepts can be determined by substituting $x = 0$ into the equation and solving for y.
Distance between two points	79	The distance between the points $A(x_1, y_1)$ and $B(x_2, y_2)$ is denoted $d(A, B)$. From the Pythagorean theorem, $$d(A, B) = \sqrt{(x_2 - x_1)^2 + (y_2 - y_1)^2}$$
Equation of a circle	80	The standard form of an equation of a circle with center at (h, k) and radius r is $$(x - h)^2 + (y - k)^2 = r^2$$ The general form of an equation of a circle is $$x^2 + y^2 + Dx + Ey + F = 0$$ where D, E, and F are real numbers.
Midpoint of a segment	82	The midpoint of the line segment with endpoints $A(x_1, y_1)$ and $B(x_2, y_2)$ has coordinates $$\left(\frac{x_1 + x_2}{2}, \frac{y_1 + y_2}{2} \right)$$

Topic	Page	Remarks
Symmetry about the *x*-axis	84	A graph is symmetric about the *x*-axis if for any point (x, y) on the graph, $(x, -y)$ is also on the graph.
Test for symmetry about the *x*-axis	84	The graph of an equation in *x* and *y* is symmetric about the *x*-axis if replacing (x, y) with $(x, -y)$ in the equation produces an equivalent equation.
Symmetry about the *y*-axis	84	A graph is symmetric about the *y*-axis if for any point (x, y) on the graph, $(-x, y)$ is also on the graph.
Test for symmetry about the *y*-axis	84	The graph of an equation in *x* and *y* is symmetric about the *y*-axis if replacing (x, y) with $(-x, y)$ in the equation produces an equivalent equation.
Symmetry about the origin	84	A graph is symmetric about the origin if for any point (x, y) on the graph, $(-x, -y)$ is also on the graph.
Test for symmetry about the origin	84	The graph of an equation in *x* and *y* is symmetric about the origin if replacing (x, y) with $(-x, -y)$ in the equation produces an equivalent equation.
Function	91	A function is a rule that assigns to each element of a set of inputs one and only one element of a set of outputs. Given a function *f*, if the input is *x*, then the output is denoted by $f(x)$.
Domain	92	The domain of a function *f* is the set of real numbers *x* for which $f(x)$ is defined.
Zero of a function	92	A zero of a function *f* is a number *a* in the domain of *f* such that $f(a) = 0$.
Graph of a function	102	The graph of a function *f* is the set of all points with coordinates (x, y) that are solutions of the equation $y = f(x)$, where *x* is in the domain of *f*.
Vertical-line test	104	A given graph is the graph of a function if and only if any vertical line intersects the graph at no more than one point.
x-Intercepts of a function	105–106	The *x*-intercepts of a function *f* are the points with coordinates $(a, f(a))$ such that *a* is a zero of *f*.
y-Intercept of a function	105	The *y*-intercept of a function is the point with coordinates $(0, f(0))$.
Increasing function	107	A function *f* is increasing over an open interval if for any two values x_1 and x_2 in the interval, $x_1 < x_2$ implies that $f(x_1) < f(x_2)$. The graph of an increasing function rises as we move from left to right.
Decreasing function	107	A function *f* is decreasing over an open interval if for any two values x_1 and x_2 in the interval, $x_1 < x_2$ implies that $f(x_1) > f(x_2)$. The graph of an increasing function falls as we move from left to right.
Turning point	107	A turning point is a point where the graph changes from increasing to decreasing, or vice versa.
Qualitative graph	111	A qualitative graph is a rough sketch that emphasizes the qualities and behavior of a function *f*, rather than precise solutions to the equation $y = f(x)$.

Continued

Topic	Page	Remarks
Viewing window	118	The rectangular portion of the coordinate plane displayed by a graphing utility. The viewing window $$[\text{Xmin, Xmax}]_{\text{Xscale}} \times [\text{Ymin, Ymax}]_{\text{Yscale}}$$ describes the set of all points (x, y) such that $\text{Xmin} \le x \le \text{Xmax}$ and $\text{Ymin} \le y \le \text{Ymax}$. The marks on the x-axis and y-axis are determined by the values of Xscale and Yscale.
Mathematical modeling	131	Describing a physical situation in mathematical terms such as functions, equations, or inequalities.
Translating a graph	151	See the box on page 155 for translating the graph of $y = f(x)$.
Expanding or compressing a graph	155	See the box on page 158 for expanding or compressing the graph of $y = f(x)$ vertically. See the box on page 159 for expanding or compressing the graph of $y = f(x)$ horizontally.
Reflecting a graph	160	See the box on page 161 for reflecting the graph of $y = f(x)$ about the x-axis or y-axis.
Composition of functions	177	Given two functions f and g, the composition of f with g, denoted $f \circ g$, is the function with output $f(g(x))$ for an input x.

◉ MISCELLANEOUS EXERCISES

1. Determine $d(A, B)$ and the midpoint M of the line segment with endpoints $A(-2, 11)$ and $B(5, -13)$.

2. Repeat Exercise 1 with $A(-6, 5)$ and $B(-4, 5)$.

In Exercises 3–6, determine the radius, center, and standard equation of the circle described.

3. The circle is the graph of $x^2 + y^2 - 4x + 4y = 0$.

4. The circle is the graph of $x^2 + y^2 - 10x + 6y - 2 = 0$.

5. The center of the circle is $(-1, 5)$, and the circle passes through $(2, 2)$.

6. The line segment with endpoints $(2, -3)$ and $(6, 1)$ is a diameter.

In Exercises 7–10, a portion of the graph of the given equation is shown. Use the symmetry tests to complete the graph. Label each intercept of the graph with its coordinates.

7. $y = x|x| - x$

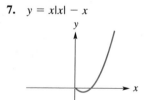

8. $(x + 2)^3 = y^2$

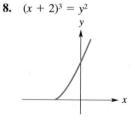

9. $y = \dfrac{4 - x^2}{x^2}$

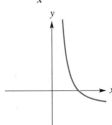

10. $|x| + 2|y| = 2$

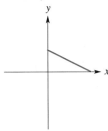

11. What point is symmetric to the point $A(-7, 3)$ with respect to the point $P(1, 2)$?

12. What point is symmetric to the point $A(2, 11)$ with respect to the horizontal line through the point $P(4, 3)$?

13. Given the function $h(x) = 3 - \sqrt{x - 4}$:
   (a) Determine the zeros and domain of h.
   (b) Evaluate $h(5)$.
   (c) Simplify $h(4t^2)$.
   (d) Simplify $6h(r^2 + 4)$.

14. Given the function $L(x) = \dfrac{24}{x} - 8$:
   (a) Determine the zeros and domain of L.
   (b) Evaluate $L(-2)$.
   (c) Simplify $L\left(\dfrac{4}{3k}\right)$.
   (d) Simplify $2L(r) - L(2r)$.

15. Given the function $f(x) = 4 - 3x$, determine:
$$\frac{f(x + h) - f(x)}{h}$$

16. Given the function $g(x) = x^2 + 5x$, determine:
$$\frac{g(x + h) - g(x)}{h}$$

17. Given the function
$$F(x) = \begin{cases} \dfrac{1}{x} & x \le -1 \\ 1 + x_2 & x > -1 \end{cases}$$
evaluate:
   (a) $F(2)$ (b) $F\left(-\frac{2}{3}\right)$ (c) $F\left(-\frac{3}{2}\right)$
   (d) $F(a_2)$

18. Given the function
$$P(x) = \begin{cases} \sqrt{x}_4 & x < 2 \\ 5 - x & x \ge 2 \end{cases}$$
evaluate:
   (a) $P(0)$ (b) $P(-3)$ (c) $P(2)$
   (d) $P(\sqrt{2})$

In Exercises 19–24, use a graphing utility to determine a complete graph of each function and approximate the x-intercepts (to the nearest 0.001).

19. $y = x^3 - 4x + 3$ **20.** $y = 12x - x^3 - 4$

21. $y = 2^x - 8x - 20$ **22.** $y = \left(\frac{1}{2}\right)^x - 8x - 20$

23. $y = 2 - \sqrt[3]{25 - x^2}$ **24.** $y = \sqrt[3]{x^2 - 20} - 3$

In Exercises 25–28:

(a) Use a graphing utility to determine the complete graph of f.
(b) Determine the x-intercepts of f exactly using algebraic methods.

25. $f(x) = -3x^2 + 15$ **26.** $f(x) = 16 - 4x^3$

27. $f(x) = x^3 - 9x^2 - 6x$ **28.** $f(x) = 5x + 2x^2 - x^3$

29. Approximate (to the nearest 0.01) the solutions of the equation $x^3 - 12x + 7 = 0$ by using the graph of an appropriate function.

30. Approximate (to the nearest 0.01) the solutions of the equation $3^x + 16x = 0$ by using the graph of an appropriate function.

In Exercises 31–36, determine the graph of each function and compare this graph to one of the graphs in the directory of graphs on page 108.

31. $y = |x + 4|$ **32.** $y = |x| - 3$

33. $y = (x - 1)^3 + 4$ **34.** $y = -x^3 - 6$

35. $y = \dfrac{-1}{x - 2}$

36. $y = \dfrac{2}{x} + 4$

In Exercises 37–40, use the graph of the function f shown.

$y = f(x)$

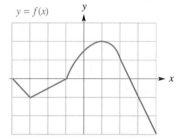

37. Determine the value of $f(-3), f(-1), f(1),$ and $f(4)$.

38. Determine the value of $f(-4), f(-2), f(2),$ and $f(3)$.

39. Determine the x-intercepts and turning points of the graph of f.

40. Determine the open intervals over which f is positive, negative, increasing, and decreasing.

In Exercises 41–50, use the graph of g shown to sketch the graph of each function:

$y = g(x)$

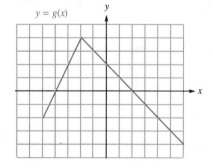

41. $y = g(x - 4) + 2$

42. $y = g(x + 3) - 4$

43. $y = 2g(x) - 2$

44. $y = \frac{1}{2}g(x) + 6$

45. $y = -g\left(\frac{1}{2}x\right)$

46. $y = g(-2x)$

47. $y = g(x) + \frac{1}{2}x$

48. $y = g(x) - x$

49. $y = g(x) + |x|$

50. $y = g(x) - |x|$

In Exercises 51 and 52, each graph is a modification of one of the functions from the directory of graphs on page 108. Determine an equation for the graph.

51.

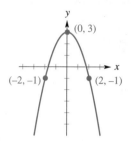

52.

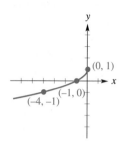

In Exercises 53–58, let $f(x) = x^2 - 6$ and $g(x) = 4x + 1$. Determine:

53. (a) $f + g$ (b) $f - 2g$

54. (a) $3f + g$ (b) $g - f$

55. (a) fg (b) $(f)^2$

56. (a) f/g (b) $(g)^3$

57. (a) $f \circ g$ (b) $f \circ f$

58. (a) $g \circ f$ (b) $g \circ g$

59. Determine two functions f and g such that

$$(f \circ g)(x) = 2^{4x - 3}$$

[Do not use $f(x) = x$ or $g(x) = x$.]

60. Determine two functions f and g such that

$$(f \circ g)(x) = \dfrac{1}{x - 3} + (x - 3)^2$$

[Do not use $f(x) = x$ or $g(x) = x$.]

61. (a) Given the points $A(0, 2), B(-3, 6), C(0, 6),$ and $D(4, 2)$, show that the midpoints of the sides of quadrilateral $ABCD$ form a parallelogram.

(b) Repeat part (a) with $A(x_1, y_1), B(x_2, y_2), C(x_3, y_3),$ and $D(x_4, y_4)$.

62. (a) Given the points $A(0, 0), B(4, 0),$ and $C(0, 6)$, show that the midpoint of the hypotenuse of the right triangle ABC is the same distance from each vertex of the triangle.

(b) Repeat part (a) with $A(0, 0), B(a, 0),$ and $C(0, b)$.

63. A function of the form $f(x) = A \cdot x^n$ was used to generate the data in the table at the top of the next page. What are the values of A and n?

x	0	-1	1	3	-2
$f(x)$	0	-4	4	108	-32

64. A function of the form $f(x) = A \cdot b^x$ was used to generate the data in the table. What are the values of A and b?

x	0	2	4	7	10
$f(x)$	$\frac{3}{4}$	3	12	96	768

65. From the following scenario, sketch a qualitative graph for the height of water in the tank as a function of time.

> *An inlet pipe begins to fill a large water tank (with vertical sides) at a constant rate. Soon after, the drain is opened and water drains out at a constant rate. (This rate is less than the rate of the inlet pipe.) Later, the inlet pipe is turned off, and the tank continues to drain until it is empty.*

66. From the following scenario, sketch a qualitative graph for the concentration of antifreeze in the radiator as a function of total amount of additive.

> *Mia notices that the radiator fluid in her truck is low and that the concentration of antifreeze in the radiator is less than 50%. She adds pure antifreeze to fill the radiator. Later, she notices that the radiator fluid is low again by the same amount, and she fills it with water. (Assume that the additives mix completely.)*

67. A weight suspended by a spring from the ceiling of a classroom is pulled down and released. The height h (in inches) of the weight (relative to its rest position) t seconds after it is released is represented by the graph.

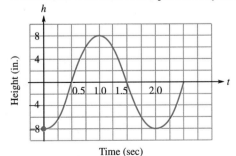

(a) Use the graph to construct a table for the function using the values $t = 0, 0.5, 1.0, 1.5, 2.0, 2.5$. (Approximate the values of h to the nearest inch.)

(b) Determine the open intervals over which the weight is moving down and over which it is moving up.

(c) What is the maximum height according to this graph? What is the minimum? For what values of t do these occur?

68. Nitrogen is used as a fertilizer in a patch of tomato plants. Past experience shows that the yield T of the patch (in pounds of tomatoes) is a function of the added nitrogen b (in bags). This function is represented by the graph.

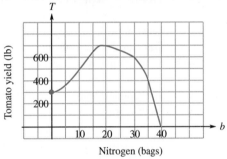

(a) Use the graph to construct a table for the function using the values $b = 10, 20, 30, 40$. (Approximate the values of T to the nearest pound.)

(b) Determine the open intervals over which the yield is increasing and the open intervals over which it is decreasing.

(c) What is the maximum yield possible according to this graph? What is the minimum? For what values of b do these occur?

69. A computer keyboard manufacturer finds that a start-up cost of $24,000 is necessary, and then each keyboard costs $35 to produce. Determine the total cost C of producing x keyboards. What is the cost of producing 2000 keyboards?

70. A rectangle has an area of 30 square feet. Determine the length of a diagonal D as a function of the width w.

71. From past experience, a manufacturer of manual lawn-mowers has found that the profit (or loss), in thousands of dollars, realized from manufacturing x thousand units annually is given by

$$P(x) = -1.2^x + 90.3x^2 - 4250$$

(a) Determine a complete graph of P.
(b) What is the minimum number of units that should be manufactured to produce a profit? What is the maximum number of units that will produce a profit?
(c) How many units should be manufactured to produce the maximum profit? What is this profit (to the nearest hundred dollars)?

72. A wire that is 40 centimeters in length is cut into two pieces of length x and $40 - x$. Each piece is bent into a square.
(a) Determine a function for the total area of both squares as a function of x. What is the domain?
(b) Determine a complete graph of the function in part (a).
(c) Use the graph to determine the minimum total area. Is there a maximum total area? Explain.

73. The points P and Q are vertices of the base angles of an isosceles triangle, and they lie on the graph of $y = 18 - \frac{1}{2}x^2$. The base PQ of the triangle is above and parallel to the x-axis, and the third vertex is at the origin, as shown in the figure.

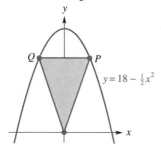

(a) Determine a function for the area of the triangle as a function of the x-coordinate of the point P. What is the domain of this function?
(b) Determine a complete graph of the function in part (a).
(c) Use the graph to determine the maximum area.

74. A rancher needs to construct a rectangular corral using 330 feet of fencing and two sides of a barn that is 60 feet by 80 feet.

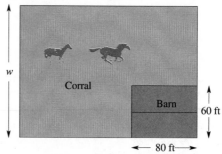

(a) Determine a function for the area of the corral as a function of its width w.
(b) Determine a complete graph of the function in part (a).
(c) Use the graph to determine the maximum area.

75. The cost of operating a delivery truck is $\$0.26 + 0.007s$ per mile, where s is the average speed. The driver is paid $18 per hour.
(a) Determine the total cost of making a delivery 40 miles away as a function of the average speed the truck is driven.
(b) Determine a graph of the function in part (a) on the interval [0, 55].
(c) Use the graph to determine the minimum cost.

■ CHAPTER TEST

1. Determine the center and radius of the graph of
$$x^2 + y^2 + x - 6y + 7 = 0$$

2. Determine the value(s) of x so that the distance between $A(2, 13)$ and $B(x, 5)$ is 10.

3. A portion of the graph of $x^3 + y^2 = 8$ is shown. Use the symmetry tests to complete the graph, and label the intercepts of the graph with its coordinates.

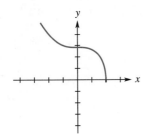

4. Given the function: $P(x) = \begin{cases} -x - 5 & x < -2 \\ -3 & -2 \le x < 1 \\ 2x & x \ge 1 \end{cases}$

(a) Evaluate $P(-7)$.
(b) Determine $P(t + 2)$, given that $t \ge 0$.
(c) Sketch the graph of P.

5. Given the function: $f(x) = \dfrac{4\sqrt{x}}{x - 2}$

(a) Evaluate $f(9)$. (b) Simplify $f(4k + 4)$.

6. Determine the zeros of $f(x) = 2x^3 - 6x$.

7. Given the function

$$g(x) = \frac{3x - 1}{2}$$

determine and simplify

$$\frac{g(x + h) - g(x)}{h}$$

8. Use a graphing utility to determine a complete graph of

$$y = \left(\tfrac{3}{2}\right)^x - 2\sqrt{x^2 + 1} + 2$$

9. Use a graphing utility to determine a complete graph of $f(x) = x^3 + x^2 - 5x - 5$. Determine the exact x-intercepts using algebraic methods.

10. Approximate (to the nearest 0.01) the solutions of the equation $x^4 - 8x^3 + 15 = 0$ by using the graph of an appropriate function.

11. From the following scenario, sketch a qualitative graph for the depth d of water in the tank as a function of time.

An inlet pipe fills an empty spherical water tank at a constant rate. When the tank is full, the inlet pipe is shut off.

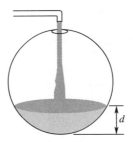

12. A real estate agent has determined that the annual cost C (per square foot) of leasing a commercial building with x square feet is given by

$$C = 0.02x^2 - 38.4x + 30{,}500$$

Use a graphing utility to determine the size of the office that minimizes the annual cost per square foot, if the available offices range from:
(a) 700 to 1200 square feet
(b) 1100 to 1600 square feet

13. A gutter is formed from a rectangular sheet of metal 8 feet long and 9 inches wide by folding up equal rectangles on each side to form right angles with the bottom (see the figure). Determine the dimensions that maximize the capacity (volume) of the gutter.

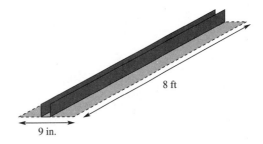

8 ft

9 in.

14. The given graph is a modification of one of the functions from the directory of graphs (page 108). Determine an equation for the graph.

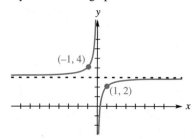

15. Use the functions $f(x) = 5x^2 + x - 1$ and $g(x) = 3 - 2x^2$ to determine and simplify each function.

   (a) $f + g$ (b) fg (c) $f \circ g$

In Exercises 16–18, use the given graphs of f and g to sketch the graph of each function.

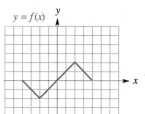

16. $y = f(x + 2) + 1$ 17. $y = -2g(x)$

18. $y = f(x) + g(x)$

19. Determine functions f and g such that $(f \circ g)(x) = (2^x + x)^3$.

20. A function of the form $f(x) = mx + b$ was used to generate the data in the table. What are the values of m and b?

x	0	2	4	10
$f(x)$	5	6.5	8	12.5

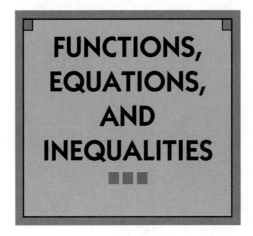

FUNCTIONS, EQUATIONS, AND INEQUALITIES

The concepts and techniques discussed in this chapter, along with those from your previous algebra courses and Chapter 1, comprise a "toolbox" of methods that can be used to solve many of the equations and inequalities that arise in mathematics and other fields.

In this chapter, we apply the results of our discussions in Chapter 1 to the functions, graphs, and equations that are used in the most basic mathematical models. Sections 2.1 and 2.2 focus on linear and quadratic functions. The next two sections provide widely used methods for solving equations and inequalities. Finally, Section 2.5 shows how these concepts are used in the application of mathematics.

LINES AND LINEAR FUNCTIONS

The simplest and most commonly used functions are *linear functions*. A linear function, as the name suggests, is one that has a *line* as a graph. These functions are also called *first-degree functions*, because they can be represented as first-degree polynomial expressions. As we shall see, sketching the graph of a linear function is rather straightforward—using a graphing utility generally is not required.

We start by discussing a concept familiar to you from previous algebra courses, the *slope of a line*.

■ **NOTE:** *The term* line *will always mean* straight line.

Slope of a Line

The direction and steepness of a nonvertical line in the coordinate plane is described by its *slope*. Simply speaking, this number is the ratio of the vertical and horizontal changes between two distinct points on the line.

Slope of a Line

The **slope** of a nonvertical line that passes through the points $P(x_1, y_1)$ and $Q(x_2, y_2)$ is the number

$$\frac{y_2 - y_1}{x_2 - x_1}$$

Using the notation $\Delta y = y_2 - y_1$ and $\Delta x = x_2 - x_1$ (from Section 1.1), the slope of the line can also be expressed as $\Delta y / \Delta x$.

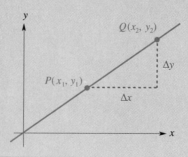

The ratio $\Delta y / \Delta x$ of a particular line does not depend upon which two points of the line are used to do this computation (see Exercise 64).

EXAMPLE 1 Determining the Slope of a Line

Determine the slope $\Delta y/\Delta x$ of each line shown.

(a)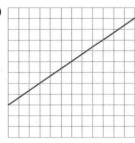

(b)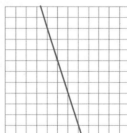

(c)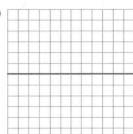

(d)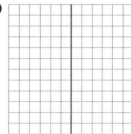

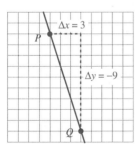

Figure 2.1

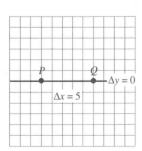

Figure 2.2

Figure 2.3

SOLUTION

In each case, we select two points on the line, determine Δx and Δy, and compute the slope $\Delta y/\Delta x$.

(a) Two points on the line, P and Q, are marked in Figure 2.1. The change in x from P to Q is 6, and the change in y is 4. The slope is

$$\frac{\Delta y}{\Delta x} = \frac{4}{6} = \frac{2}{3}$$

(b) Two points, P and Q, are marked in Figure 2.2. The change in x along the line segment is 3. The change in y is -9. (Notice that the change in y is negative because the line falls from left to right.) The slope is

$$\frac{\Delta y}{\Delta x} = \frac{-9}{3} = -3$$

(c) The change in x between the marked points in Figure 2.3 is 5. Because the line is horizontal, the change in y is 0. The slope is

$$\frac{\Delta y}{\Delta x} = \frac{0}{5} = 0$$

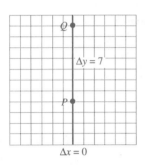

Figure 2.4

(d) The change in x is 0, since the line is vertical (Figure 2.4). The change in y is 7 for the points shown. The slope of the line is undefined, since

$$\frac{\Delta y}{\Delta x} = \frac{7}{0} \quad \text{is undefined}$$

Example 1 shows the four possible directions for a line: rising from left to right, falling from left to right, horizontal, and vertical. The slopes of these lines are generalized in Table 2.1.

TABLE 2.1

Description	Graph	Slope
Line rises from left to right		Slope of the line is positive, because $\Delta y > 0$ and $\Delta x > 0$ imply that $\Delta y / \Delta x > 0$
Line falls from left to right		Slope of the line is negative, because $\Delta y < 0$ and $\Delta x > 0$ imply that $\Delta y / \Delta x < 0$
Line is horizontal		Slope of the line is 0, because $\Delta y = 0$ implies that $\Delta y / \Delta x = 0$
Line is vertical		Slope of the line is undefined, because $\Delta x = 0$ implies that $\Delta y / \Delta x$ is undefined

EXAMPLE 2   Using Slope and a Point to Graph a Line

(a) Sketch the line with slope $\frac{3}{4}$ and y-intercept $(0, -7)$.

(b) Sketch the line passing through the point $(2, 5)$ with slope -2.

SOLUTION

(a) We start by plotting the point $(0, -7)$, as shown in Figure 2.5a. Because

$$\frac{\Delta y}{\Delta x} = \frac{3}{4}$$

we can assign $\Delta x = 4$ and $\Delta y = 3$, and use these values to locate $(4, -4)$, another point on the line (Figure 2.5b). The last step is to draw the line through these two points (Figure 2.5c).

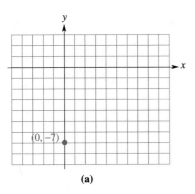

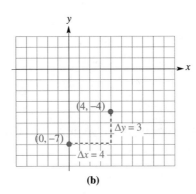

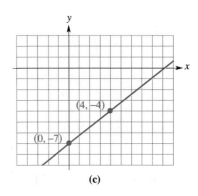

(a) **(b)** **(c)**

Figure 2.5

(b) This line can be graphed using the same three steps as in part (a). Figure 2.6a shows the point (2, 5). Since

$$\frac{\Delta y}{\Delta x} = -2 = \frac{-2}{1}$$

we use $\Delta x = 1$ and $\Delta y = -2$ to determine (3, 3), another point on the line (Figure 2.6b), and then draw the line (Figure 2.6c).

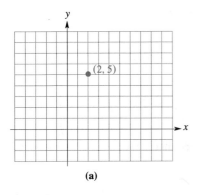

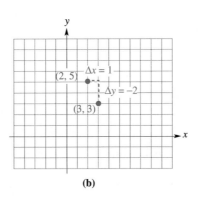

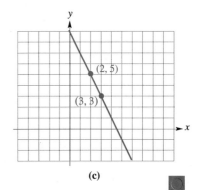

(a) **(b)** **(c)**

Figure 2.6

Slope–Intercept and Point–Slope Forms

In the directory of graphs (inside the front cover), we see that the graph of the identity function $f(x) = x$ is a line passing through the origin (the graph is repeated in Figure 2.7, p. 196). The graph of a function in the form $y = mx$ (where m is a real number) is also a line passing through the origin, because, as we saw in Section 1.6, this graph is a vertical compression, expansion, or reflection of

■ **NOTE:** *Traditionally, the letter m is used to represent the slope of a line. The reason why is not completely clear, but most agree that it is either an abbreviation of the French word* monter *(to climb), or the German word* momentagschweindekeit *(instantaneous change).*

the graph of $f(x) = x$. What is the slope $\Delta y/\Delta x$ of the graph of $y = mx$? The points $P(0, 0)$ and $Q(1, m)$ are on the graph of $y = mx$, since they make the equation true (Figure 2.8). Computing the slope using these two points, we get

$$\frac{\Delta y}{\Delta x} = \frac{m}{1} = m$$

The slope of the graph of $y = mx$ is m.

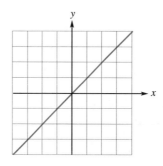

Figure 2.7
$y = x$

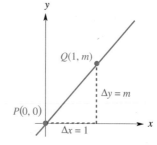

Figure 2.8
$y = mx$

◉ **EXAMPLE 3   Sketching the Graph of $y = mx$**

Sketch the graph of each function.

(a) $y = 2x$ **(b)** $y = -\frac{1}{4}x$

SOLUTION

In each case, the equation is in the form $y = mx$, so its graph is a line passing through the origin with slope m.

(a) By comparing $y = 2x$ with $y = mx$, we see that the slope m is 2. Because

$$\frac{\Delta y}{\Delta x} = 2 = \frac{2}{1}$$

we start at $(0, 0)$ and use $\Delta x = 1$ and $\Delta y = 2$ to locate $(1, 2)$, another point on the line. Drawing a line through $(0, 0)$ and $(1, 2)$ completes the sketch (Figure 2.9).

(b) The graph of $y = -\frac{1}{4}x$ has slope

$$\frac{\Delta y}{\Delta x} = \frac{-1}{4}$$

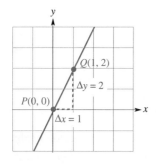

Figure 2.9
$y = 2x$

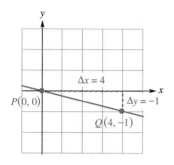

Figure 2.10
$y = -\frac{1}{4}x$

and it passes through (0, 0). With $\Delta x = 4$ and $\Delta y = -1$, we get a second point, $(4, -1)$, and then sketch the line (Figure 2.10).

In Section 1.6, we saw that, given the graph of a function $y = f(x)$, the graph of $y = f(x) + k$ can be constructed by translating this graph k units vertically. In Example 4, this concept is applied to linear functions.

● **EXAMPLE 4   Vertical Translation of the Graph of $y = mx$**
Use a vertical translation of the graph of $y = 2x$ to sketch the graph of $y = 2x + 6$.

SOLUTION
The graph of $y = 2x + 6$ is the graph of $y = 2x$ (see Figure 2.9) translated 6 units up (Figure 2.11).

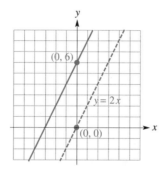

Figure 2.11
$y = 2x + 6$

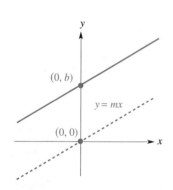

Figure 2.12
$y = mx + b$

In general, the graph of $f(x) = mx + b$ is the same as the graph of $g(x) = mx$ translated b units vertically. Because the y-intercept of $y = mx$ is the origin, the y-intercept of $y = mx + b$ is $(0, b)$ (Figure 2.12). This leads us to the following generalization:

Slope–Intercept Form

> If f is a linear function such that its graph has slope m and y-intercept $(0, b)$, then
> $$f(x) = mx + b$$
> This form is the **slope–intercept** form of f.

Every nonvertical line is the graph of a linear function. (The graph of a vertical line is not the graph of a linear function because it fails the vertical-line test.)

■ **NOTE:** *The slope–intercept form of a linear function is unique, since a nonvertical line has exactly one slope and one y-intercept.*

The discussion in Section 1.6 included horizontal translations as well as vertical translations. Specifically, if a graph of a function passing through the origin is translated h units horizontally and k units vertically, the translation passes through the point (h, k), as illustrated in Figure 2.13. Example 5 illustrates this idea applied to linear functions.

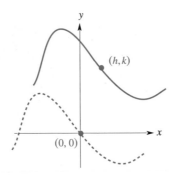

Figure 2.13

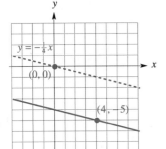

Figure 2.14
$y = -\frac{1}{4}(x - 4) - 5$

● EXAMPLE 5   The Graph of $y = m(x - h) + k$

Use a translation of the graph of $y = -\frac{1}{4}x$ (see Figure 2.10) to sketch the graph of $y = -\frac{1}{4}(x - 4) - 5$.

SOLUTION

If we define g by $g(x) = -\frac{1}{4}x$, then $y = -\frac{1}{4}(x - 4) - 5$ is equivalent to $y = g(x - 4) - 5$. Thus, the graph of $y = -\frac{1}{4}(x - 4) - 5$ is the graph of $g(x) = -\frac{1}{4}x$ translated 4 units to the right and 5 units down (Figure 2.14), so it is the line with slope $-\frac{1}{4}$ passing through the point $(4, -5)$. ■

We say that the equation $y = -\frac{1}{4}(x - 4) - 5$ in Example 5 is in *point-slope* form. This form allows us to easily determine that $(4, -5)$ is a point through which the line passes and that $-\frac{1}{4}$ is the slope of the line. Replacing $(4, -5)$ with (h, k) and $-\frac{1}{4}$ with m gives the following generalization:

Point–Slope Form

■ **NOTE:** *Unlike the slope–intercept form, a point–slope form for a given line is not unique; for each point on a given line, there exists a point–slope form for that line.*

> The graph of a linear function f with **point–slope** form
> $$f(x) = m(x - h) + k$$
> has slope m and passes through the point (h, k).

The point–slope form often allows us to construct a mathematical expression for a function from given data about the function. The data may include the slope of the graph and a point on that graph, or two points on the graph.

⬤ EXAMPLE 6 Determining a Function from Given Data

Sketch the graph of the linear function described and determine its slope–intercept form.

(a) The function f is such that $f(2) = 3$ and its graph has slope $\frac{1}{2}$.

(b) The function g is such that $g(-3) = 4$ and $g(2) = -1$.

SOLUTION

(a) Because $f(2) = 3$, the point $(2, 3)$ is on the graph of f. The graph of f is obtained by sketching the line with slope $\frac{1}{2}$ that passes through $(2, 3)$, as shown in Figure 2.15. The point–slope form $f(x) = m(x - h) + k$ with $(h, k) = (2, 3)$ and $m = \frac{1}{2}$ is

$$f(x) = \tfrac{1}{2}(x - 2) + 3$$

Simplifying this first-degree polynomial expression gives

$$f(x) = \tfrac{1}{2}(x - 2) + 3 = \tfrac{1}{2}x - 1 + 3 = \tfrac{1}{2}x + 2$$

The slope–intercept form of f is $f(x) = \frac{1}{2}x + 2$.

(b) Since $g(-3) = 4$ and $g(2) = -1$, the graph of g passes through $(-3, 4)$ and $(2, -1)$. Figure 2.16 shows that $\Delta x = 5$ and $\Delta y = -5$, so the slope is

$$m = \frac{\Delta y}{\Delta x} = \frac{-5}{5} = -1$$

The point–slope form for the function is $g(x) = m(x - h) + k$ in which $(h, k) = (-3, 4)$ and $m = -1$, so

$$g(x) = -1[x - (-3)] + 4 = -x - 3 + 4 = -x + 1$$

The slope–intercept form of g is $g(x) = -x + 1$. ⬛

In part (b) of Example 6, you may wonder why we chose $(-3, 4)$ instead of $(2, -1)$ for (h, k). Actually, either of these two points will produce the same results. If we use $(2, -1)$ for (h, k) with $m = -1$ in the point–slope form, we get the same slope–intercept form for g:

$$g(x) = m(x - h) + k = -1(x - 2) - 1 = -x + 2 - 1 = -x + 1$$

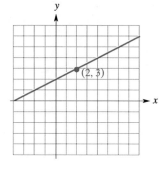

Figure 2.15

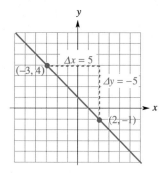

Figure 2.16

Equations of a Line

In the directory of graphs (inside the front cover), we see that for any real number k, the graph of the constant function $y = k$ is a horizontal line with y-intercept $(0, k)$ (the graph is repeated in Figure 2.17, p. 200). Each point on this graph has a y-coordinate of k. In a similar way, a vertical line with x-intercept $(h, 0)$ for some

⬛ **NOTE:** *The graph of $y = k$ can be thought of as the graph of the slope–intercept form $f(x) = 0x + k$, which has a graph with slope 0 and y-intercept $(0, k)$.*

real number h is the graph of the equation $x = h$ (Figure 2.18). Each point on this line has h as an x-coordinate.

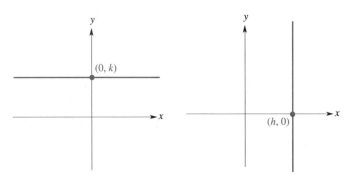

Figure 2.17
$y = k$

Figure 2.18
$x = h$

Special Cases

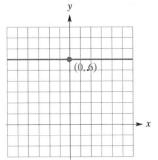

Figure 2.19
$y + 7 = 13$

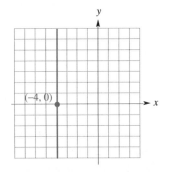

Figure 2.20
$-2x = 8$

> **1.** The graph of the equation $y = k$ is a horizontal line with y-intercept $(0, k)$.
> **2.** The graph of the equation $x = h$ is a vertical line with x-intercept $(h, 0)$.

⬤ **EXAMPLE 7   Sketching Vertical and Horizontal Lines**
Sketch each graph and determine its slope.

(a) $y + 7 = 13$ **(b)** $-2x = 8$

SOLUTION
(a) The equation $y + 7 = 13$ is equivalent to $y = 6$, so its graph is a horizontal line with y-intercept $(0, 6)$, as shown in Figure 2.19. Because the graph is a horizontal line, its slope is zero.
(b) The equation $-2x = 8$ is equivalent to $x = -4$. Its graph is a vertical line with x-intercept $(-4, 0)$, as shown in Figure 2.20. The slope of this line is undefined.   ⬤

Many equations other than those in the forms $y = mx + b$, $y = k$, or $x = h$ have graphs that are lines. For example, the graph of the equation $2x + 7y =$

14 is a line, because this equation can be rewritten in slope–intercept form as $y = -\frac{2}{7}x + 2$. (Take a moment to verify this.) We say that $2x + 7y = 14$ is the *general form* of the equation.

General Form

> An equation that can be written in the **general form**
>
> $$Ax + By = C$$
>
> for some real numbers A, B, and C (A and B not both 0) has a graph that is a line. Conversely, any line has a corresponding general form equation.

The proof of these statements is outlined in the exercises (see Exercises 65 and 66).

■ EXAMPLE 8   Determining the Graph from the General Form

Determine the slope–intercept form of $3x + 4y = 24$ and use it to sketch the graph of this equation.

SOLUTION

The slope–intercept form $y = mx + b$ is determined by solving $3x + 4y = 24$ for y:

$$3x + 4y = 24$$
$$4y = -3x + 24$$
$$\tfrac{1}{4}(4y) = \tfrac{1}{4}(-3x + 24)$$
$$y = -\tfrac{3}{4}x + 6$$

The slope of the line is $-\frac{3}{4}$, and the y-intercept is $(0, 6)$. The steps for sketching the graph using the slope and y-intercept are shown in Figure 2.21.

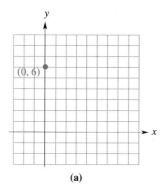

(a)

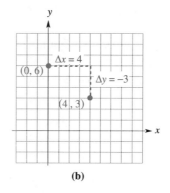

(b)

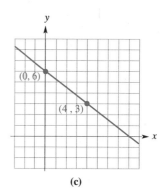

(c)

Figure 2.21

Parallel Lines and Perpendicular Lines

Consider the graphs of the functions $y = \frac{1}{3}x + 2$ and $y = \frac{1}{3}x - 4$ (Figure 2.22). Each has slope $\frac{1}{3}$, and since each is a vertical translation of $y = \frac{1}{3}x$, these lines are parallel. This observation suggests the following:

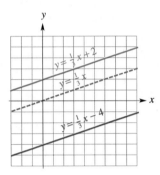

Figure 2.22

Parallel Lines and Slopes

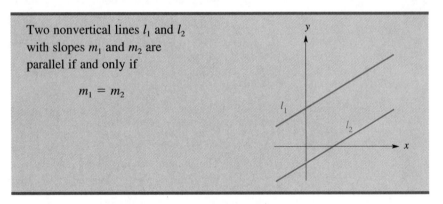

Two nonvertical lines l_1 and l_2 with slopes m_1 and m_2 are parallel if and only if

$$m_1 = m_2$$

A proof of this statement is outlined in Exercise 67.

● EXAMPLE 9   Equations of Parallel Lines

Determine the linear function f (in slope–intercept form) with a graph passing through $(4, -2)$ and parallel to the graph of $3x + 4y = 24$.

SOLUTION

Example 8 showed that the slope of the graph of $3x + 4y = 24$ is $-\frac{3}{4}$. The graph of f also has slope $-\frac{3}{4}$ because it is parallel to $3x + 4y = 24$. Using the point–slope form with $m = -\frac{3}{4}$ and $(h, k) = (4, -2)$, we get

$$f(x) = -\frac{3}{4}(x - 4) - 2 = -\frac{3}{4}x + 3 - 2 = -\frac{3}{4}x + 1$$

The slope–intercept form of f is $f(x) = -\frac{3}{4}x + 1$. The graphs of $3x + 4y = 24$ and $f(x) = -\frac{3}{4}x + 1$ are shown in Figure 2.23.

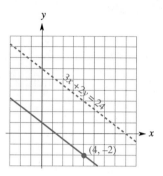

Figure 2.23
$y = -\frac{3}{4}x + 1$

There is also a connection between slope and perpendicular lines.

Perpendicular Lines and Slopes

Two nonvertical lines l_1 and l_2 with slopes m_1 and m_2 are perpendicular if and only if

$$m_1 m_2 = -1$$

A proof of this statement is outlined in Exercise 68. Note that the relation $m_1 m_2 = -1$ is equivalent to

$$m_1 = -\frac{1}{m_2} \qquad \text{and} \qquad m_2 = -\frac{1}{m_1}$$

■ **NOTE:** *Compare this example with Example 3 of Section 1.1.*

EXAMPLE 10   Slope and Perpendicular Line Segments

Determine whether the triangle with vertices $P(6, -3)$, $Q(0, 5)$, and $R(-2, 1)$ is a right triangle.

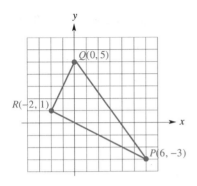

Figure 2.24

SOLUTION

The right angle appears to be $\angle PRQ$ (Figure 2.24). If the slope m_{PR} of the line passing through P and R and the slope m_{RQ} of the line passing through R and Q are such that $m_{PR}m_{RQ} = -1$, then $\angle PRQ$ is a right angle. Computing the slopes yields:

$$m_{PR} = \frac{(-3) - 1}{6 - (-2)} = \frac{-4}{8} = -\frac{1}{2}$$

$$m_{RQ} = \frac{5 - 1}{0 - (-2)} = \frac{4}{2} = 2$$

Because

$$m_{PR}m_{RQ} = \left(-\tfrac{1}{2}\right)(2) = -1$$

triangle PQR is a right triangle.

Applications

The models of many real-life situations are linear functions. The methods outlined in this section and in Section 1.6 often can be used to find a linear function that solves an applied problem.

 **EXAMPLE 11   Application: Straight-Line Depreciation**

A small accounting firm buys a personal computer for $6000. For tax purposes, the firm is allowed to depreciate the value of the computer by $1200 per year. (This method is called *straight-line depreciation.*)

(a) Determine a linear function F that gives the depreciated value of the computer after x years.

(b) Use a graphing utility to determine the graph of the function.

(c) Find the depreciated value of the computer after 2 years.

(d) After how many years is the depreciated value of the computer $0?

SOLUTION

(a) The depreciated value of the computer is

$$\begin{pmatrix} \text{Depreciated} \\ \text{value} \end{pmatrix} = \begin{pmatrix} \text{Original} \\ \text{value} \end{pmatrix} - \begin{pmatrix} \text{Amount of} \\ \text{depreciation} \end{pmatrix}$$

$$= 6000 - 1200 \begin{pmatrix} \text{Number of} \\ \text{years} \end{pmatrix}$$

$$F(x) = 6000 - 1200x$$

(b) The graph of the function is shown in Figure 2.25.

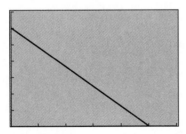

Figure 2.25
$y = 6000 - 1200x$
$[0, 6] \times [0, 7000]_{1000}$

(c) Because $F(2) = 6000 - 1200(2) = 3600$, the depreciated value of the computer after 2 years is $3600. (Use the trace feature of a graphing utility to verify this result.)

(d) The depreciated value of the computer is $0 after x years if $F(x) = 0$. Thus,

$$F(x) = 0$$
$$6000 - 1200x = 0$$
$$-1200x = -6000$$
$$x = 5$$

The depreciated value is $0 after 5 years.

Suppose that a function f is used as a model for an applied problem. A *data point* is an ordered pair $(x, f(x))$ for some x in the domain of f. Example 12 illustrates how a linear function can be found from two of its data points.

EXAMPLE 12 Application: Finding a Linear Function from Data Points

In 1780, a French balloonist by the name of Jacques Charles discovered that the volume of a fixed amount of gas held at a constant pressure is a linear function of its temperature. In a specific experiment, it is observed that if 2.000 liters of oxygen at 0°C is warmed to 100°C, the volume of the oxygen increases to 2.732 liters.

(a) Determine V, the volume of oxygen (in liters), as a function of t, its temperature (in degrees Celsius), in this experiment.

(b) Use your graphing utility to determine the graph of the function.

(c) What is the volume of oxygen at 250°C?

(d) Determine the zero of the function (to the nearest 1°C). What is its significance?

SOLUTION

(a) From the data collected in the experiment, we know that if $t = 0$, then $V = 2.000$, and if $t = 100$, then $V = 2.732$. Thus, two data points for this function are $(0, 2.000)$ and $(100, 2.732)$. The slope of the graph of the function is

$$m = \frac{\Delta V}{\Delta t} = \frac{2.732 - 2.000}{100 - 0} = 0.00732$$

and the V-intercept is $(0, 2.000)$. Thus, the function is

$$V = 0.00732t + 2.000$$

(b) The graph of the function is shown in Figure 2.26.

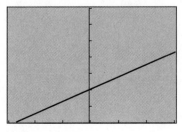

Figure 2.26
$y = 0.00732x + 2.000$
$[-300, 300]_{100} \times [0, 7]$

(c) The volume of oxygen at 250°C is the value of V for which $t = 250$:

$$V = 0.00732(250) + 2.000 = 3.83$$

The volume at 250°C is 3.83 liters. (Use the trace feature of your graphing utility to verify this result.)

(d) The zero of the function is the value of t for which $V = 0$:

$$0 = 0.00732t + 2.000$$
$$-0.00732t = 2.000$$
$$t = \frac{2.000}{-0.00732}$$
$$t \approx -273$$

This temperature, -273°C, is called *absolute zero*, since the volume of the gas is theoretically 0 at this temperature.

EXERCISE SET 2.1 ----------

A

In Exercises 1 and 2, determine the slope of the lines (a)–(d) given.

1.
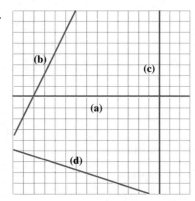

2.
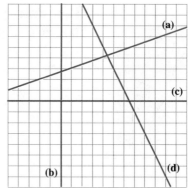

In Exercises 3–8, sketch the graph of each line described.

3. The line has slope $\Delta y/\Delta x = 1$ and y-intercept $(0, 3)$.

4. The line has slope $\Delta y/\Delta x = -2$ and y-intercept $(0, -1)$.

5. The line has slope $\Delta y/\Delta x = 2/3$ and passes through the point $(-3, 5)$.

6. The line has slope $\Delta y/\Delta x = -5/2$ and passes through the point $(4, -2)$.

7. The slope $\Delta y/\Delta x$ of the line is undefined, and the line has x-intercept $(-4, 0)$.

8. The slope $\Delta y/\Delta x$ of the line is 0, and the line passes through the origin.

In Exercises 9–14, sketch both equations on the same coordinate plane.

9. $y = -\frac{2}{3}x;\ y = -\frac{2}{3}x + 6$

10. $y = \frac{5}{2}x;\ y = \frac{5}{2}x - 3$

11. $y = -2x;\ y = -2(x - 3)$

12. $y = 3x;\ y = 3(x + 4)$

13. $y = \frac{3}{4}x;\ y = \frac{3}{4}(x + 2) - 5$

14. $y = -\frac{3}{2}x;\ y = -\frac{3}{2}(x - 1) - 4$

In Exercises 15–22, determine an equation for each line described in one of these forms: $y = mx + b$, $y = k$, or $x = h$.

15. The line has slope $m = \frac{3}{5}$ and y-intercept $(0, -2)$.

16. The line has slope $m = -2$ and y-intercept $(0, 7)$.

17. The line is parallel to the x-axis and has y-intercept $(0, 7)$.

18. The line is parallel to the y-axis and has x-intercept $\left(-\frac{7}{2}, 0\right)$.

19. The line has slope $m = -6$ and passes through the point $(3, 7)$.

20. The line has slope $m = \frac{2}{3}$ and passes through the point $(4, -1)$.

21. The line passes through the points $(-8, 2)$ and $(2, -3)$.

22. The line passes through the points $(0, -6)$ and $(3, 0)$.

In Exercises 23–30, determine the slope–intercept form of the given linear equation and then sketch its graph.

23. $2x - 4y = 16$

24. $3x + 2y = -12$

25. $3y + 2 = 14$

26. $-y + 5 = 9$

27. $x = 4y$

28. $-3y = 4x$

29. $\dfrac{2x}{3} - \dfrac{y}{4} = 2$

30. $\dfrac{x}{3} + \dfrac{y}{6} = -1$

B

In Exercises 31–38, determine a function $f(x) = mx + b$ for the graph described.

31. The line is parallel to $x + 2y = 9$ and has y-intercept $(0, 2)$.

32. The line is parallel to $x - y = 12$ and has y-intercept $(0, 6)$.

33. The line is parallel to $x + 2y = 9$ and passes through the point $(-2, 3)$.

34. The line is parallel to $x - y = 12$ and passes through the point $(5, -1)$.

35. The line is the perpendicular bisector of the line segment with endpoints $A(-3, 7)$ and $B(5, 1)$. (The *perpendicular bisector* of the line segment is the line that is perpendicular to the line segment and passes through the midpoint of the line segment.)

36. The line is the perpendicular bisector of the line segment with endpoints $P(0, 6)$ and $Q(-4, 0)$. (See Exercise 35.)

37. The line is tangent to the circle $x^2 + y^2 = 25$ at the point $P(-4, 3)$.
   HINT: This line is perpendicular to the radius of the circle at the point of tangency.

38. The line is tangent to the circle $x^2 + y^2 = 40$ at the point $P(2, 6)$. (See Exercise 37.)

In Exercises 39–44, determine a function f in slope–intercept form for each graph shown.

39.

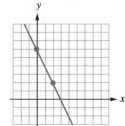

40.

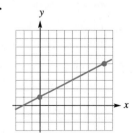

41.
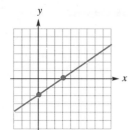

42.
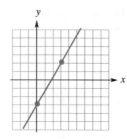

43.
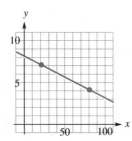

44.
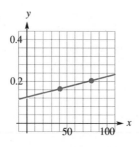

In Exercises 45–48, sketch the graph of each equation.
HINT: Use different scales on the *x* and *y* axes.

45. $2x + 30y = 150$

46. $20x - y = 60$

47. $y = -6000x + 30,000$

48. $y = 0.1x - 150$

49. Points that lie on the same line are *collinear*. Three points *A*, *B*, and *C* are collinear if and only if the slopes of the line segments *AB*, *BC*, and *CA* are all equal. Use this fact to determine whether the points $A(1, 3)$, $B(4, 7)$, and $C(10, 15)$ are collinear.

50. Determine whether the points $S(-8, 10)$, $T(-4, -5)$, and $U(-5, -1)$ are collinear. (See Exercise 49.)

51. Use slopes to determine whether the triangle with vertices $A(1, -6)$, $B(7, -8)$, and $C(3, 0)$ is a right triangle.

52. Use slopes to determine whether the triangle with vertices $H(1, 1)$, $Q(7, 0)$, and $R(8, 8)$ is a right triangle.

53. A *rhombus* is a quadrilateral with all four sides of equal length. A result from geometry states that a quadrilateral is a rhombus if and only if its two diagonals are perpendicular and have the same midpoint. Use this fact to determine whether the quadrilateral with vertices $A(-2, 3)$, $B(2, 4)$, $C(3, 8)$, and $D(-1, 7)$ is a rhombus.

54. Determine whether the quadrilateral with vertices $P(-4, -7)$, $Q(1, -6)$, $R(-2, -2)$, and $S(-7, -3)$ is a rhombus. (See Exercise 53.)

55. A moving company buys a small truck for $25,000 and plans to depreciate it using the straight-line method so that after 8 years, the truck's value is $7000.
   (a) Determine a linear function *F* that gives the depreciated value of the truck after *x* years. Sketch the graph of *F*.
   (b) What is the depreciated value of the truck after 2 years?

56. Temperature can be measured by degrees Celsius (°C) or degrees Fahrenheit (°F). Also, the Celsius measure of temperature *C* is a linear function of the corresponding Fahrenheit measure *F*. Given that ice melts at 0°C and at 32°F and that water boils at 100°C and at 212°F, find this function.

57. A biologist determines that the number of times a cricket chirps per minute is a linear function of the temperature. By changing the ambient temperature in a cricket habitat, the biologist collects the data shown in the table:

Temperature (°F), T	60	96	120
Number of chirps, n	55	64	70

Determine the function.

58. The water pressure (in pounds per square inch) on an underwater diver is a linear function of the diver's depth (in feet). During a particular dive, the following data are collected:

Depth	20	30	45
Pressure	23.4	27.7	34.2

Determine the function.

59. Nancy invests $4500 in a diesel-powered wood splitter. Experience tells her that it will cost $16 per hour in maintenance and supplies to run the splitter. She plans to charge $25 per hour to those who contract her services. Let *x* represent the number of hours that the splitter operates.

(a) Determine C, the total cost of operation, as a linear function of x.

(b) Determine R, the total revenue generated, as a linear function of x.

(c) Determine P, the total profit of operation, as a linear function of x. What is the significance of the slope of this function?

HINT Quick Review: Profit = Revenue − Cost

(d) The zero of the function P is called the *break-even point*. Find this zero and explain why this name is appropriate.

60. A car starts at an altitude of 5400 feet above sea level on a straight road that has a grade of 4%. (A grade of 4% means that the road changes 4 feet vertically over a 100 foot horizontal run.) At the end of the trip, the car is at 2000 feet.

(a) Determine the altitude of the car as a linear function of x, the horizontal distance traveled.

(b) How far did the car travel (to the nearest 1000 feet)?

C

61. Show that the intercepts of the linear equation

$$\frac{x}{a} + \frac{y}{b} = 1$$

are $(a, 0)$ and $(0, b)$. (This equation is called the *two-intercept form* of a line.)

62. Use Exercise 61 to determine the graph of the given equation. Rewrite the equation in slope–intercept form $y = mx + b$.

(a) $\dfrac{x}{2} + \dfrac{y}{8} = 1$ (b) $\dfrac{x}{-3} + \dfrac{y}{-6} = 1$

(c) $\dfrac{x}{\frac{3}{2}} + \dfrac{y}{\frac{5}{2}} = 1$

63. Suppose that f and g are linear functions. Determine which of the following are also linear functions. Support your answers.

(a) $f + g$ (b) fg (c) $f \circ g$

64. This exercise outlines the proof that the slope of a non-vertical line does not depend upon which two points of the line are used in its calculation.

(a) Show that right triangles ABC and PQR in Figure 2.27 are similar triangles.

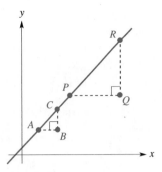

Figure 2.27

(b) Explain why

$$\frac{d(B, C)}{d(A, B)} = \frac{d(Q, R)}{d(P, Q)}$$

and how this completes the proof.

65. This exercise outlines the proof that the graph of an equation in the general form $Ax + By = C$ is a line.

(a) Show that if $B = 0$ and $A \neq 0$, the graph of $Ax + By = C$ is a vertical line.

(b) Show that if $B \neq 0$, the graph of $Ax + By = C$ is a line with slope $-A/B$ and y-intercept C/B.

(c) Explain why parts (a) and (b) complete the proof.

66. This exercise outlines the proof that every line corresponds to an equation in the general form $Ax + By = C$.

(a) Show that any vertical line corresponds to an equation in general form.

(b) Show that an equation $y = mx + b$ is equivalent to an equation in the form $Ax + By = C$.

(c) Explain why parts (a) and (b) complete the proof.

67. This exercise outlines the proof that two nonvertical lines are parallel if and only if their slopes are equal.

(a) Suppose that the graphs of $y = m_1x + b_1$ and $y = m_2x + b_2$ intersect. If $x = c$ at the point of intersection, explain why $m_1c + b_1 = m_2c + b_2$ and why it follows that

$$c = \frac{b_2 - b_1}{m_1 - m_2}$$

(b) Explain why the proof follows directly from part (a).

68. This exercise outlines the proof that two nonvertical lines are perpendicular if and only if the product of their slopes is -1. The graphs of $y = mx_1 + b_1$ and $y = mx_2 + b_2$ are shown as l_1 and l_2, respectively, in Figure 2.28.

(a) Explain why $d(P, Q) = \sqrt{1 + c^2}$, $d(Q, R) = \sqrt{1 + d^2}$, and $d(R, P) = c + d$.

(b) According to the Pythagorean theorem, $\angle PQR$ is a right angle if and only if

$$d(P, Q)^2 + d(Q, R)^2 = d(R, P)^2$$

Use this fact and part (a) to show that $\angle PQR$ is a right angle if and only if $cd = 1$.

(c) Explain why $m_1 = c$ and $m_2 = -d$, and that this completes the proof.

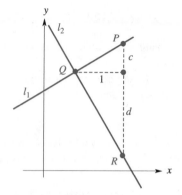

Figure 2.28

QUADRATIC FUNCTIONS

Functions that can be represented by second-degree polynomial expressions such as $f(x) = x^2 + 4x - 3$ are *quadratic functions*. Quadratic functions are used to solve a great range of problems. For example, a quadratic function can describe the path of a tossed ball, model the shape of a suspension bridge, or help determine the dimensions of a cattle pen with maximum area.

The Graph of $y = ax^2$

The graph of the squaring function $f(x) = x^2$ was featured in the directory of graphs in Section 1.3 and is repeated here in Figure 2.29. Because ax^2 is a constant multiple of x^2, the graph of $y = ax^2$ (for some real number a) is a vertical expansion, compression, or reflection of the graph of f.

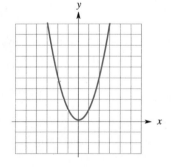

Figure 2.29
$y = x^2$

EXAMPLE 1 The Graph of $y = ax^2$

Use your graphing utility to determine the graph of each function. Describe how the graph compares to the graph of $f(x) = x^2$.

(a) $y = 2x^2$ (b) $y = \frac{1}{4}x^2$ (c) $y = -3x^2$

SOLUTION

(a) Because $y = 2x^2$ is equivalent to $y = 2f(x)$, the graph of $y = 2x^2$ is a vertical expansion of the graph of f by a factor of 2 (Figure 2.30).

(b) The graph of $y = \frac{1}{4}x^2$ is a vertical compression of the graph of f by a factor of $\frac{1}{4}$, since $y = \frac{1}{4}x^2$ is equivalent to $y = \frac{1}{4}f(x)$ (Figure 2.31).

(c) Because $y = -3x^2$ can be rewritten as $y = -3f(x)$ (Figure 2.32), the graph of $y = -3x^2$ is a vertical expansion by a factor of 3 and a reflection about the x-axis of the graph of f.

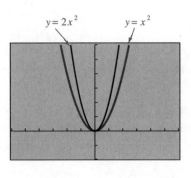

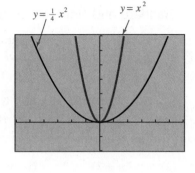

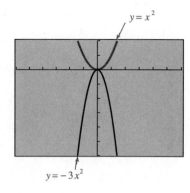

Figure 2.30
$[-6, 6] \times [-2, 6]$

Figure 2.31
$[-6, 6] \times [-2, 6]$

Figure 2.32
$[-6, 6] \times [-6, 2]$

Each of the graphs in Example 1 is similar in shape to the graph of $f(x) = x^2$. The shape of the graph of $y = ax^2$ (for any nonzero real number a) is a *parabola*. If $a > 0$, as in parts (a) and (b), then the graph of $y = ax^2$ opens upward; this parabola is *concave up*. If $a < 0$, as in part (c), then the graph of $y = ax^2$ is *concave down*.

The turning point of a parabola is its *vertex*. The vertex is the lowest point of a parabola that is concave up and the highest point of a parabola that is concave down. The graph of $y = ax^2$ is symmetric with respect to the y-axis.

The Graph of $y = ax^2$

The graph of $y = ax^2$ is a **parabola** that is symmetric with respect to the y-axis. Its **vertex** is $(0, 0)$. If $a > 0$, then the parabola is **concave up**, and if $a < 0$, then the parabola is **concave down** (Figure 2.33).

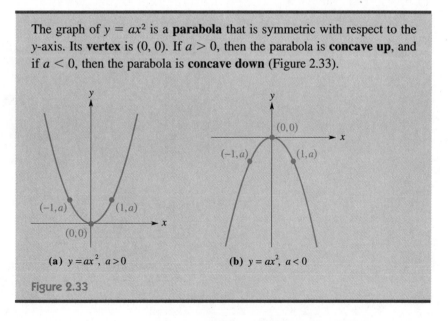

(a) $y = ax^2$, $a > 0$

(b) $y = ax^2$, $a < 0$

Figure 2.33

Standard and General Forms

In Section 1.6, we saw that a translation of a graph has the same shape as the original graph. In Example 2, we apply this idea to the graph of $y = ax^2$.

⬤ EXAMPLE 2 Sketching the Graph of $y = a(x - h)^2 + k$

Use your graphing utility to determine the graph of each function. Describe how the graph compares to the graph of $g(x) = ax^2$ (for some real number a).

(a) $y = (x - 3)^2 - 2$ **(b)** $y = 2(x - 5)^2 + 3$ **(c)** $y = -3(x + 4)^2$

SOLUTION

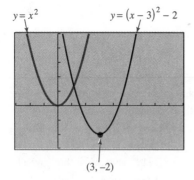

$y = x^2$ $y = (x - 3)^2 - 2$

$(3, -2)$

Figure 2.34
$[-3, 9] \times [-3, 5]$

(a) If we define g by $g(x) = x^2$, then $y = (x - 3)^2 - 2$ is equivalent to $y = g(x - 3) - 2$. The graph of $y = (x - 3)^2 - 2$ is the graph of $g(x) = x^2$ translated 3 units to the right and 2 units down (Figure 2.34). The vertex is $(3, -2)$.

(b) If $g(x) = 2x^2$, then $y = 2(x - 5)^2 + 3$ is equivalent to $y = g(x - 5) + 3$. The graph of $y = 2(x - 5)^2 + 3$ is the graph of $g(x) = 2x^2$ translated 5 units to the right and 3 units up (Figure 2.35). The vertex of the parabola is $(5, 3)$.

(c) In this case, g is defined by $g(x) = -3x^2$. The equation $y = -3(x + 4)^2$ is equivalent to $y = g(x + 4)$. Its graph is the graph of g translated 4 units to the left (Figure 2.36). The vertex is $(-4, 0)$.

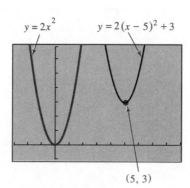

$y = 2x^2$ $y = 2(x - 5)^2 + 3$

$(5, 3)$

Figure 2.35
$[-3, 9] \times [-1, 7]$

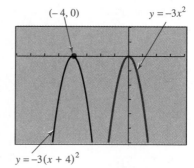

$(-4, 0)$ $y = -3x^2$

$y = -3(x + 4)^2$

Figure 2.36
$[-8, 4] \times [-6, 2]$ ⬤

In general, the graph of $f(x) = a(x - h)^2 + k$ is the graph of $y = ax^2$ translated h units horizontally and k units vertically. A quadratic function in this form is said to be in *standard form*.

Standard Form
of a Quadratic Function

A quadratic function f is in **standard form** if it is written

$$f(x) = a(x - h)^2 + k$$

where a, h, and k are real numbers ($a \neq 0$). The graph of this quadratic function is a parabola with the shape of $y = ax^2$ and vertex (h, k).

In Example 2 we saw how the graph of a quadratic function can be described from its standard form. On the other hand, Example 3 shows how we can determine the standard form of a quadratic function from its graph.

EXAMPLE 3 Finding the Function for a Parabola

The parabola in Figure 2.37 passes through $(2, -3)$ and has its vertex at $(4, 5)$. Determine a function for the parabola in the form $f(x) = a(x - h)^2 + k$.

SOLUTION

Because the vertex is $(4, 5)$, we can substitute 4 for h and 5 for k in the standard form:

$$f(x) = a(x - h)^2 + k$$
$$f(x) = a(x - 4)^2 + 5 \quad \text{Let } h = 4 \text{ and } k = 5.$$

We still need to determine the value of a. Because the point $(2, -3)$ is on the graph, we can substitute 2 for x and -3 for $f(x)$ in this last equation and solve for a:

$$f(x) = a(x - 4)^2 + 5$$
$$-3 = a(2 - 4)^2 + 5 \quad \text{Substitute } x = 2 \text{ and } f(x) = -3.$$
$$-3 = 4a + 5$$
$$-8 = 4a$$
$$-2 = a$$

The parabola is the graph of $f(x) = -2(x - 4)^2 + 5$.

A quadratic function in standard form, such as $f(x) = \frac{1}{2}(x - 2)^2 - 3$, can be simplified in the same manner as any polynomial expression:

$$f(x) = \frac{1}{2}(x - 2)^2 - 3 = \frac{1}{2}(x^2 - 4x + 4) - 3 = \frac{1}{2}x^2 - 2x - 1$$

Thus, $f(x) = \frac{1}{2}x^2 - 2x - 1$ is another way to express this function. This form of the function is its *general form*.

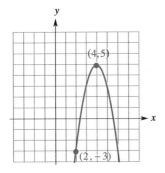

Figure 2.37

General Form
of a Quadratic
Function

> A quadratic function in the form
>
> $$f(x) = ax^2 + bx + c$$
>
> where a, b, and c are real numbers ($a \neq 0$), is in **general form**.

As we have just seen, rewriting a quadratic function from standard form to general form is rather straightforward. On the other hand, finding the standard form of a quadratic function given its general form involves a bit more thought. The following example illustrates how completing the square is used to accomplish this task.

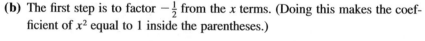 **EXAMPLE 4   Rewriting a Quadratic Function in Standard Form**
Rewrite each quadratic function in standard form, and verify your result with a graphing utility.

(a) $f(x) = x^2 + 6x - 5$ **(b)** $f(x) = -\frac{1}{2}x^2 + 5x + 4$

SOLUTION

(a) We start by grouping the x terms together:

$$f(x) = (x^2 + 6x \quad) - 5$$

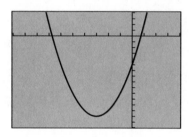

Figure 2.38
$y = x^2 + 6x - 5$ and
$y = (x + 3)^2 - 14$
$[-10, 4] \times [-16, 4]$

The number that is added to $x^2 + 6x$ to complete the square is 9, because half of 6 (the coefficient of the x term) is 3, and $3^2 = 9$. By adding and subtracting 9, the function remains unchanged:

$$f(x) = (x^2 + 6x + 9) - 5 - 9$$

We get $f(x) = (x + 3)^2 - 14$ by factoring $x^2 + 6x + 9$ and collecting terms. The graphs of $y = x^2 + 6x - 5$ and $y = (x + 3)^2 - 14$ are the same parabola (Figure 2.38).

(b) The first step is to factor $-\frac{1}{2}$ from the x terms. (Doing this makes the coefficient of x^2 equal to 1 inside the parentheses.)

$$
\begin{aligned}
f(x) &= -\tfrac{1}{2}x^2 + 5x + 4 \\
&= -\tfrac{1}{2}(x^2 - 10x \quad) + 4 \\
&= -\tfrac{1}{2}(x^2 - 10x + 25) + 4 + \tfrac{25}{2} \quad \text{Complete the square.} \\
&= -\tfrac{1}{2}(x - 5)^2 + \tfrac{33}{2} \quad\quad\quad \text{Factor and collect terms.}
\end{aligned}
$$

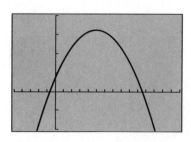

Figure 2.39
$y = -\frac{1}{2}x^2 + 5x + 4$ and
$y = -\frac{1}{2}(x - 5)^2 + \frac{33}{2}$
$[-5, 15] \times [-10, 20]_5$

Notice that by adding 25 to the x terms inside the parentheses, the entire function is decreased by $\frac{25}{2}$; to compensate, $\frac{25}{2}$ is also added to the function. The standard form of f is $f(x) = -\frac{1}{2}(x - 5)^2 + \frac{33}{2}$. The graphs of $y = -\frac{1}{2}x^2 + 5x + 4$ and $y = -\frac{1}{2}(x - 5)^2 + \frac{33}{2}$ coincide, which verifies our result (Figure 2.39).

Intercepts of a Parabola

The x- and y-intercepts are an important part of any graph. The intercepts of the graph of a quadratic function, along with its vertex, are essential features in making a quick sketch of the parabola. Finding the intercepts of a quadratic function f is the same as finding the intercepts of any function. The y-intercept is $(0, f(0))$; the x-intercepts are found by solving the quadratic equation $f(x) = 0$.

EXAMPLE 5 Finding the Intercepts of a Quadratic Function

Determine the intercepts of each function algebraically, and verify your answer with a graphing utility.

(a) $y = (x - 3)^2 - 4$ **(b)** $y = -2x^2 + 8x - 2$

SOLUTION

(a) The y-intercept is determined by letting $x = 0$ and evaluating y:

$$y = (0 - 3)^2 - 4 = 9 - 4 = 5$$

The y-intercept is $(0, 5)$. To find the x-intercepts, we solve the equation for x after substituting $y = 0$:

$$y = (x - 3)^2 - 4$$
$$0 = (x - 3)^2 - 4 \quad \text{Let } y = 0.$$
$$(x - 3)^2 = 4$$
$$x - 3 = \pm 2$$
$$x = 3 \pm 2$$

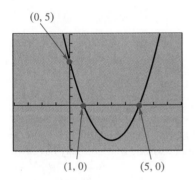

$(0, 5)$

$(1, 0)$ $(5, 0)$

Figure 2.40
$y = (x - 3)^2 - 4$
$[-4, 8] \times [-5, 8]$

The x-intercepts are $(1, 0)$ and $(5, 0)$ (Figure 2.40).

(b) We let $x = 0$ and evaluate y to find the y-intercept:

$$y = -2(0)^2 + 8(0) - 2 = -2$$

The y-intercept is $(0, -2)$. The x-intercepts are found by solving for x after substituting $y = 0$:

$$y = -2x^2 + 8x - 2$$
$$0 = -2x^2 + 8x - 2 \quad \text{Let } y = 0.$$

This last equation is best solved by the quadratic formula.

$$-(0) = -(-2x^2 + 8x - 2) \qquad \begin{array}{l}\text{Multiply each side by } -1 \text{ to} \\ \text{make the } x^2 \text{ coefficient positive.}\end{array}$$
$$0 = 2x^2 - 8x + 2$$
$$x = \frac{-(-8) \pm \sqrt{(-8)^2 - 4(2)(2)}}{2(2)} = \frac{8 \pm \sqrt{48}}{4} = \frac{8 \pm 4\sqrt{3}}{4} = 2 \pm \sqrt{3}$$

The solutions to the equation are $2 - \sqrt{3}$ and $2 + \sqrt{3}$. The x-intercepts are $(2 - \sqrt{3}, 0)$ and $(2 + \sqrt{3}, 0)$ [approximately $(0.3, 0)$ and $(3.7, 0)$]. Figure 2.41 shows the graph of $y = -2x^2 + 8x - 2$ and its intercepts.

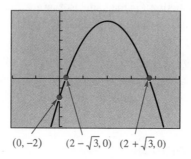

$(0, -2)$ $(2 - \sqrt{3}, 0)$ $(2 + \sqrt{3}, 0)$

Figure 2.41
$y = -2x^2 + 8x - 2$
$[-2, 5] \times [-5, 7]$

The graph of each function in Example 5 has exactly one y-intercept, as does every quadratic function. Also, each of these graphs has two x-intercepts, but, as you will see in Exercises 49–51, the graph of a quadratic function may have two, one, or zero x-intercepts.

Intercepts of a Quadratic Function

> The y-intercept of a quadratic function f is $(0, f(0))$.
>
> The x-intercepts of a quadratic function f correspond to the solutions to the quadratic equation $f(x) = 0$. If this equation has zero, one, or two solutions, then the graph of f has zero, one, or two x-intercepts, respectively.

Maximum and Minimum Values

The vertex of a parabola that is concave up has a y-coordinate that is less than that of any other point on the parabola. That is, the y-coordinate of the vertex is the minimum value of the function. Likewise, a quadratic function that has a parabola that is concave down for a graph has a maximum value equal to the y-coordinate of the vertex.

Maximum or Minimum Value of a
Quadratic Function

For the function $f(x) = a(x - h)^2 + k$ or $f(x) = ax^2 + bx + c$:

1. If $a > 0$, then the graph of f is concave up, and the function has a minimum value. This minimum value is the y-coordinate of the vertex.
2. If $a < 0$, then the graph of f is concave down, and the function has a maximum value. This maximum value is the y-coordinate of the vertex.

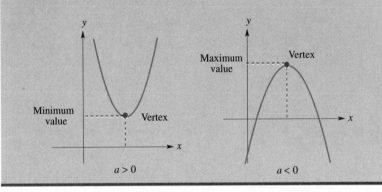

Finding the vertex of the graph of a quadratic equation in standard form $f(x) = a(x - h)^2 + k$ is rather straightforward—it is the point (h, k). On the other hand, determining the vertex of a quadratic function in general form $f(x) = ax^2 + bx + c$ is more difficult, especially if finding its standard form by completing the square becomes messy. We can avoid this sometimes difficult manipulation by deriving a general formula for the vertex of $f(x) = ax^2 + bx + c$, and using it to determine the vertex of any parabola.

$$f(x) = ax^2 + bx + c$$

$$= a\left(x^2 + \frac{b}{a}x\right) + c \qquad \text{Group the } x \text{ terms.}$$

$$= a\left(x^2 + \frac{b}{a}x + \frac{b^2}{4a^2}\right) + c - \frac{b^2}{4a} \qquad \text{Complete the square.}$$

$$f(x) = a\left(x + \frac{b}{2a}\right)^2 + \left(c - \frac{b^2}{4a}\right) \qquad \text{Factor and collect terms.}$$

Comparing this last form of the function to the standard form $f(x) = a(x - h)^2 + k$, we see that the vertex of $f(x) = ax^2 + bx + c$ is

$$(h, k) = \left(-\frac{b}{2a}, c - \frac{b^2}{4a}\right)$$

In practice, it is usually easier to compute the x-coordinate of the vertex,

$$h = -\frac{b}{2a}$$

and then evaluate $f(h)$ to find the y-coordinate, k.

▣ EXAMPLE 6 Finding the Vertex of a Parabola

Determine the vertex of the graph of the function $f(x) = -2x^2 + 8x - 2$.

SOLUTION

The graph of f was the subject of part (b) in Example 5 (see Figure 2.41). The x-coordinate of the vertex is

$$h = -\frac{b}{2a} = -\frac{8}{2(-2)} = 2$$

The y-coordinate of the vertex is

$$k = f(h) = f(2) = -2(2)^2 + 8(2) - 2 = 6$$

The vertex of the graph of $f(x) = -2x^2 + 8x - 2$ is $(2, 6)$, as shown in Figure 2.42.   ▣

A summary of how to find the vertex of the graph of a quadratic function follows.

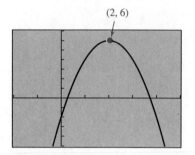

Figure 2.42
$y = -2x^2 + 8x - 2$
$[-2, 5] \times [-5, 7]$

$(2, 6)$

Finding the Vertex of a Parabola

1. *Standard form:* $f(x) = a(x - h)^2 + k$
The vertex of the graph of $f(x) = a(x - h)^2 + k$ is (h, k).
2. *General form:* $f(x) = ax^2 + bx + c$
The x-coordinate of the vertex of the graph of $f(x) = ax^2 + bx + c$ is

$$h = -\frac{b}{2a}$$

You can determine the y-coordinate by evaluating $f(h)$.

Applications

Quadratic functions serve as models for many of the applications encountered in physics, chemistry, economics, and other fields. As always, finding out information about the function—intercepts, turning points, values at particular points—helps us answer questions about the particular application.

In Section 1.5, we saw that *optimization problems* entailed finding either the minimum value or maximum value of a function over its domain. Locating the vertex of a quadratic function algebraically is an efficient way to solve these types of problems.

EXAMPLE 7 Application: Motion of a Thrown Ball

A ball is thrown directly up in the air from a height of h_0 feet with an initial velocity of v_0 feet per second. The height (in feet) t seconds after it is thrown is given by

$$H(t) = -16t^2 + v_0 t + h_0$$

From ground level, a ball is tossed up with an initial velocity of 120 feet per second.

(a) Determine the graph of H.

(b) At what time will the ball hit the ground?

(c) How long does it take for the ball to reach its maximum height? How high does the ball go?

SOLUTION

The problem tells us that $v_0 = 120$ and $h_0 = 0$. Therefore, the function that describes the height of the ball is $H(t) = -16t^2 + 120t$.

(a) The graph of H is shown in Figure 2.43.

(b) The ball is at ground level when $H(t) = 0$:

$$-16t^2 + 120t = 0$$
$$-8t(2t - 15) = 0$$
$$t(2t - 15) = 0$$

$$t = 0 \quad \bigg| \quad 2t - 15 = 0$$
$$2t = 15$$
$$t = 7.5$$

Of course, $t = 0$ is a solution to the equation because the ball is at ground level when it is thrown. The value $t = 7.5$ corresponds to 7.5 seconds, the time when the ball hits the ground.

(c) The maximum height of the ball is the maximum value of H. The x-coordinate of the vertex (h, k) is

$$h = -\frac{120}{2(-16)} = 3.75$$

The y-coordinate is

$$k = H(3.75) = -16(3.75)^2 + 120(3.75) = 225$$

The vertex is $(3.75, 225)$, as shown in Figure 2.44. Thus, the maximum height of the ball, 225 feet, is attained 3.75 seconds after it is thrown.

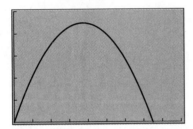

Figure 2.43
$y = -16x^2 + 120x$
$[0, 9] \times [0, 250]_{50}$

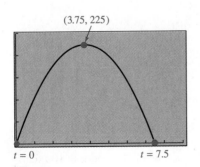

Figure 2.44
$[0, 9] \times [0, 250]_{50}$

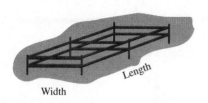

Width   Length

Figure 2.45

EXAMPLE 8 Application: Maximum Area of a Cattle Pen

Two holding pens for cattle (shown in Figure 2.45) are to be constructed with 600 feet of fencing. What is the maximum total area for the holding pens?

SOLUTION

The function we seek depends upon the relation

$$\left(\begin{array}{c}\text{Area of}\\\text{the pens}\end{array}\right) = \left(\begin{array}{c}\text{Width of}\\\text{the pens}\end{array}\right)\left(\begin{array}{c}\text{Length of}\\\text{the pens}\end{array}\right)$$

Imagine how you would construct these holding pens given 600 feet of fencing. First, you might cut three equal pieces of fence for the widths; then cut the remaining portion in half for the two lengths.

Because the length of the pens is determined by the width you choose, it follows that the area of the pens is a function of the width. Also,

$$3\left(\begin{array}{c}\text{Width}\\\text{of pens}\end{array}\right) + 2\left(\begin{array}{c}\text{Length}\\\text{of pens}\end{array}\right) = 600$$

$$2\left(\begin{array}{c}\text{Length}\\\text{of pens}\end{array}\right) = 600 - 3\left(\begin{array}{c}\text{Width}\\\text{of pens}\end{array}\right)$$

$$\left(\begin{array}{c}\text{Length}\\\text{of pens}\end{array}\right) = 300 - \tfrac{3}{2}\left(\begin{array}{c}\text{Width}\\\text{of pens}\end{array}\right)$$

Now we can construct a function A for the area of the pens in terms of the width w:

$$\left(\begin{array}{c}\text{Area of}\\\text{the pens}\end{array}\right) = \left(\begin{array}{c}\text{Width of}\\\text{the pens}\end{array}\right)\left(\begin{array}{c}\text{Length of}\\\text{the pens}\end{array}\right)$$

$$A(w) = w(300 - \tfrac{3}{2}w)$$

Writing A in general form gives

$$A(w) = -\tfrac{3}{2}w^2 + 300w$$

The graph of A is shown in Figure 2.46.

The vertex (h, k) of this parabola is such that

$$h = -\frac{b}{2a} = -\frac{300}{2\left(-\frac{3}{2}\right)} = 100$$

and

$$k = A(h) = A(100) = -\tfrac{3}{2}(100)^2 + 300(100) = 15{,}000$$

A maximum area of 15,000 square feet is attained when the width is 100 feet.

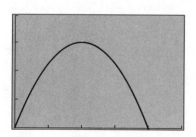

Figure 2.46
$y = -\tfrac{3}{2}x^2 + 300x$
$[0, 250]_{50} \times [0, 20{,}000]_{5000}$

EXERCISE SET 2.2

A

In Exercises 1–4, use a graphing utility to determine the graph of each function. Describe how the graph compares to the graph of $f(x) = x^2$.

1. (a) $y = 3x^2$ **(b)** $y = -x^2$

2. (a) $y = -\frac{2}{3}x^2$ **(b)** $y = -3x^2$

3. (a) $y = -2x^2$ **(b)** $y = 1.5x^2$

4. (a) $y = -2x^2$ **(b)** $y = 0.6x^2$

In Exercises 5–14, use a graphing utility to determine the graph of each function g. Describe how the graph of g compares to the graph of $f(x) = ax^2$ (for some real number a).

5. $g(x) = (x - 3)^2$ **6.** $g(x) = x^2 - 6$

7. $g(x) = (x - 2)^2 + 4$ **8.** $g(x) = (x + 5)^2 - 2$

9. $g(x) = 2(x + 3)^2 - 8$ **10.** $g(x) = \frac{1}{2}(x - 4)^2 + 2$

11. $g(x) = -(x + 4)^2 - 6$ **12.** $g(x) = -x^2 + 12$

13. $g(x) = -\frac{1}{3}(x + 3)^2$

14. $g(x) = -3(x - 6)^2 + 12$

In Exercises 15–20, determine a function in standard form for each parabola.

15.

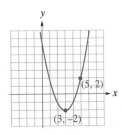

16.

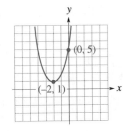

17.

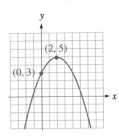

18.

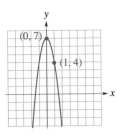

19.

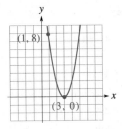

20.

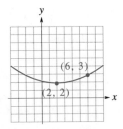

In Exercises 21–32, rewrite each quadratic function in standard form. Verify your result with a graphing utility.

21. $y = x^2 + 4x - 8$ **22.** $y = x^2 - 6x + 2$

23. $y = 2x^2 - 4x + 5$ **24.** $y = -x^2 + 8x - 1$

25. $y = -\frac{1}{2}x^2 + 2x - 2$ **26.** $y = -2x^2 - 8x - 8$

27. $y = 3x^2 + 18x$ **28.** $y = \frac{1}{4}x^2 + x$

29. $y = x^2 + 3x + \frac{1}{4}$ **30.** $y = x^2 + 5x + \frac{9}{4}$

31. $y = (x - 2)^2 + (x + 4)^2$

32. $y = (x + 1)^2 + (x - 3)^2$

In Exercises 33–40, determine the intercepts and vertex of the graph of each function.

33. $y = (x - 2)^2 - 4$ **34.** $y = (x + 1)^2 - 9$

35. $y = -2(x + 6)^2 + 8$ **36.** $y = -\frac{1}{3}(x - 4)^2 + 3$

37. $y = x^2 - 2x - 5$ **38.** $y = x^2 + 4x - 6$

39. $y = -x(x - 2) - 4$

40. $y = 2(x - 3)(x + 1) + 5$

B

41. A manufacturer of fax machines finds that the cost (in dollars) generated by manufacturing x units per week is given by the function

$$C(x) = 0.15x^2 - 39x + 4500$$

How many units should be manufactured to minimize the cost?

42. An installer of supermarket refrigeration units discovers that the profit or loss (in dollars) generated by installing x units per week is given by the function

$$P(x) = -0.1x^2 + 24.8x - 467.2$$

How many units should be installed to maximize the profit?

43. The main cable of a suspension bridge has the shape of a parabola. The cables are strung from the tops of two towers, 200 feet apart, each 55 feet high. The cable is 5 feet above the roadway at the point that is directly between the towers (see the figure).

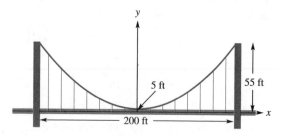

(a) Determine an equation for the parabola.
(b) How long is the vertical support cable that is 40 feet from one of the towers?

44. The arch of a doorway is in the shape of a parabola. The bottom of the arch is 4 feet above the floor, and the top of the arch is 8 feet above the floor. The doorway is 4 ft wide (see the figure).

(a) Determine an equation for the parabola.
(b) Is it possible to pass a box with dimensions $6 \times 3 \times 10$ feet through the doorway?

45. A rain gutter is constructed by folding the edges of a sheet of metal 12 inches wide so that the cross section of the gutter is a rectangle (see the figure). The capacity of the gutter is the product of the length of the gutter times the area of the cross section. How much edge should be folded up on each side to maximize the capacity?

46. What is the greatest area of a pasture in the shape of a rectangle that can be fenced with 6000 feet of fencing?

47. A rectangle is inscribed in a right isosceles triangle with hypotenuse of 18 units (see the figure).

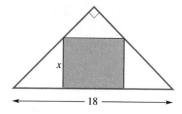

(a) Show that the area of the rectangle is given by $A(x) = x(18 - 2x)$, where x is the height of the rectangle. What is the domain of A?
(b) What is the maximum possible area of all such rectangles?

48. A rectangle is constructed so that one side is on the positive y-axis, one side is on the positive x-axis, and one corner is on the graph of $y = -\frac{2}{3}x + 4$ (see the figure).

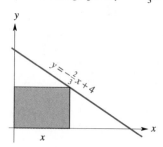

(a) Show that the area of the rectangle is given by

$$R(x) = x\left(-\tfrac{2}{3}x + 4\right)$$

where x is the width of the rectangle. What is the domain of R?

(b) What is the maximum possible area of all such rectangles?

C

49. Suppose that $f(x) = x^2 - kx + 16$. Determine a value for k such that the graph of f has:

(a) No x-intercepts

(b) Exactly one x-intercept

(c) Two x-intercepts

50. Suppose that $f(x) = x^2 - 4x + c$. Determine a value for c such that the graph of f has:

(a) No x-intercepts

(b) Exactly one x-intercept

(c) Two x-intercepts

51. Suppose that $f(x) = ax^2 + bx + c$. Explain why:

(a) The graph of f has no intercepts if and only if $b^2 < 4ac$.

(b) The graph of f has exactly one intercept if and only if $b^2 = 4ac$.

(c) The graph of f has two intercepts if and only if $b^2 > 4ac$.

52. At the same time that a car passes through an intersection traveling east at 30 miles per hour, a truck traveling south toward the intersection at 40 miles per hour is 50 miles from the intersection. Determine the time at which the car and truck are closest to each other. How far apart are they at this time?

HINT Quick Review: Use the Pythagorean theorem to express the square of the distance between the car and truck as a function of the time after the car passes through the intersection.

53. Use the distance formula to find the point on the graph of $y = -\tfrac{3}{2}x + 13$ that is closest to the origin.

54. A ball is thrown directly up in the air from a height of h_0 feet with an initial velocity of v_0 feet per second. The height (in feet) t seconds after it is thrown is given by $H(t) = -16t^2 + v_0 t + h_0$. Show that the maximum height of the ball is

$$\frac{v_0^2}{64} + h_0$$

SECTION 2.3

- Solution by Factoring
- Solution by Taking Roots
- Radical Equations
- Other Equations
- Applications

MORE EQUATIONS

The common thread running through any algebra course is that of solving equations. In the first chapter, we reviewed the algebraic methods of solving linear and quadratic equations. In Section 1.4, we investigated the graphical solution of equations. In this section, we will build on our current set of algebraic methods to solve other types of equations.

Solutions found by these algebraic methods can be anticipated graphically beforehand, or they can be verified graphically afterward. (Of course, you can still check a solution directly by substituting it back into the equation.) Using both algebraic and graphical methods to solve an equation will increase the chances that your solution is correct.

Solution by Factoring

In Section 7 (page 58), we found that quadratic equations such as

$$(x - 2)(3x + 5) = 0$$

can be solved using the zero factor property. We repeat this property here:

Zero Factor Property

> If A and B are both algebraic expressions such that $AB = 0$, then $A = 0$ or $B = 0$.

Of course, solving an equation using this property hinges on being able to write the equation in factored form. The zero factor property also applies to more than two factors; for example, if $ABC = 0$ then $A = 0$ or $B = 0$ or $C = 0$.

■ EXAMPLE 1   Solving an Equation by Factoring

Solve each equation. Verify the solutions graphically.

(a) $5x^3 - 20x = 0$ **(b)** $(x - 4)(x^2 - 4x + 2) = 0$

SOLUTION

(a) The left side of this equation can be written in factored form:

$$5x^3 - 20x = 0$$
$$5x(x^2 - 4) = 0 \quad \text{Factor out the common factor.}$$
$$5x(x - 2)(x + 2) = 0 \quad \text{Factor the difference of squares.}$$
$$x(x - 2)(x + 2) = 0 \quad \text{Divide each side by 5.}$$

This last equation is completely factored, so using the zero factor property, we get three linear equations:

$x = 0$	$x - 2 = 0$	$x + 2 = 0$
	$x = 2$	$x = -2$

The solutions are -2, 0, and 2. The graph of $y = 5x^3 - 20x$ is shown in Figure 2.47. This graph has x-intercepts that correspond to our solutions.

(b) This equation is already in a factored form, so by the zero factor property, either $x - 4 = 0$ or $x^2 - 4x + 2 = 0$. The solution of $x^2 - 4x + 2 = 0$ requires the quadratic formula because it is not factorable:

$$(x - 4)(x^2 - 4x + 2) = 0$$

$x - 4 = 0$	$x^2 - 4x + 2 = 0$
$x = 4$	$x = \dfrac{-(-4) \pm \sqrt{(-4)^2 - 4(1)(2)}}{2(1)}$
	$= \dfrac{4 \pm \sqrt{8}}{2}$
	$= 2 \pm \sqrt{2}$

The solutions are 4, $2 + \sqrt{2}$ (approximately 3.4), and $2 - \sqrt{2}$ (approximately 0.6). The graph of $y = (x - 4)(x^2 - 4x + 2)$ is shown in Figure 2.48. This graph appears to have x-intercepts that correspond to our solutions.

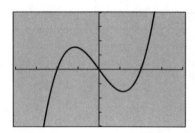

Figure 2.47
$y = 5x^3 - 20x$
$[-4, 4] \times [-40, 40]_{10}$

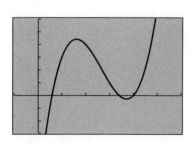

Figure 2.48
$y = (x - 4)(x^2 - 4x + 2)$
$[-1, 6] \times [-3, 6]$

Many equations can be solved by the methods used for quadratic equations by making an appropriate substitution.

EXAMPLE 2 Solving an Equation by Substitution

Solve the equation: $2x^4 - 11x^2 + 12 = 0$

SOLUTION

This equation is not a quadratic equation in x. However, by rewriting the equation as

$$2(x^2)^2 - 11(x^2) + 12 = 0$$

and substituting u for x^2, we can rewrite this equation as

$$2u^2 - 11u + 12 = 0$$

This quadratic equation in u can be solved by factoring:

$$2u^2 - 11u + 12 = 0$$
$$(2u - 3)(u - 4) = 0$$

$$2u - 3 = 0 \qquad\qquad u - 4 = 0$$

$$u = \frac{3}{2} \qquad\qquad u = 4$$

Now we replace u with x^2 and solve the resulting equations for x:

$$x^2 = \frac{3}{2} \qquad\qquad x^2 = 4$$

$$x = \pm \sqrt{\frac{3}{2}} \qquad\qquad x = \pm 2$$

$$x = \pm \frac{\sqrt{6}}{2}$$

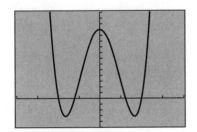

Figure 2.49
$y = 2x^4 - 11x^2 + 12$
$[-4, 4] \times [-5, 15]$

The solutions are -2, $-\sqrt{6}/2$, $\sqrt{6}/2$, and 2. We verify these solutions graphically in Figure 2.49.

Solution by Taking Roots

In Section 7 (page 58), we saw that a quadratic equation such as $(x - 2)^2 = 7$ can be solved by first taking the square root of each side:

$$(x - 2)^2 = 7$$
$$x - 2 = \pm\sqrt{7}$$
$$x = 2 \pm \sqrt{7}$$

Thus, the solutions to $(x - 2)^2 = 7$ are $2 - \sqrt{7}$ and $2 + \sqrt{7}$.

In Example 3, we extend this method of taking roots to similar equations of higher degree. An equation such as $x^4 = 16$, in which the exponent is even, has two real solutions, $x = 2$ and $x = -2$. On the other hand, an equation in which the exponent is odd, such as $t^3 = 27$, has only one real solution, $t = 3$.

EXAMPLE 3   Solving Equations by Taking Roots

Solve each equation.

(a) $(w + 2)^3 = 24$ (b) $2n^4 = 26$

(c) $x^3 = -\frac{1}{8}$ (d) $(2t - 6)^5 = 0$

(e) $-2(p^2 - 1)^6 = 4$

SOLUTION

(a) Taking the cube root of each side of this equation gives us a linear equation to solve:

$$(w + 2)^3 = 24$$
$$w + 2 = \sqrt[3]{24} \qquad \text{Take the cube root of each side.}$$
$$w + 2 = 2\sqrt[3]{3} \qquad \text{Simplify the radical.}$$
$$w = -2 + 2\sqrt[3]{3} \quad \text{Add } -2 \text{ to each side.}$$

The solution to the equation is $-2 + 2\sqrt[3]{3}$ (approximately 0.9). This equation has only one solution because the exponent 3 is odd. The graph of the corresponding function, $y = (x + 2)^3 - 24$, supports our conclusion (Figure 2.50).

(b) After dividing each side by 2, we take fourth roots of each side:

$$2n^4 = 26$$
$$n^4 = 13 \qquad \text{Divide each side by 2.}$$
$$n = \pm\sqrt[4]{13} \quad \text{Take fourth roots of each side.}$$

This equation has two solutions, $-\sqrt[4]{13}$ (approximately -1.9) and $\sqrt[4]{13}$ (approximately 1.9), because the exponent 4 is even. The graph of the corresponding function $y = 2x^4 - 26$ is shown in Figure 2.51.

(c) We proceed in the same way as we did in part (a) by taking the cube root of each side:

$$x^3 = -\frac{1}{8}$$
$$x = \sqrt[3]{-\frac{1}{8}} \quad \text{Take the cube root of each side.}$$
$$x = -\frac{1}{2}$$

The graph of $y = x^3 + \frac{1}{8}$ is shown in Figure 2.52.

(d) Taking the fifth root of each side of the equation gives a linear equation to solve:

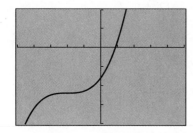

Figure 2.50
$y = (x + 2)^3 - 24$
$[-5, 5] \times [-40, 20]_{10}$

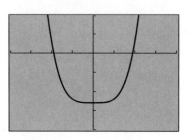

Figure 2.51
$y = 2x^4 - 26$
$[-4, 4] \times [-40, 20]_{10}$

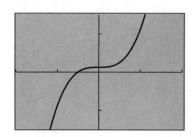

Figure 2.52
$y = x^3 + \frac{1}{8}$
$[-2, 2] \times [-1.5, 1.5]$

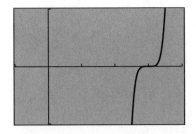

Figure 2.53
$y = (2x - 6)^5$; $[-1, 4] \times [-1, 1]$

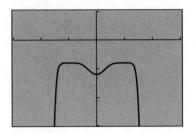

Figure 2.54
$y = -2(x^2 - 1)^6 - 4$
$[-3, 3] \times [-15, 5]_5$

$$(2t - 6)^5 = 0$$
$$2t - 6 = \sqrt[5]{0} \quad \text{Take the fifth root of each side.}$$
$$2t - 6 = 0$$
$$2t = 6$$
$$t = 3$$

The graph of $y = (2x - 6)^5$ is shown in Figure 2.53.

(e) The first step is to divide each side by -2:

$$-2(p^2 - 1)^6 = 4$$
$$(p^2 - 1)^6 = -2$$

This last equation has no real number solutions because a real number raised to an even power cannot be negative. Therefore, the graph of $y = -2(p^2 - 1)^6 - 4$ has no x-intercepts (Figure 2.54).

The results of Example 3 are generalized here:

Solving Equations
by Taking Roots

Suppose that n is a positive integer, a is a positive real number, and u is an algebraic expression.

If n is odd, then

$$u^n = a \qquad \text{implies that} \qquad u = \sqrt[n]{a}$$
$$u^n = -a \qquad \text{implies that} \qquad u = -\sqrt[n]{a}$$

If n is even, then

$$u^n = a \qquad \text{implies that} \qquad u = \sqrt[n]{a} \ \text{ or } \ u = -\sqrt[n]{a}$$
$$u^n = -a \qquad \text{has no real solutions for } u$$

If n is either even or odd, then

$$u^n = 0 \qquad \text{implies that} \qquad u = 0$$

Radical Equations

An equation with square roots, cube roots, or even higher roots can often be solved by raising each side to an appropriate power. For example, squaring each side of the equation $\sqrt{x} = 3$ gives $x = 9$ as a solution.

Unfortunately, raising each side of an equation to an even power does not necessarily give an equivalent equation. For example, squaring both sides of the equation $x = 2$, which has only 2 as a solution, gives $x^2 = 4$, which has solutions -2 and 2. The extra solution generated by squaring each side is called an *extra-*

■ **NOTE:** *Raising each side of an equation to an odd power does not generate extraneous solutions, but it is still a good idea to check your solutions to these equations.*

neous solution. Extraneous solutions may occur whenever each side of an equation is raised to an even power.

You should check a potential solution to determine whether it is an actual solution of the equation or an extraneous solution. This check can be done either graphically or by substituting it back into the equation.

■ EXAMPLE 4   Solving an Equation with a Radical

Solve the equation and check the solutions graphically.

(a) $\sqrt{x + 4} = x - 8$ **(b)** $\sqrt[3]{x^2 - 6x} = 2$

SOLUTION

(a) We start by squaring each side to eliminate the square root:

$$\sqrt{x + 4} = x - 8$$
$$(\sqrt{x + 4})^2 = (x - 8)^2 \qquad \text{Square each side.}$$
$$x + 4 = x^2 - 16x + 64$$
$$0 = x^2 - 17x + 60$$
$$0 = (x - 5)(x - 12)$$
$$x - 5 = 0 \quad \bigg| \quad x - 12 = 0$$
$$x = 5 \quad \bigg| \qquad x = 12$$

So 5 and 12 are potential solutions for the equation $\sqrt{x + 4} = x - 8$. The graph of the corresponding function $y = \sqrt{x + 4} - x + 8$ (Figure 2.55) has only one x-intercept, which appears to be 12. We conclude that one potential solution, 5, is extraneous. Also, by direct substitution, we get

$$\sqrt{5 + 4} \stackrel{?}{=} 5 - 8 \quad \bigg| \quad \sqrt{12 + 4} \stackrel{?}{=} 12 - 8$$
$$\sqrt{9} \stackrel{?}{=} -3 \quad \bigg| \qquad \sqrt{16} \stackrel{?}{=} 4$$
$$3 \neq -3 \quad \bigg| \qquad 4 = 4$$

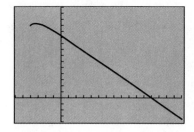

Figure 2.55
$y = \sqrt{x + 4} - x + 8$
$[-6, 16] \times [-4, 15]$

The solution to the equation $\sqrt{x + 4} = x - 8$ is 12.

(b) The cube root is eliminated by raising each side to the third power:

$$\sqrt[3]{x^2 - 6x} = 2$$
$$(\sqrt[3]{x^2 - 6x})^3 = 2^3 \quad \text{Cube each side.}$$
$$x^2 - 6x = 8$$
$$x^2 - 6x - 8 = 0$$

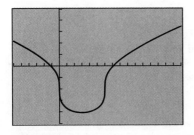

Figure 2.56
$y = \sqrt[3]{x^2 - 6x} - 2$
$[-6, 16] \times [-5, 5]$

This last equation cannot be solved by factoring, so we use the quadratic formula:

$$x = \frac{-(-6) \pm \sqrt{(-6)^2 - 4(1)(-8)}}{2(1)} = \frac{6 \pm \sqrt{68}}{2} = 3 \pm \sqrt{17}$$

The solutions of $\sqrt[3]{x^2 - 6x} = 2$ are $3 - \sqrt{17}$ (approximately -1.1) and $3 + \sqrt{17}$ (approximately 7.1). Because we raised each side of the equation to an odd power, we did not generate extraneous solutions. The graph of $y = \sqrt[3]{x^2 - 6x} - 2$ verifies our solutions (Figure 2.56).

An equation with two or more square roots is solved by isolating one square root on one side of the equation, squaring each side, and repeating this process until the square roots are eliminated.

■ EXAMPLE 5 Solving an Equation with Two Radicals

Solve: $\sqrt{6 - x} + \sqrt{x} = 3$

SOLUTION
We start by rewriting the equation so that $\sqrt{6 - x}$ is on the left side, and then squaring:

$$\sqrt{6 - x} + \sqrt{x} = 3$$
$$\sqrt{6 - x} = 3 - \sqrt{x}$$
$$(\sqrt{6 - x})^2 = (3 - \sqrt{x})^2$$
$$6 - x = 3^2 - 2(3)\sqrt{x} + (\sqrt{x})^2$$
$$6 - x = 9 - 6\sqrt{x} + x$$

The remaining radical can be eliminated in a similar way:

$$6 - x = 9 - 6\sqrt{x} + x$$
$$-3 - 2x = -6\sqrt{x}$$
$$(-3 - 2x)^2 = (-6\sqrt{x})^2$$
$$9 + 12x + 4x^2 = 36x$$
$$4x^2 - 24x + 9 = 0$$

We solve this last equation using the quadratic formula:

$$x = \frac{-(-24) \pm \sqrt{(-24)^2 - 4(4)(9)}}{2(4)} = \frac{24 \pm \sqrt{432}}{8}$$
$$= \frac{24 \pm 12\sqrt{3}}{8} = \frac{6 \pm 3\sqrt{3}}{2}$$

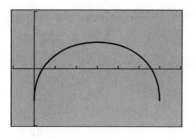

Figure 2.57
$y = \sqrt{6-x} + \sqrt{x} - 3$
$[-1, 7] \times [-1, 1]$

The potential solutions are $(6 - 3\sqrt{3})/2$ and $(6 + 3\sqrt{3})/2$ (approximately 0.4 and 5.6, respectively). The corresponding graph of $y = \sqrt{6-x} + \sqrt{x} - 3$ verifies that both of these potential solutions are indeed solutions (Figure 2.57).

Other Equations

Not every equation fits neatly into one of the categories demonstrated so far in this section. Many equations require a mixture of these and other algebraic techniques. On the other hand, many equations cannot be solved by any algebraic means, and we must settle for approximate graphical solutions. When solving an equation, try examining the graph of the corresponding function, simplifying expressions, or substituting for an expression (as we did in Example 2). The next three examples illustrate some of these suggestions.

EXAMPLE 6   Solving an Absolute Value Equation

Solve: $|x^2 - 3x| = 2$

SOLUTION

The graph of $y = |x^2 - 3x| - 2$ (Figure 2.58) suggests that there are four solutions. (Two of the solutions appear to be integers, 1 and 2.)

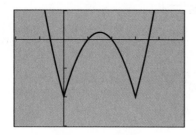

Figure 2.58
$y = |x^2 - 3x| - 2$
$[-2, 5] \times [-3, 1]$

In Section 1 (page 2), we saw that if $a > 0$, then $|u| = a$ implies that either $u = -a$ or $u = a$. Thus, a solution to $|x^2 - 3x| = 2$ is a solution to either $x^2 - 3x = -2$ or $x^2 - 3x = 2$.

The first equation, $x^2 - 3x = -2$, can be solved by factoring:

$$x^2 - 3x = -2$$
$$x^2 - 3x + 2 = 0$$
$$(x - 1)(x - 2) = 0$$
$$x - 1 = 0 \quad | \quad x - 2 = 0$$
$$x = 1 \quad | \quad x = 2$$

The second equation, $x^2 - 3x = 2$, cannot be solved by factoring, so we use the quadratic formula:

$$x^2 - 3x = 2$$
$$x^2 - 3x - 2 = 0$$
$$x = \frac{-(-3) \pm \sqrt{(-3)^2 - 4(1)(-2)}}{2(1)} = \frac{3 \pm \sqrt{17}}{2}$$

The solutions to $|x^2 - 3x| = 2$ are $(3 - \sqrt{17})/2$ (approximately -0.6), 1, 2, and $(3 + \sqrt{17})/2$ (approximately 3.6). These solutions, found by algebraic methods, agree with the solutions we anticipated by examining the graph of $y = |x^2 - 3x| - 2$ in Figure 2.58.

■ EXAMPLE 7   Solving an Equation with Rational Exponents

Solve: $x^{2/3}(2x^{2/3} - 7) = 4$

SOLUTION

The graph of $y = x^{2/3}(2x^{2/3} - 7) - 4$ (Figure 2.59) predicts two solutions to the equation $x^{2/3}(2x^{2/3} - 7) = 4$. The solutions appear to be -8 and 8.

Substituting u for $x^{2/3}$ gives us a quadratic equation to solve:

$$x^{2/3}(2x^{2/3} - 7) = 4$$
$$u(2u - 7) = 4 \quad \text{Let } u = x^{2/3}.$$

This equation in u is solved by simplifying and then factoring:

$$u(2u - 7) = 4$$
$$2u^2 - 7u = 4$$
$$2u^2 - 7u - 4 = 0$$
$$(2u + 1)(u - 4) = 0$$

$$2u + 1 = 0 \qquad \qquad u - 4 = 0$$
$$u = -\tfrac{1}{2} \qquad \qquad u = 4$$

The next step is to replace u with $x^{2/3}$ in these equations and solve for x:

$u = -\tfrac{1}{2}$	$u = 4$
$x^{2/3} = -\tfrac{1}{2}$	$x^{2/3} = 4$ $\qquad$ Replace u with $x^{2/3}$.
$(x^{2/3})^3 = \left(-\tfrac{1}{2}\right)^3$	$(x^{2/3})^3 = 4^3$ $\qquad$ Cube each side.
$x^2 = -\tfrac{1}{8}$	$x^2 = 64$
$x = \pm\sqrt{-\tfrac{1}{8}}$	$x = \pm 8$ $\qquad$ Take square roots of each side.

Because $\sqrt{-\tfrac{1}{8}}$ is undefined as a real number, the first equation has no solutions. The solutions to $x^{2/3}(2x^{2/3} - 7) = 4$ are -8 and 8, as Figure 2.59 suggests.

■

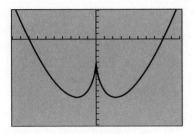

Figure 2.59
$y = x^{2/3}(2x^{2/3} - 7) - 4$
$[-10, 10] \times [-15, 5]$

■ EXAMPLE 8   Solving an Equation with Negative Exponents

Solve: $2x^{-2} - 5x^{-1} = 1$

SOLUTION

The equation is quadratic in the expression x^{-1}, so we substitute u for x^{-1} and solve for u:

$$2x^{-2} - 5x^{-1} = 1$$
$$2(x^{-1})^2 - 5x^{-1} = 1$$
$$2u^2 - 5u = 1 \quad \text{Let } u = x^{-1}.$$
$$2u^2 - 5u - 1 = 0$$

Using the quadratic formula gives

$$u = \frac{-(-5) \pm \sqrt{(-5)^2 - 4(2)(-1)}}{2(2)} = \frac{5 \pm \sqrt{33}}{4}$$

and

$$u = \frac{5 - \sqrt{33}}{4} \qquad\qquad u = \frac{5 + \sqrt{33}}{4}$$

$$x^{-1} = \frac{5 - \sqrt{33}}{4} \qquad\qquad x^{-1} = \frac{5 + \sqrt{33}}{4} \qquad \text{Replace } u \text{ with } x^{-1}.$$

$$\frac{1}{x} = \frac{5 - \sqrt{33}}{4} \qquad\qquad \frac{1}{x} = \frac{5 + \sqrt{33}}{4}$$

$$x = \frac{4}{5 - \sqrt{33}} \qquad\qquad x = \frac{4}{5 + \sqrt{33}}$$

Rationalizing the denominator, we obtain

$$\frac{4}{5 - \sqrt{33}} = \frac{-5 - \sqrt{33}}{2} \qquad \text{and} \qquad \frac{4}{5 + \sqrt{33}} = \frac{-5 + \sqrt{33}}{2}$$

■ **NOTE:** *Exercise 67 shows another way to solve the equation in Example 8.*

These are the solutions to $2x^{-2} - 5x^{-1} = 1$. We leave it to you to verify these solutions with a graphing utility.

Applications

In the next example, we use a geometric formula along with the techniques of this section to solve an applied problem.

■ **EXAMPLE 9   Application: Constructing a Survival Shelter**

A wilderness survival shelter is to be constructed from two lightweight panels, each with dimensions 5 feet $\times$ 10 feet (Figure 2.60). Furthermore, the volume of the shelter must be 100 cubic feet, and its height h must be at least 3 feet. Determine the height h and the base b of the shelter.

SOLUTION

The volume of the shelter is given by the relation

$$\begin{pmatrix} \text{Volume of} \\ \text{shelter} \end{pmatrix} = \begin{pmatrix} \text{Length of} \\ \text{shelter} \end{pmatrix} \begin{pmatrix} \text{Cross-sectional} \\ \text{area of shelter} \end{pmatrix}$$

The volume of the shelter is 100 cubic feet, and the length of the shelter is 10 feet. The area of the cross section is the area of the triangle ABC (Figure 2.60), which is $\frac{1}{2}hb$. By substituting these quantities and simplifying, we get

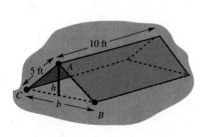

Figure 2.60

$$100 = 10 \cdot \tfrac{1}{2}hb$$
$$100 = 5hb$$
$$20 = hb$$

Can we rewrite this last equation, $20 = hb$, in terms of h only? Applying the Pythagorean theorem to the triangle ABM (Figure 2.61) yields an expression in h for b:

$$(\tfrac{1}{2}b)^2 + h^2 = 5^2$$
$$\tfrac{1}{2}b = \sqrt{25 - h^2}$$
$$b = 2\sqrt{25 - h^2}$$

Substituting this quantity into $20 = hb$ gives us an equation in h only:

$$20 = hb$$
$$20 = h(2\sqrt{25 - h^2})$$
$$10 = h\sqrt{25 - h^2}$$

Squaring each side of this radical equation eliminates the square root:

$$(10)^2 = (h\sqrt{25 - h^2})^2$$
$$100 = h^2(25 - h^2)$$
$$100 = 25h^2 - h^4$$
$$h^4 - 25h^2 + 100 = 0$$

Letting $u = h^2$ makes this equation a quadratic equation in u that can be solved by factoring:

$$h^4 - 25h^2 + 100 = 0$$
$$(h^2)^2 - 25(h^2) + 100 = 0$$
$$u^2 - 25u + 100 = 0 \qquad \text{Let } u = h^2.$$
$$(u - 20)(u - 5) = 0$$

$u - 20 = 0$	$u - 5 = 0$	
$u = 20$	$u = 5$	
$h^2 = 20$	$h^2 = 5$	Replace u with h^2.
$h = \pm\sqrt{20}$	$h = \pm\sqrt{5}$	
$h = \pm 2\sqrt{5}$		

We discard the negative solutions, $-2\sqrt{5}$ and $-\sqrt{5}$, because the height h must be positive. We can also discard the solution $\sqrt{5}$ (approximately 2.24), because the height must be at least 3 feet. Therefore, the height is $2\sqrt{5}$ feet. The base b is

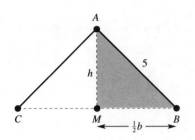

Figure 2.61

$$b = 2\sqrt{25 - h^2} = 2\sqrt{25 - (2\sqrt{5})^2} = 2\sqrt{25 - 20} = 2\sqrt{5}$$

Thus, both the height h and the base b of the shelter are $2\sqrt{5}$ (approximately 4.47) feet.

Exercise Set 2.3

A

In Exercises 1–6, solve each equation using the zero factor property. Use a graphing utility to verify your solution.

1. $x^3 - 4x^2 - 12x = 0$
2. $2x^3 - x^2 - 15x = 0$
3. $(x^2 - 1)(x + 6) = 0$
4. $(x - 5)(x^2 - 9) = 0$
5. $(x^2 - 9)(x^2 - 5) = 0$
6. $(x^2 - 16)(x^2 - 2) = 0$

In Exercises 7–16, solve each equation by taking roots. Use a graphing utility to verify your solution.

7. $(2x - 6)^3 = 8$
8. $(4 - x)^4 = 1$
9. $27(x - \frac{2}{3})^3 = 1$
10. $8(x - \frac{3}{2})^4 = \frac{1}{2}$
11. $2x^6 + 5 = 3$
12. $\frac{1}{2}(x^4 - 5) = -3$
13. $x^5 + 5 = 3$
14. $x^4 - 7 = 2$
15. $3(x^3 + 5) = 15$
16. $2(x^8 + 9) - 4 = 14$

In Exercises 17–24, solve each radical equation. Use a graphing utility to verify your solution.

17. $2\sqrt{x + 5} = 6$
18. $\frac{1}{2}\sqrt{3 - x} = 4$
19. $\sqrt[3]{2x + 12} = 4$
20. $5\sqrt[3]{4x} = 10$
21. $\sqrt{x^2 + x + 6} = x + 1$
22. $\sqrt{x^2 - 2x - 7} = x + 3$
23. $\sqrt[4]{36 - x} = 3$
24. $\sqrt[6]{2x - 7} = 1$

In Exercises 25–30, solve each absolute value equation. Use a graphing utility to verify your solution.

25. $|3x| = 12$
26. $|x + 4| = 7$
27. $|x^2 - 5| = 4$
28. $|12 - x^2| = 3$
29. $|x + 6| = 2x$
30. $|2x - 4| = x$

B

In Exercises 31–48, use a graphing utility to determine the graph of each equation and then solve the equation.

31. $(x - 5)^{3/2} = 8$
32. $(x + 2)^{2/3} = 9$
33. $2\sqrt[3]{2x + 1} - 4 = 6$
34. $\frac{1}{2}\sqrt[4]{x - 4} + 3 = 4$
35. $x^2(3x^2 - 5) = 12$
36. $\sqrt{x}(4\sqrt{x} + 1) = 39$
37. $x^4 + 5x^3 + x + 5 = 0$
38. $2x^5 - 3x^4 - 4x + 6 = 0$
39. $|x^2 - 4x| = 4$
40. $|x^2 + 3x| = 2$
41. $2x^{2/3} = 3x^{1/3} - 1$
42. $x^{2/3} + 3x^{1/3} = 10$
43. $(x^2 - 2)^2 - 4(x^2 - 2) - 21 = 0$
44. $(x + 1)^3 - 4(x + 1) = 0$
45. $\sqrt{3x + 1} - \sqrt{x - 1} = 2$
46. $\sqrt{x + 1} + \sqrt{x - 4} = 5$
47. $(x + 4) + 5\sqrt{x + 4} - 6 = 0$
48. $x^{-2} - 5x^{-1} + 6 = 0$
49. Solve: $3x - x^{3/2} = 0$.
   HINT Quick Review: First factor x from the left side.
50. Solve: $\frac{1}{2}x^2 - x^{5/2} = 0$.
   HINT Quick Review: First factor x^2 from the left side.
51. If an amount of money P is invested at an annual rate r compounded n times per year, then the value A of the investment after t years is given by

$$A = P\left(1 + \frac{r}{n}\right)^{nt}$$

(This formula will be discussed at length in Chapter 4.) At what rate (to the nearest 0.1%) must $10,000 be invested in a savings account that is compounded four times per year so that the value of the account is $12,500 after 3 years?

52. At what rate (to the nearest 0.1%) must $5000 be invested in a savings account that is compounded monthly so that

the value of the account is $8200 after 7 years? (See Exercise 51.)

53. A balance scale uses a series of brass cylinders with masses 10 grams, 15 grams, and so on. The diameter of each cylinder is equal to its height. If the height of the 10 gram cylinder is 2.7 centimeters, what is the height (to the nearest 0.1) of the 15 gram cylinder?

54. The lateral surface area S of a cone with height h and radius r is given by the formula $S = \pi r \sqrt{r^2 + h^2}$. Using an appropriate graph, determine (to the nearest 0.1 centimeter) the radius of a cone with a lateral surface area of 80 square centimeters and a height of 6.0 centimeters.

C

In Exercises 55–66, use a graphing utility to determine the graph of each equation and then solve the equation.

55. $\sqrt{x} = \sqrt[4]{x + 2}$

56. $\sqrt[3]{x + 1} = \sqrt[6]{4x}$

57. $\sqrt{x} = \sqrt{10 + 3\sqrt{x}}$

58. $\sqrt{x + 12} = \sqrt{x + 2\sqrt{x}}$

59. $|2x| = x^2 - 3x - 6$

60. $|x|^3 = (x - 4)(x^2 + 4x)$

61. $\sqrt{x} + 3\sqrt[3]{x^2 - 8} = 0$

62. $\sqrt{x^2 - 12}(x^{-1} + 2) = 0$

63. $\sqrt{x + 2} = |x - 3|$

64. $\sqrt{x} = |x - 6|$

65. $\sqrt[3]{2x} = \sqrt{x}$

66. $\sqrt[3]{x - 2} = \sqrt[4]{3x - 6}$

67. Solve $2x^{-2} - 5x^{-1} = 1$ by first multiplying each side of the equation by x^2 and then solving the resulting quadratic equation. Compare your solutions with those in Example 8.

68. Explain why each equation has no real numbers as solutions.

(a) $\sqrt{x^2 + 2x} = -1$ **(b)** $-3(x^6 - 4x)^4 = 5$

(c) $|x + 8| = -x^2$

| SECTION 2.4 | **INEQUALITIES** |

- Solving an Inequality with a Graph
- Linear and Quadratic Inequalities
- Polynomial Inequalities
- Absolute Value Inequalities
- Applications

Solving an inequality, like solving an equation, is the process of finding all the real numbers that make the inequality true. In Section 6 (page 47), we solved linear inequalities. In this section, we will solve a more general collection of inequalities. A graphing utility will play a large part in finding these solutions, as it did with equations.

Inequalities can be sorted into four groups: greater than ($>$), greater than or equal to ($\geq$), less than ($<$), and less than or equal to ($\leq$). We say that an inequality is in *zero form* if the right side of the inequality is 0. Any inequality can be rewritten in zero form by adding the same quantity to both sides of the inequality. For example, $x^2 \leq 4x - 3$ can be rewritten as $x^2 - 4x + 3 \leq 0$ by adding $-4x + 3$ to each side of the inequality. In general, any inequality in x can be rewritten in one of the forms $f(x) > 0$, $f(x) < 0$, $f(x) \geq 0$, or $f(x) \leq 0$ for some function f. We call this function the *corresponding function* of the inequality and its graph the *corresponding graph* of the inequality.

Solving an Inequality with a Graph

In Section 1.3, we saw that the graph of a function f reveals the values of x for which the function is negative, 0, or positive. Example 1 demonstrates how the graph of a function can be used to solve an inequality.

■ EXAMPLE 1 Solving an Inequality Using a Graph

The graph of the function f is shown in Figure 2.62. Determine the solutions of the following inequalities:

(a) $f(x) > 0$ **(b)** $f(x) < 0$ **(c)** $f(x) \geq 0$ **(d)** $f(x) \leq 0$

SOLUTION

(a) The graph of f is above the x-axis between -5 and 2 (not including 2), and also between 7 and 8 (not including 7). Thus, the solution to $f(x) > 0$ is

$$-5 \leq x < 2 \qquad \text{or} \qquad 7 < x \leq 8$$

These values are graphed on the x-axis in Figure 2.63, just as the solutions to linear inequalities were graphed on a number line in Section 6 (page 47).

(b) The graph of f is below the x-axis between 2 and 7 (not including either 2 or 7), so the solution to $f(x) < 0$ is

$$2 < x < 7$$

The solution is graphed on the x-axis in Figure 2.64.

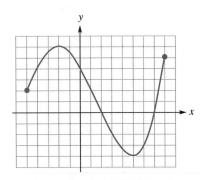

Figure 2.62
$y = f(x)$

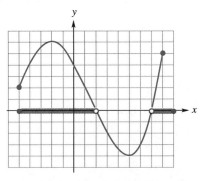

Figure 2.63

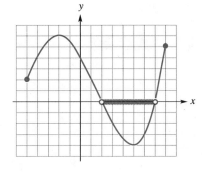

Figure 2.64

(c) The solutions to this inequality are those values of x for which the graph of f is on or above the x-axis. It follows from part (a) that the solution to $f(x) \geq 0$ is

$$-5 \leq x \leq 2 \qquad \text{or} \qquad 7 \leq x \leq 8$$

The solution is graphed on the x-axis in Figure 2.65.

(d) The set of all x for which the graph of f is on or below the x-axis is the solution to this inequality (Figure 2.66). The solution to $f(x) \leq 0$ is

$$2 \leq x \leq 7$$

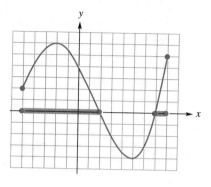

Figure 2.65   Figure 2.66

The process used in Example 1 suggests the following method for solving inequalities graphically:

Solving an Inequality Graphically

Step 1: Rewrite the inequality in zero form.

Step 2: Determine the corresponding function f for the inequality and its graph.

Step 3: Identify the points of the graph that correspond to the solution of the inequality.

$f(x) > 0$:   Points above the x-axis
$f(x) < 0$:   Points below the x-axis
$f(x) \geq 0$:   Points on or above the x-axis
$f(x) \leq 0$:   Points on or below the x-axis

Step 4: Find the values of x that correspond to these points. These values of x are the solution to the inequality.

Linear and Quadratic Inequalities

The simplest inequalities are linear inequalities and quadratic inequalities. A *linear inequality* is one in which the corresponding graph is a line. A *quadratic inequality* is one in which the corresponding graph is a parabola.

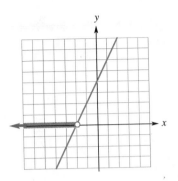

Figure 2.67
$y = 2x + 4$

■ **EXAMPLE 2** **Solving a Linear Inequality Graphically**

Solve the linear inequality $3x + 1 < x - 3$ graphically.

SOLUTION

The zero form of this inequality is $(3x + 1) - (x - 3) < 0$, or when simplified, $2x + 4 < 0$. The solution of this inequality corresponds to the set of points on the graph of $y = 2x + 4$ that are below the x-axis (Figure 2.67). (This graph is a line with slope 2 and y-intercept 4.) The solution to $3x + 1 < x - 3$ is

$$x < -2$$ ■

■ **EXAMPLE 3** **Solving Quadratic Inequalities Graphically**

Solve each quadratic inequality graphically.

(a) $-(x - 3)(x + 5) \leq 0$ **(b)** $x^2 < 4x + 6$ **(c)** $2x^2 + 3 \geq 4x$

SOLUTION

(a) The graph of $y = -(x - 3)(x + 5)$ is a parabola that is concave down and has x-intercepts at $x = -5$ and $x = 3$ (Figure 2.68). The points on the graph that are on or below the x-axis correspond to the solution of the inequality. Thus, the solution to $-(x - 3)(x + 5) \leq 0$ is

$$x \leq -5 \qquad \text{or} \qquad x \geq 3$$

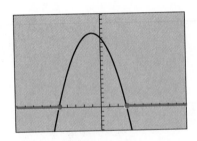

Figure 2.68
$y = -(x - 3)(x + 5)$
$[-10, 10] \times [-5, 20]$

(b) The zero form of $x^2 < 4x + 6$ is $x^2 - 4x - 6 < 0$. The solution of this inequality is the set of x values that correspond to the points below the x-axis on the graph of $y = x^2 - 4x - 6$ (Figure 2.69). These values of x are between the x-intercepts, which appear to be about $(-1, 0)$ and $(5, 0)$. We can use the quadratic formula to determine the x-intercepts of the corresponding function exactly:

$$x = \frac{-(-4) \pm \sqrt{(-4)^2 - 4(1)(-6)}}{2(1)} = \frac{4 \pm \sqrt{40}}{2} = 2 \pm \sqrt{10}$$

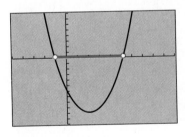

Figure 2.69
$y = x^2 - 4x - 6$
$[-5, 10] \times [-12, 8]$

The solutions are $2 - \sqrt{10}$ (approximately -1.2) and $2 + \sqrt{10}$ (approximately 5.2). The solution of $x^2 < 4x + 6$ is

$$2 - \sqrt{10} < x < 2 + \sqrt{10}$$

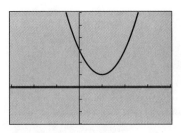

Figure 2.70
$y = 2x^2 - 4x + 3$
$[-3, 4] \times [-3, 6]$

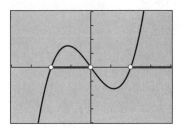

Figure 2.71
$y = 5x^3 - 20x$
$[-4, 4] \times [-40, 40]_{10}$

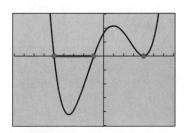

Figure 2.72
$y = (x + 1)(x + 5)(x - 4)^2$
$[-9, 7] \times [-250, 150]_{50}$

(c) The inequality $2x^2 + 3 \geq 4x$ is equivalent to $2x^2 - 4x + 3 \geq 0$. The entire graph of $y = 2x^2 - 4x + 3$ is above the x-axis, so the solution of the inequality $2x^2 + 3 \geq 4x$ is the set of all real numbers (Figure 2.70).

Polynomial Inequalities

Inequalities that involve higher-degree polynomials can be solved graphically in the same way as linear and quadratic inequalities.

EXAMPLE 4   Solving a Third-Degree Polynomial Inequality Graphically

Solve $5x^3 > 20x$ graphically.

SOLUTION
The zero form of the inequality is $5x^3 - 20x > 0$, so the corresponding function is $f(x) = 5x^3 - 20x$. The graph of f was discussed in Example 1a of Section 2.3 and is shown in Figure 2.71. The points that are above the x-axis correspond to the solution of the inequality, so the solution is

$$-2 < x < 0 \qquad \text{or} \qquad x > 2$$

EXAMPLE 5   Solving a Fourth-Degree Polynomial Inequality Graphically

Solve: $(x + 1)(x + 5)(x - 4)^2 \leq 0$

SOLUTION
The corresponding function is $f(x) = (x + 1)(x + 5)(x - 4)^2$ (Figure 2.72). The graph of this function has x-intercepts at $(-5, 0)$, $(-1, 0)$, and $(4, 0)$; the point $(4, 0)$ is also a turning point. The solution to the inequality corresponds to the points on or below the x-axis. Thus, the solution to $(x + 1)(x + 5)(x - 4)^2 \leq 0$ is

$$-5 \leq x \leq -1 \qquad \text{or} \qquad x = 4$$

Polynomial equations and inequalities are discussed more completely in the next chapter.

Absolute Value Inequalities

In Section 1 (page 2), we solved basic inequalities using algebraic methods based on facts about absolute values and inequalities. These facts are repeated here:

Absolute Value

If $a > 0$, then

$$|u| < a \quad \text{if and only if} \quad -a < u < a$$
$$|u| > a \quad \text{if and only if} \quad u < -a \quad \text{or} \quad u > a$$
$$|u| = a \quad \text{if and only if} \quad u = -a \quad \text{or} \quad u = a$$

The next example shows both an algebraic solution and a graphical solution to an absolute value inequality.

■ EXAMPLE 6   Solving an Absolute Value Inequality

Consider the absolute value inequality: $|x - 6| < 0.4$

(a) Solve the inequality by using algebraic methods.
(b) Solve the inequality graphically.

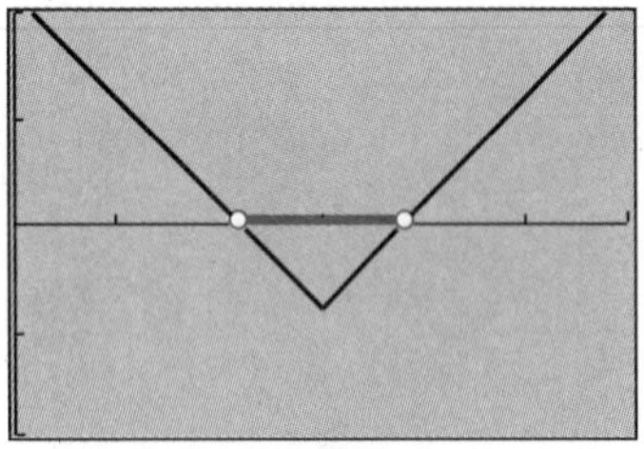

SOLUTION

(a) The inequality $|x - 6| < 0.4$ is equivalent to $-0.4 < x - 6 < 0.4$. By adding 6 to each of the three members of the compound inequality, we get

$$5.6 < x < 6.4$$

(b) The zero form of the inequality is $|x - 6| - 0.4 < 0$. The intercepts of the graph of $f(x) = |x - 6| - 0.4$ appear to be at $x = 5.6$ and $x = 6.4$ (Figure 2.73). [Pause here and verify this conjecture by showing that $f(5.6) = 0$ and $f(6.4) = 0$.] The points on the graph that are below the x-axis correspond to the values of x between these intercepts. It follows that the solution is

$$5.6 < x < 6.4 \qquad ■$$

Absolute value inequalities more complicated than the one solved in Example 6 usually are solved best by graphical methods.

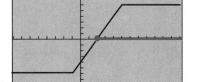

Figure 2.73
$y = |x - 6| - 0.4$
$[4.5, 7.5]_{0.5} \times [-1, 1]_{0.5}$

Figure 2.74
$y = |x + 1| - |x - 5|$
$[-8, 12] \times [-10, 10]$

■ **NOTE:** *Exercise 55 shows how this inequality can be solved by algebraic methods.*

■ EXAMPLE 7   Solving an Inequality with Two Absolute Values

Solve the absolute value inequality: $|x + 1| - |x - 5| \geq 0$

SOLUTION

The graph of $f(x) = |x + 1| - |x - 5|$ appears to have an x-intercept at $x = 2$ (Figure 2.74).

This conjecture is verified by showing that $f(2) = 0$:

$$f(2) = |2 + 1| - |2 - 5| = 3 - 3 = 0$$

The solution of $|x + 1| - |x - 5| \geq 0$ is $x \geq 2$.   ■

Applications

Inequalities play a role similar to equations in solving applied problems. For example, scientific experiments often involve numbers that are the result of measurements. As such, these numbers are approximations that are only as accurate as the measuring tool used. For example, if the length l of a metal rod is measured to be 87 centimeters by a meter stick, we know that its length is probably very close to 87 centimeters, but not exactly equal. The absolute value of the difference between the actual length and the measured length $|l - 87|$ is the *absolute error* of the measure. The *relative error* of the measurement is the ratio of the absolute error and the actual value,

$$\frac{|l - 87|}{l}$$

This quantity is usually expressed as a percent.

EXAMPLE 8   Application: Evaluating a Scientific Instrument

A manufacturer of scientific instruments tests its balance scales for accuracy by measuring an object with a mass of 20.000 grams. In order to pass the test, the relative error must be less than 0.05%. What is the acceptable range of measures for the scale?

SOLUTION

Suppose that the scale measures the mass of the object to be m grams. The relative error in this measurement is given by

$$\frac{|20.000 - m|}{20.000}$$

In order for the relative error to be less than 0.05%, m must be a solution to the inequality

$$\frac{|20.000 - m|}{20.000} < 0.0005$$

Multiplying each side of this inequality by 20.000 gives $|20.000 - m| < 0.01$. The solution to this inequality is shown with the graph of

$$y = |20 - x| - 0.01$$

in Figure 2.75. We leave it to you to show that the solution is

$$19.99 < m < 20.01$$

That is, in order for the scale to pass the test, the measured value must be between 19.99 and 20.01 grams.

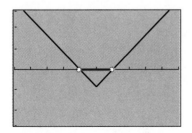

Figure 2.75
$y = |20 - x| - 0.01$
$[19.95, 20.05]_{0.01} \times [-0.03, 0.03]_{0.01}$

EXERCISE SET 2.4

A

1. Solve each inequality using the graph of f in Figure 2.76.
(a) $f(x) > 0$ (b) $f(x) < 0$
(c) $f(x) \geq 0$ (d) $f(x) \leq 0$

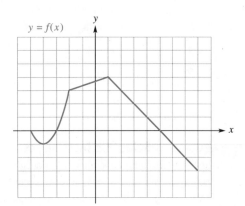

$y = f(x)$

Figure 2.76

2. Solve each inequality using the graph of g in Figure 2.77.
(a) $g(x) > 0$ (b) $g(x) < 0$
(c) $g(x) \geq 0$ (d) $g(x) \leq 0$

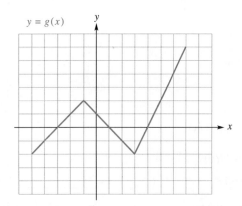

$y = g(x)$

Figure 2.77

In Exercises 3–8, solve each linear inequality graphically by sketching its corresponding graph.

3. $3x - 4 > 11$ **4.** $\frac{1}{2}x + 3 > 4$

5. $4(x - 2) + 3 \leq 3x - 8$ **6.** $3(x - 4) \leq x - 8$

7. $x(x - 3) + 5 \geq x^2 - 2x - 3$

8. $(2x + 1)(2x - 1) \geq 4(x + 3)^2 + 11$

In Exercises 9–16, solve each quadratic inequality using its corresponding graph.

9. $(x - 5)(x + 2) > 0$ **10.** $x(x + 3) \leq 0$

11. $x^2 - 2x - 3 \leq 0$ **12.** $3x^2 - x - 2 \geq 0$

13. $-\frac{1}{2}x^2 - 2x - 6 < 0$ **14.** $-2x^2 - x - 3 \leq 0$

15. $x(x - 4) \geq 21$ **16.** $x(2 - x) > 15$

In Exercises 17–24, solve each absolute value inequality by:
(a) Using algebraic methods
(b) Using its corresponding graph.

17. $|x - 3| < 0.6$ **18.** $|x + 4| < 0.5$

19. $|2x - 3| \geq 5$ **20.** $|\frac{1}{2}x + 1| \leq 3$

21. $3|x + 2| > 18$ **22.** $\frac{2}{3}|x - 4| < 8$

23. $|x - 6| \leq 0$ **24.** $|x - 5| > 0$

B

In Exercises 25–42, solve each inequality graphically.

25. $x^4 \leq 36 - 9x^2$ **26.** $x^3 > x^2 + 6x$

27. $x^5 \leq x^3$ **28.** $x^4 > -x$

29. $(2x - 1)^2 \geq (x + 3)^2$ **30.** $(x + 3)^2 \geq 2x^2$

31. $(x + 3)(x - 2)(x + 7)^2 \leq 0$

32. $(x + 1)^2(x - 6)^2(x + 4) \leq 0$

33. $\frac{3}{4}|x - 5| - 2 < 10$ **34.** $2|x - 5| + 7 < 19$

35. $|x^2 - 8| < 1$ **36.** $|x^2 - 1| < 2$

37. $|x| < |x + 5|$ **38.** $|x - 1| \geq |x + 1|$

39. $\sqrt{x - 4} \leq 3$ **40.** $\sqrt{6 - x} \geq 2$

41. $\sqrt{x^2 - 4} + x > 2$

42. $\sqrt{x^2 + 1} + 2x > 3x - 1$

In Exercises 43–48, determine the domain of each function.

43. $f(x) = \sqrt{x - 2}$ **44.** $g(x) = \sqrt{5 - x}$

45. $P(x) = \sqrt{x^2 - 14}$ **46.** $Q(x) = \sqrt{6 - x^2}$

47. $h(x) = \dfrac{2}{\sqrt{x^2 - 3x - 4}}$ **48.** $h(x) = \dfrac{-3}{\sqrt{|x| - 4}}$

49. A ball tossed up in the air is $64t - 16t^2$ feet above the ground t seconds after it is tossed. During what period of time is the ball above 48 feet?

50. A model rocket is launched upward so that t seconds after launching, the height of the rocket (in feet) is given by $h = 320t - 16t^2$. During what periods of time is the rocket less than 1200 feet above the ground?

51. Jamal finds that the profit per week earned by his word processing business is a function of the number of dollars spent on advertising per week. Specifically, if he spends x dollars per week, then his profit is given by $P = -0.2x^2 + 22.6x + 124$.

     (a) What amounts spent on advertising will generate a profit?

         HINT Quick Review: Jamal earns a profit if $P > 0$.

     (b) What should Jamal spend (to the nearest dollar) to maximize his profit?

52. Acme Car Rentals finds that the cost of maintenance over the life of an individual car is a function of its *service interval*, the distance the car travels between regular servicing. If the car's service interval is x thousand miles, then the cost of maintenance (in dollars) is

$$C = 13.5x^2 - 108x + 4094.5$$

State law requires that the car's service interval cannot exceed 15,000 miles.

     (a) For what range of service intervals does the cost of maintenance not exceed $3600, assuming the state law is not violated?

     (b) What service interval minimizes the cost of maintenance?

53. A scientific instrument uses a laser beam to measure distances. The instrument is tested for accuracy by measuring a distance of 160.000 centimeters. In order to pass the test, the measurement must have a relative error less than 0.001%. What is the acceptable range of measures for the instrument?

54. A *cesium clock* is used by physicists to measure time. A new cesium clock is tested by measuring a specified time with both the new clock and a clock that is known to be accurate. If the accurate clock measures the time period to be 1.25004 seconds, what is the acceptable range of times for the new clock so that the relative error is less than 0.005%?

C

55. **(a)** Suppose that a and b are two nonnegative real numbers. Show that $a \ge b$ if and only if $a^2 \ge b^2$.
         HINT: Rewrite $a^2 \ge b^2$ as $a^2 - b^2 \ge 0$ and factor.

     (b) Use part (a) to solve $|x + 1| - |x - 5| \ge 0$ algebraically. Compare your solution to Example 7.

56. Show that if the sum of two numbers is 1, their product cannot exceed $\frac{1}{4}$.

57. Solve each inequality by inspection. Explain your reasoning.

     (a) $\sqrt{x^3 + 2x - \dfrac{5}{x}} < 0$ **(b)** $(2x^5 - 4x^2 + 3)^4 \ge 0$

     (c) $\left| 2 + \dfrac{3 - x}{x} \right| \le -5$

58. The *arithmetic mean* of two nonnegative real numbers a and b is defined by $(a + b)/2$. The *geometric mean* of a and b is defined by $\sqrt{ab}$. This exercise outlines a proof that the arithmetic mean of two nonnegative numbers is greater than or equal to their geometric mean. (This inequality is called the *arithmetic–geometric mean inequality*.)

     (a) Explain why $(\sqrt{a} - \sqrt{b})^2 \ge 0$ with equality if and only if $a = b$.

     (b) Use part (a) to show that $a - 2\sqrt{ab} + b \ge 0$ and thus $(a + b)/2 \ge \sqrt{ab}$. Also, show that $(a + b)/2 = \sqrt{ab}$ if and only if $a = b$.

     (c) Verify the identity for $a = 12$, $b = 3$ and for $a = 8$, $b = 8$.

THE APPLICATION OF MATHEMATICS: MODELING

The primary purpose of studying algebra is to solve applied problems. The number and variety of applied problems that can be solved efficiently by constructing and solving an algebraic equation is truly amazing. Unfortunately, because these applied problems are so varied, there is no one specific process that works for solving all applied problems. In this section, we will concentrate on developing strategies to attack applied problems by solving some representative examples.

NOTE: *Although cleverness helps, the most important tools in solving applied problems are effort and persistence.*

Mathematical Models and Solving Problems

An equation that can be used to solve an applied problem is a *mathematical model* for the problem. As most experienced problem-solvers will acknowledge, developing a model for an applied problem is the most important and difficult part of its solution. In Example 1, we use a *verbal equation* to develop the model, as we have done in several previous applications.

EXAMPLE 1   Application: Dividing a Commission

Ted, Pham, and Maria form a partnership to work on a sound design project for a recording studio. Because of their different abilities and responsibilities in the project, they decide on the following method for sharing the commission. Ted receives the smallest share, Pham receives 10% more than Ted, and Maria receives 20% more than Ted. The entire commission is $2046. Find each member's share of the commission.

SOLUTION
Pause here to read the problem again. Can you see the connections between the shares of the three partners and the total commission? These quantities are such that

$$\left(\begin{array}{c}\text{Ted's share of} \\ \text{the commission}\end{array}\right) + \left(\begin{array}{c}\text{Pham's share of} \\ \text{the commission}\end{array}\right) + \left(\begin{array}{c}\text{Maria's share of} \\ \text{the commission}\end{array}\right) = \left(\begin{array}{c}\text{Total} \\ \text{commission}\end{array}\right)$$

This relation, a hybrid of an algebraic equation and an English sentence, is the verbal equation for the problem.

The next step is to translate this verbal equation to an algebraic equation with one variable. One of these quantities—the total commission—is given in the problem. The other three quantities—the shares of the three partners—are left for us to find. Pham's share and Maria's share are computed from Ted's share. Specifically, Pham's share is 10% more than Ted's, which is the same as saying that Pham's share is 110% of Ted's share. Likewise, Maria's share is 120% of Ted's. Suppose that we let

$$x = \text{Amount of Ted's share}$$

It follows that

$$1.10x = \text{Amount of Pham's share}$$
$$1.20x = \text{Amount of Maria's share}$$

We substitute these expressions, along with the given total commission, into the verbal equation to determine the mathematical model for the problem:

$$\left(\begin{array}{c}\text{Ted's share of}\\\text{the commission}\end{array}\right) + \left(\begin{array}{c}\text{Pham's share of}\\\text{the commission}\end{array}\right) + \left(\begin{array}{c}\text{Maria's share of}\\\text{the commission}\end{array}\right) = \left(\begin{array}{c}\text{Total}\\\text{commission}\end{array}\right)$$
$$x \qquad + \qquad 1.10x \qquad + \qquad 1.20x \qquad = \qquad 2046$$

The next step is to solve this algebraic equation:

$$x + 1.10x + 1.20x = 2046$$
$$3.3x = 2046$$
$$x = \frac{2046}{3.3} = 620$$

We interpret $x = 620$ as meaning Ted's share is \$620. Pham's share (110% of \$620) is \$682. Maria's share (120% of \$620) is \$744.

CHECK: We can check our results by adding the shares to see if together they equal the total commission:

$$620 + 682 + 744 = 2046$$

The sum equals the total commission, so the shares are correct as stated above.

The steps used to solve the applied problem in Example 1 illustrate the general steps for solving applied problems.

Solving Applied Problems

Step 1: *Understand the problem*
Read and reread the problem carefully. Try explaining the problem to someone else in your own words to see if you understand it. Draw a sketch, if possible, to illustrate the problem.

Step 2: *Develop an equation*
Decide which quantities are important to the problem. Write a verbal equation that relates these quantities. Fill in the quantities that you know. Choose a variable and write the unknown quantities in terms of this variable.

Step 3: *Solve the equation*
Use algebraic and/or graphing techniques to solve the equation. Interpret the solution to the equation to answer the question asked in the problem.

Step 4: *Check your answer*
Make sure your answer satisfies the requirements of the problem.

These steps provide a framework to organize your thoughts and to get you started. You might need to modify these steps to fit the particular problem and your own style.

Models Using Linear Equations

Linear equations serve as mathematical models for a large number of applied problems.

● EXAMPLE 2 Application: Evaluating Travel Times

A truck leaves a warehouse traveling 40 miles per hour. Soon after, the warehouse dispatcher notices that the truck's bill of goods was left behind. One hour after the truck left the warehouse, a courier with the bill is sent to catch the truck. How long will it take the courier to catch the truck if the courier travels 50 miles per hour?

SOLUTION

Both the truck and the courier travel from the warehouse to the point at which the courier catches the truck (Figure 2.78), so

$$\begin{pmatrix} \text{Distance traveled} \\ \text{by courier} \end{pmatrix} = \begin{pmatrix} \text{Distance traveled} \\ \text{by truck} \end{pmatrix}$$

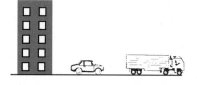

Figure 2.78

■ **NOTE:** *If an object travels a distance d at rate r for a time period t, then*

$$Distance = Rate \times Time$$

or

$$d = rt$$

The distance traveled by the truck is equal to the rate of the truck times the number of hours the truck travels. Similarly, the distance traveled by the courier is equal to the rate of the courier times the number of hours the courier travels:

$$\begin{pmatrix} \text{Distance traveled} \\ \text{by courier} \end{pmatrix} = \begin{pmatrix} \text{Distance traveled} \\ \text{by truck} \end{pmatrix}$$

$$\begin{pmatrix} \text{Rate of} \\ \text{courier} \end{pmatrix} \begin{pmatrix} \text{Time traveled} \\ \text{by courier} \end{pmatrix} = \begin{pmatrix} \text{Rate of} \\ \text{truck} \end{pmatrix} \begin{pmatrix} \text{Time traveled} \\ \text{by truck} \end{pmatrix}$$

There are four quantities in this verbal equation. The rates are given, 50 miles per hour for the courier and 40 miles per hour for the truck. The times are related because the truck travels 1 hour more than the courier. We let

$$t = \text{Number of hours traveled by the courier}$$
$$t + 1 = \text{Number of hours traveled by the truck}$$

Substituting these values and expressions into the verbal equation gives an algebraic equation to solve:

$$50t = 40(t + 1)$$
$$50t = 40t + 40$$
$$10t = 40$$
$$t = 4$$

Because $t = 4$, the courier travels 4 hours, and because $t + 1 = 5$, the truck travels 5 hours.

CHECK: To check the solution, we show that the two vehicles travel the same distance:

Distance traveled by truck: 40 miles per hour for 5 hours = 200 miles

Distance traveled by courier: 50 miles per hour for 4 hours = 200 miles

The courier catches the truck 4 hours after the courier leaves the warehouse.

■

The next example involves the computation of *simple interest*. The interest on a particular investment in 1 year is the product of the amount invested times the rate of interest.

■ **EXAMPLE 3   Application: Planning an Investment Portfolio**

A private investor divides a portfolio of $22,000 between a low-risk mutual fund paying 5% per year and a high-risk stock paying 12%. If the total interest on the portfolio in 1 year is $1800, determine the amount placed in each investment.

SOLUTION

From the statement of the problem, we get the verbal equation:

$$\left(\begin{array}{c}\text{Interest from}\\\text{mutual fund}\end{array}\right) + \left(\begin{array}{c}\text{Interest from}\\\text{stock}\end{array}\right) = \left(\begin{array}{c}\text{Total}\\\text{interest}\end{array}\right)$$

The interest in 1 year from each investment is the rate times the amount, so

$$\left(\begin{array}{c}\text{Rate of}\\\text{mutual fund}\end{array}\right)\left(\begin{array}{c}\text{Amount invested}\\\text{in mutual fund}\end{array}\right) + \left(\begin{array}{c}\text{Rate of}\\\text{stock}\end{array}\right)\left(\begin{array}{c}\text{Amount invested}\\\text{in stock}\end{array}\right) = \left(\begin{array}{c}\text{Total}\\\text{interest}\end{array}\right)$$

We let

$$m = \text{Amount invested in mutual fund}$$
$$22{,}000 - m = \text{Amount invested in stock}$$

Substituting these expressions and the given information into the last verbal equation gives this algebraic equation to solve:

$$0.05m + 0.12(22{,}000 - m) = 1800$$
$$0.05m + 2640 - 0.12m = 1800$$
$$-0.07m + 2640 = 1800$$
$$-0.07m = -840$$
$$m = 12{,}000$$

We conclude that the amount invested in the mutual fund is $12,000 and the amount invested in the stock is $10,000.

CHECK: We check the result:

$$5\% \text{ of } \$12{,}000 = \$\ 600$$
$$12\% \text{ of } \$10{,}000 = \underline{\$1200}$$
$$\$1800$$

The sum equals the total interest, so the amounts are correct.

Models Using Nonlinear Equations

Mathematical models in the form of nonlinear equations can often be solved using the techniques of Section 2.3. Using your graphing utility to verify solutions (as we did in Section 2.3) will help prevent errors in solving these equations.

EXAMPLE 4   Application: Designing a Landscape Project

The landscape plans for a backyard specify a lawn with an area of 1200 square feet surrounded by a planting strip of a constant width. The yard is a rectangle,

32 feet wide and 55 feet long. What are the dimensions of the lawn? How wide is the planting strip?

SOLUTION

A sketch of the yard is shown in Figure 2.79. Because the lawn is a rectangle,

$$\left(\begin{matrix}\text{Width of}\\\text{lawn}\end{matrix}\right)\left(\begin{matrix}\text{Length of}\\\text{lawn}\end{matrix}\right) = \left(\begin{matrix}\text{Area of}\\\text{lawn}\end{matrix}\right)$$

In Figure 2.79 we let

$$x = \text{Width of the planting strip}$$

and thus,

$$32 - 2x = \text{Width of lawn}$$
$$55 - 2x = \text{Length of lawn}$$

Substituting these values and expressions into the verbal equation gives

$$(32 - 2x)(55 - 2x) = 1200$$
$$1760 - 174x + 4x^2 = 1200$$
$$4x^2 - 174x + 560 = 0$$
$$2x^2 - 87x + 280 = 0$$
$$(x - 40)(2x - 7) = 0$$

$$x - 40 = 0 \quad\quad 2x - 7 = 0$$
$$x = 40 \quad\quad\quad x = \tfrac{7}{2}$$

We can discard 40 as an unreasonable width for the planting strip, because the yard is only 32 feet wide. Therefore, the width of the planting strip is $\frac{7}{2}$ feet.

Figure 2.80 shows the check of this solution to the problem. The dimensions of the lawn are 25 feet × 48 feet, and the planting strip is $3\frac{1}{2}$ feet wide. ■

The next example is typical of many applied problems that determine the results of mixing fluids of different concentrations.

■ EXAMPLE 5   Application: Mixing Acid Solutions

An artist needs a 30% acid solution for etching copper. How many milliliters of a 40% acid solution must be mixed with 200 milliliters of 10% acid solution to get the appropriate mixture?

SOLUTION

The statement of the problem is summarized in Figure 2.81. Notice that the figure is labeled with the concentrations and volumes of the solutions and the mixture. The volume of the 40% mixture is labeled as x milliliters because that is the quantity we are asked to find.

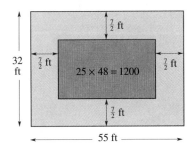

Figure 2.79

Figure 2.80

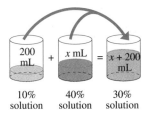

Figure 2.81

The concentration of an acid solution is the ratio of the amount of acid to the amount of solution. We write this relation for the mixture as a verbal equation:

$$\frac{\left(\begin{array}{c}\text{Amount of}\\\text{acid in mixture}\end{array}\right)}{\left(\begin{array}{c}\text{Total amount}\\\text{of mixture}\end{array}\right)} = \left(\begin{array}{c}\text{Concentration of}\\\text{acid in mixture}\end{array}\right)$$

This verbal equation is equivalent to

$$\frac{\left(\begin{array}{c}\text{Amount of acid}\\\text{in 10\% solution}\end{array}\right) + \left(\begin{array}{c}\text{Amount of acid}\\\text{in 40\% solution}\end{array}\right)}{\left(\begin{array}{c}\text{Total amount}\\\text{of 10\% solution}\end{array}\right) + \left(\begin{array}{c}\text{Total amount}\\\text{of 40\% solution}\end{array}\right)} = \left(\begin{array}{c}\text{Concentration of}\\\text{acid in mixture}\end{array}\right)$$

We are told that 10% of the 200 milliliters is acid, so

$$0.10(200) = \text{Amount of acid in 10\% solution}$$

If we let

$$x = \text{Total amount of 40\% acid solution}$$

as Figure 2.81 suggests, then

$$0.40x = \text{Amount of acid in 40\% solution}$$

The algebraic equation that results by substituting these values into the last verbal equation is

$$\frac{0.10(200) + 0.40x}{200 + x} = 0.30$$

or

$$\frac{20 + 0.40x}{200 + x} = 0.30$$

Multiplying each side by $200 + x$ gives a linear equation that can be solved by the usual methods:

$$(200 + x)\left(\frac{20 + 0.40x}{200 + x}\right) = (200 + x)0.30$$
$$20 + 0.40x = 60 + 0.30x$$
$$20 + 0.10x = 60$$
$$0.10x = 40$$
$$x = \frac{40}{0.10} = 400$$

This means that $x = 400$ milliliters of the 40% solution is needed.

CHECK: The amount of acid in the mixture is 10% of 200 milliliters plus 40% of 400 milliliters or $20 + 160 = 180$ milliliters of acid. The total amount of the mixture is $200 + 400 = 600$ milliliters of solution. The concentration of the mixture is $\frac{180}{600} = 0.30$, or 30%. This computation verifies our solution.

The equation

$$\frac{20 + 0.40x}{200 + x} = 0.30$$

in Example 5 is a *rational equation*, because the left side of the equation is a rational expression. (Rational equations are the focus of Section 3.5.) If we consider the concentration of acid in the mixture as a function C of the number of milliliters of 40% solution added, then our work in Example 5 tells us that

$$C(x) = \frac{20 + 0.40x}{200 + x}$$

The graph of C (Figure 2.82) contains the point $(400, 0.30)$, which represents the solution in Example 5. But this graph tells us much more about the problem than just its solution. Notice, for example, that as the 40% acid solution is added to the 200 milliliters of 10% acid solution, the concentration increases rapidly at first and then slows down, approaching 40%.

In the next two examples, we use the Pythagorean theorem.

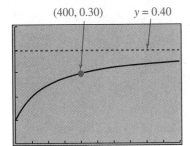

Figure 2.82
$$y = \frac{20 + 0.40x}{200 + x}$$
$[0, 1000]_{100} \times [0, 0.5]_{0.1}$

◼ EXAMPLE 6 Application: Installing a Communications Cable

A communications cable is to be installed from a broadcast station to a relay site and then to a remote receiver on an island 40 kilometers offshore from a point 80 kilometers down the shore from the broadcast station. The total length of cable can be no longer than 100 kilometers without a significant loss of signal, and the relay station is to be as far as possible from the broadcast station. How far from the broadcast station should the relay station be built?

SOLUTION
The verbal model for the problem is

$$\left(\begin{array}{c}\text{Length of cable}\\ \text{along the shore}\end{array}\right) + \left(\begin{array}{c}\text{Length of cable}\\ \text{across water}\end{array}\right) = \left(\begin{array}{c}\text{Total length}\\ \text{of cable}\end{array}\right)$$

From Figure 2.83, we let

$$x = \left(\begin{array}{c}\text{Length of cable}\\ \text{along the shore}\end{array}\right)$$

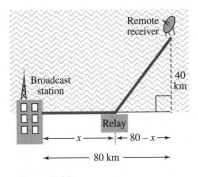

Figure 2.83

The total length of cable is 100 kilometers. The length of cable across the water is the hypotenuse of the triangle with legs 40 and $80 - x$, so by the Pythagorean theorem, we get

$$\left(\begin{array}{c}\text{Length of cable} \\ \text{across water}\end{array}\right) = \sqrt{40^2 + (80 - x)^2}$$

The verbal equation becomes

$$x + \sqrt{40^2 + (80 - x)^2} = 100$$
$$\sqrt{40^2 + (80 - x)^2} = 100 - x$$
$$40^2 + (80 - x)^2 = (100 - x)^2$$
$$1600 + 6400 - 160x + x^2 = 10{,}000 - 200x + x^2$$
$$8000 - 160x = 10{,}000 - 200x$$
$$40x = 2000$$
$$x = 50$$

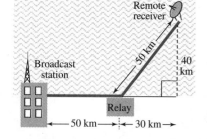

Broadcast station

Remote receiver

50 km

40 km

Relay

← 50 km → ← 30 km →

Figure 2.84

The relay station should be built 50 kilometers from the broadcast station. (The check is shown in Figure 2.84.)

The next example uses the formula for the volume of a cone.

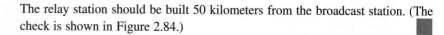 **EXAMPLE 7 Application: Constructing a Cone**

A cone is formed by cutting a quarter sector from a paper circle with radius 12 centimeters and attaching the edges formed by the cut (Figure 2.85). What is the volume of the resulting cone?

SOLUTION

The volume of a cone of radius r and height h is given by the formula $V = \frac{1}{3}\pi r^2 h$. The formula tells us that the problem of finding the volume of the cone hinges on finding its radius and height.

The circumference of the base is equal to $\frac{3}{4}$ of the circumference of the paper circle. The circumference of the paper circle is

$$2\pi\left(\begin{array}{c}\text{Radius of} \\ \text{paper circle}\end{array}\right) = 2\pi(12) = 24\pi$$

so the circumference of the base of the cone is

$$\tfrac{3}{4}(24\pi) = 18\pi$$

By solving $2\pi r = 18\pi$, we determine that $r = 9$, so the radius of the base is 9 centimeters.

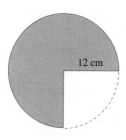

12 cm

h

r

Figure 2.85

Figure 2.85 and the Pythagorean theorem suggest that

$$h^2 + 9^2 = 12^2$$
$$h^2 + 81 = 144$$
$$h^2 = 63$$
$$h = \sqrt{63} \quad \text{or} \quad 3\sqrt{7}$$

It follows that the volume of the cone is given by

$$V = \tfrac{1}{3}\pi r^2 h = \tfrac{1}{3}\pi(9)^2(3\sqrt{7}) = 81\pi\sqrt{7}$$

The volume of the cone is $81\pi\sqrt{7}$ square centimeters (approximately 673.3 square centimeters).

EXERCISE SET 2.5

A

In Exercises 1–16, a verbal equation for each problem is given. Complete the solution of the problem by deriving and solving an equation, checking your solution, and writing a sentence that answers the question.

1. The sum of two consecutive numbers is 31. What are the numbers?

$$\begin{pmatrix} \text{First} \\ \text{number} \end{pmatrix} + \begin{pmatrix} \text{Second} \\ \text{number} \end{pmatrix} = \begin{pmatrix} \text{Sum of} \\ \text{numbers} \end{pmatrix}$$

2. The sum of three consecutive odd numbers is 57. What are the numbers?

$$\begin{pmatrix} \text{First} \\ \text{number} \end{pmatrix} + \begin{pmatrix} \text{Second} \\ \text{number} \end{pmatrix} + \begin{pmatrix} \text{Third} \\ \text{number} \end{pmatrix} = \begin{pmatrix} \text{Sum of} \\ \text{numbers} \end{pmatrix}$$

3. Hal's exam scores in his mathematics class are 81, 86, and 82. What must he score on his last exam to raise his average to 85?

$$\frac{\begin{pmatrix} \text{Sum of} \\ \text{test scores} \end{pmatrix}}{4} = \begin{pmatrix} \text{Hal's} \\ \text{average} \end{pmatrix}$$

4. Barbara's quiz scores in her physics class are 76 and 82. The final exam counts as two quizzes in determining her average. What must she score on her final exam to raise her average to 80?

$$\frac{\begin{pmatrix} \text{Sum of} \\ \text{quiz scores} \end{pmatrix} + 2\begin{pmatrix} \text{Exam} \\ \text{score} \end{pmatrix}}{4} = \begin{pmatrix} \text{Barbara's} \\ \text{average} \end{pmatrix}$$

5. Aldo's cat Max is twice as heavy as his dog Lefty. Together they weigh 48 pounds. How much does each of his animals weigh?

$$\begin{pmatrix} \text{Max's} \\ \text{weight} \end{pmatrix} + \begin{pmatrix} \text{Lefty's} \\ \text{weight} \end{pmatrix} = 48$$

6. Nguyen's regular pay rate is $16 per hour. His overtime pay rate is $24 per hour. His pay for 36 hours of work in 1 week is $700. How many overtime hours did Nguyen work?

$$\begin{pmatrix} \text{Regular} \\ \text{pay} \end{pmatrix} + \begin{pmatrix} \text{Overtime} \\ \text{pay} \end{pmatrix} = \begin{pmatrix} \text{Total} \\ \text{pay} \end{pmatrix}$$

7. The profit over the second quarter of the year for a pest control company is 10% more than its first quarter profit. The total profit over both quarters is $120,000. What is the profit in the first quarter?

$$\begin{pmatrix} \text{First quarter} \\ \text{profit} \end{pmatrix} + \begin{pmatrix} \text{Second quarter} \\ \text{profit} \end{pmatrix} = \begin{pmatrix} \text{Total} \\ \text{profit} \end{pmatrix}$$

8. Geoffrey's annual salary is 16% more than Scott's. To-
    gether, they earn $108,000. How much does each make?

    $$\left(\begin{array}{c}\text{Geoffrey's}\\\text{salary}\end{array}\right) + \left(\begin{array}{c}\text{Scott's}\\\text{salary}\end{array}\right) = \left(\begin{array}{c}\text{Total}\\\text{salary}\end{array}\right)$$

9. A rectangle is twice as long as it is wide. Its area is 128
    square meters. What are the dimensions of the rectangle?

    $$\left(\begin{array}{c}\text{Length of}\\\text{rectangle}\end{array}\right)\left(\begin{array}{c}\text{Width of}\\\text{rectangle}\end{array}\right) = \left(\begin{array}{c}\text{Area of}\\\text{rectangle}\end{array}\right)$$

10. The base of a triangle is 10 feet longer than its height. Its
    area is 28 square feet. What are the base and height of the
    triangle?

    $$\frac{1}{2}\left(\begin{array}{c}\text{Base of}\\\text{triangle}\end{array}\right)\left(\begin{array}{c}\text{Height of}\\\text{triangle}\end{array}\right) = \left(\begin{array}{c}\text{Area of}\\\text{triangle}\end{array}\right)$$

11. A water-softening unit needs a 20% salt solution to flush
    its reservoir. How much of a 10% solution must be
    mixed with 30 liters of a 40% solution to get the proper
    solution?

    $$\frac{\left(\begin{array}{c}\text{Amount of}\\\text{salt in mixture}\end{array}\right)}{\left(\begin{array}{c}\text{Total amount}\\\text{of mixture}\end{array}\right)} = \left(\begin{array}{c}\text{Concentration of}\\\text{salt in mixture}\end{array}\right)$$

12. A cooling system in an automobile requires 2.5 gallons of
    a 50% antifreeze solution. Presently, the system has a
    30% solution. How much of the present solution must be
    replaced with pure antifreeze to bring the system up to the
    recommended concentration?

    $$\frac{\left(\begin{array}{c}\text{Amount of}\\\text{antifreeze in mixture}\end{array}\right)}{\left(\begin{array}{c}\text{Total amount}\\\text{of mixture}\end{array}\right)} = \left(\begin{array}{c}\text{Concentration of}\\\text{antifreeze in mixture}\end{array}\right)$$

13. A car leaves an intersection traveling 50 miles per hour.
    Two hours later, a motorcycle leaves the same intersec-
    tion traveling 35 miles per hour in the opposite direction.
    How long after the motorcycle leaves the intersection are
    the vehicles 220 miles apart?

    $$\left(\begin{array}{c}\text{Distance traveled}\\\text{by car}\end{array}\right) + \left(\begin{array}{c}\text{Distance traveled}\\\text{by motorcycle}\end{array}\right) = \left(\begin{array}{c}\text{Total}\\\text{distance}\end{array}\right)$$

14. An airplane flies at 80 miles per hour for the first third of
    a trip and then 120 miles per hour for the rest of the trip.
    The entire trip is 400 miles. How long does the trip take?

    $$\left(\begin{array}{c}\text{Distance traveled}\\\text{at slower speed}\end{array}\right) + \left(\begin{array}{c}\text{Distance traveled}\\\text{at faster speed}\end{array}\right) = \left(\begin{array}{c}\text{Total}\\\text{distance}\end{array}\right)$$

15. An investor plans to divide a portfolio of $25,000 be-
    tween an insured savings account paying 3% per year and
    high-risk bonds paying 9%. If the total interest on the
    portfolio in 1 year is $1380, determine the amount to be
    placed in each investment.

    $$\left(\begin{array}{c}\text{Interest from}\\\text{savings account}\end{array}\right) + \left(\begin{array}{c}\text{Interest from}\\\text{bonds}\end{array}\right) = \left(\begin{array}{c}\text{Total}\\\text{interest}\end{array}\right)$$

16. An investor divides a portfolio between a low-risk mutual
    fund paying 4% per year and a high-risk stock paying
    10% such that the amount invested in the mutual fund is
    three times the amount invested in the stock. If the total
    interest on the portfolio in 1 year is $6780, determine the
    amount placed in each investment.

    $$\left(\begin{array}{c}\text{Interest from}\\\text{mutual fund}\end{array}\right) + \left(\begin{array}{c}\text{Interest from}\\\text{stock}\end{array}\right) = \left(\begin{array}{c}\text{Total}\\\text{interest}\end{array}\right)$$

B

In Exercises 17–38, solve the applied problems. (Be sure to set
up a verbal equation.)

17. An investor plans to divide a portfolio of $12,000 evenly
    between an insured savings account and high-risk bonds.
    The bonds pay at a rate that is twice the rate of the sav-
    ings account. If the total interest on the portfolio in 1 year
    is $1260, determine the rate of each investment.

18. A retiree divides his pension portfolio of $270,000 be-
    tween a conservative mutual fund paying at a rate of 4%
    and high-risk bonds paying at a rate of 8%. Both invest-
    ments pay the same amount of interest annually. How
    much is invested in each investment?

19. When the length of a rectangle is decreased by 4 meters
    and its width is increased by 3 meters the result is a
    square that has the same area as the original rectangle.
    What are the dimensions of the original rectangle?

20. A rectangle has a perimeter of 84 centimeters and an area
    of 405 square centimeters. What are its dimensions?

21. An open box 4 inches high is to be made by cutting squares from the corners of a square piece of cardboard and folding up the flaps. How large must the original piece of cardboard be if the volume of the box is to be 960 cubic inches?

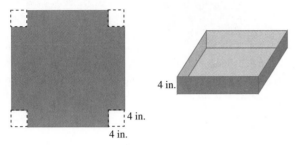

22. In a right triangle with a hypotenuse of 20 inches, the longer leg is twice the length of the shorter leg. What are the lengths of the legs?

23. A concrete walkway with a constant width is to be constructed along two adjacent sides of a rectangular corner lot that is 50 meters × 24 meters. The total area of the walkway is 408 square meters. How wide is the walkway?

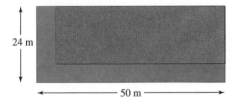

24. A rectangle of printed material with area 90 square inches is to be placed on a poster so that there are 3 inch margins at the top and bottom and 2 inch margins on the sides. The area of the entire poster is 208 square inches. What are the dimensions of the poster?

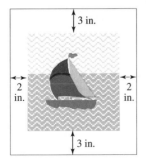

25. A plane that flies at a rate of 255 miles per hour with no wind flies 150 miles with a tailwind. When the plane turns around and flies against the wind, it travels only 105 miles in the same amount of time. What is the speed of the wind?

26. A kayaker travels upstream 12 miles and then returns. The entire trip takes 2 hours. With no current, the kayaker paddles at a rate of 15 miles per hour. How fast is the current of the stream?

27. An automobile manufacturer makes a standard-sized automobile rated at 28.5 miles per gallon and a compact rated at 34.5 miles per gallon. How many of each type should be manufactured so that 1,000,000 cars that have an average rate of 30 miles per gallon are produced?

28. Gasoline is rated by its octane content. A petroleum distributor wishes to make 200,000 gallons of 89 octane gasoline by mixing 82 octane gasoline with 90 octane gasoline. How much of each should the distributor use?

29. A stockbroker charges a fee of $50 plus 1.2% of the amount of the transaction. On a particular transaction, the fee was $138.80. What was the amount of the transaction?

30. A mortgage lender charges a fixed fee of $395 plus 2% of the amount of the loan. On a particular loan, the fee was $2945. What was the amount of the loan?

31. A train crosses a highway traveling north at 60 miles per hour. One hour later, a car traveling east at 45 miles per hour on the highway crosses the train track. To the nearest 0.1 hour, after how long will the train and car be 100 miles apart?

32. Two bicyclists are each 50 kilometers from an intersection—one directly north, the other directly south. Each rides toward the intersection at 30 kilometers per hour. After how long are they 40 kilometers apart?

33. A window is to be formed by placing an isosceles right triangle on top of a rectangle, as shown at the top of the next page. Overall, the window is to be 4 feet tall, and due to energy conservation restrictions, the area of the

window is to be 12 square feet. How wide should the window be?

4 ft

34. A farmer plans to construct two square livestock pens that share a side. The total area is to be 120 square meters. What are the dimensions of the pens?

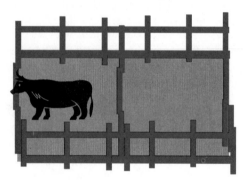

35. A cylindrical can 5 inches high is constructed such that its volume is 200 cubic inches. What is the radius of the can?

36. A balloon with a radius of 10 centimeters is inflated until its volume is doubled. What is the radius of the balloon after it is inflated?

C

37. A plane that flies at a rate of 200 miles per hour is used for a round trip between two cities that are 600 miles apart. On one particular round trip, the plane had a tailwind on the first leg of the trip that acted as a headwind on the return trip. Overall, this round trip took 24 minutes longer than a trip with no wind. What was the speed of the wind?

38. Anne drops a rock down a well and hears it splash 6.6 seconds later. Given that the rock falls $16t^2$ feet in t seconds, and that sound travels 550 feet per second, how deep is the well?

● QUICK REFERENCE

Topic	Page	Remarks
Slope of a line	192	The slope m of a nonvertical line passing through the points (x_1, y_1) and (x_2, y_2) is given by $$m = \frac{y_2 - y_1}{x_2 - x_1}$$ The slope of a vertical line is undefined.
Slope–intercept form	197	The slope–intercept form of a linear function f is $$f(x) = mx + b$$ where m and b are real numbers. The graph is a line with slope m and y-intercept $(0, b)$.

Topic	Page	Remarks
Point–slope form	198	The point–slope form of a linear function f is $$f(x) = m(x - h) + k$$ where m, h, and k are real numbers. The graph is a line with slope m, and it passes through the point (h, k).
Horizontal line	200	The graph of the equation $y = k$ is a horizontal line with y-intercept $(0, k)$.
Vertical line	200	The graph of the equation $x = h$ is a vertical line with y-intercept $(h, 0)$.
General form of an equation of a line	201	An equation in the form $$Ax + By = C$$ for some real numbers A, B, and C (A and B not both 0), is the general form of an equation of a line.
Parallel lines	202	Two nonvertical lines with slopes m_1 and m_2 are parallel if and only if $m_1 = m_2$.
Perpendicular lines	203	Two nonvertical lines with slopes m_1 and m_2 are perpendicular if and only if $m_1 m_2 = -1$.
Standard form of a quadratic function	213	The standard form of a quadratic function f is $$f(x) = a(x - h)^2 + k$$ where a, h, and k are real numbers ($a \neq 0$). The graph of f is a parabola with the shape of $y = ax^2$ and vertex (h, k).
General form of a quadratic function	214	The general form of a quadratic function f is $$f(x) = ax^2 + bx + c$$ where a, b, and c are real numbers ($a \neq 0$). The graph of f is a parabola with the shape of $y = ax^2$. The vertex of the parabola has x-coordinate $$h = -\frac{b}{2a}$$
Maximum or minimum value of a quadratic function	217	For the quadratic function $f(x) = a(x - h)^2 + k$ or $f(x) = ax^2 + bx + c$, the y-coordinate of the vertex is the: Minimum value of the function if $a > 0$ Maximum value of the function if $a < 0$
Zero factor property	224	If $AB = 0$, then $A = 0$ or $B = 0$. This can be extended to more than two factors; for example, if $ABC = 0$, then $A = 0$ or $B = 0$ or $C = 0$.

Continued

Topic	Page	Remarks
Solving an equation by taking roots	227	See Example 3 on page 226. Suppose that n is a positive integer, a is a positive number, and u is an algebraic expression.
		If n is odd, then
		$$u^n = a \quad \text{implies that} \quad u = \sqrt[n]{a}$$
		$$u^n = -a \quad \text{implies that} \quad u = -\sqrt[n]{a}$$
		If n is even, then
		$$u^n = a \quad \text{implies that} \quad u = \pm \sqrt[n]{a}$$
		$$u^n = -a \quad \text{has no real solutions for } u$$
Extraneous solutions	227–228	Raising each side of an equation to a higher power and certain other operations can lead to another equation that is not equivalent to the original equation. Solutions of the resulting equation that do not satisfy the original equation are extraneous solutions. See Example 4 on page 228.
Mathematical model	244	An equation or inequality that can be used to solve an applied problem is a mathematical model.

■ MISCELLANEOUS EXERCISES

In Exercises 1–6, sketch the graph of each line described, and determine its equation in slope–intercept form.

1. The line has slope $m = -\frac{2}{3}$ and y-intercept $(0, 7)$.

2. The line has slope $m = \frac{1}{5}$ and y-intercept $(0, -3)$.

3. The line has slope $m = 4$ and y-intercept $(0, -6)$.

4. The line has y-intercept $(0, -7)$ and passes through the point $(3, 5)$.

5. The line is parallel to the graph of $y = 2x$ and has an x-intercept $(-10, 0)$.

6. The line is perpendicular to the graph of $y = 2x$ and has an x-intercept $(-10, 0)$.

In Exercises 7–12, determine the slope–intercept form of the given linear equation and sketch its graph.

7. $2x - 6y = 18$

8. $5x + 2y = 4$

9. $\frac{x}{2} + \frac{y}{8} = 1$

10. $x - \frac{y}{4} = 1$

11. $200x = 1000 + y$

12. $y + 0.05x = 0.2$

In Exercises 13–18:

(a) Rewrite each quadratic function f in standard form.

(b) Determine the graph of f.

(c) Determine the intercepts and vertex of the graph of f.

13. $f(x) = x^2 + 4x - 21$

14. $f(x) = x^2 - 2x - 8$

15. $f(x) = x^2 + 5x + \frac{5}{2}$

16. $f(x) = x^2 + 3x - 2$

17. $f(x) = \frac{1}{4}x^2 + x - 2$

18. $f(x) = 2x^2 + 12x - 5$

In Exercises 19–34, use a graphing utility to determine the corresponding graph of each equation, and solve each equation algebraically.

19. $(x + 2)^2 = 13$

20. $(x - 4)^2 = 40$

21. $(x^2 - 4x - 8)(2x - 5) = 0$

22. $(2x^2 - 7)(x + 3) = 0$

23. $(8 - x)^3 = 16$

24. $(x^2 - 3)^3 = 125$

25. $\frac{1}{4}\sqrt{x + 24} = 2$

26. $\sqrt[3]{4x + 7} = 3$

27. $|x - 3| = 9$

28. $|x^2 - 5x| = 6$

29. $\sqrt{2x+1} + \sqrt{x} = 5$

30. $(x-2) + 6\sqrt{x-2} = 7$

31. $2x^{2/5} = 4x^{1/5} - 2$

32. $x^{-1/2} = 2x^{1/2} - 1$

33. $\sqrt{x-1} = |x-5|$

34. $\sqrt{4-x} = |x+2|$

In Exercises 35–42, solve each inequality graphically.

35. $x^2 + 2x \le 3$

36. $3x^2 - 2 > x$

37. $2|x + \frac{1}{2}| \le 5$

38. $|x^2 - 6| > 10$

39. $|x| \ge |x+4|$

40. $\sqrt{7-x} < 6$

41. $(x^2 + 7)(3-x)(x+4) > 0$

42. $(x+1)^2(x-2)(x+7) \le 0$

43. A train leaves a town that is 2800 feet above sea level, and travels on a straight track with a grade of 3%. (A grade of 3% means that the road changes 3 feet vertically over a 100 foot horizontal run.) At the end of the trip, the train is at 4000 feet above sea level. Determine the altitude of the train as a linear function of x, the horizontal distance traveled. How far did the train travel (to the nearest 0.1 mile)?

44. A publishing company buys a computer for $3800 and plans to depreciate it using the straight-line method so that after 5 years, its value is $1100.
    (a) Determine a linear function F that gives the depreciated value of the computer after x years. Sketch the graph of F.
    (b) What is the depreciated value of the computer after 1 year?

45. A manufacturer of compact disc players finds that the average cost per unit (in dollars) of making x units per week is given by the function $C(x) = 0.05x^2 - 13x + 1500$. How many units should be manufactured per week to minimize the average cost?

46. Suppose that $a + b$ is 12. What is the minimum value of $a^2 + b^2$?

47. If a ball is thrown directly up in the air from a height of h_0 feet with an initial velocity of v_0 feet per second, the height (in feet) t seconds after it is thrown is given by

$$H(t) = -16t^2 + v_0 t + h_0$$

Now, from ground level, a ball is tossed up with an initial velocity of 48 feet per second.
(a) At what time will the ball hit the ground?
(b) Over what interval of time is the ball above 27 feet?
(c) How long does it take for the ball to reach its maximum height? How high does the ball go?

48. A ball is thrown upward, staying aloft for 4 seconds. What is its initial velocity? (See Exercise 47.)

49. The sum of three consecutive odd numbers is 129. What are the numbers?

$$\left(\begin{array}{c}\text{First}\\\text{number}\end{array}\right) + \left(\begin{array}{c}\text{Second}\\\text{number}\end{array}\right) + \left(\begin{array}{c}\text{Third}\\\text{number}\end{array}\right) = \left(\begin{array}{c}\text{Sum of}\\\text{numbers}\end{array}\right)$$

50. Orlando's quiz scores in his mathematics class are 96 and 88. The final exam counts as three quizzes in determining his average. What must he score on his final exam to raise his average to 90?

$$\frac{\left(\begin{array}{c}\text{Sum of}\\\text{quiz scores}\end{array}\right) + 3\left(\begin{array}{c}\text{Exam}\\\text{score}\end{array}\right)}{5} = \left(\begin{array}{c}\text{Orlando's}\\\text{average}\end{array}\right)$$

51. A bottle of mineral water contains 24 ounces of liquid with 30% calcium content. How many ounces of this liquid should be poured out and replaced with a 55% calcium solution to obtain a drink that is 40% calcium?

52. How many quarts of acid must be added to 20 quarts of a solution that is 55% acid to obtain a new solution that is 70% acid?

53. A retiree divides his pension portfolio of $24,000 between a conservative mutual fund paying at a rate of 4% and moderate-risk bonds paying at a rate of 6%. Both investments pay the same amount of interest annually. How much is invested in each investment?

54. An investor plans to divide a portfolio of $10,000 evenly between an insured savings account and high-risk bonds. The bonds pay at a rate that is twice the rate of the savings account. If the total interest on the portfolio in 1 year is $726, determine the rate of each investment.

55. Two cars, each traveling at a constant speed, leave simultaneously from the same town. One heads south and the other heads east. The eastbound car travels 10 miles per hour faster than the southbound car. After 2 hours, the cars are 100 miles apart. How fast is each car traveling?

56. A boat, capable of traveling 40 miles per hour in still water, travels 96 miles up a river from camp and then returns downstream to camp. The entire trip takes 5 hours. What is the speed of the current of the river?

CHAPTER TEST

1. Determine the slope–intercept form of the linear function f such that $f(3) = -2$ and $f(-6) = 10$.

2. Determine the slope–intercept form of the linear equation $2x - 3y = 6$ and sketch its graph.

3. Determine the area of the triangle bounded by the graph of $3x + 22y = 33$ and the coordinate axes.

4. Rewrite the function $f(x) = x^2 - 2x - 1$ in the standard form $f(x) = a(x - h)^2 + k$. Determine the vertex and intercepts of the graph of f, and determine a graph of f.

5. Determine a function in the form $f(x) = a(x - h)^2 + k$ of the parabola with vertex $(-7, 12)$ and passing through $(-1, -6)$.

6. A manufacturer of portable radios finds that the cost of producing x units per month is given by the function

$$C(x) = 0.03x^2 - 24x + 11{,}500$$

How many units should be manufactured to minimize the cost?

7. Determine the corresponding graph of $|2x^3 - 1| = 15$ and then solve the equation algebraically.

8. Patrick Hale, a drag racer and engineer, determined that the speed s of a car at the end of a $\frac{1}{4}$ mile race is related to the weight w and power h of the car according to the equation

$$s = 234\sqrt[3]{\frac{h}{w}}$$

where s is in miles per hour, h is in horsepower, and w is in pounds. Determine the horsepower necessary for a 3438 pound Corvette to attain a speed of 91 miles per hour at the end of a $\frac{1}{4}$ mile race.

9. Solve $x + 2\sqrt{x - 1} = 9$ graphically.

10. Solve $|2x - 6| \le |x|$ graphically.

For Exercises 11 and 12, write a verbal equation that describes the problem and use it to solve the problem.

11. How many ounces of a solution that is 6% fluoride must be mixed with 12 ounces of a solution that is 11% fluoride to obtain a mixture that is 8% fluoride?

12. A person who can row 4 miles per hour in still water takes 2 hours to go a certain distance upstream and 1 hour to go the same distance downstream. Assuming the rate of the river is constant, at what rate is the river flowing?

POLYNOMIAL AND RATIONAL FUNCTIONS

In this chapter, we continue the investigation of functions and equations that began in Chapters 1 and 2. Sections 3.1 and 3.2 discuss polynomial functions of degree 3 or greater. (First-degree and second-degree functions were the subjects of Sections 2.1 and 2.2, respectively.) In Section 3.3, the number system is extended to include non-real numbers in order to allow the solution of *all* polynomial equations. The next two sections explore rational functions and equations. In Section 3.6, parametric equations are introduced, giving us a different method of describing curves in the coordinate plane.

A graphing utility removes much of the tedium found in graphing polynomial functions and solving polynomial equations by the traditional methods. As you read and work through the chapter, take note of the characteristics of these graphs and equations. By the end of this chapter, you will be able to solve many kinds of applied problems in which these graphs and equations arise—for example, maximizing the strength of a beam, designing a shipping container, and modeling the motion of a projectile.

POLYNOMIAL FUNCTIONS

Polynomial functions are the simplest and most widely used set of algebraic functions. Their values can be computed using only the basic arithmetic operations of addition, subtraction, and multiplication of real numbers. These functions serve as models for many applications of mathematics.

We start by defining a polynomial function.

Polynomial Functions

A **polynomial function** is a function P that can be expressed in the form

$$P(x) = a_n x^n + a_{n-1} x^{n-1} + a_{n-2} x^{n-2} + \cdots + a_1 x + a_0$$

The **coefficients** of P are a_n, a_{n-1}, a_{n-2}, $\cdots$, a_1, and a_0 $(a_n \neq 0)$. The **degree** of P is n. The **leading term** of P is $a_n x^n$. The domain of P is the set of all real numbers.

■ **NOTE:** *The degree of the zero function is undefined, because it has no highest-degree term.*

A function of the form $P(x) = k$ in which k is a nonzero constant is said to be a *zero-degree* polynomial function. The function $P(x) = 0$ is said to be the *zero function*.

In Chapter 2, we saw that the graph of a first-degree polynomial function $f(x) = mx + b$ is a line, and that the graph of a second-degree polynomial function $f(x) = ax^2 + bx + c$ is a parabola. What do the graphs of higher-degree polynomial functions such as $P(x) = -2x^3 + 7x^2 + 5x$ look like? What can we determine about these graphs, and what do they tell us about these functions? Answering these questions is the focus of this section.

Monomial Functions and Their Graphs

The simplest types of polynomial functions are *monomial functions*. A monomial function can be written in the form $f(x) = ax^n$ for some positive integer n and real number a. (The cases $y = ax$ and $y = ax^2$ were discussed in Sections 2.1 and 2.2.)

The graphs of $y = x^n$ for different values of n are shown in Figure 3.1. Notice that for odd values of n (Figure 3.1a), each graph passes through the points $(-1, -1)$, $(0, 0)$, and $(1, 1)$. For even values of n (Figure 3.1b), each graph passes through the points $(-1, 1)$, $(0, 0)$, and $(1, 1)$.

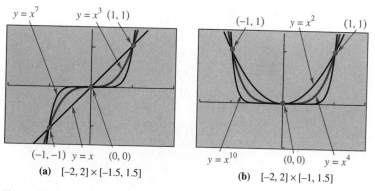

(a) [−2, 2] × [−1.5, 1.5]   (b) [−2, 2] × [−1, 1.5]

Figure 3.1

In general, the graphs of $y = x^n$ (n a positive integer) have two basic shapes (Figure 3.2), one if n is an odd number, another if n is an even number.

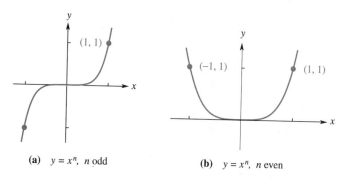

(a) $y = x^n$, n odd   (b) $y = x^n$, n even

Figure 3.2

The graph of a monomial function $y = ax^n$ is either a vertical compression or expansion of the graph of $y = x^n$.

◉ EXAMPLE 1   The Graph of $y = ax^n$

Use a graphing utility to determine the graph of each function. Describe how the graph compares to the graph of $g(x) = x^n$ (for some positive integer n).

(a) $y = \frac{1}{3}x^4$ **(b)** $y = -2x^5$

SOLUTION

(a) The graph of $y = \frac{1}{3}x^4$ is a vertical compression of the graph of $g(x) = x^4$ (Figure 3.3).

(b) The graph of $y = -2x^5$ is a vertical expansion and reflection through the x-axis of the graph of $g(x) = x^5$ (Figure 3.4).

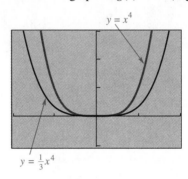

Figure 3.3
$[-2, 2] \times [-1, 3]$

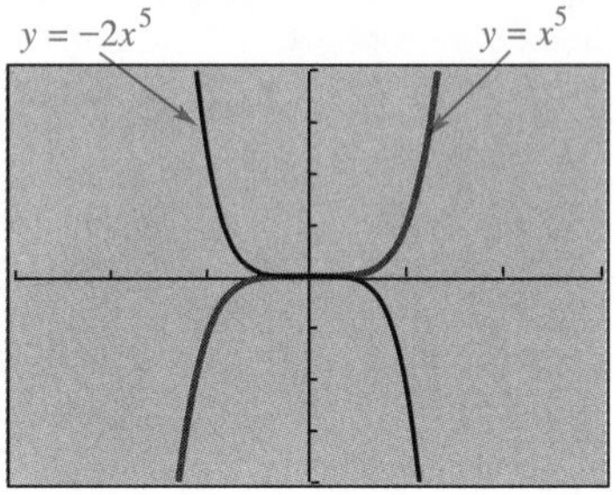

Figure 3.4
$[-3, 3] \times [-4, 4]$

We can generalize about the shape of the graph of a monomial function $f(x) = ax^n$ from Example 1. If n is an even number, the graph of f appears to be symmetric with respect to the y-axis. This conjecture can be verified by showing that f is an even function:

$$f(-x) = a(-x)^n = ax^n = f(x)$$

On the other hand, if n is an odd number, then the graph of f is symmetric with respect to the origin, because f is an odd function:

$$f(-x) = a(-x)^n = a(-x^n) = -ax^n = -f(x)$$

We can also generalize about the behavior of the graph of $y = ax^n$ for the extreme values of x. Notice, for example, that the y-coordinates of the graph of $y = \frac{1}{3}x^4$ get larger and larger in the positive direction as x gets larger and larger in the positive direction (Figure 3.5). We describe this behavior by saying that "y grows positively without bound as x grows positively without bound" and by writing

$$y \to +\infty \quad \text{as} \quad x \to +\infty$$

Furthermore, for $y = \frac{1}{3}x^4$, y also grows positively without bound as x grows negatively without bound (Figure 3.5), so we write

$$y \to +\infty \quad \text{as} \quad x \to -\infty$$

Figure 3.6 shows the four cases of what can happen with the graph of a monomial function $y = ax^n$ for extreme values of x.

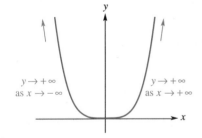

Figure 3.5
$y = \frac{1}{3}x^4$

■ **NOTE:** $y \to +\infty$ as $x \to +\infty$ *may be read as "y approaches positive infinity as x approaches positive infinity."*

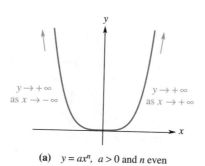

(a) $y = ax^n, \ a > 0$ and n even

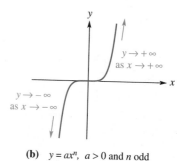

(b) $y = ax^n, \ a > 0$ and n odd

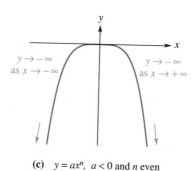

(c) $y = ax^n, \ a < 0$ and n even

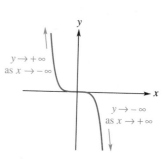

(d) $y = ax^n, \ a < 0$ and n odd

Figure 3.6
The graph of $y = ax^n$ and its behavior for extreme values of x.

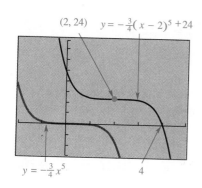

Figure 3.7
$[-2, 5] \times [-30, 80]_{10}$

⬤ **EXAMPLE 2   Graph of a Translation of a Monomial Function**
Consider the function

$$y = -\tfrac{3}{4}(x - 2)^5 + 24$$

(a) Use a graphing utility to determine the graph of this function. Compare this graph with the graph of

$$g(x) = -\tfrac{3}{4}x^5$$

(b) Describe the behavior of the graph for extreme values of x.
(c) Determine the x-intercept of the graph.

SOLUTION
(a) The graph of $y = -\tfrac{3}{4}(x - 2)^5 + 24$ is shown in Figure 3.7. This graph is a translation of the graph of $g(x) = -\tfrac{3}{4}x^5$, 2 units to the right and 24 units up, because $y = -\tfrac{3}{4}(x - 2)^5 + 24$ is equivalent to $y = g(x - 2) + 24$.

(b) The behavior of $y = -\frac{3}{4}(x - 2)^5 + 24$ for extreme values of x is the same as that of $g(x) = -\frac{3}{4}x^5$:

$$y \to +\infty \quad \text{as} \quad x \to -\infty \qquad \text{and} \qquad y \to -\infty \quad \text{as} \quad x \to +\infty$$

(c) To find the x-intercept, we let $y = 0$ and solve for x:

$$0 = -\tfrac{3}{4}(x - 2)^5 + 24$$
$$\tfrac{3}{4}(x - 2)^5 = 24$$
$$(x - 2)^5 = 32 \qquad \text{Multiply each side by } \tfrac{4}{3}.$$
$$x - 2 = 32^{1/5} \qquad \text{Take the fifth root of each side.}$$
$$x - 2 = 2$$
$$x = 4$$

The x-intercept is $(4, 0)$, which agrees with Figure 3.7.

Graphs of Polynomial Functions

We have seen that the graphs of first-degree polynomial functions, second-degree functions, and monomial functions are rather straightforward to describe. Unfortunately, this is not true for polynomial functions in general. However, certain generalizations about the graphs of polynomial functions, along with a graphing utility, will help ease the problem of determining these graphs.

■ EXAMPLE 3 The Graph of a Polynomial Function

Use a graphing utility to determine the graphs of the functions P and Q in the given viewing window. [Note $Q(x)$ is the leading term of $P(x)$.] What is the behavior of the graphs for the extreme values of x?

(a) $P(x) = x^4 - 4x^2 + 2$ and $Q(x) = x^4$ in $[-4, 4] \times [-10, 40]_5$
(b) $P(x) = -2x^3 + 2x^2 + 23x + 11$ and $Q(x) = -2x^3$ in $[-4, 8] \times [-400, 200]_{50}$

SOLUTION

(a) The graphs of P and Q are shown in Figure 3.8. The behavior of each of these functions for extreme values of x is described by

$$y \to +\infty \quad \text{as} \quad x \to -\infty \qquad \text{and} \qquad y \to +\infty \quad \text{as} \quad x \to +\infty$$

(b) The graphs of the functions are shown in Figure 3.9. The behavior of both P and Q are described by

$$y \to +\infty \quad \text{as} \quad x \to -\infty \qquad \text{and} \qquad y \to -\infty \quad \text{as} \quad x \to +\infty$$

$y = Q(x)$ $y = P(x)$ $y = Q(x)$ $y = P(x)$

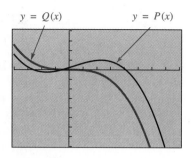

Figure 3.8   **Figure 3.9**
$[-4, 4] \times [-10, 40]_5$ $[-4, 8] \times [-400, 200]_{50}$

In Example 3, the graph of P behaves in the same manner as the graph of Q for extreme values of x. For large values of x, the value of $P(x)$ is dominated by its first term. For example, consider the function $P(x) = x^4 - 4x^2 + 2$ in part (a) of Example 3. The values of $P(20)$ and $P(-10)$ are

$$P(20) = (20)^4 - 4(20)^2 + 2 = 160,000 - 1600 + 2 = 158,402$$
$$P(-10) = (-10)^4 - 4(-10)^2 + 2 = 10,000 - 400 + 2 = 9602$$

Notice that $P(20)$ is approximately $160,000$—the value of the first term of the sum. Likewise, $P(-10)$ is approximately equal to its first term, $10,000$. Loosely speaking, for relatively large values of x, positive or negative, $x^4 - 4x^2 + 2$ is approximately x^4. In general, we call this result the *leading term property*.

Leading Term Property

> The behavior of the graph of a polynomial function for extreme values of x is the same as that of its leading term. Specifically, if
>
> $$P(x) = a_n x^n + a_{n-1} x^{n-1} + a_{n-2} x^{n-2} + \cdots + a_1 x + a_0$$
>
> then the behavior of the graph of P for extreme values of x is the same as the behavior of
>
> $$y = a_n x^n$$
>
> for extreme values of x.

The leading term property and Figure 3.6 give us a way to describe the behavior of any polynomial function P for extreme values of x. This information, along with the x-intercepts, helps significantly in determining a complete graph of P.

EXAMPLE 4 The Graph of a Polynomial Function

Let $f(x) = 3(x^2 - 2x - 3)(x^2 - 20)$.

(a) Describe the behavior of the graph of f for extreme values of x.

(b) Determine the x-intercepts of the graph algebraically.

(c) Use a graphing utility to determine the graph of f.

SOLUTION

(a) Imagine simplifying $3(x^2 - 2x - 3)(x^2 - 20)$ without actually doing so; you can see that the leading term of f is $3 \cdot x^2 \cdot x^2$, or $3x^4$. Thus, the behavior of the graph of f for extreme values of x is the same as $y = 3x^4$. Figure 3.6a shows that

$$y \to +\infty \quad \text{as} \quad x \to +\infty \qquad \text{and} \qquad y \to +\infty \quad \text{as} \quad x \to -\infty$$

(b) The x-intercepts are determined by setting $y = 0$ and solving for x:

$$0 = 3(x^2 - 2x - 3)(x^2 - 20)$$

$$0 = (x^2 - 2x - 3)(x^2 - 20) \qquad \text{Divide each side by 3.}$$

$x^2 - 2x - 3 = 0$	$x^2 - 20 = 0$
$(x + 1)(x - 3) = 0$	$x^2 = 20$
$x + 1 = 0 \qquad x - 3 = 0$	$x = \pm\sqrt{20}$
$x = -1 \qquad\quad x = 3$	$x = \pm 2\sqrt{5}$

The solutions to the equation are $-2\sqrt{5}$ (approximately -4.5), -1, 3, and $2\sqrt{5}$ (approximately 4.5). The x-intercepts are $(-2\sqrt{5}, 0)$, $(-1, 0)$, $(3, 0)$, and $(2\sqrt{5}, 0)$

(c) A complete graph of f should show its behavior for extreme values of x, its x-intercepts, and its turning points. Using the techniques of Section 1.4, we arrive at a viewing window (Figure 3.10a) that allows these features of the graph to be displayed (Figure 3.10b).

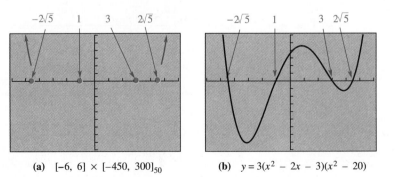

(a) $[-6, 6] \times [-450, 300]_{50}$ **(b)** $y = 3(x^2 - 2x - 3)(x^2 - 20)$

Figure 3.10

Certain traits of the graphs of polynomial functions make it fairly easy to sketch these graphs by hand or to determine them with a graphing utility. The five characteristics that follow are suggested by Examples 3 and 4, but most of their proofs actually require calculus. For now, we will assume them to be true.

Common Characteristics of Polynomial Functions

For the graph of any polynomial function P of degree n, the following statements are true:

1. The graph has no breaks or jumps. You can sketch it over any interval of real numbers without lifting your pencil. (We say that P is *continuous over the real numbers.*)
2. The graph is *smooth* everywhere; it has no "corners" or "cusps."
3. The behavior of the graph for extreme values of x is one of the four described in Figure 3.6. As x grows without bound either positively or negatively, y grows without bound either positively or negatively.
4. The graph has no more than $n - 1$ turning points. For example, a fifth-degree polynomial function can have up to four turning points, but it cannot possibly have five or more.
5. The graph has no more than n x-intercepts. (It follows directly that a polynomial function can have no more than n zeros, and that the equation $P(x) = 0$ has no more than n solutions.)

These five characteristics suggest a fairly good mental picture of a given polynomial function, even before turning to a graphing utility. Furthermore, these characteristics provide tests that can be applied to a given graph to determine whether it may be the graph of a polynomial function.

■ EXAMPLE 5   Identifying the Graph of a Polynomial Function

Determine whether each graph could be the complete graph of a polynomial function. If not, explain why. If so, describe the leading term and its possible degree.

(a)

(b)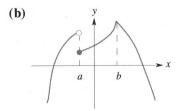

(c)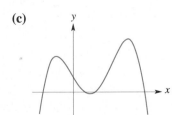

SOLUTION

(a) Because y appears to approach 0 as x grows positively without bound, this graph cannot be the complete graph of a polynomial function. If the graph were that of a polynomial function, y would grow without bound (either positively or negatively) as x grows positively without bound.

(b) The graph has both a break at $x = a$ and a corner at $x = b$. Either of these features disqualifies this graph as a graph of a polynomial function.

(c) This graph appears to satisfy all the characteristics of the graph of a polynomial function. If it is the graph of a polynomial function, then the three turning points tell us that the degree must be at least four. Also,

$$y \to -\infty \quad \text{as} \quad x \to -\infty \qquad \text{and} \qquad y \to -\infty \quad \text{as} \quad x \to +\infty$$

The leading term property and Figure 3.6c suggest that the leading term ax^n has a negative and n even.

■ **NOTE:** *Even though the graph in part (c) has the five characteristics listed above, there is no guarantee that it is the graph of a polynomial function.*

Zeros of Polynomial Functions

In Example 4, we found that the polynomial function

$$f(x) = 3(x^2 - 2x - 3)(x^2 - 20)$$

has x-intercepts $(-2\sqrt{5}, 0)$, $(-1, 0)$, $(3, 0)$, and $(2\sqrt{5}, 0)$. Factoring f over the integers gives

$$f(x) = 3\underbrace{(x + 1)(x - 3)}_{(x^2 - 2x - 3)}(x^2 - 20)$$

The factors $x + 1$ and $x - 3$ of $f(x)$ led to the zeros -1 and 3 of f in Example 4, which, in turn, led to the x-intercepts $(-1, 0)$ and $(3, 0)$.

What about the other two x-intercepts, $(-2\sqrt{5}, 0)$ and $(2\sqrt{5}, 0)$? Factoring $f(x)$ over the real numbers gives

■ **NOTE:** *Factoring over the real numbers is discussed in Section 3 (page 22).*

$$f(x) = 3(x + 1)(x - 3)\underbrace{(x + 2\sqrt{5})(x - 2\sqrt{5})}_{(x^2 - 20)}$$

In a similar way, the factors $x + 2\sqrt{5}$ and $x - 2\sqrt{5}$ correspond to these two x-intercepts.

The following statement summarizes this correspondence between linear factors and the x-intercepts of a polynomial function:

Linear Factors and x-Intercepts

> If a polynomial $P(x)$ has a factor in the form $x - c$ for some real number c, then the graph of P has an x-intercept $(c, 0)$.

The proof of this statement is detailed in Exercise 54.

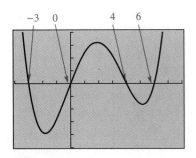

Figure 3.11
$y = 2x(x - 4)(x + 3)(x - 6)$
$[-4, 5] \times [-250, 200]_{50}$

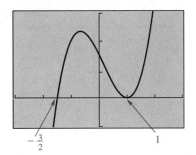

Figure 3.12
$y = (x - 1)^2(x + \frac{3}{2})$
$[-3, 3] \times [-1, 3]$

■ EXAMPLE 6   x-Intercepts of a Polynomial Function

Use a graphing utility to determine the graph of each polynomial function, and label its x-intercepts.

(a) $f(x) = 2x(x - 4)(x + 3)(x - 6)$ **(b)** $g(x) = (x - 1)^2(x + \frac{3}{2})$

SOLUTION

(a) The factors x, $x - 4$, $x + 3$, and $x - 6$ imply that the graph of f has x-intercepts at 0, 4, −3, and 6. The graph of f agrees with our findings (Figure 3.11).

(b) As both the factors and the graph (Figure 3.12) suggest, the x-intercepts of g are at 1 and $-\frac{3}{2}$. ■

The graph of $g(x) = (x - 1)^2(x + \frac{3}{2})$ in part (b) of Example 6 possesses an interesting feature; the point $(1, 0)$ serves as both an x-intercept and a turning point of the graph. The reason for this feature of the graph is that $(x - 1)^2$, a square of a linear factor, occurs in the factorization. (An explanation is outlined in Exercise 55.)

Applications

Because of their relative simplicity and variety, polynomial functions arise in many mathematical models. In Example 7, the graph of a third-degree polynomial is used to solve an engineering problem.

■ EXAMPLE 7   Application: Maximizing the Strength of a Beam

A beam is to be cut from a log 12 inches in diameter (Figure 3.13). The cross section of the beam is a rectangle. The strength of this beam is proportional to the product of its width w and the square of its depth d. More precisely, the strongest of all beams that can be milled from the log is a beam for which the quantity $S = wd^2$ is the greatest. What is the width (to the nearest 0.1 inch) of the strongest beam that can be milled from this log?

SOLUTION

Our plan of attack is to determine S as a function of w. Applying the Pythagorean theorem to the right triangle ABC shown in Figure 3.13 gives us

$$w^2 + d^2 = 12^2$$

or

$$d^2 = 144 - w^2$$

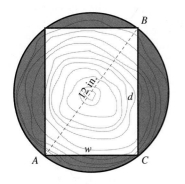

Figure 3.13

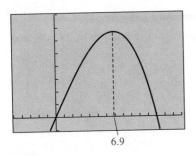

Figure 3.14
$y = -x^3 + 144x$
$[-5, 15] \times [-100, 800]_{100}$

By substitution, we get

$$S = wd^2 = w(144 - w^2) = -w^3 + 144w$$

The graph of the polynomial function $S = -w^3 + 144w$ is shown in Figure 3.14. The domain of the function is $0 < w < 12$, because the width of the beam must be positive and also less than the diameter of the beam. By using the features of a graphing utility, we approximate the value of w that yields the largest value of S (Figure 3.14). The width of the strongest beam is approximately 6.9 inches.

EXERCISE SET 3.1

A

In Exercises 1–6, use a graphing utility to determine the graphs of both functions in the same viewing window. Compare the graph of the function in part (b) with the graph of the function in part (a).

1. (a) $y = x^3$ (b) $y = \frac{1}{2}x^3$
2. (a) $y = x^4$ (b) $y = \frac{1}{2}x^4$
3. (a) $y = x^4$ (b) $y = -\frac{1}{8}x^4$
4. (a) $y = x^3$ (b) $y = -\frac{1}{2}x^3$
5. (a) $y = x^5$ (b) $y = \frac{1}{6}x^5$
6. (a) $y = x^6$ (b) $y = -x^6$

In Exercises 7–12, for each polynomial function:

(a) Use a graphing utility to determine the graph of the function. Compare this graph with the graph of $g(x) = ax^n$ (for some real number a and some positive integer n).
(b) Describe the behavior of the graph for extreme values of x.
(c) Determine the x-intercepts of the graph.

7. $y = \frac{1}{2}(x - 4)^3$ 8. $y = -x^3 + 24$
9. $y = -\frac{1}{8}x^4 + 6$ 10. $y = \frac{1}{2}(x - 2)^4 - 8$
11. $y = -\frac{1}{6}(x + 2)^5 - 16$ 12. $y = -(x - 1)^6 - 5$

In Exercises 13–18, use a graphing utility to determine the graphs of the functions P and Q in the viewing window given. [Note that $Q(x)$ is the leading term of $P(x)$.] Describe the behavior of the graphs for the extreme values of x.

13. $P(x) = x^3 - 4x$ and $Q(x) = x^3$ in $[-5, 5] \times [-30, 30]_{10}$
14. $P(x) = x^4 - 9x^2$ and $Q(x) = x^4$ in $[-6, 6] \times [-25, 200]_{25}$
15. $P(x) = -2x^3 + x^2 + 3x - 4$ and $Q(x) = -2x^3$ in $[-5, 5] \times [-10, 10]_5$
16. $P(x) = \frac{2}{3}x^5 - 4x^3 + 2x$ and $Q(x) = \frac{2}{3}x^5$ in $[-5, 5] \times [-40, 40]_{10}$
17. $P(x) = (3x^2 - 4)(5 - x^3)$ and $Q(x) = -3x^5$ in $[-4, 4] \times [-40, 40]_{10}$
18. $P(x) = (x^3 - 4x - 7)(3 - 2x)$ and $Q(x) = -2x^4$ in $[-5, 5] \times [-200, 50]_{25}$

In Exercises 19–24, determine whether each graph could be the complete graph of a polynomial function. If not, explain why. If so, describe the leading term and its possible degree.

19.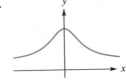

20.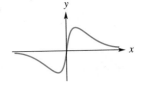

21.

22.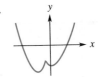

23.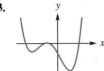

24.

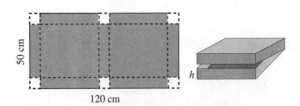

Figure 3.15

In Exercises 25–30, use a graphing utility to determine the graph of each function and label its x-intercepts.

25. $y = -(x + 4)(x - 5)(x - 1)$

26. $y = 2(x + 1)(x - 3)(x - 8)$

27. $y = \frac{1}{2}x(x - 2)(x + \pi)(x - \sqrt{5})$

28. $y = 2(x + \sqrt[3]{4})(x + \frac{9}{2})(x - 1)$

29. $y = (x + 3)^2(x - 3)^2$

30. $y = -x^2(x - 2)(x - 7)$

B

In Exercises 31–46, for each function:

(a) Describe the behavior of the graph of the function for extreme values of x.

(b) Determine the x-intercepts of the graph algebraically.

(c) Use a graphing utility to determine the graph.

31. $N(x) = x^3 - 25x$

32. $M(x) = x^3 + 8x^2$

33. $g(x) = 7x^2 - x^3$

34. $f(x) = 16x - x^3$

35. $Q(x) = x^3 - 2x^2 - 8x$

36. $P(x) = x^3 - 5x^2 + 4x$

37. $f(x) = -x^4 + 4x^3 + 21x^2$

38. $g(x) = \frac{1}{2}x^4 - x^3 - 4x^2$

39. $h(x) = x^3 - 4x^2 - x + 4$

40. $d(x) = 2x^3 - 10x^2 - 6x + 30$

41. $A(x) = x^4 - 9x^2 + 8$

42. $B(x) = x^4 - x^2 - 12$

43. $T(x) = (x^3 - 12x)(x + 7)$

44. $R(x) = (x^2 + 9x + 8)(24 - x^2)$

45. $F(x) = (x - 2)(x^2 + 5) - (x - 2)(4x + 5)$

46. $G(x) = (x^2 - 2)(4x^2 + 5) - (2x^2 + 1)(4x^2 + 5)$

47. A pizza box is made by cutting six equal-sized squares from a cardboard rectangle with dimensions 50×120 centimeters (Figure 3.15).

(a) Determine the volume V of the box as a function of its height h. What is the domain of V?

(b) Determine the graph of V and use the graph to find the height of the box that has a volume of 10,000 cubic centimeters.

48. The cone of a rocket has height 60 inches and radius 30 inches. A cylinder containing the payload is to be constructed so that it fits inside the cone (Figure 3.16).

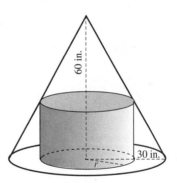

Figure 3.16

(a) Show that the volume V of the payload is given by the function

$$V = \pi r^2 (60 - 2r)$$

where r is the radius of the cylinder.

(b) Determine the graph of this function. Approximate the maximum volume of the payload.

In Exercises 49–52, explain why the graph shown is not a complete graph of the given function. Determine a complete graph of the function.

49. $f(x) = x^3 - 16x$

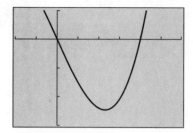

50. $f(x) = x^3 - 8x^2$

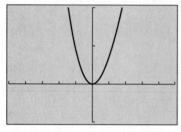

51. $f(x) = x^4 + 12x^3 - 4x^2 - 48x$

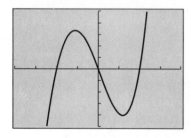

52. $f(x) = x^4 - 25x^2 + 40$

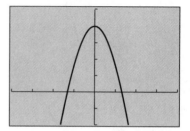

C

53. Use the characteristics of the graphs of polynomial functions listed in this section to explain why a polynomial function of odd degree must have at least one zero.

54. Suppose that a polynomial function P can be written as $P(x) = (x - c)q(x)$, where q is also a polynomial function. Show that $P(c) = 0$. Explain why this fact implies that the graph of P has an x-intercept $(c, 0)$.

55. The graph of $P(x) = \left(x + \frac{3}{2}\right)(x - 1)^2$ from part (b) of Example 6 is shown in Figure 3.12. Determine the graph of $y = \frac{5}{2}(x - 1)^2$, and explain why it appears that $P(x) \approx \frac{5}{2}(x - 1)^2$ for values of x near 1.

SECTION 3.2

- The Division Algorithm
- The Factor Theorem
- Factoring and Solving by Graphing
- Applications

POLYNOMIAL EQUATIONS

Simply stated, a polynomial equation is one that can be written in the form $P(x) = 0$ for some polynomial function P. The problem of solving these equations was a driving force in the development of modern mathematics. The methods of solving first-degree and second-degree equations have been known since antiquity. In the fourteenth century, formulas for the solutions of third-degree and fourth-degree equations were discovered. These formulas are much like the quadratic formula used for solving second-degree equations, but so much more complicated and unwieldy that today they provide little more than historical interest. In the 1830's, a French mathematician named Evariste Galois proved that there are no such general formulas for polynomial equations of degree 5 or greater. Since then, mathematicians have developed methods to solve certain polynomial equations. If these methods fail on a particular polynomial equation, then its solutions can be approximated using the techniques of Section 1.4. In this section, we investigate these methods for solving polynomial equations of degree 3 or greater.

We start by discussing again the division of polynomials, which we introduced in the first chapter (Section 4).

■ **NOTE:** *Evariste Galois (1811–1832), while a teenager, solved one of the great mathematical problems of his day with his work on the solutions of polynomial equations. Tragically, he was killed in a duel at the age of 20.*

The Division Algorithm

An improper fraction such as $\frac{75}{13}$ can be expressed as the sum of an integer and a proper fraction by performing long division:

$$\begin{array}{r} 5 \\ 13\overline{)75} \\ \underline{65} \\ 10 \end{array}$$

We can write this result as

or

$$\frac{75}{13} = 5 + \frac{10}{13}$$

$$75 = 13 \cdot 5 + 10$$

The division of polynomials can be handled in the same way, as we see in Example 1.

▢ EXAMPLE 1 Using the Division Algorithm for Polynomials

Divide the polynomial $P(x) = x^3 - 2x + 21$ by $d(x)$.

(a) $d(x) = x^2 - 2x + 4$ **(b)** $d(x) = x + 3$

SOLUTION
(a) We perform the long division in the manner described in Section 4:

$$\begin{array}{r} x + 2 \\ x^2 - 2x + 4\overline{)x^3 + 0x^2 - 2x + 21} \\ \underline{x^3 - 2x^2 + 4x} \\ 2x^2 - 6x + 21 \\ \underline{2x^2 - 4x + 8} \\ -2x + 13 \end{array}$$

Thus

$$\frac{x^3 - 2x + 21}{x^2 - 2x + 4} = (x + 2) + \frac{-2x + 13}{x^2 - 2x + 4}$$

or

$$x^3 - 2x + 21 = (x^2 - 2x + 4)(x + 2) + (-2x + 13)$$

(b) By the division algorithm, we get

$$
\begin{array}{r}
x^2 - 3x + 7 \\
x + 3 \overline{)x^3 + 0x^2 - 2x + 21} \\
\underline{x^3 + 3x^2} \\
-3x^2 - 2x \\
\underline{-3x^2 - 9x} \\
7x + 21 \\
\underline{7x + 21}
\end{array}
$$

The remainder is 0, so

$$\frac{x^3 - 2x + 21}{x + 3} = x^2 - 3x + 7$$

or

$$x^3 - 2x + 21 = (x + 3)(x^2 - 3x + 7)$$

The result of the division algorithm can be summarized as follows:

Division Algorithm
for Polynomials

If $P(x)$ and $d(x)$ are polynomials of positive degree, and the degree of d is less than or equal to the degree of P, then there exist unique polynomials $q(x)$ and $r(x)$ such that

$$\frac{P(x)}{d(x)} = q(x) + \frac{r(x)}{d(x)}$$

or, equivalently,

$$P(x) = d(x)q(x) + r(x)$$

where $r(x) = 0$ or the degree of r is less than the degree of d, and the degree of q is less than the degree of P.

In particular, if the divisor $d(x)$ is of the form $x - c$ for some number c, then the remainder is a number r, and $P(x) = (x - c)q(x) + r$.

Specifically, in Example 1 we found that

$$\underbrace{x^3 - 2x + 21}_{P(x)} = \underbrace{(x^2 - 2x + 4)}_{d(x)}\underbrace{(x + 2)}_{q(x)} + \underbrace{(-2x + 13)}_{r(x)}$$

■ **NOTE:** *The division algorithm also plays an important role in Section 3.4.*

and that

$$\underbrace{x^3 - 2x + 21}_{P(x)} = \underbrace{(x + 3)}_{d(x)}\underbrace{(x^2 - 3x + 7)}_{q(x)} + \underbrace{0}_{r}$$

The Factor Theorem

Example 1b suggests that because the remainder is 0, the divisor $x + 3$ is a factor of $P(x) = x^3 - 2x + 21$, and that

$$P(x) = (x + 3)(x^2 - 3x + 7)$$

Notice that the factor $x + 3$ is equal to 0 if x is -3. Moreover,

$$P(-3) = [(-3) + 3][(-3)^2 - 3(-3) + 7] = (0)(25) = 0$$

This suggests the following theorem:

Factor Theorem

> Suppose that P is a polynomial function. A number c is a zero of P if and only if $x - c$ is a factor of $P(x)$.

To prove the factor theorem, we use the division algorithm to write P as

$$P(x) = (x - c)q(x) + r$$

for some polynomial function q and some number r. Evaluating $P(c)$ gives

$$P(c) = (c - c)q(c) + r = (0)q(c) + r = 0 + r = r$$

Because $P(c)$ is equal to r, $P(c)$ can be 0 if and only if r is 0. Thus, $P(c)$ is 0 if and only if

$$P(x) = (x - c)q(x)$$

The conclusion we draw from this last statement is that c is a zero of P if and only if $x - c$ is a factor of $P(x)$. This proves the theorem.

The following list of four equivalent statements is a direct result of the factor theorem:

Corollary to the Factor Theorem

> Suppose that P is a polynomial function and that c is a real number. If any one of these four statements is true, then so are the other three:
>
> **1.** The real number c is a zero of the function P.
> **2.** The real number c is a solution to the equation $P(x) = 0$.
> **3.** The point $(c, 0)$ is an x-intercept of the graph of P.
> **4.** The first-degree binomial $x - c$ is a factor of the polynomial $P(x)$.

In Chapter 2, we saw that statements 1, 2, and 3 are either all true or all false for any function P and real number c. The factor theorem implies that statement 4 is true if and only if statement 1 is true. Therefore, all four statements are equivalent.

◼ EXAMPLE 2 Using the Factor Theorem to Solve Polynomial Equations

Solve $x^3 - 7x^2 - 21x + 27 = 0$, given that 9 is one of the solutions of the equation.

SOLUTION

The factor theorem suggests that since 9 is a solution of the equation $x^3 - 7x^2 - 21x + 27 = 0$, $x - 9$ is a factor of $x^3 - 7x^2 - 21x + 27$. More precisely, there exists a polynomial q such that

$$x^3 - 7x^2 - 21x + 27 = (x - 9)q(x)$$

We can find q by using the division algorithm:

$$
\begin{array}{r}
x^2 + 2x - 3 \\
x - 9\overline{)x^3 - 7x^2 - 21x + 27} \\
\underline{x^3 - 9x^2} \\
2x^2 - 21x \\
\underline{2x^2 - 18x} \\
-3x + 27 \\
\underline{-3x + 27}
\end{array}
$$

Thus,

$$q(x) = x^2 + 2x - 3$$

and

$$x^3 - 7x^2 - 21x + 27 = (x - 9)(x^2 + 2x - 3)$$

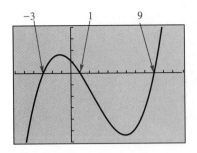

Figure 3.17
$y = x^3 - 7x^2 - 21x + 27$
$[-6, 12] \times [-150, 100]_{25}$

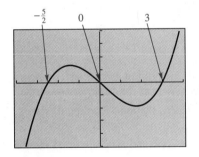

Figure 3.18
$y = x^3 - \frac{1}{2}x^2 - \frac{15}{2}x$
$[-4, 4] \times [-25, 20]_5$

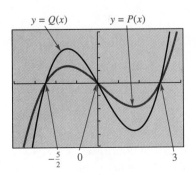

Figure 3.19
$[-4, 4] \times [-25, 20]_5$

This factorization allows us to solve the equation:

$$x^3 - 7x^2 - 21x + 27 = 0$$
$$(x - 9)(x^2 + 2x - 3) = 0 \qquad \text{Rewrite the polynomial as } (x - 9)q(x).$$
$$(x - 9)(x + 3)(x - 1) = 0 \qquad \text{Factor } x^2 + 2x - 3 \text{ using FOIL.}$$

$$\begin{array}{c|c|c} x - 9 = 0 & x + 3 = 0 & x - 1 = 0 \\ x = 9 & x = -3 & x = 1 \end{array}$$

The solutions to the equation are -3, 1, and 9. These solutions are verified by the graph of $y = x^3 - 7x^2 - 21x + 27$ in Figure 3.17.

The factor theorem also gives us a way to construct a polynomial function from its zeros.

EXAMPLE 3   Constructing a Polynomial Function with Given Zeros

Determine a polynomial function P that has the zeros 3, 0, and $-\frac{5}{2}$.

SOLUTION

By the factor theorem, a polynomial function that has 3, 0, and $-\frac{5}{2}$ as zeros also has linear factors $x - 3$, $x - 0$, and $x - \left(-\frac{5}{2}\right)$. We can construct P by using these factors:

$$P(x) = (x - 3)(x - 0)[x - \left(-\frac{5}{2}\right)] = (x - 3)x\left(x + \frac{5}{2}\right) = x^3 - \frac{1}{2}x^2 - \frac{15}{2}x$$

The graph of

$$P(x) = x^3 - \frac{1}{2}x^2 - \frac{15}{2}x$$

has $(3, 0)$, $(0, 0)$, and $\left(-\frac{5}{2}, 0\right)$ as x-intercepts, as shown in Figure 3.18.

The function $P(x) = x^3 - \frac{1}{2}x^2 - \frac{15}{2}x$ is not a unique answer for Example 3. For instance, the polynomial function Q defined by

$$Q(x) = 2P(x) = 2\left(x^3 - \frac{1}{2}x^2 - \frac{15}{2}x\right) = 2x^3 - x^2 - 15x$$

also has the same x-intercepts as P (Figure 3.19), and therefore has the same zeros. The function Q has an advantage over the function P in that it has integer coefficients.

Factoring and Solving by Graphing

In Section 3.1, we saw that an nth-degree polynomial function P can have at most n x-intercepts. From the corollary to the factor theorem, we know that for each x-intercept $(c, 0)$ of its graph, P has a linear factor $x - c$. This gives us a way to find linear factors of a polynomial function by using its graph.

EXAMPLE 4   Using a Graph to Factor a Polynomial

Factor $P(x) = 2x^3 - 8x^2 - 40x + 96$ using the graph of P.

SOLUTION

The x-intercepts of the graph of P shown in Figure 3.20 appear to be $(-4, 0)$, $(2, 0)$, and $(6, 0)$. We can verify this by checking that -4, 2, and 6 are zeros of P:

$$P(-4) = 2(-4)^3 - 8(-4)^2 - 40(-4) + 96 = 0$$
$$P(2) = 2(2)^3 - 8(2)^2 - 40(2) + 96 = 0$$
$$P(6) = 2(6)^3 - 8(6)^2 - 40(6) + 96 = 0$$

Because -4, 2, and 6 are zeros of P, $x + 4$, $x - 2$, and $x - 6$ are linear factors of P. Simplifying $(x + 4)(x - 2)(x - 6)$ gives

$$x^3 - 4x^2 - 20x + 48$$

so it follows that the factorization of P is

$$2x^3 - 8x^2 - 40x + 96 = 2(x + 4)(x - 2)(x - 6)$$

You can check the result of Example 4 by determining the graphs of both $y = 2x^3 - 8x^2 - 40x + 96$ and $y = 2(x + 4)(x - 2)(x - 6)$ in the same viewing window. (Take a moment to do this.) If the factorization is correct, then these two graphs are identical.

In the next example, we see how the factor theorem and the graph of a function help to solve a polynomial equation.

EXAMPLE 5   Using a Graph to Solve a Polynomial Equation

Use an appropriate graph to solve the equation

$$2x^4 + 38x + 24 = 8x^3 + 11x^2$$

SOLUTION

The equation is equivalent to

$$2x^4 - 8x^3 - 11x^2 + 38x + 24 = 0$$

The solutions to this equation are the x-intercepts of the graph of

$$P(x) = 2x^4 - 8x^3 - 11x^2 + 38x + 24$$

(Figure 3.21). Two of the x-intercepts are $(-2, 0)$ and $(4, 0)$, so -2 and 4 are solutions to the equation. (Pause here to verify this assertion.) From the graph, there appear to be two other x-intercepts of the graph of P, so there are two other real number solutions to the equation.

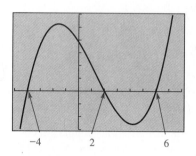

Figure 3.20

$y = 2x^3 - 8x^2 - 40x + 96$

$[-5, 8] \times [-75, 150]_{25}$

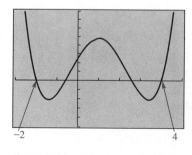

Figure 3.21

$y = 2x^4 - 8x^3 - 11x^2 + 38x + 24$

$[-3, 5] \times [-50, 80]_{10}$

The factor theorem predicts that $x + 2$ and $x - 4$ are linear factors of P. For some second-degree polynomial $q(x)$,

$$P(x) = (x + 2)(x - 4)q(x)$$

If we can determine the zeros of q, we will also find the two remaining solutions of the equation. By long division, we get

$$P(x) = (x + 2)\underbrace{(2x^3 - 12x^3 + 13x + 12)}_{(x - 4)q(x)}$$

Since $(x - 4)q(x) = 2x^3 - 12x^3 + 13x + 12$, we can find q by dividing $2x^3 - 12x^3 + 13x + 12$ by $x - 4$. Because

$$\frac{2x^3 - 12x^3 + 13x + 12}{x - 4} = 2x^2 - 4x - 3$$

we can write

$$2x^3 - 12x^3 + 13x + 12 = (x - 4)(2x^2 - 4x - 3)$$

and so,

$$P(x) = (x + 2)(x - 4)(2x^2 - 4x - 3)$$

This factorization of the polynomial allows us to solve the equation:

$$2x^4 - 8x^3 - 11x^2 + 38x + 24 = 0$$
$$(x + 2)(x - 4)(2x^2 - 4x - 3) = 0$$

$x + 2 = 0$	$x - 4 = 0$	$2x^2 - 4x - 3 = 0$
$x = -2$	$x = 4$	$x = \dfrac{-(-4) \pm \sqrt{(-4)^2 - 4(2)(-3)}}{2(2)}$
		$= \dfrac{4 \pm \sqrt{40}}{4}$
		$= \dfrac{2 \pm \sqrt{10}}{2}$

The solutions to the equation are -2, 4, $(2 + \sqrt{10})/2$ (approximately 2.6), and $(2 - \sqrt{10})/2$ (approximately -0.6). These solutions coincide with the x-intercepts of the graph of $y = 2x^4 - 8x^3 - 11x^2 + 38x + 24$ shown in Figure 3.21.

Note the strategy applied in Example 5. We used long division and the obvious x-intercepts to factor the equation into linear factors and a quadratic factor, which allowed us to solve the equation.

This strategy will not work on all polynomial equations, especially if there are no obvious solutions; in these cases we must settle for approximations. Also, exact answers such as $(2 + \sqrt{10})/2$ are not always preferable to approximations (such as 2.6) in certain applications. Quite often, using a graphing utility to approximate solutions as we did in Section 1.4 is the easiest way (or only way) to satisfactorily solve a polynomial equation.

Applications

Polynomial equations of third-degree or higher arise frequently in applications of mathematics. We will see Example 6 that once an applied problem is modeled by a polynomial equation, the methods of this section can be applied to solve that problem.

⬤ EXAMPLE 6 Application: Designing a Shipping Container

A whoesale supplier of umbrellas uses a cardboard box with a volume of 14,000 square inches for shipping. The base of the box must be square, and for reasons of cost, the sum of the girth and height of the package should be as large as the supplier's parcel service will accept, which is 180 inches. (The *girth* of a package is the perimeter of its base as shown in red in Figure 3.22.) Also, the side of the base must be no longer than 25 inches. What size box should the supplier use?

SOLUTION

The parcel service's restriction tells us that

$$\left(\begin{matrix}\text{Height}\\\text{of box}\end{matrix}\right) + 4 \left(\begin{matrix}\text{Side of}\\\text{base}\end{matrix}\right) = 180$$

or

$$\left(\begin{matrix}\text{Height}\\\text{of box}\end{matrix}\right) = 180 - 4 \left(\begin{matrix}\text{Side of}\\\text{base}\end{matrix}\right)$$

Using s for the side of the base and h for the height of the box, we get

$$h = 180 - 4s$$

The volume of the box is given by

$$\left(\begin{matrix}\text{Side of}\\\text{base}\end{matrix}\right)^2 \left(\begin{matrix}\text{Height}\\\text{of box}\end{matrix}\right) = \left(\begin{matrix}\text{Volume}\\\text{of box}\end{matrix}\right)$$

so it follows that

$$s^2 h = 14{,}000$$
$$s^2(180 - 4s) = 14{,}000 \qquad \text{Substitute } h = 180 - 4s.$$
$$-4s^3 + 180s^2 = 14{,}000$$
$$-4s^3 + 180s^2 - 14{,}000 = 0$$

The graph of $y = -4x^3 + 180x^2 - 14{,}000$ is shown in Figure 3.23a.

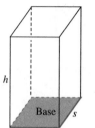

Figure 3.22

h

Base s

s

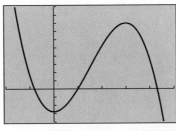
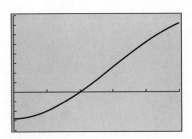

(a) $y = -4x^3 + 180x^2 - 14{,}000$
$[-20, 50]_{10} \times [-20{,}000, 50{,}000]_{5000}$

(b) $y = -4x^3 + 180x^2 - 14{,}000$
$[0, 25]_5 \times [-20{,}000, 40{,}000]_{5000}$

Figure 3.23

The only x-intercept in the interval $0 < x < 25$ appears to be $(10, 0)$ (Figure 3.23b). This conjecture can be verified by substituting $s = 10$ into the equation:

$$-4(10)^3 + 180(10)^2 - 14{,}000 \overset{?}{=} 0$$
$$-4000 + 18{,}000 - 14{,}000 \overset{?}{=} 0$$
$$0 = 0$$

The solution to the equation in the interval $0 < x < 25$ is $s = 10$. If $s = 10$, then

$$h = 180 - 4(10) = 140$$

The dimensions of the box are $10 \times 10 \times 140$ inches.

EXERCISE SET 3.2

A

In Exercises 1–6, divide each polynomial $P(x)$ by the divisor $d(x)$ given, and write P in the form $P(x) = d(x)q(x) + r(x)$.

1. $P(x) = 2x^3 + x^2 - 10x + 2;\ d(x) = 2x + 1$

2. $P(x) = 3x^3 - 2x^2 - 12x - 4;\ d(x) = 3x - 2$

3. $P(x) = x^4 - x^3 - 8x^2 + 6x - 4;\ d(x) = x^2 + 2x - 1$

4. $P(x) = 2x^4 + x^3 - 5x^2 + 14x + 7;\ d(x) = 2x^2 + x - 5$

5. $P(x) = x^4 - x^3 - x^2 + 8x + 4;\ d(x) = x^2 - 3$

6. $P(x) = 6x^4 - 2x^3 + 29x^2 - 7x + 28;\ d(x) = 2x^2 + 7$

In Exercises 7–12, solve each equation, given that c is one of the solutions of the equation.

7. $x^3 - 7x^2 - 20x + 96 = 0;\ c = 8$

8. $x^3 - 31x - 30 = 0;\ c = 6$

9. $x^3 - 7x - 6 = 0;\ c = 3$

10. $2x^3 - 10x^2 - 28x + 60 = 0;\ c = -3$

11. $x^4 + 7x^3 - 4x^2 = 28x;\ c = 2$

12. $6x^4 + 7x^3 = x^2 + 2x;\ c = -1$

In Exercises 13–18, solve each polynomial equation using an appropriate graph.

13. $x^3 - 2x^2 - x + 2 = 0$

14. $x^3 - 3x^2 - 4x + 12 = 0$

15. $x^3 + 3x^2 - 8x - 24 = 0$

16. $x^4 - 6x^3 - 28x^2 + 72x + 192 = 0$

17. $x^3 + 2x^2 - 15x = 36$ 18. $2x^3 + x^2 = 8x - 5$

In Exercises 19–24, determine a polynomial function P with the zeros given.

19. $-2, 2, 6$

20. $-5, -1, 3$

21. $-4, -\frac{3}{2}, 0, 4$

22. $-3, 0, \frac{5}{3}$

23. $-\sqrt{5}, -\sqrt{2}, \sqrt{2}, \sqrt{5}$

24. $-\sqrt{3}, -1, 1, \sqrt{3}$

B

In Exercises 25–36, solve each polynomial equation.

25. $x^3 = 5x^2 - 18$

26. $x^3 - 60 = -10x^2 - 14x$

27. $-2x^4 + 14x^3 + 26x^2 = 70x + 80$

28. $x^4 + 60 = -4x^3 + 17x^2 + 48x$

29. $x^4 + x^3 = 10x^2 + 16x + 96$

30. $x^4 + 2x^3 = 5x^2 + 18x + 36$

31. $2x^5 + 16x^3 + 120x = 12x^4 + 20x^2 + 160$

32. $x^4 + 12x^3 + 24x^2 = x^5 + 24x + 288$

33. $(x^2 - 20)(x^3 + 6x^2 + 6x - 8) = 0$

34. $(x^2 + 4)(x^3 + 7x^2 + 10x - 6) = 0$

35. $(x^3 - 6x^2 - 8x + 48)(x^3 + 2x^2 + 5x + 10) = 0$

36. $(x^3 + 7x^2 - 2x - 14)(x^3 + x^2 + 2x + 2) = 0$

37. Solve: $x^3 - \frac{13}{4}x^2 - \frac{59}{8}x + \frac{35}{2} = 0$
    HINT QUICK REVIEW: Multiply each side of the equation by 8.

38. Solve: $x^3 - \frac{13}{6}x^2 - \frac{20}{3}x + \frac{25}{2} = 0$
    HINT: See Exercise 37.

In Exercises 39–44, determine each polynomial function P described.

39. The degree of P is 3. The zeros of P are $-2, \frac{1}{2}$, and 4. The coefficient of the leading term is -4.

40. The degree of P is 3. The zeros of P are $-\frac{2}{3}$, 0, and 5. The coefficient of the leading term is 3.

41. The degree of P is 4. Also, $P(-2) = 0$, $P(1) = 0$, $P(\sqrt{3}) = 0$, $P(-\sqrt{3}) = 0$, and $P(-1) = 8$.

42. The degree of P is 4. P is an even function. Also, $P(3) = 0$, $P(-1) = 0$, and $P(0) = 9$.

43. P is the third-degree polynomial whose graph is shown in Figure 3.24.

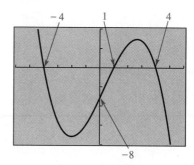

Figure 3.24
$[-6, 6] \times [-20, 10]_5$

44. P is the third-degree polynomial whose graph is shown in Figure 3.25.

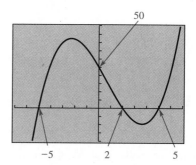

Figure 3.25
$[-7, 7] \times [-40, 100]_{10}$

45. Solve the inequality $x^3 - 5x^2 + 6x \leq 0$ graphically.

46. Solve the inequality $2x^3 + 90 \leq 11x^2 + 21x$ graphically.

47. Determine the domain of the function $f(x) = \sqrt{9x^2 - x^4}$.

48. Determine the domain of the function $f(x) = \sqrt[4]{x^3 + 7x^2}$.

49. The graph of $y = x^4 - 4x^3 - 5x^2 + 12x + 6$ has four x-intercepts. Two of these intercepts are $(-\sqrt{3}, 0)$ and $(\sqrt{3}, 0)$. What are the other two x-intercepts?

50. The graph of $y = -x^4 + 8x^3 + 22x^2 + 160x - 40$ has four x-intercepts. Two of these intercepts are $(-2\sqrt{5}, 0)$ and $(2\sqrt{5}, 0)$. What are the other two x-intercepts?

51. Factor $x^4 + 2x^3 - 23x^2 - 16x + 120$.

52. Factor $x^4 - 3x^3 - 40x^2 + 36x + 336$.

53. At the start of an experiment, a box has height, length, and width each equal to 24 centimeters. The box changes shape such that the height increases 2 centimeters per second, the length increases 1 centimeter per second, and the width decreases 3 centimeters per second.
(a) Show that the volume of this box after t seconds is given by $V = (24 + 2t)(24 + t)(24 - 3t)$.
(b) After how many seconds is the volume 8874 cubic centimeters?

54. A cone starts with radius 24 centimeters and height 36 centimeters. The cone changes shape such that the radius increases 3 centimeters per second and the height decreases 3 centimeters per second.
(a) Show that the volume of the cone after t seconds is given by $V = 9\pi(8 + t)^2(12 - t)$.
(b) After how many seconds is the volume 9000π cubic centimeters?

55. A grain silo is constructed from a cylinder and a hemisphere. The height of the silo is 12 meters. What must the radius be if the volume is 99π cubic meters?

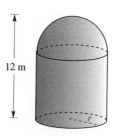

12 m

56. A propane tank is constructed from a cylinder and two hemispheres. The length of the tank is 90 inches. What must the radius of the cylinder be if the volume of the tank is 792π cubic inches?

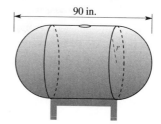

90 in.

C

57. (a) Use the factor theorem to show that $x - 1$ is a factor of $x^{56} - 6x^{43} + 3x^{12} + 2$.
(b) Use the factor theorem to show that $x + 1$ is a factor of $4x^{27} + 6x^{18} + 9x^{11} + 7$.

58. (a) Use the factor theorem to show that if n is a positive integer, then $x - a$ is a factor of $x^n - a^n$.
(b) Use the factor theorem to show that if n is a positive even integer, then $x + a$ is a factor of $x^n - a^n$.
(c) Use the factor theorem to show that if n is a positive odd integer, then $x + a$ is a factor of $x^n + a^n$.

COMPLEX NUMBERS

The set of real numbers is not sufficient to solve all the polynomial equations we have encountered. For example, if we try to solve the quadratic equation $x^2 = -4$ by taking square roots of each side, we get $x = \pm\sqrt{-4}$. Because the expression $\sqrt{-4}$ is undefined as a real number, this equation has no real number solutions.

In this section, we define a set of numbers—the *complex numbers*—with which we can solve equations such as $x^2 = -4$. Also, we need this more complete number system in order to continue our discussion of higher-order polynomial functions and equations from Sections 3.1 and 3.2.

The Imaginary Unit and Complex Numbers

Our goal is to define expressions such as $\sqrt{-4}$. The usual properties of square roots suggest that we can write

$$\sqrt{-4} = \sqrt{4(-1)} = \sqrt{4}\sqrt{-1} = 2\sqrt{-1}$$

Making sense of this last expression reduces to assigning some meaning to $\sqrt{-1}$. We define the value of $\sqrt{-1}$ as the number i, the *imaginary unit*.

Imaginary Unit

> The **imaginary unit** is $i = \sqrt{-1}$.

■ **NOTE:** *A better name for i might be the "imageless unit," since this number has no image on the real coordinate line.*

This number i is not a real number, since the square root of a negative number is not defined in the real number system, but it does give us a way to discuss expressions such as $\sqrt{-4}$:

$$\sqrt{-4} = 2\sqrt{-1} = 2i$$

We call $2i$ the *principal square root* of -4. It follows directly that the solutions to $x^2 = -4$ are $2i$ and $-2i$.

Principal Square Root of a Negative Number

> If a is a positive real number, then $\sqrt{-a}$, the **principal square root** of $-a$, is $\sqrt{a}\,i$. Furthermore, the solutions to the equation $x^2 = -a$ are $\sqrt{a}\,i$ and $-\sqrt{a}\,i$.

Astute readers will take issue with using the property

$$\sqrt{ab} = \sqrt{a}\sqrt{b}$$

from Section 5 (page 39) to say that $\sqrt{-4} = 2\sqrt{-1}$, because this property applied to nonnegative numbers a and b only. In this case, however, this apparent misuse does not cause any inconsistencies. But using this property when both numbers a and b are negative does lead to contradictions (see Exercise 68).

Example 1 shows how other expressions are simplified using the imaginary unit i.

◉ EXAMPLE 1   Using the Imaginary Unit

Rewrite each expression using the imaginary unit i.

(a) $\sqrt{-45}$ (b) $\dfrac{12 - \sqrt{-28}}{2}$

SOLUTION

In each case, we use the definition of the principal square root of a negative number.

(a) $\sqrt{-45} = \sqrt{45}\,i = 3\sqrt{5}\,i$

(b) $\dfrac{12 - \sqrt{-28}}{2} = \dfrac{12 - \sqrt{28}\,i}{2} = \dfrac{12 - 2\sqrt{7}\,i}{2} = \dfrac{12}{2} - \dfrac{2\sqrt{7}\,i}{2} = 6 - \sqrt{7}i$

From the definition of i, it follows directly that $i^2 = -1$. On occasion, higher powers of i need to be simplified. For example, using the fact that $i^2 = -1$ and the normal rules of exponents, we see that

$$i^3 = (i^2)i = (-1)i = -i \qquad i^6 = (i^2)^3 = (-1)^3 = -1$$
$$i^5 = (i^2)^2 i = (-1)^2 i = i \qquad i^{12} = (i^2)^6 = (-1)^6 = 1$$

NOTE: *A better name would be the "complete number system," since this number system contains the solution to all quadratic equations.*

In fact, any integer power of i is equal to i, $-i$, -1, or 1 (see Exercise 69).

We can use this non-real number i to construct an extended number system that contains the solutions of all quadratic equations. We call this new system of numbers the *complex number system.*

Complex Numbers

> The **complex number system** is the set of all numbers $a + bi$, in which a and b are real numbers, and i is the imaginary unit. The term a is the **real part** of $a + bi$, and the term bi is the **imaginary part** of $a + bi$. Also, two complex numbers $a + bi$ and $c + di$ are **equal** if $a = c$ and $b = d$.

Any real number is also a complex number. For example, the real number -2 can be written as $-2 + 0i$, so it qualifies as a complex number according to the definition. Thus, the set of real numbers is a subset of the set of complex numbers (Figure 3.26).

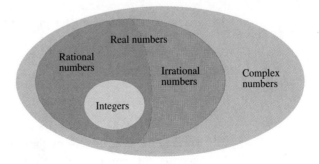

Figure 3.26

■ **NOTE:** *Some graphing calculators perform operations on complex numbers. For many of these calculators, a complex number such as 2 + 4i is represented as (2, 4).*

Operations on Complex Numbers

Addition, subtraction, multiplication, and division are performed in a similar way on complex numbers as on first-degree polynomials. The one exception is that i^2, when it occurs in an expression, is replaced by -1. Example 2 illustrates how each of these operations is performed.

◉ EXAMPLE 2   Operations on Complex Numbers

Evaluate each complex expression.

(a) $(4 + 2i) + (3 - i)$ **(b)** $\left(\frac{3}{2} - 6i\right) - \left(\frac{1}{2} + 4i\right)$

(c) $(3 + 4i)(1 - 2i)$ **(d)** $\dfrac{7 + 4i}{4 - 2i}$

SOLUTION

(a) The plan of attack is to simplify the expression by collecting similar terms:

$$(4 + 2i) + (3 - i) = 4 + 2i + 3 - i = 7 + i$$

(b) We again collect similar terms:

$$\left(\tfrac{3}{2} - 6i\right) - \left(\tfrac{1}{2} + 4i\right) = \tfrac{3}{2} - 6i - \tfrac{1}{2} - 4i = 1 - 10i$$

(c) The FOIL method from Section 2 (page 12) allows us to simplify the product of two complex numbers:

■ **NOTE:** *This computation is similar to simplifying $(3 + 4x)(1 - 2x)$ by FOIL to get $3 - 2x - 8x^2$.*

$$
\begin{aligned}
(3 + 4i)(1 - 2i) &= 3 - 2i - 8i^2 && \text{Use FOIL.}\\
&= 3 - 2i - 8(-1) && \text{Replace } i^2 \text{ with } -1.\\
&= 3 - 2i + 8\\
&= 11 - 2i
\end{aligned}
$$

(d) Using the definition of i, the given expression can be rewritten as

$$\frac{7 + 4i}{4 - 2i} = \frac{7 + 4\sqrt{-1}}{4 - 2\sqrt{-1}}$$

In Section 5 (page 39), we saw that an expression with a square root in one of the terms of the denominator can be simplified by multiplying the numerator and the denominator by the conjugate of the denominator. In this expression, the conjugate of the denominator $4 - 2i$ is the complex number $4 + 2i$.

$$\frac{7 + 4i}{4 - 2i} = \frac{7 + 4i}{4 - 2i} \cdot \frac{4 + 2i}{4 + 2i} \qquad \text{Multiply numerator and denominator by the conjugate of } 4 - 2i.$$

$$= \frac{28 + 30i + 8i^2}{16 - 4i^2} \qquad \text{Simplify numerator and denominator with FOIL.}$$

$$= \frac{28 + 30i + 8(-1)}{16 - 4(-1)} \qquad \text{Replace } i^2 \text{ with } -1.$$

$$= \frac{28 + 30i - 8}{16 + 4}$$

$$= \frac{20 + 30i}{20}$$

$$= 1 + \tfrac{3}{2}i$$

These four operations may be generalized as follows:

Operations on Complex Numbers

Suppose that $a + bi$ and $c + di$ are complex numbers.

Addition:

$$(a + bi) + (c + di) = (a + c) + (b + d)i$$

Subtraction:

$$(a + bi) - (c + di) = (a - c) + (b - d)i$$

Multiplication:

$$(a + bi)(c + di) = ac + (ad + bc)i + (bd)i^2$$
$$= (ac - bd) + (ad + bc)i$$

Division:

$$\frac{a + bi}{c + di} = \frac{a + bi}{c + di} \cdot \frac{c - di}{c - di} = \frac{(ac + bd) + (-ad + bc)i}{c^2 - d^2i^2}$$

$$= \frac{ac + bd}{c^2 + d^2} + \frac{-ad + bc}{c^2 + d^2}i$$

As we saw in part (d) of Example 2, the *conjugate of a complex number* plays a role in computing the quotient of two complex numbers.

Conjugate of a Complex Number

Let $z = a + bi$. The **conjugate** of z is

$$\bar{z} = a - bi$$

For example,

$$\overline{3 + 4i} = 3 - 4i \qquad\qquad \overline{-\sqrt{2} - i} = -\sqrt{2} + i$$

$$\overline{2i} = \overline{0 + 2i} = 0 - 2i = -2i \qquad \overline{6} = \overline{6 + 0i} = 6 - 0i = 6$$

The importance of the conjugate of a complex number (or *complex conjugate*) lies in the fact that the product of a complex number z and its conjugate $\bar{z}$ is a real number:

$$z\bar{z} = (a + bi)\overline{(a + bi)} = (a + bi)(a - bi) = a^2 - b^2i^2 = a^2 + b^2$$

Example 3 shows more computations involving complex numbers.

⬤ EXAMPLE 3   Operations on Complex Numbers

Let $w = 3 + i$ and $z = -1 + 2i$. Evaluate:

(a) $2z - 3w$ **(b)** $i^4\bar{w}$ **(c)** z^{-1}

SOLUTION

(a)
$$2z - 3w = 2(-1 + 2i) - 3(3 + i)$$
$$= -2 + 4i - 9 - 3i$$
$$= -11 + i$$

(b)
$$i^4\bar{w} = i^4\overline{(3 + i)}$$
$$= i^4(3 - i)$$
$$= (i^2)^2(3 - i)$$
$$= (-1)^2(3 - i) \quad \text{Replace } i^2 \text{ with } -1.$$
$$= 1(3 - i)$$
$$= 3 - i$$

(c)
$$z^{-1} = (-1 + 2i)^{-1}$$

$$= \frac{1}{-1 + 2i}$$

$$= \frac{1}{-1 + 2i} \cdot \frac{-1 - 2i}{-1 - 2i} \quad \begin{array}{l}\text{Multiply numerator and denominator}\\ \text{by the conjugate of } -1 + 2i.\end{array}$$

$$= \frac{-1 - 2i}{(-1)^2 - 4i^2}$$

$$= \frac{-1 - 2i}{1 - 4(-1)} \qquad \text{Replace } i^2 \text{ with } -1.$$

$$= \frac{-1 - 2i}{5}$$

$$= -\frac{1}{5} - \frac{2}{5}i$$

Quadratic Equations

Up to this point, solving an equation has meant finding all the real numbers that make the equation true. An equation such as $x^2 = 2$ has real number solutions, namely $-\sqrt{2}$ and $\sqrt{2}$. As we saw earlier, the equation $x^2 = -4$ has no real solutions, but it does have complex number solutions, $-2i$ and $2i$. We say that $x^2 = -4$ has no solutions when *solved over the real numbers,* but it has solutions $-2i$ and $2i$ when *solved over the complex numbers.*

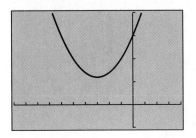

Figure 3.27
$y = (x + 3)^2 + 6$
$[-10, 4] \times [-5, 20]_5$

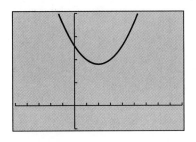

Figure 3.28
$x^2 - 4x + 13 = 0$
$[-5, 9] \times [-5, 20]_5$

 EXAMPLE 4 Solving Quadratic Equations Over the Complex Numbers

Solve each quadratic equation over the complex numbers.

(a) $(x + 3)^2 = -6$ **(b)** $x^2 - 4x + 13 = 0$

SOLUTION

(a) This equation can be solved by the method of taking square roots of each side:

$$(x + 3)^2 = -6$$
$$x + 3 = \pm\sqrt{-6}$$
$$x = -3 \pm \sqrt{6}\,i$$

The complex number solutions are $-3 - \sqrt{6}\,i$ and $-3 + \sqrt{6}\,i$.

(b) We use the quadratic formula to solve the equation $x^2 - 4x + 13 = 0$:

$$x = \frac{-(-4) \pm \sqrt{(-4)^2 - 4(1)(13)}}{2(1)} = \frac{4 \pm \sqrt{-36}}{2} = \frac{4 \pm 6i}{2} = 2 \pm 3i$$

The solutions are $2 - 3i$ and $2 + 3i$.

Neither of the equations solved in Example 4 has real number solutions, so it is not a surprise that their graphs have no x-intercepts (Figures 3.27 and 3.28).

Higher-Order Polynomial Equations

Higher-order polynomial equations may also have solutions that are not real numbers, as we see in Example 5.

EXAMPLE 5 Solving a Cubic Equation Over the Complex Numbers

Solve $x^3 + 8 = 0$ over the complex numbers.

SOLUTION

We start by factoring:

$$x^3 + 8 = 0$$
$$x^3 + 2^3 = 0 \qquad \text{Rewrite as a sum of cubes.}$$
$$(x + 2)(x^2 - 2x + 4) = 0 \qquad \text{Factor using the sum of cubes formula.}$$
$$x + 2 = 0 \quad | \quad x^2 - 2x + 4 = 0$$

The solution to the equation $x + 2 = 0$ is -2. We solve $x^2 - 2x + 4 = 0$ with the quadratic formula:

$$x = \frac{-(-2) \pm \sqrt{(-2)^2 - 4(1)(4)}}{2(1)} = \frac{2 \pm \sqrt{-12}}{2} = \frac{2 \pm 2\sqrt{3}\,i}{2} = 1 \pm \sqrt{3}\,i$$

The complex solutions to the equation are -2, $1 + \sqrt{3}\,i$, and $1 - \sqrt{3}\,i$.

Notice that if we had used the method of taking cube roots in Example 5 (as outlined in Section 2.3), we would have found only the real number solution, as follows:

$$x^3 + 8 = 0$$
$$x^3 = -8$$
$$x = \sqrt[3]{-8} \qquad \text{Take the cube root of each side.}$$
$$x = -2$$

In general, if we try to solve a polynomial equation of degree greater than 2 by taking roots, we do not find all the complex number solutions.

EXAMPLE 6   Finding the Complex Zeros of a Function

Find the complex zeros of $f(x) = x^4 - 13x^2 - 48$.

SOLUTION

The complex zeros of the function are the complex number solutions of the equation $f(x) = 0$, which we can solve by factoring:

$$x^4 - 13x^2 - 48 = 0$$
$$(x^2)^2 - 13(x^2) - 48 = 0$$
$$(x^2 - 16)(x^2 + 3) = 0$$
$$x^2 - 16 = 0 \quad \Big| \quad x^2 + 3 = 0$$
$$x^2 = 16 \quad \Big| \quad x^2 = -3$$
$$x = \pm 4 \quad \Big| \quad x = \pm \sqrt{3}\,i$$

The complex number solutions are -4, 4, $\sqrt{3}\,i$, and $-\sqrt{3}\,i$.

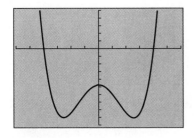

Figure 3.29
$y = x^4 - 13x^2 - 48$
$[-6, 6] \times [-100, 50]_{10}$

Suppose that we try to find the zeros of the function in Example 6 graphically. The graph of $f(x) = x^4 - 13x^2 - 48$ shows two x-intercepts, $(-4, 0)$ and $(4, 0)$, that correspond to the real zeros of the function (Figure 3.29). The non-real zeros, $\sqrt{3}\,i$ and $-\sqrt{3}\,i$, are not represented by x-intercepts.

Fundamental Theorem of Algebra

One of the problems in solving a polynomial equation is determining how many solutions to expect. The number of real solutions of an equation $P(x) = 0$ is easy to determine. Each real solution corresponds to an x-intercept of the graph of the function P, so counting the x-intercepts of the graph tells you how many real solutions there are.

What about the number of complex solutions of a polynomial equation, or equivalently, what about the number of complex zeros of a polynomial function? For example, the equation $x^3 + 8 = 0$ in Example 5 has three complex solutions, one of which is a real number. Likewise, the function $f(x) = x^4 - 13x^2 - 48$ in Example 6 has four complex zeros, two of which are real numbers. The theorems discussed in the remainder of this section address this question.

The *fundamental theorem of algebra,* as the name suggests, is of significant importance in mathematics.

Fundamental Theorem of Algebra

> Every polynomial function of degree $n > 0$ has at least one complex zero.

NOTE: *The theorem does not tell us how to find the zero — just that it exists. This type of theorem is called an* existence *theorem.*

The proof of the theorem entails calculus and advanced complex number theory, so we will be content to state it without proof.

The fundamental theorem of algebra allows us to say something about the number of complex zeros of a polynomial function. Suppose that we have the nth-degree polynomial equation

$$P(x) = a_n x^n + a_{n-1} x^{n-1} + a_{n-2} x^{n-2} + \cdots + a_1 x + a_0$$

and that c_1 is a zero of P. By the factor theorem of Section 3.2, we know that there exists a polynomial function of degree $n - 1$, say $q_{n-1}(x)$, such that

$$P(x) = (x - c_1)q_{n-1}(x)$$

By the same reasoning, there exists $q_{n-2}(x)$, a polynomial function of degree $n - 2$ such that

$$q_{n-1}(x) = (x - c_2)q_{n-2}(x)$$

so

$$P(x) = (x - c_1)q_{n-1}(x)$$
$$= (x - c_1)[(x - c_2)q_{n-2}(x)]$$
$$= (x - c_1)(x - c_2)q_{n-2}(x)$$

Continuing this process eventually gives us

$$P(x) = (x - c_1)(x - c_2) \cdot \cdots \cdot (x - c_n)q$$

in which q is a constant. This number q is equal to a_n, the leading coefficient of P (see Exercise 70). Thus,

$$P(x) = a_n(x - c_1)(x - c_2) \cdot \cdots \cdot (x - c_n)$$

and, by the factor theorem, the complex numbers $c_1, c_2, \cdots, c_n$ are zeros of P.

Are these numbers $c_1, c_2, \cdots, c_n$ the only zeros of P? Suppose that

$$a_n(k - c_1)(k - c_2) \cdot \cdots \cdot (k - c_n) = 0$$

for some complex number k. The zero factor property implies that one of these factors must be 0, say $k - c_j = 0$ for some j; and it follows that $k = c_j$. Thus, the only zeros of P are the numbers $c_1, c_2, \cdots, c_n$.

Our discussion above is the proof of the *complete factorization theorem.*

Complete Factorization Theorem

If P is a polynomial function of degree $n > 0$ and with leading coefficient a_n, then
$$P(x) = a_n(x - c_1)(x - c_2) \cdot \cdots \cdot (x - c_n)$$
where $c_1, c_2, \cdots, c_n$ are the complex zeros of P.

Nothing in our proof of this theorem guarantees that the zeros of a polynomial function are distinct. Indeed, some of the zeros of a polynomial function may appear in more than one of its linear factors.

For example, the polynomial function

$$P(x) = 2(x - 5)^3\left(x + \tfrac{3}{2}\right)(x + 7)^2$$

has three factors $x - 5$, one factor $x + \tfrac{3}{2}$, and two factors $x + 7$. Because the factor $x - 5$ occurs three times, we say that 5 is *a zero of multiplicity* 3. Likewise, $-\tfrac{3}{2}$ is a root of multiplicity 1, and -7 is a zero of multiplicity 2. Even though this function has only three distinct zeros, 5, $-\tfrac{3}{2}$, and -7, we get six zeros if we list each zero according to its multiplicity:

$$5, 5, 5, -\tfrac{3}{2}, -7, -7$$

In general, if the zeros of a polynomial function P are counted according to multiplicity, then this number is equal to the degree of the function. This fact follows from the complete factorization theorem, since each of these zeros corresponds to one of the linear factors of P, and the number of linear factors is equal to the degree of P.

Number of Complex Zeros
of a Polynomial Function

> If P is a polynomial function of degree n, then P has exactly n zeros (counting multiplicity). Equivalently, the polynomial equation $P(x) = 0$ has n solutions (counting multiplicity).

Our findings in Examples 5 and 6 agree with this statement. The equation $x^3 + 8 = 0$ has three solutions, and the function $f(x) = x^4 - 13x^2 - 48$ has four zeros.

● EXAMPLE 7 Finding the Complex Zeros of a Function Graphically

Let $g(x) = x^3 + 11x^2 + 23x + 45$. Use the graph of g to determine its complex zeros and its linear factored form.

SOLUTION

Since g is a polynomial function of degree 3 with leading coefficient 1, its linear factored form is $g(x) = (x - c_1)(x - c_2)(x - c_3)$ in which c_1, c_2, and c_3 are the zeros of the function. The graph of g appears to have an x-intercept at $(-9, 0)$, implying that $x + 9$ is a linear factor of g (Figure 3.30). By the factor theorem of Section 3.2, we can write g as $g(x) = (x + 9)q(x)$ for some polynomial function $q(x)$. We can find $q(x)$ by long division:

$$
\begin{array}{r}
x^2 + 2x + 5 \\
x + 9 \overline{)x^3 + 11x^2 + 23x + 45} \\
\underline{x^3 + 9x^2} \\
2x^2 + 23x \\
\underline{2x^2 + 18x} \\
5x + 45 \\
\underline{5x + 45}
\end{array}
$$

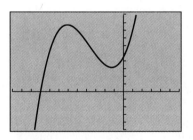

Figure 3.30
$y = x^3 + 11x^2 + 23x + 45$
$[-12, 6] \times [-50, 100]_{10}$

Thus,

$$g(x) = (x + 9)(x^2 + 2x + 5)$$

The remaining two zeros of g are the solutions of $x^2 + 2x + 5 = 0$:

$$x = \frac{-2 \pm \sqrt{2^2 - 4(1)(5)}}{2(1)} = \frac{-2 \pm \sqrt{-16}}{2} = \frac{-2 \pm 4i}{2} = -1 \pm 2i$$

The linear factored form of g is

$$g(x) = (x + 9)[x - (-1 + 2i)][x - (-1 - 2i)]$$

or

$$g(x) = (x + 9)(x + 1 - 2i)(x + 1 + 2i)$$

In Example 7, the two non-real zeros are a pair of complex conjugates. This occurrence is not coincidental.

Complex Conjugate Theorem

> Suppose that P is a polynomial function with real number coefficients. If $a + bi$ is a zero of P, then so is its complex conjugate $a - bi$.

The proof of this theorem is left as an exercise (see Exercise 72).

EXAMPLE 8 Determining a Polynomial Function with Given Zeros

Determine the polynomial function P of degree 4 with real number coefficients, with leading coefficient 2, and with zeros 3 (multiplicity 2) and $1 - i$ (multiplicity 1).

SOLUTION

The function P is of the form

$$P(x) = 2(x - c_1)(x - c_2)(x - c_3)(x - c_4)$$

in which c_1, c_2, c_3, and c_4 are the four zeros of P. Two of these zeros are 3, and one is $1 - i$. By the complex conjugate theorem, $1 + i$ is also a zero. Thus,

$$P(x) = 2(x - 3)(x - 3)[x - (1 - i)][x - (1 + i)]$$

Multiplying these linear factors gives us a simplified form of P:

$$P(x) = 2\underbrace{(x - 3)(x - 3)}_{x^2 - 6x + 9}\underbrace{[x - (1 - i)][x - (1 + i)]}_{x^2 - 2x + 2}$$
$$= 2(x^2 - 6x + 9)(x^2 - 2x + 2)$$
$$= 2(x^4 - 8x^3 + 23x^2 - 30x + 18)$$
$$= 2x^4 - 16x^3 + 46x^2 - 60x + 36$$

The function is

$$P(x) = 2x^4 - 16x^3 + 46x^2 - 60x + 36$$

EXERCISE SET 3.3

A

In Exercises 1–8, rewrite each expression using the imaginary unit i, and simplify.

1. (a) $\sqrt{-9}$ (b) $\sqrt{-13}$
2. (a) $\sqrt{-25}$ (b) $\sqrt{-7}$
3. (a) $\sqrt{-12}$ (b) $-\sqrt{-8}$
4. (a) $\sqrt{-24}$ (b) $-\sqrt{-27}$
5. (a) $7 - \sqrt{-3}$ (b) $-\sqrt{4} + \sqrt{-4}$
6. (a) $9 - \sqrt{-7}$ (b) $-\sqrt{16} + \sqrt{-9}$
7. (a) $\dfrac{8 + \sqrt{-16}}{4}$ (b) $\dfrac{14 - \sqrt{-12}}{2}$
8. (a) $\dfrac{6 + \sqrt{-9}}{3}$ (b) $\dfrac{4 - \sqrt{-32}}{4}$

In Exercises 9–16, evaluate each expression. Write your answer in the form $a + bi$.

9. (a) $(2 + i) + (4 - 7i)$ (b) $(4 + 10i) - (12 - 2i)$
10. (a) $(8 - 2i) + (10 - i)$ (b) $(4 - i) - (-8 + 5i)$
11. (a) $(2 + 2i)(1 + 3i)$ (b) $(3 + i)(4 - 5i)$
12. (a) $(-4 + i)(3 - 2i)$ (b) $(4 + 2i)(5 - i)$
13. (a) $\left(-\frac{5}{2} + 4i\right)\left(2 - \frac{1}{2}i\right)$ (b) $(2 + 3i)^2$
14. (a) $\left(3 + \frac{1}{2}i\right)\left(\frac{1}{3} - 2i\right)$ (b) $(6 - i)^2$
15. (a) $\dfrac{2 - i}{1 + i}$ (b) $\dfrac{5 + i}{4 - 3i}$
16. (a) $\dfrac{3 + i}{1 - 2i}$ (b) $\dfrac{-2 - 2i}{4 + 2i}$

In Exercises 17–28, solve each quadratic equation over the complex numbers.

17. $4x^2 = -12$ 18. $2x^2 = -18$
19. $(x - 4)^2 = -16$ 20. $(x + 5)^2 = -36$
21. $9(x - 5)^2 + 16 = 0$ 22. $12(x + 2)^2 + 3 = 0$
23. $x^2 - 2x + 6 = 0$ 24. $x^2 + 4x + 12 = 0$
25. $x^2 + 5x - 6 = 0$ 26. $2x^2 - 7x + 6 = 0$
27. $3x^2 = x - 1$ 28. $2x^2 = 6x - 7$

B

In Exercises 29–40, let $w = 2 - i$ and $z = 3 - 7i$. Simplify each expression, and write your answers in the form $a + bi$.

29. $2wz - z^2$ 30. $3z + 2w^2$ 31. w^3
32. z^3 33. $\overline{w}w$ 34. $\overline{z}z$
35. $z(2 + w)$ 36. $w(3 - 4z)$ 37. $\dfrac{w}{iz}$
38. $\dfrac{2z}{w + i}$ 39. w^{-2} 40. z^{-2}

In Exercises 41–52, solve each polynomial equation over the complex numbers.

41. $(x^2 + 12)(x^2 - 6x + 8) = 0$
42. $(x^2 - 5x - 6)(x^2 + 8) = 0$
43. $x^3 - 64 = 0$ 44. $x^3 + 27 = 0$
45. $x^4 + 13x^2 + 36 = 0$ 46. $x^4 + 10x^2 + 9 = 0$
47. $x^3 + 3x^2 + 8x + 24 = 0$
48. $x^3 - 2x^2 + x - 2 = 0$
49. $x^3 + 6x^2 - x - 30 = 0$
50. $x^3 + 2x^2 - 41x - 42 = 0$
51. $x^4 - 6x - 105 = 2x^3 + 32x^2$
52. $2x^4 - 5x - 15 = x^3 - 7x^2$

In Exercises 53–60, write each polynomial function f in linear factored form.

53. $f(x) = 2x^2 + 4x - 30$ 54. $f(x) = -5x^2 + 20$
55. $f(x) = \frac{1}{2}x^2 + 8$ 56. $f(x) = -x^2 - 9$
57. $f(x) = x^3 - 2x^2 + 25x - 50$
58. $f(x) = x^3 + 3x^2 + 16x + 48$
59. $f(x) = 2x^4 + 8x^3 + 8x^2 - 32x - 64$
60. $f(x) = -x^4 - 2x^3 + 14x^2 + 32x + 32$

In Exercises 61–66, determine each polynomial function P (with real number coefficients) that is described.

61. The leading coefficient is -2, the degree is 3, zeros are 1 (multiplicity 2) and -3.

62. The leading coefficient is 5, the degree is 3, zeros are $\frac{1}{5}$ and 7 (multiplicity 2).

63. The leading coefficient is 1, the degree is 4, zeros are -4 (multiplicity 2) and $\sqrt{3}\,i$.

64. The leading coefficient is 2, the degree is 4, zeros are -1 (multiplicity 2) and $2i$.

65. The degree is 3, zeros are -2 (multiplicity 2) and 1, and $P(-4) = -5$.

66. The degree is 4, zeros are 1 (multiplicity 2) and -1 (multiplicity 2), and $P(2) = 4$.

C

67. Solve $x^2 + 2ix + 3 = 0$ over the complex numbers. Does this solution violate the complex conjugate theorem?

68. Evalute $\sqrt{-4}\sqrt{-9}$ and $\sqrt{(-4)(-9)}$. Is it true that for all real numbers x and y, $\sqrt{x}\sqrt{y} = \sqrt{xy}$?

69. **(a)** Explain why any integer n can be written in the form $n = 4k + r$ for some positive integer k and $r = 0, 1, 2,$ or 3.

(b) Show that for any integer n, $i^n = i^{4k+r} = i^r$. Explain why this result implies that $i^n = 1, i, -1,$ or $-i$.

70. Given a polynomial function

$$P(x) = a_n x^n + a_{n-1} x^{n-1} + a_{n-2} x^{n-2} + \cdots + a_1 x + a_0$$

with a linear factored form

$$P(x) = q(x - c_1)(x - c_2) \cdot \cdots \cdot (x - c_n)$$

show that $q = a_n$.

HINT: Consider the simplified form of the linear factored form.

71. Suppose that $z_1 = a_1 + b_1 i$, $z_2 = a_2 + b_2 i$, and n is a positive integer. Show that:

(a) $z_1 + \overline{z_1} = 2a_1$ **(b)** $\overline{z_1 + z_2} = \overline{z_1} + \overline{z_2}$
(c) $\overline{z_1 \cdot z_2} = \overline{z_1} \cdot \overline{z_2}$ **(d)** $z_1 \overline{z_1} = a_1^2 + b_1^2$
(e) $(\overline{z_1})^n = \overline{z_1^n}$

72. Suppose that

$$P(x) = a_n x^n + a_{n-1} x^{n-1} + a_{n-2} x^{n-2} + \cdots + a_1 x + a_0$$

is a polynomial function with real number coefficients. Use the results of Exercise 71 to show that if z is a complex number, then $P(\overline{z}) = \overline{P(z)}$. Explain why this fact proves the complex conjugate theorem.

RATIONAL FUNCTIONS

In Section 1.7, we saw that the sum, difference, product, or ratio of any two functions is a function. Furthermore, if $n(x)$ and $d(x)$ are polynomial functions, then $n(x) + d(x)$, $n(x) - d(x)$, and $n(x)d(x)$ are polynomial functions. However, the ratio $n(x)/d(x)$ is not a polynomial function in general; therefore, we define a *rational function* as follows:

Rational Function

A **rational function** f is a function that can be written in the form

$$f(x) = \frac{n(x)}{d(x)}$$

where $n(x)$ and $d(x)$ are polynomials.

A few examples of rational functions are

$$f(x) = \frac{x - 6}{x + 2} \qquad g(x) = \frac{-8}{x^2 + 3} \qquad h(x) = \frac{x + 2}{x^2 - 4}$$

NOTE: *The relationship between rational functions and polynomials is similar to the relationship between rational numbers and integers. Just as a rational number is the ratio of two integers, a rational function is the ratio of two polynomials.*

As with the graph of a polynomial function, the graph of a rational function is smooth (without corners). However, unlike a polynomial function, the graph of a rational function may have breaks, called *discontinuities*. For this reason, knowing the domain is an important part of determining the graph of a rational function.

Domain and Zeros

The domain of a rational function is the set of all real numbers such that its denominator is not 0, since division by 0 is undefined. For example, the rational function

$$f(x) = \frac{x - 6}{x + 2}$$

is undefined at $x = -2$. Consequently, the domain of f is all real numbers except -2.

A rational function is *reduced* if its numerator and denominator have no common factors. We will be concerned primarily with reduced rational functions. The zeros of a reduced rational function

$$f(x) = \frac{n(x)}{d(x)}$$

occur at the zeros of $n(x)$, since

$$\frac{n(x)}{d(x)} = 0 \qquad \text{if and only if} \qquad n(x) = 0$$

● **EXAMPLE 1   Finding the Domain and Zeros**
Determine the domain and zeros of each function.

(a) $f(x) = \dfrac{3x + 4}{x^2 - 9}$ **(b)** $g(x) = \dfrac{-5}{x^2 + 2}$ **(c)** $h(x) = \dfrac{x^2 + x}{2x + 2}$

SOLUTION

(a) The domain of f is the set of all real numbers except for the zeros of the denominator. Setting the denominator equal to 0 and solving, we get

$$x^2 - 9 = 0$$
$$(x - 3)(x + 3) = 0$$

$x - 3 = 0$	$x + 3 = 0$
$x = 3$	$x = -3$

The domain of f is the set of all real numbers except 3 and -3. Now, before we determine the zeros, we note that the rational function

$$f(x) = \frac{3x + 4}{x^2 - 9}$$

is reduced. The zeros of f correspond to the zeros of the numerator:

$$3x + 4 = 0$$
$$x = -\tfrac{4}{3}$$

(b) Since $x^2 + 2$ is not equal to 0 for any real value of x, the domain is the set of all real numbers. The rational function

$$g(x) = \frac{-5}{x^2 + 2}$$

is reduced, so the zeros of g correspond to the zeros of the numerator. Since the numerator, -5, is not equal to 0 for any real value of x, g has no zeros.

(c) Since the only zero of the denominator is -1, the domain of h is the set of all real numbers except -1. To determine the zeros, we first reduce $h(x)$:

$$h(x) = \frac{x^2 + x}{2x + 2}$$
$$= \frac{x(x + 1)}{2(x + 1)}$$
$$= \frac{x}{2} \qquad x \neq -1$$

The only zero of the numerator of $x/2$ is 0. Therefore, the only zero of h is 0.

Vertical Asymptotes

Consider the rational function

$$f(x) = \frac{1}{x - 2}$$

We can graph f by starting with the graph of $y = 1/x$ from the directory of graphs in Section 1.3 (repeated here in Figure 3.31). The graph of f is a translation of $y = 1/x$ to the right 2 units (Figure 3.32).

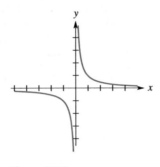

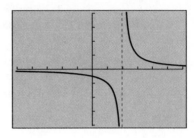

Figure 3.31

$$y = \frac{1}{x}$$

Figure 3.32

$$f(x) = \frac{1}{x - 2}$$

$$[-5, 6] \times [-4, 4]$$

Notice the behavior of $f(x) = 1/(x - 2)$ near $x = 2$. For values of x to the right of $x = 2$, as x approaches 2, the corresponding values of y increase without bound (Figure 3.33). Similarly, as x approaches 2 from the left, y decreases without bound (Figure 3.34).

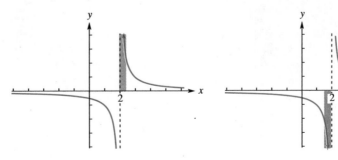

Figure 3.33   **Figure 3.34**

■ **GRAPHING NOTE:**

*Asymptotes are shown as
dashed lines because they are
not part of the actual graph.*

Vertical Asymptotes

The line $x = 2$ is called a *vertical asymptote* of f. Notice that 2 is a zero of the denominator of $f(x) = 1/(x - 2)$. In general, if a is a zero of the denominator of a reduced rational function $f(x)$, then the line $x = a$ is a vertical asymptote.

Let $f(x) = \dfrac{n(x)}{d(x)}$ be a reduced rational function:

If $d(a) = 0$, then the line $x = a$ is a vertical asymptote of the graph of f.

Since vertical asymptotes occur at values of x that are excluded from the domain, *the graph of a rational function never crosses its vertical asymptote.* This is important to remember when you use a graphing utility, because it may display a graph that erroneously intersects a vertical asymptote.

⬤ **EXAMPLE 2   Finding the Vertical Asymptotes**

Determine the vertical asymptotes of each function. Use a graphing utility to verify your results.

(a) $f(x) = \dfrac{2x^3}{6x^2 + 3x - 9}$ **(b)** $g(x) = \dfrac{7x}{x^2 + 4}$

SOLUTION

(a) First note that

$$f(x) = \frac{2x^3}{6x^2 + 3x - 9} = \frac{2x^3}{3(x - 1)(2x + 3)}$$

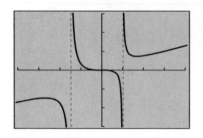

Figure 3.35

$y = \dfrac{2x^3}{6x^2 + 3x - 9}$

$[-4, 4] \times [-3, 3]$

is in reduced form. The vertical asymptotes correspond to the zeros of the denominator:

$$6x^2 + 3x - 9 = 0$$
$$3(x - 1)(2x + 3) = 0$$

$$x - 1 = 0 \quad \big| \quad 2x + 3 = 0$$
$$x = 1 \quad \big| \quad x = -\tfrac{3}{2}$$

Therefore, the lines $x = 1$ and $x = -\tfrac{3}{2}$ are the vertical asymptotes. The graph of f is shown in Figure 3.35.

(b) Since the numerator and denominator of

$$g(x) = \frac{7x}{x^2 + 4}$$

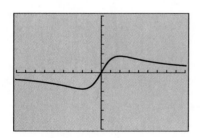

Figure 3.36

$y = \dfrac{7x}{x^2 + 4}$

$[-9, 9] \times [-6, 6]$

have no common factors, the rational function is reduced. The denominator has no real zeros. Thus, there are no vertical asymptotes. The graph of g is shown in Figure 3.36. ⬤

Behavior for Extreme Values of x

Consider the rational function

$$f(x) = \frac{3x + 7}{x + 2}$$

Performing polynomial division, we get

$$\begin{array}{r} 3 \\ x + 2 \overline{)\,3x + 7} \\ \underline{3x + 6} \\ 1 \end{array}$$

Thus, the function can also be expressed as

$$f(x) = 3 + \frac{1}{x + 2}$$

In this form we can sketch the graph of f by translating the graph of $y = 1/x$ 2 units to the left and 3 units up (Figure 3.37).

Notice the behavior of f for large values of x. As x increases without bound, y approaches 3 (Figure 3.38). Similarly, as x decreases without bound, y approaches 3 (Figure 3.39).

The behavior of f for values of x that are to the far right or far left of the graph is denoted as follows:

$$y \to 3 \quad \text{as} \quad x \to \pm\infty$$

The line $y = 3$ is called a *horizontal asymptote*. The key to finding the behavior of f for extreme values of x is to perform the division and write the function in the form

$$f(x) = 3 + \frac{1}{x + 2}$$

Recall from Section 3.2 that any reduced rational function $f(x) = n(x)/d(x)$ can be written in the form

$$f(x) = q(x) + \frac{r(x)}{d(x)}$$

where $r(x) = 0$, or the degree of $r(x)$ is less than the degree of $d(x)$. The purpose of expressing $f(x)$ in this form is revealed in the next example.

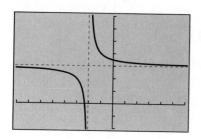

Figure 3.37

$$y = 3 + \frac{1}{x + 2}$$

$$[-8, 6] \times [-2, 7]$$

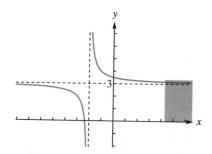

Figure 3.38

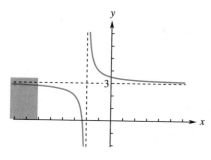

Figure 3.39

◉ EXAMPLE 3   Behavior for Extreme Values of x

Express $f(x)$ in the form

$$f(x) = q(x) + \frac{r(x)}{d(x)}$$

where the degree of $r(x)$ is less than the degree of $d(x)$. Graph the functions f and q in the given viewing window, and compare their behavior for extreme values of x.

(a) $f(x) = \dfrac{5 - 4x}{2x + 1}$; over $[-15, 15] \times [-10, 10]$

(b) $f(x) = \dfrac{10}{4x^2 - 9}$; over $[-9, 9] \times [-6, 6]$

(c) $f(x) = \dfrac{4x^2 - 17x - 8}{x - 5}$; over $[-10, 20]_5 \times [-35, 70]_{10}$

SOLUTION

(a) Performing the division, we have

$$\begin{array}{r} -2 \\ 2x + 1{\overline{\smash{\big)}\,-4x + 5}} \\ \underline{-4x - 2} \\ 7 \end{array}$$

Therefore,

$$f(x) = -2 + \frac{7}{2x + 1}$$

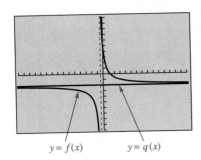

$y = f(x)$ $y = q(x)$

Figure 3.40
$[-15, 15] \times [-10, 10]$

The quotient is

$$q(x) = -2$$

The graphs of f and q are shown in Figure 3.40. For extreme values of x, the graph of f approaches the graph of q. The line $y = -2$ is a horizontal asymptote of the graph of f.

(b) The degree of the numerator is less than the degree of the denominator, so we simply write

$$f(x) = \frac{10}{4x^2 - 9} = 0 + \frac{10}{4x^2 - 9}$$

The graph of the quotient

$$q(x) = 0$$

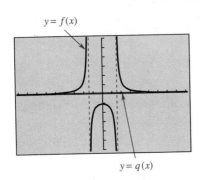

$y = f(x)$

$y = q(x)$

Figure 3.41
$[-9, 9] \times [-6, 6]$

is the x-axis. The graphs of f and q are shown in Figure 3.41. The graph of f approaches that of q for extreme values of x. The line $y = 0$ is a horizontal asymptote of the graph of f.

(c) Dividing $x - 5$ into $4x^2 - 17x - 8$, we have

$$
\begin{array}{r}
4x + 3 \\
x - 5 \overline{)4x^2 - 17x - 8} \\
\underline{4x^2 - 20x} \\
3x - 8 \\
\underline{3x - 15} \\
7
\end{array}
$$

Thus, $f(x)$ can be written as

$$f(x) = (4x + 3) + \frac{7}{x - 5}$$

The quotient is

$$q(x) = 4x + 3$$

Figure 3.42 shows that the graph of f approaches the graph of q at the extreme values of x. The line $y = 4x + 3$ is a *slant asymptote* of the graph of f. ■

Example 3 shows us that for extreme values of x, the graph of a rational function approaches the graph of the quotient $q(x)$ of the numerator and denominator. The following summarizes how to determine the behavior of the graph of a rational function for extreme values of x:

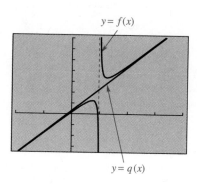

$y = f(x)$

$y = q(x)$

Figure 3.42
$[-10, 20]_5 \times [-35, 70]_{10}$

Behavior of the Graph of a
Rational Function for
Extreme Values of x

■ **NOTE:** *We have noted that the graph of a rational function cannot cross a vertical asymptote. The graph may, however, cross a horizontal or slant asymptote.*

If f is a reduced rational function written in the form

$$f(x) = q(x) + \frac{r(x)}{d(x)}$$

where the degree of $r(x)$ is less than the degree of $d(x)$, then the graph of f approaches the graph of $y = q(x)$ for extreme values of x.
  In particular:

If $q(x) = k$, where k is a constant, then $y = k$ is a **horizontal asymptote.**
If $q(x) = mx + b$, then $y = mx + b$ is a **slant asymptote.**

Graphs of Rational Functions

We are now ready to sketch the complete graph of a rational function. Our approach will be to first determine the domain, intercepts, vertical asymptotes, and behavior for extreme values of x. After we sketch the intercepts and asymptotes, we will use a graphing utility to determine the graph.

Graphing a Reduced
Rational Function

To sketch the graph of a reduced rational function $f(x) = \dfrac{n(x)}{d(x)}$:

Step 1: Determine and plot the intercepts. The x-intercepts correspond to the zeros of $n(x)$; the y-intercept corresponds to $f(0)$.

Step 2: Determine and sketch the vertical asymptotes. The vertical asymptotes correspond to the zeros of $d(x)$.

Step 3: Determine the behavior for extreme values of x by writing $f(x)$ in the form

$$f(x) = q(x) + \frac{r(x)}{d(x)}$$

where the degree of $r(x)$ is less than the degree of $d(x)$. The graph of f approaches the graph of $y = q(x)$ for extreme values of x.

Step 4: Use a graphing utility to complete the graph. The viewing window should contain the intercepts and asymptotes.

⬤ EXAMPLE 4   Graphing a Rational Function

List the intercepts and asymptotes of the rational function

$$f(x) = \frac{2x}{x + 1}$$

and graph the function.

SOLUTION

First, notice that

$$\frac{2x}{x + 1}$$

is reduced. The x-intercepts correspond to the zeros of the numerator. The only zero of $2x$ is $x = 0$; therefore, the only x-intercept is $(0, 0)$. The y-intercept is $(0, f(0)) = (0, 0)$.

The vertical asymptotes correspond to the zeros of the denominator. Since the only zero of $x + 1$ is $x = -1$, the only vertical asymptote is $x = -1$.

The behavior for extreme values of x can be determined by dividing $x + 1$ into $2x$:

$$
\begin{array}{r}
2 \\
x + 1 \overline{)\, 2x + 0} \\
\underline{2x + 2} \\
-2
\end{array}
$$

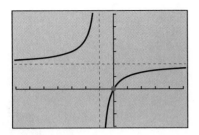

Figure 3.43
$$y = \frac{2x}{x + 1}$$
$[-7, 5] \times [-3, 6]$

This gives

$$f(x) = \frac{2x}{x + 1} = 2 + \frac{-2}{x + 1}$$

Since the quotient is $q(x) = 2$, $y = 2$ is a horizontal asymptote.

We set the window of our graphing utility so that it contains the intercepts and asymptotes (Figure 3.43).

x-intercept: $(0, 0)$

y-intercept: $(0, 0)$

Vertical asymptote: $x = -1$

Horizontal asymptote: $y = 2$

EXAMPLE 5   Graphing a Rational Function

List the intercepts and asymptotes of the rational function

$$f(x) = \frac{x - 1}{x^2 - 2x - 1}$$

and graph the function.

SOLUTION

We first note that

$$\frac{x - 1}{x^2 - 2x - 1}$$

is reduced. The zero of the numerator is 1, so the x-intercept is $(1, 0)$. Since $f(0) = 1$, the y-intercept is $(0, 1)$.

By the quadratic formula, the zeros of the denominator $x^2 - 2x - 1$ are

$$x = \frac{-(-2) \pm \sqrt{(-2)^2 - 4(1)(-1)}}{2(1)} = 1 \pm \sqrt{2}$$

Therefore, the vertical asymptotes are $x = 1 + \sqrt{2}$ and $x = 1 - \sqrt{2}$.

To determine the behavior of f for extreme values of x, we divide the denominator $x^2 - 2x - 1$ into $x - 1$; since the degree of the numerator is less than the degree of the denominator, we have

$$f(x) = \frac{x - 1}{x^2 - 2x - 1} = 0 + \frac{x - 1}{x^2 - 2x - 1}$$

Thus, $y = 0$ is a horizontal asymptote.

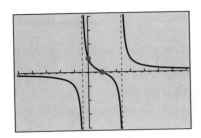

Figure 3.44
$$y = \frac{x - 1}{x^2 - 2x - 1}$$
$[-5, 7] \times [-4, 4]$

Figure 3.44 shows the graph in a viewing window containing the intercepts and asymptotes.

x-intercept: $(1, 0)$

y-intercept: $(0, 1)$

Vertical asymptotes: $x = 1 + \sqrt{2}$ and $x = 1 - \sqrt{2}$

Horizontal asymptote: $y = 0$

◼ EXAMPLE 6 Graphing a Rational Function

List the intercepts and asymptotes of the rational function

$$f(x) = \frac{2x^2 - x - 3}{3x - 6}$$

and graph the function.

SOLUTION

The rational function

$$\frac{2x^2 - x - 3}{3x - 6} = \frac{(2x - 3)(x + 1)}{3(x - 2)}$$

is reduced. Since the zeros of the numerator are $\frac{3}{2}$ and -1, the x-intercepts are $\left(\frac{3}{2}, 0\right)$ and $(-1, 0)$. The y-intercept is $\left(0, \frac{1}{2}\right)$.

The zero of the denominator is 2, so the vertical asymptote is $x = 2$.

We determine the behavior for extreme values of x by dividing $3x - 6$ into $2x^2 - x - 3$:

$$
\begin{array}{r}
\frac{2}{3}x + 1 \\
3x - 6 \overline{)2x^2 - x - 3} \\
\underline{2x^2 - 4x } \\
3x - 3 \\
\underline{3x - 6} \\
3
\end{array}
$$

Thus,

$$f(x) = \frac{2x^2 - x - 3}{3x - 6} = \left(\frac{2}{3}x + 1\right) + \frac{3}{3x - 6}$$

The quotient is $q(x) = \frac{2}{3}x + 1$; therefore, $y = \frac{2}{3}x + 1$ is a slant asymptote.

Using a graphing utility with an appropriate viewing window, we get Figure 3.45.

x-intercepts: $\left(\frac{3}{2}, 0\right)$ and $(-1, 0)$

y-intercept: $\left(0, \frac{1}{2}\right)$

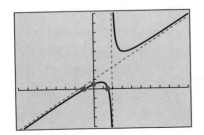

Figure 3.45
$$y = \frac{2x^2 - x - 3}{3x - 6}$$
$[-8, 10] \times [-4, 8]$

Vertical asymptote: $x = 2$

Slant asymptote: $y = \frac{2}{3}x + 1$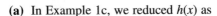

Most of the rational functions discussed so far have been reduced. One exception is

$$h(x) = \frac{x^2 + x}{2x + 2} = \frac{x(x + 1)}{2(x + 1)}$$

in Example 1c. When graphing functions such as these it is important to note the domain before reducing.

EXAMPLE 7 Graphing a Nonreduced Rational Function

Graph each function.

(a) $h(x) = \dfrac{x^2 + x}{2x + 2}$ **(b)** $g(x) = \dfrac{x + 2}{x^2 - 4}$

SOLUTION

(a) In Example 1c, we reduced $h(x)$ as

$$h(x) = \frac{x}{2} \qquad x \neq -1$$

For all values of x except -1, $h(x) = x/2$. At $x = -1$, $h(x)$ is not defined. Therefore, the graph of h is the same as the graph of $y = x/2$, except that there is a hole at $x = -1$ (Figure 3.46).

(b) The domain is all real numbers except 2 or -2. To sketch the graph we reduce $g(x)$:

$$\frac{x + 2}{x^2 - 4} = \frac{\cancel{x + 2}}{\cancel{(x + 2)}(x - 2)}$$

$$= \frac{1}{x - 2} \qquad x \neq -2$$

The graph of g is the same as the graph of $y = 1/(x - 2)$, except g is undefined at $x = -2$. The graph of $y = 1/(x - 2)$ was given earlier in Figure 3.32. The graph of g is the same, except there is a hole at $x = -2$ (Figure 3.47).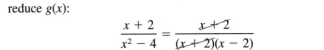

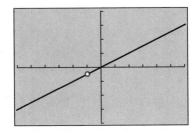

Figure 3.46
$y = \dfrac{x^2 + x}{2x + 2}$
$[-6, 6] \times [-4, 4]$

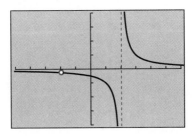

Figure 3.47
$y = \dfrac{x + 2}{x^2 - 4}$
$[-5, 6] \times [-4, 4]$

Applications

The graphs of rational functions are very useful for visualizing fundamental concepts in calculus. Rational functions are also used to model many other real-life situations, especially when a situation exhibits asymptotic behavior.

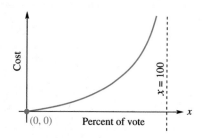

Cost

$x = 100$

(0, 0)   Percent of vote x

Figure 3.48

In many endeavors there is a trade-off between the cost and the corresponding benefit. For example, a political candidate compares the cost of a campaign against the percentage of voters she or he expects to win. Suppose the candidate will win 0% of the vote if nothing is spent. The percentage of the vote increases as the amount spent increases. However, the cost of securing 60% of the vote is more than twice the cost of securing 30% of the vote. Furthermore, the cost of securing a percentage of the vote near 100% would be unreasonably high. If C is the cost of securing x percent of the vote, a qualitative graph of this situation might look like Figure 3.48. Notice the intercept $(0, 0)$ and the vertical asymptote at $x = 100$. A rational function that fits this *cost–benefit model* is

$$C = \frac{kx}{100 - x} \qquad 0 \le x < 100$$

where k is a constant.

■ **EXAMPLE 8   Application: Cost–Benefit Model for Campaign Expenditures**

A political candidate estimates that spending $140,000 will secure 28% of the vote. Let C represent the cost (in dollars) of securing x percent of the vote in the cost–benefit model

$$C = \frac{kx}{100 - x} \qquad 0 \le x < 100$$

(a) Determine the constant k.
(b) Use a graphing utility to graph the model.
(c) Use the model to predict the cost of securing 40% of the vote.

SOLUTION

(a) We are given that $C = 140{,}000$ when $x = 28$. If we substitute these values into the model, we can solve for the constant k:

$$140{,}000 = \frac{k(28)}{100 - 28}$$

$$140{,}000 = \frac{k(7)}{18}$$

$$360{,}000 = k$$

Substituting 360,000 for k in the model

$$C = \frac{kx}{100 - x}$$

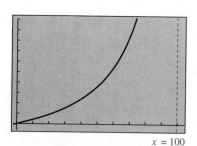

Figure 3.49
$$y = \frac{360,000x}{100 - x}$$
$$[-10,100]_{10} \times$$
$$[-50,000, 1,000,000]_{100,000}$$

we get

$$C = \frac{360,000x}{100 - x}$$

(b) The only intercept is (0, 0). The vertical asymptote is $x = 100$. A graph is shown in Figure 3.49.

(c) Substituting $x = 40$ into our model gives

$$C = \frac{360,000x}{100 - x} = \frac{360,000(40)}{100 - 40} = 240,000$$

The cost of securing 40% of the vote is $240,000.

EXAMPLE 9 Application: Velocity of a Skydiver

A skydiver who jumps from an airplane is influenced by gravity and a drag force that increases as her velocity increases. As a result, her velocity increases rapidly at first, and then increases more slowly, approaching 60 meters per second (called the *limiting velocity*). Figure 3.50 shows a graph of the skydiver's velocity v (in meters per second) t seconds after jumping.

(a) If the skydiver's velocity is 48 meters per second 5 seconds after she jumps, determine a model for this graph of the form

$$v = \frac{at}{t + b}$$

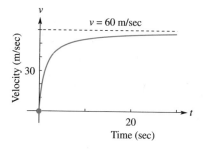

Figure 3.50

where a and b are constants.

(b) Use the model to predict the skydiver's velocity 10 seconds after jumping.

SOLUTION

(a) Dividing at by $t + b$ gives

$$
\require{enclose}
\begin{array}{r}
a \\
t + b \overline{\smash{)}at + 0} \\
\underline{at + ab} \\
- ab
\end{array}
$$

Thus,

$$v = a - \frac{ab}{t + b}$$

Since

$$v \rightarrow 60 \quad \text{as} \quad t \rightarrow +\infty$$

the graph has a horizontal asymptote $v = 60$. It follows that $a = 60$. Thus, our model has the form

$$v = \frac{at}{t + b} = \frac{60t}{t + b}$$

We are given that $v = 48$ when $t = 5$. Substituting these values into the model, we can solve for the constant b:

$$v = \frac{60t}{t + b}$$

$$48 = \frac{60(5)}{5 + b}$$

$$5 + b = \frac{60(5)}{48}$$

$$b = \frac{60(5)}{48} - 5 = 1.25$$

Therefore, our model is

$$v = \frac{60t}{t + 1.25}$$

(b) According to the model, the velocity 10 seconds after the skydiver jumps is

$$v = \frac{60(10)}{10 + 1.25} = 53\tfrac{1}{3}$$

The skydiver's velocity after 10 seconds is $53\tfrac{1}{3}$ meters per second.

EXERCISE SET 3.4

A

In Exercises 1–6, determine the domain and zeros of each function.

1. $f(x) = \dfrac{2x - 5}{x^2 - 4}$

2. $g(x) = \dfrac{4x - 1}{x^2 - 16}$

3. $F(x) = \dfrac{4}{2x + 1}$

4. $G(x) = \dfrac{2}{x - 6}$

5. $f(x) = \dfrac{3x^2 - 6x}{2x}$

6. $h(x) = \dfrac{x^2 + 4x}{2x^2 - 2x}$

In Exercises 7–12, determine the vertical asymptote(s) of each function. Use a graphing utility to verify your results.

7. $f(x) = \dfrac{x}{3x + 4}$

8. $g(x) = \dfrac{x + 2}{3x - 8}$

9. $F(x) = \dfrac{2x + 1}{x^2 - 9}$

10. $G(x) = \dfrac{1 - 2x}{x^2 + x - 6}$

11. $h(x) = \dfrac{5x + 1}{x^2 + 3}$

12. $H(x) = \dfrac{x^2 - 2x}{3x^2 + 15}$

In Exercises 13–18, express $f(x)$ in the form

$$f(x) = q(x) + \frac{r(x)}{d(x)}$$

where the degree of $r(x)$ is less than the degree of $d(x)$. Graph the functions f and q in the given viewing window, and compare their behavior for extreme values of x.

13. $f(x) = \dfrac{3x - 1}{x + 2}$; $[-17, 13] \times [-10, 10]$

14. $f(x) = \dfrac{2x + 9}{x + 1}$; $[-15, 15] \times [-10, 10]$

15. $f(x) = \dfrac{5x - 1}{x^2 + 4}$; $[-10, 10] \times [-6, 6]$

16. $f(x) = \dfrac{4}{2x - 5}$; $[-8, 10] \times [-6, 6]$

17. $f(x) = \dfrac{2x^2 + 5}{x + 1}$; $[-10, 10] \times [-20, 15]_5$

18. $f(x) = \dfrac{4x^2 + 2x - 1}{2x}$; $[-5, 5] \times [-6, 10]$

B

In Exercises 19–38, list the intercepts and asymptotes of each rational function, and graph the function.

19. $f(x) = \dfrac{2}{2x + 7}$

20. $g(x) = \dfrac{3}{2x - 5}$

21. $F(x) = \dfrac{2x}{x - 4}$

22. $G(x) = \dfrac{3x}{x + 2}$

23. $h(x) = \dfrac{5 - 2x}{x - 2}$

24. $H(x) = \dfrac{6 - x}{x + 3}$

25. $f(x) = \dfrac{x - 3}{(x + 2)^2}$

26. $F(x) = \dfrac{x - 6}{(2x - 3)^2}$

27. $g(x) = \dfrac{3x}{2x^2 - 8}$

28. $G(x) = \dfrac{-5x}{2x^2 - 18}$

29. $r(x) = \dfrac{4x - 6}{x^2 + 3}$

30. $R(x) = \dfrac{3x + 6}{x^2 + 4}$

31. $f(x) = \dfrac{x^2 + 2x + 1}{2x}$

32. $g(x) = \dfrac{2x^2 + 10x + 12}{3x}$

33. $h(x) = \dfrac{2x^2 + 3}{2x^2 - 7x}$

34. $R(x) = \dfrac{x^2 - 5}{x^2 + 4x}$

35. $S(x) = \dfrac{3x^2 - 2x - 1}{3x - 5}$

36. $T(x) = \dfrac{2x^2 - 5x - 3}{2x + 5}$

37. $F(x) = \dfrac{x^2 - 4}{x^2 + 1}$

38. $G(x) = \dfrac{4x^2 - 9}{x^2 + 3}$

In Exercises 39–44, graph each nonreduced rational function.

39. $f(x) = \dfrac{2x^2 + 4x}{3x + 6}$

40. $g(x) = \dfrac{3x^2 + 4x}{2x}$

41. $G(x) = \dfrac{x^3 - x^2}{4x - 4}$

42. $H(x) = \dfrac{x^3 + 2x^2}{2x + 4}$

43. $h(x) = \dfrac{2x^2 - 6x}{x^3 - 6x^2 + 9x}$

44. $F(x) = \dfrac{x^2 - x}{x^3 - x^2}$

45. Use compressions, expansions, translations, or reflections of the graph of $f(x) = 1/x$ to find an equation for the graph in Figure 3.51.

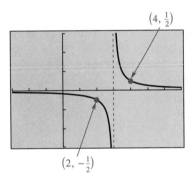

$\left(4, \frac{1}{2}\right)$

$\left(2, -\frac{1}{2}\right)$

Figure 3.51
$[-3, 7] \times [-3, 3]$

46. An oil company estimates that spending $400,000 will clean up 70% of an oil spill. Let C represent the cost (in dollars) of cleaning up x percent of the spill in the cost–benefit model

$$C = \frac{kx}{100 - x} \qquad 0 \le x < 100$$

   (a) Determine the constant k.
   (b) Use a graphing utility to graph the model
   (c) Use the model to predict the cost of cleaning up 90% of the spill.

47. A biochemical company currently spends $320,000 to remove 60% of the air pollutants in its factory emissions. The company anticipates a new federal regulation that will require a higher removal rate. Let C represent the cost (in dollars) of cleaning up x percent of the pollutants in the cost–benefit model

$$C = \frac{kx}{100 - x} \qquad 0 \le x < 100$$

(a) Determine the constant k.
(b) Use a graphing utility to graph the model.
(c) Use the model to predict the cost of removing 90% of the pollutants.

48. The velocity of a lightweight ball that has been dropped from a very tall bridge increases rapidly at first, but then increases more slowly as a result of wind resistance. Eventually, the velocity approaches 120 feet per second. A graph of the velocity v (in feet per second), t seconds after release is shown in Figure 3.52.

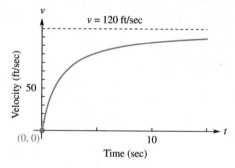

Figure 3.52

(a) If the velocity is 80 feet per second 3 seconds after the ball is released, determine a model of the form

$$v = \frac{at}{t + b}$$

where a and b are constants.
(b) Use the model to predict the velocity 6 seconds after release.

49. A landscaper's design calls for a dwarf evergreen tree that typically grows rapidly at first, and then more slowly, to a maximum height of 6 feet. A graph of the height h (in feet) t years after sprouting is shown in Figure 3.53.

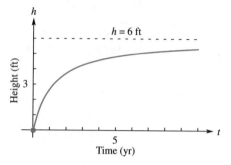

Figure 3.53

(a) Assume the height approaches 6 feet. If the height of a 2 year old tree is $3\frac{3}{4}$ feet, determine a model of the form

$$h = \frac{at}{t + b}$$

where a and b are constants.
(b) Use the model to determine the age of a tree that is 5 feet tall.

50. An artist creates a copper-etching solution by mixing a solution that is 40% acid with 200 milliliters of a solution that is 10% acid.
(a) Let x represent the amount (in milliliters) of 40% solution that is added. Show that the concentration $C(x)$ of acid in the mixture is

$$C(x) = \frac{20 + 0.4x}{200 + x}$$

HINT Quick Review: See Example 5 in Section 2.5.

(b) Graph the model for $x \ge 0$.
(c) Explain the significance of the y-intercept and the horizontal asymptote.

51. A pharmacist creates an antiseptic solution by mixing a solution that is 30% alcohol with 50 milliliters of a solution that is 10% alcohol.

(a) Let x represent the amount (in milliliters) of 30% solution that is added. Show that the concentration $C(x)$ of alcohol in the mixture is

$$C(x) = \frac{5 + 0.3x}{50 + x}$$

HINT: See Exercise 50.

(b) Graph the model for $x \geq 0$.

(c) Explain the significance of the y-intercept and the horizontal asymptote.

52. The weight $w(h)$, in pounds, of an object h miles above sea level can be approximated by

$$w(h) = \left(\frac{4000}{h + 4000}\right)^2 S$$

where S is the weight of the object at sea level. Suppose a person weighs 140 pounds at sea level.

(a) Graph the function w for $0 \leq w \leq 5000$.

(b) Determine the y-intercept and the horizontal asymptote, and explain their significance.

53. A company that manufactures squirt guns quotes a variable cost of $4.20 per unit and a fixed cost of $6000. Thus, a model for the cost C of producing x units is $C = 4.20x + 6000$. The average cost per unit is

$$\overline{C} = \frac{C}{x} = \frac{4.20x + 6000}{x}$$

(a) Determine the average cost per unit for a level of production of $x = 10{,}000$ units.

(b) Graph the model for $\overline{C}$ for $x > 0$. What is the horizontal asymptote? What does the horizontal asymptote represent?

C

In Exercises 54 and 55, express $f(x)$ in the form

$$f(x) = q(x) + \frac{r(x)}{d(x)}$$

where the degree of $r(x)$ is less than the degree of $d(x)$. Graph the functions f and q in the given viewing window, and compare their behavior for extreme values of x.

54. $f(x) = \dfrac{x^3 + 4}{4x}$; $[-6, 6] \times [-7, 10]$

55. $f(x) = \dfrac{2x^3 + 5}{10x}$; $[-5, 5] \times [-6, 8]$

56. Find an equation for the graph in Figure 3.54.

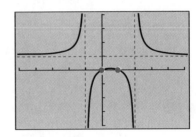

Figure 3.54
$[-5, 5] \times [-4, 4]$

RATIONAL EQUATIONS AND INEQUALITIES

An equation (or inequality) that contains a rational function is a *rational equation* (or *rational inequality*). This section develops techniques for solving rational equations and inequalities. Our approach will be similar to our approach in solving polynomial equations in Section 3.2.

Solving Rational Equations

From Section 3.4 we know that graphing a rational function requires finding the domain. Similarly, when solving a rational equation we must determine those values that cause division by 0 in any of the terms of the equation and exclude them from the set of solutions.

⬤ EXAMPLE 1   Solving a Rational Equation

Solve

$$\frac{3x - 1}{2x} = x - \frac{3}{4}$$

and check the solutions with an appropriate graph.

SOLUTION

Notice that $x = 0$ is not a possible solution because the expression

$$\frac{3x - 1}{2x}$$

is undefined at $x = 0$.

We multiply each side of the given equation by the least common denominator of all the terms in the equation to eliminate the denominators:

$$4x\left(\frac{3x - 1}{2x}\right) = 4x\left(x - \frac{3}{4}\right) \qquad \text{The least common demoninator is } 4x.$$

$$4x\left(\frac{3x - 1}{2x}\right) = 4x(x) - 4x\left(\frac{3}{4}\right)$$

$$2(3x - 1) = 4x^2 - 3x$$

Solving the resulting equation, we get

$$6x - 2 = 4x^2 - 3x$$

$$0 = 4x^2 - 9x + 2$$

$$0 = (4x - 1)(x - 2)$$

$$4x - 1 = 0 \quad \bigg| \quad x - 2 = 0$$

$$x = \tfrac{1}{4} \quad \bigg| \quad x = 2$$

CHECK: To check our solutions, we graph the corresponding function,

$$y = \frac{3x - 1}{2x} - x + \frac{3}{4}$$

(Figure 3.55). The graph has x-intercepts at $\left(\frac{1}{4}, 0\right)$ and $(2, 0)$, which verifies our solutions.   ⬤

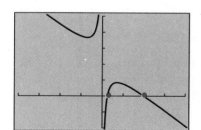

Figure 3.55
$$y = \frac{3x - 1}{2x} - x + \frac{3}{4}$$
$$[-4, 4] \times [-2, 5]$$

◼ EXAMPLE 2   A Rational Equation with No Solutions

Solve

$$\frac{2x+1}{x+5} + 1 = \frac{x-4}{x+5}$$

and check the solutions with an appropriate graph.

SOLUTION

First, we note that $x = -5$ is not a possible solution since it causes division by 0 in the terms

$$\frac{2x+1}{x+5} \quad \text{and} \quad \frac{x-4}{x+5}$$

Next, we multiply each side of the equation by the least common denominator, $x + 5$:

$$(x+5)\left(\frac{2x+1}{x+5} + 1\right) = (x+5)\left(\frac{x-4}{x+5}\right)$$

$$(x+5)\left(\frac{2x+1}{x+5}\right) + (x+5) \cdot 1 = (x+5)\left(\frac{x-4}{x+5}\right)$$

$$(2x+1) + (x+5) = x - 4$$

$$3x + 6 = x - 4$$

$$2x = -10$$

$$x = -5$$

Since we excluded $x = -5$ as a possible solution, this rational equation has no solutions.

CHECK: The graph of the corresponding function,

$$y = \frac{2x+1}{x+5} + 1 - \frac{x-4}{x+5}$$

(Figure 3.56) has no x-intercepts, verifying our result.

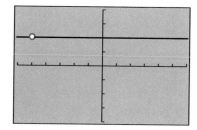

Figure 3.56
$$y = \frac{2x+1}{x+5} + 1 - \frac{x-4}{x+5}$$
$[-6, 6] \times [-4, 4]$

◼ **NOTE:** *Recall from Section 2.3 that squaring both sides of an equation sometimes produces false solutions called* extraneous *solutions. The value* $x = -5$ *in this example is also an extraneous solution.*

◼ EXAMPLE 3   A Rational Equation with an Extraneous Solution

Solve

$$\frac{x}{x+1} = \frac{2}{x^2 - 1}$$

and check the solutions with an appropriate graph.

SOLUTION

The value of x that causes division by 0 in the term

$$\frac{x}{x + 1}$$

is -1, and the values that cause division by 0 in the term

$$\frac{2}{x^2 - 1}$$

are -1 and 1. Therefore, the values -1 and 1 are not possible solutions.

The denominator $x^2 - 1$ factors as $(x + 1)(x - 1)$, so the least common denominator of

$$\frac{x}{x + 1} \quad \text{and} \quad \frac{2}{x^2 - 1}$$

is $(x + 1)(x - 1)$. Multiplying each side by the least common denominator eliminates the denominators in the equation:

$$\frac{x}{x + 1} = \frac{2}{x^2 - 1}$$

$$\frac{x}{x + 1} = \frac{2}{(x + 1)(x - 1)}$$

$$(x + 1)(x - 1)\left(\frac{x}{x + 1}\right) = (x + 1)(x - 1)\left[\frac{2}{(x + 1)(x - 1)}\right]$$

$$(x - 1)x = 2$$

Solving for x, we get

$$x^2 - x - 2 = 0$$

$$(x - 2)(x + 1) = 0$$

$$x - 2 = 0 \quad | \quad x + 1 = 0$$

$$x = 2 \quad | \quad x = -1$$

The potential solutions are $x = 2$ and $x = -1$. Since we first noted that $x = -1$ cannot be a solution, the solution to the original rational equation is $x = 2$.

CHECK: The graph of the corresponding function,

$$y = \frac{x}{x + 1} - \frac{2}{x^2 - 1}$$

in Figure 3.57 has one x-intercept at $x = 2$, verifying our result.

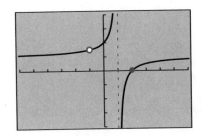

Figure 3.57

$$y = \frac{x}{x + 1} - \frac{2}{x^2 - 1}$$

$[-6, 6] \times [-4, 4]$

Now is a good time to summarize the method for solving equations with rational expressions:

Solving Rational Equations

> **To solve an equation containing rational expressions:**
>
> **Step 1:** Determine all values of the variable that make any denominator equal to 0.
>
> **Step 2:** Determine the least common denominator of all terms in the equation. Eliminate the denominators by multiplying each side of the equation by the least common denominator.
>
> **Step 3:** Solve the resulting equation.
>
> **Step 4:** Check the solutions against the values in step 1. Any value that causes division by 0 must be excluded from the set of solutions.

In addition to the procedure listed above, you should graph the corresponding function on a graphing utility and check that the *x*-intercepts coincide with your solutions. One final note on this procedure: After eliminating the denominators, the resulting equation in step 3 may be a polynomial equation of degree 3 or higher. In this case, use the methods discussed in Section 3.2.

Solving Rational Inequalities

Consider the rational inequality

$$\frac{2x - 5}{x + 1} \le 0$$

If we multiply both sides by the least common denominator $x + 1$, we may need to reverse the inequality:

$$\text{If } x + 1 \text{ is positive, then: } \quad (x + 1)\frac{2x - 5}{x + 1} \le (x + 1)0$$

$$\text{If } x + 1 \text{ is negative, then: } \quad (x + 1)\frac{2x - 5}{x + 1} \ge (x + 1)0$$

In general, eliminating the denominators in a rational inequality can be more involved than simply multiplying each side by the least common denominator. Instead, we will solve rational inequalities using a graphic approach that is similar to the way we solved inequalities in Section 2.4.

EXAMPLE 4 Solving a Rational Inequality

Use an appropriate graph to solve each inequality.

(a) $\dfrac{2x - 5}{x + 1} \le 0$ (b) $\dfrac{x^3 - 4x^2 + 1}{x^2 - 1} > -1$

SOLUTION

(a) Using the techniques for graphing rational functions in Section 3.4, we graph

$$y = \frac{2x - 5}{x + 1}$$

(Figure 3.58). The solution to

$$\frac{2x - 5}{x + 1} \le 0$$

corresponds to those values of x for which the graph is on or below the x-axis. The graph is below the x-axis for all values of x between -1 and $\frac{5}{2}$. Therefore, the solution to the inequality is

$$-1 < x \le \tfrac{5}{2}$$

(b) We add 1 to each side to get the equivalent inequality

$$\frac{x^3 - 4x^2 + 1}{x^2 - 1} + 1 > 0$$

Combining the two terms on the left side, we get

$$\frac{x^3 - 4x^2 + 1}{x^2 - 1} + \frac{x^2 - 1}{x^2 - 1} > 0$$

$$\frac{x^3 - 3x^2}{x^2 - 1} > 0$$

$$\frac{x^2(x - 3)}{(x + 1)(x - 1)} > 0$$

The graph of

$$y = \frac{x^2(x - 3)}{(x + 1)(x - 1)}$$

is shown in Figure 3.59. The solution to

$$\frac{x^3 - 4x^2 + 1}{x^2 - 1} > -1$$

corresponds to those values of x for which the graph is above the x-axis. Thus, the solution is $-1 < x < 0$ or $0 < x < 1$ or $x > 3$.

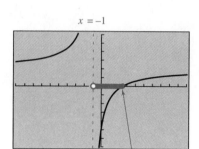

Figure 3.58

$y = \dfrac{2x - 5}{x + 1}$

$[-10, 10] \times [-7, 6]$

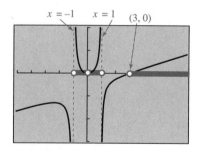

Figure 3.59

$y = \dfrac{x^2(x - 3)}{(x + 1)(x - 1)}$

$[-5, 7] \times [-15, 10]_5$

Applications

Many applications call for solving a rational equation or inequality. The next example and the exercises that follow should give you a better idea of the variety of applications that require solving rational equations or inequalities.

Figure 3.60

EXAMPLE 5   Application: Designing a Cost-Effective Poster

A print shop gets an order for a poster with 2.5 inch margins at the top and bottom and 2 inch margins at the sides (Figure 3.60). The area of the printed material is to be 200 square inches.

(a) Determine the area of the poster as a function of the vertical length of the printed region.

(b) Based on the cost of materials, filling the order is cost-effective if the total area of the poster is 350 square inches. Determine the dimensions of the printed portion.

(c) Suppose filling the order is cost-effective if the total area of the poster is less than 350 square inches. Determine an interval for the possible values of the length of the poster. Give an example, based on this interval, of the possible dimensions of the printed region.

SOLUTION

(a) Since the area of the printed portion is to be 200 square inches,

$$\left(\begin{array}{c}\text{Width of printed}\\ \text{portion}\end{array}\right)\left(\begin{array}{c}\text{Length of printed}\\ \text{portion}\end{array}\right) = 200$$

or

$$\left(\begin{array}{c}\text{Width of printed}\\ \text{portion}\end{array}\right) = \dfrac{200}{\left(\begin{array}{c}\text{Length of printed}\\ \text{portion}\end{array}\right)}$$

Letting w and l represent the width and length of the printed portion, respectively, we have

$$w = \frac{200}{l}$$

Now consider the area of the entire poster.

$$\left(\begin{array}{c}\text{Area of}\\ \text{poster}\end{array}\right) = \left(\begin{array}{c}\text{Width of}\\ \text{poster}\end{array}\right)\left(\begin{array}{c}\text{Length of}\\ \text{poster}\end{array}\right)$$

$$= \left(\begin{array}{c}\text{Width of printed}\\ \text{portion plus margins}\end{array}\right)\left(\begin{array}{c}\text{Length of printed}\\ \text{portion plus margins}\end{array}\right)$$

Replacing the width and length of the printed portion with w and l, respectively, we have

$$\begin{pmatrix} \text{Area of} \\ \text{poster} \end{pmatrix} = (w + 2 + 2)(l + 2.5 + 2.5)$$

$$= \left(\frac{200}{l} + 4 \right)(l + 5) \qquad \text{Substitute } \frac{200}{l} \text{ for } w.$$

If we expand and simplify this last expression, the area of the poster is

$$A(l) = 220 + \frac{1000}{l} + 4l$$

(b) Since the area of the poster must be 350 square inches, we have

$$220 + \frac{1000}{l} + 4l = 350$$

$$220l + 1000 + 4l^2 = 350l \qquad \text{Multiply each side by } l.$$

$$4l^2 - 130l + 1000 = 0$$

$$2l^2 - 65l + 500 = 0 \qquad \text{Divide each side by 2.}$$

$$(l - 20)(2l - 25) = 0$$

$$l - 20 = 0 \quad \Big| \quad 2l - 25 = 0$$

$$l = 20 \quad \Big| \quad l = \tfrac{25}{2} \quad \text{or} \quad 12\tfrac{1}{2}$$

If the length l is 20, then the width is

$$w = \frac{200}{l} = \frac{200}{20} = 10$$

If the length l is $12\tfrac{1}{2}$, then the width is

$$w = \frac{200}{l} = \frac{200}{12\tfrac{1}{2}} = 16$$

The poster can have a printed portion that is either 20 by 10 inches or $12\tfrac{1}{2}$ by 16 inches.

(c) If the area of the poster must be less than 350 square inches, we have

$$220 + \frac{1000}{l} + 4l < 350$$

Subtracting 350 from each side and combining terms, we get

$$-130 + \frac{1000}{l} + 4l < 0$$

$$\frac{-130l}{l} + \frac{1000}{l} + \frac{4l^2}{l} < 0$$

$$\frac{2(2l^2 - 65l + 500)}{l} < 0$$

$$\frac{2(l - 20)(2l - 25)}{l} < 0$$

Since l represents a length, we seek only positive solutions to the inequality. The graph of

$$y = \frac{2(x - 20)(2x - 25)}{x} \qquad \text{for } x > 0$$

is shown in Figure 3.61. The solution of the inequality corresponds to the values of x for which the graph is below the x-axis. Therefore, the solution is the interval

$$12\tfrac{1}{2} < l < 20$$

Based on this interval, there are an infinite number of choices for l. For example, if we choose the length of the printed area to be, say $l = 15$ inches, then the width must be

$$w = \frac{200}{l} = \frac{200}{15} = 13\tfrac{1}{3} \text{ inches}$$

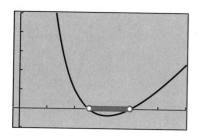

Figure 3.61
$$y = \frac{2(x - 20)(2x - 25)}{x}$$
$[-1, 30]_5 \times [-10, 50]_{10}$

EXERCISE SET 3.5

A

In Exercises 1–22, solve each equation, and check the solutions with an appropriate graph.

1. $\dfrac{x}{3} = \dfrac{1}{2} + 2x$

2. $\dfrac{x}{4} = \dfrac{2}{3} - x$

3. $\dfrac{x + 3}{x} = \dfrac{2}{5}$

4. $\dfrac{x + 2}{x} = \dfrac{4}{3}$

5. $\dfrac{5}{6x} = \dfrac{2}{3}$

6. $\dfrac{1}{4x} = \dfrac{5}{2}$

7. $\dfrac{10}{x} + 3 = \dfrac{1}{2x}$

8. $\dfrac{5}{x} = \dfrac{2}{3x} + 4$

9. $\dfrac{2}{5w} - \dfrac{3}{10} = \dfrac{1}{2w}$

10. $\dfrac{6}{7k} + \dfrac{5}{14} = \dfrac{1}{2k}$

11. $\dfrac{2x - 1}{3x} = x - \dfrac{3}{2}$

12. $\dfrac{x + 1}{4x} = x - \dfrac{8}{3}$

13. $\dfrac{5}{a + 2} + \dfrac{2}{2 - a} = \dfrac{a}{a^2 - 4}$

14. $\dfrac{2}{p+3} = \dfrac{p-3}{p^2-9} + \dfrac{4}{3-p}$

15. $\dfrac{2x-1}{x+4} = \dfrac{x-5}{x+4} - 1$

16. $\dfrac{2x+3}{x+6} + 1 = \dfrac{x-3}{x+6}$

17. $\dfrac{3}{x+1} - \dfrac{1}{2x} = 1$

18. $\dfrac{6}{x+1} + \dfrac{1}{x-1} = 1$

19. $\dfrac{x-4}{2x+1} + \dfrac{1}{x} = \dfrac{x+1}{2x^2+x}$

20. $\dfrac{x}{x+1} + \dfrac{1}{x} = \dfrac{x+2}{x^2+x}$

21. $\dfrac{0.10(50)+0.60x}{50+x} = 0.40$

22. $\dfrac{0.20(100)+0.50x}{100+x} = 0.30$

In Exercises 23–34, use an appropriate graph to solve each inequality.

23. $\dfrac{3}{2-x} < 0$

24. $\dfrac{1}{6-2x} < 0$

25. $\dfrac{x+3}{2x} \geq 0$

26. $\dfrac{x-1}{3x} \leq 0$

27. $\dfrac{2x-7}{x+4} \geq 0$

28. $\dfrac{4x-9}{x-3} \geq 0$

29. $\dfrac{x^2-3x-1}{x-1} \leq 1$

30. $\dfrac{x^2-2x+2}{x+2} \geq 1$

31. $x - \dfrac{3}{x} \leq \dfrac{1}{2}$

32. $x - \dfrac{4}{3x} \leq \dfrac{4}{3}$

33. $\dfrac{4-3x}{x^2-4} > -1$

34. $\dfrac{-3x^2+15x-6}{2x^2-7x} > -2$

B

In Exercises 35–37, multiply each side by the least common denominator, and solve the resulting polynomial equation using the methods of Section 3.2.

35. $\dfrac{x-5}{3x} + \dfrac{x}{3} = \dfrac{-1}{x^2}$

36. $\dfrac{x-3}{2x} + \dfrac{x}{2} = \dfrac{1}{x^2}$

37. $\dfrac{1}{4x^2} - x = \dfrac{2x-3}{4x}$

38. Suppose a traveler averages x miles per hour to a destination, and then returns averaging y miles per hour. The average speed A for the round trip is

$$A = \dfrac{2xy}{x+y}$$

(This is called the *harmonic mean* of x and y.)
(a) Hal drives from Wasco to Duarte averaging 40 miles per hour. At what speed must he return if his average speed for the round trip must be 48 miles per hour?
(b) Solve the equation

$$A = \dfrac{2xy}{x+y}$$

for x.

39. The resistance R (in ohms) of the circuit in Figure 3.62 is

$$R = \dfrac{r_1 r_2}{r_1 + r_2}$$

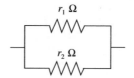

Figure 3.62

(a) The resistance across one resistor is $r_1 = 9$ ohms, and the second resistor is yet to be installed. If the resistance of the circuit must be $R = 3$ ohms, determine the resistance r_2.
(b) Solve the equation

$$R = \dfrac{r_1 r_2}{r_1 + r_2}$$

for r_1.

40. The average monthly cost (in dollars) of ordering and maintaining x units of a particular item in a store is modeled by

$$A(x) = \dfrac{300}{x} + 150 + \dfrac{3x}{2}$$

(a) Determine the number of units to order so that the average monthly cost is $300.
(b) Determine an interval for the number of units to order so that the average monthly cost is less than $300.

41. A manufacturer of computer hard disks has found that the cost (in dollars) of manufacturing x units per week is modeled by the function $C(x) = 0.1x^2 - 6x + 84$. The average weekly cost (in dollars per unit) is

$$\overline{C}(x) = \frac{C(x)}{x} = \frac{x}{10} - 6 + \frac{84}{x}$$

(a) Determine the number of units to manufacture so that the average weekly cost is $40 per unit.

(b) Determine an interval for the number of units to manufacture so that the average weekly cost is less than $40 per unit.

42. A rectangular pen is to be constructed that will enclose 396 square feet.

(a) Determine the length $L(x)$ of the fence needed as a function of the width x of the pen.

(b) Determine the dimensions of the pen if 100 feet of fencing material is to be used.

(c) Determine an interval for the possible width of the pen if less than 100 feet of fencing material is to be used.

43. A glass manufacturer gets an order for a rectangular window with 3 inch margins at the top and bottom and 2 inch margins at the sides (Figure 3.63). The interior area for an etched design requires 48 square inches.

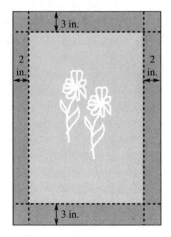

Figure 3.63

(a) Determine the area of the window as a function of the vertical length of the etched portion.

(b) Based on the cost of materials, filling the order will be cost-effective if the total area of the window is 160 square inches. Determine the dimensions of the etched portion.

(c) Suppose filling the order is cost-effective if the total area of the window is less than 160 square inches. Determine an interval for the length of the window. Based on this interval, give an example of the possible dimensions of the etched region.

44. A supplier of shipping boxes receives an order for an open-top box with a square base and a capacity (volume) of 240 cubic inches (Figure 3.64).

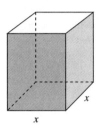

Figure 3.64

(a) Let x represent the length of a side of the square base. Determine the number of square inches of material used to construct each box as a function of x.

(b) Suppose that filling the order will be cost-effective if the amount of material used to construct each box is 196 square inches. Determine the length of the square base.

HINT Quick Review: After eliminating the denominator, use the methods for solving polynomial equations in Section 3.2.

(c) Suppose filling the order is cost-effective if the amount of material used is less than 196 square inches. Determine an interval for the length of the square base. Based on this interval, give an example of the possible dimensions of the box.

45. The controller for a rapid transit rail system must determine the schedule for two trains. One train leaves the station and travels south to a destination 12 miles away. The other train leaves the station and travels north to a destination 18 miles away, leaving 5 minutes before the southbound train. If both trains travel at the same rate and arrive at their destination at the same time, how long will it take each train to reach its destination?

46. In still water, a boat has a cruising speed of 24 miles per hour. If the boat can travel 60 miles downstream in the same amount of time that it takes to travel 45 miles upstream, what is the speed of the current?

47. An airplane flies 384 miles against the wind and returns with the wind in a total of 5 hours. In calm conditions the airplane flies 160 miles per hour. What is the speed of the wind?

48. A pump can fill a pool in $\frac{3}{4}$ of a day, and another pump can fill the same pool in 1 day. Working together, how long will it take both pumps to fill the pool?

49. One pipe can fill a tank in $1\frac{1}{2}$ hours. If the tank is full, the drain for the tank can empty it in 2 hours. How long will it take the pipe to fill the empty tank if the drain is accidentally left open?

C

50. A package supplier must design a container in the shape of a right circular cylinder with an open top. The container must hold 40 cubic inches.
 (a) Let r be the radius of the circular base. Show that the amount of material (in square inches) required to construct the container is
$$A(r) = \frac{80}{r} + \pi r^2$$

 (b) Suppose that manufacturing the container will be cost-effective if the amount of material used to construct each container is less than 121 square inches. To the nearest 0.01 inch, determine an interval for the radius of the circular base.
 HINT Quick Review: After eliminating the denominator, use the methods for solving polynomial equations in Section 3.2.

51. Let x be a positive integer greater than 1. Show that
$$\frac{1}{x} + \frac{1}{x+1} + \frac{1}{x+2} + \frac{1}{x+3} + \cdots + \frac{1}{x^2} > 1$$

| SECTION 3.6 | **PARAMETRIC EQUATIONS** |

- Sketching Plane Curves
- Parametric Equations and Graphing Utilities
- Circles and Circular Functions
- Common Parametric Representations
- Applications

There are many curves in the coordinate plane that are not graphs of functions. For example, the graph of $y^2 - x = 6$ (Figure 3.65) is not the graph of a function since it violates the vertical-line test of Section 1.3. The same is true for the graph of $x^2 + y^2 = 1$, the circle of radius 1 centered at the origin (Figure 3.66). In this

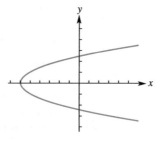

Figure 3.65
$y^2 - x = 6$

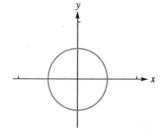

Figure 3.66
$x^2 + y^2 = 1$

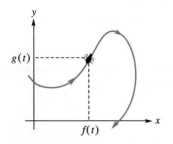

Figure 3.67

section, we investigate a method used to describe curves that are not necessarily the graphs of functions.

Consider, for a moment, a bug crawling on a coordinate plane (Figure 3.67). The position (x, y) of the bug at a particular time t can be described by a pair of equations $x = f(t)$ and $y = g(t)$. These equations are *parametric equations* (or a *parametric representation*) for this path. The variable t is called a *parameter*. Curves described by parametric equations are called *plane curves*.

Sketching Plane Curves

A plane curve that is described by parametric equations $x = f(t)$ and $y = g(t)$ can be sketched by first evaluating $f(t)$ and $g(t)$ for certain values of t, plotting the points $(f(t), g(t))$ for these values of t, and then drawing a curve through these points.

🔵 EXAMPLE 1   Sketching a Plane Curve

Sketch the plane curve described by $x = \frac{1}{2}t^2$, $y = t + 2$ for $-2 \le t \le 2$.

SOLUTION

Table 3.1 shows x and y values for the integer values of t over the interval $-2 \le t \le 2$. These points (x, y) are plotted in Figure 3.68a. Figure 3.68b shows the smooth curve that passes through these points. This curve is the plane curve described by $x = \frac{1}{2}t^2$, $y = t + 2$ over $-2 \le t \le 2$. The arrows show the direction of the curves as the value of t increases.

TABLE 3.1

t	$x = \frac{1}{2}t^2$	$y = t + 2$
-2	2	0
-1	$\frac{1}{2}$	1
0	0	2
1	$\frac{1}{2}$	3
2	2	4

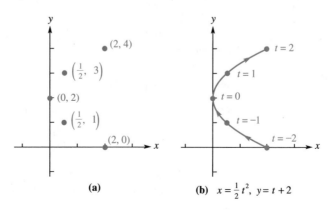

Figure 3.68

Our discussion of plane curves and parametric equations so far is summarized in the following definitions:

Plane Curves

> Suppose that f and g are functions. A **plane curve** C that is described by
>
> $$x = f(t), \quad y = g(t)$$
>
> is the set of points $(f(t), g(t))$; see Figure 3.69. The equations are **parametric equations** for C, and t is the **parameter**. The set of values of t for which C is defined is the **domain** of C. The direction along C for which t increases from a to b is the **orientation** of C.

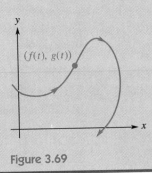

Figure 3.69

In addition to simply plotting points as we did in Example 1, a plane curve can also be determined by using its parametric equations to find an equation in x and y only. Example 2 illustrates this process.

EXAMPLE 2   Eliminating the Parameter for a Plane Curve

Find an equation in x and y only for the plane curve described by each given parametric representation, and use this equation to sketch the graph. Label the orientation.

(a) $x = t - 2, y = t^2 - 5, \quad -3 \le t \le 3$ **(b)** $x = 4 - 2\sqrt{t}, y = \sqrt{t} + 2$

SOLUTION

In each case, the plan of attack is to manipulate the parametric equations to derive an equation in which the variable t does not occur.

(a) Solving the first equation $x = t - 2$ for t gives $t = x + 2$. Substituting $t = x + 2$ into the second equation, we get

$$y = t^2 - 5 = (x + 2)^2 - 5$$

The graph of $y = (x + 2)^2 - 5$ is a parabola that is concave up with vertex $(-2, -5)$ (Figure 3.70a). The initial point of the curve, corresponding to $t = -3$, is $(-5, 4)$ since

$$x = t - 2 = -3 - 2 = -5$$

and

$$y = t^2 - 5 = (-3)^2 - 5 = 4$$

Likewise, the terminal point of the curve, corresponding to $t = 3$, is $(1, 4)$ because

$$x = t - 2 = 3 - 2 = 1$$

and

$$y = t^2 - 5 = (3)^2 - 5 = 4$$

The graph and its orientation are shown in Figure 3.70b.

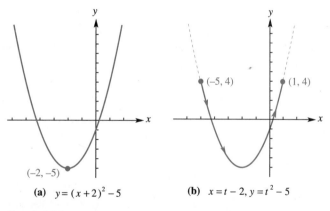

(a) $y = (x + 2)^2 - 5$ **(b)** $x = t - 2, y = t^2 - 5$

Figure 3.70

■ NOTE: *Because there is no interval for t given in part (b), we will assume the domain for t to be the largest interval possible for which the equations are defined.*

(b) We use $t \geq 0$ for the domain of t because of the square root terms in both x and y. Instead of solving one of the equations for t as we did in part (a), we can solve one of the equations, say $y = \sqrt{t} + 2$, for $\sqrt{t}$ and then use this expression in $x = 4 - 2\sqrt{t}$ to obtain an equation in x and y only. Thus,

$$y = \sqrt{t} + 2$$
$$y - 2 = \sqrt{t}$$

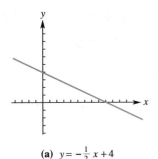

(a) $y = -\frac{1}{2}x + 4$

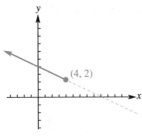

(b) $x = 4 - 2\sqrt{t},\ y = \sqrt{t} + 2$

Figure 3.71

■ GRAPHING NOTE: *Not every graphing utility implements parametric equations in the same way. You should pause here to learn how your utility graphs parametric equations.*

■ GRAPHING NOTE: *If you decrease the value of tStep, your graphing utility will plot more points.*

Then

$$x = 4 - 2\sqrt{t}$$
$$x = 4 - 2(y - 2) \qquad \text{Substitute } y - 2 \text{ for } \sqrt{t}.$$
$$x = 8 - 2y$$
$$y = -\tfrac{1}{2}x + 4 \qquad\qquad \text{Rewrite in slope–intercept form.}$$

The graph of $y = -\frac{1}{2}x + 4$ is a line with slope $-\frac{1}{2}$ and y-intercept $(0, 4)$, as shown in Figure 3.71a. The initial point of the curve, $(4, 2)$, corresponds to $t = 0$. Plotting a few more points shows the orientation and the extent of the graph of $x = 4 - 2\sqrt{t}$, $y = \sqrt{t} + 2$ for $t \geq 0$ (Figure 3.71b). ◖

The equation in rectangular coordinates x and y only that results from eliminating the parameter t from a parametric representation of a plane curve is called a *rectangular equation* of the curve. Finding a rectangular equation of a parametric representation often gives you a more familiar equation with which to work. However, the orientation of the curve is not evident from the rectangular equation.

Parametric Equations and Graphing Utilities

A graphing utility makes quick work of graphing plane curves from a set of parametric equations. The graphing utility uses the same method we used in Example 1 — that is, plotting and connecting points for particular values of the parameter t. The difference is that a graphing utility will plot many more points than you would probably be willing to plot. The selection of points to be plotted is determined by the minimum value of t (tMin), the maximum value of t (tMax), and the increment between two successive values of t (tStep). For example, if the domain of t for a particular set of parametric equations is $2 \leq t \leq 4$ and the increment is 0.2, then tMin = 2, tMax = 4, and tStep = 0.2. The points plotted and connected correspond to the following values for t:

$$2, 2.2, 2.4, 2.6, 2.8, 3.0, 3.2, 3.4, 3.6, 3.8, 4.0$$

These values are determined by starting with tMin (2) and then repeatedly adding tStep (0.2) until reaching tMax (4).

◖ EXAMPLE 3   Using a Graphing Utility to Determine a Plane Curve

Determine the graph of $x = t^2$, $y = \frac{1}{4}t^3 - t$ for $-3 \leq t \leq 3$ using a graphing utility.

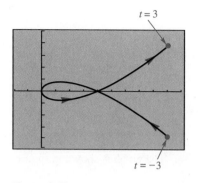

Figure 3.72
$x = t^2, y = \frac{1}{4}t^3 - t, -3 \le t \le 3$
$[-2, 10] \times [-5, 5]$

SOLUTION

Because the domain of t is $-3 \le t \le 3$, we set tMin $= -3$ and tMax $= 3$. A suitable value for tStep is 0.1. (You should experiment with different values of tStep to see if another value of tStep gives you a better sense of the curve.) The graph is shown in Figure 3.72.

Circles and Circular Functions

Circles occur frequently in the study of mathematics and science, so being able to determine them by a parametric representation using a graphing utility is often useful. Two functions, the *sine* and the *cosine*, are defined especially to determine the graph of a circle using parametric equations. Appropriately, these two functions are called *circular functions*. The following summary and two examples explain how they are used to determine the graphs of circles.

Circular Functions

> The two functions, **sine** and **cosine**, are **circular functions.** The sine of a real number t is written sin t, and the cosine of a real number t is written cos t. A parametric representation of the graph of $x^2 + y^2 = 1$, the circle centered at the origin with radius 1, is given by $x = \cos t, y = \sin t$, for $0 \le t \le 2\pi$ (Figure 3.73). The orientation of the curve is counterclockwise, and the curve both starts and stops at the point $(1, 0)$.

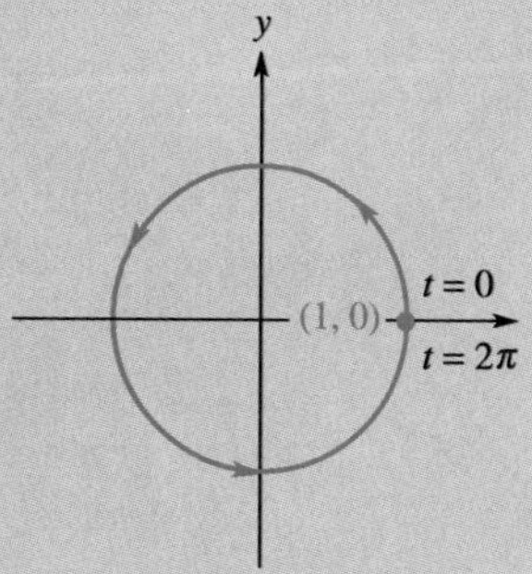

Figure 3.73
$x = \cos t, y = \sin t, 0 \le t \le 2\pi$

■ **GRAPHING NOTE:** *All graphics calculators are capable of computing the values of circular functions. Be sure that your calculator is set in* radian mode *when you use these functions.*

Circular functions and their many applications are the focus of the field of mathematics called *trigonometry.*

EXAMPLE 4   Using Circular Functions

Use a graphing utility to determine the graph of $x = \cos t$, $y = \sin t$ over each given domain of t.

(a) $0 \leq t \leq \pi$ **(b)** $\dfrac{\pi}{2} \leq t \leq 2\pi$

SOLUTION

(a) Because $0 \leq t \leq \pi$ is a subset of the interval $0 \leq t \leq 2\pi$, the resulting plane curve is a portion of the graph of $x^2 + y^2 = 1$. The graph is shown in Figure 3.74.

(b) The interval $\pi/2 \leq t \leq 2\pi$ is also a subset of $0 \leq t \leq 2\pi$, so this graph is also a part of $x^2 + y^2 = 1$ (Figure 3.75).

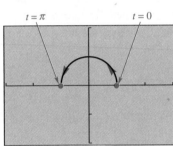

Figure 3.74

$x = \cos t$, $y = \sin t$, $0 \leq t \leq \pi$

$[-3, 3] \times [-2, 2]$

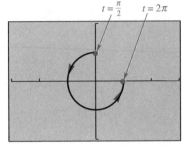

Figure 3.75

$x = \cos t$, $y = \sin t$, $\dfrac{\pi}{2} \leq t \leq 2\pi$

$[-3, 3] \times [-2, 2]$

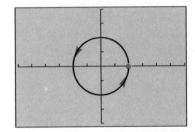

Figure 3.76

$x = 2 \cos t$, $y = 2 \sin t$, $0 \leq t \leq 2\pi$

$[-6, 6] \times [-4, 4]$

Example 5 suggests a method for graphing a circle centered at the origin with a radius other than 1.

EXAMPLE 5   Determining the Graphs of Circles

Use a graphing utility to determine each plane curve described for $0 \leq t \leq 2\pi$, and find a rectangular equation for its graph.

(a) $x = 2 \cos t$, $y = 2 \sin t$ **(b)** $x = 6 \cos t$, $y = 6 \sin t$

SOLUTION

(a) The graph of $x = 2 \cos t$, $y = 2 \sin t$, a circle centered at the origin with radius 2, is shown in Figure 3.76. The rectangular equation for this circle is $x^2 + y^2 = 4$.

(b) The graph of $x = 6 \cos t$, $y = 6 \sin t$ is a circle of radius 6 (Figure 3.77). The rectangular equation is $x^2 + y^2 = 36$.

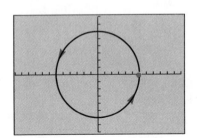

Figure 3.77

$x = 6 \cos t$, $y = 6 \sin t$, $0 \leq t \leq 2\pi$

$[-12, 12] \times [-8, 8]$

The generalization of Example 5 is that the graph of $x^2 + y^2 = r^2$, a circle centered at the origin with radius r, is given by the parametric equations

$$x = r \cos t, \quad y = r \sin t \qquad \text{for } 0 \le t \le 2\pi$$

Common Parametric Representations

Table 3.2 shows typical parametric representations for four types of plane curves that occur frequently in mathematics and science. Example 6 also demonstrates how this table is used.

TABLE 3.2

Parametric Representations of Common Plane Curves

Description of curve	Parametric representation	Curve and orientation
Graph of $y = f(x)$ from $P(a, b)$ to $Q(c, d)$	$x = t, y = f(t), a \le t \le c$	
Graph of $x = g(y)$ from $P(a, b)$ to $Q(c, d)$	$x = g(t), y = t, b \le t \le d$	
Circle centered at the origin with radius r	$x = r \cos t, y = r \sin t,$ $0 \le t \le 2\pi$	
Line segment from $P(a, b)$ to $Q(c, d)$	$x = a + (c - a)t,$ $y = b + (d - b)t,$ $0 \le t \le 1$	

■ EXAMPLE 6 Finding Parametric Representations

Use Table 3.2 to determine a parametric representation for each graph with the given rectangular equation.

(a)

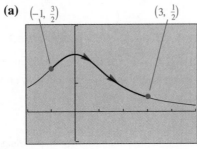

$$y = \frac{6}{x^2 + 3}$$
$$[-2, 5] \times [-1, 3]$$

(b)

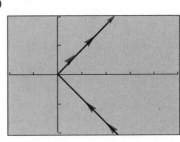

$$x = |y|$$
$$[-2, 5] \times [-2, 2]$$

(c)

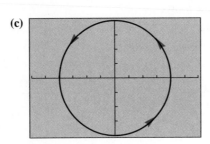

$$x^2 + y^2 = 16$$
$$[-6, 6] \times [-4, 4]$$

(d)

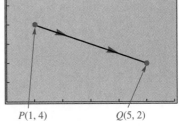

$P(1, 4)$ $Q(5, 2)$

Line segment PQ
$$[0, 6] \times [0, 6]$$

SOLUTION

(a) This curve is the graph of $y = f(x)$, where

$$f(x) = \frac{6}{x^2 + 3}$$

By Table 3.2, a parametric representation for the curve is

$$x = t, \quad y = \frac{6}{t^2 + 3}$$

Because the curve goes from $P\left(-1, \frac{3}{2}\right)$ to $Q\left(3, \frac{1}{2}\right)$, the domain for t is $-1 \le t \le 3$.

(b) By Table 3.2, a parametric representation for the curve is

$$x = |t|, \quad y = t$$

Because the curve has no endpoints, the domain for t is the set of all real numbers.

(c) The graph is a circle centered at the origin with radius 4. According to Table 3.2, a parametric representation for the curve is

$$x = 4 \cos t, \quad y = 4 \sin t \qquad \text{for } 0 \le t \le 2\pi$$

■ **GRAPHING NOTE:** *You can check each answer in Example 6 with a graphing utility.*

(d) Table 3.2 gives a parametric representation of the line segment from $P(a, b)$ to $Q(c, d)$ as $x = a + (c - a)t, y = b + (d - b)t$, with domain $0 \le t \le 1$. The given line segment goes from $P(1, 4)$ to $Q(5, 2)$, so we get

$$x = a + (c - a)t = 1 + (5 - 1)t = 1 + 4t$$
$$y = b + (d - b)t = 4 + (2 - 4)t = 4 - 2t$$

The parametric representation is

$$x = 1 + 4t, \quad y = 4 - 2t \qquad \text{for } 0 \le t \le 1$$

Applications

Parametric representations are also useful in modeling applications where the position of an object is a function of time. In the next example, we will see how parametric equations allow us to connect an object's horizontal position x and its vertical position y with the elapsed time t since the object began to move.

■ **EXAMPLE 7   Application: Projectile Motion**

A projectile is launched from ground level at a $30°$ angle with the horizontal (Figure 3.78). The position of the projectile t seconds after launch is given by the parametric equations

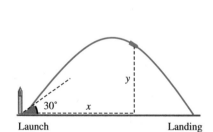

Figure 3.78

$$x = \frac{\sqrt{3}}{2}v_0 t, \quad y = \frac{1}{2}v_0 t - 16t^2$$

where v_0 is the initial velocity of the projectile (in feet per second). For a particular launch, the initial velocity is 96 feet per second.

(a) Use a graphing utility to determine the graph of the path of the projectile.

(b) How long after launch does the projectile land, and how far from the launch site (to the nearest foot) does the projectile land?

SOLUTION

(a) Given that $v_0 = 96$, the parametric representation is

$$x = \frac{\sqrt{3}}{2}(96)t = 48\sqrt{3}\, t, \quad y = \frac{1}{2}(96)t - 16t^2 = 48t - 16t^2$$

Figure 3.79 shows the path of the projectile.

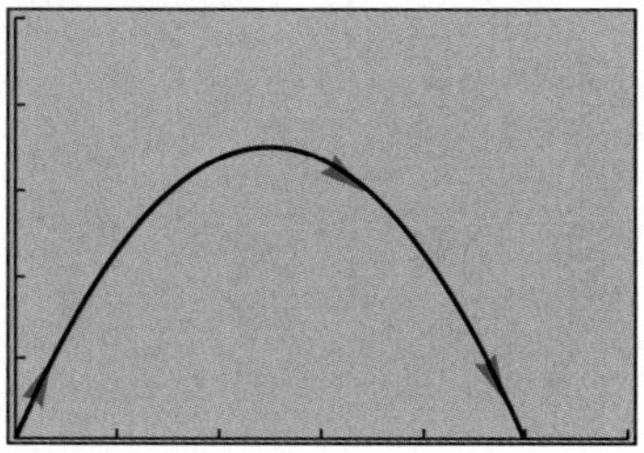

Figure 3.79
$x = 48\sqrt{3}\,t, y = 48t - 16t^2$
$[0, 300]_{50} \times [-0, 50]_{10}$

(b) When the projectile lands, the height y of the projectile is 0. Thus,

$$y = 48t - 16t^2$$
$$0 = 48t - 16t^2 \qquad \text{Let } y = 0.$$
$$0 = 16t(3 - t)$$

$$16t = 0 \quad \bigg| \quad 3 - t = 0$$
$$t = 0 \quad \bigg| \quad t = 3$$

These two answers are the values of t for which the height of the projectile is 0. At $t = 0$, the projectile is launched, and at $t = 3$, the projectile lands. The distance the projectile lands from the launch site is the value of x when $t = 3$:

$$x = 48\sqrt{3}\,t = 48\sqrt{3}\,(3) = 144\sqrt{3}$$

This distance is 249 feet (to the nearest foot).

EXERCISE SET 3.6

A

In Exercises 1–8, sketch each plane curve for the given interval of t by building a table of values, plotting the corresponding points, and drawing a smooth curve through them. Show the orientation of the curve.

1. $x = t - 1, y = 5 - t, -2 \le t \le 4$

2. $x = 6 - t, y = t - 4, -3 \le t \le 5$

3. $x = 1 + 2t, y = 4 - 2t, 0 \le t \le 4$

4. $x = \frac{1}{2}t, y = 2 + \frac{3}{2}t, -6 \le t \le 8$

5. $x = 2t^2 + 2, y = \frac{1}{2}t$

6. $x = 2t - 1, y = t^2 + 2$

7. $x = t - 1, y = \dfrac{t}{t - 1}, t \ge 2$

8. $x = \dfrac{2t}{t + 1}, y = 2 - t, t \ge 0$

In Exercises 9–14, find a rectangular equation in x and y for each plane curve described, and use the equation to sketch the graph. (Be sure to show the orientation of the curve.)

9. $x = t - 2, y = t^2, -2 \le t \le 2$

10. $x = 4 - t, y = \sqrt{t}, 0 \le t \le 9$

11. $x = t^3, y = 2t^3 + 1$

12. $x = \dfrac{1}{t^2}, y = \dfrac{2}{t^2} - 1$

13. $x = t - 2, y = \sqrt{t}, t \ge 0$

14. $x = 2\sqrt{t}, y = 4t, t \ge 0$

In Exercises 15–22, use a graphing utility to determine each plane curve described.

15. $x = t^2 - 2, y = t^3, -2 \le t \le 2$

16. $x = t^3 + 1, y = t^2, -2 \le t \le 3$

17. $x = 2t^3, y = t - 2, -1 \le t \le 1$

18. $x = \sqrt[3]{t}, y = t - 2, -1 \le t \le 8$

19. $x = \dfrac{6}{t + 1}, y = t - 2, 1 \le t \le 5$

20. $x = \dfrac{10}{t^2 + 1}$, $y = t^2 + t$, $-1 \le t \le 2$

21. $x = 3^t + 2$, $y = 3^{-t}$, $-1 \le t \le 2$

22. $x = 2^t$, $y = 2^{-t} - 1$, $-2 \le t \le 2$

In Exercises 23–30, determine a parametric representation for each curve shown.

23.

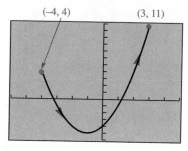

$y = x^2 + 2x - 4$
$[-6, 5] \times [-6, 11]$

24.

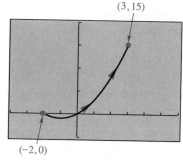

$y = x^2 + 2x$
$[-4, 6] \times [-5, 20]_5$

25.

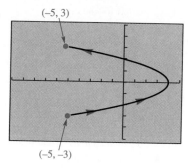

$x = 4 - y^2$
$[-10, 5] \times [-5, 5]$

26.

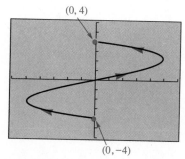

$x = 16y - y^3$
$[-30, 30]_5 \times [-6, 6]$

27.

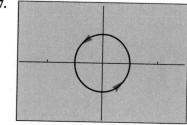

$x^2 + y^2 = \frac{1}{4}$
$[-1.5, 1.5] \times [-1, 1]$

28.

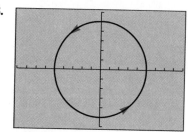

$x^2 + y^2 = 25$
$[-9, 9] \times [-6, 6]$

29.

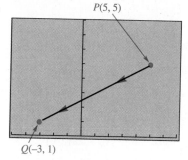

Line segment PQ
$[-5, 7] \times [0, 8]$

30.

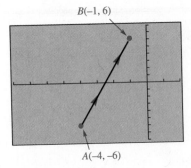

Line segment AB
$[-8, 2] \times [-8, 8]$

B

In Exercises 31–36, sketch the curve described by each parametric representation and determine a rectangular equation for the curve.

31. $x = 2 \cos t,\ y = 2 \sin t,\ 0 \leq t \leq 2\pi$

32. $x = 4 \cos t,\ y = 4 \sin t,\ 0 \leq t \leq 2\pi$

33. $x = \cos t,\ y = \sin t,\ 0 \leq t \leq \dfrac{3\pi}{2}$

34. $x = \cos t,\ y = \sin t,\ \dfrac{\pi}{2} \leq t \leq 2\pi$

35. $x = 5 \cos t,\ y = 5 \sin t,\ -\dfrac{\pi}{2} \leq t \leq \dfrac{\pi}{2}$

36. $x = 6 \cos t,\ y = 6 \sin t,\ -\pi \leq t \leq \pi$

In Exercises 37–44, determine a parametric representation for each curve described.

37. The line segment from $(2, 4)$ to $(-3, 20)$

38. The line segment from $(-1, -6)$ to $(11, 21)$

39. The graph of $y = x^2 - 4$ from $(-2, 0)$ to $(2, 0)$

40. The graph of $y = x^3 - 4x$ from $(0, 0)$ to $(4, 48)$

41. The graph of $2x + y = 4$ from $(0, 4)$ to $(2, 0)$

42. The graph of $4x - 3y = 12$ from $(3, 0)$ to $(0, -4)$

43. The arc of the circle shown in Figure 3.80

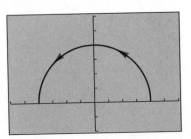

Figure 3.80
$[-6, 6] \times [-2, 6]$

44. The arc of the circle shown in Figure 3.81

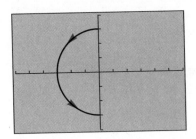

Figure 3.81
$[-6, 6] \times [-4, 4]$

In Exercises 45 and 46:

(a) Use a graphing utility to determine the curve described.
(b) Determine a rectangular equation for the curve.

45. $x = 2 + 2 \cos t,\ y = -1 + 2 \sin t,\ 0 \leq t \leq 2\pi$

46. $x = -3 + 3 \cos t,\ y = 2 + 3 \sin t,\ 0 \leq t \leq 2\pi$

47. The graphs of $x = f(t)$ and $y = g(t)$ are shown in Figure 3.82 (at the top of the next page).
(a) Use these graphs to build a table of values of x and y for $t = -4, -2, 0, 2, 4$.
(b) Use the table in part (a) to sketch the graph of the curve with parametric representation $x = f(t),\ y = g(t)$ for $-4 \leq t \leq 4$.

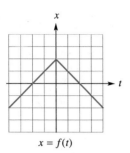

$x = f(t)$

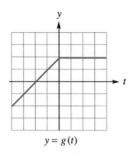

$y = g(t)$

Figure 3.82

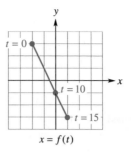

$x = f(t)$

Figure 3.85

48. Repeat Exercise 47 using Figure 3.83.

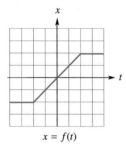

$x = f(t)$

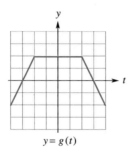

$y = g(t)$

Figure 3.83

49. The graph of $x = f(t)$, $y = g(t)$, $-2 \le t \le 2$ is shown in Figure 3.84. Determine $f(t)$ and $g(t)$.

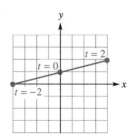

Figure 3.84

50. The graph of $x = f(t)$, $y = g(t)$, $-2 \le t \le 2$ is shown in Figure 3.85. Determine $f(t)$ and $g(t)$.

51. Suppose that the projectile in Example 7 has an initial velocity of 120 feet per second.
   (a) Use a graphing utility to determine the graph of the path of the projectile.
   (b) How long after launch does the projectile land, and how far from the launch site (to the nearest foot) does the projectile land?

52. What is the maximum height of the projectile in Exercise 51? How long after launch does this maximum occur?

C

The curve C described by $x = f(t)$, $y = g(t)$ is shown in Figure 3.86. For Exercises 53–56, sketch the graph of each parametric representation and explain the relationship between this curve and C.

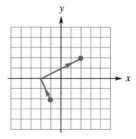

Figure 3.86

53. **(a)** $x = f(t) + 2, y = g(t) + 3$
   (b) $x = f(t) - 1, y = g(t) - 4$
   (c) $x = f(t) + 4, y = g(t) - 2$

54. **(a)** $x = f(t)$, $y = 2g(t)$ **(b)** $x = 3f(t)$, $y = g(t)$
   (c) $x = 3f(t)$, $y = 2g(t)$

55. **(a)** $x = -f(t)$, $y = g(t)$ **(b)** $x = f(t)$, $y = -g(t)$
   (c) $x = -f(t)$, $y = -g(t)$

56. **(a)** $x = g(t)$, $y = f(t)$ **(b)** $x = -g(t)$, $y = -f(t)$

57. Show that the curve C described by $x = a + (c - a)t$, $y = b + (d - b)t$ for $0 \le t \le 1$ is the line segment from $P(a, b)$ to $Q(c, d)$.
   HINT: First find the rectangular equation for C and show that the graph of this equation passes through P and Q.

58. *Predator–prey models* are used to mathematically model the relationship between the populations of a predator and its prey in a closed biological system. For example, the populations of foxes (the predators) and hares (the prey) in an isolated valley over a 60 month period are given by the graph in Figure 3.87. (The value of t is the month of the period.)
   (a) Describe the populations of foxes and hares over each of the following periods: 0–15 months, 15–30 months, 30–45 months, 45–60 months.

(b) Sketch the graph of $y = f(t)$, where $f(t)$ is the number of foxes in the valley at month t.

(c) Sketch the graph of $y = h(t)$, where $h(t)$ is the number of hares in the valley at month t.

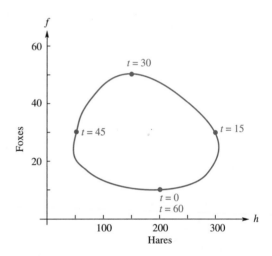

Figure 3.87

QUICK REFERENCE

Topic	Page	Remarks
Polynomial function	262	A polynomial function is a function that can be expressed in the form $$f(x) = a_n x^n + a_{n-1} x^{n-1} + \cdots + a_2 x^2 + a_1 x + a_0$$ where $a_n \ne 0$. See the box on page 262.
Leading term property	267	The behavior of the graph of a polynomial function $$f(x) = a_n x^n + a_{n-1} x^{n-1} + \cdots + a_2 x^2 + a_1 x + a_0$$ is the same as that of the graph of $y = a_n x^n$ for extreme values of x. See Figure 3.6 for the behavior of $y = a_n x^n$ for extreme values of x.
Factor theorem	277	For a polynomial function P, a number c is a zero of P if and only if $x - c$ is a factor of $P(x)$.

Topic	Page	Remarks
Corollary to the factor theorem	278	Suppose that P is a polynomial function and c is a real number. If any of the following four statements is true, then they are all true: **1.** The real number c is a zero of the function P. **2.** The real number c is a solution to the equation $P(x) = 0$. **3.** The point $(c, 0)$ is an x-intercept of the graph of P. **4.** The binomial $x - c$ is a factor of $P(x)$.
Imaginary unit, i	286	The imaginary unit is $i = \sqrt{-1}$. The imaginary unit is not a real number.
Complex number	287	A complex number is a number that can be written in the form $a + bi$, where a and b are real numbers, and i is the imaginary unit. Since a real number a can be rewritten as $a = a + 0i$, any real number is also a complex number.
Fundamental theorem of algebra	293	Every polynomial function of positive degree has at least one complex zero.
Complete factorization theorem	294	If P is a polynomial function of degree $n > 0$, $$P(x) = a_n x^n + a_{n-1} x^{n-1} + \cdots + a_2 x^2 + a_1 x + a_0$$ then $P(x)$ factors as $$P(x) = a_n(x - c_1)(x - c_2) \cdot \cdots \cdot (x - c_n)$$ where $c_1, c_2, \ldots, c_n$ are the complex zeros of P.
Complex conjugate theorem	296	Suppose P is a polynomial function with real number coefficients. If the complex number $a + bi$ is a zero of P, then its complex conjugate $a - bi$ is also a zero of P.
Rational function	298	A rational function f is a function that can be written in the form $$f(x) = \frac{n(x)}{d(x)}$$ where $n(x)$ and $d(x)$ are polynomials.
Vertical asymptote	302	A line $x = a$ is a vertical asymptote of the graph of a function f if for values of x near a, the graph of f approaches the line. If f is a reduced rational function, $$f(x) = \frac{n(x)}{d(x)}$$ and $d(a) = 0$ for some real number a, then the line $x = a$ is vertical asymptote of the graph of f.

Continued

Topic	Page	Remarks
Horizontal and slant asymptotes	305	If f is a reduced rational function written in the form $$f(x) = q(x) + \frac{r(x)}{d(x)}$$ where the degree of $r(x)$ is less than the degree of $d(x)$, then the graph of f approaches the graph of $y = q(x)$ for extreme values of x. In particular, if $q(x) = k$, where k is a constant, then $y = k$ is a horizontal asymptote. If $q(x) = mx + b$, where m and b are constants, then $y = mx + b$ is a slant asymptote.
Extraneous solutions	317	Multiplying each side of an equation by an expression, and certain other operations, can lead to another equation that is not equivalent to the original equation. Solutions of the resulting equation that do not satisfy the original equation are extraneous solutions. See Example 3 on page 317.
Parametric equations	327	Parametric equations are equations in which coordinates are expressed in terms of an arbitrary constant called a parameter. The set of points described by the parametric equations $$x = f(t), y = g(t)$$ where f and g are functions, is a plane curve; the variable t is the parameter. The direction along the curve for increasing values of t is the orientation of the curve. See Table 3.2 for parametric representations of common plane curves, such as a line or a circle.

◼ MISCELLANEOUS EXERCISES

In Exercises 1–6, for each function:

(a) Describe the behavior of the graph of the function for extreme values of x.
(b) Determine the x-intercepts of the graph algebraically.
(c) Use a graphing utility to determine the graph.

1. $y = \frac{1}{4}(x + 4)^5 - 24$ **2.** $y = -(x - 3)^6 + 27$

3. $y = x^4 - 4x^3 - x + 4$

4. $y = -x^3 + 5x^2 + 6x - 30$

5. $y = (x^3 - 8x)(x + 7)$ **6.** $(x^2 - 9)(x^2 + 4x - 4)$

In Exercises 7–14, determine the graph of each function. Determine intercepts and asymptotes, and describe the behavior of the graph for extreme values of x.

7. $y = \frac{-4}{x^5}$ **8.** $y = \frac{1}{2x^4}$

9. $y = \frac{4x - 8}{x^2 + x - 6}$ **10.** $y = \frac{2x - 2}{x^2 - 5x + 4}$

11. $f(x) = \frac{3x - 4}{x - 2}$ **12.** $f(x) = \frac{5x}{2x + 4}$

13. $y = \frac{1}{2}x + \frac{2}{x^2 - 16}$ **14.** $y = -x + \frac{2}{x(x - 4)}$

In Exercises 15–20, solve each equation over the real numbers.

15. $x^3 + 3x^2 - 22x = 60$ **16.** $2x^3 + 2x^2 = 36x - 36$

17. $\frac{5}{3x} + \frac{5}{6} = \frac{1}{2x}$ **18.** $\frac{5}{x} = \frac{1}{4x} - \frac{3}{2}$

19. $1 + \dfrac{x+1}{x+3} = \dfrac{x}{2x+7}$

20. $\dfrac{x+5}{x-2} - \dfrac{5}{x+2} = \dfrac{28}{x^2-4}$

21. Determine a fourth-degree polynomial function P with zeros -3, -1, $\frac{1}{2}$, and 2, and y-intercept $(0, 6)$.

22. Determine a third-degree polynomial function Q such that $Q(2) = 0$, $Q(5) = 0$, $Q(-2) = 0$, and $Q(0) = 10$.

23. Write $x^4 + 4x^3 - 10x^2 - 64x - 96$ as the product of linear factors.

24. Write $x^4 + 2x^3 + x^2 + 32x - 240$ as the product of linear factors.

In Exercises 25–28, evaluate each expression. Write your answer in the form $a + bi$.

25. (a) $(6 + 4i) + (-5 + 2i)$
  (b) $(3 - 2i)(-1 + 4i)$

26. (a) $(1 + 2i) - (-2 - 3i)$
  (b) $(7 + i)(-5 - 2i)$

27. (a) $\dfrac{2 + 6i}{4 - i}$
  (b) $\dfrac{i}{(1 + 2i)^2}$

28. (a) $\dfrac{3 - i}{4 + 5i}$
  (b) $\dfrac{(1 + \sqrt{3}\,i)^2}{i}$

In Exercises 29–32, let $w = 2 - i$ and $z = 3 + 7i$. Simplify each expression, and write the answer in the form $a + bi$.

29. (a) $wz + z$
  (b) w^{-1}

30. (a) $4\bar{z}$
  (b) $\bar{z}^2 - z^2$

31. (a) $z(w + z)$
  (b) z^{-1}

32. (a) $-(w + \bar{w})$
  (b) $\bar{w}^2 + w^2$

In Exercises 33–38, solve each equation.

33. $x^2 + 2x + 4 = 0$

34. $x^2 - 6x + 10 = 0$

35. $x^4 = 9$

36. $x^4 + 7x^2 = 18$

37. $x^3 + 2x^2 = 8x + 16$

38. $x^4 - 2x^3 + 11x^2 = 28x + 42$

In Exercises 39–46, solve each inequality.

39. $x^4 - 20 > x^2$

40. $x^3 + 18 < 5x^2$

41. $\dfrac{3x - 12}{x - 2} \le 0$

42. $\dfrac{x^2 - 5x - 6}{x - 9} \ge 0$

43. $2x - 1 \ge \dfrac{6}{x}$

44. $3x + \dfrac{4}{x} \ge 4$

45. $\dfrac{x^2 - 3x - 14}{x - 2} \ge 1$

46. $\dfrac{3 - 2x}{x^2 - 1} \ge -1$

In Exercises 47–52, use a graphing utility to determine the graph of each plane curve described and determine a rectangular equation for the curve.

47. $x = 4 - t$, $y = 2t - 1$, $-4 \le t \le 3$

48. $x = \frac{1}{2}t$, $y = 2 - \frac{5}{2}t$, $t \ge 0$

49. $x = \dfrac{t}{t + 1}$, $y = t + 1$, $t > -1$

50. $x = e^{-t}$, $y = -e^t$

51. $x = 3\cos t$, $y = 3\sin t$, $0 \le t \le 2\pi$

52. $x = \cos t$, $y = \sin t$, $0 \le t \le \pi$

In Exercises 53–56, determine the parametric representation of each curve described.

53. The line segment from $(2, 0)$ to $(0, -5)$

54. The graph of $y = 16 - x^2$ from $(-3, 7)$ to $(3, 7)$

55. The graph of $x = y^3 - 4y$ from $(0, 0)$ to $(48, 4)$

56. The graph of $2x + y = 4$ from $(0, 4)$ to $(2, 0)$

57. The graph of $x = f(t)$ is shown below.

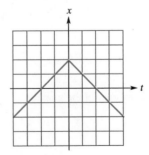

(a) Build a table of values for the parametric representation

$$x = f(t), \quad y = t + 1$$

of x for $t = -4, -2, 0, 2, 4$.

(b) Use the table in part (a) to sketch the graph of the curve with this parametric representation.

58. Repeat Exercise 57 using $y = 5 - t$.

59. A truck travels 60 miles on a rural road, after which it travels 45 miles on a highway. The speed of the truck on the highway is 20 miles per hour faster than on the road, and the entire trip takes $2\frac{1}{4}$ hours. What are the traveling speeds of the truck on the road and the highway?

60. An inlet pipe of a swimming pool can fill the pool in 12 hours. The pool's drain can empty it in 16 hours. After how many hours will the inlet pipe fill the pool if the drain is left open accidently?

◼ CHAPTER TEST

1. Let $f(x) = (5x - 6)(x^2 - 11)$.
   (a) Describe the behavior of the graph of f for extreme values of x.
   (b) Determine the x-intercepts of the graph algebraically.

2. Let $f(x) = -x^3 + 6x$.
   (a) Determine the x-intercepts of the graph of f algebraically.
   (b) Use a graphing utility to determine the graph.

3. Divide $P(x) = 3x^3 - 10x^2 + 17x - 19$ by $d(x) = x^2 - x + 3$ and write P as $P(x) = d(x)q(x) + r(x)$.

4. Solve the polynomial equation $7x^3 + 3x^2 = 98x + 42$.

5. Solve the inequality $2x^3 + 3x^2 \le 23x + 12$ graphically.

6. Determine a polynomial function P with zeros $-1, \frac{1}{2}$, and 3, such that the graph of P has the same behavior as the graph of $y = 4x^3$ for extreme values of x.

7. Let $w = 4 - 3i$ and $z = 2 + i$. Evaluate:
   (a) $\overline{w}z$ **(b)** $\dfrac{10}{z}$

8. Solve $8x^3 - 3x^2 + 40x - 15 = 0$ over the complex numbers.

9. List the intercepts and asymptotes of the graph of
$$f(x) = \frac{-3x + 12}{2x - 4}$$

10. List the intercepts and asymptotes of the graph of
$$f(x) = \frac{x + 1}{x^2 - 9}$$
and graph f.

11. Solve: $\dfrac{x + 1}{x + 2} = \dfrac{2}{x^2 + 2x}$

12. A manufacturer of carrying cases for laptop computers has found that the cost of manufacturing x units per week is modeled by the function $C(x) = 0.5x^2 - x + 72$. The average weekly cost (in dollars per unit) is
$$\overline{C}(x) = \frac{C(x)}{x} = \frac{x}{2} - 1 + \frac{72}{x}$$

Determine an interval for the number of units to manufacture so that the average weekly cost is less than \$19 per unit.

13. Use a graphing utility to determine the plane curve described by
$$x = \frac{t^4 + 3}{2}, \quad y = t + 2 \qquad -1 \le t \le 2$$

Be sure to show the orientation of the curve.

14. Determine a parametric representation of the curve shown.

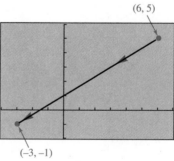

$(6, 5)$

$(-3, -1)$

$[-4, 7] \times [-2, 6]$

CHAPTER FOUR

EXPONENTIAL AND LOGARITHMIC FUNCTIONS

This chapter introduces two sets of functions called *exponential* and *logarithmic functions*. These functions are widely used to develop mathematical models.

Sections 4.1 and 4.2 discuss exponential functions and their applications. In Section 4.3, we see how the process of one function can be "undone" by another function—its inverse. The concepts of this section are then applied in Section 4.4 to define a logarithmic function as the inverse function of an exponential function. In Section 4.5, the common techniques used to solve equations involving exponential and logarithmic functions are demonstrated.

When you complete this chapter, you will be able to use these functions, along with their graphs, to solve important real-life problems in fields such as science, engineering, business, psychology, and economics.

EXPONENTIAL FUNCTIONS

Most of the functions we have encountered so far in this text are *algebraic functions*. These functions are constructed from constants and variables by using the basic operations of adding, subtracting, multiplying, dividing, and taking roots. Linear functions, quadratic functions, polynomial functions, and rational functions are all examples of algebraic functions. Many useful functions are not algebraic, however. These nonalgebraic functions are called *transcendental functions*, because they transcend the basic operations.

In this section, we introduce a set of transcendental functions called *exponential functions*. These nonalgebraic functions are instrumental in mathematical models for applications as diverse as compound interest on an investment, the spread of a disease through a society, and the decay of a dangerous radioactive isotope.

Definition of an Exponential Function

Simply stated, an *exponential function* with *base b* is of the form $f(x) = b^x$. For example, the function $f(x) = 4^x$ is an exponential function with base 4.

What is the domain of the function $f(x) = 4^x$? That is, for what values of x does the expression 4^x make sense? In Section 5 (page 39), expressions such as 4^x were defined for any rational number x. For example:

$$4^3 = 64$$
$$4^{-2} = \frac{1}{4^2} = \frac{1}{16}$$
$$4^{3/2} = \sqrt{4^3} = \sqrt{64} = 8$$

NOTE: *You can assume that the properties of rational exponents from Section 5 (page 39) also hold for real number exponents.*

The expression 4^x can also be defined for irrational values of x. We assign a value to an expression such as $4^{\sqrt{3}}$ by using a decimal representation of an irrational number. For example, saying that $\sqrt{3} = 1.732\,050\,807\,568\,8...$ means that $\sqrt{3}$ can be approximated with increasing precision by the rational numbers 1.7, 1.73, 1.732, 1.7320, 1.73205, and so on. In the same manner, $4^{\sqrt{3}}$ can be approximated to any degree of precision by raising 4 to these rational numbers. This scheme is demonstrated in Table 4.1:

TABLE 4.1

Expression	$4^{1.7}$	$4^{1.73}$	$4^{1.732}$	$4^{1.7320}$	$4^{1.73205}$
Value (to the nearest 0.0001)	10.5561	11.0043	11.0349	11.0349	11.0357

More computations would show that $4^{\sqrt{3}} = 11.035\,664\,635\,963...$. This approach allows us to say that the domain of a function such as $f(x) = 4^x$ is the entire set of real numbers, both rational and irrational.

■ **NOTE:** *If b is negative, then we encounter expressions like* $(-3)^{1/2}$ *or* $(-0.7)^{3/4}$*, which are undefined as real numbers.*

In order for b^x to be defined for all real numbers x, the base b must be positive. We also rule out $b = 1$, because $f(x) = 1^x$ is really the constant function $f(x) = 1$.

The outputs of an exponential function $f(x) = b^x$ are positive, since a positive number b raised to any real number is positive.

These results are summarized by the following:

Definition of Exponential Function

> If f is a function of the form $f(x) = b^x$ such that $b > 0$ and $b \neq 1$, then f is an **exponential function**. The domain of f is the set of all real numbers, and the range is the set of all positive real numbers.

In the past, when computing the values of exponential functions was done by hand, it was a tedious, time-consuming task. Now, a calculator can approximate the values of exponential functions quickly and accurately.

■ EXAMPLE 1   Evaluating Exponential Functions

Given that $f(x) = 4^x$ and $g(x) = \left(\frac{2}{3}\right)^x$, use a calculator to approximate each value (to the nearest 0.0001).

(a) $f(1 + \sqrt{2})$ **(b)** $g(-\pi)$

SOLUTION

The keystroke sequence and output for most graphics calculators is shown (yours may vary somewhat).

(a) The value of $f(1 + \sqrt{2})$ is $4^{1+\sqrt{2}}$. The keystroke sequence

$$4 \;\boxed{\wedge}\;\boxed{(}\; 1 \;\boxed{+}\;\boxed{\sqrt{}}\; 2 \;\boxed{)}$$

yields the display 28.411 973 205, or 28.4120 (to the nearest 0.0001).

(b) The value of $g(-\pi)$ is $\left(\frac{2}{3}\right)^{-\pi}$. The keystroke sequence is

$$\boxed{(}\; 2 \;\boxed{\div}\; 3 \;\boxed{)}\;\boxed{\wedge}\;\boxed{(-)}\;\boxed{\pi}$$

The value displayed is 3.574 431 723, or 3.5744 (to the nearest 0.0001).

■

Graphs of Exponential Functions

We can get a sense of the graphs of $y = b^x$ by looking at some examples, first for $b > 1$ and then for $0 < b < 1$.

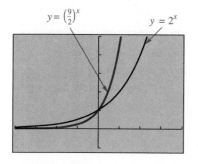

Figure 4.1
$[-4, 4] \times [-1, 5]$

TABLE 4.2

x	-2	-1	0	1	2
$y = \left(\frac{9}{2}\right)^x$	$\frac{4}{81}$	$\frac{2}{9}$	1	$\frac{9}{2}$	$\frac{81}{4}$
$y = 2^x$	$\frac{1}{4}$	$\frac{1}{2}$	1	2	4

■ **GRAPHING NOTE:** *It appears that* $y = \left(\frac{1}{2}\right)^x$ *is a reflection of* $y = 2^x$ *through the y-axis, and that* $y = \left(\frac{2}{9}\right)^x$ *is a reflection of* $y = \left(\frac{9}{2}\right)^x$ *through the y-axis as well. This is indeed the case. (See Exercise 53.)*

The graphs of $y = 2^x$ and $y = \left(\frac{9}{2}\right)^x$ are shown in Figure 4.1. Table 4.2 lists values for each function. The range for these functions is the set of positive real numbers, so the graphs are completely above the x-axis, and there are no x-intercepts. Notice also that the y-intercept of each graph is (0, 1).

The graphs of $y = 2^x$ and $y = \left(\frac{9}{2}\right)^x$ behave similarly for extreme values of x. Figure 4.1 suggests that these functions are described by

$$y \rightarrow 0 \quad \text{as} \quad x \rightarrow -\infty \quad \text{and} \quad y \rightarrow +\infty \quad \text{as} \quad x \rightarrow +\infty$$

The graphs of $y = \left(\frac{1}{2}\right)^x$ and $y = \left(\frac{2}{9}\right)^x$ are shown in Figure 4.2, and Table 4.3 lists a few values for each function. Like the functions with graphs shown in Figure 4.1, these functions are defined for all real numbers, and their values are strictly positive. Each graph has a y-intercept (0, 1), but no x-intercepts. For extreme values of x, these graphs are described by

$$y \rightarrow +\infty \quad \text{as} \quad x \rightarrow -\infty \quad \text{and} \quad y \rightarrow 0 \quad \text{as} \quad x \rightarrow +\infty$$

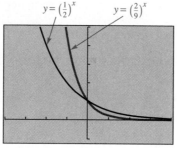

Figure 4.2
$[-4, 4] \times [-1, 5]$

TABLE 4.3

x	-2	-1	0	1	2
$y = \left(\frac{2}{9}\right)^x$	$\frac{81}{4}$	$\frac{9}{2}$	1	$\frac{2}{9}$	$\frac{4}{81}$
$y = \left(\frac{1}{2}\right)^x$	4	2	1	$\frac{1}{2}$	$\frac{1}{4}$

Graph of an Exponential Function

These graphs suggest some generalizations about exponential functions and their graphs:

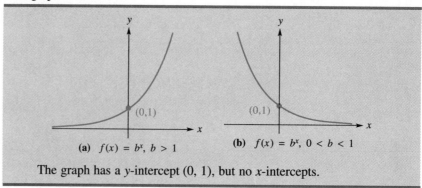

(a) $f(x) = b^x$, $b > 1$ (b) $f(x) = b^x$, $0 < b < 1$

The graph has a y-intercept (0, 1), but no x-intercepts.

Continued

> The x-axis ($y = 0$) is a horizontal asymptote.
>
> The graph of f is smooth, and it has no breaks.
>
> The function f is increasing over its domain if $b > 1$, and decreasing over its domain if $0 < b < 1$.
>
> The behavior of f for extreme values of x is described by
>
> $$y \to 0 \quad \text{as} \quad x \to -\infty \quad \text{and} \quad y \to +\infty \quad \text{as} \quad x \to +\infty$$
>
> if $b > 1$, and
>
> $$y \to +\infty \quad \text{as} \quad x \to -\infty \quad \text{and} \quad y \to 0 \quad \text{as} \quad x \to +\infty$$
>
> if $0 < b < 1$.

As we see in Examples 2 and 3, graphs of other functions can be constructed from these graphs using the translations, expansions, compressions, and reflections discussed in Section 1.6.

EXAMPLE 2 Translations of Exponential Functions

Use a graphing utility to determine the graph of $y = 2^x - 6$. Compare this graph with the graph of $f(x) = 2^x$. Determine the intercepts of $y = 2^x - 6$.

SOLUTION

The graph of $y = 2^x - 6$ (Figure 4.3) is a vertical translation, 6 units down, of the graph of $f(x) = 2^x$, because $y = 2^x - 6$ is equivalent to $y = f(x) - 6$. The horizontal asymptote of the graph is $y = -6$.

The y-intercept is found by letting $x = 0$ and solving for y:

$$y = 2^0 - 6 = 1 - 6 = -5$$

Thus, the y-intercept is $(0, -5)$. The x-intercept is determined by letting $y = 0$ and solving for x:

$$0 = 2^x - 6$$
$$2^x = 6$$

We can approximate this solution graphically as $(2.5850, 0)$, to the nearest 0.0001, using the techniques of Section 1.4.

EXAMPLE 3 Compressions and Expansions of an Exponential Function

Use a graphing utility to determine the graph of each function. Compare the graph with the graph of $g(x) = (0.6)^x$.

(a) $y = -3(0.6)^x$ **(b)** $y = (0.6)^{2x}$

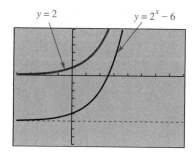

$y = 2$ $y = 2^x - 6$

Figure 4.3
$[-4, 8] \times [-9, 6]$

■ **NOTE:** *In Section 4.4, we will see how to determine this x-intercept algebraically.*

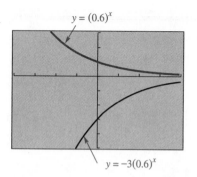

Figure 4.4
$[-4, 4] \times [-5, 3]$

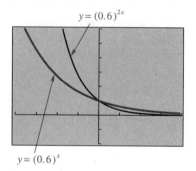

Figure 4.5
$[-4, 4] \times [-2, 6]$

SOLUTION

(a) The graph of $y = -3(0.6)^x$ is a vertical expansion and a reflection through the x-axis of g, since $y = -3(0.6)^x$ is equivalent to $y = -3g(x)$ (Figure 4.4). Like g, this graph has no x-intercepts. The y-intercept of $y = -3(0.6)^x$ is $(0, -3)$, because when $x = 0$,

$$y = -3(0.6)^0 = -3(1) = -3$$

(b) The function $y = (0.6)^{2x}$ can be written as $y = g(2x)$. The graph of this equation is a horizontal compression, by a factor of $\frac{1}{2}$, of the graph of g. Both the graphs of g and $y = (0.6)^{2x}$ are shown in Figure 4.5. The graph of $y = (0.6)^{2x}$ has no x-intercepts; the y-intercept is $(0, 1)$. ∎

Compound Interest

Money deposited in an investment, such as a savings account, earns interest according to the interest rate of the investment. After a fixed interval of time, the interest earned during that time is added to the amount of the investment and, in turn, earns interest over the next period of time. This process is called *compounding*, and the fixed interval of time is called the *compounding period*.

Suppose, for example, that $2000 is invested for 2 years at 6% compounded annually. This means that at the end of each year, the amount of the investment increases by 6%. Each time a compounding period ends, the investment becomes 106% of the previous amount. This amount is computed by multiplying the amount from the previous period by 1.06 (Table 4.4):

TABLE 4.4

	Computation	Amount
Initial amount	2000	$2000.00
Amount after one period	2000(1.06)	$2120.00
Amount after two periods	$[2000(1.06)](1.06) = 2000(1.06)^2$	$2247.20

Thus, the accumulated amount of the investment after 2 years is $2247.20.

Now suppose that the interest is compounded twice annually. The interest earned each $\frac{1}{2}$ year is 3% (since half of 6% is 3%). There are four compounding periods in 2 years. At the end of each compounding period, the investment grows by 3%, or becomes 1.03 times the amount from the previous period. Table 4.5 shows the amount of the investment after four compounding periods:

TABLE 4.5

	Computation	Amount
Initial amount	2000	$2000.00
Amount after one period	2000(1.03)	$2060.00
Amount after two periods	[2000(1.03)](1.03) = 2000(1.03)^2	$2121.80
Amount after three periods	[2000(1.03)^2](1.03) = 2000(1.03)^3	$2185.45
Amount after four periods	[2000(1.03)^3](1.03) = 2000(1.03)^4	$2251.02

The accumulated amount of this investment after 2 years is $2251.02.

What if interest is compounded monthly on this investment? Over 2 years, there are 24 compounding periods. The rate per compounding period is 0.5% (since one-twelfth of 6% is 0.5%). At the end of each compounding period, the amount of the investment is 1.005 times the previous amount. Instead of compiling another table showing each of the 24 compounding periods, we can adapt the last line of the previous computations to find the accumulated amount, using 24 instead of 4 and 1.005 instead of 1.03:

$$2000(1.005)^{24} = 2254.319\,552\,41...$$

The accumulated amount of the investment at the end of 2 years is $2254.32.

The generalization of this result provides a model for compound interest applications.

Compound Interest Model

The model for **compound interest** is:

$$\begin{pmatrix} \text{Accumulated} \\ \text{amount} \end{pmatrix} = \begin{pmatrix} \text{Initial} \\ \text{amount} \end{pmatrix} \left[1 + \begin{pmatrix} \text{Rate of growth} \\ \text{per period} \end{pmatrix} \right]^{\text{(Number of periods)}}$$

Specifically, if P dollars are invested at an annual interest rate r compounded n times annually, then the amount of the investment A after t years is given by the formula

$$A = P\left(1 + \frac{r}{n}\right)^{nt}$$

■ EXAMPLE 4 Application: Using the Compound Interest Model

Suppose that you have won a modest fortune in the state lottery, and you would like to invest some of your money. A bank has an account available with a 4.5% rate compounded monthly.

(a) If you deposit $5000 with the bank, what is the accumulated amount after 3 years?

(b) How much should you deposit now, so that the accumulated amount is $15,000 after 5 years?

SOLUTION

We use the compound interest formula

$$A = P\left(1 + \frac{r}{n}\right)^{nt}$$

in both computations.

■ **NOTE:** *The decimal representation of 4.5% is 0.045.*

(a) To determine A, the accumulated amount after 3 years, we use the compound interest formula with $P = 5000$, $r = 0.045$, $n = 12$, and $t = 3$:

$$A = P\left(1 + \frac{r}{n}\right)^{nt} = 5000\left(1 + \frac{0.045}{12}\right)^{12 \cdot 3} = 5721.239\ 161\ 024...$$

The accumulated amount after 3 years is $5721.24.

(b) The initial amount P can be determined by using the compound interest formula with $A = 15,000$, $r = 0.045$, $n = 12$, and $t = 5$:

$$A = P\left(1 + \frac{r}{n}\right)^{nt}$$

$$15,000 = P\left(1 + \frac{0.045}{12}\right)^{12 \cdot 5}$$

$$15,000 = P(1.00375)^{60} \qquad \text{Simplify the numerical expressions.}$$

$$\frac{15,000}{(1.00375)^{60}} = P \qquad \text{Divide each side by } (1.00375)^{60}.$$

Evaluating P on a calculator yields

$$P = \frac{15,000}{(1.00375)^{60}} = 11,982.784\ 86...$$

In order for the accumulated amount to be $15,000 after 5 years, $11,982.78 must be deposited now.

Other Applications

A large number of the mathematical models using exponential functions are derived from the compound interest application. In these models, the rate at which a quantity changes at a particular moment is proportional to its value at that moment. A model of this type in which the quantity increases is an *exponential growth model*; one in which the quantity decreases is an *exponential decay model*.

One method of gauging the exponential growth of a particular quantity is by its *doubling time*. The doubling time of a quantity is the period over which the quantity doubles in size. It follows that the shorter this time interval, the faster the rate of growth.

EXAMPLE 5   Application: Exponential Growth of Bacteria

Small rural water systems are often contaminated with bacteria by animals. Suppose that a water tank is infested with a colony of 100,000 *E. coli* bacteria. In this tank the colony doubles in number every 4 days.

Let $A(t)$ represent the number of bacteria in the colony after t days.

(a) Determine the number present after 4, 8, and 12 days.
(b) Determine a formula for $A(t)$.
(c) To the nearest 1000, how many bacteria are present after 11 days?

SOLUTION

(a) The initial amount in the tank doubles by the end of day 4, doubles again by the end of day 8, and doubles a third time by the end of day 12. Thus,

$$A(4) = 200,000$$
$$A(8) = 400,000$$
$$A(12) = 800,000$$

The bacteria in the tank after 4, 8, and 12 days number 200,000, 400,000 and 800,000, respectively.

(b) The situation can be described in terms of the compound interest model. The initial amount is 100,000. The rate of growth for each 4 day compounding period is 100%. The number of 4 day compounding periods in t days is $t/4$. According to the compound interest model,

$$\begin{pmatrix} \text{Accumulated} \\ \text{amount} \end{pmatrix} = \begin{pmatrix} \text{Initial} \\ \text{amount} \end{pmatrix} \left[1 + \begin{pmatrix} \text{Rate of growth} \\ \text{per period} \end{pmatrix} \right]^{\text{Number of periods}}$$

we get

■ **NOTE:** *100% is equivalent to 1.*

$$A(t) = 100,000(1 + 1)^{t/4}$$

or

$$A(t) = (100,000)2^{t/4}$$

(c) The number of bacteria after 11 days is $A(11)$:

$$A(11) = (100,000)2^{11/4} = 672,717.132\ 203...$$

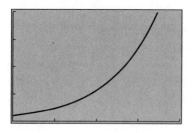

Figure 4.6
$y = (100,000)2^{x/4}$
$[0, 20]_5 \times [0, 2,000,000]_{500,000}$

Doubling Time Formula

■ **NOTE:** *The time t and doubling time d should be measured in the same units — seconds, days, etc.*

After 11 days, there are 673,000 bacteria (to the nearest 1000) present in the tank. ■

The graph of the function A from Example 5 is shown in Figure 4.6. Notice that the graph increases slowly for small values of x, and then increases more rapidly as x becomes larger. This behavior is characteristic for the graphs of exponential growth models.

The result of Example 5 can be generalized. If the initial amount is A_0 instead of 100,000, and if the doubling time is d units of time instead of 4 days, then we get the following formula:

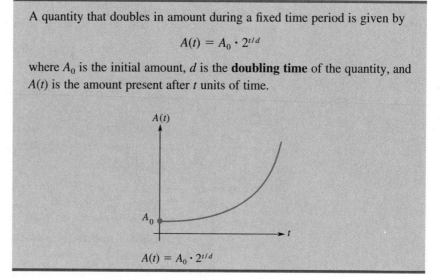

A quantity that doubles in amount during a fixed time period is given by

$$A(t) = A_0 \cdot 2^{t/d}$$

where A_0 is the initial amount, d is the **doubling time** of the quantity, and $A(t)$ is the amount present after t units of time.

$$A(t) = A_0 \cdot 2^{t/d}$$

Just as doubling time is used to gauge the exponential growth of an increasing quantity, the exponential decay of a decreasing quantity is measured by its *half-life*. The half-life of a decreasing quantity is the period over which the quantity decreases by half. The shorter this time interval, the faster the rate of decay.

■ **EXAMPLE 6 Application: Exponential Decay of Radioactive Elements**

Lead-209, a radioactive isotope, decays to nonradioactive lead over time. The half-life of lead-209 is 3.3 hours. Suppose that 20 milligrams of lead-209 are created by a particle physics experiment. Let $A(t)$ be the amount of lead-209 present t hours after the experiment.

(a) Determine the amount of lead-209 present 3.3, 6.6, and 9.9 hours after the experiment.

(b) Determine a formula for $A(t)$.

(c) How much lead-209 is left 8 hours after the experiment (to the nearest 0.01 milligram)?

SOLUTION

(a) The initial amount of lead-209 is reduced by half after 3.3 hours, reduced by half again after 6.6 hours, and reduced by half a third time after 9.9 hours:

$$A(3.3) = 10$$
$$A(6.6) = 5$$
$$A(9.9) = 2.5$$

The amount of lead-209 after 3.3 hours is 10 milligrams, after 6.6 hours is 5 milligrams, and after 9.9 hours is 2.5 milligrams.

(b) The compound interest model can be used to describe this situation. Each 3.3 hours, the amount decreases by 50%, so we consider the growth rate as -50% per 3.3 hour period. The number of periods in t hours is $t/3.3$. Using the model

$$\begin{pmatrix}\text{Accumulated}\\ \text{amount}\end{pmatrix} = \begin{pmatrix}\text{Initial}\\ \text{amount}\end{pmatrix}\left[1 + \begin{pmatrix}\text{Rate of growth}\\ \text{per period}\end{pmatrix}\right]^{\text{Number of periods}}$$

we get

■ **NOTE:** *-50% is equivalent to $-\frac{1}{2}$.*

$$A(t) = 20\left[1 + \left(-\tfrac{1}{2}\right)\right]^{t/3.3}$$

or

$$A(t) = 20\left(\tfrac{1}{2}\right)^{t/3.3}$$

(c) The amount of lead-209 after 8 hours is $A(8)$:

$$A(8) = 20\left(\tfrac{1}{2}\right)^{8/3.3} = 3.726\ 149\ 788...$$

There are 3.73 milligrams after 8 hours (to the nearest 0.01 milligram). ■

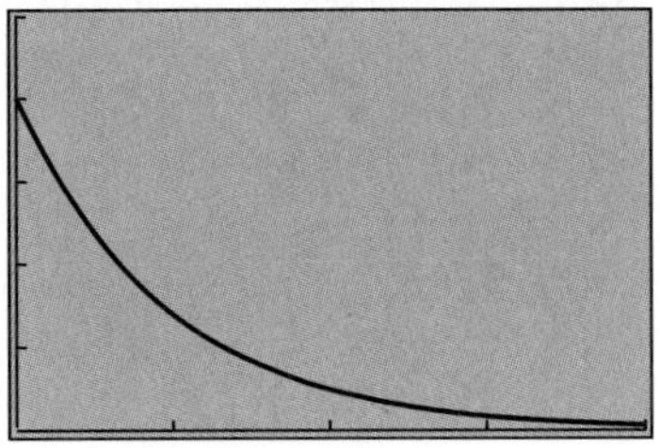

Figure 4.7
$y = 20\left(\frac{1}{2}\right)^{x/3.3}$
$[0, 20]_5 \times [0, 25]_5$

The graph of the function A from Example 6 is shown in Figure 4.7. Notice that the graph decreases rapidly for small values of x, and "flattens out" as x becomes larger. This behavior is true in general for the graphs of exponential decay models.

We can generalize from Example 6. If the initial amount is A_0 instead of 20, and if the half-life is h units of time instead of 3.3 hours, then the following formula results:

Half-Life Model

The model for a quantity that halves in amount in a fixed time period is

$$A(t) = A_0 \cdot \left(\tfrac{1}{2}\right)^{t/h}$$

NOTE: *The time t and half-life h should be measured in the same units.*

where A_0 is the initial amount, h is the **half-life** of the quantity, and $A(t)$ is the amount present after t units of time.

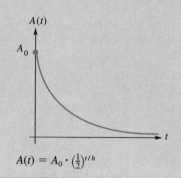

$$A(t) = A_0 \cdot \left(\tfrac{1}{2}\right)^{t/h}$$

EXERCISE SET 4.1

A

In Exercises 1–6, use a calculator to approximate (to the nearest 0.0001) the values given for the function.

1. $f(x) = 2^x$
   (a) $f(1.25)$ (b) $f(\sqrt{5})$ (c) $f(3 - \sqrt{6})$
   (d) $f\left(\dfrac{1}{\pi}\right)$

2. $g(x) = \left(\tfrac{2}{5}\right)^x$
   (a) $g(0.45)$ (b) $g(-\sqrt[3]{10})$ (c) $g\left(\dfrac{1}{\sqrt{3}}\right)$
   (d) $g(1 + \pi)$

3. $A(t) = 1000(1.015)^t$
   (a) $A(0.25)$ (b) $A(4)$ (c) $A(2)$
   (d) $A(30)$

4. $P(t) = 200(0.985)^{t/3}$
   (a) $P(3)$ (b) $P(10)$ (c) $P(0.05)$
   (d) $P(1.11)$

5. $Q(t) = 14\left(\tfrac{1}{2}\right)^{t/5}$
   (a) $Q(5)$ (b) $Q(7)$ (c) $Q(0.04)$
   (d) $Q(25)$

6. $R(t) = 0.25(2)^{12t}$
   (a) $R(1)$ (b) $R(1.05)$ (c) $R(0.5)$
   (d) $R(1.5)$

In Exercises 7–18, use a graphing utility to determine the graph of each function. Compare the graph in part (b) with the graph in part (a). Also, for the graph in part (b), determine the y-intercept and horizontal asymptote, and approximate the x-intercept (to the nearest 0.0001).

7. (a) $y = 3^x$ (b) $y = 3^x - 5$

8. (a) $y = 5^x$ (b) $y = 5^x - 12$

9. (a) $y = -\left(\tfrac{2}{3}\right)^x$ (b) $y = -\left(\tfrac{2}{3}\right)^x + 3$

10. (a) $y = -\left(\tfrac{4}{5}\right)^x$ (b) $y = -\left(\tfrac{4}{5}\right)^x + 2$

11. (a) $y = -2^x$ (b) $y = -2^{x+1}$

12. (a) $y = -\left(\tfrac{1}{4}\right)^x$ (b) $y = -\left(\tfrac{1}{4}\right)^{x-3}$

13. (a) $y = (0.8)^x$ (b) $y = 2(0.8)^x$

14. (a) $y = (4.8)^x$ (b) $y = \tfrac{1}{2}(4.8)^x$

15. (a) $y = (1.2)^x$ (b) $y = (1.2)^{6x}$

16. **(a)** $y = 10^x$ **(b)** $y = 10^{x/2}$

17. **(a)** $y = 4^x$ **(b)** $y = -4^{-x}$

18. **(a)** $y = \left(\frac{1}{2}\right)^x$ **(b)** $y = -\left(\frac{1}{2}\right)^{-x}$

In Exercises 19–24, use the compound interest formula to compute the accumulated amount for each investment described.

19. $400 at 6% compounded annually for 4 years

20. $1575 at 12% compounded annually for 2 years

21. $1500 at $3\frac{1}{2}$% compounded monthly for 7 years

22. $350 at $7\frac{1}{2}$% compounded monthly for 20 years

23. $1800 at 10% compounded 360 times annually for 2 years

24. $10 at 12% compounded 360 times annually for 80 years

B

In Exercises 25–28, use the compound interest formula to determine how much should be invested under the given conditions to obtain the given accumulated amount *A*.

25. 5% compounded 4 times a year for 8 years; $A = \$10,000$

26. 8% compounded 4 times a year for 3 years; $A = \$20,000$

27. $7\frac{1}{2}$% compounded monthly for 4 years; $A = \$5250$

28. $6\frac{1}{4}$% compounded monthly for 36 years; $A = \$120,000$

29. At what interest rate (to the nearest 0.01%) compounded annually must $1000 be invested for 2 years so that the accumulated amount is $1200?

30. At what interest rate (to the nearest 0.01%) compounded twice a year must $2500 be invested for 1 year so that the accumulated amount is $2650?

In Exercises 31–36, the graph of $y = k \cdot b^x$ is shown. Determine the values of *k* and *b*.

31.

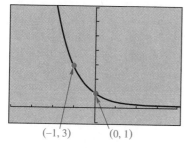

$(-1, 3)$ $(0, 1)$

32.

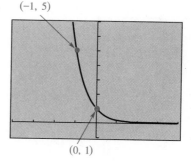

$(-1, 5)$

$(0, 1)$

33.

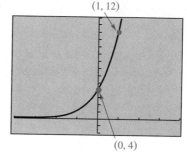

$(1, 12)$

$(0, 4)$

34.

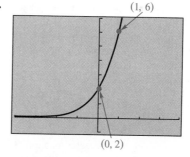

$(1, 6)$

$(0, 2)$

35.

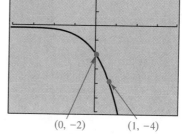

$(0, -2)$ $(1, -4)$

36.

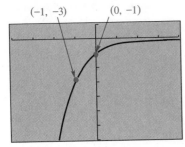

$(-1, -3)$ $(0, -1)$

37. The population of the world doubles roughly every 38 years. In 1995, the world population was 5.8 billion. Let $A(t)$ represent the world population t years after 1995.
   (a) Determine the world population 38 years and 76 years after 1995.
   (b) Determine a formula for A.
   (c) What will the world population be in the year 2000 (to the nearest 0.1 billion)?

38. The population of an endangered species of an insect is being depleted by the destruction of its habitat. At the current rate of decrease, the half-life of the population is 10 years. The present population is estimated to be 122 million.
   (a) Estimate the insect population after 10 years and 20 years.
   (b) Determine a function A such that $A(t)$ is the insect population in t years.
   (c) To the nearest million, what will the insect population be in 14 years?

39. The half-life of the radioactive isotope gallium-70 is 20 minutes.
   (a) Use the half-life model to construct a function Q for the amount of gallium-70 present after t minutes if the initial amount is 25 milligrams.
   (b) How much is left after 5 minutes? 1 hour (to the nearest 0.1 milligram)?

40. The half-life of uranium-238 is 4.5×10^9 years.
   (a) Use the half-life model to construct a function Q for the amount of uranium-238 present after t years if the initial amount is Q_0.
   (b) What percentage of the initial amount is left after 10^{10} years?

41. The World Health Organization estimated that in 1995, 18 million persons were infected by HIV, the virus that leads to AIDS. At the current rate of infection, the number of infections doubles every 8.4 years.
   (a) Use the doubling time model to construct a function P for the number of infections (in millions) t years after 1995.
   (b) Use the function from part (a) to predict the number of persons infected by 2010 (to the nearest million).

42. Tuberculosis (TB) again has become a serious health problem. In 1995, 3.0 million persons died of TB worldwide. At its present rate of increase, this number will double every 23 years.
   (a) Use the doubling time model to construct a function P for the number of deaths t years after 1995.
   (b) Use the function from part (a) to predict the number of deaths in 2010.

43. An initial amount of $10,000 is invested for 4 years at 6% compounded n times a year. Determine the accumulated amounts for $n = 1, 4, 12, 500, 10,000$. Comment on the results.

44. An initial amount of $6000 is invested for 12 years at 8% compounded n times a year. Determine the accumulated amounts for $n = 1, 4, 12, 360, 20,000$. Comment on the results.

45. A common method for financing retirement is to invest a large amount of money at a given rate r compounded annually and then withdraw a fixed amount of money from this investment each year. The maximum amount of withdrawal W possible if an investment P is to last n years is given by the formula

$$(1 + r)^n = \frac{W}{W - Pr}$$

   (a) Solve this equation for W.
   (b) If $300,000 is invested in an account at a rate of 6% with the intention of making withdrawals for 15 years, how much can be withdrawn each year?

46. The monthly payment for a home mortgage or car loan is computed by the formula

$$m = \frac{Pr}{1 - (1 + r)^{-n}}$$

where P is the initial amount borrowed, r is the monthly interest rate (this is the annual rate divided by 12), and n is the number of monthly payments (this is the number of years of the loan times 12). Suppose that the annual rate of a particular 30 year mortgage is 8%.

(a) Compute the monthly payment for a mortgage of $100,000. What is the total amount paid over the 30 years?

(b) How much could be borrowed initially if the monthly payment cannot exceed $1000?

47. Repeat Exercise 46 for a 15 year mortgage at 8%.

48. Repeat Exercise 46 for a 30 year mortgage at 12%.

C

49. A particular type of tinted glass reduces the intensity of sunlight by 20% per millimeter of thickness.

(a) Write a function I that models the intensity of sunlight passing through a pane of this glass that is x millimeters thick, given that I_0 is the intensity of the sunlight entering the glass.

(b) An energy-efficient window using this glass is 15 millimeters thick. If the intensity of sunlight entering the window is 6000 watts per square meter, what is the intensity of the light passing through the window?

50. Caffeine is a chemical stimulant that is found in coffee and cola. A typical human body eliminates 10% of this compound each hour after ingestion.

(a) Write a function Q that models the amount of caffeine present in the body t hours after the ingestion, given that Q_0 is the amount ingested.

(b) Suppose that Maria has a double espresso coffee (60 milligrams of caffeine) at 8 AM—just before physics lab. How much caffeine remains in her system when the lab is over at 11:30 AM?

51. The *Gompertz model* is used to describe a quantity that increases rapidly at first, and then increases slowly to its maximum amount. This model is used, for example, for the sales of a new product, or the introduction of a new species into a favorable environment.

(a) Sketch a qualitative graph of this model.

(b) The formula used for this model is of the form $y = c \cdot a^{b^x}$, where $c > 0$, $0 < a < 1$, and $0 < b < 1$. Suppose that $a = 0.2$ and $b = 0.4$. Use your graphing utility to determine the graphs of this equation for $c = 1, 2, 4$, and 8. What is the significance of c?

52. Adapt the formula given in Exercise 46 to compute the payment for a home mortgage of $100,000 for 30 years at an annual rate of 8% if the payments are made bimonthly (24 payments per year). Explain the consequence of making bimonthly payments instead of monthly payments.

53. Show algebraically that the graph of $y = (1/b)^x$ is the reflection of the graph of $y = b^x$ through the y-axis.

54. Use a graphing utility to graph the equations $y = 2^x$, $y = x^2$, and $y = x^4$ together in each of the windows $[0, 10] \times [0, 100]_{10}$, $[0, 20] \times [0, 20{,}000]_{1000}$, and $[0, 40]_{10} \times [0, 10^6]_{10^5}$. Can you draw a conclusion about the relative rate of increase of these functions?

THE NATURAL EXPONENTIAL FUNCTION

Quite often, a mathematical model that solves one problem can be used to attack another. In the last section, we saw that the compound interest model not only helps solve compound interest problems, but also serves as a model for exponential growth and decay applications.

In this section, the compound interest model leads us to a very useful function—the *natural exponential function*. We start by defining a most important real number, the number e.

The Number *e*

Consider the simple transaction of investing $1 for 1 year at 100% interest compounded k times during the year. According to the compound interest formula, the accumulated amount of this investment at the end of the year is

$$A = P\left(1 + \frac{r}{n}\right)^{nt} = \left(1 + \frac{1}{k}\right)^{k \cdot 1} = \left(1 + \frac{1}{k}\right)^{k}$$

In the discussion in Section 4.1, we saw that as the number of compounding periods increases, the accumulated amount also increases. However, the graph of

$$y = \left(1 + \frac{1}{x}\right)^{x}$$

shown in Figure 4.8 suggests that no matter how large the value of k becomes in the function

$$A = \left(1 + \frac{1}{k}\right)^{k}$$

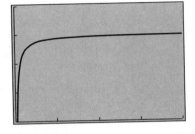

Figure 4.8

$$y = \left(1 + \frac{1}{x}\right)^{x}$$

$[0, 40]_{10} \times [0, 4]$

the value of A never exceeds 2.72. (Pause here to use the trace feature of a graphing utility to verify this assertion.) We interpret this to mean that no matter how often interest is compounded, the accumulated value will never be more than $2.72.

A more detailed investigation shows that as k grows without bound, this amount grows to approximately 2.718 281 828 459 045. This irrational number rivals the number π in its application to mathematical models. Because of its importance, it is rewarded with its own symbol, the letter e.

The Number *e*

> As $k \rightarrow +\infty$,
>
> $$\left(1 + \frac{1}{k}\right)^{k} \rightarrow 2.718\,281\,828\,459\,045...$$
>
> This number is denoted by the letter e. In other words,
>
> $$e = 2.718\,281\,828\,459\,045...$$

■ **NOTE:** *The Swiss mathematician Leonhard Euler introduced the number e in his 1748 publication* Introductio in Analysin Infinitorium.

With the key marked e^x on your calculator, you can evaluate expressions directly that involve the number e.

■ EXAMPLE 1 Evaluating Expressions with *e*

Write each number in the form e^k or ce^k for real numbers c and k, and use your calculator to evaluate the expression (to the nearest 0.0001).

(a) e (b) $\dfrac{1}{e^2}$ (c) $4\sqrt{e}$ (d) $\dfrac{6e^{1/3}}{2e^{-5/3}}$

SOLUTION

The keystroke sequences for most graphics calculators are shown.

(a) $e = e^1 = 2.7183$ Keystroke sequence: $\boxed{e^x}$ 1

(b) $\dfrac{1}{e^2} = e^{-2} = 0.1353$ Keystroke sequence: $\boxed{e^x}$ $\boxed{(-)}$ 2

(c) $4\sqrt{e} = 4e^{1/2} = 6.5949$ Keystroke sequence: 4 $\boxed{e^x}$ $\boxed{(}$ 1 $\boxed{\div}$ 2 $\boxed{)}$

(d) $\dfrac{6e^{1/3}}{2e^{-5/3}} = \dfrac{6}{2} \cdot \dfrac{e^{1/3}}{e^{-5/3}}$

$\qquad\qquad = 3e^{(1/3)-(-5/3)}$

$\qquad\qquad = 3e^2 = 22.1672$ Keystroke sequence: 3 $\boxed{e^x}$ 2   ■

The Natural Exponential Function

The exponential function with base e is the most important of all exponential functions, playing a significant role in the study of calculus and physics.

The Natural Exponential Function

> The function $f(x) = e^x$ is the **natural exponential function**. Its domain is the set of all real numbers, and its range is the set of all positive real numbers.

The domain and range of this function follow directly from the discussion of exponential functions in the last section. Because $e > 1$, the graph of the natural exponential function has the general shape of $y = b^x$ with $b > 1$. The graph of $y = e^x$ is shown in Figure 4.9, along with the graphs of $y = 2^x$ and $y = 3^x$. Notice that because $2 < e < 3$, the graph of $y = e^x$ is between the graphs of $y = 2^x$ and $y = 3^x$.

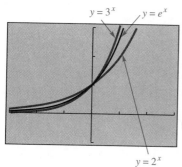

Figure 4.9
$[-3, 3] \times [-1, 3]$

■ **EXAMPLE 2**   Using the Graph of $y = e^x$

Use a graphing utility to determine the graph of each function. Compare the graph with the graph of $f(x) = e^x$. Determine the horizontal asymptote and the intercepts.

(a) $y = e^x - 4$ (b) $y = e^{-x}$ (c) $y = 150e^{0.2x}$

SOLUTION

(a) Because the function $y = e^x - 4$ can be written as $y = f(x) - 4$, the graph of $y = e^x - 4$ is a translation, 4 units down, of the graph of f (Figure 4.10).

The horizontal asymptote is $y = -4$. The y-intercept is $(0, -3)$, because if $x = 0$, then

$$y = e^0 - 4 = 1 - 4 = -3$$

Letting $y = 0$ yields

$$0 = e^x - 4$$
$$e^x = 4$$

NOTE: *The equation $e^x = 4$ is solved algebraically in Section 4.4, Example 4.*

We can use the techniques of Section 1.4 to approximate this solution graphically, and find the x-intercept $(1.3863, 0)$, to the nearest 0.0001.

(b) The function $y = e^{-x}$ is equivalent to $y = f(-x)$, so its graph (Figure 4.11) is a reflection about the y-axis of the graph of f. The horizontal asymptote of the graph is the x-axis ($y = 0$). The y-intercept is $(0, 1)$, the same as the y-intercept of f. This graph has no x-intercepts because

$$0 = e^{-x}$$

has no solutions, since the exponential function has strictly positive values.

(c) The function $y = 150e^{0.2x}$ is equivalent to $y = 150f(0.2x,)$ or

$$y = 150f\left(\tfrac{1}{5}x\right)$$

The discussion in Section 1.6 tells us that the graph of this function is a vertical expansion and a horizontal expansion, by factors of 150 and 5, respectively, of the graph of f (Figure 4.12). The horizontal asymptote is the x-axis ($y = 0$). The y-intercept is $(0, 150)$. It has no x-intercepts.

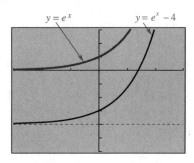

Figure 4.10
$[-3, 3] \times [-6, 3]$

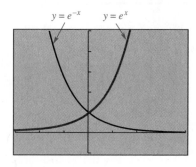

Figure 4.11
$[-3, 4] \times [-1, 5]$

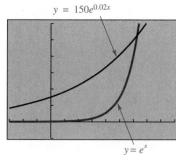

Figure 4.12
$[-3, 9] \times [-100, 600]_{100}$

Exponential functions are often combined by the arithmetic operations and function composition described in Section 1.7 to form other useful functions. In Example 3, two exponential functions are added to create an important engineering model.

EXAMPLE 3 Application: The Graph of a Catenary
The graph of

$$c(x) = \tfrac{1}{2} e^x + \tfrac{1}{2} e^{-x}$$

is an example of a *catenary*. This curve is the model for a cable hanging from two fixed points, or a beam spanning an expanse between two walls. Determine the graph of this function and its behavior for extreme values of x.

SOLUTION
The graph of c is shown in Figure 4.13.

As x grows without bound positively, the value of y grows large, but more importantly, the graph of c approaches the graph of $y = \tfrac{1}{2} e^x$. This happens because the graph of c is the sum of the graphs of $y = \tfrac{1}{2} e^x$ and $y = \tfrac{1}{2} e^{-x}$, and, as x grows without bound positively, the term $\tfrac{1}{2} e^{-x}$ tends to 0. We describe this behavior by

$$c(x) \approx \tfrac{1}{2} e^x \quad \text{as} \quad x \to +\infty$$

Likewise, we can also describe c by

$$c(x) \approx \tfrac{1}{2} e^{-x} \quad \text{as} \quad x \to -\infty$$

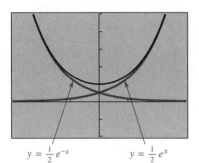

$y = \tfrac{1}{2} e^{-x}$ $\qquad$ $y = \tfrac{1}{2} e^x$

Figure 4.13
$y = \tfrac{1}{2} e^x + \tfrac{1}{2} e^{-x}$
$[-3, 3] \times [-2, 5]$

Continuous Compounding

In Section 4.1, the accumulated amount of a $2000 investment for 2 years with an annual interest rate of 6% was computed for different compounding periods. These values can be determined using the compound interest formula

$$A = P \left(1 + \frac{r}{n} \right)^{nt}$$

with $P = 2000$, $r = 0.06$, and $t = 2$. Table 4.6 shows the results for $n = 1, 2, 12, 120,$ and 1000.

TABLE 4.6

n	1	2	12	120	1000
$2000 \left(1 + \dfrac{0.06}{n} \right)^{n \cdot 2}$	\$2247.20	\$2251.02	\$2254.32	\$2254.93	\$2254.99

The graph of

$$y = 2000 \left(1 + \frac{0.06}{x} \right)^{x \cdot 2}$$

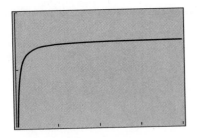

Figure 4.14

$$y = 2000\left(1 + \frac{0.06}{x}\right)^{x \cdot 2}$$

$[0, 40]_{10} \times [2240, 2260]_{10}$

■ **NOTE:** *By letting $n = 0.06k$, we can change $0.06/n$ to $1/k$.*

is shown in Figure 4.14. Both the table and graph suggest that even though the accumulated amount of the investment increases as n increases, it seems to level off, never exceeding \$2255. (Take a moment to use a graphing utility with the trace feature to investigate this claim.)

So, what does become of the accumulated amount of this investment

$$A = 2000\left(1 + \frac{0.06}{n}\right)^{n \cdot 2}$$

as n grows without bound positively? We can answer this question by substituting $0.06k$ for n in this equation:

$$A = 2000\left(1 + \frac{0.06}{n}\right)^{n \cdot 2}$$

$$= 2000\left(1 + \frac{0.06}{0.06k}\right)^{(0.06k)(2)} \qquad \text{Replace } n \text{ with } 0.06k.$$

$$= 2000\left(1 + \frac{1}{k}\right)^{0.12k} \qquad \text{Simplify.}$$

$$= 2000\left[\left(1 + \frac{1}{k}\right)^{k}\right]^{0.12} \qquad \text{Rewrite, using exponent rules.}$$

As n grows very large, so does k. From the definition of e, we know that

$$\left(1 + \frac{1}{k}\right)^{k} \to e \quad \text{as} \quad k \to +\infty$$

Thus,

$$2000\left[\left(1 + \frac{1}{k}\right)^{k}\right]^{0.12} \to 2000e^{0.12} \quad \text{as} \quad n \to +\infty \quad \text{and} \quad k \to +\infty$$

No matter how many compounding periods per year are allowed, A, the accumulated value of the investment, can never exceed $2000e^{0.12}$, or \$2254.99 (to the nearest 0.01).

As the number of compounding periods per year grows larger, the length of each compounding period becomes smaller. Computing interest in this manner is referred to as *continuous compounding*.

In general, if the initial amount is P instead of \$2000, the interest rate is r instead of 6%, the investment is for t years instead of 2 years, and n is the number of compounding periods, then following the same argument leads to

$$P\left(1 + \frac{r}{n}\right)^{nt} \to Pe^{rt} \quad \text{as} \quad n \to +\infty$$

Continuous Compounding Formula

> If P dollars are invested for t years at an annual interest rate r **compounded continuously**, then A, the accumulated amount of the investment, is given by the formula
>
> $$A = Pe^{rt}$$

The continuous compounding formula can be used to approximate compound interest in problems where the compounding period is very short compared to the term of the investment. The advantage is that the continuous model gives a computation that is easier to implement on a calculator.

EXAMPLE 4 Application: Using the Continuous Compounding Model

NOTE: Many banks describe this compounding as "daily compounding."

An investment of $250 for 2 years offers 4% annual interest compounded 360 times per year.

(a) Use the compound interest formula to compute the accumulated amount.

(b) Use the continuous compounding formula to approximate the accumulated amount.

SOLUTION

(a) We use $P = 250$, $t = 2$, $r = 0.04$, and $n = 360$ in the formula

$$A = P\left(1 + \frac{r}{n}\right)^{nt}$$

to compute A:

$$A = P\left(1 + \frac{r}{n}\right)^{nt} = 250\left(1 + \frac{0.04}{360}\right)^{360 \cdot 2} = 270.820\,563...$$

To the nearest cent, the accumulated amount is $270.82.

(b) We use $P = 250$, $t = 2$, $r = 0.04$ in the formula $A = Pe^{rt}$ to compute A:

$$A = Pe^{rt} = 250e^{(0.04)(2)} = 270.821\,767...$$

To the nearest cent, the approximation of the accumulated amount is $270.82. This approximation is virtually equal to the actual value found in part (a).

Other Applications

The natural exponential function helps us to develop another model for the continuous growth and decay applications that were introduced in the last section.

Exponential Growth
or Decay Model

> If the rate at which a quantity changes at a particular moment is proportional to the amount of the quantity at that moment, then a model for this quantity is
>
> $$A = Pe^{rt}$$
>
> where A is the accumulated amount, P is the initial amount, r is the rate of change, and t is the time elapsed.
>
> If $r > 0$, then this is an **exponential growth model**, and the quantity increases with time. If $r < 0$, then this is an **exponential decay model**, and the quantity decreases with time.

The next two examples show how this model can be used in two common applications.

⬤ EXAMPLE 5 Application: Population Growth

The population of the western United States (the states in the Pacific and Rocky Mountain time zones) can be closely approximated (to the nearest 0.01 million) by the exponential growth model $A = Pe^{rt}$. The rate of change r is 0.0247 when t is measured in years, and A is measured in millions of people. The population in 1930 was 12.32 million.

(a) Determine the population as a function A of the number of years t since 1930, and use a graphing utility to determine the graph of A.

(b) Use A to approximate the populations in 1950 and 1980 (the actual populations were 20.19 million and 42.41 million, respectively).

SOLUTION

(a) The initial population is 12.32 million, so P is 12.32 in this exponential growth formula. Also, r is 0.0247, so the function is

$$A(t) = 12.32e^{0.0247t}$$

where t is the number of years after 1930. The graph is shown in Figure 4.15.

(b) The populations in 1950 ($t = 20$) and in 1980 ($t = 50$) are given by

$$A(20) = 12.32e^{0.0247(20)} = 20.190\ 737\ 5...$$
$$A(50) = 12.32e^{0.0247(50)} = 42.360\ 823\ 4...$$

Note that these approximations, 20.19 million in 1950 and 42.36 million in 1980, are quite close to the actual populations. ⬤

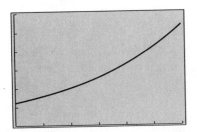

Figure 4.15
$y = 12.32e^{0.0247x}$
$[0, 60]_{10} \times [0, 60]_{10}$

EXAMPLE 6 Application: Decay of Nuclear Material

The components of an atomic weapon may prove dangerous to humans in the future even if the weapon is not used. One of the components of such a device is plutonium-239, an artificially created radioactive element so lethal to humans that a speck too small to be seen with the unaided eye can be fatal. The radioactive decay of this element obeys an exponential decay model $A = Pe^{rt}$ in which the rate of change r is -0.0289 when t is measured in thousands of years. Suppose that a small tactical nuclear weapon is to be disassembled in accordance with a recent disarmament treaty. The weapon contains 12.00 kilograms of plutonium-239 that must be safely stored until it is harmless.

(a) Determine the amount of plutonium-239 left as a function A of the time t since the disassembly of the weapon.

(b) Use A to approximate (to the nearest 0.01 kilogram) how much plutonium-239 from this amount will be left after 10,000 years.

(c) Use the trace feature of a graphing utility to determine when the amount will be 3.00 kilograms (to the nearest thousand years).

SOLUTION

(a) Because r is -0.0289 and P is 12, the exponential decay model $A = Pe^{rt}$ becomes

$$A(t) = 12e^{-0.0289t}$$

where t is measured in thousands of years. The graph of A is shown in Figure 4.16.

(b) The value of $A(10)$ is given by

$$A(10) = 12e^{-0.0289(10)} = 8.988\,146\,465...$$

The amount left after 10,000 years is 8.99 kilograms (to the nearest 0.01 kilogram).

(c) The trace feature of your calculator should show you a point on the graph of A at about $(48, 3)$. Thus, the initial amount of 12 kilograms decays to 3 kilograms in about 48,000 years (to the nearest thousand years).

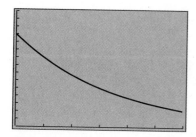

Figure 4.16
$y = 12e^{-0.0289x}$
$[0, 60]_{10} \times [0, 15]$

How quickly a hot object cools (or a cold object warms) depends upon the difference between the temperatures of the object and its surrounding environment. More precisely, the difference in the temperatures of the object and the surrounding environment decays at a rate that is proportional to that difference. Therefore, the exponential growth and decay model is applicable:

$$\left(\begin{array}{c}\text{Present difference}\\ \text{in temperature}\end{array}\right) = \left(\begin{array}{c}\text{Initial difference}\\ \text{in temperature}\end{array}\right)e^{rt}$$

NOTE: *Because the difference in temperature of the object and its environment is decreasing, the rate of change r is always negative.*

If $A(t)$ is the object's temperature at time t, A_0 is the initial temperature of the object, C is the temperature of the environment, and r is the rate of change, then we get

$$A(t) - C = (A_0 - C)e^{rt}$$

or

$$A(t) = C + (A_0 - C)e^{rt}$$

This result is called *Newton's law of cooling* [after Isaac Newton (1642–1727), English mathematician and scientist].

Newton's Law of Cooling Formula

An object initially at temperature A_0 is placed in an environment of constant temperature C. The temperature of the object at time t is given by

$$A(t) = C + (A_0 - C)e^{rt}$$

where r is the rate of change.

The physical setting and the units of measure (minutes or hours, Fahrenheit or Celsius) determine the rate of change.

EXAMPLE 7 Application: Newton's Law of Cooling

A roast is removed from a freezer (25°F) and placed in a room (75°F). From past experience in identical conditions, the rate of change has been determined to be $r = -0.03$ if t is measured in minutes.

(a) Write a function A for the temperature using Newton's law of cooling. Use a graphing utility to determine the graph of A and its horizontal asymptote.

(b) What is the temperature (to the nearest 1°F) of the roast 1 hour after it is removed from the freezer?

SOLUTION

(a) We use $A_0 = 25$, $C = 75$, and $r = -0.03$ in the formula:

$$A(t) = C + (A_0 - C)e^{rt}$$
$$A(t) = 75 + (25 - 75)e^{-0.03t}$$
$$A(t) = 75 - 50e^{-0.03t}$$

The graph of $y = 75 - 50e^{-0.03x}$ is shown in Figure 4.17. This graph shows that A is an increasing function, and it has a horizontal asymptote at $y = 75$. This agrees with the situation because the roast warms quickly at first, then warms more slowly, eventually approaching room temperature at 75°F.

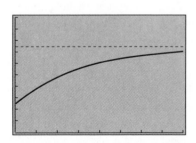

Figure 4.17
$y = 75 - 50e^{-0.03x}$
$[0, 80]_{10} \times [0, 100]_{10}$

(b) The temperature after 1 hour, or 60 minutes, is

$$A(60) = 75 - 50e^{-0.03(60)} = 66.735\ 005\ 6...$$

The temperature of the roast 1 hour after it is removed from the freezer is 67°F (to the nearest 1°F).

EXERCISE SET 4.2

A

In Exercises 1–4, write the number in the form e^k or ce^k for real numbers c and k, and use a calculator to approximate the expression (to the nearest 0.0001).

1. **(a)** $\dfrac{5}{e}$ **(b)** $\sqrt[3]{e}$

2. **(a)** $\dfrac{1}{\sqrt{e}}$ **(b)** $\sqrt{e^5}$

3. **(a)** $3(e^3)^2$ **(b)** $\dfrac{14e^6}{7e^3}$

4. **(a)** $(6e)(2e^{-4})$ **(b)** $\dfrac{\sqrt[3]{e^2}}{\sqrt{e}}$

In Exercises 5–18, use a graphing utility to determine the graph of each function. Compare the graph with the graph of $f(x) = e^x$. Also, determine the horizontal asymptote and y-intercept, and approximate the x-intercept (to the nearest 0.001).

5. $y = e^x + 2$ **6.** $y = e^x + 7$

7. $y = e^x - 6$ **8.** $y = e^{x-2}$

9. $y = e^{x+3}$ **10.** $y = e^{x-5}$

11. $y = e^{x+4} - 32$ **12.** $y = e^{x+2} - 13$

13. $y = \frac{1}{2}e^x$ **14.** $y = 2e^x$

15. $y = e^{-x/3}$ **16.** $y = \frac{1}{2}e^{-x}$

17. $y = 200e^{-0.5x}$ **18.** $y = 360e^{-0.1x}$

In Exercises 19–24, use the continuous compounding formula to approximate the accumulated amount for each investment described. (Compare your answers to your answers to Exercises 19–24 in Section 4.1.)

19. \$400 at 6% compounded annually for 4 years

20. \$1575 at 12% compounded annually for 2 years

21. \$1500 at $3\frac{1}{2}$% compounded monthly for 7 years

22. \$350 at $7\frac{1}{2}$% compounded monthly for 20 years

23. \$1800 at 10% compounded 360 times annually for 2 years

24. \$10 at 12% compounded 360 times annually for 80 years

B

In Exercises 25–28, use a graphing utility to determine the graph of each function. Determine the behavior of the graph for extreme values of x.

25. $y = \frac{1}{2}e^x - \frac{1}{2}e^{-x}$ **26.** $y = e^{2x} + e^{-2x}$

27. $y = e^x - x$ **28.** $y = e^{-x} + x$

In Exercises 29–32, find two functions f and g such that $f \circ g = h$ for the given function h. [Don't use $f(x) = x$ or $g(x) = x$.]

29. $h(x) = \frac{1}{3}e^{4x-1}$ **30.** $h(x) = 3e^{2x+4}$

31. $h(x) = e^{2x} - 4e^x + 7$ **32.** $h(x) = e^x - e^{-x}$

33. Air pressure is a function of altitude above sea level. The pressure (in pounds per square inch) at an altitude of h miles above sea level is given by the function $P(h) = 14.7e^{-0.203h}$. Approximate the air pressure at:
  (a) 6 miles above sea level
  (b) Sea level
  (c) 14,000 feet above sea level

34. The population (in thousands) of a city can be approximated by the average walking speed s (in feet per second) of persons in the downtown area during lunch hour using the function $P(s) = 0.87\ e^{2.7s-0.14}$. Estimate the population (to the nearest ten thousand) of a city where it has been observed that this speed is:

(a) 1.5 feet per second   (b) 2.8 feet per second

35. One unit of carbon monoxide (CO) and one unit of nitrogen dioxide (NO_2) react to produce one unit of carbon dioxide (CO_2) and one unit of nitrous oxide (NO). Under certain conditions, the number of units of both CO_2 and NO produced t seconds after the start of the reaction of three units of CO and one unit of NO_2 is given by

$$A(t) = \frac{3e^{1.2t} - 3}{3e^{1.2t} - 1}$$

Determine $A(t)$ to the nearest 0.0001 for:

(a) $t = 0.5$ (b) $t = 1$ (c) $t = 12$

36. Suppose that n is a positive integer. The notation $n!$ is used to express the product of all positive integers less than or equal to n. (Read $n!$ as n *factorial*.) For example, $4! = 4 \times 3 \times 2 \times 1 = 24$. For large values of n, n factorial can be approximated by *Stirling's formula*:

$$n! \approx e^{-n}n^n\sqrt{2\pi n}$$

Use your calculator first to compute $n!$ directly and then to approximate $n!$ using Stirling's formula for:

(a) $n = 5$ (b) $n = 9$ (c) $n = 12$

37. The population of the southern United States can be closely approximated (to the nearest 0.01 million) by the exponential growth model $A = Pe^{rt}$, where $r = 0.0139$ when t is measured in years and A is measured in millions. The population in 1940 was 41.67 million.

(a) Determine a model for the population as a function of the number of years since 1940.

(b) Use the function from part (a) to approximate the population in 1960 (the actual population was 54.97 million) and in 1980 (the actual population was 75.37 million).

38. One of the lethal by-products of a thermonuclear explosion is the radioactive element strontium-90. The radioactive decay of this element adheres to an exponential decay model $A = Pe^{rt}$, where r is -0.0248 when t is measured in years. A small underground nuclear test releases 220 grams of this element.

(a) Determine a model for the amount of strontium-90 remaining as a function of the time since the test.

(b) How much strontium-90 from this amount will be left 10 years after the test?

(c) Use the graph of the function to determine when the amount will be 50 grams.

39. After decreasing for years, the number of California gray whales is now growing. Their population can be approximated by the exponential growth model $A = Pe^{rt}$, where r is 0.015 when t is measured in years. In 1994, there were approximately 22,000 whales.

(a) Determine the number of whales as a function of the number of years since 1994.

(b) Approximate the number of California gray whales in the year 2000.

(c) Use the graph of the function to determine in what year the number will be 30,000.

40. The concentration of carbon dioxide (CO_2) in the atmosphere increases according to the exponential growth model $A = Pe^{rt}$, where r is 0.0025 when time is measured in years. In 1900, the concentration was 281 parts per million.

(a) Determine a model for the concentration of CO_2 as a function of the number of years since 1900.

(b) Use the function from part (a) to approximate the concentration of CO_2 in the year 1997.

(c) Use the graph of the function to determine in what year the concentration will be 400 parts per million.

41. The *declining balance method of depreciation* is a method of accounting in which the value of an item follows the exponential decay model $A = Pe^{rt}$. Suppose that a company uses this method to depreciate a truck that originally cost $54,000. When time is measured in years, the rate of change r is -0.2.

(a) Determine a model for the value of the truck as a function of the number of years since its purchase.

(b) Use the function from part (a) to determine the value of the truck 3 years after its purchase.

(c) The truck has an effective life of 8 years. What is its scrap value (its value at the end of its effective life)?

42. When a company stops advertising a product, it expects that the number N of sales per week of the product will eventually decrease to a minimum level N_0. The quantity $N - N_0$ decreases according to the exponential decay

model $A = Pe^{rt}$. For a particular product, $N = 100,000$, $N_0 = 72,000$, and r is -0.03 when t is measured in weeks.

(a) Determine the number of sales per week as a function of the number of weeks since the company stops advertising.

(b) Use the function in part (a) to approximate the number of sales during the tenth week after the company stops advertising.

(c) Use the graph of the function to determine in what week the number of sales will be 80,000.

Newton's Law of Cooling

Exercises 43–46 involve Newton's law of cooling.

43. A cup of coffee initially at a temperature of 155°F is allowed to cool in a room where the temperature is 70°F. The rate of change r is -0.045 when t is measured in minutes.

(a) Write a function for the temperature t minutes after the coffee starts to cool.

(b) What is the temperature of the coffee 15 minutes after it is allowed to cool?

(c) Use the graph of the function to determine when the temperature is 120°F.

44. The dinosaur embryo freezer at a popular wild game park has been sabotaged. The freezer, initially at $-10°C$, begins to warm to the room temperature of 25°C. The rate of change r is -0.16 when t is measured in hours. The embryos will be unusable once their temperature becomes 17°C.

(a) Write a function for the temperature t hours after the sabotage.

(b) What is the temperature of the embryos 3 hours after the sabotage?

(c) Use the graph of this function to determine when the embryos become unusable.

45. A light bulb operates at a temperature of 90°C. The bulb is turned off and allowed to cool in a room at 20°C. The rate of change r is -0.18 when t is measured in minutes. Soon, the bulb is 44°C.

(a) Write a function for the temperature t minutes after the bulb starts to cool.

(b) Use the graph of this function to determine when the bulb was turned off.

46. A coroner arrives at exactly midnight at the scene of a recent murder, an apartment in the fashionable east side of town. He ascertains that since death the corpse has cooled to 28°C from its live temperature of 37°C. The apartment is maintained at a constant temperature of 21°C. The coroner knows the rate of change r in this situation is -0.41 when t is measured in hours.

(a) Write a function for the temperature of the corpse t hours after the murder.

(b) Use the graph of this function to determine when the murder was committed.

Exercises 47 and 48 are applications of the *limited growth model*. This model applies when a quantity A is initially 0 and changes such that the difference between it and a maximum amount C decays exponentially. A formula for this model is

$$A = C - Ce^{rt}$$

where $r < 0$. The qualitative graph of this model is shown in Figure 4.18.

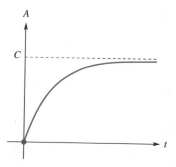

Figure 4.18
$A = C - Ce^{rt}$

47. Eric, wishing to speak the universal language Esperanto, sets a goal to learn 3000 vocabulary words. The number he learns in t weeks is given by $A = 3000 - 3000e^{-0.05t}$. (In this context, the limited growth model is called the *learning curve*.)

(a) How many words will he know in 6 weeks?

(b) Use the graph of this function to determine when Eric will know 2000 words. What happens as t grows very large?

48. The cost of cleaning up a polluted lake is such that the cost of removing $A\%$ is given by $A = 100 - 100e^{-0.12t}$, where t is measured in thousands of dollars.

(a) What percentage of the pollution is eliminated for $10,000?

(b) Use the graph of this function to determine the cost of removing half the pollution. What happens as t grows very large?

Exercises 49 and 50 are applications of the *logistic growth model*. This model applies when a quantity A is initially small and changes at a rate that is proportional to both its present amount and the difference between it and a maximum amount C. A formula for this model (obtained using calculus) is

$$A = \frac{C}{1 + ke^{rt}}$$

where $k > 0$ and $r < 0$ are constants. The qualitative graph of this model is shown in Figure 4.19.

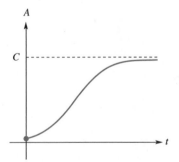

Figure 4.19

$$A = \frac{C}{1 + ke^{rt}}$$

49. A disease spreads through an isolated community. The number of persons infected t days after the infection is approximated by the function

$$f(t) = \frac{2000}{1 + 199e^{-0.12t}}$$

(a) How many will be infected in a week?

(b) Use the graph of f to determine when 1000 persons will be infected. What happens as t grows very large?

50. The spread of a rumor through a particular social group is such that the number of people who have heard the rumor t days after its inception is approximated by the function

$$f(t) = \frac{244}{1 + 121e^{-0.42t}}$$

(a) How many will have heard the rumor in 3 days?

(b) Use the graph of f to determine when 80 persons will have heard the rumor. What happens as t grows very large?

C

51. **(a)** Use a graphing utility to determine the graph of $f(x) = x^{1/x}$. For what value of x does the maximum value of f occur?

(b) Use part (a) to decide the larger of e^{π} and π^{e} without actually calculating their values.

HINT: First decide on the larger of $e^{1/e}$ and $\pi^{1/\pi}$.

52. Values of a transcendental function can be computed by *approximating polynomials*. The polynomials given here are used to approximate e^x.

$$P_1(x) = 1 + x \qquad\qquad P_2(x) = 1 + x + \frac{x^2}{2}$$

$$P_3(x) = 1 + x + \frac{x^2}{2} + \frac{x^3}{6}$$

$$P_4(x) = 1 + x + \frac{x^2}{2} + \frac{x^3}{6} + \frac{x^4}{24}$$

(a) Use a graphing utility to graph these polynomials and $y = e^x$.

(b) Compute $e^{0.6}$, $P_1(0.6)$, $P_2(0.6)$, $P_3(0.6)$, and $P_4(0.6)$. Compare the results.

53. A mug of coffee initially at 140°F cools to 122°F after 10 minutes in a room where the temperature is 72°F. Use

Newton's law of cooling to predict the temperature of the coffee after another 10 minutes.

54. **(a)** Show that for $b > 0$, b^x can be written as e^{Bx} for some real number B.

(b) Explain why part (a) shows that the graph of any exponential function $f(x) = b^x$ is a horizontal expansion, a horizontal compression, or a reflection through the y-axis of the graph of $y = e^x$.

INVERSE FUNCTIONS

A simple method of encoding a message is to assign each letter of the message with a positive integer so that the message becomes a sequence of numbers. For example, the message

MATHEMATICS

is encoded by the sender of the message, using Table 4.7, as

$$13-1-20-8-5-13-1-20-9-3-19$$

This encoding scheme is actually a function f, where

$$f\left(\begin{array}{c}\text{Text}\\\text{letter}\end{array}\right) = \left(\begin{array}{c}\text{Code}\\\text{value}\end{array}\right)$$

For example, we can say that

$$f(A) = 1 \qquad f(M) = 13 \qquad f(Y) = 25$$

TABLE 4.7

Text letter	Code value	Text letter	Code value	Text letter	Code value
A	1	J	10	S	19
B	2	K	11	T	20
C	3	L	12	U	21
D	4	M	13	V	22
E	5	N	14	W	23
F	6	O	15	X	24
G	7	P	16	Y	25
H	8	Q	17	Z	26
I	9	R	18		

The domain of f is the set

$$A, B, C, D, E, \ldots, Y, Z$$

and its range is the set

$$1, 2, 3, 4, 5, \ldots, 25, 26$$

The recipient of the message also uses Table 4.7, but in a different way. In decoding the message, each code value is assigned a text letter. If we name this function g, then

$$g\binom{\text{Code}}{\text{value}} = \binom{\text{Text}}{\text{letter}}$$

For example, we can say that

$$g(1) = A \qquad g(13) = M \qquad g(25) = Y$$

The decoding function g can be described as "undoing" what the encoding function f does. We call g the *inverse function* of f. Notice that the domain of g is the range of f, and the range of g is the domain of f (Figure 4.20).

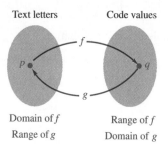

Text letters   Code values

p q

Domain of f Range of f
Range of g Domain of g

Figure 4.20

Another way to describe the relationship between f and g is to say that for any text letter p, and for any code value q,

$$g[f(p)] = p \qquad f[g(q)] = q$$

For example, from Table 4.7, we have

$$g[f(R)] = g(18) = R \quad \text{and} \quad f[g(10)] = f(J) = 10$$

In this section, we develop the general concept of the inverse function. In Section 4.4, we tackle the problem of defining inverse functions for the exponential functions.

Definition of an Inverse Function

NOTE: *Recall from Section 1.7 that* $(f \circ g)(x) = f[g(x)]$.

For any two functions f and g, we say that g *is the inverse function of* f (and f *is the inverse function of* g) if both $g \circ f$ and $f \circ g$ are the identity function $h(x) = x$.

Inverse Functions

Two functions f and g are **inverses** of each other if

$$g[f(p)] = p \quad \text{for every } p \text{ in the domain of } f$$

and

$$f[g(q)] = q \quad \text{for every } q \text{ in the domain of } g$$

p

f

$g \circ f$

$f(p)$

g

p

q

g

$f \circ g$

$g(q)$

f

q

EXAMPLE 1 Verifying Inverse Functions

Show that f and g are inverse functions. Use a graphing utility to determine their graphs.

(a) $f(x) = \frac{1}{3}x + 2$; $g(x) = 3x - 6$ **(b)** $f(x) = \sqrt[3]{x + 7}$; $g(x) = x^3 - 7$

SOLUTION

We need to verify that $g[f(p)] = p$ for all p in the domain of f, and $f[g(q)] = q$ for all q in the domain of g.

(a) The domain for both f and g is the set of real numbers.

$$g[f(p)] = g(\tfrac{1}{3}p + 2) = 3(\tfrac{1}{3}p + 2) - 6 = p + 6 - 6 = p$$
$$f[g(q)] = f(3q - 6) = \tfrac{1}{3}(3q - 6) + 2 = q - 2 + 2 = q$$

Because $g[f(p)] = p$ and $f[g(q)] = q$, f and g are inverse functions. The graph of f is a line with slope $\frac{1}{3}$ and y-intercept 2; the graph of g is a line with slope 3 and y-intercept -6 (Figure 4.21).

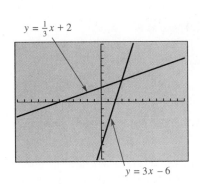

$y = \frac{1}{3}x + 2$

$y = 3x - 6$

Figure 4.21
$[-12, 12] \times [-8, 8]$

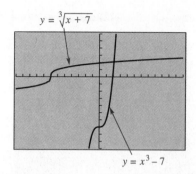

Figure 4.22
$[-12, 12] \times [-10, 6]$

(b) The domain for both f and g is the set of real numbers.

$$g[f(p)] = g(\sqrt[3]{p + 7}) = (\sqrt[3]{p + 7})^3 - 7 = p + 7 - 7 = p$$
$$f[g(q)] = f(q^3 - 7) = \sqrt[3]{(q^3 - 7) + 7} = \sqrt[3]{q^3} = q$$

The functions f and g are inverse functions, since $g[f(p)] = p$ and $f[g(q)] = q$. The graphs of f and g can be determined by using a graphing utility (Figure 4.22).

Example 1 shows an important connection between the graphs of two functions f and g that are inverses. If $f(p) = q$, then the point (p, q) is on the graph of f. Since g is the inverse of f, we have $g(q) = p$, and the point (q, p) is on the graph of g. These two points are symmetric with respect to the line $y = x$ (Figure 4.23), which gives us the following result:

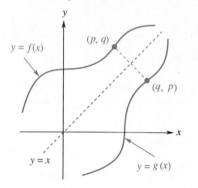

Figure 4.23

> Suppose that the function g is the inverse of the function f. The graph of g is the reflection of the graph of f through the line $y = x$.

Often, a pair of functions may appear to be inverses of each other, even though they are not.

◉ EXAMPLE 2   Functions That Are Not Inverses
Show that f and g are not inverse functions. Use a graphing utility to determine their graphs.

(a) $f(x) = 2x + 3$; $g(x) = \frac{1}{2}x - 3$ **(b)** $f(x) = \sqrt{x}$; $g(x) = x^2$

SOLUTION
In each case, we first check to see if $g[f(p)] = p$ and $f[g(q)] = q$, and then we verify our conclusion by examining the graphs of the functions.

(a) $g[f(p)] = g(2p + 3) = \frac{1}{2}(2p + 3) - 3 = p + \frac{3}{2} - 3 = p - \frac{3}{2}$

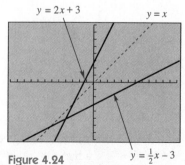

Figure 4.24
$[-12, 12] \times [-8, 8]$

Because $g[f(p)] \neq p$, f and g are not inverses. There is no need to check $f[g(q)]$. The graphs of f and g also confirm that they are not inverse functions (Figure 4.24). The graph of g is not the reflection of the graph of f through the line $y = x$.

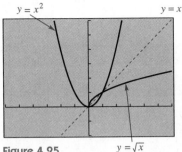

Figure 4.25
$[-6, 6] \times [-2, 6]$

(b)
$$g[f(p)] = g(\sqrt{p}) = (\sqrt{p})^2 = p$$

However,

$$f[g(q)] = f(q^2) = \sqrt{q^2} = |q|$$

Because $f[g(q)] \neq q$, f and g not inverse functions. Figure 4.25 also shows that the functions are not symmetric with respect to the line $y = x$.

Part (b) of Example 2 suggests an interesting question. Can the function $g(x) = x^2$ have an inverse function f? If it did, then f would have to "undo" g for each number in the domain of g. Specifically, since $g(2) = 4$ and $g(-2) = 4$, the function f would have to be such that $f(4) = 2$ and $f(4) = -2$. This clearly violates the definition for a function, which mandates that $f(4)$ should have only one value. Thus, no inverse function can exist for g.

The graph of $g(x) = x^2$ also shows the difficulty in finding its inverse. The graph of g and its reflection through the line $y = x$ are shown in Figure 4.26. The reflection is not the graph of a function, since it violates the vertical-line test.

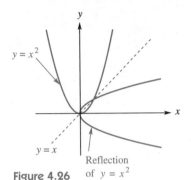

Figure 4.26

One-to-One Functions

Examples 1 and 2 suggest two questions. First, how can we tell if a given function has an inverse function? And second, if a function does have an inverse, how do we construct it? We start to answer the first question by defining a property of functions.

One-to-One Functions

> A function f is **one-to-one** if, for each y in the range of f, there is only one x in the domain of f such that $y = f(x)$. In other words, f is one-to-one if
>
> $$f(c_1) = f(c_2) \quad \text{implies that} \quad c_1 = c_2$$

Another way to define this property is to say that a function f is one-to-one if no two elements of its domain are assigned by f to the same element of its range.

If f is a one-to-one function, then each point $(x, f(x))$ on the graph of f has a y-coordinate that is shared by no other point on the graph (Figure 4.27). If this is true for the graph of f, then no two points on the graph lie on the same horizontal line.

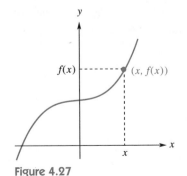

Figure 4.27

Horizontal-Line Test

> A function is a one-to-one function if and only if no horizontal line intersects its graph at more than one point.

Figure 4.28 shows three possibilities for the graph of an equation.

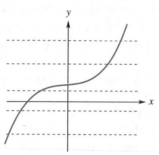

(a) This is the graph of a one-to-one function. It passes the horizontal-line test.

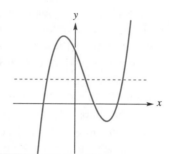

(b) This is the graph of a function that is not one-to-one. It fails the horizontal-line test.

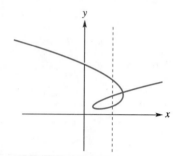

(c) This is not the graph of a function. It fails the vertical-line test.

Figure 4.28

⬤ **EXAMPLE 3   Determining Graphically Whether a Function is One-to-One**

Use the graph of the given function to determine whether the function is one-to-one, and then verify this conjecture algebraically.

(a) $f(x) = x^3 - 8$ **(b)** $T(x) = (x - 6)^2$

SOLUTION

(a) The graph of f is shown in Figure 4.29. It appears to pass the horizontal-line test, which suggests that h is a one-to-one function. We can verify that f is a one-to-one function by showing that $f(c_1) = f(c_2)$ implies $c_1 = c_2$:

$$f(c_1) = f(c_2)$$
$$c_1{}^3 - 8 = c_2{}^3 - 8$$
$$c_1{}^3 = c_2{}^3$$
$$c_1 = c_2 \qquad \text{Take the cube root of each side.}$$

This verifies that f is a one-to-one function.

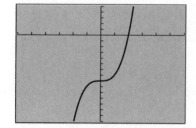

Figure 4.29
$y = x^3 - 8$
$[-6, 6] \times [-15, 5]$

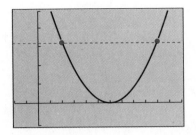

Figure 4.30
$y = (x - 6)^2$
$[-2, 12] \times [-5, 25]_5$

(b) The graph of T is shown in Figure 4.30. It apparently does not pass the horizontal-line test (one line is shown), so T is not one-to-one. We proceed in the same way as in part (a) to verify this conclusion:

$$T(c_1) = T(c_2)$$

$$(c_1 - 6)^2 = (c_2 - 6)^2$$

$$c_1 - 6 = \pm(c_2 - 6) \quad \text{Take the square root of each side.}$$

There are two possibilities, either $c_1 - 6 = c_2 - 6$ or $c_1 - 6 = -c_2 + 6$. From the second possibility, we get $c_1 = -c_2 + 12$. For example, if $c_2 = 2$, then $c_1 = 10$, and

$$T(2) = (2 - 6)^2 = 16$$

$$T(10) = (10 - 6)^2 = 16$$

Because 2 and 10 are two numbers such that $T(2) = T(10)$, T is not one-to-one.

Finding Inverse Functions

We are now ready to answer the question of which functions have inverses. The reflection through the line $y = x$ of the graph of a given function passes the vertical-line test if and only if the graph of the given function passes the horizontal-line test. This connection justifies the following conclusion:

Existence of an Inverse Function

> The **inverse of a function** f, denoted by f^{-1}, exists if and only if f is a one-to-one function.

◼ **NOTE:** *Don't confuse the inverse f^{-1} with the reciprocal of the function f.*

In Exercises 61 and 62, you are asked to show that exponential functions are one-to-one and therefore have inverses.

Now that we have a way of deciding whether an inverse function exists for a given function, how do we construct it if it does?

◉ EXAMPLE 4   Finding the Inverse of a Polynomial Function

Find the inverse of the function $f(x) = x^3 - 8$. (This function was shown to be one-to-one in Example 3.)

SOLUTION
The definition of an inverse function tells us that

$$f[f^{-1}(x)] = x$$

for any x in the domain of f^{-1}. If we let

$$y = f^{-1}(x)$$

then

$$f(y) = x$$

or

$$y^3 - 8 = x$$

Solving this equation for y gives us $f^{-1}(x)$:

$$y^3 - 8 = x$$
$$y^3 = x + 8$$
$$y = \sqrt[3]{x + 8} \qquad \text{Take the cube root of each side.}$$

Because $y = f^{-1}(x)$, the inverse of f is $f^{-1}(x) = \sqrt[3]{x + 8}$.

CHECK:

$$f^{-1}[f(p)] = f^{-1}(p^3 - 8) = \sqrt[3]{(p^3 - 8) + 8} = \sqrt[3]{p^3} = p$$
$$f[f^{-1}(q)] = f(\sqrt[3]{q + 8}) = (\sqrt[3]{q + 8})^3 - 8 = q + 8 - 8 = q$$

This verifies that

$$f^{-1}(x) = \sqrt[3]{x + 8}$$

The steps of the last example are generalized here:

Finding the Inverse of a Function

> Suppose that f is a one-to-one function. To determine f^{-1}:
>
> **Step 1:** Set $f(y) = x$.
> **Step 2:** Solve this equation for y as an expression in x. This expression is $f^{-1}(x)$.
> **Step 3:** Verify that $f^{-1}[f(p)] = p$ for each p in the domain of f, and $f[f^{-1}(q)] = q$ for each q in the domain of f^{-1}.

EXAMPLE 5   Finding the Inverse of a Rational Function

The function

$$C(x) = \frac{3x}{x + 1}$$

is a one-to-one function. Find C^{-1}.

SOLUTION

We set $C(y) = x$ and solve the resulting equation for y:

$$\frac{3y}{y + 1} = x$$

$$3y = x(y + 1) \quad \text{Multiply each side by } y + 1.$$

$$3y = xy + x$$

$$3y - xy = x \qquad \text{Collect } y \text{ terms on the left side.}$$

$$y(3 - x) = x \qquad \text{Factor } y \text{ from the left side.}$$

$$y = \frac{x}{3 - x} \qquad \text{Divide each side by } 3 - x.$$

Thus,

$$C^{-1}(x) = \frac{x}{3 - x}$$

CHECK:

$$C^{-1}[C(p)] = C^{-1}\left(\frac{3p}{p + 1}\right) = \frac{\dfrac{3p}{p + 1}}{3 - \dfrac{3p}{p + 1}} = \frac{\dfrac{3p}{p + 1}}{3 - \dfrac{3p}{p + 1}} \cdot \frac{p + 1}{p + 1}$$

$$= \frac{3p}{3(p + 1) - 3p} = \frac{3p}{3} = p$$

$$C[C^{-1}(q)] = C\left(\frac{q}{3 - q}\right) = \frac{3\left(\dfrac{q}{3 - q}\right)}{\dfrac{q}{3 - q} + 1} = \frac{\dfrac{3q}{3 - q}}{\dfrac{q}{3 - q} + 1} \cdot \frac{3 - q}{3 - q}$$

$$= \frac{3q}{q + (3 - q)} = \frac{3q}{3} = q$$

This verifies that

$$C^{-1}(x) = \frac{x}{3 - x}$$

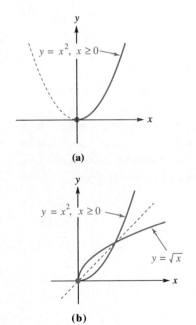

(a)

(b)

Figure 4.31

Unfortunately, not all functions for which we would like to find inverse functions are one-to-one. For example, we saw earlier that $g(x) = x^2$, the squaring function, is not a one-to-one function. However, by restricting the domain of g to the interval $x \geq 0$, this function becomes one-to-one. Over this restricted domain, the inverse of g is $g^{-1}(x) = \sqrt{x}$ (Figure 4.31).

EXAMPLE 6 Finding the Inverse of a Function with a Restricted Domain

Show graphically that $F(x) = x^2 - 4x - 5$ is a one-to-one function over the domain $x \geq 2$. Find F^{-1} and use a graphing utility to graph both F and F^{-1}.

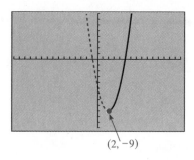

$(2, -9)$

Figure 4.32
$y = x^2 - 4x - 5$
$[-15, 15] \times [-12, 8]$

$y = 2 + \sqrt{x + 9}$

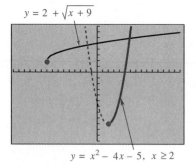

$y = x^2 - 4x - 5, \ x \geq 2$

Figure 4.33
$[-15, 15] \times [-12, 8]$

SOLUTION

First, we rewrite the quadratic function in standard form by completing the square:

$$F(x) = (x^2 - 4x \qquad) - 5 = (x^2 - 4x + 4) - 5 - 4 = (x - 2)^2 - 9$$

This tells us that the graph of F is a parabola that is concave up with vertex $(2, -9)$.

Over the domain $x \geq 2$, this is a one-to-one function (Figure 4.32). Next, we solve $F(y) = x$ to determine F^{-1}:

$$(y - 2)^2 - 9 = x$$
$$(y - 2)^2 = x + 9$$
$$y - 2 = \pm\sqrt{x + 9} \qquad \text{Take the square root of each side.}$$
$$y = 2 \pm \sqrt{x + 9}$$

This gives us two candidates for the inverse, $y = 2 + \sqrt{x + 9}$ and $y = 2 - \sqrt{x + 9}$. Because the range of F^{-1} is identical to the domain of F, namely $[2, +\infty)$, we choose $F^{-1}(x) = 2 + \sqrt{x + 9}$, since this is the solution such that $y \geq 2$.

The check is left to you, but the graphs of F and F^{-1} (Figure 4.33) strongly suggest that

$$F^{-1}(x) = 2 + \sqrt{x + 9}$$

Parametric Representation of an Inverse Function

In the last three examples, we have determined the graph of the inverse of a given function f by solving the equation $f(y) = x$. For some functions, solving this equation is not easy, or even possible. There is another way, however, to determine the graph of the inverse of a given function—using parametric equations.

Suppose that f is a one-to-one function. Using Table 3.2 of Section 3.6, we determine a parametric representation of the graph of $y = f(x)$ to be

$$x = t, \quad y = f(t)$$

The graph of $y = f^{-1}(x)$ is the same as the graph of $x = f(y)$, so, by Table 3.2, this graph has a parametric representation

$$x = f(t), \quad y = t$$

This curve is the graph of f^{-1}.

◉ EXAMPLE 7 Parametric Representation of an Inverse Function

Use a graphing utility in parametric mode to determine the graph of the one-to-one function $f(x) = x^3 + \frac{1}{3}x - 2$ and its inverse.

SOLUTION

A parametric representation for the graph of f is given by

$$x = t, \quad y = t^3 + \frac{1}{3}t - 2 \qquad -6 \leq t \leq 6$$

The graph of f^{-1} has a parametric representation

$$x = t^3 + \frac{1}{3}t - 2, \quad y = t \qquad -6 \leq t \leq 6$$

Both graphs are shown in Figure 4.34.

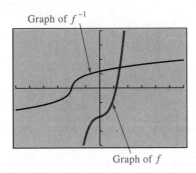

Graph of f^{-1}

Graph of f

Figure 4.34
$[-6, 6] \times [-4, 4]$

Applications

Many important mathematical relationships can be expressed as functions in different ways. For example, consider a square with side length s and perimeter P. The perimeter P is a function of s, namely $P = 4s$. On the other hand, we can also express s as a function of P, $s = \frac{1}{4}P$. It is no coincidence that these two functions, $f(s) = 4s$ and $g(P) = \frac{1}{4}P$, are inverses.

Frequently, the relationship between two quantities can be expressed by two different functions, where one is the inverse of the other.

◉ EXAMPLE 8 Application: Period of a Pendulum

The pendulum of a clock is a staff with a weighted bob near the end of the staff. The time required for the pendulum to swing back and forth is the *period* of the

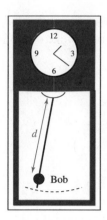

Figure 4.35

pendulum. The period t is a function of the distance d from the pivot point to the bob along the staff (Figure 4.35). Specifically, $t = f(d)$, where

$$f(d) = 2\pi\sqrt{\frac{d}{32}}$$

Determine d as a function of t. What is the application of this function?

SOLUTION

We need to find d by solving the equation $f(d) = t$:

$$2\pi\sqrt{\frac{d}{32}} = t$$

$$\sqrt{\frac{d}{32}} = \frac{t}{2\pi} \qquad \text{Divide each side by } 2\pi.$$

$$\frac{d}{32} = \frac{t^2}{4\pi^2} \qquad \text{Square each side.}$$

$$d = \frac{8t^2}{\pi^2} \qquad \text{Multiply each side by 32.}$$

It follows that the function

$$g(t) = \frac{8t^2}{\pi^2} \qquad t \geq 0$$

is the inverse of f. The function g can be used to compute the distance d from the pivot to the bob necessary for the period to be t seconds. Thus, the bob can be adjusted so that the clock maintains the correct time.

 EXERCISE SET 4.3

A

In Exercises 1–4, show that f and g are inverse functions. Use a graphing utility to determine their graphs.

1. $f(x) = 4x - 5$; $g(x) = \frac{1}{4}(x + 5)$

2. $f(x) = \frac{1}{3}x + 1$; $g(x) = 3x - 3$

3. $f(x) = \sqrt[3]{x + 2}$; $g(x) = x^3 - 2$

4. $f(x) = \dfrac{1}{x - 2}$; $g(x) = \dfrac{1}{x} + 2$

In Exercises 5–8, show that f and g are not inverse functions. Use a graphing utility to determine their graphs.

5. $f(x) = 6x + 2$; $g(x) = \frac{1}{6}x - 2$

6. $f(x) = \frac{1}{3}x + 1$; $g(x) = 3x - 1$

7. $f(x) = \sqrt{x} + 9$; $g(x) = (x - 9)^2$

8. $f(x) = \dfrac{1}{x + 4}$; $g(x) = x + 4$

In Exercises 9–16, use a graphing utility to determine the graph of each function f, and use the graph to decide whether f is one-to-one.

9. $f(x) = 2x - 7$

10. $f(x) = 4 - \frac{1}{2}x^2$

11. $f(x) = \frac{1}{2}\sqrt{x} + 6$

12. $f(x) = x + \sqrt{4 - x^2}$

13. $f(x) = x^3 + 3x - 2$

14. $f(x) = \frac{2}{5}x^3 - 2x + 8$

15. $f(x) = 2^x$

16. $f(x) = e^x$

In Exercises 17–24, each function is one-to-one. Find the inverse function.

17. $f(x) = 3x - 4$

18. $f(x) = 12 + \frac{3}{5}x$

19. $R(x) = x^3 + 1$

20. $P(x) = (x - 6)^3$

21. $f(x) = \dfrac{1}{x - 5}$

22. $L(x) = \dfrac{1}{2x}$

23. $q(x) = \sqrt[3]{x - 5}$

24. $T(x) = \sqrt[3]{x} - 2$

B

In Exercises 25–30, use the graph of each function to determine whether the function is one-to-one, and then verify your conclusion algebraically (see Example 3).

25. $f(x) = (x + 2)^2$

26. $f(x) = x^2 - 3$

27. $f(x) = \sqrt[3]{x} - 8$

28. $f(x) = (x + 1)^3$

29. $f(x) = \dfrac{2}{x - 3}$

30. $f(x) = \dfrac{12}{x^2 + 4}$

In Exercises 31–36, given each one-to-one function f, determine f^{-1}.

31. $f(x) = \dfrac{2x - 1}{x + 4}$

32. $f(x) = \dfrac{7 - 3x}{5 - 4x}$

33. $f(x) = \dfrac{4}{\sqrt{x}}$

34. $f(x) = \dfrac{1}{\sqrt{x - 3}}$

35. $f(x) = 3 + \sqrt{x - 4}$

36. $f(x) = 4 - \sqrt{x + 2}$

In Exercises 37–40, the graph of a one-to-one function f is shown. Sketch the graph of the inverse of f.

37.

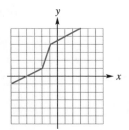

38.

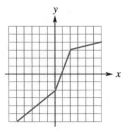

39.

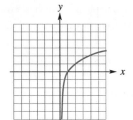

40.

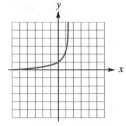

In Exercises 41–44, show graphically that the function f over the given domain is a one-to-one function. Find the inverse function f^{-1} algebraically.

41. $f(x) = x^2 - 4, \, x \le 0$

42. $f(x) = |x + 4|^2, \, x \ge -4$

43. $f(x) = \sqrt{36 - x^2}, \, x \le 0$

44. $f(x) = -\sqrt{16 - x^2}, \, x \ge 0$

In Exercises 45–48, write the given function f in the form $f(x) = a(x - h)^2 + k$. Sketch the graph of the function for $x \ge h$. Find the inverse g of this one-to-one function, and sketch its graph.

45. $f(x) = x^2 - 4x + 5$

46. $f(x) = x^2 + 6x + 1$

47. $f(x) = -\frac{1}{2}x^2 - 3x + 4$

48. $f(x) = -2x^2 - 4x + 8$

49. An intelligence quotient (IQ) is the ratio of a person's mental age and chronological age times 100. The function

$$I(m) = \frac{100m}{13}$$

gives the IQ of a person who is 13 years old with a mental age m. Find the inverse of I and explain its application.

50. From past experience, a manufacturer of bicycle helmets has found that the cost C (in dollars) of manufacturing n helmets per month is given by

$$C(n) = 28n + 5700$$

Find the inverse of C and explain its application.

51. The function

$$F(c) = \tfrac{9}{5}c + 32$$

is used to determine the corresponding Fahrenheit measure of a temperature with Celsius measure c. Find the inverse of F and explain its application.

52. The distance (in feet) that an automobile driver needs to brake from a velocity of v miles per hour under ideal conditions is given by the function

$$d(v) = v + \frac{v^2}{25}$$

Find the inverse of d and explain its application.

53. Consider: $f(x) = \begin{cases} -x^2 & x < 0 \\ x^2 & x \geq 0 \end{cases}$

  (a) Sketch the graph of f and f^{-1}.
  (b) From the graph of f^{-1} in part (a), determine f^{-1}.

54. Repeat Exercise 53 for: $f(x) = \begin{cases} \tfrac{1}{2}x & x < 0 \\ x + 2 & 0 \leq x \leq 4 \\ \tfrac{1}{2}x + 4 & x > 4 \end{cases}$

In Exercises 55–60, determine the graph of f and f^{-1} by using the parametric mode on a graphing utility.

55. $f(x) = x^3 + 2x - 6$

56. $f(x) = -\tfrac{1}{2}x^3 - 3x + 5$

57. $f(x) = e^{2x} - 4$

58. $f(x) = 2^{x-3}$

59. $f(x) = e^{-x} + 2$

60. $f(x) = 3(\tfrac{1}{2})^x$

C

61. (a) Show that a function that is increasing over its entire domain is one-to-one.
  (b) Show that a function that is decreasing over its entire domain is one-to-one.

62. Use Exercise 61 to show that $f(x) = b^x$, $b > 0$ and $b \neq 1$, is a one-to-one function.

63. Suppose that three one-to-one functions f, g, and h are such that $h = f \circ g$. What can be said about h^{-1} in terms of f^{-1} and g^{-1}?

64. Suppose that f and g are one-to-one functions. Does it follow that $f + g$ is a one-to-one function?

65. Given that

$$f(x) = \frac{2x}{x - 3}$$

determine $f^{-1}(x)$, $f(x^{-1})$, and $[f(x)]^{-1}$.

66. Suppose that f and g are inverse functions. Describe the inverse of the given function in terms of g.
  (a) $f(x - 2)$
  (b) $f(x) - 5$
  (c) $f(3x)$
  (d) $2f(\tfrac{1}{3}x)$

LOGARITHMIC FUNCTIONS

In Section 4.3, we saw that for any one-to-one function f, there exists another function that serves as the inverse function for f. An exponential function $f(x) = b^x$ certainly is one-to-one, so it has an inverse. In this section, we investigate *logarithmic functions*, which are the inverses of exponential functions.

In Example 2 of Section 4.1, we could not find the x-intercept of the graph of $y = 2^x - 6$ exactly because we could not solve the equation $2^x = 6$ algebraically. We did, however, approximate the solution graphically as 2.5850.

The solution to $2^x = 6$ is an exponent; it is the number to which 2 must be raised to get 6. We call this number $\log_2 6$, the *logarithm base 2 of 6*. We say that the value of $\log_2 6$ is 2.5850 (to the nearest 0.0001).

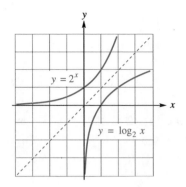

Figure 4.36

Logarithms and Logarithmic
Functions

○ **NOTE:** *Simply speaking, a*
logarithm is an exponent.

More generally, suppose we define $f(x) = 2^x$. Because f is one-to-one, there exists a function g that is the inverse of f. We define this inverse function as

$$g(x) = \log_2 x$$

The graph of $y = \log_2 x$ is the reflection of the graph of $y = 2^x$ through the line $y = x$ (Figure 4.36). Because g is the inverse of the exponential function f, the domain of g is the set of positive real numbers (the range of f), and the range of g is the set of all real numbers (the domain of f).

Definition of a Logarithmic Function

We generalize from our discussion so far:

> Suppose that $N > 0$ and that $b > 0$, $b \neq 1$. The solution to the equation $b^x = N$ is $x = \log_b N$, the **logarithm base b of N**.
>
> Furthermore, the inverse of the exponential function $f(x) = b^x$, where $b > 0$ and $b \neq 1$, is the **logarithmic function**
>
> $$g(x) = \log_b x$$
>
> The domain of g is the set of positive real numbers, and the range of g is the set of all real numbers.

We say that the *exponential equation $b^x = N$* is equivalent to the *logarithmic equation $x = \log_b N$*. Evaluating logarithms often depends upon solving an equivalent exponential equation by inspection.

○ **EXAMPLE 1 Determining the Values of Logarithms**
Determine the value of each logarithm exactly.

(a) $\log_3 9$ (b) $\log_2 \frac{1}{16}$ (c) $\log_{13} 1$ (d) $\log_7 7$
(e) $\log_2 4\sqrt{2}$ (f) $\log_{(1/5)} 25$ (g) $\log_{10}(10^{-134})$ (h) $\log_6(-36)$

SOLUTION
The plan here calls for writing $x = \log_b N$ as its equivalent exponential equation, $b^x = N$, and solving this equation by inspection. The solution x to this exponential equation is the value of $\log_b N$.

$x = \log_b N$	$b^x = N$	*Value of $\log_b N$*
(a) $x = \log_3 9$	$3^x = 9$	Because $3^2 = 9$, the value of $\log_3 9$ is 2.
(b) $x = \log_2 \frac{1}{16}$	$2^x = \frac{1}{16}$	Because $2^{-4} = \frac{1}{16}$, the value of $\log_2 \frac{1}{16}$ is -4.

$x = \log_b N$	$b^x = N$	*Value of* $\log_b N$
(c) $x = \log_{13} 1$	$13^x = 1$	Because $13^0 = 1$, the value of $\log_{13} 1$ is 0.
(d) $x = \log_7 7$	$7^x = 7$	Because $7^1 = 7$, the value of $\log_7 7$ is 1.
(e) $x = \log_2 4\sqrt{2}$	$2^x = 4\sqrt{2}$	Because $4\sqrt{2} = 2^2 2^{1/2} = 2^{5/2}$, the value of $\log_2 4\sqrt{2}$ is $\frac{5}{2}$.
(f) $x = \log_{(1/5)} 25$	$\left(\frac{1}{5}\right)^x = 25$	Because $\left(\frac{1}{5}\right)^{-2} = 25$, the value of $\log_{(1/5)} 25$ is -2.
(g) $x = \log_{10}(10^{-134})$	$10^x = 10^{-134}$	Because $10^{-134} = 10^{-134}$, the value of $\log_{10}(10^{-134})$ is -134.
(h) $x = \log_6(-36)$	$6^x = -36$	This equation has no solution because $6^x > 0$ for all real numbers x. The expression $\log_6(-36)$ is undefined.

Most logarithms cannot be evaluated by inspection as in Example 1. For example, trying to evaluate $\log_2 6$ in the manner of this example reduces to solving the equivalent exponential equation $2^x = 6$. This has no obvious solution because 6 is not a recognizable power of 2.

In Example 2, we see how being able to change an exponential or logarithmic equation into an equivalent form allows us to solve formulas that involve these transcendental functions.

EXAMPLE 2   Solving Exponential and Logarithmic Formulas

Solve each formula for t.

(a) $A = 2^{6t}$ **(b)** $r = \log_3(t + 5)$

SOLUTION

(a) The equivalent logarithmic equation for the exponential equation $A = 2^{6t}$ is

$$6t = \log_2 A$$

or

$$t = \tfrac{1}{6} \log_2 A$$

(b) The equivalent exponential equation for $r = \log_3(t + 5)$, a logarithmic equation, is

$$t + 5 = 3^r$$

which we can solve for t:

$$t = 3^r - 5$$

Because the domain of a logarithmic function is the set of positive real numbers, the value of $\log_b N$ is defined only if the expression N is greater than 0.

EXAMPLE 3 Domains of Logarithmic Functions

Determine the domain of the given function.

(a) $L(x) = \log_2(x - 4)$ **(b)** $w(t) = \log_{10}|t|$

SOLUTION

(a) The domain of L is the set of all x such that the expression $x - 4$ is positive. Thus, the domain of L is the set of all x such that

$$x - 4 > 0$$

or

$$x > 4$$

(b) The domain of w is such that $|t|$ is positive. Because $|t|$ is positive for all real numbers t except 0, the domain of w is the set of all nonzero real numbers, which is described by

$$t < 0 \quad \text{or} \quad t > 0$$

Natural and Common Logarithms

Any positive real number b (except 1) can serve as the base for a logarithm. In practice, however, two numbers are used far more than all others.

One of these bases is e, the base of the natural exponential function.

Natural Logarithm

> The logarithm base e is the **natural logarithm;** the natural logarithm of a positive real number x is denoted **ln** x. Thus, **ln** $x = \log_e x$.

The other important base is 10, because of its pivotal role as the base of the decimal number system.

Common Logarithm

> The logarithm base 10 is the **common logarithm;** the common logarithm of a positive real number x is denoted **log** x. Thus, **log** $x = \log_{10} x$.

Your calculator has two keys, $\boxed{\text{LN}}$ and $\boxed{\text{LOG}}$, used to compute the logarithms for these two bases directly.

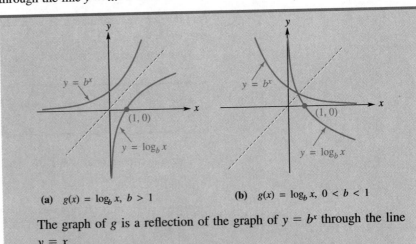 **EXAMPLE 4   Approximating Natural and Common Logarithms**
Approximate each value (to the nearest 0.0001) with your calculator.

(a) $\ln 4$ **(b)** $\log 0.00219$ **(c)** $\ln(-4)$

GRAPHING NOTE: *Compare the answer to part (a) of Example 4 to the graphical approximation of the x-intercept in part (a) of Example 2, Section 4.2.*

SOLUTION
The keystrokes shown are for most graphics calculators.

(a) The keystrokes for this expression are $\boxed{\text{LN}}$ 4. This yields $1.386\,294\,361...$. Thus, to the nearest 0.0001,

$$\ln 4 = 1.3863$$

(b) The keystrokes are $\boxed{\text{LOG}}$ 0.00219. This gives $-2.659\,555\,885...$. Thus, to the nearest 0.0001,

$$\log 0.00219 = -2.6596$$

NOTE: *Take a moment to see how your calculator handles the computation of* $\ln(-4)$.

(c) Because -4 is not positive, $\ln(-4)$ is undefined.

Graphs of Logarithmic Functions

Graphing utilities do not determine the graphs of logarithmic functions directly, except for the natural logarithmic function $y = \ln x$ and the common logarithmic function $y = \log x$. We can, however, sketch the graph of a logarithmic function $g(x) = \log_b x$ by reflecting the graph of the exponential function $f(x) = b^x$ through the line $y = x$.

Graph of a Logarithmic Function

(a) $g(x) = \log_b x,\ b > 1$ **(b)** $g(x) = \log_b x,\ 0 < b < 1$

The graph of g is a reflection of the graph of $y = b^x$ through the line $y = x$.
The graph of g has an x-intercept $(1, 0)$, but no y-intercept.

Continued

The y-axis ($x = 0$) is a vertical asymptote.

The graph of g is smooth, and it has no breaks.

The function g is increasing over its domain if $b > 1$, and decreasing over its domain if $0 < b < 1$.

The behavior of g is described by

$$y \rightarrow -\infty \quad \text{as} \quad x \rightarrow 0 \qquad \text{and} \qquad y \rightarrow +\infty \quad \text{as} \quad x \rightarrow +\infty$$

if $b > 1$, and by

$$y \rightarrow +\infty \quad \text{as} \quad x \rightarrow 0 \qquad \text{and} \qquad y \rightarrow -\infty \quad \text{as} \quad x \rightarrow +\infty$$

if $0 < b < 1$.

The graphs of logarithmic functions can be used to construct the graphs of other functions using translations, compressions, expansions, and reflections.

EXAMPLE 5   Graphing a Logarithmic Function with a Graphing Utility

Use a graphing utility to determine the graph of $y = \ln x - 2$, and compare this graph with the graph of $f(x) = \ln x$. Determine the vertical asymptote and approximate the x-intercept (to the nearest 0.0001).

SOLUTION

The graph of $y = \ln x - 2$ is a translation 2 units down of the graph of $y = \ln x$ (Figure 4.37). The vertical asymptote of the graph is the y-axis ($x = 0$). Replacing y with 0 and solving for x gives us the x-intercept:

$$0 = \ln x - 2$$
$$\ln x = 2$$
$$x = e^2 \qquad \text{Rewrite as an equivalent exponential equation.}$$

The x-intercept is (e^2, 0), or (7.3891, 0), to the nearest 0.0001.

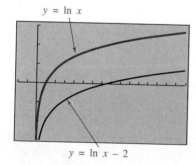

$y = \ln x$

$y = \ln x - 2$

Figure 4.37
$[-2, 16] \times [-3, 3]$

Properties of Logarithms

Before the advent of computing machinery, the significance of logarithms was in simplifying tedious numerical computations. This use of logarithms has become unimportant because of the advances in technology, but the three properties that supported this historical use are still helpful in manipulating logarithmic expressions.

Properties of Logarithms

If P and Q are positive expressions and r is a real number, then

1. $\log_b(PQ) = \log_b P + \log_b Q$

2. $\log_b\left(\dfrac{P}{Q}\right) = \log_b P - \log_b Q$

3. $\log_b(P^r) = r\log_b P$

Using one of these properties to rewrite a logarithmic expression from the form on the left side of the property into the form on the right side of the property is called **expanding** the expression.

Each of these properties is tied to an exponent law from Section 2 (page 12). Property 1 is proved here. Properties 2 and 3 are left to you (Exercise 62).

To prove property 1, we start by letting

$$m = \log_b P \qquad \text{and} \qquad n = \log_b Q$$

The equivalent exponential equations of these two logarithmic equations are

$$P = b^m \qquad \text{and} \qquad Q = b^n$$

Thus,

$$PQ = (b^m)(b^n) = b^{m+n}$$

The logarithmic form of this last equation is

$$\log_b(PQ) = m + n$$

After substituting $m = \log_b P$ and $n = \log_b Q$, we get

$$\log_b(PQ) = \log_b P + \log_b Q$$

This proves property 1.

Logarithm Identities

These identities follow directly from the definition of logarithms and their properties ($b > 0$, $b \neq 1$, and $x > 0$).

1. $\log_b 1 = 0$ **2.** $\log_b b = 1$ **3.** $\log_b\left(\dfrac{1}{x}\right) = -\log_b x$

4. $\log_b(b^x) = x$ **5.** $b^{\log_b x} = x$

Identities 1 and 5 are proved here. Identities 2, 3, and 4 are left to you (Exercise 63).

For identity 1, $\log_b 1 = 0$, the equivalent exponential equation is

$$b^0 = 1$$

This is true for all positive numbers b. This proves identity 1.

The equivalent logarithmic equation for $b^{\log_b x} = x$, identity 5, is

$$\log_b x = \log_b x$$

This is obviously true for all positive numbers b and x. This proves identity 5.

The next two examples show these properties and identities in action.

EXAMPLE 6 Expanding Logarithmic Expressions

Expand each logarithmic expression completely.

(a) $\log_5(kN)$ **(b)** $\log_a\left(\dfrac{y}{2}\right)$ **(c)** $\log_{1.02}(a^7)$

(d) $\ln\left(\dfrac{PVT}{\sqrt{n}}\right)$ **(e)** $\log\left(\dfrac{xy}{az}\right)$ **(f)** $\log_2(x^2 - y^2)$

SOLUTION

(a) $\log_5(kN) = \log_5 k + \log_5 N$ Use property 1.

(b) $\log_a\left(\dfrac{y}{2}\right) = \log_a y - \log_a 2$ Use property 2.

(c) $\log_{1.02}(a^7) = 7\log_{1.02} a$ Use property 3.

(d) $\ln\left(\dfrac{PVT}{\sqrt{n}}\right) = \ln(PVT) - \ln(\sqrt{n})$ Use property 2.

$\qquad\qquad = (\ln P + \ln V + \ln T) - \ln(n^{1/2})$ Use property 1.

$\qquad\qquad = \ln P + \ln V + \ln T - \frac{1}{2}\ln n$ Use property 3.

(e) $\log\left(\dfrac{xy}{az}\right) = \log(xy) - \log(az)$ Use property 2.

$\qquad\qquad = (\log x + \log y) - (\log a + \log z)$ Use property 1.

$\qquad\qquad = \log x + \log y - \log a - \log z$

(f) By factoring $x^2 - y^2$, we get $(x - y)(x + y)$. Then

$$\log_2(x^2 - y^2) = \log_2[(x - y)(x + y)]$$
$$= \log_2(x - y) + \log_2(x + y) \qquad \text{Use property 1.}$$

> **NOTE:** *In part (f) of Example 6, you may be tempted to rewrite* $\log_2(x + y)$ *as* $\log_2 x + \log_2 y$, *or* $\log_2(x - y)$ *as* $\log_2 x - \log_2 y$. *These "simplifications" are not supported by the properties above. Avoid these temptations.*

EXAMPLE 7 Writing Logarithmic Expressions As One Logarithm

Write each quantity in the form $\log_b P$, in which P is an algebraic expression.

(a) $3\log_b(2x) - \frac{1}{3}\log_b z$ **(b)** $\log_b(a + 3) - \log_b(2a + 1) + \log_b a$

SOLUTION

(a) $3 \log_b(2x) - \frac{1}{3} \log_b z = \log_b[(2x)^3] - \log_b(z^{1/3})$ Use property 3.

$$= \log_b\left(\frac{(2x)^3}{z^{1/3}}\right) \qquad \text{Use property 2.}$$

$$= \log_b\left(\frac{8x^3}{z^{1/3}}\right)$$

(b) $\log_b(a + 3) - \log_b(2a + 1) + \log_b a$

$$= \log_b\left(\frac{a + 3}{2a + 1}\right) + \log_b a \quad \text{Use property 2.}$$

$$= \log_b\left(\frac{a + 3}{2a + 1} \cdot a\right) \qquad \text{Use property 1.}$$

$$= \log_b\left(\frac{a^2 + 3a}{2a + 1}\right)$$

An expression that includes exponential or logarithmic expressions is a *transcendental expression*. Quite often, a transcendental expression can be simplified to an algebraic expression using properties of logarithms.

EXAMPLE 8   Simplifying Transcendental Expressions

Simplify each transcendental expression to an algebraic expression.

(a) $\ln(e^{\sqrt{2}+x})$ **(b)** $\log(10^{-2x})$ **(c)** $e^{3\ln x}, x > 0$

(d) $10^{\log 5 + \log x}, x > 0$

SOLUTION

(a) $\ln(e^{\sqrt{2}+x}) = \sqrt{2} + x$ Use identity 4.

(b) $\log(10^{-2x}) = -2x$ Use identity 4.

(c) $e^{3\ln x} = e^{\ln(x^3)} = x^3$ Use property 3 and identity 5.

(d) $10^{\log 5 + \log x} = 10^{\log(5x)} = 5x$ Use property 1 and identity 5.

Applications

Logarithmic functions assume a role almost as important as the role of exponential functions in the application of mathematics. This importance is due in part to their being the inverse functions of exponential functions, and in part to their properties and characteristics.

EXAMPLE 9   Application: Earthquakes and the Richter Scale

The *Richter scale* assigns a magnitude number M to the amount of energy E released by an earthquake. (The unit of measure for E is an erg.) These numbers are related by the formula

$$M = \frac{\log E - 11.4}{1.5}$$

In 1964, a severe earthquake in Alaska, killing hundreds and destroying millions of dollars worth of property, released 10^{24} ergs. What was the magnitude of this earthquake (to the nearest 0.1)?

SOLUTION

We substitute 10^{24} for E and evaluate M:

$$M = \frac{\log E - 11.4}{1.5}$$

$$= \frac{\log(10^{24}) - 11.4}{1.5} \qquad \text{Replace } E \text{ with } 10^{24}.$$

$$= \frac{24 - 11.4}{1.5} \qquad \text{Use identity 4: } \log(10^{24}) = 24.$$

$$= 8.4$$

The magnitude of the 1964 Alaska earthquake was 8.4.

Many scales, like the Richter scale, use logarithmic functions to measure human sensation to a stimulus. There are two reasons for this choice. First, the stimulus is usually measured over a large range of numbers, and it is difficult to measure both large and small numbers easily. The second reason is psychological; if a stimulus increases at a constant rate, then a human's physiological response increases, but at a diminishing rate. This phenomenon is called the *Weber–Fechner law of stimulus and sensation*. In practice, a logarithmic function is a good model for this phenomenon. More examples are discussed in Exercises 53–58.

EXAMPLE 10   Application: Velocity of a Rocket

Jet propulsion is the principle behind the locomotion of jet airplanes, rockets, and squids. In the case of the rocket, fuel on board burns and expels exhaust, which propels the rocket. As the fuel burns, the rocket loses mass. If k (in kilometers per second) is the velocity of the exhaust, m is the present mass of the rocket (in kilograms), m_0 is the initial mass of the rocket (in kilograms), and t is the number of seconds after launch, then the velocity v (in kilometers per second) is given by

$$v(t) = k \ln\left(\frac{m_0}{m}\right) - 0.00981t$$

The Saturn V rocket was used in the Apollo program that landed humans on the moon. Suppose the exhaust velocity of the Saturn V is 2.46 kilometers per second. On a particular mission, the fuel is expended completely after 150 seconds. At this time, the mass of the rocket and its payload is 27% of their initial mass. Determine the velocity of the rocket at this time (to the nearest 0.01).

SOLUTION

The present and final mass are not given specifically, but we are told that $m = 0.27m_0$. Using this and $k = 2.46$, we get

$$v = 2.46 \ln\left(\frac{m_0}{m}\right) - 0.00981t = 2.46 \ln\left(\frac{m_0}{0.27m_0}\right) - 0.00981(150)$$

$$= 2.46 \ln\left(\frac{1}{0.27}\right) - 0.00981(150)$$

$$= 1.749\,459\,97\ldots$$

The velocity at the time the fuel is expended is 1.75 kilometers per second (to the nearest 0.01).

 EXERCISE SET 4.4

A

In Exercises 1–4, determine the value of each logarithm exactly.

1. **(a)** $\log_2 8$ **(b)** $\log_3\left(\frac{1}{9}\right)$
   (c) $\log_7 1$ **(d)** $\log_{12} 12$

2. **(a)** $\log_5 125$ **(b)** $\log_4\left(\frac{1}{16}\right)$
   (c) $\log_{(2/3)} 1$ **(d)** $\log_9 9$

3. **(a)** $\log_3 3\sqrt{3}$ **(b)** $\log_{(1/4)} 16$
   (c) $\log 10$ **(d)** $\log_3(-9)$

4. **(a)** $\log_2 8\sqrt{2}$ **(b)** $\log_{(2/5)}\left(\frac{4}{25}\right)$
   (c) $\ln e$ **(d)** $\log_4(-2)$

In Exercises 5 and 6, write each exponential equation as an equivalent logarithmic equation.

5. **(a)** $4^t = 2P$ **(b)** $(xy)^t = 8$
6. **(a)** $10^k = 7.5$ **(b)** $e^{-2t} = m$

In Exercises 7 and 8, write each logarithmic equation as an equivalent exponential equation.

7. **(a)** $\log_5(2t) = M$ **(b)** $\log_{(x/2)}(4t) = 5$
8. **(a)** $\ln\left(\frac{2}{t}\right) = 12$ **(b)** $\log_t(2x) = 3$

In Exercises 9 and 10, approximate the value of each logarithm (to the nearest 0.0001) with a calculator.

9. **(a)** $\log 13.4$ **(b)** $\log 0.0135$
   (c) $\ln 8.4$ **(d)** $\ln 0.075$

10. (a) $\log 127.8$ **(b)** $\log(1.2 \times 10^{14})$
     (c) $\ln 127.8$ **(d)** $\ln(1.2 \times 10^{14})$

In Exercises 11–16, expand each logarithm completely.

11. (a) $\log_3(11x)$ **(b)** $\log_9\left(\dfrac{13s}{t}\right)$

12. (a) $\log_2(7k)$ **(b)** $\log_9\left(\dfrac{5a}{b}\right)$

13. (a) $\log(xyz)$ **(b)** $\log_7\left(\dfrac{3P^2}{Q}\right)$

14. (a) $\ln(2x^3)$ **(b)** $\log\left(\dfrac{3v}{w^2}\right)$

15. (a) $\ln(\sqrt{1-x})$ **(b)** $\log_2\left(\dfrac{1}{\sqrt{x^2-9}}\right)$

16. (a) $\log_{11}\left(\sqrt[3]{\dfrac{x}{5}}\right)$ **(b)** $\ln\left(\dfrac{x}{\sqrt{x^2-25}}\right)$

In Exercises 17–22, write each expression in the form $\log_b P$, in which P is a simplified expression.

17. (a) $\log 3 + \log y$ **(b)** $\ln(5a) - \ln(2b)$
18. (a) $\log v + \log(2w)$ **(b)** $\ln x - \ln(4z)$
19. (a) $2\log(a-3) - \log(a+4)$
     (b) $\log_2(x^2-3) - \log_2 y - \log_2 z$
20. (a) $\log_2(k-3) - 3\log_2(k+5)$
     (b) $\log_5(z-3) - \log_5 x + \log_5(2y)$
21. (a) $\log x + \frac{1}{2}\log(x^2-6)$
     (b) $\ln(6x) + \frac{1}{2}\ln x - \ln(2x)$
22. (a) $\frac{1}{2}\ln(2y) + \ln(x^4+4x)$
     (b) $\log(12z) + \frac{1}{3}\log z - \log(4z)$

B

In Exercises 23–26, simplify the transcendental expression to an algebraic expression.

23. (a) $e^{\ln(5y)}$ **(b)** $10^{1+\log x}$ **(c)** $\left(\frac{1}{2}\right)^{\log_2(4x)}$
24. (a) $10^{\log(2z)}$ **(b)** $e^{\ln t + \ln 2}$ **(c)** $(0.1)^{\log(ab)}$
25. (a) $\log(10^{2.54w})$ **(b)** $\ln\left(\dfrac{1}{e^{4x}}\right)$
     (c) $\log_{(1/10)}(10^x)$

26. (a) $\ln(e^{2+t^3})$ **(b)** $-\log(0.001^{2x})$
     (c) $\dfrac{\ln(e^{6x})}{\ln(e^{3x})}$

In Exercises 27–32, determine the domain of the given function f.

27. $f(x) = \ln(x^2 - 16)$ **28.** $f(x) = \log_9(25 - x^2)$
29. $f(x) = \log_2|x - 3|$ **30.** $f(x) = \log|4 + x|$
31. $f(x) = \ln\left(\dfrac{\sqrt{x+1}}{x}\right)$ **32.** $f(x) = \ln\left(\dfrac{x}{\sqrt{x-3}}\right)$

In Exercises 33–38, determine the graph of the function f and its asymptote using a graphing utility. Approximate the x-intercept to the nearest 0.0001.

33. $f(x) = \ln(x + 2)$ **34.** $f(x) = -4 + \log x$
35. $f(x) = 5 - 4\log x$ **36.** $f(x) = 2 - \ln(x - 3)$
37. $f(x) = \frac{1}{2}\log(x + 5)$ **38.** $f(x) = \ln(\frac{1}{2}x) - 1$

In Exercises 39–42, decide if each statement is always true. If so, label it TRUE; otherwise label it FALSE.

39. (a) $\log_b(2x + y) \overset{?}{=} \log_b(2x) + \log_b y$
     (b) $\ln(a - b^2) \overset{?}{=} \ln a - \ln(b^2)$
     (c) $\log\left(\dfrac{1}{5d}\right) \overset{?}{=} -\log(5d)$
     (d) $\log_k 0 \overset{?}{=} 1$

40. (a) $\log_2(a + 4c) \overset{?}{=} \log_2 a + \log_2(4c)$
     (b) $\log_5(y - 7z) \overset{?}{=} \log_5 y - \log_5(7z)$
     (c) $\dfrac{1}{\log_4 x} \overset{?}{=} -\log_4 x$
     (d) $\log_k 1 \overset{?}{=} 0$

41. (a) $\dfrac{\log_b y}{\log_b x} \overset{?}{=} \log_b\left(\dfrac{y}{x}\right)$
     (b) $(\ln x)(\ln y) \overset{?}{=} \ln(x + y)$
     (c) $\log_{(1/b)}(b^2) \overset{?}{=} \frac{1}{2}$
     (d) $\ln(3c) \overset{?}{=} 3\ln c$

42. (a) $\dfrac{\log_b y}{\log_b x} \overset{?}{=} \log_b(y - x)$
     (b) $(\ln x)(\ln y) \overset{?}{=} \ln(xy)$
     (c) $\log_{(1/b)} b \overset{?}{=} -1$
     (d) $\ln(3c) \overset{?}{=} (\ln c)^3$

43. Given $h(x) = \ln(2x^3 - 4)$, find two functions f and g such that $f \circ g = h$. [Don't use $f(x) = x$ or $g(x) = x$.]

44. Repeat Exercise 43, using $h(x) = \dfrac{1}{2 \ln x} + \ln x$.

In Exercises 45–48, approximate (to the nearest 0.0001) the value of the logarithm $\log_b N$ by approximating the x-intercept of the graph of $y = b^x - N$.

45. $\log_2 12$ **46.** $\log_3 10$ **47.** $\log_{(1/2)} 6$ **48.** $\log_{(2/3)} 5$

49. Two modern dialects that independently evolved from a common ancestral tongue share a number of words. The number of years since the dialects split is estimated by the function $N(r) = -5000 \ln r$, where r is the proportion of words from the ancestral tongue in both dialects.
  (a) Two South Pacific societies share 32% of a common ancestral tongue. Use the function to estimate how long ago the dialects split.
  (b) It has been approximately 1300 years since a French tongue split into north and south dialects. Use the graph of N to estimate the percentage of words common to both dialects.

50. A barometer is an instrument that measures atmospheric pressure in millimeters of mercury (mm Hg). If the atmospheric pressure at sea level is h_0, then one can determine the altitude (in meters above sea level) by the formula

$$A(h) = 22{,}860 \ln\left(\frac{h_0}{h}\right)$$

where h is the atmospheric pressure at that altitude. Suppose that $h_0 = 760$ mm Hg.
  (a) At a mountain resort, the pressure is measured as 700 mm Hg. What is the altitude according to the function?
  (b) A hang glider soars unexpectedly to 3000 meters. Use the graph of A to estimate the atmospheric pressure on the hang glider.

51. The 1992 earthquake in Southern California released 3.2×10^{22} ergs; its aftershock released one-tenth of this energy. What are the magnitudes of this earthquake and its aftershock? (See Example 9.)

52. On another mission of the Saturn V rocket discussed in Example 10, the fuel is expended completely after 120 seconds, at which time the mass of the rocket and its payload is 32% of its initial mass. Determine its velocity at this time (to the nearest 0.01 kilometer per second).

Sound Intensity

(Exercises 53–56) The intensity of what you hear depends upon the density of energy hitting your eardrum, which is released by the object producing the sound. The *decibel scale* assigns a magnitude number dB to the intensity I (measured in watts per square meter) according to the formula

$$dB = 10 \log I + 120$$

53. **(a)** The sound intensity in a library quiet-reading room is 10^{-10} watts per square meter. What is the decibel magnitude?
  (b) The sound intensity from a nearby jet at takeoff is 10^3 watts per square meter. What is the decibel magnitude?

54. Use a graph of dB in terms of I to answer the following:
  (a) What is the sound intensity of a soft whisper (28 dB)?
  (b) What is the sound intensity of a nearby train (110 dB)?

55. The minimum intensity of sound that is detectable by the typical human ear is I_0, the *threshold of hearing*. If $I_0 = 10^{-12}$ watts per square meter, what is the decibel magnitude of I_0? Show that

$$dB = 10 \log\left(\frac{I}{I_0}\right)$$

56. The intensity of sound I at a distance m meters from a source that emits P watts of sound energy is given by

$$I = \frac{P}{4\pi m^2}$$

  (a) Find a formula for dB in terms of P and m.
  (b) What is the decibel level 5 meters from a kazoo emitting 10^{-3} watts?

Magnitudes of Astronomical Bodies

(Exercises 57 and 58) In the second century BC, Hipparchus, an astronomer born in present-day Turkey, divided the distant stars that could be seen with the unaided eye into six groups according to their relative brightness. He labeled the group he perceived as brightest as "first magnitude," and he labeled the group he perceived as dimmest as "sixth magnitude." Nearly 2000 years later, a mathematical model was imposed on this system such that the magnitude m of a star is given by

$$m = -19 - 2.5 \log l$$

where l is the measure of brightness (in watts per square meter) from the star at the point of observation.

57. **(a)** The brightness of Rigel, a star in the constellation Orion, is 4.8×10^{-4} watts per square meter. What is its magnitude?
   (b) The brightness of the star Proxima Centauri is 9.5×10^{-13} watts per square meter. What is its magnitude?

58. What is the magnitude of a star that is 1000 times brighter than a star with magnitude 3.2?

C

59. Show that: $\ln(x + \sqrt{x^2 - 1}) = -\ln(x - \sqrt{x^2 - 1})$

60. Determine the graph of $f(x) = 2^x$ and its inverse by using the parametric mode of a graphing utility.

61. Determine the graph of $f(x) = \left(\frac{1}{3}\right)^x$ and its inverse by using the parametric mode of a graphing utility.

62. **(a)** Prove property 2: $\log_b\left(\dfrac{P}{Q}\right) = \log_b P - \log_b Q$
   (b) Prove property 3: $\log_b(P^r) = r \log_b P$

63. **(a)** Prove identity 2: $\log_b b = 1$
   (b) Prove identity 3: $\log_b\left(\dfrac{1}{x}\right) = -\log_b x$
   (c) Prove identity 4: $\log_b(b^x) = x$

64. Values of a transcendental function can be computed with *approximating polynomials*. The polynomials given here are used to approximate the function $f(x) = \ln(1 + x)$.

$$P_1(x) = x \qquad\qquad P_2(x) = x - \frac{x^2}{2}$$

$$P_3(x) = x - \frac{x^2}{2} + \frac{x^3}{3} \qquad P_4(x) = x - \frac{x^2}{2} + \frac{x^3}{3} - \frac{x^4}{4}$$

   (a) Use a graphing utility to graph these polynomials and f.
   (b) Compute $f(0.6)$, $P_1(0.6)$, $P_2(0.6)$, $P_3(0.6)$, and $P_4(0.6)$. Compare the results.

EXPONENTIAL AND LOGARITHMIC EQUATIONS

Because of the many applications of exponential and logarithmic functions, developing tools to solve equations involving these functions is very important.

Most methods of solving these transcendental equations are underpinned by two facts. First, exponential and logarithmic functions are one-to-one functions. Second, an exponential equation can be written as an equivalent logarithmic equation, and vice versa.

Change-of-Base Formula

In Section 4.4, we saw that the solution to the equation $2^x = 6$ is $\log_2 6$, since the equivalent logarithmic equation is $x = \log_2 6$. Because calculators only evaluate natural and common logarithms, we could not find the value of this logarithm directly. But there is a way we can evaluate $\log_2 6$ on a calculator. Using the natural logarithmic function to solve the equation $2^x = 6$, we get

$$2^x = 6$$

$$\ln(2^x) = \ln 6 \quad \text{If } a = b, \text{ then } \ln a = \ln b.$$

$$x \ln 2 = \ln 6 \quad \text{Use property 3 of logarithms.}$$

$$x = \frac{\ln 6}{\ln 2} \quad \text{Divide each side by } \ln 2.$$

NOTE: *The keystrokes to*
evaluate

$$\frac{\ln 6}{\ln 2}$$

on most graphics calculators
are

$$\boxed{\text{LN}}\, 6 \,\boxed{\div}\, \boxed{\text{LN}}\, 2$$

Approximating $(\ln 6)/(\ln 2)$ by calculator yields 2.5850 (to the nearest 0.001), the same value we found when we solved the equation $2^x = 6$ graphically in Section 4.1.

Likewise, we could have applied the common logarithm in the same way to get

$$x = \frac{\log 6}{\log 2}$$

(Take a moment to evaluate this expression.) In fact, we could use any base b. The obvious advantage of using the natural logarithm or the common logarithm is that these expressions can be evaluated on most calculators.

This idea is generalized in the *change-of-base formula*.

Change-of-Base Formula

If a, b, and x are positive real numbers, and a and b are not equal to 1, then

$$\log_a x = \frac{\log_b x}{\log_b a}$$

Specifically:

1. If $b = e$, then

$$\log_a x = \frac{\ln x}{\ln a}$$

2. If $b = 10$, then

$$\log_a x = \frac{\log x}{\log a}$$

The next example shows how the change-of-base formula is used to approximate values of logarithms.

EXAMPLE 1 · Using the Change-of-Base Formula

Approximate each value (to the nearest 0.0001), using the change-of-base formula and natural logarithm key on a calculator.

(a) $\log_5 86$ **(b)** $\log_{(1/3)} 54$

SOLUTION

(a) $\log_5 86 = \dfrac{\ln 86}{\ln 5} = 2.7676$ (to the nearest 0.0001)

(b) $\log_{(1/3)} 54 = \dfrac{\ln 54}{\ln\left(\frac{1}{3}\right)} = -3.6309$ (to the nearest 0.0001)

Using the change-of-base formula, you can determine the graphs of logarithmic functions directly with a graphing utility.

EXAMPLE 2 · Graphing Logarithmic Functions

Use a graphing utility to determine the graph of $y = \log_2 x$.

SOLUTION

Because

$$\log_2 x = \frac{\ln x}{\ln 2}$$

by the change-of-base formula, we can determine the graph of $y = \log_2 x$ with a graphing utility by graphing

$$y = \frac{\ln x}{\ln 2}$$

This graph is shown in Figure 4.38.

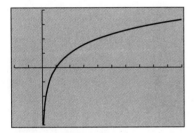

Figure 4.38
$y = \log_2 x$
$[-2, 10] \times [-4, 4]$

Solving Exponential Equations

Simply put, an exponential equation is an equation in which the variable is an exponent, or part of an expression that is an exponent. The next three examples illustrate general methods used for solving exponential equations.

EXAMPLE 3 · Solving an Exponential Equation Exactly

Solve each exponential equation exactly.

(a) $3^x = 81$ **(b)** $\left(\frac{1}{2}\right)^{3x} = 8\sqrt{2}$

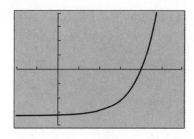

Figure 4.39
$y = 3^x - 81$
$[-2, 6] \times [-100, 100]_{25}$

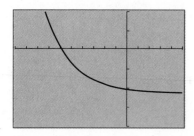

Figure 4.40
$y = \left(\frac{1}{2}\right)^{3x} - 8\sqrt{2}$
$[-2, 1]_{0.2} \times [-20, 10]_5$

SOLUTION

(a) Because 81 is a recognizable power of 3, namely 3^4, we get

$$3^x = 81$$
$$3^x = 3^4 \qquad \text{Replace 81 with } 3^4.$$
$$x = 4 \qquad \text{If } b^x = b^N, \text{ then } x = N.$$

The solution to the equation $3^x = 81$ is 4. The graph of $y = 3^x - 81$ has an x-intercept $(4, 0)$, which supports our solution (Figure 4.39).

(b) Both the left and right sides of the equation can be written as powers of 2:

$$\left(\tfrac{1}{2}\right)^{3x} = (2^{-1})^{3x} = 2^{-3x}$$

and

$$8\sqrt{2} = 2^3 2^{1/2} = 2^{7/2}$$

Thus,

$$\left(\tfrac{1}{2}\right)^{3x} = 8\sqrt{2}$$
$$2^{-3x} = 2^{7/2} \qquad \text{Replace } \left(\tfrac{1}{2}\right)^{3x} \text{ with } 2^{-3x} \text{ and } 8\sqrt{2} \text{ with } 2^{7/2}.$$
$$-3x = \tfrac{7}{2} \qquad \text{If } b^x = b^N, \text{ then } x = N.$$
$$x = -\tfrac{7}{6}$$

The solution to the equation is $-\tfrac{7}{6}$. Our answer is confirmed by the graph of $y = \left(\tfrac{1}{2}\right)^{3x} - 8\sqrt{2}$ shown in Figure 4.40.

The solutions of the equations in Example 3 depended upon our being able to write the equation in the form $b^x = b^N$, and then drawing the conclusion that $x = N$. This conclusion is valid because exponential functions are one-to-one.

Not all exponential equations can be solved in this way, however. Example 4 shows a method of solution for an exponential equation using its equivalent logarithmic equation.

⬤ **EXAMPLE 4   Solving Exponential Equations by Using Common and Natural Logarithms**

Solve each equation exactly, and approximate the solution (to the nearest 0.0001).

(a) $10^x = 24$ (b) $4e^{-x} = 38$

SOLUTION

(a) The right side of the equation cannot be written easily in the form 10^N, so the plan of attack is to solve the equivalent logarithmic equation. The equivalent logarithmic equation of $10^x = 24$ is

$$x = \log 24 = 1.3802 \text{ (to the nearest 0.0001)}$$

The graph of $y = 10^x - 24$ supports our solution (Figure 4.41).

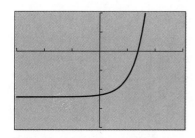

Figure 4.41
$y = 10^x - 24$
$[-3, 3] \times [-40, 20]_{10}$

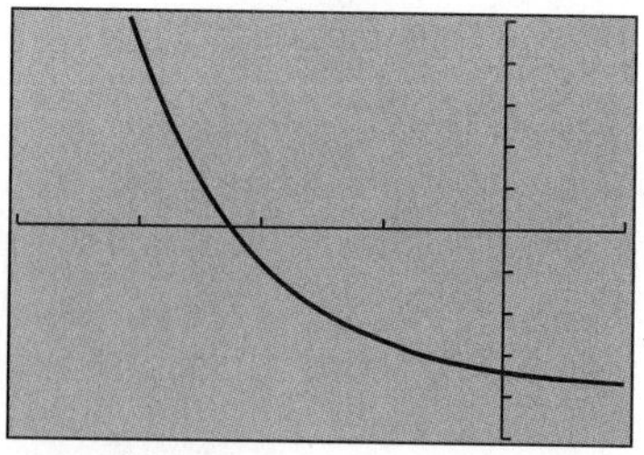

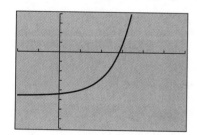

Figure 4.42
$y = 4e^{-x} - 38$
$[-4, 1] \times [-50, 30]_{10}$

(b) Dividing each side of this equation by 4 gives

$$e^{-x} = \tfrac{19}{2}$$

The equivalent logarithmic equation is

$$-x = \ln\left(\tfrac{19}{2}\right)$$

or

$$x = -\ln\left(\tfrac{19}{2}\right) = -2.2513 \text{ (to the nearest 0.0001)}$$

The graph of $y = 4e^{-x} - 38$ (Figure 4.42) verifies our solution.

■ EXAMPLE 5 Approximating the Solutions to an Exponential Equation

Solve each equation exactly, and approximate the solution (to the nearest 0.0001).

(a) $2(3^x) - 6 = 40$ **(b)** $e^{2x} - 4e^x = 5$

SOLUTION

(a) We first solve for 3^x, and then find the corresponding logarithmic equation.

$$2(3^x) - 6 = 40$$
$$2(3^x) = 46$$
$$3^x = 23 \qquad \text{Divide each side by 2.}$$
$$x = \log_3 23 \qquad \text{Rewrite as an equivalent logarithmic equation.}$$

The exact solution of the equation $2(3^x) - 6 = 40$ is $\log_3 23$. The change-of-base formula is used to approximate this value:

$$\log_3 23 = \frac{\ln 23}{\ln 3} = 2.8540 \text{ (to the nearest 0.0001)}$$

The graph of $y = 2(3^x) - 46$ (Figure 4.43) confirms the solution.

(b) This is a quadratic equation that can be solved for e^x:

$$e^{2x} - 4e^x = 5$$
$$(e^x)^2 - 4e^x - 5 = 0$$
$$u^2 - 4u - 5 = 0 \quad \text{Let } u = e^x.$$
$$(u + 1)(u - 5) = 0$$

$$u + 1 = 0 \quad | \quad u - 5 = 0$$
$$u = -1 \quad | \quad u = 5$$
$$e^x = -1 \quad | \quad e^x = 5 \quad \text{Substitute } e^x \text{ for } u.$$

Because e^x is strictly positive for all real numbers x, $e^x = -1$ has no solution.

Figure 4.43
$y = 2(3^x) - 46$
$[-2, 6] \times [-80, 40]_{10}$

The logarithmic equation equivalent to the second equation, $e^x = 5$, is

$$x = \ln 5 = 1.6094 \text{ (to the nearest 0.0001)}$$

The graph of $y = e^{2x} - 4e^x - 5$ verifies this answer (Figure 4.44).

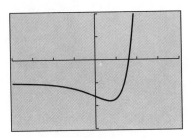

Figure 4.44
$y = e^{2x} - 4e^x - 5$
$[-4, 4] \times [-15, 10]_5$

The strategies of these last three examples are summarized here:

Strategies for Solving Exponential Equations

Suppose that $f(x)$ and $g(x)$ are expressions in terms of x.

1. If the equation can be written as $b^{f(x)} = b^N$, then the equation is equivalent to $f(x) = N$. Solve this equation for x. (See Example 3.)
2. The equation $b^{f(x)} = a$ is equivalent to $f(x) = \log_b a$. Use the change-of-base formula with natural logarithms or common logarithms to approximate $\log_b a$ with your calculator, if necessary. (See Examples 4 and 5.)
3. If the equation can be written as $b^{f(x)} = a^{g(x)}$, taking natural logarithms or common logarithms of both sides yields $f(x) \log b = g(x) \log a$, which can be solved for x. (See Exercises 49–52.)

Solving Logarithmic Equations

A *logarithmic equation* is one that involves a logarithmic function. The general methods of solving these equations are illustrated in the next three examples.

EXAMPLE 6   Solving Logarithmic Equations Directly

Solve each logarithmic equation exactly.

(a) $\log_6(2x) = 3$ **(b)** $3 + 2 \log_4 x = 8$

SOLUTION

(a) The equivalent exponential equation of $\log_6(2x) = 3$ is $2x = 6^3$. Solving this equation for x yields:

$$2x = 6^3$$
$$2x = 216$$
$$x = 108$$

The solution to the equation is 108.

(b) First, we solve for $\log_4 x$:

$$3 + 2\log_4 x = 8$$
$$2\log_4 x = 5$$
$$\log_4 x = \tfrac{5}{2}$$

The equivalent exponential equation of $\log_4 x = \tfrac{5}{2}$ is $x = 4^{5/2}$. This number can be simplified:

$$x = 4^{5/2} = (4^{1/2})^5 = (2)^5 = 32$$

The solution to the equation is 32.

The properties and identities of logarithms from the last section can be used to solve logarithmic equations. Care must be taken, however, because the use of these properties and identities may generate extraneous roots.

● **EXAMPLE 7 Solving Logarithmic Equations by Using Logarithmic Properties**

Solve

$$\log x + \log(x - 3) = 1$$

Check your solutions graphically.

SOLUTION

The left side of the equation can be rewritten using property 1 of logarithms from Section 4.4:

$$\log x + \log(x - 3) = 1$$

$\log[x(x - 3)] = 1$ Use property 1 of logarithms: $\log_b(PQ) = \log_b P + \log_b Q$.

$x(x - 3) = 10^1$ Rewrite as equivalent exponential equation.

$$x^2 - 3x = 10$$
$$x^2 - 3x - 10 = 0$$
$$(x - 5)(x + 2) = 0$$

$x - 5 = 0 \qquad x + 2 = 0$

$x = 5 \qquad\qquad x = -2$

The potential solutions are 5 and -2.

The solutions to the equation $\log x + \log(x - 3) = 1$ correspond to the x-intercepts of the graph of $y = \log x + \log(x - 3) - 1$. Figure 4.45 shows only one x-intercept, at $x = 5$, so the only true solution is 5. The other apparent solution, -2, is extraneous, since there is no x-intercept at -2. (Try checking -2 in the original equation.) Thus, the only solution of the equation is 5.

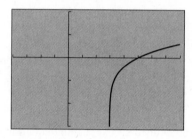

Figure 4.45
$y = \log x + \log(x - 3) - 1$
$[-4, 8] \times [-3, 2]$

EXAMPLE 8 Using the One-to-One Property to Solve Logarithmic Equations

Solve $\ln(x + 2) = -\ln x$ exactly. Approximate the solution (to the nearest 0.0001), and check the solution graphically.

SOLUTION

The right side of the equation can be rewritten using identity 3 of logarithms from Section 4.4. So,

$$\ln(x + 2) = -\ln x$$

$$\ln(x + 2) = \ln\left(\frac{1}{x}\right) \quad \text{Use identity 3 of logarithms: } \log_b\left(\frac{1}{x}\right) = -\log_b x.$$

$$x + 2 = \frac{1}{x} \quad \log_b x = \log_b N \text{ implies } x = N.$$

$$x^2 + 2x = 1 \quad \text{Multiply each side by } x.$$

$$x^2 + 2x - 1 = 0$$

This last equation can be solved by the quadratic formula:

$$x = \frac{-2 \pm \sqrt{2^2 - 4(1)(-1)}}{2(1)} = \frac{-2 \pm \sqrt{8}}{2} = \frac{-2 \pm 2\sqrt{2}}{2} = -1 \pm \sqrt{2}$$

The apparent solutions are $-1 + \sqrt{2}$ (approximately 0.4142) and $-1 - \sqrt{2}$ (approximately -2.4142).

The solutions to the equation $\ln(x + 2) = -\ln x$ are the x-intercepts of the graph of $y = \ln(x + 2) + \ln x$. This graph (Figure 4.46) has an x-intercept between 0 and 1, so there is one solution, $x = -1 + \sqrt{2}$. Because there is no x-intercept between -3 and -2, the other apparent solution, $-1 - \sqrt{2}$, is extraneous.

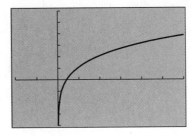

Figure 4.46
$y = \ln(x + 2) + \ln x$
$[-2, 6] \times [-4, 6]$

The strategies of these last three examples are summarized here:

Strategies for Solving
Logarithmic Equations

Suppose that $f(x)$ and $g(x)$ are expressions in terms of x.

1. If the equation can be written as $\log_b f(x) = N$, then it is equivalent to $f(x) = b^N$. Solve this equation for x. (See Examples 6 and 7.)

2. If the equation can be written as $\log_b f(x) = \log_b g(x)$, then it is equivalent to $f(x) = g(x)$. Solve this equation for x. (See Example 8.)

In both strategies, it is important to check for extraneous solutions.

Applications

Solving exponential and logarithmic equations is important in many of the applications we have discussed in the previous sections of this chapter.

EXAMPLE 9   Application: Doubling Time of an Investment

An investment offers an interest rate of 6% compounded monthly. After how long will the amount of the investment double in value?

SOLUTION
To answer this question, we use the compound interest formula

$$A = P\left(1 + \frac{r}{n}\right)^{nt}$$

with $r = 0.06$ and $n = 12$. If the initial amount is P, then the accumulated amount is $2P$, because the accumulated amount is double the initial amount. The problem reduces to solving the equation

$$2P = P\left(1 + \frac{0.06}{12}\right)^{12t}$$

for t.

$$2P = P\left(1 + \frac{0.06}{12}\right)^{12t}$$

$2P = P(1.005)^{12t}$ Simplify numerically.

$2 = (1.005)^{12t}$ Divide each side by P.

$12t = \log_{1.005} 2$ Rewrite as an equivalent logarithmic equation.

$t = \frac{1}{12} \log_{1.005} 2$

The amount of time required is $\frac{1}{12} \log_{1.005} 2$, which, using the change-of-base formula, is

$$\frac{1}{12}\left(\frac{\ln 2}{\ln 1.005}\right) \quad \text{or} \quad \frac{\ln 2}{12 \ln 1.005} = 11.581\,310\,13...$$

This value is 11.58 to the nearest 0.01. The investment will double in approximately 11 years, 7 months.

EXAMPLE 10 Application: The Richter Scale
In Example 9 of Section 4.4, a formula for computing the magnitude M of an earthquake releasing E ergs of energy was given as

$$M = \frac{\log E - 11.4}{1.5}$$

Solve the formula for E. How much energy was released by the 1992 San Francisco earthquake, which had a Richter number of 7.1?

SOLUTION

First, we solve the equation for log E:

$$M = \frac{\log E - 11.4}{1.5}$$

$$1.5M = \log E - 11.4$$

$$1.5M + 11.4 = \log E$$

This logarithmic equation is equivalent to the exponential equation

$$E = 10^{1.5M+11.4}$$

Substituting 7.1 for M and evaluating by calculator gives

$$E = 10^{1.5(7.1)+11.4} \approx 1.1 \times 10^{22}$$

The earthquake released approximately 1.1×10^{22} ergs.

EXERCISE SET 4.5

A

In Exercises 1–6, use the change-of-base formula to approximate each value (to the nearest 0.0001) using a calculator.

1. $\log_2 12$

2. $\log_3 10$

3. $\log_{(1/2)} 6$

4. $\log_{(2/3)} 5$

5. $\log_{(1/2)} 7$

6. $\log_{(1/3)} 12$

In Exercises 7–12, use a graphing utility to determine the graph of each logarithmic equation.

7. $y = \log_3 x$

8. $y = \log_5 x$

9. $y = 5 - \log_2 x$

10. $y = -2 + \log_{(1/2)} x$

11. $y = \log_3(x-4)$

12. $y = \log(x + 6)$

In Exercises 13–18, solve each exponential equation exactly without using a calculator.

13. $4^x = 64$

14. $2^x = 32$

15. $\left(\frac{2}{3}\right)^x = \frac{9}{4}$

16. $\left(\frac{3}{5}\right)^x = \frac{5}{3}$

17. $\left(\frac{1}{3}\right)^x = 9\sqrt{3}$

18. $6^x = \dfrac{\sqrt{6}}{36}$

In Exercises 19–24, solve each logarithmic equation exactly without using a calculator.

19. $\log_3 x = 4$

20. $\log_2 x = 6$

21. $\log_6 x = -2$

22. $\log_2 x = -3$

23. $\log_{(1/2)} x = 4$

24. $\log_{(1/3)} x = -2$

B

In Exercises 25–36, solve each exponential equation exactly, and approximate the solution (to the nearest 0.0001) using a calculator.

25. $5^{2x} = 74$

26. $3^{x-4} = 48$

27. $e^{2x-1} = 123$

28. $10^{x/2} = 1.8$

29. $75(2)^{t/12} = 300$

30. $140\left(\frac{1}{2}\right)^{t/4} = 350$

31. $2000 = 540e^{0.06t}$

32. $10 = 28e^{-0.04t}$

33. $2^{x+2} - 2^x = 48$

34. $2e^x - 9 = 2$

35. $e^{2x} - 4(e^x) = 0$

36. $4^{2x} - 7(4^x) = 8$

In Exercises 37–48, solve each logarithmic equation exactly, and approximate the solutions (to the nearest 0.0001). Use a graphing utility to check for extraneous solutions.

37. $\log_2(3x) = 5$

38. $\log_7(5x) = 2$

39. $3 \ln(4x) = 8$

40. $3 \log(x - 7) = 6$

41. $\log x + \log(x - 15) = 2$

42. $\log x + \log(x + 3) = 1$

43. $\ln x = 2x - 3 - \ln\left(\dfrac{1}{x}\right)$

44. $3x + 5 = \log x - \log\left(\dfrac{x}{10}\right)$

45. $\log(x + 1) + \log(x - 1) = \log(x + 5)$

46. $\ln(2x^2 + x) = 2 \ln(x + 2)$

47. $\log(x^2 - 1) - \log(x + 4) = \log x$

48. $\log_2 x + \log_2(x^2 - 8) = \log_2(8x)$

49. Consider the equation $18^x = 4^{2x+1}$.

    (a) Show that this equation is equivalent to

$$x \ln 18 = (2x + 1) \ln 4$$

    (b) Solve the equation in part (a) to show that

$$x = \frac{\ln 4}{\ln 18 - 2 \ln 4}$$

    (c) Use a calculator to approximate x (to the nearest 0.0001), and check the solution with an appropriate graph.

50. Solve the equation $3^{x+2} = 10^{x-1}$ in the manner outlined in Exercise 49.

51. Solve the equation $4(2^x) = 5^x$ in the manner outlined in Exercise 49.

52. Solve the equation $\frac{1}{2}(6^x) = 2^x$ in the manner outlined in Exercise 49.

In Exercises 53–56, find the inverse function f^{-1} for the given function f.

53. $f(x) = 2^x - 5$

54. $f(x) = 5^{x+3}$

55. $f(x) = 4e^{x-2} - 11$

56. $f(x) = 10^{2x-4} + 5$

57. The population of the western United States is approximated (in millions) by the function

$$P(t) = 12.32e^{0.0247t}$$

where t is the number of years after 1930 (see Example 5, Section 4.2). Determine when the population was 30 million (to the nearest year) by solving an appropriate equation.

58. An investment offers 4.5% annual interest compounded monthly. How many months will it take to double the present amount in this investment?

59. An investment offers 8% annual interest compounded quarterly (four times per year). How many quarters will it take to triple the present amount in this investment?

60. The radioactive decay of plutonium-239 obeys an exponential decay model

$$A = Pe^{-0.0289t}$$

where t is measured in thousands of years (see Example 6, Section 4.2). To the nearest thousand years, what is the half-life of this element?

HINT: Solve the equation $\frac{1}{2}P = Pe^{-0.0289t}$ for t.

61. In 1988, an earthquake of magnitude 6.8 in Armenia killed 55,000 persons. How much energy was released by this earthquake? (See Example 9, Section 4.4.)

62. Solve for t: $A = C + (A_0 - C)e^{rt}$

63. Solve for t: $I = \dfrac{E}{R}(1 - e^{-Rt/L})$

C

64. Solve the equation: $\frac{1}{2}(e^x - e^{-x}) = 4$

65. Solve the equation: $\frac{1}{2}(e^x - e^{-x}) = 3$

66. In this chapter, there are two formulas for the exponential growth model, $P(t) = P_0 e^{rt}$ (where r is the rate of change) and $P(t) = P_0(2)^{t/d}$ (where d is the doubling time).

    (a) Show that $rd = \ln 2$.

    (b) The *Rule of 70* is a computation used by financiers to roughly approximate the doubling time of an investment at $R\%$. This is done by dividing 70 by R; the quotient is the doubling time in years. For example, the doubling time of an investment paying 8% annual interest is about 9 years because $\frac{70}{8} \approx 9$. Explain why this approximation works using part (a).

QUICK REFERENCE

Topic	Page	Remarks
Exponential function	347	An exponential function f is a function of the form $$f(x) = b^x$$ with $b > 0$ and $b \neq 1$.
Compound interest model	351	The amount A of an investment after t years when P dollars are invested at an annual interest rate r compounded n times per year is given by $$A = P\left(1 + \frac{r}{n}\right)^{nt}$$
Doubling time	354	The doubling time of a quantity is the time required for the quantity to double in size. If an initial amount A_0 of a quantity with doubling time d grows exponentially, then the amount $A(t)$ after t units of time is given by $$A(t) = A_0 \cdot 2^{t/d}$$
Half-life	356	The half-life of a quantity is the time required for the quantity to decay to half its size. If an initial amount A_0 of a quantity with half-life h decays exponentially, then the amount $A(t)$ after t units of time is given by $$A(t) = A_0 \cdot \left(\tfrac{1}{2}\right)^{t/h}$$
The number e	360	As $k \to +\infty$, $$\left(1 + \frac{1}{k}\right)^k \to e = 2.718\,281\,828\,459\,045...$$ The irrational number e is one of the most important constants in mathematics.
The natural exponential function	361	The natural exponential function is $$f(x) = e^x$$
Continuous compounding formula	365	The amount A of an investment after t years when P dollars are invested at an annual interest rate r compounded continuously is given by $$A = Pe^{rt}$$
Exponential growth and decay model	366	If the rate of change of a quantity is proportional to the amount of the quantity present at that moment, then a model for this quantity is also $$A = Pe^{rt}$$ where A is the accumulated amount, P is the initial amount, r is the rate of change, and t is the elapsed time. If $r > 0$, this model is called the exponential growth model. If $r < 0$, this model is called the exponential decay model.

Continued

Topic	Page	Remarks
Newton's law of cooling	368	If an object with initial temperature A_0 is placed in an environment of constant temperature C, then the temperature $A(t)$ at time t is given by $$A(t) = C + (A_0 - C)e^{rt}$$ where r is the rate of change of the difference in temperature of the object and the environment.
Inverse functions	375	Two functions f and g are inverses of each other if $$f[g(a)] = a \quad \text{for every } a \text{ in the domain of } g$$ and $$g[f(b)] = b \quad \text{for every } b \text{ in the domain of } f$$ The inverse of a function f is denoted by f^{-1}.
One-to-one function	377	A function f is one-to-one if $$f(x_1) = f(x_2) \quad \text{implies that} \quad x_1 = x_2$$ A function f is one-to-one if and only if no horizontal line crosses the graph at more than one point.
Logarithm	387	The logarithm base b of N, denoted $\log_b N$, is the exponent to which b must be raised to get N. In other words, the solution of the equation $$b^x = N$$ is $$x = \log_b N$$
Logarithmic function	387	A logarithmic function is a function of the form $$g(x) = \log_b x$$ where $b > 0$ and $b \neq 1$. It is the inverse of the exponential function $f(x) = b^x$.
Natural logarithm	389	The natural logarithm of a number x, denoted $\ln x$, is the logarithm base e of x: $$\ln x = \log_e x$$
Common logarithm	389	The common logarithm of a number x, denoted $\log x$, is the logarithm base 10 of x: $$\log x = \log_{10} x$$
Properties of logarithms	392	If P and Q are positive and r is a real number, then **1.** $\log_b(PQ) = \log_b P + \log_b Q$ **2.** $\log_b\left(\dfrac{P}{Q}\right) = \log_b P - \log_b Q$ **3.** $\log_b(P^r) = r \log_b P$

Topic	Page	Remarks
Logarithm identities	392	For $b > 0$, $b \neq 1$, and $x > 0$:

1. $\log_b 1 = 0$ **2.** $\log_b b = 1$ **3.** $\log_b\left(\dfrac{1}{x}\right) = -\log_b x$

4. $\log_b(b^x) = x$ **5.** $b^{\log_b x} = x$

Topic	Page	Remarks
Change-of-base formula	400	For $a > 0$, $b > 0$, $a \neq 1$, $b \neq 1$, and $x > 0$:

$$\log_a x = \frac{\log_b x}{\log_b a}$$

Particularly useful cases are for $b = e$ and $b = 10$:

1. $\log_a x = \dfrac{\ln x}{\ln a}$ **2.** $\log_a x = \dfrac{\log x}{\log a}$

◉ MISCELLANEOUS EXERCISES

In Exercises 1–10, determine the graph of each function. Determine the intercepts and asymptote of the graph.

1. $y = 4^x - 9$

2. $y = 12 - 3^x$

3. $y = 4\left(\frac{1}{3}\right)^x$

4. $y = -\left(\frac{2}{5}\right)^{x-1}$

5. $y = 50e^{-0.5x}$

6. $y = 2e^{0.2x} - 4$

7. $y = 3 - \ln x$

8. $y = \ln(6 - x)$

9. $y = \log_2(-2x)$

10. $y = \log_2 x + 2$

In Exercises 11 and 12, determine the value of each logarithm exactly.

11. (a) $\log_2 16$ (b) $\log_{(1/2)} 16$ (c) $\log_7(\sqrt{7})$

12. (a) $\log_6 36$ (b) $\log_{(1/3)} 27$ (c) $\log_{(1/2)}\left(\frac{1}{8}\right)$

13. Determine the domain of $\ln(\sqrt{4 - x^2})$.

14. Determine the domain of $\log_3 |x^2 - 8|$.

15. Use the compound interest formula to compute the accumulated amount after 2 years for an investment of $600 at 7.5% compounded monthly.

16. Repeat Exercise 15 for an investment of $1200 at 5% compounded four times per year.

17. Use the compound interest formula to compute the principal invested at 4% compounded two times per year, given the accumulated amount of the investment after 3 years is $24,000.

18. Repeat Exercise 17, given the accumulated amount is $15,000 after 5 years.

In Exercises 19–26, for each function f, determine f^{-1}, the inverse of f.

19. $f(x) = 2x - 4$

20. $f(x) = 7 - \frac{2}{3}x$

21. $f(x) = \sqrt[3]{8 - x}$

22. $f(x) = x^3 + 12$

23. $f(x) = \dfrac{2x}{x - 1}$

24. $f(x) = 3 - \dfrac{6}{x}$

25. $f(x) = \frac{1}{2} e^{2x-8}$

26. $f(x) = 3 + \ln(x + 4)$

In Exercises 27 and 28, rewrite the transcendental expression as an algebraic expression, and simplify.

27. (a) $e^{\ln(2t)}$ (b) $10^{\log w - 1}$ (c) $\left(\frac{1}{2}\right)^{-\log_2 t}$

28. (a) $10^{\log(4x)}$ (b) $e^{\ln 4 + \ln x}$ (c) $(0.1)^{\log(2+y)}$

In Exercises 29 and 30, expand the logarithm $\log_b N$ completely.

29. (a) $\log\left(\dfrac{7x}{10}\right)$ (b) $\ln(2xy)$ (c) $\ln(\sqrt{1 - x})$

30. (a) $\ln\left(\dfrac{5}{2t}\right)$ (b) $\log(3c^4)$

(c) $\log_{11}\left(\sqrt[3]{\dfrac{x}{5}}\right)$

In Exercises 31 and 32, write the expression in the form $\log_b P$, in which P is a simplified expression.

31. (a) $3\log(m-3) + \log(m+2)$
   (b) $\ln(x^2 - 4) + \ln x - \ln(x-2)$

32. (a) $\log(x-3) - 2\log x$
   (b) $\ln(x^2 - 1) + \ln\left[\dfrac{1}{2}(x-1)\right] - \ln x$

In Exercises 33–44, solve each equation exactly, and approximate the solutions (to the nearest 0.0001).

33. $7^{3x} = 24$ **34.** $3^{2+x} = 72$

35. $120e^{-0.4t} = 2400$ **36.** $2e^x - 9 = 0$

37. $e^{2x} - 3e^x = -2$ **38.** $4^{-x} + 8(4^x) = 6$

39. $\log_2(x-4) = 3$ **40.** $\log_6(x+14) = 2$

41. $\log(x-2) - 1 = \log(x+1)$

42. $\ln(2x^2) = 2\ln(x-2)$

43. $\ln(1-x) = 2 + \ln x$ **44.** $\log(3x) - \log 4 = 2$

In Exercises 45–48, solve each formula for the given variable.

45. $A = Pe^{rt}$; for t **46.** $A = 100 \cdot 2^{t/h}$; for h

47. $d = 10\log\left(\dfrac{I}{I_0}\right)$; for I

48. $r = \dfrac{1}{t}[\ln(C - A) - \ln A]$; for C

In Exercises 49 and 50, the graph of $y = k \cdot b^x + c$ is shown. Determine the values of k, b, and c.

49.

(−1, 5) (0, 3)

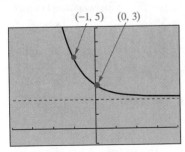

50.

(1, 1)

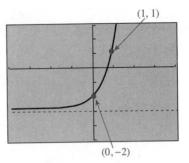

(0, −2)

51. The vertical motion of a shock absorber on an automobile is described by the function

$$s(t) = 5(0.12 - t)26.4^{-t}$$

where $s(t)$ is the displacement (in feet) of the top of the shock absorber from its rest position t seconds after the automobile hits a speed bump.
(a) Determine the complete graph of s.
(b) At what times (to the nearest 0.01 second) is the top of the shock absorber 0.2 feet above its rest position?

52. The charge (in coulombs) on a capacitor in an electronic circuit is given by the function

$$q(t) = 0.00021(1 - e^{-1.5t})$$

(a) Determine the complete graph of q.
(b) What is the maximum charge on the capacitor?

53. The population of an endangered species of rodent is being depleted by encroachment on its habitat. At the current rate of decrease, the half-life of the population is 4 years. The present population is estimated to be 1.75 million.
(a) Determine a function A such that $A(t)$ is the rodent population in t years.
(b) Estimate the population after 6 years and 10 years.

54. Air pressure is a function of altitude above sea level. The pressure P (in pounds per square inch) at an altitude of h miles above sea level is given by $P = 14.7e^{-0.203h}$.
(a) Solve this equation for h.
(b) At what altitude is the pressure 6 pounds per square inch?

55. An investment offers 5.5% annual interest compounded quarterly. How many quarters will it take to double the present amount in this investment?

56. An investment offers 4% annual interest compounded daily (360 times per year). How many days will it take to triple the present amount in this investment?

57. The proportion of carbon-14, an isotope of carbon, in living plant matter is constant. Once a plant dies, the carbon-14 in it begins to decay with a half-life of 5570 years. An archaeologist measures the amount of carbon-14 in the remains of a post in a prehistoric hut in Kenya to be one-tenth of the amount of carbon-14 as would be in living wood. How old is the hut?

58. Repeat Exercise 57, given that the proportion of carbon-14 is one-twentieth.

59. If 30% of a radioactive substance decays in 4 years, what is its half-life (to the nearest 0.01 year)?

60. An investment that earns interest continuously at a constant rate grows by 20% in 2 years. What is this interest rate?

CHAPTER TEST

1. Determine the graph of $f(x) = \ln x - 4$.

2. Let $f(x) = 2e^{x-1}$. Determine $f^{-1}(x)$, the inverse of f.

3. Write the expression

$$3 \log(x + 2) - \log(x^2 - 4)$$

in the form $\log_b P$, in which P is a simplified expression.

4. Simplify the expression: $e^{\ln b} + \log\left(\dfrac{1}{10^a}\right)$

5. Solve the equation

$$17(3^{2x}) = 250$$

exactly, and appropriate the solution (to the nearest 0.0001).

6. Solve the equation

$$\ln(5x) = 3 + \ln 2$$

exactly, and approximate the solution (to the nearest 0.0001).

7. The graph of $y = k \cdot b^x + c$ is shown. Determine the values of k, b, and c.

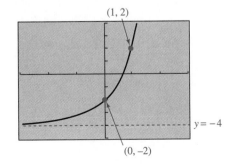

8. Use the compound interest formula to compute the accumulated amount after 6 years for an investment of $2000 at 6.5% compounded monthly.

9. The half-life of thorium-228 is 1.9 years. How much of an initial sample of 50 grams will remain after 4 years?

10. Solve the formula for r: $A = C + Be^{rt}$

11. An investment offers 7% annual interest compounded weekly (52 times per year). How long will it take to double the present amount of the investment?

12. The Richter number M of an earthquake releasing E ergs of energy is given by

$$M = \frac{\log E - 11.4}{1.5}$$

Determine the energy released by the 1993 Agana, Guam earthquake, which had a Richter number 8.0.

13. The spread of a virus in an isolated community is modeled by

$$N(t) = \frac{500}{1 + 49e^{-0.2t}}$$

where $N(t)$ is the number of people infected after t days.
(a) How many will be infected in 6 days?
(b) How many days until 100 people are infected?
(c) Use a graphing utility to determine the number of people infected after a very long time.

14. A bowl of soup initially at a temperature of 150°F cools in a room in which the temperature is a constant 68°F. Use Newton's law of cooling to determine the temperature after 10 minutes if the rate of change is $r = -0.06$.

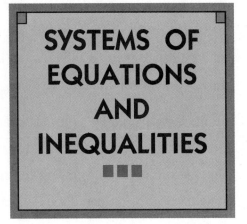

SYSTEMS OF EQUATIONS AND INEQUALITIES

So far, each of the problems we have encountered can be solved or modeled by one equation or inequality. Many situations, however, cannot be adequately modeled by just one equation. For example, predicting prices based on the forces of supply and demand, analyzing mixtures in food processing, and comparing simple interest to compound interest are applications that involve at least two equations in two variables. In the first two sections of this chapter, we consider these applications and discuss graphical and algebraic methods of solving systems of two equations in two variables.

In Sections 5.3 and 5.4, we encounter systems of three equations in three variables in problems such as analyzing electrical networks, preparing food mixtures subject to dietary and cost requirements, and modeling the temperature distribution in a plate. We solve these systems using algebraic methods that are well-suited to graphing utilities, and we introduce a powerful tool in mathematics—the matrix. In Sections 5.5 and 5.6, we move from using matrices to solve linear systems of equations to other applications, such as determining the area of a polygon and predicting market share trends. Matrix addition, subtraction, and multiplication, and the determinant are motivated by these applications. With these matrix operations in place, we define the multiplicative inverse and develop another method of solving linear systems of equations.

In the last section, we consider systems of inequalities in two variables. Systems of inequalities arise when modeling situations that are subject to constraints; for example, a food supplement may need to be prepared with at least a certain amount of protein, but no more than a certain amount of fat.

- Solving Systems by Graphing
- Nature of Solutions
- Applications

SYSTEMS OF EQUATIONS

Many applications, such as comparing an investment that pays compound interest to one that pays simple interest, or analyzing an electrical network, require two or more equations considered together, called a *system of equations*. Four examples of systems of equations are shown below.

$$\begin{cases} x + y = 9 \\ y = \frac{1}{2}x + 3 \end{cases} \qquad \text{Two equations in two variables, } x \text{ and } y$$

$$\begin{cases} y = x^2 - 3 \\ y = 2x \end{cases} \qquad \text{Two equations in two variables, } x \text{ and } y$$

$$\begin{cases} 3x + y - 3z = 6 \\ x - y - 2z = 13 \end{cases} \qquad \text{Two equations in three variables, } x, y, \text{ and } z$$

$$\begin{cases} x - 2y + 4z = 0 \\ -x - 3y - 2z = 1 \\ 5x \quad\;\; + \;\; z = 3 \end{cases} \qquad \text{Three equations in three variables, } x, y, \text{ and } z$$

The brace on the left of each system indicates that we are considering the equations simultaneously. If each equation in a system is linear (first-degree), then the system is a *linear system*. If any equation in a system is not a linear equation, then the system is a *nonlinear system*. Three of the four systems above are linear systems; the second system is nonlinear.

In this section we will consider systems of two equations in two variables, such as

$$\begin{cases} x + y = 9 \\ y = \frac{1}{2}x + 3 \end{cases}$$

A *solution* to this system is an *ordered pair* (x, y) of numbers that makes *both* equations true. A solution to the above system is $(4, 5)$, since this ordered pair satisfies each of the equations:

$$\begin{cases} (4) + (5) \overset{?}{=} 9 \quad \text{True} \\ (5) \overset{?}{=} \frac{1}{2}(4) + 3 \quad \text{True} \end{cases}$$

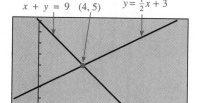

Figure 5.1
$[-2, 13] \times [-1, 9]$

Solving Systems by Graphing

The *graph of a system of equations* is the set of graphs of both equations in the system on the same coordinate plane. For example, the graph of the system

$$\begin{cases} x + y = 9 \\ y = \frac{1}{2}x + 3 \end{cases}$$

is shown in Figure 5.1. The solution to the system is the point $(4, 5)$ where the graphs intersect.

We can use the graph of a system to find the solutions of the system.

EXAMPLE 1   A System with Integer Solutions
The system

$$\begin{cases} y = x^2 - 3 \\ y = 2x \end{cases}$$

has integer solutions. Use a graphing utility to graph the system, and determine the solutions of the system.

SOLUTION

A complete graph of each equation in the system is shown in Figure 5.2.

The graph of the first equation represents all ordered pairs (x, y) that satisfy $y = x^2 - 3$. Similarly, the graph of the second equation represents all ordered pairs (x, y) that satisfy $y = 2x$. The two points where the graphs intersect correspond to the ordered pairs (x, y) that make both equations in the system true. Therefore, the solutions of the system are apparently $(-1, -2)$ and $(3, 6)$. To be sure, we check our solutions.

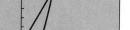
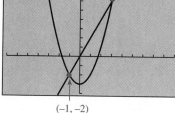

Figure 5.2
$[-7, 9] \times [-4, 8]$

CHECK: For $(-1, -2)$:

$$\begin{cases} (-2) \stackrel{?}{=} (-1)^2 - 3 & \text{True} \\ (-2) \stackrel{?}{=} 2(-1) & \text{True} \end{cases}$$

For $(3, 6)$:

$$\begin{cases} (6) \stackrel{?}{=} (3)^2 - 3 & \text{True} \\ (6) \stackrel{?}{=} 2(3) & \text{True} \end{cases}$$

The solutions to the system are $(-1, -2)$ and $(3, 6)$.

Not all systems have integer solutions as in Example 1. If a system of equations models a problem from real life, we would expect realistic, noninteger answers. We often must settle for approximations to the actual answers.

EXAMPLE 2 Solving a System

Use a graphing utility to graph the system

$$\begin{cases} y = e^x \\ y = \sqrt{x + 1} \end{cases}$$

and determine the solutions to the system. Approximate all nonexact answers to the nearest 0.001.

SOLUTION

A graph of the system is shown in Figure 5.3.

Since the graphs intersect at two points, there are two solutions to the system. One of the solutions appears to be $(0, 1)$. We check this result by substituting $x = 0$ and $y = 1$ into each of the equations:

$$\begin{cases} 1 \stackrel{?}{=} e^0 & \text{True} \\ 1 \stackrel{?}{=} \sqrt{0 + 1} & \text{True} \end{cases}$$

The other point of intersection of the two graphs appears to be approximately $(-0.8, 0.5)$. Using the zoom and trace features, we can improve the precision (Figure 5.4).

The graphs intersect at the point $(-0.797, 0.451)$, to the nearest 0.001. Our solution is not exact, so we must settle for approximations when checking:

$$\begin{cases} 0.451 \stackrel{?}{\approx} e^{-0.797} = 0.450678... & \text{True} \\ 0.451 \stackrel{?}{\approx} \sqrt{-0.797 + 1} = 0.450555... & \text{True} \end{cases}$$

The solutions are $(0, 1)$ and $(-0.797, 0.451)$.

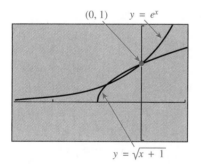

$(0, 1)$ $y = e^x$

$y = \sqrt{x + 1}$

Figure 5.3
$[-2.7, 1] \times [-1, 2]$

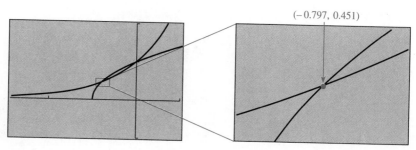

(−0.797, 0.451)

Figure 5.4

EXAMPLE 3   A System with Equations That Are Not All Functions

Use a graphing utility to graph the system

$$\begin{cases} x^2 + 4y^2 = 16 \\ y = x^3 - 3x + 3 \end{cases}$$

and determine the solutions to the system. Approximate all nonexact answers to the nearest 0.001.

SOLUTION

In order to graph the equation $x^2 + 4y^2 = 16$ with a graphing utility, we first solve for y:

$$x^2 + 4y^2 = 16$$
$$4y^2 = 16 - x^2$$
$$y^2 = \frac{16 - x^2}{4}$$
$$y = \pm \frac{\sqrt{16 - x^2}}{2}$$

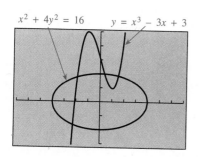

$x^2 + 4y^2 = 16 \qquad y = x^3 - 3x + 3$

Figure 5.5
$[-7, 7] \times [-3, 5]$

Thus, to graph the system, we graph the three equations

$$y = \frac{\sqrt{16 - x^2}}{2}, \quad y = -\frac{\sqrt{16 - x^2}}{2}, \quad \text{and} \quad y = x^3 - 3x + 3$$

The graph of the system is shown in Figure 5.5.

Since there are four points of intersection, the system has four solutions. The point of intersection in the third quadrant is approximately $(-2, -2)$. Using the zoom and trace features (Figure 5.6), we find that the graphs intersect at the point $(-2.251, -1.653)$, to the nearest 0.001.

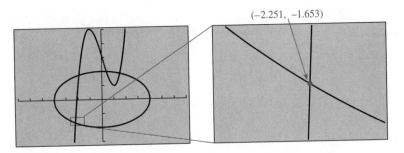

Figure 5.6

In the second quadrant, the two graphs intersect at about $(-2, 2)$. Using the zoom and trace features (Figure 5.7), we find that the graphs intersect at the point $(-1.911, 1.757)$, to the nearest 0.001.

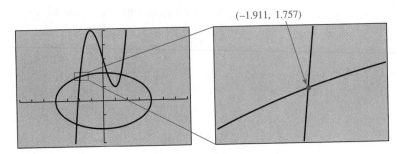

Figure 5.7

The two points of intersection in the first quadrant are approximately $(0.5, 2)$ and $(1.5, 2)$. Using the same techniques as with the first two points of intersection, we determine that these two solutions are $(0.350, 1.992)$ and $(1.495, 1.855)$, to the nearest 0.001. Thus, the solutions to the system are $(-1.911, 1.757)$, $(-2.251, -1.653)$, $(0.350, 1.992)$, and $(1.495, 1.855)$. ■

Nature of Solutions

Graphing a system of equations in two variables tells us how many solutions to expect. *Nonlinear* systems can have any number of solutions. On the other hand, a *linear* system of two equations has only three possibilities. The graphs in Figure 5.8 are typical.

A system is *consistent* if it has at least one solution. A consistent system that has exactly one solution is called an *independent* system, whereas a consistent system that has an infinite number of solutions is called a *dependent* system. A system with no solutions is an *inconsistent* system. The following box summarizes these ideas.

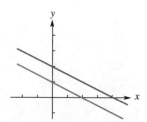

(a) One solution:
The two graphs intersect
at exactly one point.

(b) Infinite number of solutions:
The two graphs are the
same line.

(c) No solutions:
The two graphs do not
intersect.

Figure 5.8

The Nature of Solutions
for Linear Systems

A linear system of equations in two variables can have one of three possible outcomes:

1. The graphs of the two equations intersect at exactly one point, so the system has exactly one solution. A system that has exactly one solution is an **independent (consistent)** system.
2. The graphs of the two equations coincide (they are actually the same line), so the system has an infinite number of solutions. A system that has an infinite number of solutions is a **dependent (consistent)** system.
3. The graphs of the two equations are parallel and distinct, so the system has no solutions. A system that has no solutions is an **inconsistent** system.

● **EXAMPLE 4 Solving a Linear System**

Solve each linear system by graphing the system.

(a) $\begin{cases} x = 3y \\ 2x + y = 14 \end{cases}$ (b) $\begin{cases} 2x + 4y = 9 \\ x + 2y = -2 \end{cases}$ (c) $\begin{cases} y = \frac{2}{3}x - 1 \\ 2x - 3y = 3 \end{cases}$

SOLUTION

(a) Each equation in the system

$$\begin{cases} x = 3y \\ 2x + y = 14 \end{cases}$$

can be written in slope–intercept form:

$$\begin{cases} y = \frac{1}{3}x \\ y = -2x + 14 \end{cases}$$

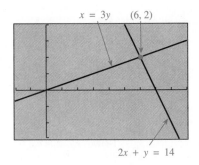

Figure 5.9
$[-2, 9] \times [-3, 4]$

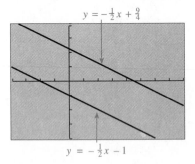

Figure 5.10
$[-4, 8] \times [-4, 4]$

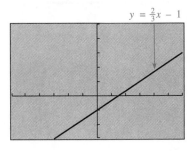

Figure 5.11
$[-6, 6] \times [-3, 5]$

From this form, we know that the slope of the first line is $\frac{1}{3}$ and the slope of the second is -2. Since the slopes are not the same, the lines are not parallel and they do not coincide, so they must intersect at exactly one point. The graphs of these equations are shown in Figure 5.9. The solution is $(6, 2)$. (This should be verified by substituting $x = 6$ and $y = 2$ into each of the equations of the system.)

(b) First, we rewrite each equation in the system

$$\begin{cases} 2x + 4y = 9 \\ \ \ x + 2y = -2 \end{cases}$$

in slope–intercept form:

$$\begin{cases} y = -\frac{1}{2}x + \frac{9}{4} \\ y = -\frac{1}{2}x - 1 \end{cases}$$

The slope of each graph is $-\frac{1}{2}$, but the y-intercepts are different. Therefore, the graphs are parallel lines, as shown in Figure 5.10. This system is inconsistent.

(c) The first equation is already in slope–intercept form. Rewriting the second equation, $2x - 3y = 3$, in slope–intercept form, we get

$$\begin{cases} y = \frac{2}{3}x - 1 \\ y = \frac{2}{3}x - 1 \end{cases}$$

In this form, it is apparent that the two equations represent the same line (Figure 5.11). This system is dependent. Any point on the line

$$y = \frac{2}{3}x - 1$$

is a solution to the system. As a way of representing the solutions, let $x = t$, where t is any real number. Substituting in the last equation gives

$$y = \frac{2}{3}t - 1$$

Thus, the solutions (x, y) to the system can be described as

$$\left(t, \frac{2}{3}t - 1\right)$$

where t is any real number.

The solution $\left(t, \frac{2}{3}t - 1\right)$ in Example 4c is called a *parametric* representation, and t is called a *parameter*. Recall from Section 3.6 that the graph of the parametric equations

$$x = t, \quad y = \frac{2}{3}t - 1$$

is a line. The parametric representation $\left(t, \frac{2}{3}t - 1\right)$ is equivalent to these parametric equations. In other words, the graph of the solutions generated by either of these two representations is the same line.

Applications

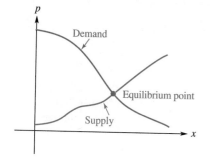

Figure 5.12

A manufacturer produces x units of an item for a selling price p per unit. Consider the relationship between x and p from the manufacturer's point of view. If the selling price p increases, then the manufacturer is willing to increase the number x of units produced. Therefore, the graph of the ordered pairs (x, p), called the *supply,* is increasing.

Now think about the relationship between x and p from the consumer's perspective. If the selling price p decreases, then consumers will buy more units x. Hence, the graph of the ordered pairs (x, p), called the *demand,* is decreasing.

If the supply and demand graphs are drawn in the same coordinate plane, the *equilibrium point* is where the two graphs intersect. A typical situation is shown in Figure 5.12. The price p at the equilibrium point is the *equilibrium price.* According to economists, this is the only price that can persist in the long run.

EXAMPLE 5 Application: Finding the Equilibrium Price

Equations for the supply and demand for a portable cassette player are given by:

Supply
$$p = 0.04x^2 + 11$$

Demand
$$p = 31 - 0.7x$$

where x is in hundreds of units and p is price in dollars. Use a graphing utility to determine the equilibrium price (to the nearest $0.01).

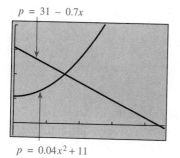

Figure 5.13
$[0, 45]_{10} \times [-5, 40]_{10}$

SOLUTION

We graph both equations in the same viewing window and locate the point where the graphs intersect. The situation calls for $x \geq 0$ and $p \geq 0$, so we limit our attention to the first quadrant (Figure 5.13).

The point of intersection in the first quadrant is approximately (15, 20). Using the zoom and trace features, we find that the graphs intersect at the point (15.26, 20.32), to the nearest 0.01 (Figure 5.14).

The equilibrium price is $20.32.

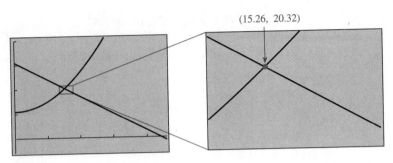

(15.26, 20.32)

Figure 5.14

● EXAMPLE 6 Application: Modeling Supply and Demand

A new type of tennis racket is introduced into a test market. After a few months, 1100 are sold at \$65 and 600 are sold at \$75. The supplier will furnish 800 rackets at \$65. At \$75, 1200 rackets can be supplied. Assume that supply and demand can be modeled by linear equations.

(a) Determine models for the demand and supply.

(b) Use the models to determine the equilibrium price (to the nearest \$0.10).

SOLUTION

(a) In order to determine a linear model, we need two points (x_1, p_1) and (x_2, p_2). Recall from Section 2.1 that an equation of the line passing through (x_1, p_1) with slope m is

$$p = m(x - x_1) + p_1$$

The linear model for demand is an equation of the line containing (1100, 65) and (600, 75). The slope of the line through these data points is

$$m = \frac{p_2 - p_1}{x_2 - x_1} = \frac{75 - 65}{600 - 1100} = -\frac{1}{50}$$

In point–slope form, the demand can be expressed as

$$p = -\tfrac{1}{50}(x - 1100) + 65$$
$$= -\tfrac{1}{50}x + 87$$

On the other hand, the supply model is an equation of the line containing the data points (800, 65) and (1200, 75). The slope of the line containing these points is

$$m = \frac{p_2 - p_1}{x_2 - x_1} = \frac{75 - 65}{1200 - 800} = \frac{1}{40}$$

Thus, the supply can be expressed in point–slope form as

■ **NOTE:** *In Section 2.1, the point–slope form is stated as* $y = m(x - h) + k.$

$$p = \tfrac{1}{40}(x - 800) + 65$$
$$= \tfrac{1}{40}x + 45$$

The demand model is $p = -\tfrac{1}{50}x + 87$, and the supply model is $p = \tfrac{1}{40}x + 45$.

(b) To determine the equilibrium price, we solve the system

$$\begin{cases} p = -\tfrac{1}{50}x + 87 \\ p = \tfrac{1}{40}x + 45 \end{cases}$$

A graph of the system is shown in Figure 5.15. The point of intersection is approximately (930, 70). Using the zoom and trace features (Figure 5.16), the point of intersection of the two graphs (to the nearest 0.01) is (933.33, 68.33):

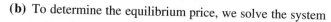

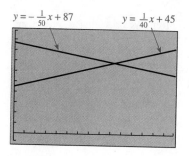

Figure 5.15
$[0, 1500]_{100} \times [-10, 100]_{10}$

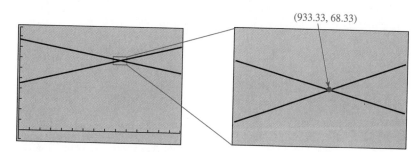

(933.33, 68.33)

Figure 5.16

The equilibrium price is $68.30 (to the nearest $0.10).

If P dollars are invested for t years at an annual simple interest rate r, the accumulated amount A is

$$A = P + Prt$$

Recall from Section 4.1 that if P dollars are invested for t years at an interest rate r compounded n times per year, the accumulated amount A is

$$A = P\left(1 + \frac{r}{n}\right)^{nt}$$

We can compare an investment that pays simple interest to one that pays compound interest by using a system of equations.

EXAMPLE 7 Application: Comparing Investments

Consider an investment of $1000 at a simple interest rate of 10% and an investment of $1000 invested at 6% compounded daily (use $n = 365$).

(a) When will the accumulated amounts for each investment be the same?

(b) If you plan to invest for 8 years or less, which of the two situations pays more?

SOLUTION

(a) The future amount A at 10% simple interest after t years is

$$A = 1000 + 1000(0.10)t = 1000 + 100t$$

The future amount A at 6% interest compounded daily for t years is

$$A = 1000 \left(1 + \frac{0.06}{365} \right)^{365t}$$

We want to determine values of $t \geq 0$ such that the future amount A is the same for both investments. In other words, we want to solve the nonlinear system

$$\begin{cases} A = 1000 + 100t \\ A = 1000 \left(1 + \dfrac{0.06}{365} \right)^{365t} \end{cases}$$

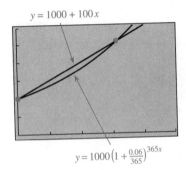

$y = 1000 + 100x$

$y = 1000\left(1 + \frac{0.06}{365}\right)^{365x}$

Figure 5.17

$[0, 25]_5 \times [-100, 3000]_{500}$

The graphs of both equations for $t \geq 0$ are shown in Figure 5.17. There are two solutions for $t \geq 0$. One solution is $(0, 1000)$, corresponding to each investment of $1000 before any time has passed (and before any interest is earned). The other solution appears to be about $(16, 2600)$. We leave it to you to verify (with a graphing utility) that this point of intersection is $(15.794, 2579.375)$, to the nearest 0.001. Thus, both investments will grow to $2579.38 in 15.794 years, or 15 years and 290 days.

(b) For $0 < t < 8$, the graph of $A = 1000 + 100t$ is above the graph of

$$A = 1000\left(1 + \frac{0.06}{365} \right)^{365t}$$

Therefore, for the first 8 years, the accumulated amount A will be greater at 10% simple interest than at 6% compounded daily. You should invest at 10% simple interest.

EXERCISE SET 5.1

A

In Exercises 1–6, each system of equations has integer solutions. Use a graphing utility to graph each system, and determine the solutions of the system.

1. $\begin{cases} y = \dfrac{x-7}{2} \\ y = 1 - x \end{cases}$

2. $\begin{cases} y = 6 - x \\ y = \frac{2}{3}x - 4 \end{cases}$

3. $\begin{cases} y = x^2 \\ y = 2 - x \end{cases}$

4. $\begin{cases} y = 3 - x^2 \\ y = x + 1 \end{cases}$

5. $\begin{cases} y = |x| + 1 \\ y = \sqrt{25 - x^2} \end{cases}$

6. $\begin{cases} y = |x - 2| - 2 \\ y = \sqrt{x + 2} \end{cases}$

In Exercises 7–14, use a graphing utility to graph each system, and determine the solutions to the system. Approximate all noninteger solutions to the nearest 0.001.

7. $\begin{cases} y = x^2 \\ y = 2\sqrt{2 - x} \end{cases}$

8. $\begin{cases} y = |x| - 1 \\ y = \ln(x + 3) + 1 \end{cases}$

9. $\begin{cases} y = |x| - 1 \\ y = (x - 2)^3 + 1 \end{cases}$

10. $\begin{cases} y = |x - 1| \\ y = (x - 2)^3 \end{cases}$

11. $\begin{cases} y = \sqrt{|x|} \\ y = e^x \end{cases}$

12. $\begin{cases} y = x - 2 \\ y = \ln x \end{cases}$

13. $\begin{cases} y = x^3 \\ y = e^x \end{cases}$

14. $\begin{cases} y = x^2 \\ y = 2^x \end{cases}$

In Exercises 15–24, solve each linear system by graphing the system.

15. $\begin{cases} x - 5y = 5 \\ y = x - 3 \end{cases}$

16. $\begin{cases} 2y - x = 4 \\ 2x + y = 9 \end{cases}$

17. $\begin{cases} y = 2x - 1 \\ 2x + 8y = 19 \end{cases}$

18. $\begin{cases} 5y = 6x + 6 \\ 2x + 5y = 10 \end{cases}$

19. $\begin{cases} 2x + y = 0 \\ 4x + 2y = 13 \end{cases}$

20. $\begin{cases} 2x - 3y = 10 \\ -x + \frac{3}{2}y = 12 \end{cases}$

21. $\begin{cases} 2x + \frac{2}{3}y = 0 \\ 9x + 3y = 0 \end{cases}$

22. $\begin{cases} x - 4y = 0 \\ -\frac{5}{2}x + 2y = 0 \end{cases}$

23. $\begin{cases} -2x + \frac{8}{3}y = 10 \\ 3x - 4y = -15 \end{cases}$

24. $\begin{cases} \frac{5}{2}x - 3y = \frac{3}{2} \\ 10x - 12y = 6 \end{cases}$

B

In Exercises 25–40, use a graphing utility to solve each system. Approximate all noninteger solutions to the nearest 0.001.

25. $\begin{cases} x^2 + y^2 = 16 \\ 5y = 5x + 4 \end{cases}$

26. $\begin{cases} 3x^2 + 4y^2 = 36 \\ 2y = 4x - 9 \end{cases}$

27. $\begin{cases} y^2 = 2x \\ x + y = 6 \end{cases}$

28. $\begin{cases} y^2 = x + 3 \\ y - 2x = 1 \end{cases}$

29. $\begin{cases} y = x^3 - 5x + 2 \\ x^2 + y^2 = 25 \end{cases}$

30. $\begin{cases} y = x^3 - 2x^2 - 2x - 1 \\ x^2 + y^2 = 16 \end{cases}$

31. $\begin{cases} x^3 + y^2 = 12 \\ y = e^x - 1 \end{cases}$

32. $\begin{cases} x^5 + y^2 = 30 \\ y = 1.7^x \end{cases}$

33. $\begin{cases} x^4 + y^2 = 10 \\ y - x^2 = 4 \end{cases}$

34. $\begin{cases} 2x^4 + y^2 = 6 \\ 2y = x^2 - 7 \end{cases}$

35. $\begin{cases} 0.03a + 0.11b = 1.155 \\ 3.8a - 2.2b = 25.3 \end{cases}$

36. $\begin{cases} 1.61a + 0.28b = -4.62 \\ -0.3a + 0.4b = 3.8 \end{cases}$

37. $\begin{cases} 0.24u^2 + 1.8v = 15.6 \\ 3.1u^2 - 7.6v = 4.9u \end{cases}$

38. $\begin{cases} 0.16t^2 + w = 1.28t + 2.5 \\ w = (1.05)^{4t} \end{cases}$

39. $\begin{cases} A = 1000(1.06)^t \\ A = \dfrac{100(1.06^{t+1} - 1)}{0.06} \end{cases}$

40.
$$\begin{cases} A = 2000\left(1 + \dfrac{r}{100}\right)^5 \\[4mm] A = \dfrac{400\left[\left(1 + \dfrac{r}{100}\right)^6 - 1\right]}{\dfrac{r}{100}} \end{cases}$$

41. A parametric representation of the solution to

$$\begin{cases} 9x + 4y = 12 \\ 6x + \frac{8}{3}y = 8 \end{cases}$$

is $(4t, 3 - 9t)$. Generate three solutions.

42. A parametric representation of the solution to

$$\begin{cases} -7x + 2y = 6 \\ \frac{1}{2}x - \frac{1}{7}y = -\frac{3}{7} \end{cases}$$

is $(2t, 7t + 3)$. Generate three solutions.

In Exercises 43–46, use a graphing utility to determine the equilibrium price (to the nearest $0.01) for each pair of supply and demand curves.

43. Supply: $p = 45 + 0.05x$; Demand: $p = 92 - 0.1x$

44. Supply: $p = 12 + 0.005x$; Demand: $p = 35 - 0.001x$

45. Supply: $p = 0.01x^2 + 40$; Demand: $p = 65 - 0.5x$

46. Supply: $p = 0.005x^2 + 15$; Demand: $p = 80 - 0.01x^2$

47. A new type of rowing machine is introduced into a test market. After 1 month, 40 are sold at a price of $150, and 24 are sold at $230. The manufacturer will supply only 30 rowing machines per month at a selling price of $150. At $230, 40 machines can be supplied per month. Assume that supply and demand can be modeled by linear equations.
   (a) Determine models for the demand and supply.
   (b) Use the models to determine the equilibrium price (to the nearest $0.10).

48. A supermarket chain wants to sell its own brand of multi-vitamins. After several weeks of test marketing in two of the supermarkets, 710 bottles are sold per week at a price of $3.50, and 520 bottles are sold per week at a price of $4.00. At a selling price of $3.50, the supplier will furnish only 580 bottles per week. At $4.00, the supplier agrees to supply 650 bottles per week. Assume that supply and demand can be modeled by linear equations.
   (a) Determine models for the demand and supply.
   (b) Use the models to determine the equilibrium price (to the nearest $0.10).

49. Consider an investment of $2500 paying a simple interest rate of 9% and an investment of $2500 invested at 5.5% compounded daily (use $n = 365$).
   (a) When will the accumulated amounts for each investment be the same?
   (b) If you plan to invest for 10 years or less, which of the two situations pays more?

50. Suppose you want to deposit $4000 into an account at your credit union. The credit union offers two options: 9% compounded daily or a simple interest rate of 12%.
   (a) When will the accumulated amounts for each investment be the same?
   (b) If you plan to close the account in less than 6 years, should you invest at 12% simple interest or 9% compounded daily?

C

51. A jet plane takes 5 hours to fly 2700 miles from San Francisco to Boston. The return flight takes 6 hours. The jet flies at a constant airspeed and assume that the wind blows at a constant speed. What is the airspeed of the jet and the speed of the wind?

52. An open-topped box with a square base has a volume of 800 cubic inches and a surface area of 550 square inches. Determine the dimensions of the box (to the nearest 0.1 inch).

53. A container is to be designed so that it is a right circular cylinder with volume 58 cubic inches and surface area 100 square inches. Determine the height and radius of the container (to the nearest 0.1 inch).

54. Use a graphing utility to solve the system

$$\begin{cases} \dfrac{5}{x} + \dfrac{2}{y} = 3 \\[2mm] \dfrac{2}{x} - \dfrac{1}{y} = \dfrac{3}{10} \end{cases}$$

55. (a) Determine the value of $b > 1$ such that the system

$$\begin{cases} y = b^x \\ y = \log_b x \end{cases}$$

has exactly one solution.

(b) What is the solution to this system?

- Elimination Method
- Substitution Method
- Applications

LINEAR SYSTEMS IN TWO VARIABLES ————————————

In Section 5.1 we solved linear and nonlinear systems of equations using a graphing utility. A graphical approach is the best way to solve most nonlinear systems in two variables. However, any linear system can be solved without graphing. In this section we discuss two algebraic methods for solving linear systems: the *elimination method* and the *substitution method*.

Elimination Method

Loosely speaking, the strategy of the elimination method is to combine the equations of a system so that one of the variables is eliminated.

● EXAMPLE 1   The Elimination Method

Solve the system using the elimination method.

$$\begin{cases} 2x + 3y = 8 \\ 4x - 9y = 1 \end{cases}$$

SOLUTION

If we multiply each side of the first equation by 3, the coefficients of the y terms will be opposites:

$$\begin{cases} 2x + 3y = 8 \\ 4x - 9y = 1 \end{cases} \quad \xrightarrow[\text{No change}]{\text{Multiply each side by 3}} \quad \begin{cases} 6x + 9y = 24 \\ 4x - 9y = 1 \end{cases}$$

Now if we add the two equations in the resulting system, we obtain an equation in one variable:

$$\begin{cases} 6x + 9y = 24 \\ 4x - 9y = \;\;1 \end{cases}$$

$$\overline{10x = 25}$$

$$x = \tfrac{5}{2}$$

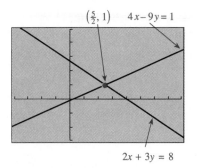

$\left(\frac{5}{2}, 1\right)$ $4x - 9y = 1$

$2x + 3y = 8$

Figure 5.18
$[-4, 8] \times [-3, 5]$

We can substitute $x = \frac{5}{2}$ into any of the equations containing both the variables x and y. Substituting $x = \frac{5}{2}$ into the first equation of the original system and solving for y, we get

$$2\left(\tfrac{5}{2}\right) + 3y = 8$$
$$5 + 3y = 8$$
$$3y = 3$$
$$y = 1$$

The solution to the system is $\left(\frac{5}{2}, 1\right)$. A graph of the system verifies our solution (Figure 5.18).

The method of elimination can be outlined in four steps:

Method of Elimination

Step 1: Write each equation in the general form $Ax + By = C$.

Step 2: Multiply one or both equations by appropriate numbers so that the coefficients of one of the variables are opposites.

Step 3: Add the resulting equations to eliminate one variable, and solve the resulting equation for the remaining variable.

Step 4: Substitute the value obtained in step 3 into either equation in the original system, and solve this equation for the second variable.

Recall from the previous section that for a linear system of two equations in two variables there are three possible outcomes: exactly one solution, an infinite number of solutions, or no solutions. We must keep these possibilities in mind when solving a linear system, regardless of the method used.

● **EXAMPLE 2   Solving Inconsistent or Dependent Systems by the Elimination Method**

Solve each system using the elimination method.

(a) $\begin{cases} -2x + 4y = 3 \\ 3x = 6y \end{cases}$ (b) $\begin{cases} -x + 4y = 3 \\ 2x - 8y = -6 \end{cases}$

SOLUTION

(a) First, we write each equation in the form $Ax + By = C$:

$$\begin{cases} -2x + 4y = 3 \\ 3x - 6y = 0 \end{cases}$$

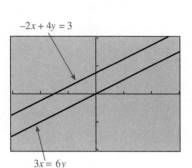

$-2x + 4y = 3$

$3x = 6y$

Figure 5.19
$[-3, 3] \times [-2, 2]$

The coefficients of the x terms will be opposites if we multiply each side of the first equation by 3 and each side of the second equation by 2:

$$\begin{cases} -2x + 4y = 3 \\ 3x - 6y = 0 \end{cases} \quad \xrightarrow{\text{Multiply each side by 3}} \quad \begin{cases} -6x + 12y = 9 \\ 6x - 12y = 0 \end{cases}$$

Adding the two equations in the resulting system, we get

$$\begin{cases} -6x + 12y = 9 \\ \underline{6x - 12y = 0} \\ \quad 0 = 9 \quad \text{False} \end{cases}$$

There are no values for x and y that make the equation $0 = 9$ a true statement. The system is inconsistent. To check our conclusion, a graph of the system is shown in Figure 5.19.

(b) Each equation in the system is in the form $Ax + By = C$. Multiplying each side of the first equation by 2 gives

$$\begin{cases} -x + 4y = 3 \\ 2x - 8y = -6 \end{cases} \quad \xrightarrow{\text{Multiply each side by 2}} \quad \begin{cases} -2x + 8y = 6 \\ 2x - 8y = -6 \end{cases}$$

Adding, we get

$$\begin{cases} -2x + 8y = 6 \\ \underline{2x - 8y = -6} \\ \quad 0 = 0 \end{cases}$$

The equation $0 = 0$ is certainly true, but what does it imply about the original system? Consider the graph of the original system. If we write each equation in slope–intercept form, the result in both cases is

$$y = \tfrac{1}{4}x + \tfrac{3}{4}$$

The graph of each equation is the same line; the system is dependent. A graph of the system (Figure 5.20) verifies our result. To determine a parametric representation of the solutions, we can replace x or y with t (where t is any real number). Substituting $y = t$ in either of the equations and solving for x gives

$$x = 4t - 3$$

Thus, the solutions (x, y) can be described parametrically as $(4t - 3, t)$.

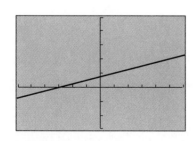

Figure 5.20
$\begin{cases} -x + 4y = 3 \\ 2x - 8y = -6 \end{cases}$
$[-6, 6] \times [-3, 5]$

Substitution Method

Another method for solving a system of equations is to use one of the equations to express a variable in terms of the other variable, and then substitute the resulting expression in the other equation.

● EXAMPLE 3   The Substitution Method

Solve the system using the substitution method: $\begin{cases} x + 2y = 1 \\ 5x - 8y = 2 \end{cases}$

SOLUTION

We can express x in terms of y by solving for x in the first equation:

$$x + 2y = 1$$
$$x = 1 - 2y$$

Substituting the expression $1 - 2y$ for x in the second equation of the system, we have

$$5(1 - 2y) - 8y = 2$$

Solving this equation for y, we get

$$5 - 18y = 2$$
$$-18y = -3$$
$$y = \tfrac{1}{6}$$

If we substitute $\tfrac{1}{6}$ for y into the equation

$$x = 1 - 2y$$

we get

$$x = 1 - 2(\tfrac{1}{6}) = \tfrac{2}{3}$$

The solution to the system is $(\tfrac{2}{3}, \tfrac{1}{6})$. A graph of the system serves as a check (Figure 5.21). ●

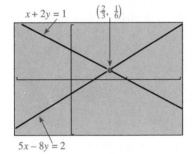

Figure 5.21
$[-1, 2] \times [-1, 1]$

The substitution method can be summarized in three steps:

Method of Substitution

Step 1: Solve one of the equations for one variable in terms of the other variable.

Step 2: Substitute the resulting expression into the other equation to get an equation in one variable. Solve this equation.

Step 3: Substitute the solution obtained in step 2 into the equation obtained in step 1. Solve the resulting equation for the second variable.

● EXAMPLE 4   Solving Inconsistent or Dependent Systems by Substitution

Solve each system using the substitution method.

(a) $\begin{cases} 9x - 3y = 8 \\ -6x + 2y = 1 \end{cases}$ **(b)** $\begin{cases} y = \frac{2}{3}x - 1 \\ 2x - 3y = 3 \end{cases}$

SOLUTION

(a) Our first step is to solve one of the equations for one variable in terms of the other variable. Solving for y in the second equation, we get

$$-6x + 2y = 1$$
$$2y = 1 + 6x$$
$$y = \tfrac{1}{2} + 3x$$

Substituting $\frac{1}{2} + 3x$ for y in the first equation and solving the resulting equation gives

$$9x - 3(\tfrac{1}{2} + 3x) = 8$$
$$9x - \tfrac{3}{2} - 9x = 8$$
$$-\tfrac{3}{2} = 8 \quad \text{False}$$

There are no values for x and y that make the equation $-\frac{3}{2} = 8$ a true statement. The system is inconsistent and has no solution. To check our conclusion, a graph of the system is shown in Figure 5.22.

(b) The first equation in the system, $y = \frac{2}{3}x - 1$, expresses y in terms of x. Substituting $\frac{2}{3}x - 1$ for y in the second equation, we get

$$2x - 3(\tfrac{2}{3}x - 1) = 3$$
$$2x - 2x + 3 = 3$$
$$3 = 3$$

The equation $3 = 3$ implies that the system is dependent. In fact, we solved this system in Example 4c in Section 5.1, and determined the solutions parametrically as $\left(t, \frac{2}{3}t - 1\right)$.

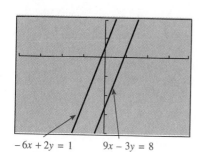

$-6x + 2y = 1 \qquad 9x - 3y = 8$

Figure 5.22
$[-4, 4] \times [-4, 2]$

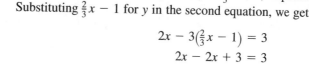

Applications

In Section 5.1, we considered applications that required solving linear as well as nonlinear systems. In this section we have limited our discussion to linear systems of two equations in two variables. Linear systems arise naturally as models for many real-world problems. There are also times when we assume that a linear system is a satisfactory model for a particular situation (as in Example 6 of Section 5.1).

We finish this section with two more applications of linear systems. The first is modeled naturally with a linear system; the second assumes that the system is linear.

● EXAMPLE 5 Application: Sugar Mixture in Food Processing

A food processing plant receives an order for 500 gallons of syrup that is 45% sugar. The plant has supplies of brand A syrup, which is 30% sugar, and brand B syrup, which is 80% sugar. How many gallons of each should be used to fill the order?

SOLUTION

Since the order consists of brand A and brand B syrup, we have

$$\left(\begin{array}{c}\text{Gallons of}\\\text{syrup of brand } A\end{array}\right) + \left(\begin{array}{c}\text{Gallons of}\\\text{syrup of brand } B\end{array}\right) = \left(\begin{array}{c}\text{Gallons of}\\\text{syrup ordered}\end{array}\right)$$

The order calls for 500 gallons. If we let a represent the number of gallons of syrup that is 30% sugar, and let b represent the number of gallons of syrup that is 80% sugar, we get

$$a + b = 500$$

Now consider the amount of sugar in each syrup:

$$\left(\begin{array}{c}\text{Gallons of sugar}\\\text{from brand } A\end{array}\right) + \left(\begin{array}{c}\text{Gallons of sugar}\\\text{from brand } B\end{array}\right) = \left(\begin{array}{c}\text{Gallons of sugar}\\\text{in syrup ordered}\end{array}\right)$$

$$\downarrow \qquad\qquad \downarrow \qquad\qquad \downarrow$$

$$0.30\left(\begin{array}{c}\text{Gallons of}\\\text{brand } A\end{array}\right) + 0.80\left(\begin{array}{c}\text{Gallons of}\\\text{brand } B\end{array}\right) = 0.45\left(\begin{array}{c}\text{Gallons of}\\\text{syrup needed}\end{array}\right)$$

or,

$$0.30a + 0.80b = 0.45(500)$$

Therefore, we have the system

$$\begin{cases} a + b = 500 \\ 0.30a + 0.80b = 225 \end{cases}$$

We will solve the system using the elimination method. Multiplying the first equation by -0.30 and then adding, we get

$$\begin{cases} -0.30a - 0.30b = -150 \\ 0.30a + 0.80b = 225 \end{cases}$$
$$\overline{ 0.50b = 75}$$
$$b = 150$$

Substituting 150 for b into either equation of the system and solving gives

$$a = 350$$

The plant must mix 350 gallons of brand A with 150 gallons of brand B to fill the order.

In Section 1.5 we determined two mathematical models for a business that manufactures a product selling for p dollars per unit. The revenue function R and the cost function C are defined as follows:

Revenue function: $R(x) = xp = $ Amount collected from selling x units

Cost function: $C(x) = $ Cost of producing and selling x units

If we assume the cost function is linear, the cost of producing x units can be modeled by

$$C(x) = mx + b$$

where m represents the variable cost per unit and b represents the fixed cost.

In order for the business to make a profit, the revenue must exceed the cost. The *break-even point* is the level of production for which revenue equals cost. In other words, the break-even point is the value for x such that

$$R(x) = C(x)$$

EXAMPLE 6   Application: Break-Even Analysis

A winery invests $100,000 in equipment to produce sparkling wine. Each case costs $220 to produce and will be sold at $310. Determine the break-even point.

SOLUTION

The cost $C(x)$ of producing x units is

$$C(x) = (\text{Cost per unit})x + (\text{Fixed cost})$$
$$= 220x + 100,000$$

The revenue $R(x)$ is

$$R(x) = (\text{Selling price per unit})(\text{Number of units})$$
$$= 310x$$

The break-even point is illustrated in Figure 5.23 (page 438).

■ **GRAPHING NOTE:** *Figure 5.23 shows that a loss occurs when sales are less than the break-even point, and that sales greater than the break-even point correspond to a profit.*

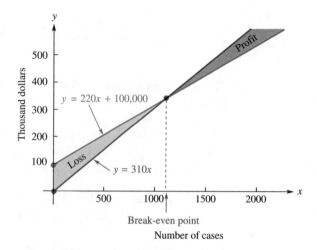

Figure 5.23

Although we could locate this point using the zoom and trace features on a graphing utility, we can easily locate it algebraically. Setting $R(x) = C(x)$, we can solve for x:

$$310x = 220x + 100{,}000$$
$$90x = 100{,}000$$
$$x = \frac{100{,}000}{90} \approx 1111$$

The break-even point is 1111 cases.

■ EXERCISE SET 5.2

A

In Exercises 1–12, solve each system of equations using the elimination method.

1. $\begin{cases} 4x - 3y = 3 \\ x + 3y = 27 \end{cases}$

2. $\begin{cases} -2x + y = 0 \\ 2x + 3y = 16 \end{cases}$

3. $\begin{cases} 5x - y = 12 \\ 8x + 7y = 2 \end{cases}$

4. $\begin{cases} x + 4y = 11 \\ 7x + 3y = 2 \end{cases}$

5. $\begin{cases} -3x + 4y = 20 \\ 2x + 3y = 15 \end{cases}$

6. $\begin{cases} 4x + 5y = 18 \\ -6x + 2y = 11 \end{cases}$

7. $\begin{cases} 5x - 9y = 4 \\ 2x - 3y = 2 \end{cases}$

8. $\begin{cases} -3x + 8y = 9 \\ -7x + 4y = 10 \end{cases}$

9. $\begin{cases} 4x - 10y = 3 \\ -6x + 15y = 10 \end{cases}$

10. $\begin{cases} 8x + 2y = 11 \\ -12x - 3y = 5 \end{cases}$

11. $\begin{cases} -6x + 10y = 4 \\ 9x - 15y = -6 \end{cases}$

12. $\begin{cases} 2x - 4y = -8 \\ -3x + 6y = 12 \end{cases}$

In Exercises 13–22, solve each system of equations using the substitution method.

13. $\begin{cases} x = 2 - 2y \\ 3x + 2y = -6 \end{cases}$

14. $\begin{cases} y = 3 - 4x \\ 7x + 2y = 4 \end{cases}$

15. $\begin{cases} 2x - y = 1 \\ -3x + 2y = -4 \end{cases}$

16. $\begin{cases} -x + 3y = 14 \\ 9x + 5y = 2 \end{cases}$

17. $\begin{cases} 10x - 3y = -13 \\ 14x + 2y = 19 \end{cases}$

18. $\begin{cases} 8x - 3y = -3 \\ -6x + 2y = 1 \end{cases}$

19. $\begin{cases} -6x + 9y = 10 \\ 8x - 12y = 5 \end{cases}$

20. $\begin{cases} 15x - 9y = 5 \\ -10x + 6y = 3 \end{cases}$

21. $\begin{cases} 3y = 3 - 4x \\ 8x + 6y = 6 \end{cases}$

22. $\begin{cases} 4y = 10 + 6x \\ 3x - 2y = -5 \end{cases}$

B

In Exercises 23–32, solve each system of equations using either the substitution or elimination method.

23. $\begin{cases} 4x - 10y = 0 \\ -6x + 15y = 0 \end{cases}$

24. $\begin{cases} 8x + 2y = 0 \\ -12x - 3y = 0 \end{cases}$

25. $\begin{cases} -x + 6y = \frac{3}{2} \\ \frac{2}{3}x - y = 0 \end{cases}$

26. $\begin{cases} 2x + y = 0 \\ \frac{4}{3}x + y = \frac{1}{2} \end{cases}$

27. $\begin{cases} \frac{1}{6}x + \frac{5}{4}y = 0 \\ x + 2 = -\frac{5}{2}y \end{cases}$

28. $\begin{cases} \frac{2}{5}x + \frac{1}{6}y = 0 \\ y + 3 = \frac{6}{5}x \end{cases}$

29. $\begin{cases} 0.5x + 0.7y = 12.6 \\ x + y = 22 \end{cases}$

30. $\begin{cases} 0.4x + 0.7y = 7.1 \\ x + y = 11 \end{cases}$

31. $\begin{cases} 0.2x + 0.6y = 4.6 \\ x + 3y = 23 \end{cases}$

32. $\begin{cases} 0.4x + 0.5y = 7.2 \\ 8x + 2y = 144 \end{cases}$

33. The solutions to the system

$$\begin{cases} -x + 4y = 3 \\ 2x - 8y = -6 \end{cases}$$

in part (b) of Example 2 are given as $(4t - 3, t)$. Another parametric representation of the solutions is $\left(T, \frac{1}{4}T + \frac{3}{4}\right)$.
   (a) Let $t = 0$, 1, and 2 in $(4t - 3, t)$ to generate three solutions.
   (b) Let $T = -3$, 1, and 5 in $\left(T, \frac{1}{4}T + \frac{3}{4}\right)$ to generate three solutions.

34. The solutions to the system

$$\begin{cases} x + 3y = 6 \\ \frac{2}{3}x + 2y = 4 \end{cases}$$

are $(6 - 3t, t)$. Another parametric representation of the solutions is $(3s, 2 - s)$.
   (a) Let $t = 0$, 1, and $\frac{1}{3}$ in $(6 - 3t, t)$ to generate three solutions.

   (b) Let $s = 2$, 1, and $\frac{5}{3}$ in $(3s, 2 - s)$ to generate three solutions.

35. A refinery receives an order for 1400 gallons of gasoline that is 89% isooctane. The refinery has a supply of gasoline that is 87% isooctane and a supply of gasoline that is 92% isooctane. How many gallons of each should be used to fill the order?

36. A goldsmith receives an order from a jeweler that requires 8 grams of 60% gold alloy. The goldsmith has a supply of 45% gold alloy and a supply of 65% gold alloy. How many grams of each should be melted together to fill the order?

37. A chemist needs 40 liters of a 35% hydrochloric acid solution. The stockroom has a supply of 20% hydrochloric acid solution and 60% hydrochloric acid solution. How many liters of each should the chemist mix to obtain the 35% solution needed?

38. A 6 quart radiator is filled with a 40% antifreeze solution. How many quarts need to be replaced with pure (100%) antifreeze so that the radiator will have a 50% antifreeze solution?

39. Refer to Exercise 38. Suppose that instead of a 50% antifreeze solution, we want the radiator to have a 60% solution. How many quarts need to be replaced with pure (100%) antifreeze?

40. A fitness expert invests $16,000 in studio equipment to produce a series of exercise videos. Production costs will be $14 per video cassette, and each cassette will be sold for $29. Determine the break-even point.

41. A communications company plans to produce and market a new type of cellular phone. An investment of $80,000 is required to begin production. Each phone costs $160 to produce and will be sold for $210. Determine the break-even point.

42. One car rental agency charges $29 a day plus $0.26 per mile. Another rental agency charges $34 a day plus $0.19 per mile. For a 1 day rental, how many miles must be driven for the costs to be equal?

43. A new breakfast cereal combines crispy rice and wheat flakes to provide 3 grams of protein per cup. A cup of crispy rice cereal provides 2 grams of protein, and a cup of wheat flakes provides 4.5 grams of protein. How much of each cereal is included in 1 cup of the mix?

44. A coffee company decides to blend two coffees: Kona ($3.40 per pound) and Colombian ($2.60 per pound). The blend is to cost $2.90 per pound. How much of each coffee is used to make 20 pounds of the blend?

45. A horse breeder prepares the daily diet for her pygmy ponies, requiring 30 ounces of protein and 5.5 ounces of fat. The diet is a mix of two types of food; one is 20% protein and 4% fat, and the other is 15% protein and 2% fat. How many ounces of each type of food is included in the mix?

46. An individual invests a total of $15,000 in two stocks, part in a high-risk stock that pays a 10% dividend and the rest in a low-risk stock that pays a 6% dividend. If the total annual return on the investment is $1120, find the amount invested in each stock.

C

47. Find an equation of the form $y = ax^2 + bx$ whose graph passes through (2, 8) and (4, 20).

48. A projectile is fired upward at time $t = 0$ from the top of a tall platform. The distance s of the projectile above the ground (in feet) is related to the time t (in seconds) by the equation

$$s = a + bt - 16t^2$$

where a and b are constants. Suppose the projectile is 334 feet above ground 1 second after it is fired, and it is 586 feet above ground 2 seconds after it is fired.
(a) Determine the constants a and b.
(b) How high is the object 4 seconds after it is fired?
(c) When does the projectile hit the ground?

49. Solve the system:
$$\begin{cases} \dfrac{5}{x} + \dfrac{2}{y} = 3 \\ \dfrac{2}{x} - \dfrac{1}{y} = \dfrac{3}{10} \end{cases}$$

HINT: Even though this is not a linear system, you can solve it using the techniques of this section. Let $u = 1/x$ and $v = 1/y$. Solve the resulting linear system and then determine the values of x and y. (This is the same system as Exercise 54 in Section 5.1.)

50. Solve the system:
$$\begin{cases} x^2 + 2y^2 = 17 \\ x^2 - y^2 = 5 \end{cases}$$

HINT: See the hint in Exercise 49.

LINEAR SYSTEMS IN MORE THAN TWO VARIABLES

Many modeling situations lead to more than one equation with several variables. In this section we will investigate linear systems with three variables, such as

$$\begin{cases} x + y + 2z = 1 \\ -2x + 2y + z = 8 \\ -x - y + z = 5 \end{cases}$$

A *solution* to such a system is an *ordered triple* (x, y, z) that satisfies all three equations. A solution to the system above is $(-3, 0, 2)$, since it satisfies each equation in the system:

$$\begin{cases} (-3) + (0) + 2(2) \overset{?}{=} 1 \quad \text{True} \\ -2(-3) + 2(0) + (2) \overset{?}{=} 8 \quad \text{True} \\ -(-3) - (0) + (2) \overset{?}{=} 5 \quad \text{True} \end{cases}$$

Row-Echelon Form and Back-Substitution

The elimination method can be extended to systems containing three or more variables. Before we discuss the method, we investigate a special set of linear systems.

⬤ EXAMPLE 1   Back-Substitution

Solve the system of equations:
$$\begin{cases} x - 2y + 4z = 18 \\ \quad\ \ y - \ z = -5 \\ \qquad\qquad z = 2 \end{cases}$$

SOLUTION

We know the value of z is 2 from the last equation in the system. If we substitute 2 for z in the second equation, we can solve for y:

$$y - (2) = -5$$
$$y = -3$$

Finally, if we substitute $z = 2$ and $y = -3$ in the first equation, we can solve for x:

$$x - 2(-3) + 4(2) = 18$$
$$x + 14 = 18$$
$$x = 4$$

The solution to the system is $(4, -3, 2)$. You should check this solution in the original system. ⬤

The process used in Example 1 is called *back-substitution,* because we start with the last equation and work backwards toward the first equation by substitution.

The system in Example 1 was easy to solve because it is written in a special form called *row-echelon form.* Loosely speaking, row-echelon form means that the system is in a steplike formation, and the leading coefficient of each equation is 1.

Row-Echelon Form

> A linear system of equations in x, y, and z is in **row-echelon form** if it satisfies the following conditions:
>
> **1.** The variable x appears in no equation after the first, and y appears in no equation after the second.
> **2.** In the first equation, the coefficient of x is 1. In any succeeding equations, the coefficient of the first (leftmost) variable is 1.

The following systems are in row-echelon form:

$$\begin{cases} x - 3y + 2z = 13 \\ \quad\;\; y - 5z = 0 \\ \qquad\qquad z = 4 \end{cases} \qquad \begin{cases} x - 3y + 2z = 13 \\ \quad\;\; y - 5z = 0 \end{cases} \qquad \begin{cases} x + y + 3z = 11 \\ \qquad\qquad z = 5 \end{cases}$$

Two examples of systems that are *not* in row-echelon form are:

$$\begin{cases} \qquad\qquad z = -6 \\ x - 3y + 2z = 13 \\ \quad\;\; y - 5z = 0 \end{cases}$$ Violates condition 1; x appears in an equation after the first.

$$\begin{cases} x - 3y + 2z = 8 \\ \quad\; 2y - 3z = 5 \\ \qquad\qquad z = -1 \end{cases}$$ Violates condition 2. In the second equation, the coefficient of the first (leftmost) variable is 2.

Any system in row-echelon form can be solved by the method of back-substitution used in Example 1.

Operations for Equivalent Systems

Now we consider solving a general linear system. Our strategy is to convert a given system into an equivalent system that is in row-echelon form. (Two systems are equivalent if they have the same solution.) To do this, we will use the three basic operations, called *elementary operations*, listed below.

Elementary Operations for Systems

> Each of the **elementary operations** on a system of equations yields an equivalent system of equations:
>
> **1.** Interchange any two equations.
> **2.** Replace an equation with a nonzero multiple of itself. (A **multiple** of an equation is the result of multiplying each side by the same real number.)
> **3.** Replace an equation with the sum of that equation and a multiple of another equation of the system.

● EXAMPLE 2   Converting into Row-Echelon Form

Solve the system of equations

$$\begin{cases} 2x - 3y + 7z = 31 \\ \;\; x - 2y + 4z = 18 \\ \;\; x - \;\; y + 2z = 11 \end{cases}$$

by finding an equivalent system that is in row-echelon form and then back-substituting.

SOLUTION

To convert this system into row-echelon form, we start with the upper left corner. To obtain a coefficient of 1 for x in the first equation, we exchange the first and second equations:

$$\begin{cases} x - 2y + 4z = 18 \\ 2x - 3y + 7z = 31 \\ x - y + 2z = 11 \end{cases}$$ Exchange the first and second equations.

We will use the top equation to eliminate the x terms below it:

$$\begin{cases} x - 2y + 4z = 18 \\ y - z = -5 \\ x - y + 2z = 11 \end{cases}$$ Replace the second equation with the sum of the second equation and -2 times the first equation.

$$\begin{cases} x - 2y + 4z = 18 \\ y - z = -5 \\ y - 2z = -7 \end{cases}$$ Replace the third equation with the sum of the third equation and -1 times the first equation.

Next, consider the y terms. We eliminate the y term in the third equation using the second equation:

$$\begin{cases} x - 2y + 4z = 18 \\ y - z = -5 \\ -1z = -2 \end{cases}$$ Replace the third equation with the sum of the third equation and -1 times the second equation.

Finally, the system will be in row-echelon form once the coefficient for z in the third equation is 1:

$$\begin{cases} x - 2y + 4z = 18 \\ y - z = -5 \\ z = 2 \end{cases}$$ Replace the third equation with -1 times the third equation.

This is the system that we solved in Example 1; the solution is $(4, -3, 2)$. ■

It is convenient to use an abbreviated notation for the elementary operations. The first, second, and third equations of a system are represented by E_1, E_2, and E_3, respectively. An example of this notation for each of the three elementary operations follows.

Notation	*Explanation*
$E_3 \rightarrow E_1$	Interchange the first and third equations.
$E_1 \rightarrow E_3$	
$-4E_3 \rightarrow E_3$	Replace the third equation with -4 times the third equation.
$5E_1 + E_2 \rightarrow E_2$	Replace the second equation with the sum of 5 times the first equation and the second equation.

EXAMPLE 3   Using Elementary Operation Notation

Solve the system of equations

$$\begin{cases} 5x + 6y + 4z = 10 \\ \tfrac{1}{2}x + y + 2z = -2 \\ 3x - 4y + z = 5 \end{cases}$$

by finding an equivalent system that is in row-echelon form and then back-substituting. Use the notation for elementary operations to show the steps.

SOLUTION

We eliminate fractional coefficients by multiplying the second equation by 2:

NOTE: *The notation is written next to the equation that is being replaced.*

$$2E_2 \rightarrow E_2 \quad \begin{cases} 5x + 6y + 4z = 10 \\ x + 2y + 4z = -4 \\ 3x - 4y + z = 5 \end{cases}$$

To obtain a coefficient of 1 for x in the first equation, we exchange the first and second equations:

$$\begin{aligned} E_2 \rightarrow E_1 \\ E_1 \rightarrow E_2 \end{aligned} \quad \begin{cases} x + 2y + 4z = -4 \\ 5x + 6y + 4z = 10 \\ 3x - 4y + z = 5 \end{cases}$$

Now we use the top equation to eliminate the x terms below it (in the first column):

$$\begin{aligned} -5E_1 + E_2 \rightarrow E_2 \\ -3E_1 + E_3 \rightarrow E_3 \end{aligned} \quad \begin{cases} x + 2y + 4z = -4 \\ -4y - 16z = 30 \\ -10y - 11z = 17 \end{cases}$$

Next, we consider the y terms; the y term in the second equation needs to be 1:

$$-\tfrac{1}{4}E_2 \rightarrow E_2 \quad \begin{cases} x + 2y + 4z = -4 \\ \quad\;\; y + 4z = -\tfrac{15}{2} \\ -10y - 11z = 17 \end{cases}$$

Using the second equation to eliminate the y term in the third equation, we get

$$10E_2 + E_3 \rightarrow E_3 \quad \begin{cases} x + 2y + 4z = -4 \\ \quad\;\; y + 4z = -\tfrac{15}{2} \\ \quad\quad 29z = -58 \end{cases}$$

Finally, the system will be in row-echelon form once the coefficient for z in the third equation is 1:

$$\tfrac{1}{29}E_3 \rightarrow E_3 \quad \begin{cases} x + 2y + 4z = -4 \\ \quad\;\; y + 4z = -\tfrac{15}{2} \\ \quad\quad\quad z = -2 \end{cases}$$

Back-substituting $z = -2$ into the second equation and solving for y gives

$$y + 4(-2) = -\tfrac{15}{2}$$
$$y = \tfrac{1}{2}$$

Back-substituting $z = -2$ and $y = \tfrac{1}{2}$ into the first equation and solving for x, we have

$$x + 2\left(\tfrac{1}{2}\right) + 4(-2) = -4$$
$$x = 3$$

The solution is $\left(3, \tfrac{1}{2}, -2\right)$. The check is left to you.

We know that a linear system of two equations in two variables has either no solutions, exactly one solution, or an infinite number of solutions. This is true for a linear system containing three or more variables.

The Nature of Solutions
for Linear Systems

A linear system of equations in three or more variables can have one of three possible outcomes:

1. The system is (**consistent** and) **independent**—there is exactly one solution.
2. The system is (**consistent** and) **dependent**—there are an infinite number of solutions.
3. The system is **inconsistent**—there are no solutions.

So far, each system in this section has had exactly one solution. The next two examples deal with the other two possibilities.

EXAMPLE 4   A System with an Infinite Number of Solutions

Solve the system of equations

$$\begin{cases} x + y + 2z = 3 \\ 2x + y + z = 5 \\ x - z = 2 \end{cases}$$

by finding an equivalent system that is in row-echelon form and then back-substituting.

SOLUTION

We proceed to determine an equivalent system in row-echelon form.

$$\begin{array}{c} -2E_1 + E_2 \rightarrow E_2 \\ -1E_1 + E_3 \rightarrow E_3 \end{array} \quad \begin{cases} x + y + 2z = 3 \\ -y - 3z = -1 \\ -y - 3z = -1 \end{cases}$$

$$-1E_2 \rightarrow E_2 \quad \begin{cases} x + y + 2z = 3 \\ y + 3z = 1 \\ -y - 3z = -1 \end{cases}$$

$$E_2 + E_3 \rightarrow E_3 \quad \begin{cases} x + y + 2z = 3 \\ y + 3z = 1 \\ 0 = 0 \end{cases}$$

The last equation, $0 = 0$, is true for any values of x, y, and z. The system is dependent.

To determine a parametric representation, we let $z = t$ in the second equation and solve for y:

$$y + 3t = 1$$
$$y = 1 - 3t$$

Now we back-substitute $z = t$ and $y = 1 - 3t$ into the first equation:

$$x + (1 - 3t) + 2t = 3$$
$$x + 1 - t = 3$$
$$x = 2 + t$$

The solutions can be represented by $(2 + t, 1 - 3t, t)$.

NOTE: *Recall seeing the situation $0 = 0$ in Example 2b of Section 5.2*

● EXAMPLE 5 A System with No Solutions

Solve the system of equations

$$\begin{cases} x + 3y - z = 5 \\ 2x + 4y - 5z = 8 \\ 3x + 9y - 3z = 20 \end{cases}$$

by finding an equivalent system that is in row-echelon form and then back-substituting.

SOLUTION

NOTE: *Since the first system in the solution has the contradiction* $0 = 5$, *we know the system has no solutions. We continued only to write the system in row-echelon form.*

Using the elementary operations, we have

$$\begin{matrix} -2E_1 + E_2 \rightarrow E_2 \\ -3E_1 + E_3 \rightarrow E_3 \end{matrix} \quad \begin{cases} x + 3y - z = 5 \\ -2y - 3z = -2 \\ 0 = 5 \end{cases}$$

$$\begin{matrix} -\frac{1}{2}E_2 \rightarrow E_2 \\ \frac{1}{5}E_3 \rightarrow E_3 \end{matrix} \quad \begin{cases} x + 3y - z = 5 \\ y + \frac{3}{2}z = 1 \\ 0 = 1 \end{cases}$$

Regardless of the values we might choose for x, y, and z, the last equation, $0 = 1$, is false. The system is inconsistent. ●

Underdetermined Systems

A system with fewer equations than variables is called an *underdetermined* system. An underdetermined linear system cannot have a unique solution; it is always dependent or inconsistent.

● EXAMPLE 6 An Underdetermined System

Solve the system of equations

$$\begin{cases} -4x - y - 2z = 7 \\ x + 2y - 3z = 0 \end{cases}$$

by finding an equivalent system that is in row-echelon form and then back-substituting.

SOLUTION

We proceed to determine an equivalent system in row-echelon form.

$$\begin{matrix} E_2 \rightarrow E_1 \\ E_1 \rightarrow E_2 \end{matrix} \quad \begin{cases} x + 2y - 3z = 0 \\ -4x - y - 2z = 7 \end{cases}$$

$$4E_1 + E_2 \rightarrow E_2 \quad \begin{cases} x + 2y - 3z = 0 \\ \qquad 7y - 14z = 7 \end{cases}$$

$$\tfrac{1}{7}E_2 \rightarrow E_2 \quad \begin{cases} x + 2y - 3z = 0 \\ \qquad y - 2z = 1 \end{cases}$$

This system is in row-echelon form. Replacing z with t in the second equation and solving for y, we have

$$y - 2t = 1$$
$$y = 1 + 2t$$

Now we back-substitute $z = t$ and $y = 1 + 2t$ in the first equation:

$$x + 2(1 + 2t) - 3t = 0$$
$$x = -2 - t$$

The solutions can be represented parametrically by $(-2 - t, 1 + 2t, t)$.

Applications

Given two points in the coordinate plane, we can determine a linear function whose graph (a line) contains them (see Section 2.1). If we are given three non-collinear points in the coordinate plane, we can determine a quadratic function whose graph (a parabola) contains them.

EXAMPLE 7 Application: Curve Fitting

Determine a quadratic function

$$f(x) = ax^2 + bx + c$$

whose graph passes through the points $(-2, 14)$, $(2, 2)$, and $(1, -1)$.

SOLUTION

Since the graph of f passes through the point $(-2, 14)$, we have

$$f(-2) = 14$$
$$a(-2)^2 + b(-2) + c = 14$$
$$4a - 2b + c = 14$$

Similarly, since the graph of f contains $(2, 2)$, and $(1, -1)$, we get

$$f(2) = 2 \qquad\qquad f(1) = -1$$
$$a(2)^2 + b(2) + c = 2 \qquad a(1)^2 + b(1) + c = -1$$
$$4a + 2b + c = 2 \qquad\qquad a + b + c = -1$$

We seek values of a, b, and c that satisfy the system

$$\begin{cases} 4a - 2b + c = 14 \\ 4a + 2b + c = 2 \\ a + b + c = -1 \end{cases}$$

To solve the system, we use the elementary operations to determine an equivalent system in row-echelon form:

$$\begin{matrix} E_3 \to E_1 \\ \\ E_1 \to E_3 \end{matrix} \quad \begin{cases} a + b + c = -1 \\ 4a + 2b + c = 2 \\ 4a - 2b + c = 14 \end{cases}$$

$$\begin{matrix} -4E_1 + E_2 \to E_2 \\ -4E_1 + E_3 \to E_3 \end{matrix} \quad \begin{cases} a + b + c = -1 \\ -2b - 3c = 6 \\ -6b - 3c = 18 \end{cases}$$

$$-\tfrac{1}{2}E_2 \to E_2 \quad \begin{cases} a + b + c = -1 \\ b + \tfrac{3}{2}c = -3 \\ -6b - 3c = 18 \end{cases}$$

$$6E_2 + E_3 \to E_3 \quad \begin{cases} a + b + c = -1 \\ b + \tfrac{3}{2}c = -3 \\ 6c = 0 \end{cases}$$

$$\tfrac{1}{6}E_3 \to E_3 \quad \begin{cases} a + b + c = -1 \\ b + \tfrac{3}{2}c = -3 \\ c = 0 \end{cases}$$

This system is in row-echelon form. Back-substitution reveals the solution to be $a = 2$, $b = -3$, and $c = 0$. Therefore, we find that the quadratic function $f(x) = ax^2 + bx + c$ passing through the three given points is

$$f(x) = 2x^2 - 3x$$

A graph of f is shown in Figure 5.24.

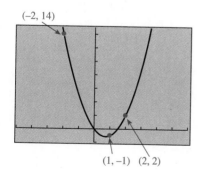

(−2, 14)

(1, −1) (2, 2)

Figure 5.24
$y = 2x^2 - 3x$
$[-5, 6] \times [-2, 15]_2$

■ EXERCISE SET 5.3

A

In Exercises 1–8, solve each system by back-substitution. Note that each system is in row-echelon form.

1. $\begin{cases} x + 4y = 9 \\ y = 3 \end{cases}$

2. $\begin{cases} x - 3y = 7 \\ y = -2 \end{cases}$

3. $\begin{cases} x + 2y - z = 11 \\ y + 2z = 11 \\ z = 3 \end{cases}$

4. $\begin{cases} x - 3y + z = 8 \\ y - 3z = 5 \\ z = 2 \end{cases}$

5. $\begin{cases} x - 2y - 4z = 4 \\ y + z = 1 \\ z = \frac{1}{3} \end{cases}$

6. $\begin{cases} x + 6y - z = 4 \\ y + z = 2 \\ z = \frac{3}{2} \end{cases}$

7. $\begin{cases} x + 2z = 4 \\ y + z = -1 \\ z = 2 \end{cases}$

8. $\begin{cases} x - 6z = 10 \\ y - z = -3 \\ z = 5 \end{cases}$

In Exercises 9–16, solve each system of equations by finding an equivalent system that is in row-echelon form and then back-substituting.

9. $\begin{cases} x + 2y - 3z = 5 \\ y + 3z = 5 \\ -2y + z = -3 \end{cases}$

10. $\begin{cases} x + y + z = 0 \\ y + 2z = 7 \\ -3y - 5z = 11 \end{cases}$

11. $\begin{cases} x + y + z = 7 \\ y - z = -1 \\ x + 2y + z = 9 \end{cases}$

12. $\begin{cases} x + y + z = 11 \\ y + 2z = 14 \\ x - 3y + 2z = 0 \end{cases}$

13. $\begin{cases} x - y + z = 4 \\ 2x + y - z = -1 \\ y + z = -1 \end{cases}$

14. $\begin{cases} x + y + 2z = 1 \\ 2x - y + z = 2 \\ y - 2z = 3 \end{cases}$

15. $\begin{cases} x + 2y - 2z = 8 \\ 3x - y + 5z = 6 \\ x + 2y - z = 7 \end{cases}$

16. $\begin{cases} x - y + 2z = -2 \\ 4x + y + 3z = 7 \\ x - y + z = 0 \end{cases}$

B

In Exercises 17–40, solve each system of equations by finding an equivalent system that is row-echelon form and then back-substituting. Use the notation for elementary operations introduced in this section to show your steps.

17. $\begin{cases} x + y + 2z = 1 \\ x + 3y - 2z = 7 \\ 2x - y + z = 2 \end{cases}$

18. $\begin{cases} x + y + 2z = 2 \\ 3x + 2y + z = 3 \\ 2x - 2y + z = 9 \end{cases}$

19. $\begin{cases} x + 2y + 3z = 3 \\ 2x + 3y + 10z = 18 \\ 3x + 9y + 10z = -1 \end{cases}$

20. $\begin{cases} x + 2y - z = 10 \\ 2x - y + 2z = -3 \\ 3x + 2y + 3z = 6 \end{cases}$

21. $\begin{cases} x + y + z = 4 \\ x - y - z = -2 \\ x + 2y - z = 1 \end{cases}$

22. $\begin{cases} x + y = -2 \\ 6x + y - z = -20 \\ 4x - 6y + z = 0 \end{cases}$

23. $\begin{cases} 2x - 3y + 2z = 4 \\ 2x + 4y + 5z = 7 \\ 3x + 2y + 4z = 7 \end{cases}$

24. $\begin{cases} 3x - 2y - 5z = 5 \\ 2x + 3y - 2z = -4 \\ 5x + 2y + 3z = 9 \end{cases}$

25. $\begin{cases} x + y + 3z = 0 \\ x - y + z = 0 \\ 2x - 3y + z = 0 \end{cases}$

26. $\begin{cases} x + y + z = 0 \\ x - y + 5z = 0 \\ 3x + 4y + z = 0 \end{cases}$

27. $\begin{cases} 2x - y - 3z = -1 \\ x + 2y - 4z = 7 \\ 3x - 5y - z = -12 \end{cases}$

28. $\begin{cases} 2x - y + 5z = 0 \\ -x + y - 4z = 2 \\ 5x - 2y + 11z = 2 \end{cases}$

29. $\begin{cases} x + 3y + z = 1 \\ x + 4y - z = 10 \\ 2x + 9y - 4z = 26 \end{cases}$

30. $\begin{cases} x + y = 6 \\ 2x + y + z = 5 \\ 3x + y + 2z = 6 \end{cases}$

31. $\begin{cases} x + y - 3z = 9 \\ 2x + 7y + 2z = 1 \end{cases}$

32. $\begin{cases} x + y + z = 1 \\ 5x + y + 4z = 0 \end{cases}$

33. $\begin{cases} -x + y + 2z = 0 \\ x - 2y + 2z = 2 \\ 3x + 2y - 2z = -1 \end{cases}$

34. $\begin{cases} 4x - y + 6z = -1 \\ -x + y = 1 \\ 3x + y - 2z = 2 \end{cases}$

35. $\begin{cases} 0.5x - 0.25y + z = 0 \\ 0.2x + 0.2y + 0.6z = 2.2 \\ 5.5x - 1.1y + 1.1z = 6.6 \end{cases}$

36. $\begin{cases} 0.1x + 0.1y - 0.1z = 0.4 \\ -2x + 0.5y + 0.5z = 2 \\ -0.5x + y - z = 2.2 \end{cases}$

37. $\begin{cases} 2x - y + \frac{4}{5}z = 1 \\ x + 5y + z = 3 \\ \frac{1}{5}x + y + z = 1 \end{cases}$ **38.** $\begin{cases} 2x + y + \frac{3}{2}z = 3 \\ x + 2y - 2z = -1 \\ \frac{1}{4}x - y + z = 1 \end{cases}$

39. $\begin{cases} x + 2y + z - w = -1 \\ 2x + z + 3w = 13 \\ -x + 4y - z + 2w = 8 \\ 3x + 2y + z + w = 5 \end{cases}$

40. $\begin{cases} x + y + z - 4w = 3 \\ 3x - y + z + w = -1 \\ 2x + y - 3w = 6 \\ x + z + w = 2 \end{cases}$

41. The solutions to the system in Example 4 are given as $(2 + t, 1 - 3t, t)$.
 (a) Let $t = 0$ and $t = 1$ in $(2 + t, 1 - 3t, t)$ to determine two of the solutions.
 (b) Another parametric representation of the solutions is $(T, 7 - 3T, T - 2)$. Let $T = 2$ and $T = 3$ in $(T, 7 - 3T, T - 2)$ to generate two solutions. Compare with the two solutions generated in part (a).
 (c) Use either $(2 + t, 1 - 3t, t)$ or $(T, 7 - 3T, T - 2)$ to generate three additional solutions to the system.

42. A parametric representation of the solution to

$$\begin{cases} 4x + 3y + 3z = 21 \\ 2x + 3y + 2z = 15 \\ 2x + 6y + 3z = 24 \end{cases}$$

is $\left(3 - \frac{1}{2}t, 3 - \frac{1}{3}t, t\right)$.
 (a) Let $t = 0$ and 6 in $\left(3 - \frac{1}{2}t, 3 - \frac{1}{3}t, t\right)$ to generate two solutions.
 (b) Another parametric representation of the solutions is $(3 - 3T, 3 - 2T, 6T)$. Let $T = 0$ and $T = 1$ in $(3 - 3T, 3 - 2T, 6T)$ to generate two solutions. Compare with the two solutions generated in part (a).
 (c) Use either the solution $\left(3 - \frac{1}{2}t, 3 - \frac{1}{3}t, t\right)$ or the solution $(3 - 3T, 3 - 2T, 6T)$ to generate four additional solutions.

43. Determine a quadratic function $f(x) = ax^2 + bx + c$ whose graph passes through the points $(1, 1)$, $(-1, 0)$, and $(3, -6)$.

44. Determine a quadratic function $f(x) = ax^2 + bx + c$

whose graph passes through the points $(-2, 4)$, $(2, 2)$ and $(4, -19)$.

45. Determine an equation of the form $x = ay^2 + by + c$ whose graph passes through the points $(-2, 1)$, $(-4, 2)$, and $(16, -2)$.

46. Determine an equation of the form $x = ay^2 + by + c$ whose graph passes through the points $(0, -1)$, $(-6, 1)$, and $(6, -2)$.

47. Determine an equation for a circle of the form $x^2 + y^2 + ax + by + c = 0$ whose graph passes through the points $(4, 2)$, $(-2, 2)$, and $(5, 1)$.

48. Determine an equation for a circle of the form $x^2 + y^2 + ax + by + c = 0$ whose graph passes through the points $(9, 10)$, $(-5, -4)$, and $(13, 2)$.

49. A total of $15,000 is divided among three investments, a certificate of deposit paying 6% annually, stocks paying 8% annually, and municipal bonds paying 9% annually. The total annual return on the investments is $1140, and the amount invested in the certificate of deposit is the same as the amount invested in municipal bonds. Find the amount placed in each of the three investments.

50. An inheritance of $24,000 is invested in three stocks that pay annual dividends of 5%, 7%, and 8%. The total annual return on the investments is $1730, and the amount invested at 7% is $1000 more than the amount invested at 5%. Find the amount invested in each stock.

51. A party mix is to be made of almonds, peanuts, and walnuts, which cost $2.10 per pound, $1.60 per pound, and $1.80 per pound, respectively. The total amount of mix is to weigh 20 pounds and cost $1.90 per pound. If the weight of the almonds in the mix must be twice that of the walnuts in the mix, find the amount of each type of nut to be mixed.

52. A furniture store sells three types of desk lamps for $65, $40, and $25 per lamp. Last year, four times as many $40 lamps were sold as $65 lamps, and the combined number of lamps sold was 334. If the total annual revenue for these three lamps was $14,550, find the number of each type of lamp that was sold.

53. A distribution center ships a total of 1200 cameras to three retail outlets, one in Los Angeles, one in New York, and one in Denver. The shipping cost per camera is $15 to Los Angeles, $18 to New York, and $12 to Denver. There are 200 more cameras shipped to New York than to Denver, and the total cost of shipping is $18,600. Find the number shipped to each outlet. If there is more than one way, give three examples.

54. A total of $30,000 is to be divided among three investments, a certificate of deposit paying 5% annually, stocks paying 6% annually, and municipal bonds paying 8% annually. The total annual return on the investments is $1841. Determine how much should be allocated to each type of investment. If there is more than one way, give three examples.

55. A coffee wholesaler blends three coffees, French roast ($2.20 per pound), Kona ($3.60 per pound), and Colombian ($2.80 per pound). The blend costs $108 for a 40 pound sack. Find the amount of each coffee to be blended

for each 40 pound sack. If there is more than one way, give three examples.

C

56. A water tank has two inlet pipes and a drain. The two inlet pipes working together can fill a tank with the drain closed in 18 minutes. With the smaller inlet pipe closed and the drain open, the larger inlet pipe can fill the tank in 60 minutes. With the larger inlet pipe closed and the drain open, the smaller inlet pipe can fill the tank in 180 minutes. How long does it take the open drain to empty the tank if both inlet pipes are closed?

57. Solve: $\begin{cases} yz + xz + xy = 6xyz \\ yz + xz - xy = 3xyz \\ 3yz - 2xz - 4xy = 5xyz \end{cases}$

SECTION 5.4

• Matrices
• Elementary Row Operations
• Reduced Row-Echelon Form

MATRICES AND LINEAR SYSTEMS

In Section 5.3 we used the systematic method of elimination and back-substitution to solve linear systems of equations. In this section we will streamline the notation and extend the method.

Matrices

Consider the system of equations

$$\begin{cases} 2x + 4y - z = 10 \\ x - 3y = -6 \\ 7x + y + 5z = 9 \end{cases}$$

The essential elements in the solution are the coefficients of the variables and the constants on the right side of the equations. We can highlight this information by representing the parts of the system in a rectangular array of numbers called a *matrix*. The *coefficient matrix* for the system above is

NOTE: *Notice that* 0 *is the coefficient of the missing z term in the second equation.*

$$\begin{bmatrix} 2 & 4 & -1 \\ 1 & -3 & 0 \\ 7 & 1 & 5 \end{bmatrix}$$

A matrix that represents the entire system is

$$\begin{bmatrix} 2 & 4 & -1 & | & 10 \\ 1 & -3 & 0 & | & -6 \\ 7 & 1 & 5 & | & 9 \end{bmatrix}$$

This matrix is called the *augmented matrix* for the system. The vertical line between the third and fourth columns serves to remind us where the equal sign belongs.

Each horizontal alignment of numbers in a matrix is called a *row,* and each vertical alignment of numbers is called a *column:*

$$\begin{array}{c} & \text{Column} & \text{Column} & \text{Column} & \text{Column} \\ & 1 & 2 & 3 & 4 \\ \begin{array}{c} \text{Row 1} \\ \text{Row 2} \\ \text{Row 3} \end{array} & \begin{bmatrix} 2 & 4 & -1 & | & 10 \\ 1 & -3 & 0 & | & -6 \\ 7 & 1 & 5 & | & 9 \end{bmatrix} \end{array}$$

A matrix with m rows and n columns is an $m \times n$ matrix, and the expression $m \times n$ is called the *dimension* of the matrix. The dimension of the above matrix, for example, is 3×4. A matrix with the same number of rows as columns is called a *square* matrix; the 3×3 matrix

$$\begin{bmatrix} 2 & 4 & -1 \\ 1 & -3 & 0 \\ 7 & 1 & 5 \end{bmatrix}$$

is an example.

■ **NOTE:** *Read $m \times n$ as "m by n." The dimension of a matrix is also known as the order of a matrix.*

Each of the numbers in a matrix is called an *entry.* In general, the entry in the ith row and jth column is denoted as a_{ij}. For example, a 2×5 matrix can be written in general as

$$\begin{bmatrix} a_{11} & a_{12} & a_{13} & a_{14} & a_{15} \\ a_{21} & a_{22} & a_{23} & a_{24} & a_{25} \end{bmatrix}$$

■ **NOTE:** *"Matrices" is the plural of matrix.*

Matrices play an important role in mathematics. Using them to solve systems of equations is only one of their many applications.

EXAMPLE 1 Augmented Matrices

Write the augmented matrix for each system of equations, and state the dimension.

(a) $\begin{cases} 3x + y = -1 \\ 2x - 5y = 4 \end{cases}$ **(b)** $\begin{cases} x + 8z = 6 \\ 3y - 5z = 1 \\ 2x + 9y = 7 \end{cases}$

SOLUTION

(a) The augmented matrix is

$$\left[\begin{array}{cc|c} 3 & 1 & -1 \\ 2 & -5 & 4 \end{array}\right]$$

Since the matrix has 2 rows and 3 columns, the dimension is 2×3.

(b) The system can be written as

$$\begin{cases} x + 0y + 8z = 6 \\ 0x + 3y - 5z = 1 \\ 2x + 9y + 0z = 7 \end{cases}$$

The augmented matrix is

$$\left[\begin{array}{ccc|c} 1 & 0 & 8 & 6 \\ 0 & 3 & -5 & 1 \\ 2 & 9 & 0 & 7 \end{array}\right]$$

The dimension is 3×4.

Elementary Row Operations

In Section 5.3 we solved systems of equations using three elementary operations. Given the augmented matrix for a system, we can solve the system more efficiently by manipulating rows of numbers instead of entire equations. In terms of matrices, each of the elementary operations corresponds to an *elementary row operation.*

Elementary Row
Operations for Matrices

> Each of the following **elementary row operations** on an augmented matrix yields a matrix that represents an equivalent system of equations:
>
> **1.** Interchange two rows.
> **2.** Replace a row with a nonzero multiple of itself.
> **3.** Replace a row with the sum of that row and a multiple of another row.

The notation for elementary row operations is essentially the same, except we use R_1, R_2, and so on, instead of E_1, E_2, etc.

EXAMPLE 2   Elementary Row Operations

Perform the indicated elementary row operation on the given matrix.

(a) Interchange the second and third rows:

$$\begin{bmatrix} 2 & -3 & 1 & | & 3 \\ 1 & 7 & 0 & | & -2 \\ 5 & 1 & -4 & | & 6 \end{bmatrix}$$

(b) Replace the third row with $-\frac{1}{3}$ times the third row:

$$\begin{bmatrix} 1 & 8 & 2 & | & -3 \\ 0 & 4 & 4 & | & 9 \\ 3 & -6 & 2 & | & 9 \end{bmatrix}$$

(c) Replace the second row with the sum of -2 times the first row and the second row:

$$\begin{bmatrix} 1 & 3 & -7 & | & -2 \\ 2 & 5 & 1 & | & -1 \\ 4 & 0 & 6 & | & 5 \end{bmatrix}$$

NOTE: *We write the elementary row operation next to the row that is replaced.*

SOLUTION

(a) $\begin{bmatrix} 2 & -3 & 1 & | & 3 \\ 1 & 7 & 0 & | & -2 \\ 5 & 1 & -4 & | & 6 \end{bmatrix}$ $\begin{matrix} R_3 \to R_2 \\ R_2 \to R_3 \end{matrix}$ $\begin{bmatrix} 2 & -3 & 1 & | & 3 \\ 5 & 1 & -4 & | & 6 \\ 1 & 7 & 0 & | & -2 \end{bmatrix}$

(b) $\begin{bmatrix} 1 & 8 & 2 & | & -3 \\ 0 & 4 & 4 & | & 9 \\ 3 & -6 & 2 & | & 9 \end{bmatrix}$ $-\frac{1}{3}R_3 \to R_3$ $\begin{bmatrix} 1 & 8 & 2 & | & -3 \\ 0 & 4 & 4 & | & 9 \\ -1 & 2 & -\frac{2}{3} & | & -3 \end{bmatrix}$

(c) $\begin{bmatrix} 1 & 3 & -7 & | & -2 \\ 2 & 5 & 1 & | & -1 \\ 4 & 0 & 6 & | & 5 \end{bmatrix}$ $-2R_1 + R_2 \to R_2$ $\begin{bmatrix} 1 & 3 & -7 & | & -2 \\ 0 & -1 & 15 & | & 3 \\ 4 & 0 & 6 & | & 5 \end{bmatrix}$

We are now ready to solve a system of linear equations using elementary row operations on the augmented matrix. The process is the same as the elimination method of Section 5.3, except the notation is streamlined.

EXAMPLE 3 **Row-Echelon Form, Elementary Row Operations, and Back-Substitution**

Solve the system using elementary row operations on its augmented matrix.

$$\begin{cases} 3x + 7y + 11z = 19 \\ y + z = 5 \\ x + 2y + 3z = 4 \end{cases}$$

SOLUTION

The augmented matrix is

$$\begin{bmatrix} 3 & 7 & 11 & | & 19 \\ 0 & 1 & 1 & | & 5 \\ 1 & 2 & 3 & | & 4 \end{bmatrix}$$

Our goal is to use elementary row operations to reduce this augmented matrix to a matrix representing an equivalent system in row-echelon form. We need the first entry in the first row to be nonzero. That entry, 3, will work; however, an entry of 1 in the first column is easier to use, so we interchange the first and third rows:

$$\begin{matrix} R_3 \rightarrow R_1 \\ \\ R_1 \rightarrow R_3 \end{matrix} \begin{bmatrix} 1 & 2 & 3 & | & 4 \\ 0 & 1 & 1 & | & 5 \\ 3 & 7 & 11 & | & 19 \end{bmatrix}$$

Next, we eliminate the entries below the first entry in the first row:

$$-3R_1 + R_3 \rightarrow R_3 \begin{bmatrix} 1 & 2 & 3 & | & 4 \\ 0 & 1 & 1 & | & 5 \\ 0 & 1 & 2 & | & 7 \end{bmatrix}$$

To make the matrix represent a system in row-echelon form, we eliminate the second entry in the third row:

$$-1R_2 + R_3 \rightarrow R_3 \begin{bmatrix} 1 & 2 & 3 & | & 4 \\ 0 & 1 & 1 & | & 5 \\ 0 & 0 & 1 & | & 2 \end{bmatrix}$$

This matrix represents the system of equations

$$\begin{cases} x + 2y + 3z = 4 \\ \quad\quad y + z = 5 \\ \quad\quad\quad z = 2 \end{cases}$$

We leave it to you to back-substitute and find that the solution is $(-8, 3, 2)$.

Reduced Row-Echelon Form

Suppose that instead of back-substituting in the row-echelon system of Example 3, we continue to perform the elementary row operations as follows:

$$\begin{bmatrix} 1 & 2 & 3 & | & 4 \\ 0 & 1 & 1 & | & 5 \\ 0 & 0 & 1 & | & 2 \end{bmatrix}$$

$$-2R_2 + R_1 \rightarrow R_1 \quad \begin{bmatrix} 1 & 0 & 1 & | & -6 \\ 0 & 1 & 1 & | & 5 \\ 0 & 0 & 1 & | & 2 \end{bmatrix}$$

$$\begin{aligned} -1R_3 + R_1 \rightarrow R_1 \\ -1R_3 + R_2 \rightarrow R_2 \end{aligned} \quad \begin{bmatrix} 1 & 0 & 0 & | & -8 \\ 0 & 1 & 0 & | & 3 \\ 0 & 0 & 1 & | & 2 \end{bmatrix}$$

This matrix represents the system

$$\begin{cases} x & = -8 \\ y & = 3 \\ z = 2 \end{cases}$$

The solution to this system is $(-8, 3, 2)$.

The last matrix,

$$\begin{bmatrix} 1 & 0 & 0 & | & -8 \\ 0 & 1 & 0 & | & 3 \\ 0 & 0 & 1 & | & 2 \end{bmatrix}$$

is in *reduced row-echelon form.* The advantage of this form is that we can read the solution of the system directly from the matrix.

Conditions for Reduced
Row-Echelon Form

A matrix is in **reduced row-echelon form** if it satisfies the following conditions:

1. The first nonzero entry of each row is 1 (called a **leading** 1).
2. The other entries in a column with a leading 1 are 0's.
3. Each leading 1 is in a column to the right of all the leading 1's above it.
4. Any rows containing only 0's appear at the bottom of the matrix.

NOTE: *No matter what row operations are used to reduce a given matrix to reduced row-echelon form, this reduced row-echelon form is always the same. In other words, every matrix is equivalent to a unique reduced row-echelon form.*

EXAMPLE 4 Reduced Row-Echelon Form

For each matrix, determine whether it is in reduced row-echelon form. If not, perform one or more elementary row operations to transform it to reduced row-echelon form.

(a) $\begin{bmatrix} 0 & 1 & 7 \\ 1 & 0 & -3 \end{bmatrix}$

(b) $\begin{bmatrix} 1 & 0 & -5 & -4 \\ 0 & 0 & 0 & 0 \\ 0 & 0 & 1 & 2 \end{bmatrix}$

$$\text{(c)} \begin{bmatrix} 1 & 0 & 0 & 0 & 0 \\ 0 & 1 & 0 & 1 & 6 \\ 0 & 0 & 2 & -3 & 0 \end{bmatrix} \qquad \text{(d)} \begin{bmatrix} 1 & 0 & 0 & 8 \\ 0 & 1 & 6 & 1 \\ 0 & 0 & 0 & 0 \end{bmatrix}$$

SOLUTION

(a) The leading 1 in the second row is not to the right of the leading 1 above it. Interchanging the rows will produce the reduced row-echelon form:

$$\begin{matrix} R_2 \rightarrow R_1 \\ R_1 \rightarrow R_2 \end{matrix} \begin{bmatrix} 1 & 0 & -3 \\ 0 & 1 & 7 \end{bmatrix}$$

(b) The leading 1 in the third row has a nonzero entry (-5) in its column, so we eliminate the -5:

$$5R_3 + R_1 \rightarrow R_1 \begin{bmatrix} 1 & 0 & 0 & 6 \\ 0 & 0 & 0 & 0 \\ 0 & 0 & 1 & 2 \end{bmatrix}$$

Also, the second row contains only 0's, so this row should be moved to the bottom of the matrix:

$$\begin{matrix} R_3 \rightarrow R_2 \\ R_2 \rightarrow R_3 \end{matrix} \begin{bmatrix} 1 & 0 & 0 & 6 \\ 0 & 0 & 1 & 2 \\ 0 & 0 & 0 & 0 \end{bmatrix}$$

(c) The first entry in the third row is not a 1. This can be remedied by multiplying the third row by $\frac{1}{2}$:

$$\frac{1}{2}R_3 \rightarrow R_3 \begin{bmatrix} 1 & 0 & 0 & 0 & 0 \\ 0 & 1 & 0 & 1 & 6 \\ 0 & 0 & 1 & -\frac{3}{2} & 0 \end{bmatrix}$$

(d) The matrix is in reduced row-echelon form.

EXAMPLE 5 Solving a System Using Reduced Row-Echelon Form

Solve the system

$$\begin{cases} 2x + 8y + 3z = 14 \\ x + 5y + 2z = 5 \\ 2x + 10y - 3z = 38 \end{cases}$$

by finding the reduced row-echelon form of its augmented matrix.

SOLUTION

The augmented matrix is

$$\begin{bmatrix} 2 & 8 & 3 & | & 14 \\ 1 & 5 & 2 & | & 5 \\ 2 & 10 & -3 & | & 38 \end{bmatrix}$$

We begin with the first column. The first step is to place a 1 in the first row and first column:

$$\begin{matrix} R_2 \to R_1 \\ R_1 \to R_2 \end{matrix} \begin{bmatrix} 1 & 5 & 2 & | & 5 \\ 2 & 8 & 3 & | & 14 \\ 2 & 10 & -3 & | & 38 \end{bmatrix}$$

Now we use this 1 to create 0's in all other entries of the first column:

$$\begin{matrix} \\ -2R_1 + R_2 \to R_2 \\ -2R_1 + R_3 \to R_3 \end{matrix} \begin{bmatrix} 1 & 5 & 2 & | & 5 \\ 0 & -2 & -1 & | & 4 \\ 0 & 0 & -7 & | & 28 \end{bmatrix}$$

We are done with the first column; now we concentrate on the second column. We need a 1 as the entry in the second row and second column:

$$-\tfrac{1}{2}R_2 \to R_2 \begin{bmatrix} 1 & 5 & 2 & | & 5 \\ 0 & 1 & \tfrac{1}{2} & | & -2 \\ 0 & 0 & -7 & | & 28 \end{bmatrix}$$

We use this 1 to create 0's in all other entries of the second column:

$$-5R_2 + R_1 \to R_1 \begin{bmatrix} 1 & 0 & -\tfrac{1}{2} & | & 15 \\ 0 & 1 & \tfrac{1}{2} & | & -2 \\ 0 & 0 & -7 & | & 28 \end{bmatrix}$$

Finally, we focus on the third column. We create an entry of 1 in the third row and third column, and use it to create 0's in all other entries of the third column:

$$-\tfrac{1}{7}R_3 \to R_3 \begin{bmatrix} 1 & 0 & -\tfrac{1}{2} & | & 15 \\ 0 & 1 & \tfrac{1}{2} & | & -2 \\ 0 & 0 & 1 & | & -4 \end{bmatrix}$$

$$\begin{matrix} \tfrac{1}{2}R_3 + R_1 \to R_1 \\ -\tfrac{1}{2}R_3 + R_2 \to R_2 \end{matrix} \begin{bmatrix} 1 & 0 & 0 & | & 13 \\ 0 & 1 & 0 & | & 0 \\ 0 & 0 & 1 & | & -4 \end{bmatrix}$$

This final matrix is in reduced row-echelon form. The solution to the system is $(13, 0, -4)$.

Some graphing utilities have a reduced row-echelon form (RREF) feature that will determine the reduced row-echelon form of a matrix. Most graphing utilities can perform elementary row operations on a matrix to find its reduced row-echelon form. Using the RREF feature or performing these operations with a graphing utility (instead of with pencil and paper) saves time and eliminates computational errors that frequently occur. Try entering the augmented matrix in Example 5 and finding its reduced row-echelon form on a graphing utility. Do the same for the examples that follow. If the graphing utility does not store the matrix prior to a row operation, consider storing some of the intermediate results.

Recall that a system of linear equations may have exactly one solution, no solution, or an infinite number of solutions. The system in Example 5 has exactly one solution. The next two examples deal with the other two cases.

EXAMPLE 6   Systems with an Infinite Number of Solutions

Solve each system of equations by finding the reduced row-echelon form of its augmented matrix.

(a) $\begin{cases} 3x + y + z = 3 \\ x + z = 1 \end{cases}$ (b) $\begin{cases} x - y + z = -1 \\ x + 2y - 5z = 2 \\ 2x + y - 4z = 1 \end{cases}$

SOLUTION

(a) The augmented matrix is

$$\left[\begin{array}{ccc|c} 3 & 1 & 1 & 3 \\ 1 & 0 & 1 & 1 \end{array}\right]$$

We determine its reduced row-echelon form as follows:

$$\begin{matrix} R_2 \to R_1 \\ R_1 \to R_2 \end{matrix} \left[\begin{array}{ccc|c} 1 & 0 & 1 & 1 \\ 3 & 1 & 1 & 3 \end{array}\right]$$

$$-3R_1 + R_2 \to R_2 \left[\begin{array}{ccc|c} 1 & 0 & 1 & 1 \\ 0 & 1 & -2 & 0 \end{array}\right]$$

The last matrix is in reduced row-echelon form and represents the linear system

$$\begin{cases} x + z = 1 \\ y - 2z = 0 \end{cases}$$

If we let $z = t$ and then back-substitute into the second equation, we get

$$y - 2t = 0$$
$$y = 2t$$

Back-substituting $z = t$ into the first equation gives

$$x + t = 1$$
$$x = 1 - t$$

Thus, the solution can be written parametrically as $(1 - t, 2t, t)$.

(b) The augmented matrix

$$\begin{bmatrix} 1 & -1 & 1 & | & -1 \\ 1 & 2 & -5 & | & 2 \\ 2 & 1 & -4 & | & 1 \end{bmatrix}$$

has a reduced row-echelon form that can be determined as follows:

$$\begin{matrix} -1R_1 + R_2 \rightarrow R_2 \\ -2R_1 + R_3 \rightarrow R_3 \end{matrix} \begin{bmatrix} 1 & -1 & 1 & | & -1 \\ 0 & 3 & -6 & | & 3 \\ 0 & 3 & -6 & | & 3 \end{bmatrix}$$

$$\tfrac{1}{3}R_2 \rightarrow R_2 \begin{bmatrix} 1 & -1 & 1 & | & -1 \\ 0 & 1 & -2 & | & 1 \\ 0 & 3 & -6 & | & 3 \end{bmatrix}$$

$$\begin{matrix} 1R_2 + R_1 \rightarrow R_1 \\ \\ -3R_2 + R_3 \rightarrow R_3 \end{matrix} \begin{bmatrix} 1 & 0 & -1 & | & 0 \\ 0 & 1 & -2 & | & 1 \\ 0 & 0 & 0 & | & 0 \end{bmatrix}$$

The last matrix is in reduced row-echelon form, representing the dependent linear system

$$\begin{cases} x - z = 0 \\ y - 2z = 1 \\ \quad\quad 0 = 0 \end{cases}$$

The last equation, $0 = 0$, is true for any values of x, y, and z, and implies that the system is dependent. We represent the solutions parametrically by letting $z = t$ and back-substituting into the second equation:

$$y - 2t = 1$$
$$y = 1 + 2t$$

Letting $z = t$ in the first equation gives

$$x - t = 0$$
$$x = t$$

The solution can be expressed parametrically as $(t, 1 + 2t, t)$.

EXAMPLE 7   A System with No Solutions

Solve the system of equations

$$\begin{cases} 2x + 6y = 5 \\ 3x + 9y = -2 \end{cases}$$

by finding the reduced row-echelon form of its augmented matrix.

SOLUTION

Starting with the augmented matrix

$$\left[\begin{array}{cc|c} 2 & 6 & 5 \\ 3 & 9 & -2 \end{array}\right]$$

we determine the reduced row-echelon form:

$$\frac{1}{2}R_1 \rightarrow R_1 \quad \left[\begin{array}{cc|c} 1 & 3 & \frac{5}{2} \\ 3 & 9 & -2 \end{array}\right]$$

$$-3R_1 + R_2 \rightarrow R_2 \quad \left[\begin{array}{cc|c} 1 & 3 & \frac{5}{2} \\ 0 & 0 & -\frac{19}{2} \end{array}\right]$$

$$-\frac{2}{19}R_2 \rightarrow R_2 \quad \left[\begin{array}{cc|c} 1 & 3 & \frac{5}{2} \\ 0 & 0 & 1 \end{array}\right]$$

$$-\frac{5}{2}R_2 + R_1 \rightarrow R_1 \quad \left[\begin{array}{cc|c} 1 & 3 & 0 \\ 0 & 0 & 1 \end{array}\right]$$

This matrix represents the system

$$\begin{cases} x + 3y = 0 \\ 0x + 0y = 1 \end{cases}$$

The last equation is not true for any values of x and y. The system is inconsistent. ∎

EXERCISE SET 5.4

A

In Exercises 1–4, write the augmented matrix for each system and state the dimension.

1. $\begin{cases} 5x + y = 6 \\ -x + 3y = 1 \end{cases}$

2. $\begin{cases} 2x - y = 0 \\ x + 7y = -8 \end{cases}$

3. $\begin{cases} 2x + y = 6 \\ x - 2y + 4z = 5 \\ 6x - 2z = -1 \end{cases}$

4. $\begin{cases} -x + 8y + 4z = 11 \\ x - y = 5 \\ 7y + z = -1 \end{cases}$

In Exercises 5–10, perform the indicated elementary row operation on each matrix. Use the notation for elementary row operations introduced in this section (for example, $3R_2 + R_1 \rightarrow R_1$).

5. Interchange the first and third rows:

$$\left[\begin{array}{cccc} 4 & 1 & -9 & 6 \\ 0 & 12 & \frac{1}{2} & -5 \\ -1 & 1 & 4 & 7 \end{array}\right]$$

6. Interchange the first and second rows:

$$\begin{bmatrix} 0 & -2 & 0 & 13 \\ 7 & \frac{2}{5} & -1 & -3 \\ 4 & 6 & 8 & \frac{3}{2} \end{bmatrix}$$

7. Replace the second row with $\frac{2}{3}$ times the second row:

$$\begin{bmatrix} 1 & -8 & 2 & 10 \\ 0 & \frac{3}{2} & 1 & -3 \\ -1 & 1 & 5 & 14 \end{bmatrix}$$

8. Replace the first row with -4 times the first row:

$$\begin{bmatrix} -\frac{1}{4} & 3 & 0 & \frac{5}{2} \\ 2 & 3 & 5 & 7 \\ -6 & 1 & 0 & 9 \end{bmatrix}$$

9. Replace the third row with the sum of -3 times the first row and the third row:

$$\begin{bmatrix} 1 & -1 & -5 & 2 \\ 2 & 3 & 5 & 6 \\ 3 & 1 & 0 & -4 \end{bmatrix}$$

10. Replace the first row with the sum of 2 times the second row and the first row:

$$\begin{bmatrix} 1 & -2 & -5 & -1 \\ 0 & 1 & 5 & 7 \\ 0 & 3 & 1 & -9 \end{bmatrix}$$

In Exercises 11–16, determine whether each matrix is in reduced row-echelon form. If not, perform one or more elementary row operations to determine the reduced row-echelon form. Use the notation for elementary row operations introduced in this section.

11. $\begin{bmatrix} 1 & 0 & 0 \\ 0 & 0 & 1 \end{bmatrix}$
12. $\begin{bmatrix} 1 & 2 & 0 \\ 0 & 0 & 0 \end{bmatrix}$

13. $\begin{bmatrix} 1 & 0 & -7 & 4 \\ 0 & 1 & 0 & 2 \\ 0 & 0 & 1 & -3 \end{bmatrix}$
14. $\begin{bmatrix} 1 & 0 & 4 & -8 \\ 0 & 0 & 0 & 0 \\ 0 & 1 & -1 & 9 \end{bmatrix}$

15. $\begin{bmatrix} 1 & 5 & 0 & 0 & -3 \\ 0 & 0 & 1 & 0 & 2 \\ 0 & 0 & 0 & 1 & 6 \end{bmatrix}$
16. $\begin{bmatrix} 0 & 0 & 0 & 0 \\ 1 & 3 & -7 & 0 \\ 0 & 0 & 0 & 1 \end{bmatrix}$

In Exercises 17–22, state the row operations used to transform the first matrix into the second matrix. State your answer in the notation used in this section.

17. $\begin{bmatrix} 2 & -4 & 3 & 18 \\ -3 & -7 & 0 & 5 \\ 0 & 2 & -9 & 1 \end{bmatrix}; \begin{bmatrix} 1 & -2 & \frac{3}{2} & 9 \\ -3 & -7 & 0 & 5 \\ 0 & 2 & -9 & 1 \end{bmatrix}$

18. $\begin{bmatrix} 1 & 7 & -2 & 11 \\ 0 & 1 & -1 & -3 \\ 0 & 0 & -2 & 0 \end{bmatrix}; \begin{bmatrix} 1 & 0 & 5 & 32 \\ 0 & 1 & -1 & -3 \\ 0 & 0 & -2 & 0 \end{bmatrix}$

19. $\begin{bmatrix} 1 & 2 & -3 & -1 \\ 0 & 4 & -1 & 1 \\ -5 & 0 & 1 & 2 \end{bmatrix}; \begin{bmatrix} 1 & 2 & -3 & -1 \\ 0 & 4 & -1 & 1 \\ 0 & 10 & -14 & -3 \end{bmatrix}$

20. $\begin{bmatrix} 1 & 3 & -6 & 5 \\ 0 & 4 & 2 & 12 \\ 0 & 0 & 7 & 11 \end{bmatrix}; \begin{bmatrix} 1 & 3 & -6 & 5 \\ 0 & 1 & \frac{1}{2} & 3 \\ 0 & 0 & 7 & 11 \end{bmatrix}$

21. $\begin{bmatrix} 1 & 8 & 11 & 0 \\ 0 & 5 & -2 & 1 \\ 0 & 1 & -3 & -1 \end{bmatrix}; \begin{bmatrix} 1 & 8 & 11 & 0 \\ 0 & 1 & -3 & -1 \\ 0 & 5 & -2 & 1 \end{bmatrix}$

22. $\begin{bmatrix} 4 & -9 & 5 & 2 \\ 1 & 0 & -5 & 1 \\ 3 & 1 & 12 & 0 \end{bmatrix}; \begin{bmatrix} 1 & 0 & -5 & 1 \\ 4 & -9 & 5 & 2 \\ 3 & 1 & 12 & 0 \end{bmatrix}$

In Exercises 23–28, each matrix is the reduced row-echelon form of the augmented matrix representing a linear system of equations in the variables x, y, and z. State the solution to the system. If the system is dependent, state the solution parametrically.

23. $\begin{bmatrix} 1 & 0 & 0 & | & -4 \\ 0 & 1 & 0 & | & 1 \\ 0 & 0 & 1 & | & 7 \end{bmatrix}$
24. $\begin{bmatrix} 1 & 0 & 0 & | & 8 \\ 0 & 1 & 0 & | & -5 \\ 0 & 0 & 1 & | & -2 \end{bmatrix}$

25. $\begin{bmatrix} 1 & 0 & 0 & | & 0 \\ 0 & 1 & 0 & | & 0 \\ 0 & 0 & 0 & | & 1 \end{bmatrix}$
26. $\begin{bmatrix} 1 & 0 & 0 & | & 0 \\ 0 & 0 & 1 & | & 0 \\ 0 & 0 & 0 & | & 1 \end{bmatrix}$

27. $\begin{bmatrix} 1 & 0 & 7 & | & -3 \\ 0 & 1 & 5 & | & 2 \\ 0 & 0 & 0 & | & 0 \end{bmatrix}$
28. $\begin{bmatrix} 1 & 0 & 0 & | & 5 \\ 0 & 0 & 1 & | & -7 \\ 0 & 0 & 0 & | & 0 \end{bmatrix}$

In Exercises 29 and 30, solve each system of equations by performing the given set of elementary row operations on the augmented matrix. The result is a matrix in reduced row-echelon form.

29. $\begin{cases} 2x + 3y + z = 0 \\ x + y + z = 1 \\ -x + y + z = 11 \end{cases}$

(a) $R_2 \rightarrow R_1$ and $R_1 \rightarrow R_2$

(b) $-2R_1 + R_2 \rightarrow R_2$ and $1R_1 + R_3 \rightarrow R_3$

(c) $-1R_2 + R_1 \rightarrow R_1$ and $-2R_2 + R_3 \rightarrow R_3$

(d) $\frac{1}{4} R_3 \rightarrow R_3$

(e) $-2R_3 + R_1 \rightarrow R_1$ and $1R_3 + R_2 \rightarrow R_2$

30. $\begin{cases} 3x + 4y - 2z = 8 \\ -2x - y - z = 4 \\ x + y + 2z = 3 \end{cases}$

(a) $R_3 \rightarrow R_1$ and $R_1 \rightarrow R_3$

(b) $2R_1 + R_2 \rightarrow R_2$ and $-3R_1 + R_3 \rightarrow R_3$

(c) $-1R_2 + R_1 \rightarrow R_1$ and $-1R_2 + R_3 \rightarrow R_3$

(d) $-\frac{1}{11} R_3 \rightarrow R_3$

(e) $1R_3 + R_1 \rightarrow R_1$ and $-3R_3 + R_2 \rightarrow R_2$

In Exercises 31–38, solve each system of equations by finding the reduced row-echelon form of the augmented matrix. (Note that these are the same systems as in Exercises 15–22 in Section 5.3.) Use the notation for elementary row operations introduced in this section to show your steps.

31. $\begin{cases} x + 2y - 2z = 8 \\ 3x - y + 5z = 6 \\ x + 2y - z = 7 \end{cases}$

32. $\begin{cases} x - y + 2z = -2 \\ 4x + y + 3z = 7 \\ x - y + z = 0 \end{cases}$

33. $\begin{cases} x + y + 2z = 1 \\ x + 3y - 2z = 7 \\ 2x - y + z = 2 \end{cases}$

34. $\begin{cases} x + y + 2z = 2 \\ 3x + 2y + z = 3 \\ 2x - 2y + z = 9 \end{cases}$

35. $\begin{cases} x + 2y + 3z = 3 \\ 2x + 3y + 10z = 18 \\ 3x + 9y + 10z = -1 \end{cases}$

36. $\begin{cases} x + 2y - z = 10 \\ 2x - y + 2z = -3 \\ 3x + 2y + 3z = 6 \end{cases}$

37. $\begin{cases} x + y + z = 4 \\ x - y - z = -2 \\ x + 2y - z = 1 \end{cases}$

38. $\begin{cases} x + y = -2 \\ 6x + y - z = -20 \\ 4x - 6y + z = 0 \end{cases}$

B

In Exercises 39–55, solve each system of equations by finding the reduced row-echelon form of the augmented matrix. Use the notation for elementary row operations introduced in this section to show your steps. If the system is dependent, state the solution parametrically.

39. $\begin{cases} 2x - 3y = -11 \\ 3x + y - 4z = 8 \\ 2x - 4z = 6 \end{cases}$

40. $\begin{cases} x - y + 3z = 6 \\ 3x + 2y - z = 8 \\ x + 2y + z = 8 \end{cases}$

41. $\begin{cases} x + z = 5 \\ 4x - y + 3z = 7 \\ x + 2y = 8 \end{cases}$

42. $\begin{cases} x + 3y - z = 5 \\ x + 3y + 2z = -1 \\ x + 2y = 1 \end{cases}$

43. $\begin{cases} x + y + z = 0 \\ 2x - y - z = 2 \\ 3x - 3y - 2z = 3 \end{cases}$

44. $\begin{cases} 2x + 4y - z = 2 \\ x + 3y + z = 12 \\ 5x + y - z = 11 \end{cases}$

45. $\begin{cases} 2x + y = -5 \\ x + y - z = -2 \\ 3x + 2y + z = -7 \end{cases}$

46. $\begin{cases} x - 3y + 2z = -12 \\ x + y - 6z = 4 \\ 2x - 5y - 5z = -20 \end{cases}$

47 $\begin{cases} x + 2y + z = 5 \\ x - y + 4z = -1 \\ 2x + 3y - 4z = 4 \end{cases}$

48. $\begin{cases} x + y = -2 \\ 3x + y - z = 1 \\ 2x - y + z = 3 \end{cases}$

49. $\begin{cases} 2x - y - z = 3 \\ x + 7y + z = 9 \\ 3x + y - 4z = 7 \end{cases}$

50. $\begin{cases} x + y - z = 0 \\ 2x + 5y + 4z = 0 \\ 3x - y - 11z = 0 \end{cases}$

51. $\begin{cases} 2x - 3y - z = 8 \\ x - y + 4z = 16 \end{cases}$

52. $\begin{cases} 2x + 3y - z = 1 \\ x + y + 5z = 5 \end{cases}$

53. $\begin{cases} 3x + 7y - 6z = 23 \\ x + 2y - z = 7 \end{cases}$

54. $\begin{cases} x - y + z + w = 8 \\ x - y + 2z = 10 \\ y - 3z + w = -11 \\ x + 2z - w = 7 \end{cases}$

55. $\begin{cases} 2x + y + z + 2w = -15 \\ x + 5y - 2z + 6w = -7 \\ 5x + 2y + z - w = -37 \\ 3x + 4y + w = 0 \end{cases}$

In Exercises 56–65 solve the system that arises by finding the reduced row-echelon form of the augmented matrix.

56. Determine a quadratic function $f(x) = ax^2 + bx + c$ whose graph passes through the points $(1, -1)$, $(2, 2)$, and $(-1, 7)$.

57. Determine a quadratic function $f(x) = ax^2 + bx + c$ whose graph passes through the points $(-1, -5)$, $(1, 9)$, and $(2, 13)$.

58. Determine a cubic function $f(x) = ax^3 + bx^2 + cx + d$ whose graph passes through the points $(1, 1)$, $(2, 7)$, $(-2, -5)$, and $(-1, 1)$.

59. Determine a cubic function $f(x) = ax^3 + bx^2 + cx + d$ whose graph passes through the points $(1, 7)$, $(-1, 5)$, $(2, 11)$ and $(-2, 31)$.

60. The equation for a circle can be written in the form $x^2 + y^2 + ax + by + c = 0$.
   (a) Determine the values of a, b, and c so that the circle passes through $(2, 4)$, $(-4, 4)$, and $(3, 3)$.
   (b) Find the center and radius of the circle.

61. A total of $43,000 is placed in three investments that pay annual interest rates of 6%, 8%, and 10%. The total annual return on the investments is $3350, and the amount invested at 8% is $7000 more than the amount invested at 10%. Find the amount placed in each investment.

62. A new fruit juice is made of orange juice, banana juice, and apple juice, which cost $0.30 per ounce, $0.15 per ounce, and $0.20 per ounce, respectively. The blend is to cost $0.22 per ounce, and will be sold in 16 ounce bottles. To satisfy the demand for vitamin requirements, there must be twice as much orange juice in the blend as banana juice. Find the amount of each type of juice in each bottle.

63. A camping supply store sells three types of sleeping bags for $220, $170, and $140 per bag. Last year, three times as many $140 bags were sold as $220 bags, and the combined number of bags sold was 156. If the total annual revenue for these three bags was $25,480, find the number of each type of bag that was sold.

64. The compositions of three types of steel are given in the table below. How many tons of each type should be melted and mixed together (to the nearest 0.01 ton) to get 15 tons of alloy that is 73% iron and 6% nickel?

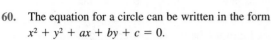

	Iron	Nickel
Type A	75%	5%
Type B	70%	7%
Type C	71%	8%

65. There are three currents I_1, I_2, and I_3 (measured in amperes) through the electrical network pictured:

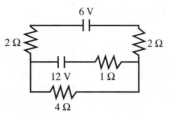

According to laws of physics, these currents satisfy the system

$$\begin{cases} I_1 + I_2 - I_3 = 0 \\ \quad\quad 4I_2 + I_3 = 12 \\ 4I_1 \quad\quad + I_3 = 6 \end{cases}$$

Solve this system to find the three currents.

C

66. The average of three numbers is 60. One of the numbers is the sum of the other two. The difference between the other two is 4. What are the three numbers?

67. The approximate temperature distribution throughout a triangular plate can be modeled using the following diagram:

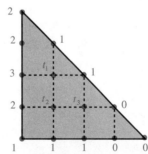

The temperature (in degrees Fahrenheit) at each indicated point along the edge is given. The temperature of each of the three interior points is to be determined. For each of these points there is a neighboring point above, below, to the right, and to the left. Assume that the temperature at each point is the average of the four temperatures at the four neighboring points; for example,

$$t_2 = \tfrac{1}{4}(t_1 + 1 + t_3 + 2)$$

Determine t_1, t_2, and t_3.

DETERMINANTS AND CRAMER'S RULE

The topics in this section are motivated by two problems: developing a formula for the solution of a system of equations, and calculating the area of a polygon in the coordinate plane. These problems, as well as other important calculations, lead to expressions that share certain mathematical patterns. These patterns can be organized with a matrix function called the *determinant*.

2×2 Determinants

Consider the following system:

$$\begin{cases} ax + by = m \\ cx + dy = n \end{cases}$$

We can solve this system for x by multiplying each side of the first equation by d, multiplying each side of the second equation by $-b$, and adding the resulting equations:

$$\begin{cases} ax + by = m & \xrightarrow{\text{Multiply by } d} & adx + bdy = md \\ cx + dy = n & \xrightarrow{\text{Multiply by } -b} & \dfrac{-bcx - bdy = -bn}{(ad - bc)x = md - bn} \end{cases}$$

If $ad - bc \neq 0$, we get

$$x = \frac{md - bn}{ad - bc}$$

In a similar way, we can solve for y to obtain

$$y = \frac{an - mc}{ad - bc}$$

The denominator for each solution is the same value,

$$ad - bc$$

This value is associated with the coefficient matrix

$$\begin{bmatrix} a & b \\ c & d \end{bmatrix}$$

of the original system according to the following definition:

2 × 2 Determinant

Given the 2 × 2 matrix

$$A = \begin{bmatrix} a & b \\ c & d \end{bmatrix}$$

the **determinant** of A is

$$ad - bc$$

The determinant may be denoted as

$$\det a \qquad \text{or} \qquad \begin{vmatrix} a & b \\ c & d \end{vmatrix}$$

An easy way to remember the determinant is to multiply along the diagonals of the matrix,

$$\begin{bmatrix} a & b \\ c & d \end{bmatrix}$$

and subtract the products in the correct order.

◼ EXAMPLE 1   Evaluating a 2 × 2 Determinant

Evaluate the determinant of each matrix.

$$A = \begin{bmatrix} 7 & 2 \\ 3 & 4 \end{bmatrix} \qquad B = \begin{bmatrix} 5 & -3 \\ 6 & 2 \end{bmatrix} \qquad C = \begin{bmatrix} r-1 & -4 \\ -2 & r+2 \end{bmatrix}$$

SOLUTION

$$\det A = \begin{vmatrix} 7 & 2 \\ 3 & 4 \end{vmatrix} = (7)(4) - (2)(3) = 22$$

$$\det B = \begin{vmatrix} 5 & -3 \\ 6 & 2 \end{vmatrix} = (5)(2) - (-3)(6) = 28$$

$$\det C = \begin{vmatrix} r-1 & -4 \\ -2 & r+2 \end{vmatrix} = (r-1)(r+2) - (-4)(-2)$$

$$= r^2 + r - 10 \qquad ◼$$

3 × 3 Determinants

Determinants of 3 × 3 matrices have important applications in engineering and can be useful for solving systems of equations. The determinant of the 3 × 3 matrix

$$\begin{bmatrix} a_1 & b_1 & c_1 \\ a_2 & b_2 & c_2 \\ a_3 & b_3 & c_3 \end{bmatrix}$$

is defined as

$$
\begin{vmatrix} a_1 & b_1 & c_1 \\ a_2 & b_2 & c_2 \\ a_3 & b_3 & c_3 \end{vmatrix} = a_1 b_2 c_3 - a_1 b_3 c_2 - a_2 b_1 c_3 + a_3 b_1 c_2 + a_2 b_3 c_1 - a_3 b_2 c_1
$$

This can be expressed in terms of 2×2 determinants as follows:

3×3 Determinant

Given the 3×3 matrix

$$
A = \begin{bmatrix} a_1 & b_1 & c_1 \\ a_2 & b_2 & c_2 \\ a_3 & b_3 & c_3 \end{bmatrix}
$$

the determinant of A is

$$
\det A = \begin{vmatrix} a_1 & b_1 & c_1 \\ a_2 & b_2 & c_2 \\ a_3 & b_3 & c_3 \end{vmatrix} = a_1 \begin{vmatrix} b_2 & c_2 \\ b_3 & c_3 \end{vmatrix} - b_1 \begin{vmatrix} a_2 & c_2 \\ a_3 & c_3 \end{vmatrix} + c_1 \begin{vmatrix} a_2 & b_2 \\ a_3 & b_3 \end{vmatrix}
$$

There is an easy way to remember this. Notice the 2×2 matrix that remains if we remove the row and column of a_1:

$$
\begin{bmatrix} a_1 & b_1 & c_1 \\ a_2 & b_2 & c_2 \\ a_3 & b_3 & c_3 \end{bmatrix} \longrightarrow \begin{bmatrix} b_2 & c_2 \\ b_3 & c_3 \end{bmatrix}
$$

Consider the 2×2 matrix that remains if we remove the row and column that b_1 is in:

$$
\begin{bmatrix} a_1 & b_1 & c_1 \\ a_2 & b_2 & c_2 \\ a_3 & b_3 & c_3 \end{bmatrix} \longrightarrow \begin{bmatrix} a_2 & c_2 \\ a_3 & c_3 \end{bmatrix}
$$

Similarly, if we remove the row and column of c_1, we get

$$
\begin{bmatrix} a_1 & b_1 & c_1 \\ a_2 & b_2 & c_2 \\ a_3 & b_3 & c_3 \end{bmatrix} \longrightarrow \begin{bmatrix} a_2 & b_2 \\ a_3 & b_3 \end{bmatrix}
$$

The determinant of a 3 × 3 matrix can be remembered in terms of these three 2 × 2 determinants as follows:

$$\begin{vmatrix} a_1 & b_1 & c_1 \\ a_2 & b_2 & c_2 \\ a_3 & b_3 & c_3 \end{vmatrix} = a_1 \begin{vmatrix} b_2 & c_2 \\ b_3 & c_3 \end{vmatrix} - b_1 \begin{vmatrix} a_2 & c_2 \\ a_3 & c_3 \end{vmatrix} + c_1 \begin{vmatrix} a_2 & b_2 \\ a_3 & b_3 \end{vmatrix}$$

$$= a_1 \begin{vmatrix} 2 \times 2 \text{ matrix remaining} \\ \text{after removing the row} \\ \text{and column of } a_1 \end{vmatrix} - b_1 \begin{vmatrix} 2 \times 2 \text{ matrix remaining} \\ \text{after removing the row} \\ \text{and column of } b_1 \end{vmatrix} + c_1 \begin{vmatrix} 2 \times 2 \text{ matrix remaining} \\ \text{after removing the row} \\ \text{and column of } c_1 \end{vmatrix}$$

EXAMPLE 2 Evaluating a 3 × 3 Determinant

Evaluate the 3 × 3 determinant: $\begin{vmatrix} 2 & 3 & -4 \\ 1 & 6 & -2 \\ -5 & 1 & 7 \end{vmatrix}$

SOLUTION

$$\begin{vmatrix} 2 & 3 & -4 \\ 1 & 6 & -2 \\ -5 & 1 & 7 \end{vmatrix} = 2 \begin{vmatrix} 2 \times 2 \text{ matrix remaining} \\ \text{after removing the row} \\ \text{and column of } 2 \end{vmatrix} - 3 \begin{vmatrix} 2 \times 2 \text{ matrix remaining} \\ \text{after removing the row} \\ \text{and column of } 3 \end{vmatrix} + (-4) \begin{vmatrix} 2 \times 2 \text{ matrix remaining} \\ \text{after removing the row} \\ \text{and column of } -4 \end{vmatrix}$$

$$= 2 \begin{vmatrix} 6 & -2 \\ 1 & 7 \end{vmatrix} - 3 \begin{vmatrix} 1 & -2 \\ -5 & 7 \end{vmatrix} + (-4) \begin{vmatrix} 1 & 6 \\ -5 & 1 \end{vmatrix}$$

$$= 2(44) - 3(-3) + (-4)(31)$$

$$= -27$$

Our method for evaluating a 3 × 3 determinant uses the first row for the coefficients (a_1, b_1, and c_1). This process is called *expanding along the first row*. We can expand along any row or column in a similar way and get the same result. For example, consider the second row of

$$\begin{bmatrix} a_1 & b_1 & c_1 \\ a_2 & b_2 & c_2 \\ a_3 & b_3 & c_3 \end{bmatrix}$$

It can be shown that

$$\begin{vmatrix} a_1 & b_1 & c_1 \\ a_2 & b_2 & c_2 \\ a_3 & b_3 & c_3 \end{vmatrix} = -a_2 \begin{vmatrix} 2 \times 2 \text{ matrix left after} \\ \text{removing the row} \\ \text{and column of } a_2 \end{vmatrix} + b_2 \begin{vmatrix} 2 \times 2 \text{ matrix left after} \\ \text{removing the row} \\ \text{and column of } b_2 \end{vmatrix} - c_2 \begin{vmatrix} 2 \times 2 \text{ matrix left after} \\ \text{removing the row} \\ \text{and column of } c_2 \end{vmatrix}$$

$$= -a_2 \begin{vmatrix} b_1 & c_1 \\ b_3 & c_3 \end{vmatrix} + b_2 \begin{vmatrix} a_1 & c_1 \\ a_3 & c_3 \end{vmatrix} - c_2 \begin{vmatrix} a_1 & b_1 \\ a_3 & b_3 \end{vmatrix}$$

Notice that the scheme is similar to expanding along the first row, but the signs of the coefficients a_2, b_2, and c_2 are different. When expanding along any row or column, you can determine the sign of the three coefficients according to the following array of signs:

$$\begin{array}{ccc} + & - & + \\ - & + & - \\ + & - & + \end{array}$$

For example, if we choose to expand along the second column, we use b_1, b_2, and b_3 as coefficients and add or subtract according to the pattern of signs:

$$\begin{vmatrix} a_1 & b_1 & c_1 \\ a_2 & b_2 & c_2 \\ a_3 & b_3 & c_3 \end{vmatrix} \qquad \begin{array}{ccc} + & - & + \\ - & + & - \\ + & - & + \end{array}$$

Specifically,

$$\begin{vmatrix} a_1 & b_1 & c_1 \\ a_2 & b_2 & c_2 \\ a_3 & b_3 & c_3 \end{vmatrix} = -b_1 \begin{vmatrix} a_2 & c_2 \\ a_3 & c_3 \end{vmatrix} + b_2 \begin{vmatrix} a_1 & c_1 \\ a_3 & c_3 \end{vmatrix} - b_3 \begin{vmatrix} a_1 & c_1 \\ a_2 & c_2 \end{vmatrix}$$

● EXAMPLE 3   Expanding a Determinant Along a Column

Evaluate the 3×3 determinant in Example 2 by expanding along the third column.

SOLUTION

$$\begin{vmatrix} 2 & 3 & -4 \\ 1 & 6 & -2 \\ -5 & 1 & 7 \end{vmatrix} = -4 \begin{vmatrix} 2 \times 2 \text{ matrix left after} \\ \text{removing the row} \\ \text{and column of } -4 \end{vmatrix} - (-2) \begin{vmatrix} 2 \times 2 \text{ matrix left after} \\ \text{removing the row} \\ \text{and column of } -2 \end{vmatrix} + 7 \begin{vmatrix} 2 \times 2 \text{ matrix left after} \\ \text{removing the row} \\ \text{and column of } 7 \end{vmatrix}$$

$$= -4 \begin{vmatrix} 1 & 6 \\ -5 & 1 \end{vmatrix} - (-2) \begin{vmatrix} 2 & 3 \\ -5 & 1 \end{vmatrix} + 7 \begin{vmatrix} 2 & 3 \\ 1 & 6 \end{vmatrix}$$

$$= -4(31) + 2(17) + 7(9)$$

$$= -27$$

Notice that the result agrees with that of Example 2.   ●

Many graphing utilities can evaluate the determinant of 2×2, 3×3, and larger square matrices. The scheme for evaluating a 4×4 determinant with pencil and paper requires expanding along a row or column, which involves four 3×3 determinants. This should motivate you to learn how to evaluate determinants with a graphing utility!

Not all graphing utilities can evaluate determinants of matrices containing nonconstant entries, such as the determinant of the matrix C in Example 1. That determinant as well as the 3×3 determinant in the next example have applications in science, mathematics, and engineering.

■ EXAMPLE 4   A 3×3 Determinant with Variables

Evaluate

$$\begin{vmatrix} i & j & k \\ -3 & 4 & 0 \\ 1 & -2 & -5 \end{vmatrix}$$

in terms of the variables i, j, and k.

SOLUTION

Expanding along the first row, we have

$$\begin{vmatrix} i & j & k \\ -3 & 4 & 0 \\ 1 & -2 & -5 \end{vmatrix} = i \begin{vmatrix} 4 & 0 \\ -2 & -5 \end{vmatrix} - j \begin{vmatrix} -3 & 0 \\ 1 & -5 \end{vmatrix} + k \begin{vmatrix} -3 & 4 \\ 1 & -2 \end{vmatrix}$$

$$= -20i - 15j + 2k$$

Cramer's Rule

In the beginning of this section, we derived a solution to the system

$$\begin{cases} ax + by = m \\ cx + dy = n \end{cases}$$

and found that

$$x = \frac{md - bn}{ad - bc} \quad \text{and} \quad y = \frac{an - mc}{ad - bc}$$

We already noted that in both cases the denominator D is the determinant of the coefficient matrix of the system of equations:

$$D = \begin{vmatrix} a & b \\ c & d \end{vmatrix}$$

Notice also that in the solution for x, the numerator is

$$\begin{vmatrix} m & b \\ n & d \end{vmatrix}$$

which is the same as D, except the first column is replaced by m and n. Similarly, the numerator for y is the same as D, except the second column is replaced by m and n. These results are known as *Cramer's rule*.

Cramer's Rule for Two
Equations in Two Variables

NOTE: *Although the essence of Cramer's rule was published by Maclaurin in 1729, the Swiss mathematician Gabriel Cramer (1704–1752) popularized the formula using determinants in his 1750 book on plane curves.*

Given the system

$$\begin{cases} ax + by = m \\ cx + dy = n \end{cases}$$

let D be the determinant of the 2×2 coefficient matrix:

$$D = \begin{vmatrix} a & b \\ c & d \end{vmatrix}$$

Let D_x be the determinant formed by replacing the entries in the column of coefficients for x with m and n, respectively:

$$D_x = \begin{vmatrix} m & b \\ n & d \end{vmatrix}$$

Similarly, let D_y be the determinant formed by replacing the entries in the column of coefficients for y with m and n, respectively:

$$D_y = \begin{vmatrix} a & m \\ c & n \end{vmatrix}$$

If $D \neq 0$, then the solution to the system is

$$x = \frac{D_x}{D} \quad \text{and} \quad y = \frac{D_y}{D}$$

A system has a unique solution if and only if $D \neq 0$. It can be shown that if $D = 0$ and $D_x = 0$, then the system is dependent. It can also be shown that if $D = 0$ and $D_x \neq 0$, then the system is inconsistent (see Exercise 52).

EXAMPLE 5 Cramer's Rule for Two Equations in Two Varaibles

Use Cramer's rule to solve: $\begin{cases} 2x + 2y = -1 \\ -4x + 3y = 9 \end{cases}$

SOLUTION
The determinant of the coefficient matrix is

$$D = \begin{vmatrix} 2 & 2 \\ -4 & 3 \end{vmatrix} = 2(3) - 2(-4) = 14$$

Since $D \neq 0$, we determine D_x and D_y:

$$D_x = \begin{vmatrix} -1 & 2 \\ 9 & 3 \end{vmatrix} = -1(3) - 9(2) = -21$$

$$D_y = \begin{vmatrix} 2 & -1 \\ -4 & 9 \end{vmatrix} = 2(9) - (-4)(-1) = 14$$

Thus,

$$x = \frac{D_x}{D} = \frac{-21}{14} = -\frac{3}{2} \quad \text{and} \quad y = \frac{D_y}{D} = \frac{14}{14} = 1$$

The solution is $\left(-\frac{3}{2}, 1\right)$.

EXAMPLE 6   Solving a Literal System

Use Cramer's rule to solve for x and y: $\quad \begin{cases} 3ux + vy = 0 \\ u^2x + 2y = f \end{cases}$

SOLUTION

The determinant of the coefficient matrix is

$$D = \begin{vmatrix} 3u & v \\ u^2 & 2 \end{vmatrix} = 3u(2) - v(u^2) = 6u - u^2v$$

Assuming that $D \neq 0$, we have

$$D_x = \begin{vmatrix} 0 & v \\ f & 2 \end{vmatrix} = 0(2) - vf = -vf$$

$$D_y = \begin{vmatrix} 3u & 0 \\ u^2 & f \end{vmatrix} = 3uf - 0(u^2) = 3uf$$

Applying Cramer's rule, we have

$$x = \frac{D_x}{D} = \frac{-vf}{6u - u^2v} = \frac{-vf}{-(u^2v - 6u)} = \frac{vf}{u^2v - 6u}$$

$$y = \frac{D_y}{D} = \frac{3uf}{6u - u^2v} = \frac{u(3f)}{u(6 - uv)} = \frac{3f}{6 - uv}$$

Assuming that $6u - u^2v \neq 0$, the solution is

$$\left(\frac{vf}{u^2v - 6u}, \frac{3f}{6 - uv}\right)$$

Cramer's rule generalizes naturally to systems with n equations in n variables. It is particularly convenient when solving systems with two equations in two variables, but for most systems with more than three variables the matrix method discussed in Section 5.4 should be used. In the box that follows we state Cramer's rule for three linear equations in three variables.

Cramer's Rule for Three Equations
in Three Variables

Given the system

$$\begin{cases} a_1x + b_1y + c_1z = k_1 \\ a_2x + b_2y + c_2z = k_2 \\ a_3x + b_3y + c_3z = k_3 \end{cases}$$

let

$$D = \begin{vmatrix} a_1 & b_1 & c_1 \\ a_2 & b_2 & c_2 \\ a_3 & b_3 & c_3 \end{vmatrix} \qquad D_x = \begin{vmatrix} k_1 & b_1 & c_1 \\ k_2 & b_2 & c_2 \\ k_3 & b_3 & c_3 \end{vmatrix}$$

$$D_y = \begin{vmatrix} a_1 & k_1 & c_1 \\ a_2 & k_2 & c_2 \\ a_3 & k_3 & c_3 \end{vmatrix} \qquad D_z = \begin{vmatrix} a_1 & b_1 & k_1 \\ a_2 & b_2 & k_2 \\ a_3 & b_3 & k_3 \end{vmatrix}$$

If $D \neq 0$, then the solution to the system is

$$x = \frac{D_x}{D}, \quad y = \frac{D_y}{D}, \quad \text{and} \quad z = \frac{D_z}{D}$$

EXAMPLE 7 Cramer's Rule for Three Equations in Three Variables

Use Cramer's rule to solve: $\begin{cases} x + y + z = 5 \\ 2x - 4y - 3z = -5 \\ x - y = 4 \end{cases}$

SOLUTION

The determinant of the coefficient matrix is

$$D = \begin{vmatrix} 1 & 1 & 1 \\ 2 & -4 & -3 \\ 1 & -1 & 0 \end{vmatrix}$$

Using the determinant feature of a graphing utility, we get

$$D = -4$$

Since $D \neq 0$, we proceed with the calculation of D_x, D_y, and D_z. Using a graphing utility, we get

$$D_x = \begin{vmatrix} 5 & 1 & 1 \\ -5 & -4 & -3 \\ 4 & -1 & 0 \end{vmatrix} = -6 \qquad D_y = \begin{vmatrix} 1 & 5 & 1 \\ 2 & -5 & -3 \\ 1 & 4 & 0 \end{vmatrix} = 10$$

$$D_z = \begin{vmatrix} 1 & 1 & 5 \\ 2 & -4 & -5 \\ 1 & -1 & 4 \end{vmatrix} = -24$$

By Cramer's rule:

$$x = \frac{D_x}{D} = \frac{-6}{-4} = \frac{3}{2}, \quad y = \frac{D_y}{D} = \frac{10}{-4} = -\frac{5}{2}, \quad \text{and} \quad z = \frac{D_z}{D} = \frac{-24}{-4} = 6$$

The solution is $\left(\frac{3}{2}, -\frac{5}{2}, 6\right)$. ■

Applications

Consider the following problem:

Determine the area of the parallelogram in Figure 5.25 (in terms of the given coordinates).

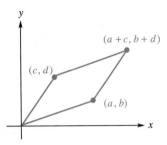

Figure 5.25

If we frame the parallelogram with a rectangle, we can partition the resulting region as shown in Figure 5.26. The triangles labeled R_1 are congruent, the triangles labeled R_2 are congruent, and the rectangles labeled R_3 are congruent. The areas of these regions are

$$R_1 = \tfrac{1}{2}ab \qquad R_2 = \tfrac{1}{2}cd \qquad R_3 = bc$$

The area P of the parallelogram is

$$\begin{aligned} P &= (\text{Area of framing rectangle}) - (2R_1 + 2R_2 + 2R_3) \\ &= (a + c)(b + d) - (ab + cd + 2bc) \\ &= ad - bc \end{aligned}$$

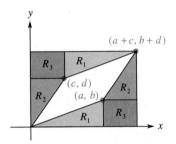

Figure 5.26

Consider the 2×2 matrix with rows consisting of the coordinates (a, b) and (c, d):

$$\begin{bmatrix} a & b \\ c & d \end{bmatrix}$$

In terms of determinants, the area of the parallelogram is

$$\begin{vmatrix} a & b \\ c & d \end{vmatrix} = ad - bc$$

This can be extended to determine the area of any polygon. The formula that follows is known as the *surveyor's formula* (see Exercise 51).

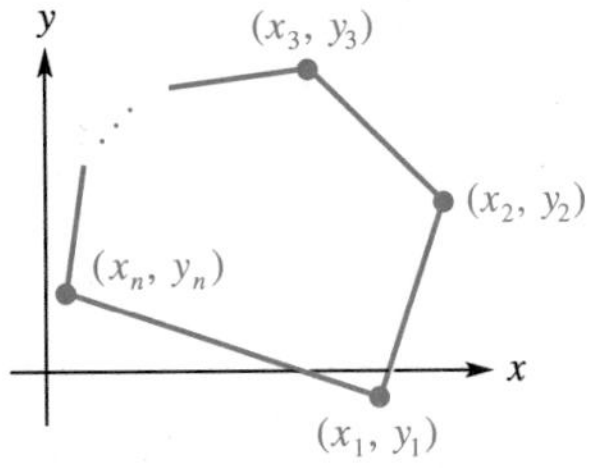

Surveyor's Formula

Let the vertices of a polygon be (x_1, y_1), (x_2, y_2), (x_3, y_3), ... ,(x_n, y_n), listed consecutively in a counterclockwise direction (see Figure 5.27, in the margin). The area A of the polygon is

$$A = \frac{1}{2}\left(\begin{vmatrix} x_1 & y_1 \\ x_2 & y_2 \end{vmatrix} + \begin{vmatrix} x_2 & y_2 \\ x_3 & y_3 \end{vmatrix} + \begin{vmatrix} x_3 & y_3 \\ x_4 & y_4 \end{vmatrix} + \cdots + \begin{vmatrix} x_n & y_n \\ x_1 & y_1 \end{vmatrix} \right)$$

Figure 5.27

EXAMPLE 8   Application: Finding the Area of a Polygon

Determine the area of the quadrilateral $PQRS$ with vertices $P(-2, 1)$, $Q(3, 0)$, $R(4, 2)$, and $S(2, 5)$.

SOLUTION

We sketch the vertices to check that they are listed consecutively in a counterclockwise direction (Figure 5.28). By the surveyor's formula, we have

$$A = \frac{1}{2}\left(\begin{vmatrix} -2 & 1 \\ 3 & 0 \end{vmatrix} + \begin{vmatrix} 3 & 0 \\ 4 & 2 \end{vmatrix} + \begin{vmatrix} 4 & 2 \\ 2 & 5 \end{vmatrix} + \begin{vmatrix} 2 & 5 \\ -2 & 1 \end{vmatrix} \right)$$

$$= \frac{1}{2}(-3 + 6 + 16 + 12)$$

$$= 15\tfrac{1}{2}$$

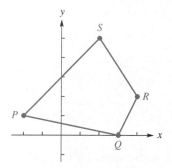

Figure 5.28

The area is $15\tfrac{1}{2}$ square units.

EXERCISE SET 5.5

A

In Exercises 1–6, evaluate each determinant.

1. $\begin{vmatrix} 2 & 3 \\ 5 & 6 \end{vmatrix}$ **2.** $\begin{vmatrix} 4 & 3 \\ 6 & 2 \end{vmatrix}$

3. $\begin{vmatrix} 8 & -5 \\ 11 & -7 \end{vmatrix}$ **4.** $\begin{vmatrix} -6 & -3 \\ 10 & 4 \end{vmatrix}$

5. $\begin{vmatrix} r+1 & 2 \\ 3 & r-1 \end{vmatrix}$ **6.** $\begin{vmatrix} r-2 & 4 \\ 2 & r+1 \end{vmatrix}$

In Exercises 7–10, determine the area of each polygon.

7. Triangle PQR with vertices $P(-3, 2)$, $Q(3, -1)$, and $R(2, 4)$.

8. Quadrilateral $ABCD$ with vertices $A(-4, 1)$, $B(-1, -2)$, $C(3, 0)$, and $D(1, 5)$.

9. Polygon $PQRST$ with vertices $P(1, 1)$, $Q(5, 0)$, $R(7, 2)$, $S(3, 6)$, and $T(-1, 3)$.

10. Polygon $ABCDE$ with vertices $A(2, 1)$, $B(8, 2)$, $C(8, 5)$, $D(0, 8)$, and $E(0, 3)$.

In Exercises 11–14, evaluate each determinant by hand. If your graphing utility evaluates determinants, use it to check your results.

11. $\begin{vmatrix} 1 & 2 & -2 \\ 5 & 0 & 4 \\ 3 & 1 & 3 \end{vmatrix}$ **12.** $\begin{vmatrix} 4 & 1 & 0 \\ -3 & 2 & 3 \\ 6 & 1 & 2 \end{vmatrix}$

13. $\begin{vmatrix} 3 & 2 & -4 \\ -2 & 3 & 1 \\ 3 & 7 & 1 \end{vmatrix}$ **14.** $\begin{vmatrix} 8 & 3 & 5 \\ 0 & 1 & -2 \\ 3 & 4 & 4 \end{vmatrix}$

15. (a) Evaluate

$$\begin{vmatrix} i & j & k \\ 1 & -4 & 4 \\ 2 & 1 & -3 \end{vmatrix}$$

by expanding along the first row.

(b) Evaluate the determinant in part (a) by expanding along the second row.

16. (a) Evaluate

$$\begin{vmatrix} i & j & k \\ 5 & 3 & 1 \\ -4 & 6 & -2 \end{vmatrix}$$

by expanding along the first row.

(b) Evaluate the determinant in part (a) by expanding along the second row.

In Exercises 17–22, use Cramer's rule to solve each system for x and y.

17. $\begin{cases} -x + 4y = 10 \\ 2x - 7y = 6 \end{cases}$ **18.** $\begin{cases} 3x + 4y = 0 \\ 4x + 5y = 3 \end{cases}$

19. $\begin{cases} 5x + 6y = 4 \\ 2x + 3y = 7 \end{cases}$ **20.** $\begin{cases} 4x + 9y = 1 \\ 3x + 6y = 5 \end{cases}$

21. $\begin{cases} 4x - y = 0 \\ 5x + 2y = 3 \end{cases}$ **22.** $\begin{cases} 3x - y = 11 \\ -4x + 5y = 8 \end{cases}$

B

In Exercises 23–26, use a graphing utility to evaluate each determinant.

23. $\begin{vmatrix} 0.5 & 1.2 & -3 \\ 2.4 & 3.1 & 0 \\ -0.7 & 5.4 & -6 \end{vmatrix}$ **24.** $\begin{vmatrix} -2.8 & 4.1 & 0 \\ 5.6 & -0.2 & 4 \\ 2 & 3.7 & 0.5 \end{vmatrix}$

25. $\begin{vmatrix} 2 & 1 & 3 & -1 \\ 6 & -2 & 0 & 5 \\ 4 & -3 & -1 & -2 \\ 7 & 11 & 3 & 8 \end{vmatrix}$ **26.** $\begin{vmatrix} 6 & -4 & 1 & 3 \\ 5 & 2 & 9 & -1 \\ 0 & -2 & 4 & 13 \\ -3 & 7 & 1 & 8 \end{vmatrix}$

In Exercises 27–34, use Cramer's rule to solve each system for x, y, and z.

27. $\begin{cases} x - y + z = 4 \\ 2x + y - z = -1 \\ y + z = -1 \end{cases}$ **28.** $\begin{cases} x + y + 2z = 1 \\ 2x - y + z = 2 \\ y - 2z = 3 \end{cases}$

29. $\begin{cases} x + y + 2z = 1 \\ 3x + y - z = 2 \\ x + z = 3 \end{cases}$ **30.** $\begin{cases} x + 2y + z = 7 \\ 2x + y - 2z = -3 \\ y + z = 4 \end{cases}$

31. $\begin{cases} 3x + y - z = 6 \\ x - y + z = -4 \\ 5x - 3y + z = -5 \end{cases}$

32. $\begin{cases} 3x - 2y + z = -2 \\ x + 4y - 3z = 6 \\ -2x + y + z = 2 \end{cases}$

33. $\begin{cases} x + y + 2z = k \\ 2x - y = 0 \\ 3y + z = 1 \end{cases}$ **34.** $\begin{cases} x + 2y + z = k \\ x + 4z = 1 \\ 2x + 2y - z = 0 \end{cases}$

In Exercises 35–38 use Cramer's rule to solve each system. Use a graphing utility to evaluate the determinants (to the nearest 0.01).

35. $\begin{cases} 2.31x + 1.87y = 12.3 \\ 0.14x - 5.60y = 4.29 \end{cases}$

36. $\begin{cases} -5.43x + 2.04y = 16.8 \\ 1.77x + 3.90y = 2.73 \end{cases}$

37. $\begin{cases} 3.14x + 1.59y - 2.65z = 1.61 \\ 8.03y + 1.90z = 2.34 \\ 3.01x - 4.77y + 6.93z = 5.51 \end{cases}$

38. $\begin{cases} 2.33x - 1.01y + 0.74z = 14.05 \\ -6.02x + 8.10z = 9.55 \\ 2.71x - 8.28y + 0.25z = 8.21 \end{cases}$

In Exercises 39 and 40, use Cramer's rule to solve each system for x and y.

39. $\begin{cases} ux + 2vy = 0 \\ -3ux + 4vy = f \end{cases}$ **40.** $\begin{cases} 8ux - vy = 0 \\ 5ux + 2vy = g \end{cases}$

In Exercises 41 and 42, use Cramer's rule to solve each system for x only.

41. $\begin{cases} x + y + 3z = -1 \\ x - 5y - 4z = 5 \\ x + 7y = 3 \end{cases}$ **42.** $\begin{cases} x + y + z = 1 \\ 2x - 3y + 7z = 2 \\ x + 2z = 1 \end{cases}$

In Exercises 43 and 44, use Cramer's rule to solve each system for y only.

43. $\begin{cases} 2x + 2y + 3z = 2 \\ x + y + z = 1 \\ 2y - z = 0 \end{cases}$

44. $\begin{cases} 3x + y + 4z = 1 \\ -x + 3y + 10z = 2 \\ x - 5y = -2 \end{cases}$

In Exercises 45 and 46, use Cramer's rule to solve each system for z only.

45. $\begin{cases} x - y + z = 1 \\ -x + 3y + z = 0 \\ 2x - 5y + 3z = 3 \end{cases}$ **46.** $\begin{cases} 2x + y + z = 1 \\ -4x + 3z = 0 \\ 3x + 5y + 2z = 2 \end{cases}$

47. Find the values of r that satisfy

$$\begin{vmatrix} r - 1 & 8 \\ 3 & r + 1 \end{vmatrix} = 0$$

48. Find the values of r that satisfy

$$\begin{vmatrix} r - 3 & 5 \\ 7 & r - 1 \end{vmatrix} = 0$$

49. Find the values of r that satisfy

$$\begin{vmatrix} r + 2 & -2 & 3 \\ -2 & r - 1 & 6 \\ 1 & 2 & r \end{vmatrix} = 0$$

C

50. Use the surveyor's formula to find an equation of the line passing through $(4, 8)$ and $(-2, -13)$.

HINT: The point (x, y) lies on the line through $(4, 8)$ and $(-2, -13)$ if and only if the area of the "triangle" formed by these three points is 0.

51. This exercise shows the idea of a derivation of the surveyor's formula.

(a) Show that the area of a triangle with vertices $(0, 0)$, (a, b), and (c, d) is

$$\frac{1}{2}\begin{vmatrix} a & b \\ c & d \end{vmatrix}$$

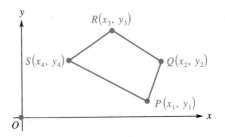

Figure 5.29

(b) Show that the area of the polygon in Figure 5.29 is

$$A = \frac{1}{2}\left(\begin{vmatrix} x_1 & y_1 \\ x_2 & y_2 \end{vmatrix} + \begin{vmatrix} x_2 & y_2 \\ x_3 & y_3 \end{vmatrix} + \begin{vmatrix} x_3 & y_3 \\ x_4 & y_4 \end{vmatrix} - \begin{vmatrix} x_1 & y_1 \\ x_4 & y_4 \end{vmatrix} \right)$$

HINT: Consider the areas of triangle OPQ, triangle OQR, triangle ORS, and triangle OPS.

(c) Show that the expression in part (b) is equal to

$$A = \frac{1}{2}\left(\begin{vmatrix} x_1 & y_1 \\ x_2 & y_2 \end{vmatrix} + \begin{vmatrix} x_2 & y_2 \\ x_3 & y_3 \end{vmatrix} + \begin{vmatrix} x_3 & y_3 \\ x_4 & y_4 \end{vmatrix} + \begin{vmatrix} x_4 & y_4 \\ x_1 & y_1 \end{vmatrix} \right)$$

52. For the system

$$\begin{cases} ax + by = m \\ cx + dy = n \end{cases}$$

we multiplied each side of the first equation by d, the second equation by $-b$, and added the resulting equations to get

$$(ad - bc)x = md - nb$$

(a) Explain why the system is dependent if $D = 0$ and $D_x = 0$.

(b) Explain why the system is inconsistent if $D = 0$ and $D_x \neq 0$.

SECTION 5.6

- Equality, Addition, and Subtraction
- Multiplying a Number and a Matrix
- Matrix Multiplication
- Applications

MATRIX OPERATIONS

In Section 5.4 we used matrices to solve systems of linear equations. Matrices have many other applications. They are especially useful in organizing and manipulating the large amounts of data that are inherent in many real-world problems. To this end, we will discuss the fundamental operations of matrices.

Equality, Addition, and Subtraction

Recall that a matrix with m rows and n columns is an $m \times n$ matrix, and the expression $m \times n$ is the *dimension* of the matrix. Two matrices are *equal* if and only if they have the same dimension and their corresponding entries are equal. For example,

$$\begin{bmatrix} -2 & 1 \\ 3 & 4 \end{bmatrix} = \begin{bmatrix} -2 & 1 \\ x & 4 \end{bmatrix} \quad \text{if and only if} \quad x = 3$$

If two matrices have the same dimension, their *sum* is the matrix that results when the corresponding entries are added. Similarly, the *difference* of two matrices with the same dimension is the matrix that results when the corresponding entries are subtracted. Addition or subtraction is not defined for matrices with different dimensions.

■ EXAMPLE 1   Adding and Subtracting Matrices

Let

$$A = \begin{bmatrix} 4 & 0 \\ -3 & 1 \\ 2 & -6 \end{bmatrix} \qquad B = \begin{bmatrix} 8 & 11 \\ 2 & -4 \\ 7 & 6 \end{bmatrix} \qquad C = \begin{bmatrix} -9 & -3 \\ 5 & 1 \end{bmatrix}$$

Determine (if possible):

(a) $A + B$ **(b)** $A + C$ **(c)** $B - A$

SOLUTION

(a) $A + B = \begin{bmatrix} 4 & 0 \\ -3 & 1 \\ 2 & -6 \end{bmatrix} + \begin{bmatrix} 8 & 11 \\ 2 & -4 \\ 7 & 6 \end{bmatrix}$

$$= \begin{bmatrix} 4+8 & 0+11 \\ -3+2 & 1+(-4) \\ 2+7 & -6+6 \end{bmatrix} = \begin{bmatrix} 12 & 11 \\ -1 & -3 \\ 9 & 0 \end{bmatrix}$$

(b) The dimension of A (3×2) does not equal the dimension of C (2×2), so these matrices cannot be added.

(c) $B - A = \begin{bmatrix} 8 & 11 \\ 2 & -4 \\ 7 & 6 \end{bmatrix} - \begin{bmatrix} 4 & 0 \\ -3 & 1 \\ 2 & -6 \end{bmatrix}$

$$= \begin{bmatrix} 8-4 & 11-0 \\ 2-(-3) & -4-1 \\ 7-2 & 6-(-6) \end{bmatrix} = \begin{bmatrix} 4 & 11 \\ 5 & -5 \\ 5 & 12 \end{bmatrix}$$

Multiplying a Number and a Matrix

Consider the repeated addition of a matrix. For example,

$$\begin{bmatrix} 2 & -7 \\ 0 & 4 \end{bmatrix} + \begin{bmatrix} 2 & -7 \\ 0 & 4 \end{bmatrix} + \begin{bmatrix} 2 & -7 \\ 0 & 4 \end{bmatrix} = \begin{bmatrix} 6 & -21 \\ 0 & 12 \end{bmatrix}$$

The notion of multiplication as repeated addition suggests that

$$\begin{bmatrix} 2 & -7 \\ 0 & 4 \end{bmatrix} + \begin{bmatrix} 2 & -7 \\ 0 & 4 \end{bmatrix} + \begin{bmatrix} 2 & -7 \\ 0 & 4 \end{bmatrix} = 3\begin{bmatrix} 2 & -7 \\ 0 & 4 \end{bmatrix}$$

We use this idea to define the *product of the real number k and the matrix A*, denoted *kA*, as the matrix that results from multiplying each entry of *A* by *k*. For example,

$$3\begin{bmatrix} 2 & -7 \\ 0 & 4 \end{bmatrix} = \begin{bmatrix} 3(2) & 3(-7) \\ 3(0) & 3(4) \end{bmatrix} = \begin{bmatrix} 6 & -21 \\ 0 & 12 \end{bmatrix}$$

We use $-A$ to represent $(-1)A$. For example,

$$-\begin{bmatrix} 5 & 2 & -2 \\ -1 & 0 & 6 \end{bmatrix} = \begin{bmatrix} -5 & -2 & 2 \\ 1 & 0 & -6 \end{bmatrix}$$

EXAMPLE 2 Combining Multiplication with Subtraction

Let

$$A = \begin{bmatrix} 2 & \frac{3}{2} \\ -6 & 1 \\ -5 & -1 \end{bmatrix} \quad \text{and} \quad B = \begin{bmatrix} -4 & -3 \\ 0 & 1 \\ 2 & 4 \end{bmatrix}$$

Determine $4A - 0.5B$.

SOLUTION

NOTE: *Most graphing utilities can perform these matrix operations. Enter A and B in a graphing utility and check the results of this example.*

$$4A - 0.5B = 4\begin{bmatrix} 2 & \frac{3}{2} \\ -6 & 1 \\ -5 & -1 \end{bmatrix} - 0.5\begin{bmatrix} -4 & -3 \\ 0 & 1 \\ 2 & 4 \end{bmatrix}$$

$$= \begin{bmatrix} 8 & 6 \\ -24 & 4 \\ -20 & -4 \end{bmatrix} - \begin{bmatrix} -2 & -1.5 \\ 0 & 0.5 \\ 1 & 2 \end{bmatrix} = \begin{bmatrix} 10 & 7.5 \\ -24 & 3.5 \\ -21 & -6 \end{bmatrix}$$

EXAMPLE 3 Factoring a Number Out of a Matrix

Factor $\frac{1}{2}$ out of the matrix

$$A = \begin{bmatrix} 3 & \frac{1}{2} \\ -4 & 1 \end{bmatrix}$$

In other words, determine B so that $A = \frac{1}{2}B$.

SOLUTION

$$\begin{bmatrix} 3 & \frac{1}{2} \\ -4 & 1 \end{bmatrix} = \frac{1}{2} \cdot 2 \begin{bmatrix} 3 & \frac{1}{2} \\ -4 & 1 \end{bmatrix} = \frac{1}{2} \begin{bmatrix} 6 & 1 \\ -8 & 2 \end{bmatrix}$$

Matrix Multiplication

Suppose a class is graded such that 25% of the grade is based on homework, 35% of the grade is based on the midterm exam, and 40% of the grade is based on the final exam:

Homework	Midterm	Final
[0.25	0.35	0.40]

To keep it simple, suppose there are only four people in the class and their scores are organized in the following matrix:

	Ann	Bill	Julio	Robert	
⎡	80	92	90	71 ⎤	Homework
	85	88	76	84	Midterm
⎣	82	80	94	88 ⎦	Final

The scores for course grades are computed as follows:

Ann: $(0.25)80 + (0.35)85 + (0.40)82 = 82.55$

Bill: $(0.25)92 + (0.35)88 + (0.40)80 = 85.8$

Julio: $(0.25)90 + (0.35)76 + (0.40)94 = 86.7$

Robert: $(0.25)71 + (0.35)84 + (0.40)88 = 82.35$

These scores can be calculated with a matrix operation called *matrix multiplication*. Place the two matrices from above side by side:

$$[0.25 \quad 0.35 \quad 0.40] \begin{bmatrix} 80 & 92 & 90 & 71 \\ 85 & 88 & 76 & 84 \\ 82 & 80 & 94 & 88 \end{bmatrix}$$

Ann's grade is computed by multiplying entries in the *row* of the left-hand matrix by corresponding entries in the first *column* of the right-hand matrix, and then adding the products:

$$\begin{bmatrix} 0.25 & 0.35 & 0.40 \end{bmatrix} \begin{bmatrix} 80 & 92 & 90 & 71 \\ 85 & 88 & 76 & 84 \\ 82 & 80 & 94 & 88 \end{bmatrix}$$

$$= \begin{bmatrix} (0.25)80 + (0.35)85 + (0.40)82 & & & \end{bmatrix}$$

$$= \begin{bmatrix} 82.55 & & & \end{bmatrix}$$

Bill's grade is computed by multiplying entries in the *row* of the left-hand matrix by corresponding entries in the second *column* of the right-hand matrix, and then adding the products:

$$\begin{bmatrix} 0.25 & 0.35 & 0.40 \end{bmatrix} \begin{bmatrix} 80 & 92 & 90 & 71 \\ 85 & 88 & 76 & 84 \\ 82 & 80 & 94 & 88 \end{bmatrix}$$

$$= \begin{bmatrix} 82.55 & (0.25)92 + (0.35)88 + (0.40)80 & & \end{bmatrix}$$

$$= \begin{bmatrix} 82.55 & 85.8 & & \end{bmatrix}$$

The grades for Julio and Robert are calculated the same way:

$$\begin{bmatrix} 0.25 & 0.35 & 0.40 \end{bmatrix} \begin{bmatrix} 80 & 92 & 90 & 71 \\ 85 & 88 & 76 & 84 \\ 82 & 80 & 94 & 88 \end{bmatrix}$$

$$= \begin{bmatrix} 82.55 & 85.8 & (0.25)90 + (0.35)76 + (0.40)94 & \end{bmatrix}$$

$$= \begin{bmatrix} 82.55 & 85.8 & 86.7 & \end{bmatrix}$$

$$\begin{bmatrix} 0.25 & 0.35 & 0.40 \end{bmatrix} \begin{bmatrix} 80 & 92 & 90 & 71 \\ 85 & 88 & 76 & 84 \\ 82 & 80 & 94 & 88 \end{bmatrix}$$

$$= \begin{bmatrix} 82.55 & 85.8 & 86.7 & (0.25)71 + (0.35)84 + (0.40)88 \end{bmatrix}$$

$$= \begin{bmatrix} 82.55 & 85.8 & 86.7 & 82.35 \end{bmatrix}$$

The calculations for all four grades can be summarized as

$$\begin{bmatrix} 0.25 & 0.35 & 0.40 \end{bmatrix} \begin{bmatrix} 80 & 92 & 90 & 71 \\ 85 & 88 & 76 & 84 \\ 82 & 80 & 94 & 88 \end{bmatrix} = \begin{bmatrix} 82.55 & 85.8 & 86.7 & 82.35 \end{bmatrix}$$

The definition of matrix multiplication relies on the ideas in this example:

Matrix Multiplication

> Let A be an $m \times n$ matrix and let B be an $n \times p$ matrix. The product AB is an $m \times p$ matrix. The entry in the ith row and jth column of AB is determined by multiplying each entry in the ith row of A by its corresponding entry in the jth column of B, and then adding the products.

Notice the dimensions of A and B in this definition. The number of columns in A must be equal to the number of rows in B:

$$
\begin{array}{cc}
A & B \\
m \times n & n \times p
\end{array}
$$

These values
must be equal

Notice also that the product AB has the same number of rows as A and the same number of columns as B:

$$
\begin{array}{cc}
A & B \\
m \times n & n \times p
\end{array}
$$

The dimension of
AB is $m \times p$

Now we will look at several examples.

■ EXAMPLE 4   Multiplying Two Matrices

Let

$$
A = \begin{bmatrix} 1 & 3 \\ -2 & 4 \\ 0 & 5 \end{bmatrix} \quad \text{and} \quad B = \begin{bmatrix} 6 & 7 \\ 8 & -9 \end{bmatrix}
$$

Determine AB.

SOLUTION

First check that the number of columns of A equals the number of rows of B:

$$
\begin{array}{cc}
A & B \\
3 \times 2 & 2 \times 2
\end{array}
$$

Equal

The dimension of
AB is 3×2

Recall from Section 5.4 that the entry in the ith row and jth column can be denoted as c_{ij}. The product AB is defined and has the form

$$\begin{bmatrix} 1 & 3 \\ -2 & 4 \\ 0 & 5 \end{bmatrix} \begin{bmatrix} 6 & 7 \\ 8 & -9 \end{bmatrix} = \begin{bmatrix} c_{11} & c_{12} \\ c_{21} & c_{22} \\ c_{31} & c_{32} \end{bmatrix}$$

The entry in the first row and first column, c_{11}, is determined by multiplying each entry in the first row of A by the corresponding entry in the first column of B, and then adding the products:

$$\begin{bmatrix} 1 & 3 \\ -2 & 4 \\ 0 & 5 \end{bmatrix} \begin{bmatrix} 6 & 7 \\ 8 & -9 \end{bmatrix} = \begin{bmatrix} 1(6) + 3(8) & c_{12} \\ c_{21} & c_{22} \\ c_{31} & c_{32} \end{bmatrix} = \begin{bmatrix} 30 & c_{12} \\ c_{21} & c_{22} \\ c_{31} & c_{32} \end{bmatrix}$$

We determine the entry in the first row and second column, c_{12}, by multiplying each entry in the first row of A by the corresponding entry in the second column of B, and then adding the products:

$$\begin{bmatrix} 1 & 3 \\ -2 & 4 \\ 0 & 5 \end{bmatrix} \begin{bmatrix} 6 & 7 \\ 8 & -9 \end{bmatrix} = \begin{bmatrix} 30 & 1(7) + 3(-9) \\ c_{21} & c_{22} \\ c_{31} & c_{32} \end{bmatrix} = \begin{bmatrix} 30 & -20 \\ c_{21} & c_{22} \\ c_{31} & c_{32} \end{bmatrix}$$

Similarly, the entry in the second row and first column, c_{21}, is determined from the second row of A and the first column of B:

$$\begin{bmatrix} 1 & 3 \\ -2 & 4 \\ 0 & 5 \end{bmatrix} \begin{bmatrix} 6 & 7 \\ 8 & -9 \end{bmatrix} = \begin{bmatrix} 30 & -20 \\ (-2)6 + 4(8) & c_{22} \\ c_{31} & c_{32} \end{bmatrix} = \begin{bmatrix} 30 & -20 \\ 20 & c_{22} \\ c_{31} & c_{32} \end{bmatrix}$$

Continuing in this way, we get

$$\begin{bmatrix} 1 & 3 \\ -2 & 4 \\ 0 & 5 \end{bmatrix} \begin{bmatrix} 6 & 7 \\ 8 & -9 \end{bmatrix} = \begin{bmatrix} 30 & -20 \\ 20 & -2(7) + 4(-9) \\ 0(6) + 5(8) & 0(7) + 5(-9) \end{bmatrix} = \begin{bmatrix} 30 & -20 \\ 20 & -50 \\ 40 & -45 \end{bmatrix}$$

● EXAMPLE 5 More Matrix Multiplication

Let

$$M = \begin{bmatrix} 2 & -1 & 5 \\ -3 & 4 & 6 \end{bmatrix} \qquad N = \begin{bmatrix} x \\ 8 \\ -2 \end{bmatrix} \qquad P = \begin{bmatrix} 11 & 0 \\ -7 & 7 \\ 12 & 9 \end{bmatrix}$$

Determine each product, if possible.

(a) *MN*　　**(b)** *MP*　　**(c)** *PM*　　**(d)** *NM*

SOLUTION

(a) First check that the number of columns of *M* equals the number of rows of *N*:

$$\begin{array}{cc} M & N \\ 2 \times 3 & 3 \times 1 \end{array}$$

Equal

The dimension of
MN is 2×1

The product *MN* is

$$\begin{bmatrix} 2 & -1 & 5 \\ -3 & 4 & 6 \end{bmatrix} \begin{bmatrix} x \\ 8 \\ -2 \end{bmatrix} = \begin{bmatrix} 2x + (-1)8 + 5(-2) \\ -3x + (4)8 + 6(-2) \end{bmatrix}$$

$$= \begin{bmatrix} 2x - 18 \\ -3x + 20 \end{bmatrix}$$

(b) The dimensions of *M*, *P*, and *MP* are

$$\begin{array}{cc} M & P \\ 2 \times 3 & 3 \times 2 \end{array}$$

Equal

The dimension of
MP is 2×2

The product of these two matrices is

$$MP = \begin{bmatrix} 2 & -1 & 5 \\ -3 & 4 & 6 \end{bmatrix} \begin{bmatrix} 11 & 0 \\ -7 & 7 \\ 12 & 9 \end{bmatrix}$$

$$= \begin{bmatrix} 2(11) + (-1)(-7) + 5(12) & 2(0) + (-1)(7) + 5(9) \\ -3(11) + 4(-7) + 6(12) & -3(0) + 4(7) + 6(9) \end{bmatrix}$$

$$= \begin{bmatrix} 89 & 38 \\ 11 & 82 \end{bmatrix}$$

(c) First, note the dimensions:

$$\begin{array}{cc} P & M \\ 3 \times 2 & 2 \times 3 \end{array}$$

Equal

The dimension of
PM is 3×3

NOTE: *The results of parts (b) and (c) point out that MP ≠ PM. In general, matrix multiplication is not commutative.*

NOTE: *Most graphing utilities will calculate the product of two matrices. Try entering matrices M and P on a graphing utility and verifying the results.*

The product is

$$PM = \begin{bmatrix} 11 & 0 \\ -7 & 7 \\ 12 & 9 \end{bmatrix} \begin{bmatrix} 2 & -1 & 5 \\ -3 & 4 & 6 \end{bmatrix}$$

$$= \begin{bmatrix} 11(2) + 0(-3) & 11(-1) + 0(4) & 11(5) + 0(6) \\ (-7)(2) + 7(-3) & (-7)(-1) + 7(4) & (-7)(5) + 7(6) \\ 12(2) + 9(-3) & 12(-1) + 9(4) & 12(5) + 9(6) \end{bmatrix}$$

$$= \begin{bmatrix} 22 & -11 & 55 \\ -35 & 35 & 7 \\ -3 & 24 & 114 \end{bmatrix}$$

(d) First, note the dimensions:

$$\begin{array}{cc} N & M \\ 3 \times 1 & 2 \times 3 \end{array}$$

Not equal

The product NM is not defined.

Applications

Matrix operations have important applications in engineering, economics, advanced mathematics, and business. We conclude this section with two business applications of matrix operations.

EXAMPLE 6   Application: Sales Commission

A chain of retail stores is divided into three regions: California, Oregon, and Washington. Each store sells video and stereo equipment in one department, and appliances in another department. The gross sales (in dollars) for a particular month in each department for each sales region are given by the following matrix:

	CA	OR	WA	
	40,000	15,000	32,000	Video/stereo
	110,000	85,000	104,000	Appliance

SECTION 5.6 • MATRIX OPERATIONS **487**

If there is a 6% commission on gross sales, determine the commission paid to each department in each region.

SOLUTION

The commission is 6% of each of the entries in the given matrix:

$$0.06 \begin{bmatrix} 40{,}000 & 15{,}000 & 32{,}000 \\ 110{,}000 & 85{,}000 & 104{,}000 \end{bmatrix} = \begin{bmatrix} 2400 & 900 & 1920 \\ 6600 & 5100 & 6240 \end{bmatrix}$$

The commissions (in dollars) by department and region are

$$\begin{array}{ccc} \text{CA} & \text{OR} & \text{WA} \end{array}$$
$$\begin{bmatrix} 2400 & 900 & 1920 \\ 6600 & 5100 & 6240 \end{bmatrix} \begin{array}{l} \text{Video/stereo} \\ \text{Appliance} \end{array}$$

EXAMPLE 7 Application: Labor Costs of Production

A cellular phone manufacturer produces two models, a standard model and a deluxe model. The production of each unit requires assembly and testing. The numbers of hours required for each model are given in the following matrix:

$$\begin{array}{cc} \text{Assembly} & \text{Testing} \end{array}$$
$$\begin{bmatrix} 2 & 1 \\ 2.5 & 1.5 \end{bmatrix} \begin{array}{l} \text{Standard} \\ \text{Deluxe} \end{array}$$

The company has two manufacturing plants, one in New Jersey and one in Mexico. The hourly labor rates (in dollars) for each plant are given by the following matrix:

$$\begin{array}{cc} \text{NJ} & \text{Mexico} \end{array}$$
$$\begin{bmatrix} 10 & 8.5 \\ 11 & 12 \end{bmatrix} \begin{array}{l} \text{Assembly} \\ \text{Testing} \end{array}$$

Use matrix multiplication to determine the total labor cost of each unit for each plant.

SOLUTION

Consider the labor costs of a particular model at a particular plant, say the standard model from New Jersey. For assembly, each standard model from New Jersey requires 2 hours at $10 per hour. For testing, each standard model from New Jersey requires 1 hour at $11 per hour. Therefore, the total labor cost of the standard model at the New Jersey plant is

$$(2 \text{ hours})(\$10 \text{ per hour}) + (1 \text{ hour})(\$11 \text{ per hour}) = \$31$$

Notice that this is the entry in the first row and first column of the following product:

$$\begin{bmatrix} 2 & 1 \\ 2.5 & 1.5 \end{bmatrix} \begin{bmatrix} 10 & 8.5 \\ 11 & 12 \end{bmatrix} = \begin{bmatrix} 31 & \\ & \end{bmatrix}$$

The total labor cost of each unit for each plant can be found with the following matrix multiplication:

$$\begin{bmatrix} 2 & 1 \\ 2.5 & 1.5 \end{bmatrix} \begin{bmatrix} 10 & 8.5 \\ 11 & 12 \end{bmatrix} = \begin{matrix} & \text{NJ} & \text{Mexico} \\ \begin{bmatrix} 31.00 & 29.00 \\ 41.50 & 39.25 \end{bmatrix} & \begin{matrix} \text{Standard} \\ \text{Deluxe} \end{matrix} \end{matrix}$$

EXERCISE SET 5.6

A

In Exercises 1–14, let

$$A = \begin{bmatrix} 3 & 11 & -7 \\ 8 & -1 & 6 \end{bmatrix} \qquad B = \begin{bmatrix} -2 & 0 & 14 \\ 9 & -10 & -5 \end{bmatrix}$$

$$C = \begin{bmatrix} -4 & 12 \\ 16 & 1 \end{bmatrix} \qquad D = \begin{bmatrix} 2.5 & 8 \\ 7 & 4 \end{bmatrix}$$

Determine (if possible):

1. $A + B$

2. $C + D$

3. $B - A$

4. $C - D$

5. $A - C$

6. $D - A$

7. $B + B$

8. $C + C$

9. $-2C$

10. $\frac{1}{2}B$

11. $2A - 5B$

12. $2D - 3C$

13. $3A + 2C$

14. $6B - 2C$

15. Factor $\frac{1}{3}$ out of the matrix

$$A = \begin{bmatrix} 2 & -1 \\ 3 & 0 \end{bmatrix}$$

16. Factor $\frac{1}{4}$ out of the matrix

$$B = \begin{bmatrix} 2 & -5 \\ 6 & 0 \end{bmatrix}$$

17. Factor 2 out of the matrix

$$C = \begin{bmatrix} -8 & 3 \\ 4 & 0 \end{bmatrix}$$

18. Factor 3 out of the matrix

$$D = \begin{bmatrix} 6 & 0 \\ 2 & -3 \end{bmatrix}$$

19. Let

$$A = \begin{bmatrix} 2 & 4 \\ 0 & -3 \\ 1 & 5 \end{bmatrix} \quad \text{and} \quad B = \begin{bmatrix} 3 & 2 \\ -1 & 6 \end{bmatrix}$$

Determine AB.

20. Let

$$A = \begin{bmatrix} 3 & 5 \\ -1 & -4 \\ 0 & 4 \end{bmatrix} \quad \text{and} \quad B = \begin{bmatrix} 2 & 3 \\ -2 & 5 \end{bmatrix}$$

Determine AB.

21. Let

$$M = \begin{bmatrix} 3 & 7 \\ 1 & -8 \end{bmatrix} \quad \text{and} \quad N = \begin{bmatrix} -5 & 2 \\ -1 & 6 \end{bmatrix}$$

Determine MN and NM.

22. Let

$$C = \begin{bmatrix} 9 & -2 \\ 5 & 1 \end{bmatrix} \quad \text{and} \quad D = \begin{bmatrix} 4 & 3 \\ -8 & 1 \end{bmatrix}$$

Determine CD and DC.

In Exercises 23–28, let

$$M = \begin{bmatrix} -3 & 0 \\ 1 & -4 \\ 11 & 6 \end{bmatrix} \quad N = \begin{bmatrix} 3 \\ x \\ -4 \end{bmatrix} \quad P = \begin{bmatrix} 3 & 0 & 6 \\ -2 & 5 & -1 \end{bmatrix}$$

Determine each product (if possible).

23. PM **24.** PN **25.** MP **26.** MN

27. NP **28.** NM

B

In Exercises 29–37, let

$$A = \begin{bmatrix} 4 & 1 & 2 \\ -1 & 0 & -3 \end{bmatrix} \quad B = \begin{bmatrix} 4 & -1 \\ 6 & -2 \\ 3 & 5 \end{bmatrix} \quad C = \begin{bmatrix} 2 & 7 \\ -1 & 0 \end{bmatrix}$$

Determine each product (if possible).

HINT: If A is a matrix, then A^2 is AA.

29. AB **30.** BA **31.** $(AB)C$ **32.** $(CA)B$

33. C^2 **34.** A^2 **35.** B^2 **36.** $(AB)^2$

37. $(BA)^2$

38. Let

$$A = \begin{bmatrix} 0.2 & 0.3 & 0.5 \\ 0.3 & 0.4 & 0.3 \\ 0.4 & 0.5 & 0.1 \end{bmatrix}$$

Use a graphing utility to evaluate each expression (to the nearest 0.001).

(a) A^2 (b) A^4 (c) A^8

39. Repeat Exercise 38 with

$$A = \begin{bmatrix} 0.2 & 0.4 & 0.4 \\ 0.7 & 0.1 & 0.2 \\ 0.6 & 0.1 & 0.3 \end{bmatrix}$$

40. Let

$$A = \begin{bmatrix} 2 & 1 & 3 \\ 4 & 8 & 7 \\ 1 & -3 & 6 \end{bmatrix} \quad X = \begin{bmatrix} x \\ y \\ z \end{bmatrix} \quad B = \begin{bmatrix} 23 \\ 11 \\ 19 \end{bmatrix}$$

(a) Determine AX.

(b) Suppose $AX = B$. What system of equations does this represent?

HINT: Recall that two matrices are equal if and only if they have the same dimension and corresponding entries are equal.

41. Repeat Exercise 40 with

$$A = \begin{bmatrix} 3 & 2 & 5 \\ 5 & 9 & 8 \\ -1 & -5 & 4 \end{bmatrix} \quad X = \begin{bmatrix} x \\ y \\ z \end{bmatrix} \quad B = \begin{bmatrix} 16 \\ 21 \\ 15 \end{bmatrix}$$

42. A drug store chain is divided into four regions: north, east, west, and south. The stores sell pharmaceuticals in one department and groceries in another department. The revenues (in dollars) for the previous month in each department for each region are given by the following matrix:

North	East	West	South	
310,000	408,000	344,000	285,000	Pharmaceuticals
560,000	731,000	586,000	320,000	Groceries

If there is a 5% tax on revenue, determine a matrix indicating the tax owed by each department in each region.

43. A salary schedule (in dollars) for the faculty at a public college depends on rank—lecturer, assistant professor, associate professor, or professor—and number of years of experience according to the following matrix:

Lecturer	Assist. Prof.	Assoc. Prof.	Prof.	
28,200	35,500	42,000	51,400	0–3 years
32,500	39,100	46,700	57,200	4–7 years
36,000	41,800	52,600	62,300	8 or more years

If the faculty receives a 4.5% raise, determine a matrix for the resulting salary schedule.

44. A company that manufactures compressors for air-conditioning units has two factories, one on the east coast and one on the west coast. Each factory makes three types of compressors, one for automobile units, one for residential units, and one for commercial units. The numbers of compressors produced per week at each factory are given by the following matrix:

East	West	
240	226	Auto
152	142	Residential
45	44	Commercial

Find a matrix that gives the weekly production levels if production is increased by 10%.

45. A furniture supplier produces two types of dining tables, a standard table and a deluxe table. The production of each table involves an assembly division and a finishing division. The numbers of hours required for each table are given in the following matrix:

$$
\begin{array}{cc}
\text{Assembly} & \text{Finishing} \\
\begin{bmatrix} 3 & 2 \\ 4 & 3.5 \end{bmatrix} & \begin{array}{l} \text{Standard} \\ \text{Deluxe} \end{array}
\end{array}
$$

The company has three manufacturing plants: one in Portland, one in Atlanta, and one in Boston. Their hourly labor rates (in dollars) for each division are given by the following matrix:

$$
\begin{array}{ccc}
\text{Portland} & \text{Atlanta} & \text{Boston} \\
\begin{bmatrix} 10 & 9.5 & 11 \\ 12 & 11 & 12.5 \end{bmatrix} & & \begin{array}{l} \text{Assembly} \\ \text{Finishing} \end{array}
\end{array}
$$

Use matrix multiplication to determine the total labor cost of each table for each plant.

46. A car stereo retailer sells three models of car stereos, standard (AM/FM radio only), preferred (AM/FM radio and cassette player), and deluxe (AM/FM radio, cassette, and CD player). The retailer sells at two retail outlets, Warehouse Sound #1 and Warehouse Sound #2. The inventory at each location is given in the following matrix:

$$
\begin{array}{ccc}
\text{Standard} & \text{Preferred} & \text{Deluxe} \\
\begin{bmatrix} 45 & 32 & 20 \\ 38 & 30 & 22 \end{bmatrix} & & \begin{array}{l} \text{Warehouse Sound \#1} \\ \text{Warehouse Sound \#2} \end{array}
\end{array}
$$

The wholesale and retail values (in dollars) for each model are given by the following matrix:

$$
\begin{array}{cc}
\text{Wholesale} & \text{Retail} \\
\begin{bmatrix} 78 & 120 \\ 94 & 185 \\ 154 & 260 \end{bmatrix} & \begin{array}{l} \text{Standard} \\ \text{Preferred} \\ \text{Deluxe} \end{array}
\end{array}
$$

(a) What is the wholesale value of the inventory at Warehouse Sound #1?

(b) Use matrix multiplication to determine a matrix that gives the wholesale and retail values for each inventory at each outlet.

C

47. Suppose that Sharp Utilities, Tech Instruments, and Kasia Calculators each have about one-third of the market for a certain type of calculator. This can be represented by the following 1×3 matrix:

$$
\begin{array}{ccc}
\text{Sharp} & \text{T.I.} & \text{Kasia} \\
\begin{bmatrix} \frac{1}{3} & \frac{1}{3} & \frac{1}{3} \end{bmatrix}
\end{array}
$$

Market research has determined that during the previous year Sharp Utilities kept 80% of its customers, lost 10% to Tech Instruments, and lost 10% to Kasia Calculators. Tech Instruments kept 85% of its customers, lost 10% to Sharp Utilities, and lost 5% to Kasia Calculators. Kasia Calculators kept 75% of its customers, lost 15% to Tech Instruments, and lost 10% to Sharp Utilities. This trend can be represented by the matrix

$$
\begin{array}{cccc}
& & \text{To:} & \\
& \text{Sharp} & \text{T.I.} & \text{Kasia} \\
\text{Sharp} & \begin{bmatrix} 0.80 & 0.10 & 0.10 \\ 0.10 & 0.85 & 0.05 \\ 0.10 & 0.15 & 0.75 \end{bmatrix}
\end{array}
$$

From: T.I.
Kasia

Use matrix multiplication to determine the share of the market each company will have this year if this trend continues.

48. Let

$$
M = \begin{bmatrix} 5 & 1 \\ 2 & 3 \end{bmatrix}
$$

Determine a 2×2 matrix X such that $XM = M$.

HINT Quick Review: Recall that two matrices are equal if and only if they have the same dimension and corresponding entries are equal.

49. Repeat Exercise 48 for: $M = \begin{bmatrix} 6 & 1 \\ 3 & 4 \end{bmatrix}$

50. Let

$$
A = \begin{bmatrix} a & b \\ c & d \end{bmatrix} \qquad B = \begin{bmatrix} x & y \\ z & w \end{bmatrix} \qquad C = \begin{bmatrix} p & q \\ r & s \end{bmatrix}
$$

(a) Determine AB. **(b)** Determine $(AB)C$.

(c) Determine BC. **(d)** Determine $A(BC)$.

(e) Compare the results of parts (b) and (d). What does this prove for all 2×2 matrices?

51. Let

$$A = \begin{bmatrix} a & b \\ c & d \end{bmatrix} \quad B = \begin{bmatrix} x & y \\ z & w \end{bmatrix} \quad C = \begin{bmatrix} p & q \\ r & s \end{bmatrix}$$

(a) Determine AB.   (b) Determine $AB + AC$.
(c) Determine $B + C$.   (d) Determine $A(B + C)$.
(e) Compare the results of parts (b) and (d). What does
     this prove for all 2×2 matrices?

52. Let

$$M = \begin{bmatrix} 4 & 1 \\ 3 & 2 \end{bmatrix}$$

Determine a 2×2 matrix X such that

$$XM = \begin{bmatrix} 1 & 0 \\ 0 & 1 \end{bmatrix}$$

SECTION 5.7

- Multiplicative Identity and Inverse
- Matrix Equations and Solving Systems
- Applications

ALGEBRA OF MATRICES

The linear equation $ax = b$ $(a \neq 0)$ can be solved by multiplying each side by the multiplicative inverse of a, giving

$$x = \left(\frac{1}{a}\right) b$$

Analogously, we can solve the equation

$$AX = B$$

where A, X, and B are matrices. To do so, we need a definition for the multiplicative inverse of a matrix. Once we can solve the matrix equation $AX = B$, we will have a different method for solving linear systems of equations.

Throughout this section we will be concerned primarily with 2×2 and 3×3 matrices. However, the discussion extends naturally to larger square matrices ($n \times n$ matrices).

Multiplicative Identity and Inverse

We begin with two special matrices,

$$I_2 = \begin{bmatrix} 1 & 0 \\ 0 & 1 \end{bmatrix} \quad \text{and} \quad I_3 = \begin{bmatrix} 1 & 0 & 0 \\ 0 & 1 & 0 \\ 0 & 0 & 1 \end{bmatrix}$$

The matrix I_2 has the property that for any 2×1 matrix $X = \begin{bmatrix} x \\ y \end{bmatrix}$,

$$\underset{I_2}{\begin{bmatrix} 1 & 0 \\ 0 & 1 \end{bmatrix}} \underset{X}{\begin{bmatrix} x \\ y \end{bmatrix}} = \underset{X}{\begin{bmatrix} x \\ y \end{bmatrix}}$$

Similarly, the matrix I_3 has the property that for any 3×1 matrix $X = \begin{bmatrix} x \\ y \\ z \end{bmatrix}$,

$$\underbrace{\begin{bmatrix} 1 & 0 & 0 \\ 0 & 1 & 0 \\ 0 & 0 & 1 \end{bmatrix}}_{I_3} \underbrace{\begin{bmatrix} x \\ y \\ z \end{bmatrix}}_{X} = \underbrace{\begin{bmatrix} x \\ y \\ z \end{bmatrix}}_{X}$$

For matrix multiplication, I_2 and I_3 play roles similar to the role that 1 plays for multiplying real numbers. For this reason, I_2 and I_3 are called *identity matrices*.

We can take this analogy between matrix multiplication and real number multiplication a step further. We know that any nonzero real number a has a multiplicative inverse a^{-1}, such that

$$a^{-1}a = 1$$

The definition of the *multiplicative inverse of a matrix* is similar.

Inverse of a Matrix

> Let A be a 2×2 matrix. If there exists a 2×2 matrix A^{-1} such that
>
> $$A^{-1}A = I_2$$
>
> then A^{-1} is the **inverse of A.**
>
> Let A be a 3×3 matrix. If there exists a 3×3 matrix A^{-1} such that
>
> $$A^{-1}A = I_3$$
>
> then A^{-1} is the **inverse of A.**

We will see that not all square matrices have inverses. It can be shown that if a square matrix does have an inverse, then the inverse is unique. It also can be shown that

$$A^{-1}A = AA^{-1}$$

■ **NOTE:** *Recall from Section 5.6 that, in general, matrix multiplication is not commutative ($AB \neq BA$). Thus, $A^{-1}A = AA^{-1}$ is an exception.*

The fact that $A^{-1}A$ equals AA^{-1} can be useful when verifying that two matrices are inverses of each other.

● **EXAMPLE 1   Verifying Inverses**

Verify that the inverse of $M = \begin{bmatrix} -2 & 1 \\ 3 & -1 \end{bmatrix}$ is $M^{-1} = \begin{bmatrix} 1 & 1 \\ 3 & 2 \end{bmatrix}$.

SOLUTION

By the definition of an inverse matrix, we need to show that $M^{-1}M = I_2$:

■ **NOTE:** *Check these results on a graphing utility by entering*

$$A = \begin{bmatrix} -2 & 1 \\ 3 & -1 \end{bmatrix}$$

$$B = \begin{bmatrix} 1 & 1 \\ 3 & 2 \end{bmatrix}$$

and calculating BA and AB.

$$M^{-1}M = \begin{bmatrix} 1 & 1 \\ 3 & 2 \end{bmatrix} \begin{bmatrix} -2 & 1 \\ 3 & -1 \end{bmatrix}$$

$$= \begin{bmatrix} 1(-2) + 1(3) & 1(1) + 1(-1) \\ 3(-2) + 2(3) & 3(1) + 2(-1) \end{bmatrix} = \begin{bmatrix} 1 & 0 \\ 0 & 1 \end{bmatrix}$$

Since $M^{-1}M = MM^{-1}$, another way to show that M and M^{-1} are inverses is to show that $MM^{-1} = I_2$:

$$MM^{-1} = \begin{bmatrix} -2 & 1 \\ 3 & -1 \end{bmatrix} \begin{bmatrix} 1 & 1 \\ 3 & 2 \end{bmatrix}$$

$$= \begin{bmatrix} (-2)1 + (1)3 & (-2)1 + (1)2 \\ (3)1 + (-1)3 & (3)1 + (-1)2 \end{bmatrix} = \begin{bmatrix} 1 & 0 \\ 0 & 1 \end{bmatrix} \quad ■$$

There is a simple formula for the inverse of a 2×2 matrix. Let

$$A = \begin{bmatrix} a & b \\ c & d \end{bmatrix} \quad \text{and} \quad A^{-1} = \begin{bmatrix} x & y \\ z & w \end{bmatrix}$$

By definition, $A^{-1}A = I_2$:

$$\begin{bmatrix} x & y \\ z & w \end{bmatrix} \begin{bmatrix} a & b \\ c & d \end{bmatrix} = \begin{bmatrix} 1 & 0 \\ 0 & 1 \end{bmatrix}$$

Multiplying the matrices on the left side of this equation gives

$$\begin{bmatrix} ax + cy & bx + dy \\ az + cw & bz + dw \end{bmatrix} = \begin{bmatrix} 1 & 0 \\ 0 & 1 \end{bmatrix}$$

If two matrices are equal, each corresponding entry must be equal. Looking at the first row of each matrix, we have

$$\begin{cases} ax + cy = 1 \\ bx + dy = 0 \end{cases}$$

and considering the second row of each matrix, we get

$$\begin{cases} az + cw = 0 \\ bz + dw = 1 \end{cases}$$

We solve the first system for x and y using Cramer's rule:

$$x = \frac{\begin{vmatrix} 1 & c \\ 0 & d \end{vmatrix}}{\begin{vmatrix} a & c \\ b & d \end{vmatrix}} = \frac{d}{ad - bc} \qquad y = \frac{\begin{vmatrix} a & 1 \\ b & 0 \end{vmatrix}}{\begin{vmatrix} a & c \\ b & d \end{vmatrix}} = \frac{-b}{ad - bc}$$

Similarly, we solve for z and w in the second system:

$$z = \frac{\begin{vmatrix} 0 & c \\ 1 & d \end{vmatrix}}{\begin{vmatrix} a & c \\ b & d \end{vmatrix}} = \frac{-c}{ad - bc} \qquad w = \frac{\begin{vmatrix} a & 0 \\ b & 1 \end{vmatrix}}{\begin{vmatrix} a & c \\ b & d \end{vmatrix}} = \frac{a}{ad - bc}$$

Thus,

$$A^{-1} = \begin{bmatrix} x & y \\ z & w \end{bmatrix} = \begin{bmatrix} \dfrac{d}{ad - bc} & \dfrac{-b}{ad - bc} \\ \dfrac{-c}{ad - bc} & \dfrac{a}{ad - bc} \end{bmatrix} = \frac{1}{ad - bc} \begin{bmatrix} d & -b \\ -c & a \end{bmatrix}$$

Therefore, if

$$A = \begin{bmatrix} a & b \\ c & d \end{bmatrix}$$

then

$$A^{-1} = \frac{1}{\det A} \begin{bmatrix} d & -b \\ -c & a \end{bmatrix}$$

This is a convenient formula for calculating the inverse of a 2×2 matrix, provided $\det A \neq 0$. It can be shown that an $n \times n$ matrix A has an inverse if and only if $\det A \neq 0$. Thus, many square matrices do not have an inverse.

Most graphing utilities can determine the inverse, if it exists, of a matrix. You should use a graphing utility to evaluate the inverse of a matrix of dimension 3×3 or larger. If the inverse does not exist, an error message will be displayed.

■ EXAMPLE 2 Finding the Inverse

Determine the inverse, if it exists, of each matrix.

(a) $A = \begin{bmatrix} -3 & 4 \\ 1 & 2 \end{bmatrix}$ **(b)** $B = \begin{bmatrix} 2 & -3 & 1 \\ 0 & 1 & -1 \\ 4 & -4 & 1 \end{bmatrix}$

(c) $C = \begin{bmatrix} 2 & -3 & 1 \\ 0 & 1 & -1 \\ 4 & -8 & 4 \end{bmatrix}$

SOLUTION

(a) If $A = \begin{bmatrix} a & b \\ c & d \end{bmatrix}$, then

$$A^{-1} = \frac{1}{\det A} \begin{bmatrix} d & -b \\ -c & a \end{bmatrix}$$

First, we evaluate the determinant of A:

$$\det A = \begin{vmatrix} -3 & 4 \\ 1 & 2 \end{vmatrix} = -10$$

Since the determinant is nonzero, the inverse exists. Letting $a = -3$, $b = 4$, $c = 1$, and $d = 2$, we have

$$A^{-1} = \frac{1}{\det A} \begin{bmatrix} d & -b \\ -c & a \end{bmatrix} = \frac{1}{-10} \begin{bmatrix} 2 & -4 \\ -1 & -3 \end{bmatrix} = \begin{bmatrix} -\frac{1}{5} & \frac{2}{5} \\ \frac{1}{10} & \frac{3}{10} \end{bmatrix}$$

(b) We enter the matrix B into a graphing utility. Using the inverse feature, we get

$$B^{-1} = \begin{bmatrix} -1.5 & -0.5 & 1 \\ -2 & -1 & 1 \\ -2 & -2 & 1 \end{bmatrix}$$

You can check this by multiplying this result with B.

(c) Entering the matrix C into a graphing utility and using the inverse feature will cause an error message. Further investigation will reveal that $\det C = 0$. Thus, C^{-1} does not exist.

Matrix Equations and Solving Systems

A linear system can be written as an equation of the form

$$AX = B$$

where A, X, and B are matrices. For example, consider the system

$$\begin{cases} a_1 x + b_1 y + c_1 z = k_1 \\ a_2 x + b_2 y + c_2 z = k_2 \\ a_3 x + b_3 y + c_3 z = k_3 \end{cases}$$

If

$$A = \begin{bmatrix} a_1 & b_1 & c_1 \\ a_2 & b_2 & c_2 \\ a_3 & b_3 & c_3 \end{bmatrix} \qquad X = \begin{bmatrix} x \\ y \\ z \end{bmatrix} \qquad B = \begin{bmatrix} k_1 \\ k_2 \\ k_3 \end{bmatrix}$$

the system can be written in the form $AX = B$:

$$\begin{bmatrix} a_1 & b_1 & c_1 \\ a_2 & b_2 & c_2 \\ a_3 & b_3 & c_3 \end{bmatrix} \begin{bmatrix} x \\ y \\ z \end{bmatrix} = \begin{bmatrix} k_1 \\ k_2 \\ k_3 \end{bmatrix}$$

If the matrix A has an inverse, we can solve the matrix equation as follows:

$$AX = B$$

$A^{-1}(AX) = A^{-1}B$ Multiply both sides on the left by A^{-1}.

$(A^{-1}A)X = A^{-1}B$ Matrix multiplication is associative.

$X = A^{-1}B$ Since $(A^{-1}A)X = IX = X$.

EXAMPLE 3 Solving a System with an Inverse Matrix

Solve each system using the inverse of the coefficient matrix.

(a) $\begin{cases} x + 2y - z = -2 \\ 3x + 4z = 8 \\ 4x + 4y + z = 5 \end{cases}$ (b) $\begin{cases} x + 2y - z = 4 \\ 3x + 4z = 1 \\ 4x + 4y + z = 6 \end{cases}$

SOLUTION

(a) First, write the system in matrix form:

$$\underset{A}{\begin{bmatrix} 1 & 2 & -1 \\ 3 & 0 & 4 \\ 4 & 4 & 1 \end{bmatrix}} \underset{X}{\begin{bmatrix} x \\ y \\ z \end{bmatrix}} = \underset{B}{\begin{bmatrix} -2 \\ 8 \\ 5 \end{bmatrix}}$$

The solution to this matrix equation is

$$X = A^{-1}B$$

Entering A and B into a graphing utility, and then calculating $A^{-1}B$, we get

NOTE: *It is important that the product $A^{-1}B$ is calculated in this order, with A^{-1} on the left and B on the right.*

$$X = A^{-1}B = \begin{bmatrix} -12 \\ 10.5 \\ 11 \end{bmatrix}$$

Therefore,

$$\begin{bmatrix} x \\ y \\ z \end{bmatrix} = \begin{bmatrix} -12 \\ 10.5 \\ 11 \end{bmatrix}$$

The solution to the system is $(-12, 10.5, 11)$.

(b) Notice that this system is identical to the first system, except that the three numbers to the right of the equal signs are different. Writing the system in matrix form, we have

$$\underbrace{\begin{bmatrix} 1 & 2 & -1 \\ 3 & 0 & 4 \\ 4 & 4 & 1 \end{bmatrix}}_{A} \underbrace{\begin{bmatrix} x \\ y \\ z \end{bmatrix}}_{X} = \underbrace{\begin{bmatrix} 4 \\ 1 \\ 6 \end{bmatrix}}_{C}$$

Since we have already entered A, we only need to enter C into a graphing utility and then calculate $A^{-1}C$. The solution to this matrix equation is

$$X = A^{-1}C = \begin{bmatrix} 11 \\ -7.5 \\ -8 \end{bmatrix}$$

The solution to the system is $(11, -7.5, -8)$.

Example 3 points out an advantage of using the inverse of the coefficient matrix to solve certain systems of linear equations. Using the inverse is an efficient method for solving two or more systems that are identical, except for the constants to the right of the equal signs.

Applications

Matrices allow us to solve many applied problems easily and efficiently, especially when the computations can be done on a graphing utility.

EXAMPLE 4   Application: Nutritional Analysis

A dog breeder feeds his award-winning retrievers a mix of Gaines Cycle dog food and Pedigree dog food. The table shows the percentages of protein and fat for each food. The breeder prepares a special diet for the puppies and another for the adults. The daily requirement for each puppy is 2.5 ounces of protein and 1 ounce of fat; each adult needs 3 ounces of protein and 1.5 ounces of fat per day. How many ounces of each food should be mixed for each diet (to the nearest 0.1 ounce)?

	Gaines Cycle	Pedigree
Protein	22%	8%
Fat	8%	6%

SOLUTION

First, consider the diet for the puppies:

$$0.22\begin{pmatrix} \text{Number of ounces} \\ \text{of Gaines Cycle} \end{pmatrix} + 0.08\begin{pmatrix} \text{Number of ounces} \\ \text{of Pedigree} \end{pmatrix} = 2.5 \text{ ounces protein}$$

and

$$0.08\left(\begin{array}{c}\text{Number of ounces}\\ \text{of Gaines Cycle}\end{array}\right) + 0.06\left(\begin{array}{c}\text{Number of ounces}\\ \text{of Pedigree}\end{array}\right) = 1 \text{ ounce fat}$$

Letting

$$x = \text{Number of ounces of Gaines Cycle}$$
$$y = \text{Number of ounces of Pedigree}$$

we have the system

$$\begin{cases} 0.22x + 0.08y = 2.5 \\ 0.08x + 0.06y = 1 \end{cases}$$

A matrix equation for the system is

$$\underset{A}{\begin{bmatrix} 0.22 & 0.08 \\ 0.08 & 0.06 \end{bmatrix}} \underset{X}{\begin{bmatrix} x \\ y \end{bmatrix}} = \underset{B}{\begin{bmatrix} 2.5 \\ 1 \end{bmatrix}}$$

The solution to this matrix equation is

$$X = A^{-1}B \approx \begin{bmatrix} 10.3 \\ 2.9 \end{bmatrix}$$

The puppies' mix should consist of 10.3 ounces of Gaines Cycle and 2.9 ounces of Pedigree (to the nearest 0.1 ounce).

Now consider the diet for the adult retrievers. The system of equations is very similar. A matrix equation for this situation is

$$\underset{A}{\begin{bmatrix} 0.22 & 0.08 \\ 0.08 & 0.06 \end{bmatrix}} \underset{X}{\begin{bmatrix} x \\ y \end{bmatrix}} = \underset{C}{\begin{bmatrix} 3 \\ 1.5 \end{bmatrix}}$$

The solution to this matrix equation is

$$X = A^{-1}C \approx \begin{bmatrix} 8.8 \\ 13.2 \end{bmatrix}$$

The mix for the adults should consist of 8.8 ounces of Gaines Cycle and 13.2 ounces of Pedigree.

● EXERCISE SET 5.7

A

In Exercises 1–4, verify that M and M^{-1} are inverses of each other.

1. $M = \begin{bmatrix} -1 & 1 \\ -2 & 3 \end{bmatrix}$; $M^{-1} = \begin{bmatrix} -3 & 1 \\ -2 & 1 \end{bmatrix}$

2. $M = \begin{bmatrix} 1 & 2 \\ 4 & 7 \end{bmatrix}$; $M^{-1} = \begin{bmatrix} -7 & 2 \\ 4 & -1 \end{bmatrix}$

3. $M = \begin{bmatrix} 1 & -1 & 2 \\ 2 & -3 & 4 \\ 1 & 0 & 3 \end{bmatrix}$; $M^{-1} = \begin{bmatrix} 9 & -3 & -2 \\ 2 & -1 & 0 \\ -3 & 1 & 1 \end{bmatrix}$

4. $M = \begin{bmatrix} 1 & 3 & 0 \\ -2 & -5 & 1 \\ 1 & 4 & 0 \end{bmatrix}$; $M^{-1} = \begin{bmatrix} 4 & 0 & -3 \\ -1 & 0 & 1 \\ 3 & 1 & -1 \end{bmatrix}$

In Exercises 5–16, determine A^{-1}, if it exists.

5. $A = \begin{bmatrix} 1 & 2 \\ 3 & 5 \end{bmatrix}$

6. $A = \begin{bmatrix} -1 & -1 \\ 2 & 1 \end{bmatrix}$

7. $A = \begin{bmatrix} -1 & 1 \\ -2 & 4 \end{bmatrix}$

8. $A = \begin{bmatrix} 1 & 1 \\ -2 & 0 \end{bmatrix}$

9. $A = \begin{bmatrix} 4 & 2 \\ 6 & 3 \end{bmatrix}$

10. $A = \begin{bmatrix} 8 & 4 \\ 6 & 3 \end{bmatrix}$

11. $A = \begin{bmatrix} -1 & 2 & -2 \\ 1 & -1 & 3 \\ 2 & -4 & 5 \end{bmatrix}$

12. $A = \begin{bmatrix} 3 & 4 & 3 \\ -1 & 3 & 1 \\ 0 & -2 & -1 \end{bmatrix}$

13. $A = \begin{bmatrix} 1 & 1 & -1 \\ 1 & 2 & -2 \\ -2 & -2 & 0 \end{bmatrix}$

14. $A = \begin{bmatrix} 1 & 1 & -1 \\ 1 & 3 & -3 \\ -2 & -2 & 1 \end{bmatrix}$

15. $A = \begin{bmatrix} 2 & 1 & -1 \\ 3 & 0 & 2 \\ -5 & -1 & -1 \end{bmatrix}$

16. $A = \begin{bmatrix} -1 & 2 & 0 \\ 1 & 0 & 2 \\ 2 & 3 & 7 \end{bmatrix}$

In Exercises 17–20, write each system as a matrix equation in the form $AX = B$.

17. $\begin{cases} 5x - 2y = 7 \\ 4x + 3y = -1 \end{cases}$

18. $\begin{cases} 6x + 7y = -5 \\ 3x - 8y = 0 \end{cases}$

19. $\begin{cases} x - 2y + 3z = 14 \\ 5x + 4y - 4z = -1 \\ 2x \qquad + 9z = 13 \end{cases}$

20. $\begin{cases} -3x + y - 7z = 31 \\ 2x - 10y + 5z = 17 \\ 2x \qquad + 9y = 20 \end{cases}$

In Exercises 21 and 22, determine x and y.

21. $\begin{bmatrix} x \\ y \end{bmatrix} = \begin{bmatrix} -2 & 3 \\ 1 & 4 \end{bmatrix} \begin{bmatrix} 6 \\ -5 \end{bmatrix}$

22. $\begin{bmatrix} x \\ y \end{bmatrix} = \begin{bmatrix} 4 & -1 \\ 7 & 2 \end{bmatrix} \begin{bmatrix} 3 \\ -6 \end{bmatrix}$

In Exercises 23 and 24, determine x, y, and z.

23. $\begin{bmatrix} x \\ y \\ z \end{bmatrix} = \begin{bmatrix} -2 & 6 & 5 \\ 3 & 1 & 2 \\ 8 & -5 & 0 \end{bmatrix} \begin{bmatrix} 4 \\ -1 \\ 7 \end{bmatrix}$

24. $\begin{bmatrix} x \\ y \\ z \end{bmatrix} = \begin{bmatrix} 4 & -5 & 1 \\ 3 & 6 & 0 \\ -2 & 9 & 8 \end{bmatrix} \begin{bmatrix} 1 \\ -2 \\ -3 \end{bmatrix}$

In Exercises 25–27, use the inverse of the coefficient matrix given in Exercise 1 to solve each system.

25. $\begin{cases} -x + y = -7 \\ -2x + 3y = 12 \end{cases}$

26. $\begin{cases} -x + y = 6 \\ -2x + 3y = -8 \end{cases}$

27. $\begin{cases} -x + y = 23 \\ -2x + 3y = 19 \end{cases}$

In Exercises 28–30, use the inverse of the coefficient matrix given in Exercise 3 to solve each system.

28. $\begin{cases} x - y + 2z = 2 \\ 2x - 3y + 4z = 0 \\ x \qquad + 3z = 5 \end{cases}$

29. $\begin{cases} x - y + 2z = -1 \\ 2x - 3y + 4z = 4 \\ x \qquad + 3z = 6 \end{cases}$

30. $\begin{cases} x - y + 2z = 11 \\ 2x - 3y + 4z = -8 \\ x \qquad + 3z = 10 \end{cases}$

B

In Exercises 31–39, solve each system using the inverse of the coefficient matrix.

31. (a) $\begin{cases} 2x + 3y = -2 \\ 3x + 5y = 6 \end{cases}$ **(b)** $\begin{cases} 2x + 3y = 11 \\ 3x + 5y = 15 \end{cases}$

(c) $\begin{cases} 2x + 3y = -30 \\ 3x + 5y = 42 \end{cases}$

32. (a) $\begin{cases} 3x + 7y = 4 \\ 2x + 5y = -9 \end{cases}$ **(b)** $\begin{cases} 3x + 7y = 0 \\ 2x + 5y = 1 \end{cases}$

(c) $\begin{cases} 3x + 7y = 20 \\ 2x + 5y = 10 \end{cases}$

33. (a) $\begin{cases} 6x + 5y = 8 \\ 2x + 2y = 11 \end{cases}$ **(b)** $\begin{cases} 6x + 5y = -5 \\ 2x + 2y = 1 \end{cases}$

(c) $\begin{cases} 6x + 5y = 21 \\ 2x + 2y = 18 \end{cases}$

34. (a) $\begin{cases} 4x + 9y = 10 \\ 2x + 5y = 6 \end{cases}$ **(b)** $\begin{cases} 4x + 9y = -14 \\ 2x + 5y = 5 \end{cases}$

(c) $\begin{cases} 4x + 9y = 34 \\ 2x + 5y = -11 \end{cases}$

35. (a) $\begin{cases} x + 2y + z = 0 \\ x + y - z = 3 \\ 2x + 5y + 3z = 2 \end{cases}$ **(b)** $\begin{cases} x + 2y + z = 4 \\ x + y - z = 7 \\ 2x + 5y + 3z = 10 \end{cases}$

(c) $\begin{cases} x + 2y + z = 20 \\ x + y - z = 32 \\ 2x + 5y + 3z = 25 \end{cases}$

36. (a) $\begin{cases} -x + y + 3z = 4 \\ 2x - y - 2z = 1 \\ -4x + 2y + 5z = -3 \end{cases}$ **(b)** $\begin{cases} -x + y + 3z = 8 \\ 2x - y - 2z = 0 \\ -4x + 2y + 5z = 9 \end{cases}$

(c) $\begin{cases} -x + y + 3z = 15 \\ 2x - y - 2z = 18 \\ -4x + 2y + 5z = 11 \end{cases}$

37. (a) $\begin{cases} 3x + 3y - z = 2 \\ 2x + 2y - z = 1 \\ -4x - 5y + 2z = 4 \end{cases}$ **(b)** $\begin{cases} 3x + 3y - z = -6 \\ 2x + 2y - z = 0 \\ -4x - 5y + 2z = 10 \end{cases}$

(c) $\begin{cases} 3x + 3y - z = 26 \\ 2x + 2y - z = 21 \\ -4x - 5y + 2z = 30 \end{cases}$

38. (a) $\begin{cases} x - y + 3z = 12 \\ 2x - y + 6z = 8 \\ x - 3y + 7z = -4 \end{cases}$ **(b)** $\begin{cases} x - y + 3z = 0 \\ 2x - y + 6z = -1 \\ x - 3y + 7z = 8 \end{cases}$

(c) $\begin{cases} x - y + 3z = 24 \\ 2x - y + 6z = 38 \\ x - 3y + 7z = 31 \end{cases}$

39. (a) $\begin{cases} -x + 2y - 3z = 4 \\ x - 4y + 7z = 8 \\ x - y + 3z = 0 \end{cases}$ **(b)** $\begin{cases} -x + 2y - 3z = 6 \\ x - 4y + 7z = -10 \\ x - y + 3z = 5 \end{cases}$

(c) $\begin{cases} -x + 2y - 3z = 23 \\ x - 4y + 7z = 28 \\ x - y + 3z = 17 \end{cases}$

40. There are three currents I_1, I_2, and I_3 (measured in amperes) through the electrical network pictured:

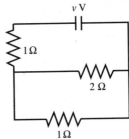

According to laws of physics, these currents satisfy the system

$$\begin{cases} I_1 - I_2 - I_3 = 0 \\ I_1 + 2I_2 = v \\ I_1 + I_3 = v \end{cases}$$

Solve this system to find the three currents for the given voltage v:

(a) $v = 9$ **(b)** $v = 12$ **(c)** $v = 16$·

41. There are three currents I_1, I_2, and I_3 (measured in amperes) through the electrical network pictured:

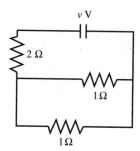

v V

2 Ω

1 Ω

1 Ω

According to laws of physics, these currents satisfy the system

$$\begin{cases} I_1 - I_2 - I_3 = 0 \\ 2I_1 + I_2 = v \\ 2I_1 + I_3 = v \end{cases}$$

Solve this system to find the three currents for the given voltage *v*:

(a) $v = 6$ **(b)** $v = 9$ **(c)** $v = 12$

42. A researcher feeds her laboratory mice a mixture of food *A* and food *B*. The following table shows the percentages of protein and fat for each food:

	Food A	Food B
Protein	18%	9%
Fat	7%	6%

The researcher prepares a diet for one group (group #1) and another for a second group (group #2). The researcher wants to supply each mouse from group #1 with 0.7 gram of protein and 0.3 gram of fat per day. The researcher wants to supply each mouse from group #2 with 1 gram of protein and 0.5 gram of fat per day. How many grams of each food should be mixed for each diet (to the nearest 0.1 gram)?

43. Repeat Exercise 42 to provide group #1 with 0.9 gram of

protein and 0.4 gram of fat and group #2 with 1.1 gram of protein and 0.6 gram of fat per day.

C

44. A tire company manufactures three models of tires: *A, B,* and *C*. Model *A* requires 3 units of steel, 3 units of rubber, and 1.5 units of fiberglass. Model *B* requires 3 units of steel, 3 units of rubber, and 3 units of fiberglass. Model *C* requires 1.5 units of steel, 4.5 units of rubber, and 3 units of fiberglass. How many model *A, B,* and *C* tires can be produced with the following amounts of materials?

(a) 510 units of steel, 510 units of rubber, 360 units of fiberglass

(b) 600 units of steel, 900 units of rubber, 600 units of fiberglass

(c) 1200 units of steel, 1400 units of rubber, 800 units of fiberglass

45. Repeat Exercise 44 using the following amounts of materials:

(a) 750 units of steel, 810 units of rubber, 720 units of fiberglass

(b) 1020 units of steel, 1020 units of rubber, 630 units of fiberglass

(c) 1450 units of steel, 1500 units of rubber, 1000 units of fiberglass

46. Let

$$P = \begin{bmatrix} 3 & 5 \\ 1 & 2 \end{bmatrix} \quad \text{and} \quad Q = \begin{bmatrix} 0 & 1 \\ 1 & -2 \end{bmatrix}$$

Determine a matrix *M* such that $MP = PQ$.

47. Suppose *A, B,* and *C* are 3 × 3 matrices with inverses A^{-1}, B^{-1}, and C^{-1}, respectively. In terms of these six matrices:

(a) Solve the equation $AMB = C$ for matrix *M*.

(b) Determine the inverse of *ABC*.

48. Show that if a matrix has an inverse, the inverse is unique.

SYSTEMS OF INEQUALITIES

In Section 6 (page 47) and Section 2.4 we discussed inequalities in one variable. Many applications require finding the solutions to *inequalities in two variables*, such as

$$x < y^2 + 3 \qquad \text{or} \qquad 4x - 5y > 10 \qquad \text{or} \qquad 2x^2 + 3y^2 \le 12$$

A *solution* to such an inequality is an ordered pair (x, y) that makes the inequality true. The best way to represent the solutions of an inequality in two variables is with a graph.

Graphing Inequalities in Two Variables

For the sake of discussion, suppose we want to graph the inequality

$$y > \tfrac{2}{3}x + 1$$

Consider first the graph of the *associated equation*

$$y = \tfrac{2}{3}x + 1$$

and any particular value for x, say $x = 6$ (Figure 5.30). We know that there is only one value for y such that $(6, y)$ is a solution to the equation $y = \tfrac{2}{3}x + 1$, namely,

$$y = \tfrac{2}{3}(6) + 1 = 5$$

For what values of y is $(6, y)$ a solution to the inequality $y > \tfrac{2}{3}x + 1$? Substituting 6 for x into the inequality gives

$$y > \tfrac{2}{3}(6) + 1$$
$$y > 5$$

Thus, $(6, y)$ is part of the graph of the inequality if $y > 5$. In other words, all points directly above $(6, 5)$ are part of the graph of $y > \tfrac{2}{3}x + 1$. This observation applies to any value of x (not just $x = 6$). The ordered pair (x, y) is part of the graph of $y > \tfrac{2}{3}x + 1$ if and only if (x, y) is above the line $y = \tfrac{2}{3}x + 1$. We indicate these points (x, y) by shading the region above the line $y = \tfrac{2}{3}x + 1$ (Figure 5.31). We use a dashed line to indicate that the points on the line are *not* part of the graph of the inequality.

The graph of the inequality $y > \tfrac{2}{3}x + 1$ is typical of the graphs of inequalities in two variables. Given an inequality in x and y, if we replace the inequality symbol ($<$, $>$, $\le$, or $\ge$) with an equal sign, then the graph of the resulting equation separates the xy-plane into two or more regions. If a point (a, b) in a region is a solution of the inequality, then every point in the region satisfies the inequality. If a point (a, b) in a region is not a solution of the inequality, then no

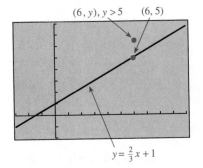

Figure 5.30
$[-3, 10] \times [-2, 8]$

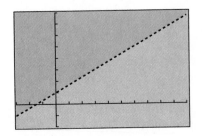

Figure 5.31
$y > \tfrac{2}{3}x + 1$

points in the region satisfy the inequality. As a consequence, we have the following method for graphing an inequality in two variables.

Graphing an Inequality in
Two Variables

An inequality in two variables can be graphed by the following procedure:

Step 1: Replace the inequality sign in the equation with an equal sign and graph the resulting **associated equation.** Graph the associated equation with a solid curve if the original inequality sign is $\leq$ or $\geq$; graph the associated equation with a dashed curve if the original inequality sign is $<$ or $>$. The resulting graph represents a boundary for the graph of the inequality.

Step 2: Select a test point (a, b) on one side of the graph of the associated equation. If the test point satisfies the original inequality, then shade the side containing (a, b) to signify that all points in that region satisfy the inequality. If the test point does not satisfy the inequality, then shade the side that does not contain (a, b) to signify that all points in that region satisfy the inequality.

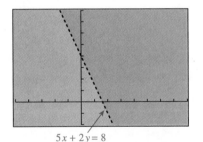

$5x + 2y = 8$

Figure 5.32
$5x + 2y > 8$
$[-5, 8] \times [-2, 8]$

■ **GRAPHING NOTE:** *Most graphing utilities are capable of shading above, below, or between graphs. Try to obtain the results of the examples in this section on a graphing utility.*

■ **EXAMPLE 1   Graphing a Linear Inequality**
Sketch the graph of the inequality: $5x + 2y > 8$

SOLUTION
The graph of the associated equation,

$$5x + 2y = 8$$

is a line, as shown in Figure 5.32. Since points on the line are not part of the graph of $5x + 2y > 8$, the line is dashed. The solution to $5x + 2y > 8$ is either the set of all points in the region below the line or the region above the line. Selecting a test point not on the line, say $(0, 0)$, we substitute into the inequality:

$$5(0) + 2(0) \overset{?}{>} 8 \quad \text{False}$$

The test point $(0, 0)$ does not satisfy the inequality, and, as can be seen in Figure 5.32, the graph of $5x + 2y > 8$ consists of all points above the line. ■

■ **EXAMPLE 2   Graphing a Nonlinear Inequality**
Sketch the graph of the inequality: $y \leq \dfrac{4}{1 + \sqrt{x}}$

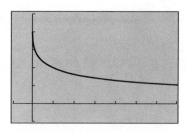

Figure 5.33
$y = \dfrac{4}{1 + \sqrt{x}}$
$[-1, 7] \times [-1, 5]$

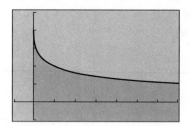

Figure 5.34
$y \leq \dfrac{4}{1 + \sqrt{x}}$

SOLUTION

First, note that because there is a square root of x, x must be greater than or equal to 0. Using a graphing utility, we graph the associated equation

$$y = \frac{4}{1 + \sqrt{x}}$$

(see Figure 5.33). Since points on the curve are part of the graph of

$$y \leq \frac{4}{1 + \sqrt{x}}$$

the curve is solid.

We choose a test point below the curve, say $(1, 1)$, and substitute the coordinates into the original inequality:

$$1 \overset{?}{\leq} \frac{4}{1 + \sqrt{1}} \quad \text{True}$$

The graph of

$$y \leq \frac{4}{1 + \sqrt{x}}$$

is the region on or below the curve (Figure 5.34). ●

Graphing Systems of Inequalities

The *graph of a system of inequalities* is the graph of the set of solutions to the system, which is the intersection of the graphs of the inequalities in the system.

▆ EXAMPLE 3 Graphing a System of Linear Inequalities

Sketch the graph of the system:
$$\begin{cases} x - 2y < 8 \\ 2x + 3y \geq 6 \end{cases}$$

SOLUTION

The first inequality has an associated equation $x - 2y = 8$. Selecting a point not on the line, say $(0, 0)$, as a test point, we have

$$(0) - 2(0) \overset{?}{<} 8 \quad \text{True}$$

Consequently, the graph of the inequality consists of all points above the line $x - 2y = 8$ (Figure 5.35).

The second inequality has an associated equation $2x + 3y = 6$. Selecting a point not on the line, say $(4, 1)$, as a test point, we have

$$2(4) + 3(1) \overset{?}{\geq} 6 \quad \text{True}$$

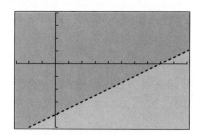

Figure 5.35
$x - 2y < 8$
$[-3, 10] \times [-5, 4]$

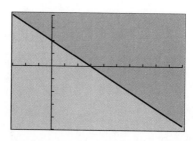

Figure 5.36
$2x + 3y \geq 6$
$[-3, 10] \times [-5, 4]$

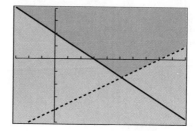

Figure 5.37
$\begin{cases} x - 2y < 8 \\ 2x + 3y \geq 6 \end{cases}$

This means the graph of the inequality consists of all points on or above the line $2x + 3y = 6$ (Figure 5.36).

The graph of the system of inequalities is the intersection of these two graphs (Figure 5.37).

EXAMPLE 4 Graphing a Nonlinear System of Inequalities

Sketch the graph of the system: $\begin{cases} y \leq 2 - x \\ y^2 \leq x + 4 \end{cases}$

SOLUTION

To graph the first inequality, we graph its associated equation $y = 2 - x$. Testing the point $(0, 0)$ in the original inequality, we get

$$(0) \stackrel{?}{\leq} 2 - (0) \quad \text{True}$$

The graph of the inequality $y \leq 2 - x$ is shown in Figure 5.38.

The associated equation of the second inequality is $y^2 = x + 4$. To graph this equation on a graphing utility, we solve for y:

$$y^2 = x + 4$$
$$y = \pm\sqrt{x + 4}$$

Thus, to graph $y^2 = x + 4$, we graph $y = \sqrt{x + 4}$ and $y = -\sqrt{x + 4}$. Selecting $(0, 0)$ as a test point, we get

$$(0)^2 \stackrel{?}{\leq} (0) + 4 \quad \text{True}$$

The graph of $y^2 \leq x + 4$ is shown in Figure 5.39.

The graph of the system of inequalities is the set of all points common to both graphs (Figure 5.40).

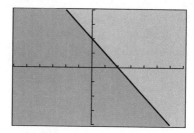

Figure 5.38
$y \leq 2 - x$
$[-6, 7] \times [-4, 4]$

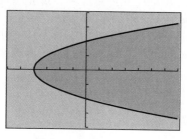

Figure 5.39
$y^2 \leq x + 4$
$[-6, 7] \times [-4, 4]$

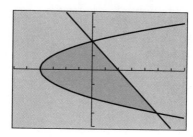

Figure 5.40
$\begin{cases} y \leq 2 - x \\ y^2 \leq x + 4 \end{cases}$

Compound Inequalities

Recall from Section 6 (page 47) that the compound inequality $a \leq t \leq b$ means that $a \leq t$ and $t \leq b$. For example,

$$-2 \leq y \leq x + 1$$

means that

$$y \geq -2 \quad \text{and} \quad y \leq x + 1$$

Thus,

$$-2 \leq y \leq x + 1$$

is equivalent to the system

$$\begin{cases} y \geq -2 \\ y \leq x + 1 \end{cases}$$

In general, a compound inequality in two variables is equivalent to a system of inequalities.

■ EXAMPLE 5   Graphing a Compound Inequality

Sketch the graph of the compound inequality: $2 \leq x + 2y \leq 9$

SOLUTION

The compound inequality $2 \leq x + 2y \leq 9$ is equivalent to the system

$$\begin{cases} x + 2y \geq 2 \\ x + 2y \leq 9 \end{cases}$$

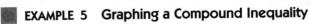

The graphs of the associated equations are shown in the same viewing window in Figure 5.41.

Selecting a test point not on the line $x + 2y = 2$, we see that the graph of the first inequality, $x + 2y \geq 2$, is the set of all points on or above the line. Selecting a test point not on the line $x + 2y = 9$ determines that the graph of the second inequality, $x + 2y \leq 9$, is the set of all points on or below the line. The graph of the compound inequality is the set of points common to both graphs (Figure 5.42). ●

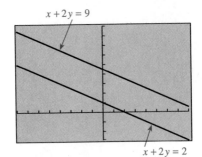

$x + 2y = 9$

Figure 5.41
$[-8, 8] \times [-3, 9]$

$x + 2y = 2$

Figure 5.42
$\begin{cases} x + 2y \geq 2 \\ x + 2y \leq 9 \end{cases}$

Applications

Many practical problems involve the graphs of systems of inequalities. Often, such problems require finding the *vertices* of the region. The vertices are the points where the boundaries of the region intersect.

◼ EXAMPLE 6 Application: Analyzing an Investment

A $15 million pension fund is to be invested in stocks, bonds, and real estate. A regulatory board requires that the amount of each type of investment is at least 20% of the total investment, and the money invested in bonds must be at least half that invested in stocks.

(a) Let x represent the amount (in millions of dollars) invested in stocks, and let y represent the amount (in millions of dollars) invested in bonds. Determine a system of inequalities that models the regulations.

(b) Graph the system of inequalities.

(c) What is the maximum amount that can be invested in stocks?

SOLUTION

(a) At least 20% of the $15 million pension fund must be invested in stocks, so

$$x \geq (0.20)(15)$$

or

$$x \geq 3$$

Similarly, 20% or more of the $15 million must be invested in bonds, giving

$$y \geq 3$$

Since the total amount invested is $15 million, the amount invested in real estate is $15 - (x + y)$. This amount must be at least 20% of the $15 million, so

$$15 - (x + y) \geq (0.20)(15)$$

which simplifies as

$$x + y \leq 12$$

The money invested in bonds must be at least half that invested in stocks, so

$$y \geq \tfrac{1}{2}x$$

Thus, a system that models the regulations is

$$\begin{cases} x \geq 3 \\ y \geq 3 \\ x + y \leq 12 \\ y \geq \tfrac{1}{2}x \end{cases}$$

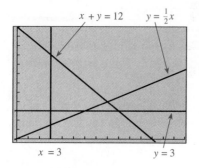

Figure 5.43

[0, 15] × [0, 12]

(b) The graphs of the associated equations are shown in Figure 5.43. The graph of $x \geq 3$ is the set of all points to the right of the vertical line $x = 3$. The graph of $y \geq 3$ is the set of all points above the horizontal line $y = 3$. Using $(0, 0)$ as a test point in the third inequality $x + y \leq 12$, we have

$$0 + 0 \stackrel{?}{\leq} 12 \quad \text{True}$$

Thus, the graph of $x + y \leq 12$ is all points below the line $x + y = 12$. Selecting $(1, 1)$ as a test point for the fourth inequality, $y \geq \frac{1}{2}x$, gives

$$1 \stackrel{?}{\geq} \tfrac{1}{2}(1) \quad \text{True}$$

So the graph of the fourth inequality of the system is the set of all points above the line $y = \frac{1}{2}x$. The graph of the system therefore is the set of all points to the right of $x = 3$, above $y = 3$, below $x + y = 12$, and above $y = \frac{1}{2}x$ (Figure 5.44).

(c) The maximum amount that can be invested in stocks corresponds to the point (x, y) in the region with the largest value of x. The point with the largest x-coordinate is the point P on the graph that is the farthest to the right, as shown in Figure 5.45. The point P is the vertex where the lines $x + y = 12$ and $y = \frac{1}{2}x$ intersect. Solving the system

$$\begin{cases} x + y = 12 \\ \quad\ y = \tfrac{1}{2}x \end{cases}$$

we determine that the coordinates of P are $(8, 4)$. The maximum amount that can be invested in stocks is $8 million.

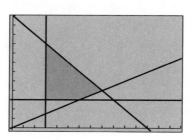

Figure 5.44

$$\begin{cases} \quad\ x \geq 3 \\ \quad\ y \geq 3 \\ x + y \leq 12 \\ \quad\ y \geq \tfrac{1}{2}x \end{cases}$$

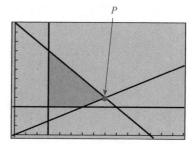

Figure 5.45

⊙ EXERCISE SET 5.8

A

In Exercises 1–16, sketch the graph of each inequality.

1. $y \geq x - 2$
2. $y \geq x + 3$
3. $x + 2y \leq 2$
4. $-x + 3y \leq 6$
5. $y > \frac{2}{3}x - 4$
6. $y > \frac{5}{4}x + 1$
7. $x \leq 5$
8. $y \geq -3$
9. $y \leq x^2 - 3x + 1$
10. $y \leq 7 + 2x - x^2$
11. $y^2 \geq 9 - x^2$
12. $y^2 \geq 4 - 4x^2$
13. $y \geq e^x$
14. $y \leq \ln x$
15. $y \leq \dfrac{4}{1 + |x|}$
16. $y \geq \dfrac{5}{2 + x^2}$

In Exercises 17–20, match each system with the corresponding region in Figure 5.46.

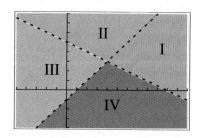

Figure 5.46
$[-5, 12] \times [-4, 9]$

17. $\begin{cases} y < x - 1 \\ y < -\frac{1}{2}x + 5 \end{cases}$
18. $\begin{cases} y > x - 1 \\ y < -\frac{1}{2}x + 5 \end{cases}$
19. $\begin{cases} y < x - 1 \\ y > -\frac{1}{2}x + 5 \end{cases}$
20. $\begin{cases} y > x - 1 \\ y > -\frac{1}{2}x + 5 \end{cases}$

In Exercises 21–32, sketch the graph of each system of inequalities.

21. $\begin{cases} y \geq 6 - x \\ y \geq 2x \end{cases}$
22. $\begin{cases} y \leq 8 - 2x \\ y \leq \frac{3}{4}x \end{cases}$

23. $\begin{cases} 3y < x - 9 \\ y > 2 \end{cases}$
24. $\begin{cases} 5y < 3x + 10 \\ x < 4 \end{cases}$
25. $\begin{cases} 2x + 3y \leq 12 \\ y - x > 0 \end{cases}$
26. $\begin{cases} 5x - 2y > 10 \\ x + y \leq 0 \end{cases}$
27. $\begin{cases} y \geq x - 2 \\ y \leq 4 - x^2 \end{cases}$
28. $\begin{cases} y \leq x + 3 \\ y \geq 1 + x^2 \end{cases}$
29. $\begin{cases} x^2 + y^2 > 4 \\ 3y - 2x \leq 12 \end{cases}$
30. $\begin{cases} x^2 + y^2 < 16 \\ 3x - 4y \leq 16 \end{cases}$
31. $\begin{cases} y \geq x^2 - 4x \\ 2y \leq 4 + 6x - x^2 \end{cases}$
32. $\begin{cases} y \geq 5x - x^2 \\ 3y \leq x^2 + 3x - 2 \end{cases}$

In Exercises 33–38, sketch the graph of each compound inequality.

33. $x \leq y \leq 5$
34. $1 \leq y \leq 2 - x$
35. $0 < 2x + 3y \leq 12$
36. $-6 \leq 5x - 3y < 15$
37. $x + 1 \leq y \leq 4 - x^2$
38. $x \leq y \leq 3x + x^2$

B

In Exercises 39–48, sketch the graph of each system of inequalities.

39. $\begin{cases} y \geq -2 \\ x \leq 4 \\ y \leq 2 - \frac{1}{2}x \end{cases}$
40. $\begin{cases} y \leq 5 \\ x \geq -3 \\ y \geq \frac{4}{3}x - 1 \end{cases}$
41. $\begin{cases} x \geq 0 \\ y \geq 0 \\ x + y \leq 8 \\ 2y - 3x \leq 6 \end{cases}$
42. $\begin{cases} x \geq 0 \\ y \geq 0 \\ y \leq 8 - \frac{4}{3}x \\ 2x - y \geq 2 \end{cases}$
43. $\begin{cases} 1 \leq x \leq 6 \\ y \geq 2 \\ 5y - 2x < 18 \end{cases}$
44. $\begin{cases} -3 \leq y \leq 5 \\ x \geq 1 \\ 2x - y < 9 \end{cases}$
45. $\begin{cases} y \geq 0 \\ 3 \leq x \leq 8 \\ y \leq \sqrt{x + 1} \end{cases}$
46. $\begin{cases} y \geq 0 \\ 0 \leq x \leq 3 \\ y \leq 2^x \end{cases}$
47. $\begin{cases} y \geq -2 \\ 0 \leq 4x + 3y \leq 12 \end{cases}$
48. $\begin{cases} y \geq 5 \\ 0 \leq 5x - 3y \leq 15 \end{cases}$

49. Determine the vertices of the region graphed in Exercise 41.

50. Determine the vertices of the region graphed in Exercise 42.

51. Determine the vertices of the region graphed in Exercise 43.

52. Determine the vertices of the region graphed in Exercise 44.

53. In Example 6, determine (to the nearest dollar) the maximum amount that can be invested in bonds.

54. A dietitian must prepare a food supplement for a gymnastics team. The dietitian has two sources, mixture I and mixture II. Each ounce of mixture I supplies 15 units of protein, 12 units of carbohydrates, and 15 units of fat. Each ounce of mixture II supplies 6 units of protein, 30 units of carbohydrates, and 8 units of fat. Each diet requires a supplement of at least 60 units of protein, at least 120 units of carbohydrates, but no more than 80 units of fat.

  (a) Let x represent the number of ounces of mixture I used in the supplement, and let y represent the number of ounces of mixture II used in the supplement. Determine a system of inequalities that models the dietary requirements.

  (b) Graph the system of inequalities.

  (c) What is the maximum amount of mixture I that can be used in the supplement? What is the minimum amount of mixture II that can be used?

55. A developer owns 20 lots of a 32 lot site, with an option to buy the remaining 12 lots. The developer plans to build either a single-family home or a duplex on each lot. The local planning commission has stipulated that the number of lots with single-family homes can be no greater than three times the number of lots with duplexes. The developer will build on every lot he owns.

  (a) Let x represent the number of lots that have single-family homes, and let y represent the number of lots with duplexes. Determine a system of inequalities that models the situation.

  (b) Graph the system of inequalities.

  (c) What is the greatest number of lots with single-family homes? What is the minimum number of lots with duplexes?

C

56. Determine a system of inequalities that describes the triangular region with vertices at $A(0, 1)$, $B(4, 1)$, and $C(0, 5)$.

57. Determine a system of inequalities that describes the triangular region with vertices at $P(2, 0)$, $Q(2, 3)$, and $R(8, 0)$.

58. **(a)** Sketch the graph of the solution of $|x| \leq y$.

  (b) Sketch the graph of the solution of $-y \leq x \leq y$.

  (c) Comparing the results of parts (a) and (b), what can you conclude?

59. For what values of x and y is the expression $\sqrt{(y - e^x)(x - y)}$ a real number?

● QUICK REFERENCE

Topic	Page	Remarks
Solution to a system of equations	419	A solution to a system of equations in two variables is an ordered pair (x, y) that satisfies all equations in the system.
	440	A solution to a system of equations in three variables is an ordered triple (x, y, z) that satisfies all the equations in the system.
Graph of a system	419	The graph of a system of equations in two variables is the set of graphs of both equations in the system on the same coordinate plane.

Topic	Page	Remarks
Consistent system	422	A system that has at least one solution is consistent.
Inconsistent system	422	A system with no solutions is inconsistent.
Independent system	422, 445	An independent system is a system that has one and only one solution.
Dependent system	422, 424	A dependent system is a system that has more than one solution. Solutions to a dependent system can be represented parametrically.
Row-echelon form	441	A system of equations in the variables x, y, and z is in row-echelon form if: **1.** The variable x appears in no equation after the first, and y appears in no equation after the second. **2.** The coefficient of the first (leftmost) variable is 1.
Elementary operations for systems	442	See the box on page 442.
Underdetermined system	447	An underdetermined system is a system with fewer equations than variables. An underdetermined linear system is either dependent or inconsistent.
Matrix	453	A matrix is a rectangular array of numbers enclosed in brackets. Each of the numbers in the matrix is called an entry.
Dimension of a matrix	453	The dimension of a matrix is $m \times n$, specifying the number of rows (m) and columns (n).
Square matrix	453	A square matrix is a matrix with the same number of rows as columns.
Elementary row operations for matrices	454	See the box on page 454. The elementary row operations for matrices correspond to the elementary operations for systems.
Reduced row-echelon form	456	See the box on page 457.
Determinant	466, 467	The value of the 2×2 determinant $$\begin{vmatrix} a & b \\ c & d \end{vmatrix}$$ is $ad - bc$. The value of a 3×3 determinant can be found by expanding along any given row or column. See the box on page 468 for expanding along the first row. See Example 3 in Section 5.5 for expanding along a column.

Continued

Topic	Page	Remarks
Cramer's rule for two equations in two variables	472	The solution (x, y) to the system $$\begin{cases} ax + by = m \\ cx + dy = n \end{cases}$$ is given by $$x = \frac{\begin{vmatrix} m & b \\ n & d \end{vmatrix}}{\begin{vmatrix} a & b \\ c & d \end{vmatrix}} \quad \text{and} \quad y = \frac{\begin{vmatrix} a & m \\ c & n \end{vmatrix}}{\begin{vmatrix} a & b \\ c & d \end{vmatrix}} \quad \text{Provided} \quad \begin{vmatrix} a & b \\ c & d \end{vmatrix} \neq 0$$
Cramer's rule for three equations in three variables	474	The solution (x, y, z) to the system $$\begin{cases} a_1x + b_1y + c_1z = k_1 \\ a_2x + b_2y + c_2z = k_2 \\ a_3x + b_3y + c_3z = k_3 \end{cases}$$ is given by $$x = \frac{\begin{vmatrix} k_1 & b_1 & c_1 \\ k_2 & b_2 & c_2 \\ k_3 & b_3 & c_3 \end{vmatrix}}{\begin{vmatrix} a_1 & b_1 & c_1 \\ a_2 & b_2 & c_2 \\ a_3 & b_3 & c_3 \end{vmatrix}}, \quad y = \frac{\begin{vmatrix} a_1 & k_1 & c_1 \\ a_2 & k_2 & c_2 \\ a_3 & k_3 & c_3 \end{vmatrix}}{\begin{vmatrix} a_1 & b_1 & c_1 \\ a_2 & b_2 & c_2 \\ a_3 & b_3 & c_3 \end{vmatrix}}, \quad z = \frac{\begin{vmatrix} a_1 & b_1 & k_1 \\ a_2 & b_2 & k_2 \\ a_3 & b_3 & k_3 \end{vmatrix}}{\begin{vmatrix} a_1 & b_1 & c_1 \\ a_2 & b_2 & c_2 \\ a_3 & b_3 & c_3 \end{vmatrix}}$$ $$\text{Provided} \quad \begin{vmatrix} a_1 & b_1 & c_1 \\ a_2 & b_2 & c_2 \\ a_3 & b_3 & c_3 \end{vmatrix} \neq 0$$
Surveyor's formula	475	The area A of the polygon with vertices $(x_1, y_1), (x_2, y_2), \ldots, (x_n, y_n)$ listed consecutively in a counterclockwise direction is given by $$A = \frac{1}{2}\left(\begin{vmatrix} x_1 & y_1 \\ x_2 & y_2 \end{vmatrix} + \begin{vmatrix} x_2 & y_2 \\ x_3 & y_3 \end{vmatrix} + \cdots + \begin{vmatrix} x_{n-1} & y_{n-1} \\ x_n & y_n \end{vmatrix} + \begin{vmatrix} x_n & y_n \\ x_1 & y_1 \end{vmatrix} \right)$$
Matrix addition and subtraction	479	Two matrices with the same dimension are added (subtracted) by adding (subtracting) corresponding entries.
Multiplying a number and a matrix	480	To multiply a number and a matrix, multiply each entry of the matrix by the number.
Matrix multiplication	481, 483	Let A and B be matrices with dimension $m \times n$ and $n \times p$, respectively. The product AB is an $m \times p$ matrix. The entry of the ith row and jth column of AB is determined by multiplying each entry in the ith row of A by its corresponding entry in the jth column of B, and adding the products.

Topic	Page	Remarks
Multiplicative identity matrices	491	The matrix $$I_2 = \begin{bmatrix} 1 & 0 \\ 0 & 1 \end{bmatrix}$$ has the property that, for any 2×1 matrix $$X = \begin{bmatrix} x \\ y \end{bmatrix}$$ $I_2 X = X$: $$\begin{bmatrix} 1 & 0 \\ 0 & 1 \end{bmatrix} \begin{bmatrix} x \\ y \end{bmatrix} = \begin{bmatrix} x \\ y \end{bmatrix}$$ The matrix $$I_3 = \begin{bmatrix} 1 & 0 & 0 \\ 0 & 1 & 0 \\ 0 & 0 & 1 \end{bmatrix}$$ has the property that, for any 3×1 matrix $$X = \begin{bmatrix} x \\ y \\ z \end{bmatrix}$$ $I_3 X = X$: $$\begin{bmatrix} 1 & 0 & 0 \\ 0 & 1 & 0 \\ 0 & 0 & 1 \end{bmatrix} \begin{bmatrix} x \\ y \\ z \end{bmatrix} = \begin{bmatrix} x \\ y \\ z \end{bmatrix}$$
Multiplicative inverse of a matrix	492	Let A be an $n \times n$ matrix. If there exists an $n \times n$ matrix A^{-1} such that $AA^{-1} = I_n$, then A^{-1} is the inverse of A. The inverse of a 2×2 matrix $$A = \begin{bmatrix} a & b \\ c & d \end{bmatrix}$$ is given by $$A^{-1} = \frac{1}{\det A} \begin{bmatrix} d & -b \\ -c & a \end{bmatrix}$$ If $\det A = 0$, the matrix A does not have an inverse.
Solution to a system of inequalities	504–505	A solution to a system of inequalities in two variables is an ordered pair (x, y) that satisfies all the inequalities in the system. A system of inequalities typically has many solutions; these solutions are generally best represented with a graph. The graph of the solutions to a system of inequalities is typically a shaded region in the coordinate plane. See Examples 3 and 4 in Section 5.8.

MISCELLANEOUS EXERCISES

In Exercises 1–6, use a graphing utility to solve each system. Approximate all noninteger solutions to the nearest 0.001.

1. $\begin{cases} y = \sqrt{25 - x^2} \\ y = (x - 1)^2 \end{cases}$

2. $\begin{cases} y = \left(\frac{3}{2}\right)^x \\ y = \sqrt[3]{x} + 1 \end{cases}$

3. $\begin{cases} xy = 6 \\ 4y = x^3 + 4 \end{cases}$

4. $\begin{cases} x^2 + y = 4x + 3 \\ x^2 + y^3 = 17 \end{cases}$

5. $\begin{cases} 16y + 1 = x \\ \sqrt{y} + 6 = \sqrt{x - 1} \end{cases}$

6. $\begin{cases} x^2 - y^2 = 9 \\ y = \dfrac{4}{x^2 + 1} \end{cases}$

In Exercises 7–10, solve each linear system by graphing the system.

7. $\begin{cases} 10x + 3y = 16 \\ 2x = y + 8 \end{cases}$

8. $\begin{cases} x + 6y = 8 \\ x = 3y + 2 \end{cases}$

9. $\begin{cases} 3x - 9y = 9 \\ \frac{2}{3}x = 2y + 2 \end{cases}$

10. $\begin{cases} 3y - 4x = 6 \\ \frac{9}{2}y = 6x - 9 \end{cases}$

In Exercises 11 and 12, use a graphing utility to determine the equilibrium price (to the nearest $0.01).

11. The supply equation is $p = 30 + 0.07x$, and the demand equation is $p = 80 - 0.05x$.

12. The supply equation is $p = 15 + 0.001x^2$, and the demand equation is $p = 45 - 0.02x$.

13. Suppose a rock is dropped from a bridge into a river, and the splash is heard 2.5 seconds later. If we let t represent the time it takes the rock to reach the water, then $2.5 - t$ represents the time it takes the sound to come back up to the bridge. According to the laws of physics, the rock falls $s = 16t^2$ feet, and the sound travels $s = 1100(2.5 - t)$ feet. How high is the bridge?

14. A purchasing agent is quoted a price of $408 for an order of envelopes. Then she negotiates a price reduction of $0.03 per envelope after the first 1000 envelopes purchased. If the negotiated cost was $336, how many envelopes were purchased?

In Exercises 15–20, solve each system of equations using the substitution or elimination method.

15. $\begin{cases} 2x - 6y = -3 \\ x + 2y = 1 \end{cases}$

16. $\begin{cases} 5x + 3y = 14 \\ 2x - 4y = 16 \end{cases}$

17. $\begin{cases} 0.4x + 0.7y = 8 \\ x + y = 16 \end{cases}$

18. $\begin{cases} 0.25x + 0.6y = 6 \\ x + y = 14 \end{cases}$

19. $\begin{cases} y - 4 = \frac{3}{5}x \\ 5y - 3x = 10 \end{cases}$

20. $\begin{cases} \dfrac{x}{3} + \dfrac{y}{2} = \dfrac{1}{2} \\ 2x + 3y = 3 \end{cases}$

In Exercises 21–24, solve each system of equations by finding an equivalent system that is in row-echelon form and then back-substituting.

21. $\begin{cases} x - 2y = 12 \\ y + z = 7 \\ 3y - z = -7 \end{cases}$

22. $\begin{cases} x - y + z = 1 \\ 2x - y + 2z = 3 \\ 4y - 5z = 14 \end{cases}$

23. $\begin{cases} x + y - z = 6 \\ 2y + 3z = 0 \\ 2x - y - z = 1 \end{cases}$

24. $\begin{cases} x + 5y + 2z = 9 \\ 3y + 3z = 21 \\ 2x - y + 3z = -19 \end{cases}$

In Exercises 25–34, solve each system of equations by finding the reduced row-echelon form of the augmented matrix. Use the notation for elementary row operations. If the system is dependent, state the solution parametrically.

25. $\begin{cases} 3x - 2y + 2z = 5 \\ -x + y - z = -3 \\ 7x + z = 3 \end{cases}$

26. $\begin{cases} 4x + 3y - z = 2 \\ x - y + 2z = -3 \\ 2x + y = -1 \end{cases}$

27. $\begin{cases} x - y + z = 3 \\ 2x + y - z = -3 \\ 3x + 3y + 4z = 9 \end{cases}$

28. $\begin{cases} x + y + z = 5 \\ x - y - 2z = 4 \\ 2x + 4y - z = 3 \end{cases}$

29. $\begin{cases} 2x - y + 2z = 8 \\ x + 3y - z = 1 \\ x - 4y + 3z = 7 \end{cases}$

30. $\begin{cases} 2x - y + z = 10 \\ 3x + 3y + 2z = 18 \\ x - 5y = 2 \end{cases}$

31. $\begin{cases} x + y + z = 0 \\ 2x + y + 2z = 3 \\ x - y + z = 10 \end{cases}$

32. $\begin{cases} x + 2y + 2z = -3 \\ 3x - 2y + z = -1 \\ -x + 6y + 3z = 4 \end{cases}$

33. $\begin{cases} x + y + z - w = 0 \\ x + 2y + 2z + w = 3 \\ y - z + 4w = 1 \\ 2x - y - z + 3w = 13 \end{cases}$

34. $\begin{cases} x + 2y - z + w = 0 \\ 2x - y + z + w = 9 \\ y + 3z - 2w = 3 \\ 3x + 4y + 2z - w = 1 \end{cases}$

35. Determine a quadratic function $f(x) = ax^2 + bx + c$ whose graph passes through the points $(-1, -2)$, $(1, 1)$, and $(3, -4)$.

36. Determine a quadratic function $f(x) = ax^2 + bx + c$ whose graph passes through the points $(-2, -1)$, $(2, 7)$, and $(4, 5)$.

37. There are three currents I_1, I_2, and I_3 (measured in amperes) through the electrical network pictured:

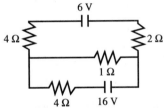

6 V

4 Ω

2 Ω

1 Ω

4 Ω   16 V

According to laws of physics, these currents satisfy the system

$\begin{cases} I_1 + I_2 - I_3 = 0 \\ 4I_2 + I_3 = 16 \\ 6I_1 + I_3 = 6 \end{cases}$

Solve this system to find the three currents.

38. A trail mix is made of nuts, raisins, and chocolate chips, which cost $0.25 per ounce, $0.10 per ounce, and $0.20 per ounce, respectively. The mix costs $0.18 per ounce and is sold in 10 ounce bags. To satisfy the demand for nutritional requirements, the number of ounces of raisins must be twice as much as chocolate chips. Find the amount of each type of ingredient in each bag.

39. Determine the area of the polygon $ABCDE$ with vertices $A(2, 0)$, $B(3, 1)$, $C(6, 3)$, $D(4, 6)$, and $E(1, 2)$.

40. Determine the area of the polygon $PQRST$ with vertices $P(-1, 1)$, $Q(2, 0)$, $R(4, 2)$, $S(4, 5)$, and $T(1, 3)$.

In Exercises 41–44, use Cramer's rule to solve each system for x and y.

41. $\begin{cases} 4x - 5y = 14 \\ -3x + 4y = 6 \end{cases}$

42. $\begin{cases} 7x - 3y = 12 \\ -2x + y = 5 \end{cases}$

43. $\begin{cases} 3px + 2qy = 0 \\ -px + py = 1 \end{cases}$

44. $\begin{cases} ax - by = 0 \\ ax + 2by = h \end{cases}$

In Exercises 45 and 46, use Cramer's rule to solve each system for x, y, and z.

45. $\begin{cases} 4x - 5y + 3z = -5 \\ -3x + 4y - 2z = 7 \\ 3x + 2y + z = 8 \end{cases}$

46. $\begin{cases} 2x - 2y + 3z = 13 \\ 5x - y + 4z = 11 \\ x + 3y - z = -10 \end{cases}$

In Exercises 47 and 48, use Cramer's rule to solve each system. Use a graphing utility to evaluate the determinants (to the nearest 0.1).

47. $\begin{cases} 0.2x + 1.5y + 3.1z = 15.4 \\ 0.6x - 1.8y + 7.2z = 11.6 \\ 3.2x - 0.7y + 4.1z = 8.3 \end{cases}$

48. $\begin{cases} 5.1x - 2.5y + 4.3z = 11.7 \\ 1.6x - 1.4y + 9.0z = 4.8 \\ 2.2x + 1.5y + 0.7z = 6.3 \end{cases}$

In Exercises 49–54, let

$$A = \begin{bmatrix} -2 & 1 \\ 3 & 4 \end{bmatrix} \qquad B = \begin{bmatrix} 0 & 8 \\ -4 & 3 \end{bmatrix} \qquad C = \begin{bmatrix} 1 & -3 & 0 \\ 2 & -5 & 6 \end{bmatrix}$$

Determine (if possible):

49. $3A + B$ **50.** $A - 2B$ **51.** AC **52.** BC

53. CA **54.** CB

55. To compare food bills at three supermarkets, the following data were collected:

	Market 1	Market 2	Market 3
Milk (per half gallon)	1.06	1.07	1.09
Bread (per loaf)	1.29	1.32	1.29
Eggs (per dozen)	1.24	1.20	1.19
Chicken (per pound)	0.89	0.93	0.91

The entries are in terms of dollars; for example, the entry in the first row and first column represents the price, $1.06, that market 1 charges for 1 half gallon of milk. Let A represent this 4×3 matrix. Suppose we purchase 3 half gallons of milk, 2 loaves of bread, 1 dozen eggs, and 1.5 pounds of chicken. Let B be the matrix [3 2 1 1.5]. Determine BA, and explain what this product represents.

56. A salary schedule for the teachers at a high school depends on the highest degree earned—bachelor of arts, master of arts, or doctorate—and number of years of experience according to the following matrix:

Bachelor's	Master's	Doctorate	
24,600	27,300	32,100	0–4 years
29,700	33,900	35,400	5–10 years
35,200	39,500	42,000	10 or more years

If the faculty receives a 3.8% raise, determine a matrix for the resulting salary schedule.

In Exercises 57–60, write each system as a matrix equation in the form $AX = B$.

57. $\begin{cases} x + 8y = 12 \\ 4x - 3y = 6 \end{cases}$

58. $\begin{cases} -x + 3y = 0 \\ 9x + 7y = -5 \end{cases}$

59. $\begin{cases} 5x + 3y - 4z = -14 \\ x + 2y - 4z = 1 \\ 2x - 5z = 10 \end{cases}$

60. $\begin{cases} x - y - 2z = 11 \\ 3x + 4y + 12z = 18 \\ -2x + 5y = 21 \end{cases}$

In Exercises 61–64, solve each system using the inverse of the coefficient matrix.

61. (a) $\begin{cases} x + 2y = -3 \\ 4x + 7y = 4 \end{cases}$ (b) $\begin{cases} x + 2y = 34 \\ 4x + 7y = 26 \end{cases}$

62. (a) $\begin{cases} 6x + 7y = 1 \\ 5x + 6y = -2 \end{cases}$ (b) $\begin{cases} 6x + 7y = 20 \\ 5x + 6y = 16 \end{cases}$

63. (a) $\begin{cases} -x - 2z = 3 \\ 3x + y + 5z = -6 \\ x - 2y + 2z = 0 \end{cases}$

(b) $\begin{cases} -x - 2z = 36 \\ 3x + y + 5z = 42 \\ x - 2y + 2z = 108 \end{cases}$

64. (a) $\begin{cases} -4x + 2y + z = 9 \\ x + y + 2z = 12 \\ 3y + 4z = 3 \end{cases}$

(b) $\begin{cases} -4x + 2y + z = 24 \\ x + y + 2z = 81 \\ 3y + 4z = 0 \end{cases}$

In Exercises 65–68, sketch the graph of each inequality.

65. $y \leq \ln(x - 3)$ 66. $y \geq |x + 2|$

67. $x^2 + 4y^2 > 36$ 68. $x^2 - y^2 < 9$

In Exercises 69 and 70, sketch the graph of each compound inequality.

69. $x^2 + 1 \leq y \leq 5$ 70. $-8 \leq y \leq (x - 1)^3$

In Exercises 71–74, sketch the graph of each system of inequalities.

71. $\begin{cases} y \geq 2 \\ x \geq 3 \\ x + y \leq 10 \end{cases}$

72. $\begin{cases} y \leq 6 \\ x \leq 4 \\ y \geq 4 - \frac{1}{2}x \end{cases}$

73. $\begin{cases} 0 \leq x \leq 4 \\ -2 \leq y \leq 1 \\ y + 1 \leq |x - 2| \end{cases}$

74. $\begin{cases} x \leq 8 \\ y^3 \leq x \\ y \geq 1 \end{cases}$

■ CHAPTER TEST

In Exercises 1 and 2, use a graphing utility to solve each system. Approximate all noninteger solutions to the nearest 0.001.

1. $\begin{cases} y = 2x - 6 \\ y = -|x| \end{cases}$

2. $\begin{cases} y = 3^x \\ x^2 + y^2 = 4 \end{cases}$

In Exercises 3 and 4, solve each linear system by graphing the system.

3. $\begin{cases} x + 2y = 10 \\ 3x - 6y = 0 \end{cases}$

4. $\begin{cases} \frac{5}{2}x - 4y = 8 \\ -15x + 24y = 20 \end{cases}$

5. Solve the system of equations using the substitution or elimination method.

$$\begin{cases} 2x + y = 1 \\ 8x + 3y = 10 \end{cases}$$

In Exercises 6 and 7, solve each system of equations by finding the reduced row-echelon form of the augmented matrix. If the system is dependent, state the solution parametrically.

6. $\begin{cases} x + y + z = -1 \\ x - 2y - z = 3 \\ 3x - z = 3 \end{cases}$

7. $\begin{cases} 7x + 4y - z = 32 \\ 3x + 2y - 3z = 18 \\ 2x + y + z = 7 \end{cases}$

8. The height of a dropped object is a function of the form $f(t) = at^2 + bt + c$, where t is the time (in seconds) and $f(t)$ is the height (in feet). After 2, 4, and 6 seconds, the corresponding heights are 29, 22, and 9 feet, respectively. Determine the constants a, b, and c. Use the result to determine the height after 5 seconds.

9. Determine the area of the polygon $ABCD$ with vertices $A(2, 3)$, $B(4, 1)$, $C(6,3)$, and $D(3, 7)$.

In Exercises 10 and 11, use Cramer's rule to solve each system for x and y.

10. $\begin{cases} 5x - 4y = 9 \\ -2x + 2y = 7 \end{cases}$

11. $\begin{cases} 3x + (L + 1)y = 1 \\ 4x + 2Ly = 0 \end{cases}$

In Exercises 12–15, let

$$A = \begin{bmatrix} 7 & 3 \\ 4 & 2 \end{bmatrix} \quad B = \begin{bmatrix} 1 & 6 \\ -2 & 3 \end{bmatrix} \quad C = \begin{bmatrix} -2 & 7 & 7 \\ 0 & 5 & 6 \\ 4 & -1 & -3 \end{bmatrix}$$

Determine (if possible):

12. AB 13. BC 14. $\det C$ 15. A^{-1}

16. For the system: $\begin{cases} -3x + y + z = 6 \\ x - y + 2z = 10 \\ 3y - 10z = 1 \end{cases}$

   (a) Write the system as a matrix equation in the form $AX = B$.
   (b) Determine the inverse of the coefficient matrix A.
   (c) Solve the system using the inverse of the coefficient matrix.

In Exercises 17 and 18, sketch the graph of each system of inequalities.

17. $\begin{cases} y \le 8 \\ x \ge 3 \\ y \ge 2x \end{cases}$

18. $\begin{cases} -2 \le x \le 4 \\ x^2 + y^2 \le 25 \end{cases}$

THE CONIC SECTIONS

■ ■ ■

This chapter is an introduction to a set of curves called *conic sections*. Conic sections (or simply *conics*) are the curves formed by a plane intersecting a *double right circular cone*. This surface is comprised of the set of all lines that intersect a fixed *axis* at the same point and angle. The lines are called *elements*, and the point of intersection is called the *vertex* of the cone (Figure 6.1).

The intersection of a cone and a plane can be one of four types of curves. If the intersecting plane is perpendicular to the axis of the cone, then the curve is a *circle* (Figure 6.2a). Tilting this plane slightly gives an *ellipse* (Figure 6.2b). By continuing to tilt this plane until it is parallel to one of the elements of the cone, the curve becomes a *parabola* (Figure 6.2c). Tilting the plane even more yields a *hyperbola* (Figure 6.2d).

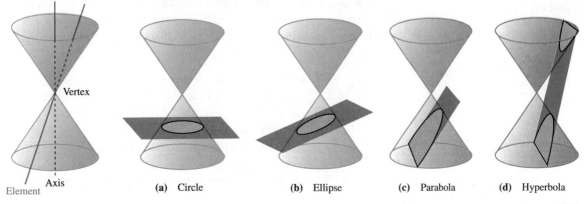

Vertex

Element · Axis

(a) Circle   **(b)** Ellipse   **(c)** Parabola   **(d)** Hyperbola

Figure 6.1   **Figure 6.2**

519

In Section 1.1, we defined a circle as the set of points in a plane that are a fixed distance from a given point. A definition such as this, in which a curve is defined as a set of points that satisfy a particular given property, is called a *locus* definition. This locus definition of a circle allowed us to derive a second-degree equation for any circle in the coordinate plane.

In Sections 6.1 and 6.2, we give locus definitions for a parabola, ellipse, and hyperbola, and we develop second-degree equations that describe them. In Section 6.3, we continue the investigation of second-degree equations, and also show that the locus definitions from Sections 6.1 and 6.2 are consistent with the geometric definitions given here.

Greek mathematicians studied conic sections as early as the fourth century BC, mainly for their aesthetic beauty and simplicity. The birth of physics and astronomy in the seventeenth century revealed that conic sections play a significant role in predicting the orbits of planets, describing the path of a falling object, and constructing telescopes. Since then, other applications of conic sections in fields such as navigation, medicine, and engineering have reinforced their importance in mathematics.

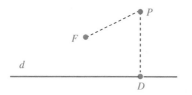

Figure 6.3
$d(P, F) = d(P, D)$

The Parabola (Locus Definition)

THE PARABOLA

As we saw in Section 2.2, the graph of a second-degree polynomial function is a parabola. In this section, we present a locus definition of parabolas, and then show that these curves are indeed the graphs of second-degree equations. Defining parabolas in this way allows us to use these graphs in describing satellite signal receivers, designing solar energy collectors, and modeling many other scientific applications.

Parabolas and Their Equations

Consider a point F and a line d marked on a sheet of paper. Each point on the paper is a certain distance from the point F and from the line d. For some points P on the paper, the two distances $d(P, F)$ and $d(P, D)$ are equal (Figure 6.3). The set of all such points P is the graph of a parabola.

> The set of all points P in a plane that are the same distance from a point F and a line d is a **parabola**. The point F is the **focus** of the parabola, and the line d is the **directrix** of the parabola. The **axis of symmetry** is the line passing through the focus that is perpendicular to the directrix. The **vertex** is the point at which the axis of symmetry intersects the parabola.

Continued

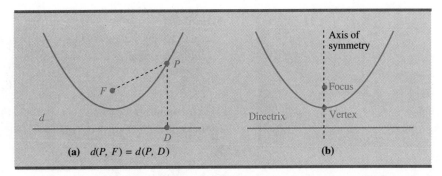

(a) $d(P, F) = d(P, D)$ **(b)**

We defined a parabola in Section 2.2 as the graph of a function $f(x) = ax^2$, for some real number a. Does this definition agree with the locus definition? The answer is yes—and here is the proof.

We place the parabola defined above on a coordinate plane, with the x-axis parallel to the directrix and through the vertex, and the y-axis falling on the axis of symmetry (Figure 6.4). The focus F is on the y-axis, say, p units above the vertex, so we label it $F(0, p)$. The directrix is a horizontal line p units below the vertex, so its equation is $y = -p$.

By the locus definition, a point $P(x, y)$ on the parabola is the same distance from the focus, $F(0, p)$, and the point closest to P on the line $y = -p$, $D(x, -p)$. This relationship allows us to determine an equation in x and y to describe this set of points P:

$$d(P, D) = d(P, F)$$
$$\sqrt{(x - x)^2 + [y - (-p)]^2} = \sqrt{(x - 0)^2 + (y - p)^2} \quad \text{Use the distance formula.}$$
$$\sqrt{(y + p)^2} = \sqrt{x^2 + (y - p)^2}$$
$$(y + p)^2 = x^2 + (y - p)^2 \quad \text{Square each side.}$$
$$y^2 + 2py + p^2 = x^2 + y^2 - 2py + p^2$$
$$4py = x^2$$
$$y = \frac{1}{4p} x^2$$

This last equation has the form $y = ax^2$, where a is $1/(4p)$, so this set of points is the graph of a second-degree polynomial function.

■ EXAMPLE 1 Verifying the Focus–Directrix Property of Parabolas

The graph of the parabola $f(x) = \frac{1}{8}x^2$ has $F(0, 2)$ as its focus and the line $y = -2$ as its directrix d. Show that the point $P(2, \frac{1}{2})$ is on the graph of f, and verify that the distance from P to the focus F is equal to the distance from P to the directrix d.

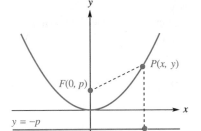

Figure 6.4

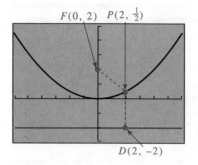

Figure 6.5
$y = \frac{1}{8}x^2$
$[-6, 6] \times [-3, 5]$

SOLUTION

The graph of f is shown in Figure 6.5. To demonstrate that $P\left(2, \frac{1}{2}\right)$ is on the parabola, we need to show that $f(2)$ is $\frac{1}{2}$:

$$f(2) = \tfrac{1}{8}(2)^2 = \tfrac{1}{2}$$

The point on the line $y = -2$ closest to P is $D(2, -2)$. We use the distance formula to evaluate $d(P, F)$ and $d(P, D)$:

$$d(P, F) = \sqrt{(0 - 2)^2 + \left(2 - \tfrac{1}{2}\right)^2} = \sqrt{(-2)^2 + \left(\tfrac{3}{2}\right)^2} = \sqrt{\tfrac{25}{4}} = \tfrac{5}{2}$$
$$d(P, D) = \sqrt{(2 - 2)^2 + \left(-2 - \tfrac{1}{2}\right)^2} = \sqrt{0 + \left(-\tfrac{5}{2}\right)^2} = \sqrt{\tfrac{25}{4}} = \tfrac{5}{2}$$

Because $d(P, F) = d(P, D)$, P is the same distance from the focus and the directrix.

If the directrix of a parabola is vertical and its focus is to the right of the directrix, then the parabola opens to the right (Figure 6.6). In this case, using $d(P, F) = d(P, D)$ leads to the equation

$$x = \frac{1}{4p}y^2$$

(See Exercise 52.)

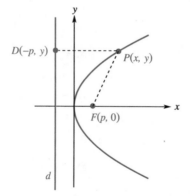

Figure 6.6
$$x = \frac{1}{4p}y^2$$

Besides opening upward or to the right, parabolas may also open downward or to the left on the coordinate plane. Figure 6.7 shows the basic equation for each of these cases.

The next two examples show how Figure 6.7 is used.

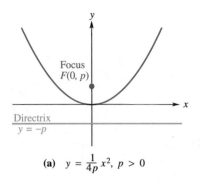

(a) $y = \dfrac{1}{4p} x^2$, $p > 0$

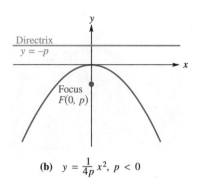

(b) $y = \dfrac{1}{4p} x^2$, $p < 0$

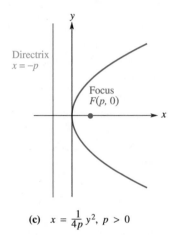

(c) $x = \dfrac{1}{4p} y^2$, $p > 0$

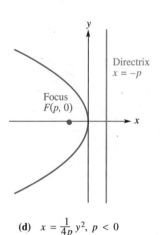

(d) $x = \dfrac{1}{4p} y^2$, $p < 0$

Figure 6.7
Basic equations for parabolas

EXAMPLE 2 Finding the Focus and the Directrix of a Parabola
The graph of the given equation is a parabola. Use a graphing utility to determine the graph of the equation, and find its focus and directrix.

(a) $y = 6x^2$ **(b)** $-12y = x^2$ **(c)** $y^2 = 4x$

SOLUTION
(a) The graph of $y = 6x^2$ is shown in Figure 6.8. Comparing this equation with

$$y = \frac{1}{4p} x^2$$

$F\left(0, \dfrac{1}{24}\right)$

$y = -\dfrac{1}{24}$

Figure 6.8
$y = 6x^2$
$[-0.6, 0.6]_{0.1} \times [-0.2, 0.6]_{0.1}$

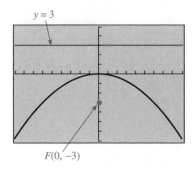

Figure 6.9
$y = -\frac{1}{12}x^2$
$[-9, 9] \times [-7, 5]$

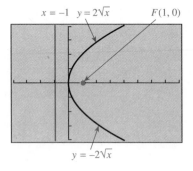

Figure 6.10
$y^2 = 4x$
$[-4, 8] \times [-4, 4]$

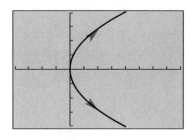

Figure 6.11
$x = \frac{1}{4}t^2, y = t, -4 \le t \le 4$
$[-4, 8] \times [-4, 4]$

in Figure 6.7, gives the value of p for this parabola:

$$6 = \frac{1}{4p}$$
$$24p = 1 \quad \text{Multiply each side by } 4p.$$
$$p = \frac{1}{24}$$

From Figure 6.7, we determine that the focus of the parabola is $\left(0, \frac{1}{24}\right)$, and its directrix is $y = -\frac{1}{24}$.

(b) Solving the equation $-12y = x^2$ for y gives $y = -\frac{1}{12}x^2$ (Figure 6.9). From

$$\frac{1}{4p} = -\frac{1}{12}$$

we get $p = -3$. The focus of the graph of $-12y = x^2$ is $(0, -3)$, and the directrix is $y = 3$.

(c) The graph of $y^2 = 4x$ can be determined with a graphing utility by first solving the equation for y:

$$y^2 = 4x$$
$$y = \pm\sqrt{4x} \quad \text{Take the square root of each side.}$$
$$y = \pm 2\sqrt{x} \quad \text{Simplify the radical.}$$

By graphing both $y = -2\sqrt{x}$ and $y = 2\sqrt{x}$, we get the graph of $y^2 = 4x$ (Figure 6.10). The equation $y^2 = 4x$ is equivalent to $x = \frac{1}{4}y^2$, so $p = 1$. The focus of this parabola is $(1, 0)$, and the directrix is $x = -1$.

The graph of $x = \frac{1}{4}y^2$ in part (c) of Example 2 also can be determined by its parametric representation from Table 3.2 in Section 3.6. Because this equation is of the form $x = g(y)$, its parametric representation is

$$x = \frac{1}{4}t^2, \quad y = t$$

Figure 6.11 shows the graph of this representation.

EXAMPLE 3 Determining an Equation of a Parabola

Determine an equation for the parabola with vertex at the origin and directrix $x = \frac{5}{2}$.

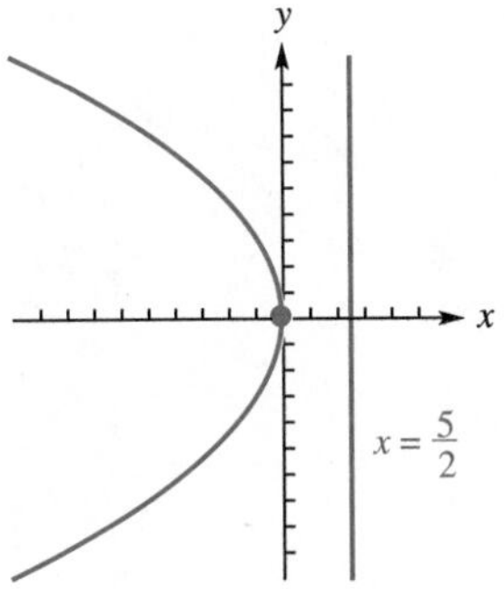

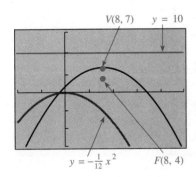

Figure 6.12

SOLUTION

The directrix is a vertical line to the right of the y-axis, so the parabola opens to the left (Figure 6.12). Figure 6.7 tells us that the equation is in the form

$$x = \frac{1}{4p} y^2 \qquad \text{for } p \text{ negative}$$

The directrix for such a parabola is $x = -p$, so it follows that p is $-\frac{5}{2}$. The equation is

$$x = \frac{1}{4\left(-\frac{5}{2}\right)} y^2$$

which is equivalent to

$$x = -\frac{1}{10} y^2$$

Translations of Parabolas

In Example 4, we determine the focus and directrix of an equation in the form $y = a(x - h)^2 + k$. As we saw in Section 2.2, the graph of this second-degree function, a parabola with vertex (h, k), is a translation of the graph of $y = ax^2$.

EXAMPLE 4   Finding the Focus and Directrix of the Graph of $y = a(x - h)^2 + k$

Determine the focus and directrix of the graph of $y = -\frac{1}{12}(x - 8)^2 + 7$.

SOLUTION

The graph of $y = -\frac{1}{12}(x - 8)^2 + 7$ is the graph of $y = -\frac{1}{12} x^2$ translated 8 units to the right and 7 units up.

In part (b) of Example 2, we found that the focus of $y = -\frac{1}{12} x^2$, $F(0, -3)$, is 3 units below the vertex of the parabola, and the directrix, $y = 3$, is 3 units above the vertex. The graph of $y = -\frac{1}{12}(x - 8)^2 + 7$ is the same parabola as the graph of $y = -\frac{1}{12} x^2$, but with vertex $V(8, 7)$ and focus $F(8, 4)$. The focus is again 3 units below the vertex, and the directrix, $y = 10$, is 3 units above the vertex (Figure 6.13).

Figure 6.13
$y = -\frac{1}{12}(x - 8)^2 + 7$
$[-10, 25]_5 \times [-15, 15]_5$

The results of Example 4 can be generalized to draw similar conclusions about the graphs of

$$y = \frac{1}{4p}(x - h)^2 + k \qquad \text{and} \qquad x = \frac{1}{4p}(y - k)^2 + h$$

Translations and the Standard
Equations of Parabolas

The graph of

$$y = \frac{1}{4p}(x - h)^2 + k$$

is a translation of the graph of

$$y = \frac{1}{4p}x^2$$

h units horizontally and k units vertically.

The graph of

$$x = \frac{1}{4p}(y - k)^2 + h$$

is a translation of the graph of

$$x = \frac{1}{4p}y^2$$

h units horizontally and k units vertically.

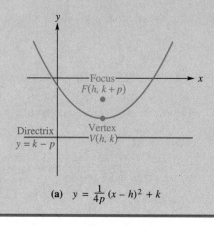

(a) $y = \frac{1}{4p}(x - h)^2 + k$

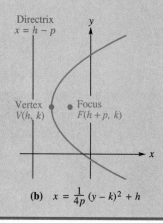

(b) $x = \frac{1}{4p}(y - k)^2 + h$

We can use the information in the above box in two ways. First, given a quadratic equation in standard form, we can determine the focus and directrix of its graph. Second, we can derive an equation for a parabola for which some information is given, such as the focus, directrix, or vertex. These uses are illustrated in Examples 5 and 6.

EXAMPLE 5   Completing the Square to Find the Focus and Directrix

Rewrite $x = \frac{1}{2}y^2 + 2y - 3$ in standard form. Use this form to determine the graph with a graphing utility, and find the focus and directrix of the graph.

SOLUTION

The equation $x = \frac{1}{2}y^2 + 2y - 3$ can be rewritten in standard form

$$x = \frac{1}{4p}(y - k)^2 + h$$

by completing the square:

$x = \frac{1}{2}y^2 + 2y - 3$

$x = \frac{1}{2}(y^2 + 4y \quad) - 3$ Group the y terms together, and factor.

$x = \frac{1}{2}(y^2 + 4y + \mathbf{4}) - 3 - \mathbf{2}$ Add $\frac{1}{2}(4)$ and subtract 2 to complete the square.

$x = \frac{1}{2}(y + 2)^2 - 5$

Solving $x = \frac{1}{2}(y + 2)^2 - 5$ for y allows us to determine the graph of this equation (just as we did in Example 2c):

$$\frac{1}{2}(y + 2)^2 - 5 = x$$

$$\frac{1}{2}(y + 2)^2 = x + 5$$

$$(y + 2)^2 = 2x + 10$$

$$y + 2 = \pm\sqrt{2x + 10}$$

$$y = -2 \pm \sqrt{2x + 10}$$

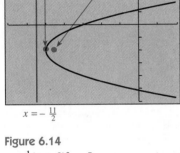

$V(-5, -2)$ $F(-\frac{9}{2}, -2)$

$x = -\frac{11}{2}$

Figure 6.14
$x = \frac{1}{2}(y + 2)^2 - 5$
$[-7, 2] \times [-6, 3]$

The graphs of $y = -2 - \sqrt{2x + 10}$ and $y = -2 + \sqrt{2x + 10}$ together give us the graph of $x = \frac{1}{2}(y + 2)^2 - 5$ (Figure 6.14).

Comparing

$$x = \frac{1}{2}(y + 2)^2 - 5 \quad \text{with} \quad x = \frac{1}{4p}(y - k)^2 + h$$

tells us that

$$\frac{1}{4p} = \frac{1}{2} \quad \text{or} \quad p = \frac{1}{2}$$

The vertex is $(-5, -2)$. Because p is positive, the parabola opens to the right. The focus is $F(-\frac{9}{2}, -2)$, $\frac{1}{2}$ unit to the right of the vertex, and the directrix is the vertical line $x = -\frac{11}{2}$, $\frac{1}{2}$ unit to the left of the vertex.

EXAMPLE 6   **Using the Focus and Directrix to Find an Equation**

Determine an equation in standard form for the parabola with focus $(6, 4)$ and directrix $y = 2$.

SOLUTION

Because the directrix is horizontal and the focus is above the directrix, the parabola opens upward (Figure 6.15). Its standard equation is in the form

$$y = \frac{1}{4p}(x - h)^2 + k$$

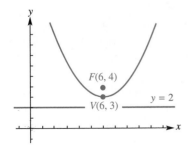

$F(6, 4)$
$V(6, 3)$ — $y = 2$

Figure 6.15

The vertex is $V(6, 3)$, since it is the point that is the same distance from the focus and directrix on the axis of symmetry. Because the focus is 1 unit above the vertex, we conclude that p is 1. Thus, the equation is

$$y = \tfrac{1}{4}(x - 6)^2 + 3$$

Applications

Applications of second-degree polynomials and their parabolic graphs were discussed in Section 2.2, but because these parabolic forms are so widely used, they merit more investigation.

The *reflective property* of parabolas, stated here, is the underlying principle behind many of the applications of this conic section.

Reflective Property of Parabolas

Light rays entering a parabola parallel to its axis of symmetry are directed by reflection to the focus. Conversely, light rays emitted from the focus are directed by reflection out of the parabola parallel to the axis of symmetry.

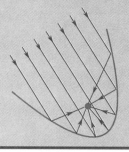

⬤ **NOTE:** *The proof of this property is part of calculus.*

This reflective property also applies for collecting or emitting sound waves and radio waves.

The television satellite dish, common to many neighborhoods, is in the shape of a parabola (Figure 6.16a). The television signal from a satellite in orbit overhead enters the parabolic dish. The rays are reflected by the dish to its receiver, mounted on a post at the focus. Based on the same concept, large reflector-type telescopes use parabolic mirrors to collect images of far-off stars and galaxies.

An automobile headlight uses the same design as the satellite dish, but with the opposite result. The filament in the headlight emits bright light in all directions. The parabolic-shaped reflector directs the light out of the lamp and ahead, onto the path of the automobile (Figure 6.16b).

Example 7 illustrates another use for the parabola based on the reflective property.

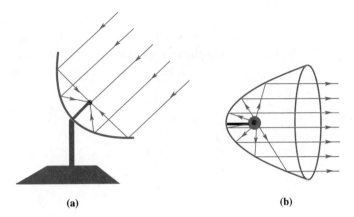

Figure 6.16

EXAMPLE 7   Application: Designing a Solar Collector

Because of the environmental and economic problems involved with burning fossil fuels such as coal and petroleum, many public utilities are investigating solar energy as a way to generate electricity. One company wants to build a prototype solar collector as shown in Figure 6.17. The overall height and width of the cross section of the trough are 1.0 meter and 1.6 meters, respectively. The inside of the parabolic trough is mirrored to reflect sunlight onto a black pipe positioned inside the trough. The heat warms a fluid in the pipe which, in turn, drives a turbine to produce electricity. Where should the pipe be placed in the trough to maximize the efficiency of the system?

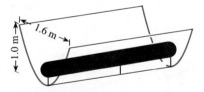

Figure 6.17

SOLUTION

The reflective property of parabolas suggests that the best position for the pipe is at the focus of the parabolic cross section. The cross section is the graph of

$$y = \frac{1}{4p}x^2$$

and the value of p will tell us where to place the pipe.

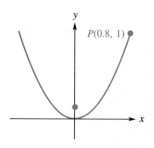

Figure 6.18

Figure 6.18 shows the cross section on the coordinate plane. Because the trough is 1.0 meter high and 1.6 meters wide, the point P on the edge of the trough has coordinates $(0.8, 1)$. We can use this fact to determine p:

$$y = \frac{1}{4p}x^2$$

$$1 = \frac{1}{4p}(0.8)^2 \quad \text{Substitute } x = 0.8 \text{ and } y = 1.$$

$$4p = (0.8)^2 \quad \text{Multiply each side by } 4p.$$

$$4p = 0.64$$

$$p = 0.16$$

The pipe should be placed 0.16 meter above the bottom of the trough.

EXERCISE SET 6.1

A

In Exercises 1–4, the focus F and directrix d of the graph of f are given. Show that each point P is on the graph of f, and verify that the distance from P to F is equal to the distance from P to d.

1. $f(x) = \frac{1}{4}x^2$; $F(0, 1)$; $d: y = -1$
  (a) $P(6, 9)$ **(b)** $P(-4, 4)$ **(c)** $P\left(1, \frac{1}{4}\right)$

2. $f(x) = -\frac{1}{8}x^2$; $F(0, -2)$; $d: y = 2$
  (a) $P(4, -2)$ **(b)** $P(-16, -32)$
  (c) $P\left(-6, -\frac{9}{2}\right)$

3. $f(x) = \frac{1}{8}(x + 4)^2 + 2$; $F(-4, 4)$; $d: y = 0$
  (a) $P(0, 4)$ **(b)** $P(-4, 2)$ **(c)** $P(4, 10)$

4. $f(x) = \frac{1}{4}x^2 + 3$; $F(0, 4)$; $d: y = 2$
  (a) $P(0, 3)$ **(b)** $P(-2, 4)$ **(c)** $P(4, 7)$

In Exercises 5–12, determine the graph of each equation, and find the focus and directrix.

5. $y = \frac{1}{12}x^2$ **6.** $y = -\frac{1}{10}x^2$

7. $x^2 + 4y = 0$ **8.** $x^2 - 16y = 0$

9. $100x = y^2$ **10.** $y^2 = 25x$

11. $6x - 3y^2 = 0$ **12.** $3y^2 + 48x = 0$

In Exercises 13–20, determine an equation in the form

$$y = \frac{1}{4p}x^2 \quad \text{or} \quad x = \frac{1}{4p}y^2$$

for each parabola described. Assume the vertex is at the origin.

13. The focus of the parabola is $(0, -5)$.

14. The focus of the parabola is $(4, 0)$.

15. The focus of the parabola is $\left(0, \frac{1}{2}\right)$.

16. The focus of the parabola is $\left(-\frac{3}{2}, 0\right)$.

17. The directrix of the parabola is $y = 3$.

18. The directrix of the parabola is $x = -\frac{1}{4}$.

19. The directrix of the parabola is $x = -2$.

20. The directrix of the parabola is $y = 1$.

In Exercises 21–28, the graph of the given equation is a parabola. Determine the graph with a graphing utility, and find the focus and directrix.

21. $y = -(x - 3)^2$ **22.** $y = x^2 + 10$

23. $x = -2(y - 4)^2$ **24.** $x = y^2 - 12$

25. $y = \frac{1}{4}(x - 3)^2 + 7$ **26.** $y = -\frac{1}{12}(x + 9)^2 - 2$

27. $x = -\frac{1}{4}(y + 1)^2 - 2$ **28.** $x = \frac{1}{6}(y - 1)^2 - \frac{3}{2}$

B

In Exercises 29–36, the graph of each equation is a parabola. Rewrite the equation in standard form, and determine the vertex, graph, focus, and directrix.

29. $(x - 2)^2 = 4(y + 1)$ **30.** $(x + 1)^2 = 4(y - 5)$

31. $y^2 - 4y - 8x - 36 = 0$ **32.** $x^2 - 10x - 8y + 49 = 0$

33. $3x^2 - 12x - 8y - 4 = 0$ **34.** $y^2 + 6y + 8x - 7 = 0$

35. $x^2 - 6x - 8y - 7 = 0$ **36.** $4y^2 - 8y + 3x = 2$

In Exercises 37–44, determine an equation in standard form for each parabola described.

37. The directrix is $y = -5$ and the focus is $(1, -3)$.

38. The directrix is $y = -4$ and the focus is $(-3, 0)$.

39. The focus is $(2, 1)$ and the vertex is $(2, 4)$.

40. The focus is $(-1, 5)$ and the vertex is $(-1, 6)$.

41. The vertex is $(2, 3)$ and the directrix passes horizontally through $(-2, 5)$.

42. The vertex is $(2, 3)$ and the directrix passes vertically through $(-2, 5)$.

43. The focus is the origin, the directrix is horizontal, and p is negative. The graph passes through $(4, 0)$.

44. The focus is the origin, the directrix is vertical, and p is negative. The graph passes through $(0, -2)$.

45. A solar cooker is constructed from a parabolic-shaped dish 2 feet in diameter and 9 inches deep. A grill is mounted at the focus of the parabolic cross section. How high above the bottom of the dish should the grill be mounted?

46. A reflecting telescope has a parabolic mirror 24 centimeters in diameter and 1.2 centimeters deep. How high above the bottom of the mirror will the reflected light collect?

47. The *focal chord* of a parabola is the line segment passing through the focus, parallel to the directrix, and with endpoints on the parabola (Figure 6.19). Show that the length of the focal chord is twice the distance from the focus F to the directrix d.

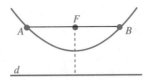

Figure 6.19

48. Determine the equations of the two parabolas that have a focal chord with endpoints $(-3, 4)$ and $(1, 4)$. (See Exercise 47.)

49. Determine the equations of the two parabolas that have a focal chord with endpoints $(-4, 5)$ and $(-4, 9)$. (See Exercise 47.)

50. Use a graphing utility to determine the graph of

$$y = \frac{1}{4p}x^2 - p$$

for $p = -3, -2, -1, 1, 2,$ and 3. Show that these parabolas share the same focus. (These curves form a *confocal family*. These families of curves have important applications in the study of electromagnetism.)

C

51. Explain how Figure 6.20 can be used to sketch a parabola with directrix d and focus F.

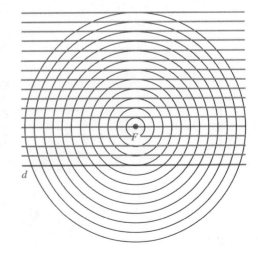

Figure 6.20

52. Use the definition of a parabola given in this section to show that the parabola with focus $(p, 0)$ and directrix $x = -p$ is the graph of

$$x = \frac{1}{4p} y^2$$

See Figure 6.21.

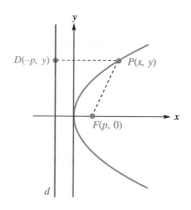

Figure 6.21

THE ELLIPSE AND HYPERBOLA

The ellipse and hyperbola have played important roles in astronomy since the late sixteenth century, when Johannes Kepler, a German astronomer, discovered that the paths of planets, moons, and comets can be modeled by these conic sections. Since then, other important applications of these curves have justified their study.

The ellipse and hyperbola can each be defined as a locus, just as the parabola was defined as a locus in the previous section. And, like the parabola, we can determine second-degree equations that describe these curves.

Ellipses and Their Equations

The following method of constructing an ellipse suggests the definition. Suppose that two thumbtacks are inserted into a drawing board, and the ends of a string are attached to the tacks. A pencil moving along a path that holds the string taut traces out an *ellipse* (Figure 6.22). Note that the sum of the distances from each of the tacks to any point on the ellipse is the same as the total length of the string.

The Ellipse (Locus Definition)

> An **ellipse** is the set of all points P in a plane such that the sum of the distances to two fixed points, F_1 and F_2, is a constant. The two points F_1 and F_2 are the **foci** of the ellipse. The midpoint O of the line segment F_1F_2 is the **center** of the ellipse. The chord of the ellipse that passes through the foci is the **major axis** of the ellipse. The endpoints of the major axis are the **major vertices** A_1 and A_2. The chord passing through the center O perpendicular to the major axis is the **minor axis**, and its endpoints B_1 and B_2 are the **minor vertices**.

Continued

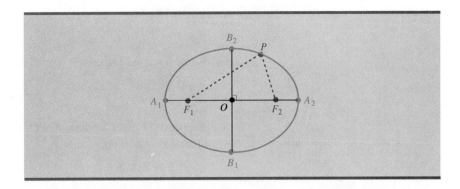

Figure 6.22

Can an ellipse be described by an equation? First, we need some preliminary results. Suppose that the distance from the center O to each of the major vertices A_1 and A_2 is a, the distance from the center O to each of the minor vertices B_1 and B_2 is b, and the distance from the center O to each of the foci F_1 and F_2 is c (Figure 6.23a). The length of the major axis is $2a$, which is also the length of the string used to construct the ellipse (see Exercise 63).

The symmetry of Figure 6.23a suggests that $d(B_2, F_2)$ is a units, and that triangle OF_2B_2 is a right triangle. Thus, by the Pythagorean theorem:

$$a^2 = b^2 + c^2$$

Next, we place the ellipse on the coordinate plane so that the center of the ellipse is at the origin, and the horizontal axis of the ellipse falls on the x-axis (Figure 6.23b). The coordinates of the foci are $F_1(-c, 0)$ and $F_2(c, 0)$. The major vertices are $A_1(-a, 0)$ and $A_2(a, 0)$, and the minor vertices are $B_1(0, -b)$ and $B_2(0, b)$. The definition of an ellipse tells us that for any point $P(x, y)$ on the ellipse, we have

$$d(P, F_1) + d(P, F_2) = 2a$$

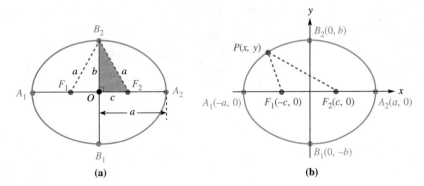

Figure 6.23

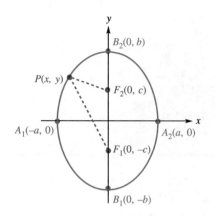

Figure 6.24

With the distance formula, this relationship becomes

$$\sqrt{(x + c)^2 + y^2} + \sqrt{(x - c)^2 + y^2} = 2a$$

After some involved algebraic manipulation, we get

$$\frac{x^2}{a^2} + \frac{y^2}{a^2 - c^2} = 1$$

(See Exercise 65.) Our finding above that $a^2 = b^2 + c^2$ means that $a^2 - c^2 = b^2$, so the last equation becomes

$$\frac{x^2}{a^2} + \frac{y^2}{b^2} = 1$$

This last equation is the *standard equation of an ellipse* with center at the origin.

If the foci are on the vertical axis of the ellipse (Figure 6.24), then the major axis is B_1B_2, the minor axis is A_1A_2, and the sum of the distances from the foci to any point on the ellipse is $2b$. Also,

$$c^2 = b^2 - a^2$$

The standard equation is again

$$\frac{x^2}{a^2} + \frac{y^2}{b^2} = 1$$

(See Exercise 67.)

⬤ EXAMPLE 1   Verifying the Focus Property of Ellipses

The graph of the ellipse

$$\frac{x^2}{25} + \frac{y^2}{9} = 1$$

has foci $F_1(-4, 0)$ and $F_2(4, 0)$ (Figure 6.25). The length of the major axis is 10. Show that the point $P\left(-3, -\frac{12}{5}\right)$ is on the ellipse, and verify that the sum of the distances from P to each of the foci is equal to the length of the major axis.

SOLUTION

First, we check whether the ordered pair $\left(-3, -\frac{12}{5}\right)$ makes the equation of the ellipse true:

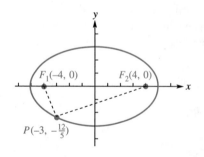

Figure 6.25
$$\frac{x^2}{25} + \frac{y^2}{9} = 1$$

$$\frac{(-3)^2}{25} + \frac{\left(-\frac{12}{5}\right)^2}{9} \overset{?}{=} 1$$

$$\frac{9}{25} + \frac{\left(\frac{144}{25}\right)}{9} \overset{?}{=} 1$$

$$\frac{9}{25} + \frac{16}{25} = 1$$

This shows that the point P is on the ellipse.

Next, we need to calculate $d(P, F_1)$ and $d(P, F_2)$, and show that their sum, $d(P, F_1) + d(P, F_2)$, is 10:

$$d(P, F_1) = \sqrt{[(-4) - (-3)]^2 + \left[0 - \left(-\frac{12}{5}\right)\right]^2}$$
$$= \sqrt{(-1)^2 + \left(\frac{12}{5}\right)^2} = \sqrt{\frac{169}{25}} = \frac{13}{5}$$
$$d(P, F_2) = \sqrt{[4 - (-3)]^2 + \left[0 - \left(-\frac{12}{5}\right)\right]^2}$$
$$= \sqrt{7^2 + \left(\frac{12}{5}\right)^2} = \sqrt{\frac{1369}{25}} = \frac{37}{5}$$

Thus,

$$d(P, F_1) + d(P, F_2) = \frac{13}{5} + \frac{37}{5} = 10$$

Sketching the graph of

$$\frac{x^2}{a^2} + \frac{y^2}{b^2} = 1$$

is not difficult once the values of a and b have been determined. Example 2 illustrates.

EXAMPLE 2   Sketching the Graph of an Ellipse
Sketch the graph of each equation. Label the foci.

(a) $\dfrac{x^2}{16} + \dfrac{y^2}{4} = 1$ (b) $\dfrac{x^2}{13} + \dfrac{y^2}{49} = 1$

SOLUTION

(a) Comparing

$$\frac{x^2}{16} + \frac{y^2}{4} = 1 \qquad \text{with} \qquad \frac{x^2}{a^2} + \frac{y^2}{b^2} = 1$$

tells us that a is 4 and b is 2 for this ellipse. The horizontal vertices are 4 units from the origin on the x-axis and the vertical vertices are 2 units from the origin on the y-axis. These vertices define a frame for the ellipse (Figure 6.26a, page 536), in which we can sketch the ellipse (Figure 6.26b). Because

$$c = \sqrt{a^2 - b^2} = \sqrt{16 - 4} = \sqrt{12} = 2\sqrt{3}$$

the foci are $(-2\sqrt{3}, 0)$ and $(2\sqrt{3}, 0)$ each about 3.5 units from the origin on the x-axis.

(b) From the equation

$$\frac{x^2}{13} + \frac{y^2}{49} = 1$$

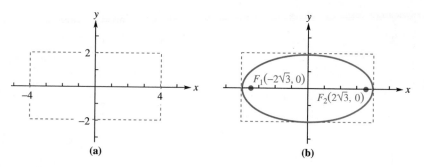

Figure 6.26

we determine that a is $\sqrt{13}$ (approximately 3.6) and b is 7. The frame of the ellipse is shown in Figure 6.27a. Since b is greater than a, the foci are on the vertical axis. Because

$$c = \sqrt{b^2 - a^2} = \sqrt{49 - 13} = \sqrt{36} = 6$$

the foci are $(0, -6)$ and $(0, 6)$. The ellipse and its foci are shown in Figure 6.27b.

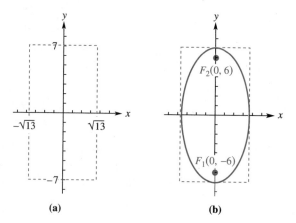

Figure 6.27

The steps for sketching an ellipse used in Example 2 are generalized below.

Sketching the Graph of an Ellipse

To sketch the graph of $\dfrac{x^2}{a^2} + \dfrac{y^2}{b^2} = 1$:

Step 1: Determine the values of a and b from the equation, and construct the frame of the ellipse.

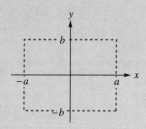

Step 2: Sketch an ellipse in the frame. The vertices of the ellipse coincide with the midpoints of the sides of the frame. The foci of the ellipse are determined by the following:

(a) If $a > b$, then

$$c = \sqrt{a^2 - b^2}$$

and the foci F_1 and F_2 are each c units from the center of the ellipse on the horizontal axis.

(b) If $b > a$, then

$$c = \sqrt{b^2 - a^2}$$

and the foci F_1 and F_2 are each c units from the center of the ellipse on the vertical axis.

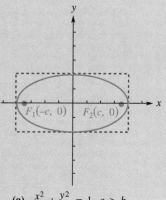
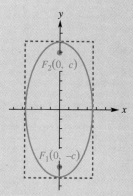

(a) $\dfrac{x^2}{a^2} + \dfrac{y^2}{b^2} = 1,\ a > b$ **(b)** $\dfrac{x^2}{a^2} + \dfrac{y^2}{b^2} = 1,\ b > a$

Example 3 shows how we can determine a standard equation for an ellipse when given certain information about the curve.

■ EXAMPLE 3   Finding the Standard Equation of an Ellipse

Determine the standard equation of the ellipse with center at the origin, foci $(-3, 0)$ and $(3, 0)$, and major axis of length 8.

SOLUTION

The equation we are to determine is of the form

$$\frac{x^2}{a^2} + \frac{y^2}{b^2} = 1$$

so the problem reduces to finding the values of a^2 and b^2. Because the foci are on the x-axis, the horizontal axis is the major axis of the ellipse. The length of the major axis, $2a$, is 8. It follows that a is 4, and a^2 is 16.

Since the foci are $(-3, 0)$ and $(3, 0)$, the value of c is 3. We can compute b^2:

$$b^2 = a^2 - c^2 = 4^2 - 3^2 = 7$$

The equation of the ellipse described is

$$\frac{x^2}{16} + \frac{y^2}{7} = 1$$

(Figure 6.28).

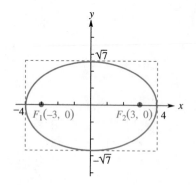

Figure 6.28

Hyperbolas and Their Equations

The fourth conic section, the *hyperbola*, is defined in a manner similar to the ellipse.

The Hyperbola (Locus Definition)

> A **hyperbola** is the set of all points P in a plane such that the difference of the distances to two fixed points, F_1 and F_2, is a constant. The two points F_1 and F_2 are the **foci** of the hyperbola. The midpoint O of the line segment F_1F_2 is the **center** of the hyperbola. The two curves that comprise the hyperbola are its **branches**. The points A_1 and A_2 of the hyperbola on the same line as the foci are the **vertices** of the hyperbola. The line segment A_1A_2 is the **transverse axis** of the hyperbola.

Continued

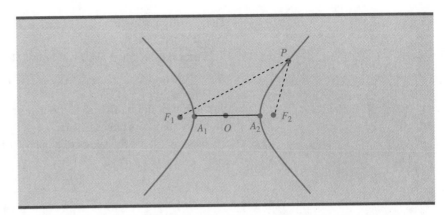

Suppose the distance from the center O to each of the vertices A_1 and A_2 is a. The length of the transverse axis, $2a$, is equal to the difference of the distances in the definition of the hyperbola given above (see Exercise 64). The branch that is closer to F_2 is the set of points P such that

$$d(P, F_1) - d(P, F_2) = 2a$$

The other branch is determined by

$$d(P, F_2) - d(P, F_1) = 2a$$

So it follows that both branches together are described by

$$d(P, F_1) - d(P, F_2) = \pm 2a$$

Next, we place a coordinate system on the hyperbola (Figure 6.29), as we did for the ellipse earlier. The vertices are $A_1(-a, 0)$ and $A_2(a, 0)$. We label the foci as $F_1(-c, 0)$ and $F_2(c, 0)$. With the distance formula, the relationship

$$d(P, F_1) - d(P, F_2) = \pm 2a$$

becomes the equation

$$\sqrt{(x + c)^2 + y^2} - \sqrt{(x - c)^2 + y^2} = \pm 2a$$

Again, after a lengthy bit of algebraic manipulation (Exercise 66), we arrive at the equation.

$$\frac{x^2}{a^2} - \frac{y^2}{c^2 - a^2} = 1$$

The experience of deriving the standard equation for an ellipse suggests that we define b such that $b^2 = c^2 - a^2$, giving us the following equation:

$$\frac{x^2}{a^2} - \frac{y^2}{b^2} = 1$$

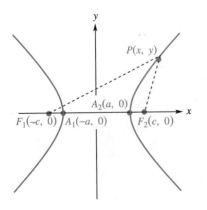

Figure 6.29

This equation is the *standard equation* of a hyperbola with a horizontal transverse axis and center at the origin.

From the sketch of the hyperbola in Figure 6.29, we can see that

$$y \to \pm\infty \quad \text{as} \quad x \to -\infty \qquad \text{and} \qquad y \to \pm\infty \quad \text{as} \quad x \to +\infty$$

Furthermore, the branches of the hyperbola seem to have asymptotes. We can show that this conjecture is true.

First, we solve the standard equation of the hyperbola for y^2:

$$\frac{x^2}{a^2} - \frac{y^2}{b^2} = 1$$

$$\frac{y^2}{b^2} = \frac{x^2}{a^2} - 1$$

$$y^2 = \frac{b^2}{a^2}x^2 - b^2$$

In the expression on the right, as x grows without bound (either positively or negatively), the term $-b^2$ becomes insignificant with respect to the size of the other term,

$$\frac{b^2}{a^2}x^2$$

Thus, the behavior of the graph for extreme values of x is described by

$$y^2 \approx \frac{b^2}{a^2}x^2 \qquad \text{as} \qquad x \to \pm\infty$$

or

$$y \approx \pm\frac{b}{a}x \qquad \text{as} \qquad x \to \pm\infty \qquad \text{(Figure 6.30)}$$

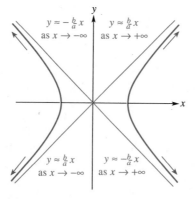

Figure 6.30

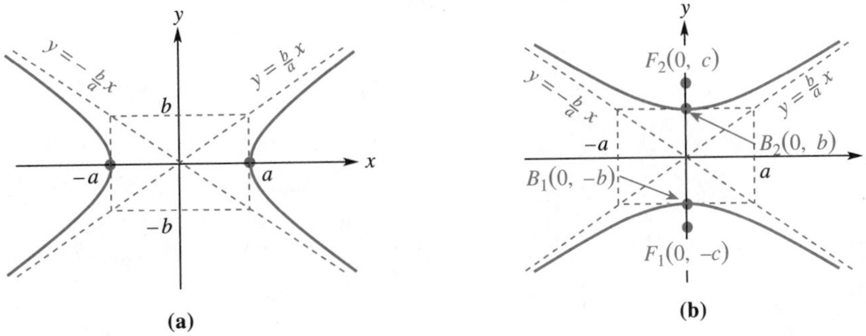

Figure 6.31

The graphs of

$$y = -\frac{b}{a}x \quad \text{and} \quad y = \frac{b}{a}x$$

are lines passing through the origin (Figure 6.31a). These lines are the *asymptotes of the hyperbola*. They can be determined by first sketching the frame, using a and b as we did for the ellipse, and then drawing the lines through the diagonal corners of the frame. Notice that the vertices of the hyperbola are on the frame.

If the transverse axis is vertical (Figure 6.31b), then the vertices are $B_1(0, -b)$ and $B_2(0, b)$, and the length of the transverse axis is $2b$. The standard equation of this hyperbola is

$$\frac{y^2}{b^2} - \frac{x^2}{a^2} = 1$$

(See Exercise 68.) The asymptotes of the graph are again

$$y = -\frac{b}{a}x \quad \text{and} \quad y = \frac{b}{a}x$$

The steps for quickly sketching the graph of a hyperbola from its equation are outlined in Example 4.

◯ EXAMPLE 4   Sketching the Graph of a Hyperbola

Sketch the graph of each equation. Label the foci.

(a) $\dfrac{x^2}{4} - \dfrac{y^2}{9} = 1$ **(b)** $\dfrac{y^2}{10} - \dfrac{x^2}{6} = 1$

SOLUTION

(a) Comparing

$$\frac{x^2}{4} - \frac{y^2}{9} = 1 \qquad \text{with} \qquad \frac{x^2}{a^2} - \frac{y^2}{b^2} = 1$$

tells us that the transverse axis is horizontal, and that a is 2 and b is 3. The frame and the asymptotes are shown in Figure 6.32a. The vertices are $(-2, 0)$ and $(2, 0)$. Because $b^2 = c^2 - a^2$, we can compute c from a and b:

$$c = \sqrt{a^2 + b^2} = \sqrt{4 + 9} = \sqrt{13}$$

The foci are $(-\sqrt{13}, 0)$ and $(\sqrt{13}, 0)$, each about 3.6 units from the origin, on the x-axis. The hyperbola and its foci are shown in Figure 6.32b.

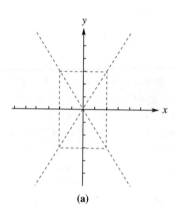

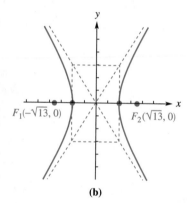

(a)   (b)

Figure 6.32

(b) The equation

$$\frac{y^2}{10} - \frac{x^2}{6} = 1$$

is of the form

$$\frac{y^2}{b^2} - \frac{x^2}{a^2} = 1$$

so the transverse axis is vertical. From the equation, we determine that b is $\sqrt{10}$ and a is $\sqrt{6}$. The frame and asymptotes are shown in Figure 6.33a. Because the transverse axis is vertical, the foci are on the vertical axis. The foci are $(0, -4)$ and $(0, 4)$ because

$$c = \sqrt{a^2 + b^2} = \sqrt{6 + 10} = \sqrt{16} = 4$$

The hyperbola and its foci are shown in Figure 6.33b.

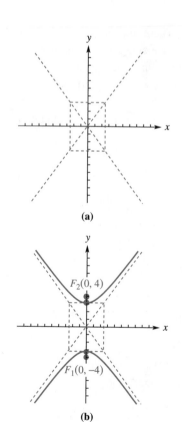

(a)

(b)

Figure 6.33

Sketching the Graph
of a Hyperbola

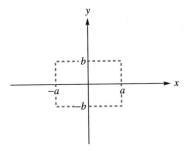

Step 1

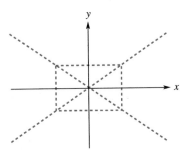

Step 2

To sketch the graph of either $\dfrac{x^2}{a^2} - \dfrac{y^2}{b^2} = 1$ or $\dfrac{y^2}{b^2} - \dfrac{x^2}{a^2} = 1$:

Step 1: Determine the values of a and b from the equation, and construct the frame of the hyperbola, as shown in the margin.

Step 2: Sketch the asymptotes of the hyperbola through the corners of the frame, as shown in the margin. The equations of the asymptotes are

$$y = \frac{b}{a}x \quad \text{and} \quad y = -\frac{b}{a}x$$

Step 3: Sketch the hyperbola using the frame and the asymptotes.

  (a) For $\dfrac{x^2}{a^2} - \dfrac{y^2}{b^2} = 1$, the vertices are $A_1(-a, 0)$ and $A_2(a, 0)$.

  (b) For $\dfrac{y^2}{b^2} - \dfrac{x^2}{a^2} = 1$, the vertices are $B_1(0, -b)$ and $B_2(0, b)$.

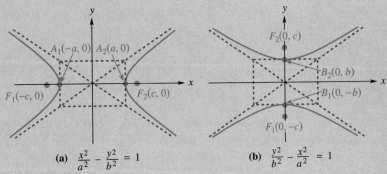

  (a) $\dfrac{x^2}{a^2} - \dfrac{y^2}{b^2} = 1$ **(b)** $\dfrac{y^2}{b^2} - \dfrac{x^2}{a^2} = 1$

The foci are both c units from the center of the hyperbola, outside the frame, on the line containing the transverse axis. In both cases, the value of c is given by $c = \sqrt{a^2 + b^2}$.

◼ EXAMPLE 5 Determining the Equation of a Hyperbola

Determine the equation of the hyperbola with center at the origin, vertices $(-6, 0)$ and $(6, 0)$, and asymptotes $y = -\frac{4}{3}x$ and $y = \frac{4}{3}x$.

SOLUTION

The vertices of the hyperbola are on the x-axis, so the transverse axis of the hyperbola is horizontal. The equation is of the form

$$\frac{x^2}{a^2} - \frac{y^2}{b^2} = 1$$

for some real numbers a and b. The length of the transverse axis is 12, so a is 6. We can find b from the equations of the asymptotes. Comparing $y = \pm\frac{4}{3}x$ with

$$y = \pm\frac{b}{a}x$$

tells us that

$$\frac{b}{a} = \frac{4}{3}$$

$$b = \frac{4}{3}a = \frac{4}{3}(6) = 8$$

The equation of the hyperbola is

$$\frac{x^2}{6^2} - \frac{y^2}{8^2} = 1$$

$$\frac{x^2}{36} - \frac{y^2}{64} = 1$$

The graph is shown in Figure 6.34.

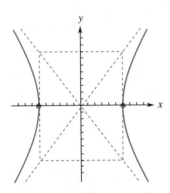

Figure 6.34
$\dfrac{x^2}{36} - \dfrac{y^2}{64} = 1$

Translations of Ellipses and Hyperbolas

In Section 1.1, we saw that the graph of $x^2 + y^2 = r^2$ is a circle of radius r with center at the origin. The graph of $(x - h)^2 + (y - k)^2 = r^2$ is also a circle of radius r, but with center (h, k). By replacing x with $x - h$ and y with $y - k$, the graph of $x^2 + y^2 = r^2$ is translated h units horizontally and k units vertically. The same is true for both ellipses and hyperbolas. By replacing x and y with $x - h$ and $y - k$ in the standard equation of an ellipse or hyperbola, we get an identical curve but with center (h, k).

EXAMPLE 6   Sketching Conic Sections Using Translations
Sketch the graph of each equation.

(a) $\dfrac{(x - 3)^2}{25} + (y + 1)^2 = 1$ (b) $\dfrac{(x + 2)^2}{9} - \dfrac{(y - 4)^2}{9} = 1$

SOLUTION

(a) The graph of

$$\frac{(x - 3)^2}{25} + (y + 1)^2 = 1$$

is an ellipse with center $(3, -1)$. Rewriting this equation as

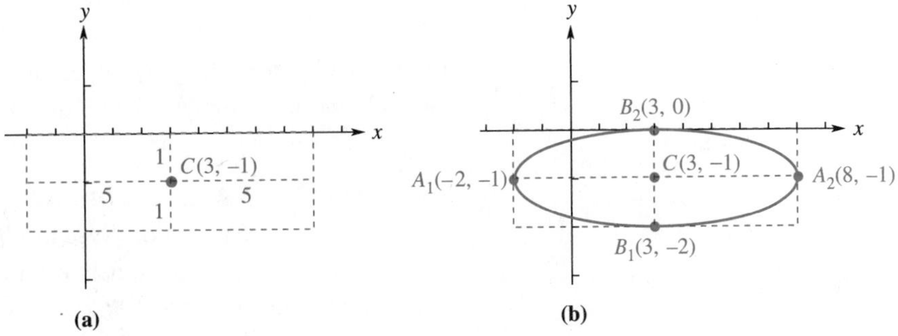

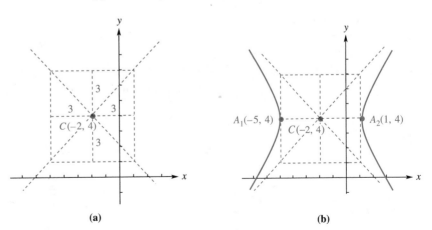

Figure 6.35

$$\frac{(x-3)^2}{5^2} + \frac{(y+1)^2}{1^2} = 1$$

tells us that a is 5 and b is 1. Starting at the center $(3, -1)$, we build the frame by moving 5 units to the right and left and 1 unit up and down (Figure 6.35a). Next, we use the frame to sketch the ellipse (Figure 6.35b).

(b) The graph of

$$\frac{(x+2)^2}{9} - \frac{(y-4)^2}{9} = 1$$

is a hyperbola with center $(-2, 4)$. The transverse axis is horizontal, and both a and b are 3. Moving 3 units up, down, right, and left from $(-2, 4)$ determines the frame. The asymptotes of the hyperbola contain the diagonals of this rectangle (Figure 6.36a). The frame and the asymptotes allow us to sketch the hyperbola (Figure 6.36b).

Figure 6.36

Applications

Ellipses and hyperbolas play important roles in the application of mathematics in the real world. Examples 7 and 8 demonstrate two such applications.

▣ EXAMPLE 7   Application: Navigating by LORAN

The *Long Range Navigation System* (*LORAN*) allows ships at sea to pinpoint their position, using radio signals from onshore transmitters. Suppose that transmitter A and transmitter B, 100 kilometers directly west of A, each send a signal simultaneously. A ship directly north of transmitter A receives the signal from A 200 microseconds before it receives the signal from B. The velocity of the signals is 300 meters per microsecond. How far (to the nearest 0.1 kilometer) is the ship from transmitter A?

SOLUTION

The signal from B takes longer to reach the ship because the ship is farther from B than it is from A. The rate of the signal is given, so we can compute this difference in distances:

$$(200 \text{ microseconds})(300 \text{ meters per microsecond}) = 60{,}000 \text{ meters}$$

The ship is 60,000 meters, or 60 kilometers, farther from B than A.

If the ship is at point $P(x, y)$ (Figure 6.37), then

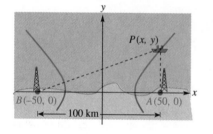

Figure 6.37

$$d(P, B) - d(P, A) = 60$$

The points P that satisfy this condition all lie on the right branch of the hyperbola shown in Figure 6.37. This hyperbola has an equation of the form

$$\frac{x^2}{a^2} - \frac{y^2}{b^2} = 1$$

where $2a = 60$, or $a = 30$. Because the transmitters are 100 kilometers apart, the foci of this hyperbola are $B(-50, 0)$ and $A(50, 0)$. The value of b^2 is given by

$$b^2 = c^2 - a^2 = 50^2 - 30^2 = 2500 - 900 = 1600$$

Thus, the equation of the hyperbola is

$$\frac{x^2}{900} - \frac{y^2}{1600} = 1$$

Since the ship is directly north of A, it follows that the x-coordinate of P is 50, and the distance from the ship to transmitter A is the y-coordinate of P. We leave it to you to verify that substituting $x = 50$ into the equation of the hyperbola and solving for y gives $y = 53\frac{1}{3}$. The ship is 53.3 kilometers (to the nearest 0.1 kilometer) from transmitter A.

Ellipses have a *reflective property* similar to that of parabolas.

Reflective Property of Ellipses

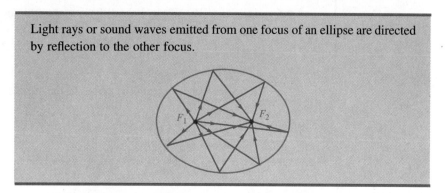

Light rays or sound waves emitted from one focus of an ellipse are directed by reflection to the other focus.

This property is exploited in the design of *whispering galleries*, elliptical-shaped rooms where two persons—each standing at a focus—can carry on a normal conversation because of the concentration of sound waves at these points. The most famous whispering gallery is in the Capitol Building in Washington, D.C. In the nineteenth century, this room was used for sessions of Congress. Records show that Martin Van Buren, the leader of the Democrats (and later President of the United States), positioned his desk at one focus and often feigned sleep to eavesdrop on his colleagues at the other focus. Later, another mathematically inclined member of Congress, Abraham Lincoln, chose a desk close to one of the minor vertices of the room, since he knew that this desk was more private than others that were available.

Example 8 shows a more modern use of this property.

EXAMPLE 8 Application: Constructing a Lithotriptor

A medical procedure called *sound wave lithotripsy* treats kidney stone conditions without surgery by using the reflective property of ellipses. The patient is placed in a lithotriptor, an elliptical tank with a seat at one end and a wave emitter at the other. The stone is positioned at one focus of the ellipse and a wave emitter is at the other focus at the same height as the stone. The waves from the emitter are reflected to the stone, which is then pulverized and passed naturally from the patient's body.

Suppose that the elliptical tank is 180 centimeters long and 90 centimeters wide. Where (to the nearest centimeter) should the emitter be placed and the stone positioned?

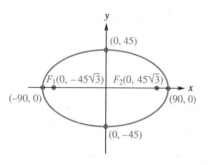

Figure 6.38

SOLUTION

The stone and the emitter are to be at the foci of the ellipse. The major axis of the ellipse is 180 centimeters, so the major vertices are 90 centimeters from the center. Likewise, the minor axis is 90 centimeters, so the minor vertices are 45 centimeters from the center (Figure 6.38).

The distance from the center to the focus of the ellipse, c, is given by

$$c = \sqrt{90^2 - 45^2} = 45\sqrt{3} \approx 78$$

The stone and the emitter should be positioned 78 centimeters (to the nearest centimeter) from the center of the tank on the major axis of the ellipse.

EXERCISE SET 6.2

A

In Exercises 1 and 2, the graph of each equation is an ellipse with the given foci and major axis. Show that each point P is on the ellipse, and verify that the sum of the distances from P to each of the foci is equal to the length of the major axis.

1. $\dfrac{x^2}{100} + \dfrac{y^2}{36} = 1$; foci: $(\pm 8, 0)$; major axis: 20

   (a) $P(0, -6)$ **(b)** $P(10, 0)$ **(c)** $P\left(6, \frac{24}{5}\right)$

2. $\dfrac{x^2}{16} + \dfrac{y^2}{25} = 1$; foci: $(0, \pm 3)$; major axis: 10

   (a) $P(4, 0)$ **(b)** $P(0, -5)$ **(c)** $P\left(\frac{-16}{5}, -3\right)$

In Exercises 3 and 4, the graph of each equation is a hyperbola with the given foci and transverse axis. Show that each point P is on the hyperbola, and verify that the difference of the distances from P to each of the foci is equal to the length of the transverse axis.

3. $\dfrac{x^2}{16} - \dfrac{y^2}{9} = 1$; foci: $(\pm 5, 0)$; transverse axis: 8

   (a) $P(-4, 0)$ **(b)** $P\left(\frac{20}{3}, -4\right)$ **(c)** $P\left(5, \frac{9}{4}\right)$

4. $\dfrac{y^2}{9} - \dfrac{x^2}{16} = 1$; foci: $(0, \pm 5)$; transverse axis: 6

   (a) $P(0, -3)$ **(b)** $P\left(-\frac{16}{3}, -5\right)$ **(c)** $\left(3, \frac{15}{4}\right)$

In Exercises 5–12, the graph of each equation is an ellipse. Sketch the graph and determine the foci and vertices of the graph.

5. $\dfrac{x^2}{100} + \dfrac{y^2}{36} = 1$ **6.** $\dfrac{x^2}{169} + \dfrac{y^2}{25} = 1$

7. $\dfrac{x^2}{144} + \dfrac{y^2}{225} = 1$ **8.** $\dfrac{x^2}{16} + \dfrac{y^2}{25} = 1$

9. $\dfrac{x^2}{50} + \dfrac{y^2}{25} = 1$ **10.** $\dfrac{x^2}{4} + \dfrac{y^2}{40} = 1$

11. $\dfrac{x^2}{4} + y^2 = 1$ **12.** $x^2 + \dfrac{y^2}{9} = 1$

In Exercises 13–20, the graph of each equation is a hyperbola. Sketch the graph and determine the foci, vertices, and asymptotes of the graph.

13. $\dfrac{x^2}{9} - \dfrac{y^2}{16} = 1$ **14.** $\dfrac{x^2}{36} - \dfrac{y^2}{36} = 1$

15. $\dfrac{y^2}{16} - \dfrac{x^2}{16} = 1$ **16.** $\dfrac{y^2}{36} - \dfrac{x^2}{64} = 1$

17. $x^2 - y^2 = 1$ **18.** $y^2 - x^2 = 1$

19. $y^2 - \dfrac{x^2}{64} = 1$ **20.** $\dfrac{x^2}{9} - y^2 = 1$

B

In Exercises 21–32, the graph of each equation is a conic section. Sketch the graph.

21. $y = \frac{1}{8}x^2 + 6$

22. $x = -\frac{1}{4}(y - 2)^2$

23. $\frac{(x - 2)^2}{12} + \frac{y^2}{3} = 1$

24. $\frac{x^2}{12} + \frac{(y + 4)^2}{16} = 1$

25. $\frac{(x - 2)^2}{4} + \frac{(y - 5)^2}{4} = 1$

26. $\frac{(x + 2)^2}{18} + \frac{(y - 4)^2}{18} = 1$

27. $\frac{(x - 4)^2}{16} - \frac{(y + 2)^2}{9} = 1$

28. $\frac{(y - 5)^2}{16} - \frac{(x - 2)^2}{16} = 1$

29. $x = -(y - 4)^2 - 5$

30. $(x - 3)^2 + (y + 3)^2 = 9$

31. $(x + 4)^2 - (y - 1)^2 = 1$

32. $\frac{1}{4}(x - 2)^2 + (y + 4)^2 = 1$

33. Determine the vertices and foci of the graph of the equation in Exercise 23.

34. Determine the vertices and foci of the graph of the equation in Exercise 24.

35. Determine the vertices, foci, and asymptotes of the graph of the equation in Exercise 27.

36. Determine the vertices, foci, and asymptotes of the graph of the equation in Exercise 28.

In Exercises 37–46, determine an equation in standard form for the conic section described.

37. An ellipse with vertices $(6, 0)$, $(-6, 0)$, $(0, -2)$, and $(0, 2)$

38. An ellipse with vertices $(5, 0)$ and $(0, 3)$, and center $(0, 0)$

39. A hyperbola with vertices $(-2, 0)$ and $(2, 0)$, and foci $(-3, 0)$ and $(3, 0)$

40. A hyperbola with a vertical transverse axis of length 6, and foci $(0, -5)$ and $(0, 5)$

41. An ellipse with vertices $(1, 2)$, $(-5, 2)$, $(-2, 0)$, and $(-2, 4)$

42. An ellipse with vertices $(5, 7)$ and $(8, 3)$, and center $(5, 3)$

43. A hyperbola with one vertex at $(-3, -1)$, and foci $(-3, 4)$ and $(-3, -2)$

44. A hyperbola with asymptotes $y = \frac{3}{2}x - 6$ and $y = -\frac{3}{2}x - 6$, and horizontal transverse axis of length 4

45. An ellipse with foci $(-8, 0)$ and $(8, 0)$, passing through the point $(-6, -2)$

46. A hyperbola with vertices $(-3, 0)$ and $(3, 0)$, passing through the point $(3, -8)$

In Exercises 47–52, sketch the graphs of each system of equations and determine their points of intersection.

47. $\begin{cases} \dfrac{x^2}{5} + \dfrac{y^2}{45} = 1 \\ x - y + 1 = 0 \end{cases}$

48. $\begin{cases} \dfrac{x^2}{48} + \dfrac{y^2}{24} = 1 \\ y = \dfrac{1}{2}x + 2 \end{cases}$

49. $\begin{cases} x^2 - y^2 = 1 \\ x^2 + y^2 = 7 \end{cases}$

50. $\begin{cases} x^2 - \dfrac{y^2}{5} = 1 \\ x^2 + y^2 = 25 \end{cases}$

51. $\begin{cases} x = y^2 + 2 \\ (x - 4)^2 + y^2 = 4 \end{cases}$

52. $\begin{cases} x = y^2 - 5 \\ \dfrac{(x + 5)^2}{54} - \dfrac{y^2}{18} = 1 \end{cases}$

In Exercises 53–58, sketch each inequality or system of inequalities.

53. $\dfrac{x^2}{16} + \dfrac{y^2}{36} \le 1$

54. $\dfrac{y^2}{25} - \dfrac{x^2}{9} \ge 1$

55. $\begin{cases} x \ge y^2 - 6 \\ y \ge \dfrac{1}{4}x \end{cases}$

56. $\begin{cases} x \ge y^2 - 6 \\ x^2 + y^2 \le 16 \end{cases}$

57. $\begin{cases} \dfrac{x^2}{36} + \dfrac{y^2}{16} \ge 1 \\ \dfrac{x^2}{16} + \dfrac{y^2}{36} \le 1 \end{cases}$

58. $\begin{cases} x^2 + y^2 \le 64 \\ \dfrac{x^2}{4} - \dfrac{y^2}{4} \ge 1 \end{cases}$

59. Rework Example 7 given that the signals received by the ship are 100 microseconds apart.

60. Rework Example 8 given that the tank is 6 feet long and 2 feet wide. (Express answer to the nearest 0.1 foot.)

61. An elliptical arch 20 feet wide is to be constructed so that a crate 10 feet wide and 6 feet high can pass under the arch. How high is the arch in the middle?

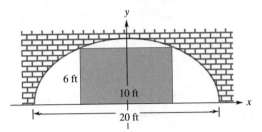

6 ft

10 ft

20 ft

62. Trinh and Pat are talking on the telephone when an explosion occurs on a road directly west of Trinh's house. Trinh hears the explosion 1.2 seconds earlier than Pat. Given that Trinh's house is 6.6 kilometers south of Pat's house, and that sound travels at 330 meters per second, how far from each house is the blast?

63. Show that for an ellipse with major axis A_1A_2,

$$d(A_1, F_1) + d(A_1, F_2) = d(A_1, A_2)$$

Explain why this fact, along with the definition of an ellipse given in this section, shows that for any point P on the ellipse, $d(P, F_1) + d(P, F_2)$ is equal to the length of the major axis.

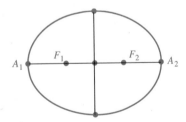

64. Show that for a hyperbola with transverse axis A_1A_2,

$$d(A_1, F_2) - d(A_1, F_1) = d(A_1, A_2)$$

Explain why this fact, along with the locus definition of a hyperbola, shows that $|d(P, F_1) - d(P, F_2)|$ is equal to the length of the transverse axis. (See the figure at the top of the next column.)

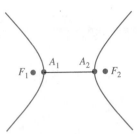

65. Show that $\sqrt{(x + c)^2 + y^2} + \sqrt{(x - c)^2 + y^2} = 2a$ is equivalent to

$$\frac{x^2}{a^2} + \frac{y^2}{b^2} = 1 \qquad \text{where } b^2 = a^2 - c^2$$

66. Show that $\sqrt{(x + c)^2 + y^2} - \sqrt{(x - c)^2 + y^2} = \pm 2a$ is equivalent to

$$\frac{x^2}{a^2} - \frac{y^2}{b^2} = 1 \qquad \text{where } b^2 = c^2 - a^2$$

67. Use the locus definition of an ellipse and Exercise 65 to show that the ellipse with foci $F_1(0, -c)$ and $F_2(0, c)$ is the graph of

$$\frac{x^2}{a^2} + \frac{y^2}{b^2} = 1 \qquad \text{where } b > a$$

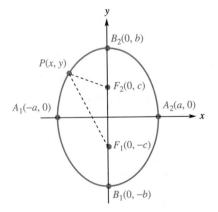

68. Use the locus definition of a hyperbola and Exercise 66 to show that the hyperbola with foci $F_1(0, -c)$ and $F_2(0, c)$ is the graph of

$$\frac{y^2}{b^2} - \frac{x^2}{a^2} = 1$$

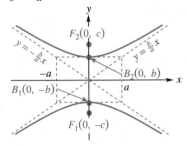

(b) Explain how the figure can be used to sketch a hyperbola with foci F_1 and F_2 and transverse axis 4.

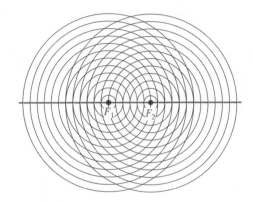

69. **(a)** Explain how the figure can be used to sketch an ellipse with foci F_1 and F_2 and major axis 10.

CONIC SECTIONS

At the beginning of this chapter, we introduced the four conic sections—circle, ellipse, parabola, and hyperbola—as intersections of a plane and a double right circular cone (Figure 6.39). In Sections 6.1 and 6.2, we gave locus definitions for these curves and developed equations for them.

In this section, we conclude the discussion of conic sections by introducing their general equations, demonstrating how conics are implemented on a graphing utility, and, finally, showing that the curves described by the locus definitions in Sections 6.1 and 6.2 are indeed the conic sections defined geometrically in Figure 6.39.

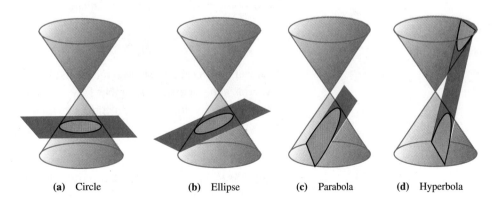

(a) Circle   **(b)** Ellipse   **(c)** Parabola   **(d)** Hyperbola

Figure 6.39

General Equations for Conic Sections

The second-degree equation

$$\frac{(x-1)^2}{4} - \frac{(y-3)^2}{9} = 1$$

is a standard equation of a hyperbola. This equation can be rewritten in the following way:

$$\frac{(x-1)^2}{4} - \frac{(y-3)^2}{9} = 1$$

$$9(x-1)^2 - 4(y-3)^2 = 36 \quad \text{Multiply each side by 36.}$$

$$9(x^2 - 2x + 1) - 4(y^2 - 6y + 9) = 36$$

$$9x^2 - 18x + 9 - 4y^2 + 24y - 36 = 36$$

$$9x^2 - 4y^2 - 18x + 24y - 63 = 0$$

The last equation, $9x^2 - 4y^2 - 18x + 24y - 63 = 0$, is the *general equation* of this hyperbola.

In a similar manner, the standard equation of any conic section can be rewritten in the general form

$$Ax^2 + Cy^2 + Dx + Ey + F = 0$$

where A, C, D, E, and F are real numbers (A and C not both 0).

Example 5 of Section 1.1 demonstrated how the second-degree equation

$$x^2 + y^2 - 6x + 12y + 32 = 0$$

can be rewritten as the standard equation of a circle:

$$(x-3)^2 + (y+6)^2 = 13$$

As you may recall, we showed this equivalence by completing the squares on the x terms and y terms. In Example 1, we use the same process of completing the squares to rewrite a second-degree equation in the standard form of a conic section.

EXAMPLE 1 Graphing a Second-Degree Equation

Rewrite the equation $x^2 + 3y^2 - 6x + 12y - 15 = 0$ in a standard form of a conic section, and sketch its graph.

SOLUTION

We complete the squares on the x terms and y terms:

$$x^2 + 3y^2 - 6x + 12y - 15 = 0$$

$$(x^2 - 6x \quad) + 3(y^2 + 4y \quad) = 15$$
Group the x terms and y terms.

$$(x^2 - 6x + 9) + 3(y^2 + 4y + 4) = 15 + 9 + 12$$
Add 9 and 12 to each side to complete the squares.

$$(x - 3)^2 + 3(y + 2)^2 = 36$$

$$\frac{(x - 3)^2}{36} + \frac{(y + 2)^2}{12} = 1$$
Divide each side by 36.

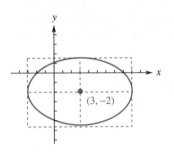

Figure 6.40

The graph of the last equation is an ellipse with center $(3, -2)$; we use 6 for a and $2\sqrt{3}$ (approximately 3.5) for b. The ellipse and its frame are shown in Figure 6.40.

The conic sections we have discussed—circles, parabolas, ellipses, and hyperbolas—have equations of the form $Ax^2 + Cy^2 + Dx + Ey + F = 0$. It is not true, however, that the graph of any equation in this general form is a conic section. There are two other possible outcomes.

First, the intersections shown in Figure 6.39 at the beginning of this section are not the only possibilities. If the plane passes through the vertex of the cone, the intersection can be a point, a line, or a pair of lines (Figure 6.41). These intersections are *degenerate conic sections*.

Second, an equation in x and y may have no real solutions. For example, the equation $x^2 + y^2 + 4 = 0$ has no real solutions. (Take a moment to confirm this statement.) We call the graph of an equation with no real solutions an *empty graph*, because there are no points on the graph.

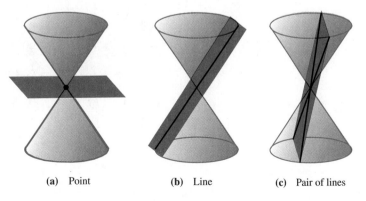

    (a) Point   **(b)** Line   **(c)** Pair of lines

Figure 6.41

The Graph of a Second-Degree
Equation

The graph of the second-degree equation

$$Ax^2 + Cy^2 + Dx + Ey + F = 0$$

(A, C, D, E, and F are real numbers; A and C not both 0) is a circle, ellipse, parabola, hyperbola, degenerate conic section, or empty graph.

⊙ EXAMPLE 2   Graphing a Second-Degree Equation

Rewrite each equation in the standard form of a conic section (if possible), and sketch its graph.

(a) $x^2 - 8x + y + 14 = 0$ **(b)** $4x^2 - y^2 - 6y - 9 = 0$

SOLUTION

(a) We can complete the square on the x terms as we did in Example 1:

$$
\begin{aligned}
x^2 - 8x + y + 14 &= 0 & \\
(x^2 - 8x \quad) + y &= -14 & \text{Group the } x \text{ terms.} \\
(x^2 + 8x + 16) + y &= -14 + 16 & \text{Add 16 to each side.} \\
(x + 4)^2 + y &= 2 & \\
y &= -(x + 4)^2 + 2 &
\end{aligned}
$$

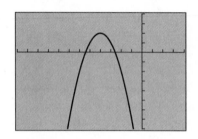

Figure 6.42
$y = -(x + 4)^2 + 2$
$[-12, 4] \times [-8, 4]$

The graph of this equation is a parabola with vertex $(-4, 2)$, as shown in Figure 6.42.

(b) Because there is no linear x term, we need to complete the square on the y terms only:

$$
\begin{aligned}
4x^2 - y^2 - 6y - 9 &= 0 & \\
4x^2 - (y^2 + 6y \quad) &= 9 & \text{Group the } y \text{ terms.} \\
4x^2 - (y^2 + 6y + 9) &= 9 - 9 & \text{Add } -9 \text{ to each side} \\
& & \text{to complete the square.} \\
4x^2 - (y + 3)^2 &= 0 & \\
(y + 3)^2 &= 4x^2 & \\
y + 3 &= \pm 2x & \text{Take square roots of each side.} \\
y &= \pm 2x - 3 &
\end{aligned}
$$

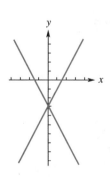

Figure 6.43
$4x^2 - y^2 - 6y - 9 = 0$

The graph of $4x^2 - y^2 - 6y - 9 = 0$ is a degenerate conic section; it is the pair of lines $y = -2x - 3$ and $y = 2x - 3$ (Figure 6.43). ⊙

Conic Sections and Graphing Utilities

Ellipses, hyperbolas, and some parabolas (those that open right or left) do not pass the vertical-line test, so in general, the graph of a conic section is not the graph of a function. However, an equation of a conic section can be solved for y, giving two functions with graphs that can be determined by a graphing utility.

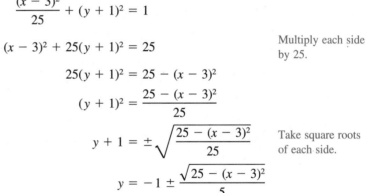

EXAMPLE 3 Using a Graphing Utility to Determine the Graph of a Conic Section

Determine the graph of the equation

$$\frac{(x - 3)^2}{25} + (y + 1)^2 = 1$$

using a graphing utility.

SOLUTION

We solve this equation for y:

$$\frac{(x - 3)^2}{25} + (y + 1)^2 = 1$$

$$(x - 3)^2 + 25(y + 1)^2 = 25 \qquad \text{Multiply each side by 25.}$$

$$25(y + 1)^2 = 25 - (x - 3)^2$$

$$(y + 1)^2 = \frac{25 - (x - 3)^2}{25}$$

$$y + 1 = \pm\sqrt{\frac{25 - (x - 3)^2}{25}} \qquad \text{Take square roots of each side.}$$

$$y = -1 \pm \frac{\sqrt{25 - (x - 3)^2}}{5}$$

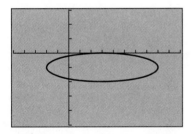

Figure 6.44

$$\frac{(x - 3)^2}{25} + (y + 1)^2 = 1$$

$[-3, 9] \times [-3, 2]$

The graphs of

$$y = -1 + \frac{\sqrt{25 - (x - 3)^2}}{5} \qquad \text{and} \qquad y = -1 - \frac{\sqrt{25 - (x - 3)^2}}{5}$$

are shown in Figure 6.44. These two graphs together give the graph of

$$\frac{(x - 3)^2}{25} + (y + 1)^2 = 1$$

A second-degree equation may have a nonzero term of the form Bxy (for some real number B). In this case, the process of rewriting the equation by completing the squares on the x terms and y terms does not work in determining a standard form of a conic section. However, the following statement, offered here without proof, is true:

General Equation of a
Conic Section

> The graph of the second-degree equation
>
> $$Ax^2 + Bxy + Cy^2 + Dx + Ey + F = 0$$
>
> (A, B, C, D, E, and F are real numbers; A and C not both 0) is a circle, ellipse, parabola, hyperbola, degenerate conic section, or empty graph.

The graph of $Ax^2 + Bxy + Cy^2 + Dx + Ey + F = 0$ (if it is a conic section) does not necessarily have axes of symmetry that are parallel to the x-axis or the y-axis. In Example 4, we show how a graphing utility helps to determine the graphs of these equations.

EXAMPLE 4 Graphing a Second-Degree Equation

Use a graphing utility to determine the graph of $x^2 - 4xy + 4y^2 - 4x = 0$.

SOLUTION

To use a graphing utility, we first solve the equation for y. Rewriting the equation as $4y^2 + (-4x)y + (x^2 - 4x) = 0$ shows it to be a quadratic equation in the form $ay^2 + by + c = 0$, where a is 4, b is $-4x$, and c is $x^2 - 4x$. We solve for y using the quadratic formula:

$$y = \frac{-b \pm \sqrt{b^2 - 4ac}}{2a}$$

$$= \frac{-(-4x) \pm \sqrt{(-4x)^2 - 4(4)(x^2 - 4x)}}{2(4)}$$

$$= \frac{4x \pm \sqrt{16x^2 - 16x^2 + 64x}}{8}$$

$$= \frac{4x \pm 8\sqrt{x}}{8}$$

$$= \tfrac{1}{2}x \pm \sqrt{x}$$

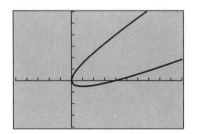

Figure 6.45
$x^2 - 4xy + 4y^2 - 4x = 0$
$[-5, 10] \times [-4, 6]$

The graphs of $y = \tfrac{1}{2}x + \sqrt{x}$ and $y = \tfrac{1}{2}x - \sqrt{x}$ together give us the graph of $x^2 - 4xy + 4y^2 - 4x = 0$, a parabola (Figure 6.45).

Parametric Representations of Conic Sections

In Section 3.6, Table 3.2 showed that a parametric representation for the graph of $x^2 + y^2 = r^2$, a circle with radius r and center at the origin, is given by

$$x = r \cos t, \quad y = r \sin t \qquad \text{for } 0 \le t \le 2\pi$$

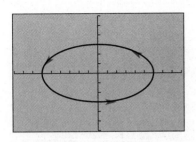

Figure 6.46
$x = 6 \cos t, y = 3 \sin t, 0 \le t \le 2\pi$
$[-9, 9] \times [-6, 6]$

The circular functions sine and cosine are also used in parametric representations of ellipses. For example, a parametric representation of the graph of

$$\frac{x^2}{36} + \frac{y^2}{9} = 1$$

is given by

$$x = 6 \cos t, \quad y = 3 \sin t \qquad \text{for } 0 \le t \le 2\pi$$

(The graph is shown in Figure 6.46. Pause here to try this representation on a graphing utility.)

Parametric representations of hyperbolas commonly use the functions

$$f(t) = \frac{e^t + e^{-t}}{2} \quad \text{and} \quad g(t) = \frac{e^t - e^{-t}}{2}$$

Because of this particular application of these functions, they are called *hyperbolic functions*.

Hyperbolic Functions

NOTE: *Hyperbolic functions are developed more completely in the study of calculus.*

Two **hyperbolic functions** are the **hyperbolic sine** of a real number t, written **sinh t**, and the hyperbolic cosine of a real number t, written **cosh t**. These functions are defined as

$$\sinh t = \frac{e^t - e^{-t}}{2} \qquad \cosh t = \frac{e^t + e^{-t}}{2}$$

Table 6.1 (page 558) shows how ellipses and hyperbolas are represented parametrically by circular and hyperbolic functions. Notice that the parametric representations for hyperbolas each describe only one branch of the curve.

EXAMPLE 5   Finding Parametric Representations of Conic Sections

Determine a parametric representation for the graph of

$$x^2 - \frac{y^2}{4} = 1 \qquad x > 0$$

using Table 6.1.

SOLUTION
The graph of

$$x^2 - \frac{y^2}{4} = 1$$

TABLE 6.1

Parametric Representations of Ellipses and Hyperbolas

Conic section and equation	Parametric representation	Curve and orientation
Ellipse $\dfrac{x^2}{a^2} + \dfrac{y^2}{b^2} = 1$	$x = a \cos t, \quad y = b \sin t$ $0 \le t \le 2\pi$	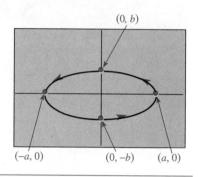
Hyperbola $\dfrac{x^2}{a^2} - \dfrac{y^2}{b^2} = 1$	$x = a \cosh t, \quad y = b \sinh t$	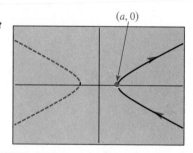
Hyperbola $\dfrac{y^2}{b^2} - \dfrac{x^2}{a^2} = 1$	$x = a \sinh t, \quad y = b \cosh t$	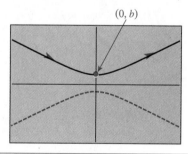

Figure 6.47
$x = \cosh t, \ y = 2 \sinh t$
$[-9, 9] \times [-6, 6]$

is a hyperbola with a horizontal transverse axis, where a is 1 and b is 2. By Table 6.1, the parametric representation is

$$x = \cosh t, \quad y = 2 \sinh t$$

The graph of this curve is shown in Figure 6.47.

The parametric representations of the graphs of parabolas with equations $y = a(x - h)^2 + k$ and $x = a(y - k)^2 + h$ can be determined with Table 3.2 in Section 3.6.

Geometry of Conic Sections

In this chapter we have defined each conic section in three ways: as an intersection of a plane and a cone, as a locus, and as a graph of a second-degree equation. Table 6.2 (page 560) gives a summary.

In Section 1.1, we developed the equation of a circle from its locus definition. We developed similar equations for parabolas, ellipses, and hyperbolas from their locus definitions in Sections 6.1 and 6.2. In Example 6, we show that the geometric definition of the ellipse is equivalent to its locus definition. (Similar connections for the parabola and hyperbola are the subjects of Exercises 55 and 56.)

◯ EXAMPLE 6   An Ellipse Is a Conic Section

Show that the geometric definition of an ellipse is consistent with its locus definition by referring to Figure 6.48.

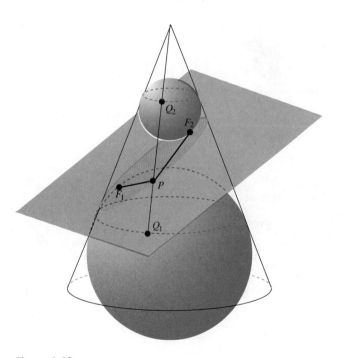

Figure 6.48

TABLE 6.2

Summary of Conic Sections

Conic section	Geometric definition	Locus definition	Equation and graph
Circle	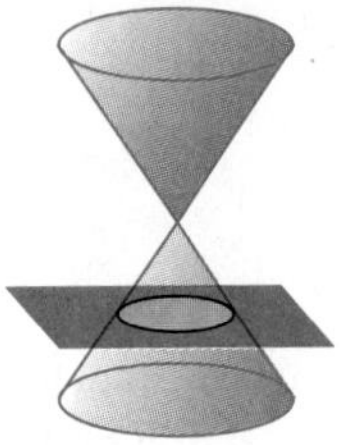	$d(P, C) = r$ 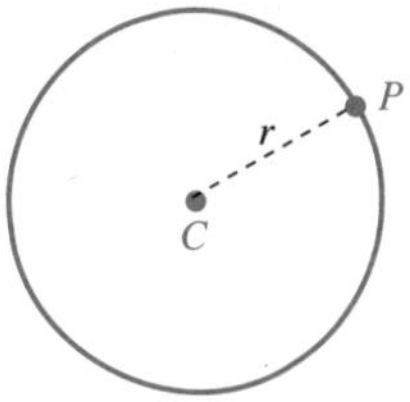	$x^2 + y^2 = r^2$ 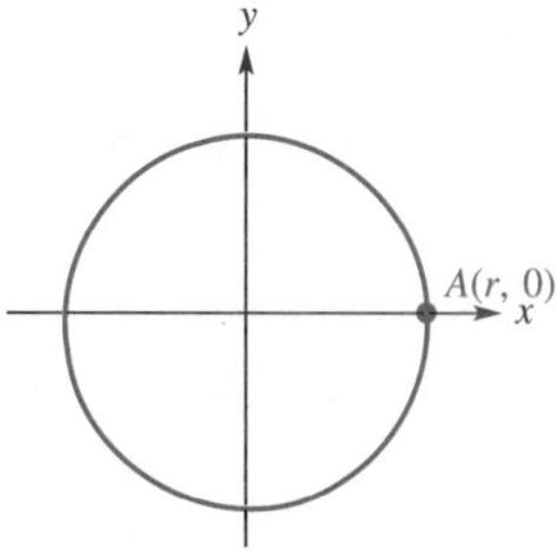
Ellipse	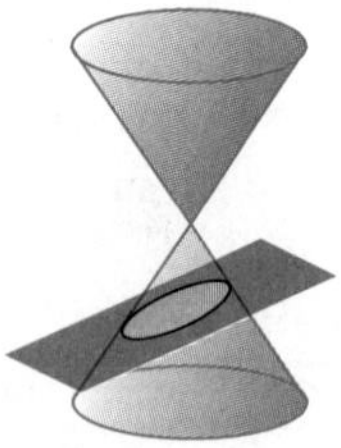	$d(P, F_1) + d(P, F_2) = 2a$ 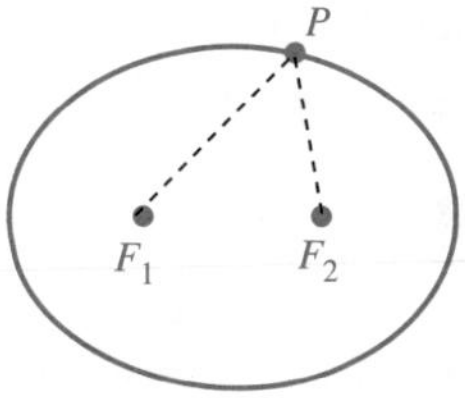	$\dfrac{x^2}{a^2} + \dfrac{y^2}{b^2} = 1$ 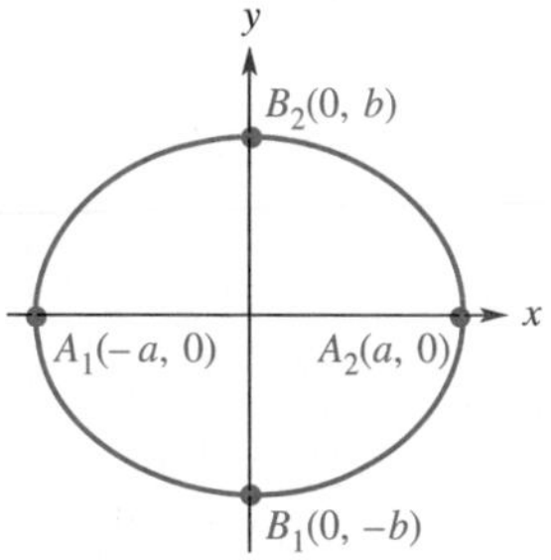
Parabola	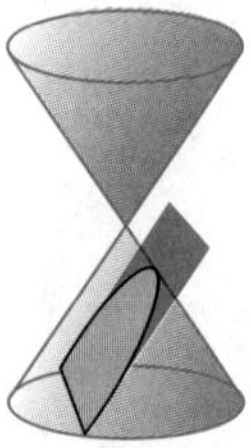	$d(P, D) = d(P, F)$ 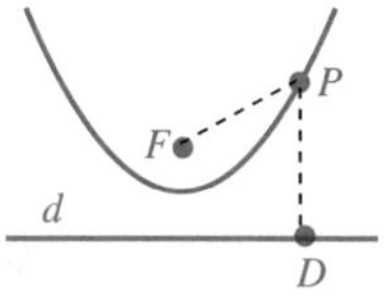	$y = \dfrac{1}{4p}\, x^2$ 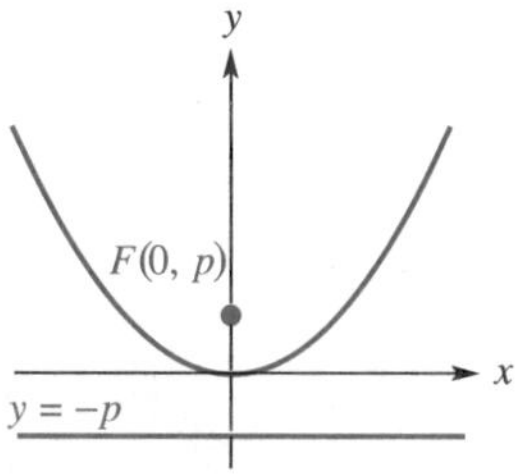
Hyperbola	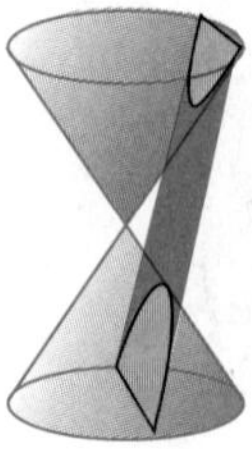	$d(P, F_1) - d(P, F_2) = \pm 2a$ 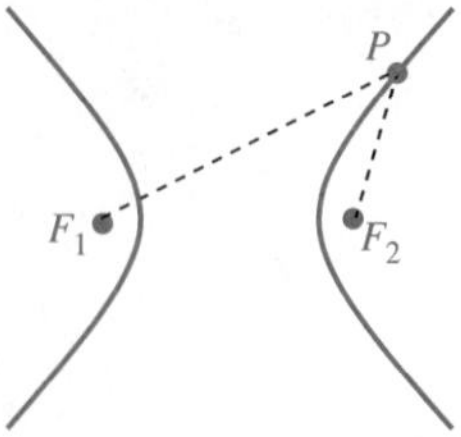	$\dfrac{x^2}{a^2} - \dfrac{y^2}{b^2} = 1$ 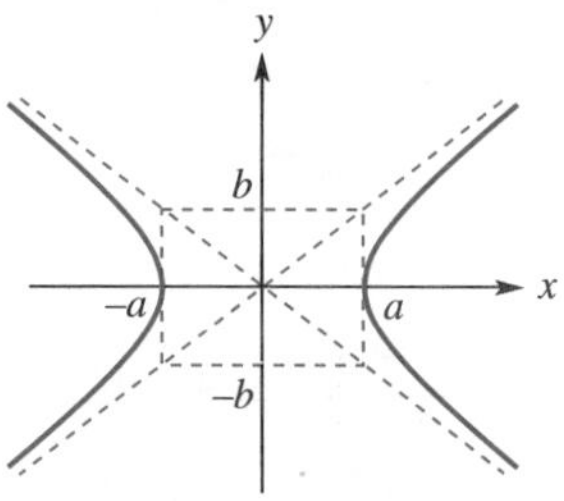

SOLUTION

Figure 6.48 shows two spheres placed inside a cone such that each sphere is tangent to the intersecting plane and touches the cone along a horizontal circle. The points of tangency with the intersecting plane are labeled F_1 and F_2.

A point P is selected on the intersection of the plane and cone, and a line is drawn from the vertex of the cone through P. The points at which this line crosses the horizontal circles are labeled Q_1 and Q_2. If we can show that $d(P, F_1) + d(P, F_2)$ is a constant for any point P on the intersection, then we have shown that the intersection is an ellipse with foci F_1 and F_2 according to the definition in Section 6.2.

Because PF_1 and PQ_1 share an endpoint and are each tangent to the same sphere, they are equal in length (see Exercise 54). The same can be said for PF_2 and PQ_2. Thus,

■ **NOTE:** *This proof is the handiwork of a Belgian mathematician, G. P. Dandelin in 1822.*

$$d(P, F_1) = d(P, Q_1) \qquad \text{and} \qquad d(P, F_2) = d(P, Q_2)$$

It follows directly that

$$d(P, F_1) + d(P, F_2) = d(P, Q_1) + d(P, Q_2) = d(Q_1, Q_2)$$

No matter what point P is selected, $d(Q_1, Q_2)$ is a constant, the distance along the cone between the two horizontal circles. This completes the proof that this intersection is an ellipse according to the locus definition in Section 6.2.

■ **EXERCISE SET 6.3** ═══

A

In Exercises 1–8, the graph of each equation is a circle, parabola, ellipse, or hyperbola. Rewrite the equation in standard form and sketch the graph.

1. $x^2 + 4y^2 - 16 = 0$

2. $x^2 - 4y^2 - 36 = 0$

3. $25x^2 + 9y^2 - 225 = 0$

4. $x^2 - 4x - y^2 - 20 = 0$

5. $2x^2 + 2y^2 - 12 = 0$

6. $2x^2 + 2y^2 + 16y + 8 = 0$

7. $x^2 - 4x - 2y - 4 = 0$

8. $4x - y^2 - 10y - 17 = 0$

In Exercises 9–16, solve each equation for y and determine the graph of the equation using a graphing utility.

9. $x = (y - 2)^2 + 3$

10. $x = \frac{1}{8}(y + 4)^2 + 2$

11. $\dfrac{x^2}{9} - \dfrac{y^2}{9} = 1$

12. $y^2 - x^2 = 1$

13. $\dfrac{x^2}{12} + \dfrac{y^2}{4} = 1$

14. $\dfrac{x^2}{36} + y^2 = 1$

15. $x^2 + y^2 = 4$

16. $x^2 + y^2 = 25$

In Exercises 17–24, find a parametric representation for each rectangular equation given, and use the representation to determine the graph of the equation using a graphing utility. Use arrows to show the orientation.

17. $x = (y - 2)^2 + 3$

18. $x = \frac{1}{8}(y + 4)^2 + 2$

19. $\dfrac{x^2}{9} - \dfrac{y^2}{9} = 1,\, x > 0$

20. $y^2 - x^2 = 1,\, y > 0$

21. $\dfrac{x^2}{12} + \dfrac{y^2}{4} = 1$

22. $\dfrac{x^2}{36} + y^2 = 1$

23. $x^2 + y^2 = 4$

24. $x^2 + y^2 = 25$

B

In Exercises 25–34, the graph of each equation is a circle, parabola, ellipse, hyperbola, degenerate conic section, or empty graph. Rewrite each equation in standard form, and sketch the graph.

25. $x^2 + 4y^2 + 24y + 32 = 0$

26. $-4x^2 + y^2 - 4y + 8 = 0$

27. $3x^2 + 3y^2 + 24x - 12y + 57 = 0$

28. $y^2 + x - 4y - 21 = 0$

29. $x^2 - y^2 - 2x + 4y + 8 = 0$

30. $x^2 + y^2 + 2x - 8y + 17 = 0$

31. $3x^2 + 6x + 2y^2 + 39 = 0$

32. $x^2 + y^2 + 2x + 21 = 0$

33. $(2x - y)^2 - (x - 2y)^2 + 3 = 0$

34. $(x - y)(x + y - 2) = 1$

In Exercises 35–42, use a graphing utility to determine the graph of each second-degree equation, and identify the graph as a parabola, ellipse, hyperbola, or degenerate conic.

35. $x^2 + 2xy + y^2 - x = 0$

36. $x^2 + 4xy + 4y^2 - 16x = 0$

37. $17x^2 - 4xy + 4y^2 - 64 = 0$

38. $65x^2 - 8xy + 16y^2 - 576 = 0$

39. $3x^2 + 2xy - y^2 - 4 = 0$

40. $2xy - y^2 - 4 = 0$

41. $3x^2 - xy - 2y^2 - 3x - 2y = 0$

42. $x^2 - 2xy + y^2 - 4x + 4y + 4 = 0$

43. Sketch the solution to the inequality:
$3x^2 + 2xy - y^2 - 4 \le 0$
HINT: See Exercise 39.

44. Sketch the solution to the inequality: $2xy - y^2 - 4 \ge 0$
HINT: See Exercise 40.

45. Sketch the graphs of $x^2 + y^2 - 6x = 0$ and $2x^2 + y^2 - 8x = 0$, and determine their points of intersection.

46. Sketch the graphs of $x^2 + y^2 + 6y = 0$ and $x^2 - y^2 + 4y = 0$, and determine their points of intersection.

47. Find a parametric representation for

$$\frac{x^2}{9} - \frac{y^2}{9} = 1 \qquad x < 0$$

and use the representation to determine the graph using a graphing utility. Use arrows to show the orientation.

48. Find a parametric representation for $y^2 - x^2 = 1,\, y < 0$, and use the representation to determine the graph using a graphing utility. Use arrows to show the orientation.

49. Suppose that the graph of $Ax^2 + Cy^2 + Dx + Ey + F = 0$ is a conic section. Determine the type of conic section given that:
(a) $AC = 0$ **(b)** $A = C$ **(c)** $AC < 0$
(d) $AC > 0$

50. Suppose that the graph of

$$Ax^2 + Bxy + Cy^2 + Dx + Ey + F = 0$$

is a conic section. It can be shown that the discriminant of the equation, $B^2 - 4AC$, is negative, 0, or positive depending on whether the conic section is an ellipse, parabola, or hyperbola, respectively. Use the discriminant to identify the graphs of the equations in Exercises 35–42.

C

Eccentricity (Exercises 51–53)

In the sixteenth century, the paths of orbiting bodies in the solar system were determined to be ellipses, with the sun at one focus of each ellipse. Some objects, such as the planet Venus, have orbits that are nearly circular, but others, such as Halley's comet, deviate greatly from a circle. Early astronomers assigned a numerical measure for the deviation of an elliptical orbit from being perfectly circular. Since then, this measure has been extended to all conic sections. The *eccentricity* of a conic section is a nonnegative number that is used to describe its relative shape. For an ellipse with foci F_1 and F_2, and major vertices A_1 and A_2, and for a hyperbola with foci F_1 and F_2, and vertices A_1 and A_2, eccentricity is defined as the ratio $d(F_1, F_2)/d(A_1, A_2)$. (The eccentricity of a circle is defined to be 0, and the eccentricity of a parabola is defined to be 1.)

51. Determine the eccentricity of the graph of:

(a) $\dfrac{x^2}{16} - \dfrac{y^2}{9} = 1$ **(b)** $\dfrac{x^2}{36} + y^2 = 1$

(c) $\dfrac{x^2}{4} + \dfrac{y^2}{2} = 1$ **(d)** $x^2 - y^2 = 1$

52. The orbit of the planet Mercury has an eccentricity of 0.3871. The semimajor axis of its orbit is 57.9 million kilometers. Determine the minimum and maximum distances of Mercury from the sun. (These points of minimum and maximum distance of an orbit are called the *perihelion* and *aphelion,* respectively. These points coincide with the major vertices of the ellipse.)

53. The orbit of Halley's comet has an eccentricity of 0.97214. The semimajor axis of its orbit is 5412.2 million kilometers. Determine the minimum and maximum distances of Halley's comet from the sun. (See Exercise 52.)

54. The lines AP_1 and AP_2 are tangent at P_1 and P_2 to a sphere with center C (Figure 6.49).

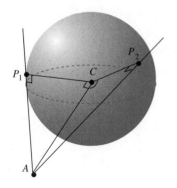

Figure 6.49

(a) Show that triangles AP_1C and AP_2C are congruent.
(b) Explain why the result of part (a) proves that $d(A, P_1) = d(A, P_2)$.

55. In Figure 6.50, a plane is parallel to an element of the cone and intersects the cone, and a sphere is tangent to this plane at F. The line PR is an element of the cone, the line segment PQ is vertical, and the line segment QD is perpendicular to the line of intersection of the two planes.

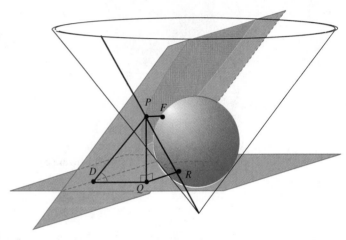

Figure 6.50

(a) Use Exercise 54 to show that $d(P, F) = d(P, R)$.
(b) Explain why angle PDQ is equal to angle PRQ, and therefore triangle PDQ is congruent to triangle PRQ.
(c) Use parts (a) and (b) to explain why $d(P, F) = d(P, D)$, and to verify that this result shows that the geometric definition and locus definition of a parabola are consistent.

56. The two spheres are tangent to the intersecting plane at F_1 and F_2 (Figure 6.51, page 564). The line Q_1Q_2 is an element of the cone.
(a) Use Exercise 54 to show that $d(P, F_1) = d(P, Q_1)$ and $d(P, F_2) = d(P, Q_2)$.
(b) Use parts (a) and (b) to explain why

$$d(P, F_2) - d(P, F_1) = d(P, Q_2) - d(P, Q_1)$$

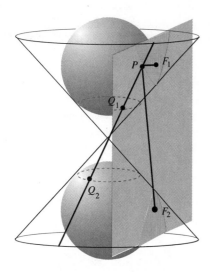

Figure 6.51

and to verify that this result shows that the geometric definition and locus definition of a hyperbola are consistent.

57. Show that

$$x = \frac{e^t + e^{-t}}{2}, \quad y = \frac{e^t - e^{-t}}{2}$$

is a parametric representation of $x^2 - y^2 = 1$.

58. Use the locus definition of an ellipse to find an equation

$$Ax^2 + Bxy + Cy^2 + Dx + Ey + F = 0$$

for the ellipse with foci $F_1(-1,-1)$ and $F_2(1, 1)$ and major axis of length 6.

59. Use the locus definition of a hyperbola to find an equation

$$Ax^2 + Bxy + Cy^2 + Dx + Ey + F = 0$$

for the hyperbola with foci $F_1(-2, 2)$ and $F_2(2,-2)$ and transverse axis of length 2.

60. Use the locus definition of a hyperbola to show that the graph of the reciprocal function $y = 1/x$ from the directory of graphs (inside the front cover) is a hyperbola with foci $F_1(-\sqrt{2}, -\sqrt{2})$ and $F_2(\sqrt{2}, \sqrt{2})$.

◉ QUICK REFERENCE

Topic	Page	Remarks
Conic sections and degenerate conic sections	519	The conic sections are the curves formed when a plane intersects a double right circular cone. The possible curves formed are a circle, parabola, ellipse, and hyperbola; the degenerate conic sections are a point, line, or pair of intersecting lines.
Parabola	520	A parabola is the set of all points in a plane that are the same distance from a fixed line (the directrix) and a fixed point (the focus) not on the line.
Axis of symmetry of a parabola	520	The axis of symmetry is the line passing through the focus that is perpendicular to the directrix of the parabola.
Vertex of a parabola	520	The vertex is the point where the axis of symmetry intersects the parabola.
Basic equations for parabolas	523	See Figure 6.7.
Ellipse	532	An ellipse is the set of all points in the plane such that the sum of the distances to two fixed points (the foci) is a constant.
Center of an ellipse	532	The center is the midpoint of the line segment joining the foci.

Topic	Page	Remarks
Major axis and major vertices	532	The major axis is the chord of the ellipse that passes through the foci. The endpoints of the major axis are the major vertices.
Minor axis and minor vertices	532	The minor axis is the chord of the ellipse passing through the center and perpendicular to the major axis. The endpoints of the minor axis are the minor vertices.
Standard equation for an ellipse	534	The standard equation for an ellipse with center at the origin is $$\frac{x^2}{a^2} + \frac{y^2}{b^2} = 1$$ If $a > b$, then the major vertices are $A_1(-a, 0)$ and $A_2(a, 0)$, the minor vertices are $B_1(0, -b)$ and $B_2(0, b)$, and the foci are $F_1(-c, 0)$ and $F_2(c, 0)$, where $c^2 = a^2 - b^2$. If $a < b$, then the major vertices are $B_1(0, -b)$ and $B_2 (0, b)$, the minor vertices are $A_1(-a, 0)$ and $A_2(a, 0)$, and the foci are $F_1(0, -c)$ and $F_2(0, c)$, where $c^2 = b^2 - a^2$.
Hyperbola	538	A hyperbola is the set of all points in the plane such that the difference of the distances to two fixed points (the foci) is a constant. The two curves that comprise a hyperbola are its branches.
Center of a hyperbola	538	The center is the midpoint of the line segment joining the foci.
Transverse axis and vertices	538	The vertices are the points of intersection of the hyperbola with the line passing through the foci. The transverse axis is the line segment joining the vertices.
Standard equations for a hyperbola	539–540	The standard equation for a hyperbola with horizontal transverse axis and center at the origin is $$\frac{x^2}{a^2} - \frac{y^2}{b^2} = 1$$ The vertices are $A_1(-a, 0)$ and $A_2(a, 0)$, the asymptotes are the graphs of $$y = \pm\frac{b}{a}x$$ and the foci are $F_1(-c, 0)$ and $F_2(c, 0)$, where $c^2 = a^2 + b^2$. The standard equation for a hyperbola with vertical transverse axis and center at the origin is $$\frac{y^2}{b^2} - \frac{x^2}{a^2} = 1$$ The vertices are $B_1(0, -b)$ and $B_2(0, b)$, the asymptotes are the graphs of $$y = \pm\frac{b}{a}x$$ and the foci are $F_1(0, -c)$ and $F_2(0, c)$, where $c^2 = a^2 + b^2$. See the box on page 543.

Topic	Page	Remarks
Graph of a second-degree equation	554	The graph of the second-degree equation

$$Ax^2 + Bxy + Cy^2 + Dx + Ey + F = 0$$

(A, B, C, D, E, and F are real numbers; A and C not both 0) is a circle, ellipse, parabola, hyperbola, degenerate conic section (point, line, or pair of lines), or empty graph.

◯ MISCELLANEOUS EXERCISES

In Exercises 1–4, determine an equation in the form

$$y = \frac{1}{4p} x^2 \qquad \text{or} \qquad x = \frac{1}{4p} y^2$$

for each parabola described. Assume the vertex is at the origin.

1. The focus of the parabola is $\left(0, \frac{5}{2}\right)$.

2. The focus of the parabola is $(-2, 0)$.

3. The directrix of the parabola is $x = -3$.

4. The directrix of the parabola is $y = -\frac{1}{2}$.

In Exercises 5–8, the graph of each equation is a parabola. Rewrite the equation in standard form, and determine the graph, focus, and directrix.

5. $x^2 - 6x - 12y - 15 = 0$

6. $y^2 - 2y - 8x - 39 = 0$

7. $2y^2 + 8y + 3x + 8 = 0$

8. $3x^2 - 24x + 4y + 48 = 0$

In Exercises 9–12, determine an equation in standard form for each parabola described.

9. The focus is $(-3, 1)$, and the directrix is $x = -6$.

10. The focus is $(0, -3)$, and the directrix is $y = 2$.

11. The focus is $(1, 2)$, the directrix is horizontal, and the graph passes through $(5, 2)$.

12. The focus is $(4, 4)$, the directrix is vertical, and the graph passes through $(4, -2)$.

13. A spotlight has a parabolic-shaped mirror 4 feet in diameter and 3 inches deep. The bulb is located at the focus of the parabolic cross section. How high above the bottom of the mirror should the bulb be placed?

14. A television satellite antenna has a parabolic-shaped dish 6 feet in diameter and 18 inches deep. The receiver is located at the focus of the parabolic cross section. How high above the bottom of the dish should the receiver be placed?

In Exercises 15–22, the graph of each equation is a conic section. Sketch the graph.

15. $\dfrac{x^2}{36} + \dfrac{y^2}{9} = 1$

16. $\dfrac{x^2}{25} + \dfrac{y^2}{49} = 1$

17. $\dfrac{y^2}{9} - \dfrac{x^2}{4} = 1$

18. $\dfrac{x^2}{16} - \dfrac{y^2}{4} = 1$

19. $\dfrac{(x-3)^2}{9} - \dfrac{(y-2)^2}{9} = 1$

20. $\dfrac{(y+4)^2}{25} - \dfrac{(x-1)^2}{36} = 1$

21. $\dfrac{(x+5)^2}{16} + \dfrac{(y-1)^2}{49} = 1$

22. $\dfrac{(x-6)^2}{25} + \dfrac{(y+2)^2}{64} = 1$

In Exercises 23–26, determine the vertices and foci of each graph.

23. $\dfrac{x^2}{11} + \dfrac{(y-4)^2}{36} = 1$

24. $\dfrac{(x-3)^2}{49} + \dfrac{y^2}{24} = 1$

25. $\dfrac{(y+2)^2}{4} - \dfrac{(x-2)^2}{5} = 1$

26. $\dfrac{(x-4)^2}{9} - \dfrac{(y+1)^2}{7} = 1$

In Exercises 27–30, determine an equation in standard form for the conic section described.

27. An ellipse with vertex $(1, 0)$, and foci $(1, 1)$ and $(1, 5)$

28. A hyperbola with vertex $(-1, -3)$, and foci $(0, -3)$ and $(-6, -3)$

29. A hyperbola with foci $(3, 0)$ and $(-3, 0)$, passing through the point $(4, 1)$

30. An ellipse with foci $(0, -4)$ and $(0, 4)$, passing through the point $(1, 3)$

In Exercises 31–34, sketch the graph of each system of equations and determine their points of intersection.

31. $\begin{cases} \dfrac{x^2}{3} + \dfrac{y^2}{6} = 1 \\ 2x - y = 0 \end{cases}$ **32.** $\begin{cases} \dfrac{x^2}{21} - \dfrac{y^2}{3} = 1 \\ x - 2y = 3 \end{cases}$

33. $\begin{cases} \dfrac{x^2}{5} - \dfrac{y^2}{5} = 1 \\ x = y^2 - 1 \end{cases}$ **34.** $\begin{cases} \dfrac{x^2}{6} + \dfrac{y^2}{12} = 1 \\ 2y = x^2 \end{cases}$

In Exercises 35–38, sketch each inequality or system of inequalities.

35. $\dfrac{x^2}{25} + \dfrac{y^2}{12} \geq 1$ **36.** $\dfrac{x^2}{20} - \dfrac{y^2}{9} \leq 1$

37. $\begin{cases} \dfrac{y^2}{4} - x^2 \geq 1 \\ y^2 \leq \frac{1}{4}(x + 4) \end{cases}$ **38.** $\begin{cases} \dfrac{x^2}{16} + \dfrac{y^2}{9} \leq 1 \\ y \leq \frac{1}{2}x^2 \end{cases}$

In Exercises 39–44, the graph of each equation is a parabola, ellipse, or hyperbola. Rewrite the equation in standard form and sketch its graph.

39. $x^2 - 2x + 4y + 1 = 0$

40. $y^2 - 8x + 6y + 9 = 0$

41. $x^2 + 4y^2 - 10x + 21 = 0$

42. $9x^2 + y^2 + 4y - 5 = 0$

43. $x^2 - y^2 - 12x + 32 = 0$

44. $y^2 - x^2 + 2y - 24 = 0$

In Exercises 45–52, the graph of each equation is a circle, parabola, ellipse, hyperbola, degenerate conic section, or empty graph. Complete the squares on the x terms and y terms to rewrite each equation in standard form, and sketch the graph.

45. $x^2 - 4y^2 - 6x - 24y - 31 = 0$

46. $x^2 + y^2 + 14x - 4y + 44 = 0$

47. $x^2 + 8x - 12y + 16 = 0$

48. $16x^2 - 9y^2 + 32x - 36y + 124 = 0$

49. $x^2 + 4y^2 + 24y + 20 = 0$

50. $9x^2 + y^2 - 18x = 0$

51. $x^2 - y^2 - 2y - 17 = 0$

52. $x^2 + y^2 - 18x + 2y + 82 = 0$

In Exercises 53 and 54, solve each equation for y and determine the graph of the equation using a graphing utility.

53. $5(x + 4)^2 + 20y^2 = 80$ **54.** $x^2 - 3y^2 = 30$

In Exercises 55–58, use a graphing utility to determine the graph of each second-degree equation, and identify the graph as a parabola, hyperbola, or ellipse.

55. $4x^2 - 4xy + y^2 - x = 0$

56. $x^2 - 4xy + 4y^2 - 8x = 0$

57. $5x^2 + 4xy + 4y^2 - 64 = 0$

58. $2xy + y^2 + 4 = 0$

In Exercises 59–62, find a parametric representation for each rectangular equation given, and use the representation to determine the graph of the equation using a graphing utility. Use arrows to show the orientation.

59. $\dfrac{x^2}{9} - \dfrac{y^2}{4} = 1,\ x > 0$ **60.** $\dfrac{y^2}{16} - \dfrac{x^2}{9} = 1,\ y > 0$

61. $\dfrac{x^2}{49} + \dfrac{y^2}{25} = 1$ **62.** $\dfrac{x^2}{16} + \dfrac{y^2}{36} = 1$

63. Use your graphing utility to determine the graph of

$$\begin{cases} x = t + \dfrac{1}{t} \\ y = t - \dfrac{1}{t} \end{cases}$$

Identify the graph as a parabola, hyperbola, or ellipse, and prove your result.

CHAPTER TEST

1. Determine an equation of the parabola with directrix $x = 3$ and focus $(-1, 3)$.

2. Sketch the graph of $x^2 = 10y$. Indicate the focus and directrix in your sketch.

3. Determine the coordinates of the focus of $y^2 - 8y = 8x + 8$.

4. Sketch the graph of

   $$\frac{x^2}{36} + \frac{y^2}{20} = 1$$

   Indicate the foci and vertices in your sketch.

5. Sketch the graph of $9(x - 1)^2 + 4(y + 2)^2 = 36$.

6. Complete the squares on the x terms and y terms, and sketch the graph of the equation

   $$4x^2 - 9y^2 - 24x - 18y + 63 = 0$$

7. Determine an equation of the hyperbola with vertices $(0, -2)$ and $(0, 2)$, and passing through $(3, 4)$.

8. A tunnel for a roadway must be built in the shape of the upper half of an ellipse. The tunnel must be 5 feet high when 2 feet from each side. If the roadway is 52 feet wide, find an equation of the ellipse, where the origin is at the center of the tunnel roadway.

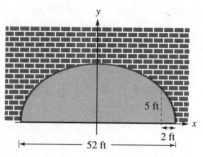

9. Sketch the graph of the system and determine the points of intersection.
   $$\begin{cases} x^2 - y^2 = 9 \\ 4(x - 1) = y^2 \end{cases}$$

10. Use a graphing utility to determine the graph of

    $$7x^2 + 4y^2 = 60$$

11. Use a graphing utility to determine the graph of

    $$x^2 + 3xy + 3y^2 + 2x = 0$$

    and identify the graph as a parabola, ellipse, or hyperbola.

12. Determine a parametric representation for the equation $x^2 - 9y^2 = 36$, $x > 0$.

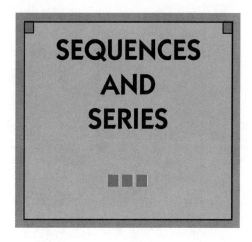

SEQUENCES AND SERIES

A schedule of monthly payments on a loan or the times for a physician's daily appointments are examples of *sequences* (ordered lists) in everyday life. Sequences of numbers, symbols, or expressions occur frequently in applied and theoretical mathematics, and we will consider applications ranging from modeling life expectancy to approximating functions with polynomials. Sequences also form the basis of pattern recognition, financial formulas, and recursive procedures in computer programming.

This chapter introduces the fundamental ideas and notation of sequences. We will investigate several special sequences and their sums. Our discussion will involve many of the functions—linear, exponential, rational—that we have already encountered. The important distinction in this chapter is that the domain of these functions is the set of natural numbers. This chapter serves both as an introduction to sequences and series, and as preparation for those of you who will go on to study calculus, where you will see these topics developed more thoroughly.

SEQUENCES

In Section 4.1 we computed the growth of $2000 invested at 6% compounded annually. We can extend this to get the amount after each of the following years:

	Computation	Amount
Original amount:		$2000.00
Amount after 1 year:	2000(1.06)	$2120.00
Amount after 2 years:	2120(1.06)	$2247.20
Amount after 3 years:	2247.20(1.06)	$2382.03

These amounts can be written as an ordered list of numbers, or *sequence*:

$$2000.00, 2120.00, 2247.20, 2382.03, \ldots$$

Sequences often arise in applied problems and play an important role in theoretical mathematics as well. In this section we will introduce the notation and some basic formulas for representing sequences.

Sequence Notation

Sequences generally are written as subscripted letters, such as

$$a_1, a_2, a_3, \ldots, a_n, \ldots$$

In particular, for the sequence 2000.00, 2120.00, 2247.20, 2382.03, ... ,

$$a_1 = 2000.00$$
$$a_2 = 2120.00$$
$$a_3 = 2247.20$$

and so on. The number a_1 is the *first term* of the sequence, a_2 is the *second term* of the sequence, and so on. In general, the *nth term* of the sequence is a_n. A sequence that has only a finite number of terms, such as

$$3, 6, 9, 12 \quad \text{or} \quad 1, \tfrac{1}{2}, \tfrac{1}{4}, \ldots, \tfrac{1}{64}$$

is called a *finite sequence*. A sequence with infinitely many terms, such as

$$2, -4, 6, -8, 10, \ldots$$

is called an *infinite sequence*.

For each positive integer *n* there corresponds exactly one value a_n. It follows that a sequence may be regarded as a function *f*:

$$a_n = f(n) \quad \text{where } n = 1, 2, 3, \ldots$$

Sequence

A **sequence** is a function whose domain is the set of positive integers.

EXAMPLE 1 Finding the Terms of a Sequence

List the first four terms of each sequence with the indicated nth term,

(a) $a_n = \dfrac{3n}{n+1}$ **(b)** $a_n = (-1)^{n+1}(2n-1)$

SOLUTION

(a) To find the first four terms of the sequence, we substitute $n = 1, 2, 3$, and 4 into the formula for a_n:

$$a_1 = \frac{3(1)}{1+1} = \frac{3}{2}$$

$$a_2 = \frac{3(2)}{2+1} = 2$$

$$a_3 = \frac{3(3)}{3+1} = \frac{9}{4}$$

$$a_4 = \frac{3(4)}{4+1} = \frac{12}{5}$$

(b) Replacing n with 1, 2, 3, and 4, we get

$$a_1 = (-1)^{1+1}[2(1) - 1] = 1$$
$$a_2 = (-1)^{2+1}[2(2) - 1] = -3$$
$$a_3 = (-1)^{3+1}[2(3) - 1] = 5$$
$$a_4 = (-1)^{4+1}[2(4) - 1] = -7$$

■ **NOTE:** *A sequence in which the terms are alternately positive and negative is called an* alternating *sequence.*

Some sequences are defined by a rule for obtaining each term from one or more preceding terms. Such a definition is called a *recursive* definition.

EXAMPLE 2 Finding the Terms of a Sequence Defined Recursively

List the first four terms of each sequence.

(a) $a_n = \begin{cases} -2 & \text{for } n = 1 \\ (a_{n-1})^2 & \text{for } n = 2, 3, 4, \ldots \end{cases}$

(b) $a_n = \begin{cases} 1 & \text{for } n = 1 \text{ and } 2 \\ a_{n-1} + a_{n-2} & \text{for } n = 3, 4, 5, \ldots \end{cases}$

SOLUTION

(a) The first term is $a_1 = -2$.

To find the second term, let $n = 2$ in $a_n = (a_{n-1})^2$:

$$a_2 = (a_1)^2$$
$$= (-2)^2 = 4 \quad \text{Substitute } a_1 = -2.$$

For the third term, let $n = 3$ in $a_n = (a_{n-1})^2$:

$$a_3 = (a_2)^2$$
$$= (4)^2 = 16 \quad \text{Substitute } a_2 = 4.$$

For the fourth term, let $n = 4$ in $a_n = (a_{n-1})^2$:

$$a_4 = (a_3)^2$$
$$= (16)^2 = 256 \quad \text{Substitute } a_3 = 16.$$

> ◉ **NOTE:** *The relationship $a_n = (a_{n-1})^2$ in part (a) can be stated in words as, "Each term is the square of the previous term." Similarly, in part (b) $a_n = a_{n-1} + a_{n-2}$ can be described as, "Each term is the sum of the previous two terms."*

Thus, the first four terms of the sequence are $-2, 4, 16, 256, \ldots$.

(b) The first two terms are $a_1 = 1$ and $a_2 = 1$.

We determine the third term by letting $n = 3$ in $a_n = a_{n-1} + a_{n-2}$:

$$a_3 = a_2 + a_1$$
$$= 1 + 1 = 2 \quad \text{Substitute } a_2 = 1 \text{ and } a_1 = 1.$$

Let $n = 4$ in $a_n = a_{n-1} + a_{n-2}$ to find the fourth term:

$$a_4 = a_3 + a_2$$
$$= 2 + 1 = 3 \quad \text{Substitute } a_3 = 2 \text{ and } a_2 = 1.$$

> ◉ **NOTE:** *The sequence in part (b) is known as the* Fibonacci sequence, *named after the mathematician Leonardo Fibonacci (circa 1175–1250), also known as Leonardo da Pisa.*

Thus, the first four terms of the sequence are $1, 1, 2, 3, \ldots$.

Finding an Expression for the *n*th Term of a Sequence

A sequence is not uniquely defined by simply listing some of the first terms; defining a sequence requires a general rule for the *n*th term, a_n. For example, what is a_5 (the fifth term) in the sequence below?

$$2, 4, 6, 8, \ldots$$

The apparent answer is $a_5 = 10$, and the corresponding *n*th term is

$$a_n = 2n$$

However, there are other possibilities. It may be that the *n*th term of the sequence is defined by

$$a_n = 2n + (n - 1)(n - 2)(n - 3)(n - 4)$$

The first five terms of this sequence are $2, 4, 6, 8, 34, \ldots$.

Even though a list of some of the first terms does not uniquely define a sequence, it is important to recognize patterns in sequences. In fact, mathematicians rely on pattern recognition to solve many problems. Given the first few terms of a sequence, you may be asked to find the *most evident* expression for the nth term. For the sequence 2, 4, 6, 8, $\ldots$, the most evident expression for the nth term is $a_n = 2n$.

Determining the nth term of a sequence based on the first few terms can be very challenging. Table 7.1 lists some common sequences.

TABLE 7.1

Common Sequences

Sequence	nth term	Sequence with nth term
2, 4, 6, 8, ...	$a_n = 2n$	2, 4, 6, 8, ... , $2n$, ...
1, 3, 5, 7, ...	$a_n = 2n - 1$	1, 3, 5, 7, ... , $2n - 1$, ...
3, 6, 9, 12, ...	$a_n = 3n$	3, 6, 9, 12, ... , $3n$, ...
4, 8, 12, 16, ...	$a_n = 4n$	4, 8, 12, 16, ... , $4n$, ...
1, 4, 9, 16, ...	$a_n = n^2$	1, 4, 9, 16, ... , n^2, ...
1, 8, 27, 64, ...	$a_n = n^3$	1, 8, 27, 64, ... , n^3, ...
2, 4, 8, 16, ...	$a_n = 2^n$	2, 4, 8, 16, ... , 2^n, ...

EXAMPLE 3 Finding the *n*th Term

Find the most evident expression for the nth term of each sequence. Then write the sequence in the form $a_1, a_2, a_3, \ldots, a_n, \ldots$.

(a) 2, 5, 8, 11, $\ldots$ **(b)** $\frac{1}{4}, \frac{2}{9}, \frac{3}{16}, \frac{4}{25}, \ldots$ **(c)** 5, -10, 15, -20, $\ldots$

SOLUTION

(a) Each term of 2, 5, 8, 11, $\ldots$ is 1 less than the sequence 3, 6, 9, 12, $\ldots$, $3n$, $\ldots$ in Table 7.1. Therefore, the most evident expression for the nth term of the sequence 2, 5, 8, 11, $\ldots$ is one less than $3n$:

$$a_n = 3n - 1$$

and the given sequence can be written as

$$2, 5, 8, \ldots, 3n - 1, \ldots$$

(b) The numerators appear to be the positive integers:

$$1, 2, 3, \ldots, n, \ldots$$

The denominators are similar to the sequence $1, 4, 9, 16, \ldots, n^2, \ldots$ in Table 7.1, except they start with the second term, 4. It follows that the denominators of the given sequence are

$$4, 9, 16, \ldots, (n + 1)^2, \ldots$$

Therefore, the nth term of the sequence $\frac{1}{4}, \frac{2}{9}, \frac{3}{16}, \frac{4}{25}, \ldots$ is

$$a_n = \frac{n}{(n + 1)^2}$$

and the given sequence can be written as

$$\frac{1}{4}, \frac{2}{9}, \frac{3}{16}, \ldots, \frac{n}{(n + 1)^2}, \ldots$$

(c) Notice from Table 7.1 that $a_n = 2n$ for the multiples of 2, $a_n = 3n$ for the multiples of 3, and $a_n = 4n$ for the multiples of 4. So the apparent rule for the nth term of the sequence

$$5, 10, 15, 20, \ldots$$

is

$$a_n = 5n$$

To make the signs of the terms alternate between positive and negative, as in the given sequence, we use the factor $(-1)^{n+1}$ (see Example 1b). Thus, the nth term of the sequence $5, -10, 15, -20, \ldots$ is

$$a_n = (-1)^{n+1}(5n)$$

and the sequence can be written as

$$5, -10, 15, \ldots, (-1)^{n+1}(5n), \ldots$$

Graph of a Sequence

We previously stated that a sequence may be regarded as a function f,

$$a_n = f(n) \qquad \text{where } n = 1, 2, 3, \ldots$$

The *ordered pairs* for the sequence $a_1, a_2, a_3, \ldots, a_n, \ldots$ are $(1, a_1), (2, a_2), (3, a_3)$, and so on. The *graph of a sequence* is the graph of these ordered pairs.

EXAMPLE 4 Graphing a Sequence

Graph f and the first five ordered pairs of the sequence a_n.

(a) $f(x) = 2x - 3$; $a_n = 2n - 3$ **(b)** $f(x) = \dfrac{2x}{x + 1}$; $a_n = \dfrac{2n}{n + 1}$

SOLUTION

(a) The graph of $f(x) = 2x - 3$ is a line with slope 2 and y-intercept $(0, -3)$, as shown in Figure 7.1. The first five ordered pairs of the sequence $a_n = 2n - 3$ are $(1, -1)$, $(2, 1)$, $(3, 3)$, $(4, 5)$, and $(5, 7)$; the graph is shown in Figure 7.2.

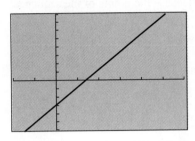 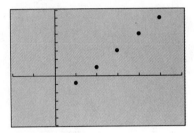

Figure 7.1
$y = 2x - 3$
$[-2, 6] \times [-6, 8]$

Figure 7.2
$a_n = 2n - 3$
$[-2, 6] \times [-6, 8]$

(b) The graph of the rational function

$$f(x) = \frac{2x}{x + 1}$$

is discussed in Example 4 of Section 3.4. The graph is shown in Figure 7.3. The first five ordered pairs of the sequence

$$a_n = \frac{2n}{n + 1}$$

are $(1, 1)$, $\left(2, \frac{4}{3}\right)$, $\left(3, \frac{3}{2}\right)$, $\left(4, \frac{8}{5}\right)$, and $\left(5, \frac{5}{3}\right)$, as shown in Figure 7.4.

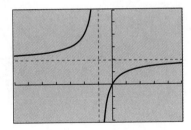 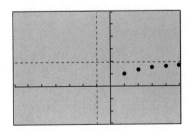

Figure 7.3
$y = \dfrac{2x}{x + 1}$
$[-7, 5] \times [-3, 6]$

Figure 7.4
$a_n = \dfrac{2n}{n + 1}$
$[-7, 5] \times [-3, 6]$

We can make an observation about the sequence in part (b) of Example 4. Since the line $y = 2$ is a horizontal asymptote for the rational function f,

$$f(x) = \frac{2x}{x + 1} \to 2 \quad \text{as} \quad x \to +\infty$$

Because the points of the sequence

$$a_n = \frac{2n}{n + 1}$$

lie on the graph of f,

$$a_n = \frac{2n}{n + 1} \to 2 \quad \text{as} \quad n \to +\infty$$

and we say the *sequence approaches* 2.

Applications

As we stated earlier, mathematicians often rely on pattern recognition to solve problems. The next example illustrates how considering smaller or simpler cases of a problem can create a sequence that leads to the solution.

◉ EXAMPLE 5   Application: Expanding Airline Operations

An airline specializing in nonstop flights wants to expand its operations to more airports. To determine the extent of expansion, the airline needs to know the number of nonstop routes. There are 0 nonstop routes when there is only 1 airport, 1 nonstop route between 2 airports, 3 nonstop routes between 3 airports, and 6 nonstop routes between 4 airports (Figure 7.5).

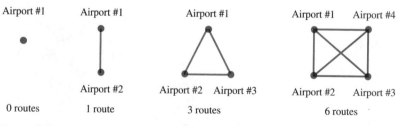

Figure 7.5

(a) Determine the number of nonstop routes between 5 airports.
(b) Use a sequence to determine the number of nonstop routes between 6 airports and the number of nonstop routes between 7 airports.

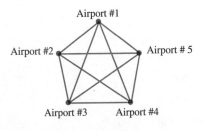

Airport #1
Airport #2
Airport # 5
Airport #3
Airport #4

Figure 7.6

SOLUTION

(a) A picture of the situation is shown in Figure 7.6. Counting the line segments, we see that there are 10 nonstop routes.

(b) The sequence for the number of nonstop routes is

$$0, 1, 3, 6, 10, \ldots$$

To determine the next two terms of the sequence, consider the increase in routes as we go from term to term. From the first term (0) to the second term (1) there is an increase of 1; from the second term (1) to the third term (3) there is an increase of 2; from the third term (3) to the fourth term (6) there is an increase of 3; and from the fourth term (6) to the fifth term (10) there is an increase of 4. The pattern suggests adding 5 to the fifth term to predict the next term:

$$0, 1, 3, 6, 10, 15, \ldots$$

Similarly, we add 6 to the sixth term to get the next term:

$$0, 1, 3, 6, 10, 15, 21, \ldots$$

Based on this sequence, the number of nonstop routes between 6 airports is 15, and the number of nonstop routes between 7 airports is 21.

The last two terms of the sequence in Example 5 were based solely on a pattern. In Exercise 60, you are asked to explain why the nth term for this sequence is

$$a_n = a_{n-1} + n - 1$$

 EXERCISE SET 7.1 ⎯⎯⎯⎯⎯⎯⎯⎯⎯⎯⎯⎯⎯⎯⎯⎯⎯⎯⎯⎯⎯⎯⎯⎯⎯⎯⎯⎯⎯⎯

A

In Exercises 1–12, list the first four terms of each sequence with the indicated nth term.

1. $a_n = 3n - 2$

2. $a_n = 2n + 5$

3. $a_n = \dfrac{n - 1}{2n}$

4. $a_n = \dfrac{n + 2}{4n - 3}$

5. $a_n = 20 - n^2$

6. $a_n = 30 - n^3$

7. $a_n = (-1)^{n+1}(n^2 - 1)$

8. $a_n = (-1)^{n+1}(4n + 1)$

9. $a_n = (-1)^n(5)$

10. $a_n = (-1)^{n+2}(4)$

11. $a_n = \left(-\frac{1}{2}\right)^{n-1}$

12. $a_n = \left(-\frac{2}{3}\right)^n$

In Exercises 13–20, list the first four terms of each sequence with the indicated nth term.

13. $a_n = \begin{cases} 3 & \text{for } n = 1 \\ 3 - a_{n-1} & \text{for } n = 2, 3, 4, \ldots \end{cases}$

14. $a_n = \begin{cases} 0 & \text{for } n = 1 \\ 6 - a_{n-1} & \text{for } n = 2, 3, 4, \ldots \end{cases}$

15. $a_n = \begin{cases} 5 & \text{for } n = 1 \\ 2a_{n-1} & \text{for } n = 2, 3, 4, \ldots \end{cases}$

16. $a_n = \begin{cases} 2 & \text{for } n = 1 \\ 3a_{n-1} & \text{for } n = 2, 3, 4, \ldots \end{cases}$

17. $a_n = \begin{cases} \frac{1}{2} & \text{for } n = 1 \\ \left(\dfrac{1}{a_{n-1}}\right)^2 & \text{for } n = 2, 3, 4, \ldots \end{cases}$

18. $a_n = \begin{cases} 256 & \text{for } n = 1 \\ \sqrt{\dfrac{1}{a_{n-1}}} & \text{for } n = 2, 3, 4, \ldots \end{cases}$

19. $a_n = \begin{cases} 1 & \text{for } n = 1 \text{ and } 2 \\ 2a_{n-1} + a_{n-2} & \text{for } n = 3, 4, \ldots \end{cases}$

20. $a_n = \begin{cases} 2 & \text{for } n = 1 \text{ and } 2 \\ a_{n-1} + 3a_{n-2} & \text{for } n = 3, 4, \ldots \end{cases}$

In Exercises 21–32, find the most evident expression for the nth term of each sequence. Write the sequence in the form $a_1, a_2, a_3, \ldots, a_n, \ldots$.

21. $7, 14, 21, 28, \ldots$ **22.** $6, 12, 18, 24, \ldots$

23. $3, 7, 11, 15, \ldots$ **24.** $1, 4, 7, 10, \ldots$

25. $2, -4, 8, -16, \ldots$ **26.** $1, -4, 9, -16, \ldots$

27. $0, 2, 4, 6, \ldots$ **28.** $0, 1, 8, 27, \ldots$

29. $\frac{1}{2}, \frac{8}{3}, \frac{27}{4}, \frac{64}{5}, \ldots$ **30.** $1, \frac{3}{4}, \frac{5}{9}, \frac{7}{16}, \ldots$

31. $\dfrac{-1}{1 \cdot 2}, \dfrac{1}{2 \cdot 3}, \dfrac{-1}{3 \cdot 4}, \dfrac{1}{4 \cdot 5}, \ldots$

32. $\dfrac{1}{1 \cdot 3}, \dfrac{-1}{2 \cdot 5}, \dfrac{1}{3 \cdot 7}, \dfrac{-1}{4 \cdot 9}, \ldots$

B

In Exercises 33–38, graph f and the first five ordered pairs of each sequence a_n.

33. $f(x) = \frac{1}{2}x + 1; \ a_n = \frac{1}{2}n + 1$

34. $f(x) = \frac{2}{3}x - 1; \ a_n = \frac{2}{3}n - 1$

35. $f(x) = 6\left(\frac{1}{2}\right)^x; \ a_n = 6\left(\frac{1}{2}\right)^n$

36. $f(x) = 15\left(\frac{2}{5}\right)^x; \ a_n = 15\left(\frac{2}{5}\right)^n$

37. $f(x) = \dfrac{3x}{x + 2}; \ a_n = \dfrac{3n}{n + 2}$

38. $f(x) = \dfrac{2x - 1}{x + 1}; \ a_n = \dfrac{2n - 1}{n + 1}$

In Exercises 39–42, determine the value that each sequence approaches as $n \to +\infty$.

39. The sequence in Exercise 35

40. The sequence in Exercise 36

41. The sequence in Exercise 37

42. The sequence in Exercise 38

In Exercises 43–50, find the sixth term of each sequence.

43. $1, 3, 7, 13, 21, \ldots$ **44.** $1, 2, 5, 10, 17, \ldots$

45. $0, 1, 5, 14, 30, \ldots$ **46.** $0, 2, 6, 14, 30, \ldots$

47. $1, 4, 10, 19, 31, \ldots$ **48.** $1, 5, 13, 25, 41, \ldots$

49. $1, 1 + 2x, 1 + 2x + 3x^2, 1 + 2x + 3x^2 + 4x^3,$
$1 + 2x + 3x^2 + 4x^3 + 5x^4, \ldots$

50. $1 + x^2, 2 + x^3, 6 + x^4, 24 + x^5, 120 + x^6, \ldots$

In Exercises 51–54, list the first four terms of each sequence.

51. $a_n = 2nx$ **52.** $a_n = n^2 x$

53. $a_n = P(1 + r)^{n-1}$ **54.** $a_n = Q\left(\frac{1}{2}\right)^{(n-1)/H}$

55. If \$1000 is deposited into an account that pays 5% interest compounded monthly, the balance A_n in the account after n months is given by

$$A_n = 1000\left(1 + \frac{0.05}{12}\right)^n$$

(a) Determine the first four terms of this sequence.
(b) Determine the balance in the account after 2 years.

56. A company depreciates its machinery according to the *constant percentage method*, which means that 4% is deducted from the value of the machine every year. The value a_n after n years of a machine that originally cost \$12,000 is given by

$$a_n = 12{,}000 \,(1 - 0.04)^n$$

(a) Determine the first three terms of this sequence.
(b) Determine the value of the machine after 10 years.

57. Each employee in a software development company has a computer that is connected to the other computers on a single circuit that can be represented by Figure 7.7:

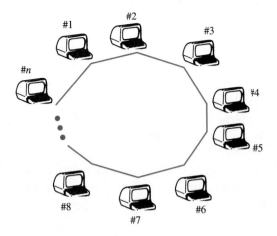

Figure 7.7

The company wants to upgrade the network so that the computers are connected directly to each other. To determine the number of connections to order, consider the following cases of smaller companies (Figure 7.8):

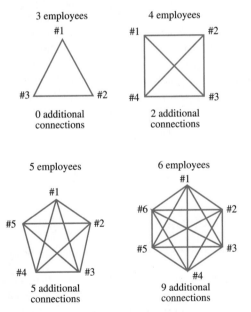

Figure 7.8

(a) Draw a picture of the case of 7 employees and determine the number of additional connections for that case. Use your result to determine the fifth term of the sequence 0, 2, 5, 9, ...

(b) Extend the sequence in part (a) to determine the number of connections to be ordered for the company if it has 10 employees.

(c) Determine an expression for the nth term of the sequence.

58. If two coins are tossed, the probability that both show heads is $\frac{1}{4}$. If three coins are tossed, the probability that exactly two show heads (and one shows tails) is $\frac{3}{8}$. If four coins are tossed, the probability that exactly two show heads (and two show tails) is $\frac{6}{16}$; and when five coins are tossed, the probability that exactly two show heads (and three show tails) is $\frac{10}{32}$. By considering the fifth term of the sequence $\frac{1}{4}, \frac{3}{8}, \frac{6}{16}, \frac{10}{32}, \dots$, determine the probability that exactly two coins show heads when six coins are tossed.

59. The typical price of a home in the United States from 1981 to 1993 can be approximated by the model

$$a_n = -0.048n^3 + 0.964n^2 - 1.908n + 67.809$$

$$n = 1, 2, 3, \dots, 13$$

where a_n is the median price of existing homes (in thousands of dollars) during the nth year, with $n = 1$ corresponding to 1981.

(a) Find the first six terms of the sequence. Compare these terms with the actual prices shown in Figure 7.9. Which is the most accurate? Which is the least accurate?

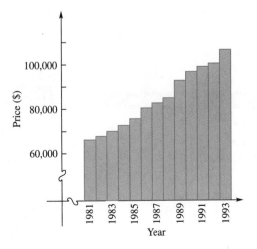

Figure 7.9

(b) Use the model to approximate the median price of a home in 1994.

C

60. Suppose that the number of nonstop routes between $n - 1$ airports in Example 5 is a_{n-1}.

(a) Explain why adding one more airport adds $n - 1$ nonstop routes. Use this to explain why the nth term for the sequence in Example 5 is

$$a_n = a_{n-1} + n - 1$$

(b) The expression for a_n in part (a) is defined recursively. Determine a nonrecursive expression for a_n.

61. For any positive number P, $\sqrt{P}$ can be approximated by the recursively defined sequence

$$a_n = \begin{cases} \dfrac{P}{2} & \text{for } n = 1 \\ \dfrac{1}{2}\,a_{n-1} + \dfrac{P}{2a_{n-1}} & \text{for } n = 2, 3, 4, \ldots \end{cases}$$

Approximate $\sqrt{6}$ by letting $P = 6$ and finding the first four terms of the sequence.

62. Find the nth term of the sequence 2, 0, 0, 0, 2, 0, 0, 0, 2, 0, 0, 0, 2, … .
HINT: Consider powers of $i = \sqrt{-1}$.

63. Let

$$r = \frac{1 + \sqrt{5}}{2} \qquad \text{and} \qquad s = \frac{1 - \sqrt{5}}{2}$$

Use your calculator to evaluate *Benet's formula*,

$$\frac{r^n - s^n}{r - s}$$

for $n = 1, 2, 3, 4, 5,$ and 6. Have you seen this sequence before?

64. Pick any positive integer as the first number of a sequence. Generate the rest of the sequence using the following rule for $n > 1$:

$$a_n = \begin{cases} \dfrac{a_{n-1}}{2} & \text{if } a_{n-1} \text{ is even} \\ 3a_{n-1} + 1 & \text{if } a_{n-1} \text{ is odd} \end{cases}$$

For example, we select any positive integer as the first term, say $a_1 = 17$. Since $a_1 = 17$ is odd, then $a_2 = 3(17) + 1 = 52$. Since $a_2 = 52$ is even, $a_3 = \frac{52}{2} = 26$. Thus, the first three terms of the sequence are 17, 52, 26, … .

(a) Generate the six sequences with first term $a_1 = 1, 2, 3, 4, 5,$ and 6. List at least the first 15 terms.

(b) Repeat part (a) by selecting another positive integer as the first term.

(c) Every sequence generated this way seems to share a property. Describe this property. (At the time of this writing, no one has been able to prove that all such sequences share this property.)

SUMMATION AND SERIES

There are many practical and important theoretical problems that require finding the sum of the terms in a sequence. For example, investigating the distance an object falls as a function of time, computing the future value of an annuity, or obtaining decimal approximations for square roots, exponentials, and logarithms are all problems that involve the sum of a finite number of terms of a sequence. In general, the sum of the first n terms of a sequence $a_1, a_2, a_3, a_4, \ldots, a_n, \ldots$ can be written as

$$a_1 + a_2 + a_3 + a_4 + \cdots + a_n$$

An expression that indicates the sum of the terms of a sequence is called a *series*.

Summation Notation

A compact notation for the sum of the first n terms of a sequence uses the symbol Σ (a stylized version of the Greek letter sigma).

Summation Notation

The sum of the first n terms of a sequence can be indicated by

$$\sum_{k=1}^{n} a_k = a_1 + a_2 + a_3 + \cdots + a_n$$

The integer k is the **index** of the summation. The equation $k = 1$ under the Σ indicates that the summation starts at 1, called the **lower limit** of the summation. The value n above the Σ indicates where to stop, and is called the **upper limit** of the summation.

■ **NOTE:** *We read* $\displaystyle\sum_{k=1}^{n} a_k$ *as*

"the summation of a_k from $k = 1$ to n."

◉ EXAMPLE 1   Expanding a Summation

Write each expression in expanded form (without summation notation) and simplify.

(a) $\displaystyle\sum_{k=1}^{4} k(2k + 1)$ **(b)** $\displaystyle\sum_{j=0}^{3}\left(-\frac{1}{2}\right)^{j}$ **(c)** $\displaystyle\sum_{k=1}^{5} 3$

SOLUTION

(a) We substitute $k = 1, 2, 3,$ and 4 successively in the expression $k(2k + 1)$ and add the resulting values:

$$\sum_{k=1}^{4} k(2k + 1) = \overbrace{1[2(1) + 1]}^{k = 1} + \overbrace{2[2(2) + 1]}^{k = 2} + \overbrace{3[2(3) + 1]}^{k = 3} + \overbrace{4[2(4) + 1]}^{k = 4}$$

$$= 3 + 10 + 21 + 36$$

$$= 70$$

(b) We replace j with 0, 1, 2, and 3 in the expression $\left(-\frac{1}{2}\right)^{j}$ and add:

$$\sum_{j=0}^{3}\left(-\frac{1}{2}\right)^{j} = \overbrace{\left(-\frac{1}{2}\right)^{0}}^{j = 0} + \overbrace{\left(-\frac{1}{2}\right)^{1}}^{j = 1} + \overbrace{\left(-\frac{1}{2}\right)^{2}}^{j = 2} + \overbrace{\left(-\frac{1}{2}\right)^{3}}^{j = 3}$$

$$= 1 - \frac{1}{2} + \frac{1}{4} - \frac{1}{8}$$

$$= \frac{5}{8}$$

(c) This expression represents the sum of the first five terms of the sequence $a_k = 3$. Specifically.

$$\sum_{k=1}^{5} 3 = \overbrace{3}^{k=1} + \overbrace{3}^{k=2} + \overbrace{3}^{k=3} + \overbrace{3}^{k=4} + \overbrace{3}^{k=5}$$

$$= 15$$

EXAMPLE 2   Writing a Sum Using Summation Notation

Express each sum in summation notation.

(a) $\dfrac{1}{1+1} + \dfrac{4}{1+2} + \dfrac{9}{1+3} + \cdots + \dfrac{225}{1+15}$

(b) $2 - 4 + 8 - 16 + \cdots - 256$

SOLUTION

(a) The numerators are the first 15 terms of the sequence 1, 4, 9, 16, ... , n^2, ... in Table 7.1 (Section 7.1). The denominators are the first 15 terms of the sequence $1 + 1, 1 + 2, 1 + 3, \ldots, 1 + n, \ldots$. Thus, the sum

$$\frac{1}{1+1} + \frac{4}{1+2} + \frac{9}{1+3} + \cdots + \frac{225}{1+15}$$

can be written in summation notation as

$$\sum_{k=1}^{15} \frac{k^2}{1+k}$$

(b) The terms are those of the sequence 2, 4, 8, 16, ... , 2^n, ... in Table 7.1. Using the factor $(-1)^{k+1}$ to make the signs alternate between positive and negative, we write the sum as

$$\sum_{k=1}^{8} (-1)^{k+1} 2^k$$

There may be more than one way to express a sum in summation notation. For example, the sum in part (a) of Example 2,

$$\frac{1}{1+1} + \frac{4}{1+2} + \frac{9}{1+3} + \cdots + \frac{225}{1+15}$$

can also be represented as

$$\sum_{k=2}^{16} \frac{(k-1)^2}{k} \qquad \text{or} \qquad \sum_{k=0}^{14} \frac{(k+1)^2}{k+2}$$

Some very important series, as well as other mathematical topics, involve special products like

$$5 \cdot 4 \cdot 3 \cdot 2 \cdot 1$$

The notation 5!, read 5 *factorial*, is used to represent this product:

$$5! = 5 \cdot 4 \cdot 3 \cdot 2 \cdot 1$$

Factorial

> If k is a positive integer, then $k!$, called **k factorial**, is defined as
>
> $$k! = k \cdot (k - 1) \cdot \cdots \cdot 3 \cdot 2 \cdot 1$$
>
> The expression 0! is defined to be 1.

For example,

$$0! = 1$$
$$1! = 1$$
$$2! = 2 \cdot 1 = 2$$
$$3! = 3 \cdot 2 \cdot 1 = 6$$
$$4! = 4 \cdot 3 \cdot 2 \cdot 1 = 24$$

EXAMPLE 3   A Series with Factorials

A polynomial used to approximate e^x is

$$P_n(x) = \sum_{k=0}^{n} \frac{x^k}{k!}$$

(a) Express $P_3(x)$ in expanded form.
(b) Compute $e^{0.5}$ and $P_3(0.5)$ to the nearest 0.001.
(c) Graph $y = e^x$ and $y = P_3(x)$ with a viewing window of $[-3, 3] \times [-1, 12]$.

SOLUTION

(a) Letting $n = 3$ and expanding the summation, we have

$$P_3(x) = \sum_{k=0}^{3} \frac{x^k}{k!}$$

$$= \frac{x^0}{0!} + \frac{x^1}{1!} + \frac{x^2}{2!} + \frac{x^3}{3!}$$

$$= \frac{1}{1} + \frac{x}{1} + \frac{x^2}{2} + \frac{x^3}{6}$$

Thus,

$$P_3(x) = 1 + x + \frac{x^2}{2} + \frac{x^3}{6}$$

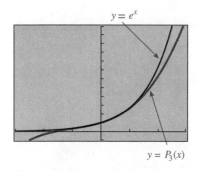

Figure 7.10
$[-3, 3] \times [-1, 12]$

(b) To the nearest 0.001, $e^{0.5} = 1.649$. To the nearest 0.001,

$$P_3(0.5) = 1 + 0.5 + \frac{0.5^2}{2} + \frac{0.5^3}{6} = 1.646$$

(c) The graphs of $y = e^x$ and $y = P_3(x)$ are shown in Figure 7.10. Notice that the graphs nearly coincide for values of x near 0.

In calculus it is shown that as the value of n increases, the polynomial

$$P_n(x) = \sum_{k=0}^{n} \frac{x^k}{k!}$$

gets closer and closer to e^x. Exercises 31–34 give you an opportunity to explore this method of approximating e^x and other important functions.

Properties of Summation

The following summation properties are used to evaluate summations and to simplify expressions containing summations.

Summation Properties

> Let a_k and b_k be sequences. For every positive integer n and any constant c, the following properties hold:
>
> **1.** $\displaystyle\sum_{k=1}^{n} c = nc$ **3.** $\displaystyle\sum_{k=1}^{n} (a_k + b_k) = \sum_{k=1}^{n} a_k + \sum_{k=1}^{n} b_k$
>
> **2.** $\displaystyle\sum_{k=1}^{n} ca_k = c \sum_{k=1}^{n} a_k$ **4.** $\displaystyle\sum_{k=1}^{n} (a_k - b_k) = \sum_{k=1}^{n} a_k - \sum_{k=1}^{n} b_k$

PROOF: We will prove properties 2 and 3. The proofs of properties 1 and 4 are left for you (Exercises 59 and 60). To prove property 2 we use the distributive property of real numbers.

$$\sum_{k=1}^{n} ca_k = ca_1 + ca_2 + ca_3 + \cdots + ca_n \qquad \text{Write the summation in expanded form.}$$

$$= c(a_1 + a_2 + a_3 + \cdots + a_n) \qquad \text{Factor } c \text{ from each term.}$$

$$= c \sum_{k=1}^{n} a_k \qquad \text{Write the expression in parentheses in summation notation.}$$

The proof of property 3 uses the commutative and associative properties of real numbers.

$$\sum_{k=1}^{n} (a_k + b_k) = (a_1 + b_1) + (a_2 + b_2) + (a_3 + b_3) + \cdots + (a_n + b_n)$$

$$= (a_1 + a_2 + a_3 + \cdots + a_n) + (b_1 + b_2 + b_3 + \cdots + b_n)$$

$$= \sum_{k=1}^{n} a_k + \sum_{k=1}^{n} b_k$$

EXAMPLE 4   Using Properties of Summation

Given that $\displaystyle\sum_{k=1}^{30} a_k = 40$ and $\displaystyle\sum_{k=1}^{30} b_k = 65$, evaluate each summation.

(a) $\displaystyle\sum_{k=1}^{30} (3a_k - b_k)$ **(b)** $\displaystyle\sum_{k=1}^{30} (8 + a_k)$

SOLUTION

(a) $\displaystyle\sum_{k=1}^{30} (3a_k - b_k) = \sum_{k=1}^{30} 3a_k - \sum_{k=1}^{30} b_k$ Use property 4.

$$= 3 \sum_{k=1}^{30} a_k - \sum_{k=1}^{30} b_k \quad \text{Use property 2.}$$

$$= 3(40) - 65 \qquad \text{Substitute } \sum_{k=1}^{30} a_k = 40 \text{ and } \sum_{k=1}^{30} b_k = 65.$$

$$= 55$$

(b) $\displaystyle\sum_{k=1}^{30} (8 + a_k) = \sum_{k=1}^{30} 8 + \sum_{k=1}^{30} a_k$ Use property 3.

$$= 30(8) + \sum_{k=1}^{30} a_k \quad \text{Use property 1.}$$

$$= 240 + 40 \qquad \text{Substitute } \sum_{k=1}^{30} a_k = 40.$$

$$= 280$$

When the great mathematician Karl F. Gauss (1777–1855) was a young boy attending primary school, his teacher gave the class the problem of finding the sum of the numbers 1, 2, 3, ... , 100. Within only a few moments, young Gauss handed the correct answer to his teacher with scarcely any work on his slate. How did he find the sum so quickly? Many believe his reasoning went something like the following.

Let $S = 1 + 2 + 3 + \cdots + 100$, the desired sum. Reversing the order of addition gives the same result, so $S = 100 + 99 + 98 + \cdots + 1$. Adding the corresponding terms of these two equations gives

$$\begin{aligned} S &= 1 + 2 + 3 + \cdots + 100 \\ S &= 100 + 99 + 98 + \cdots + 1 \\ \hline 2S &= 101 + 101 + 101 + \cdots + 101 \end{aligned}$$

The last sum can be evaluated with multiplication:

$$2S = 100(101)$$

Dividing each side by 2 gives

$$S = \frac{100(101)}{2} = 5050$$

This clever way of finding the sum is a glimpse of the mathematical talent that Gauss already possessed as a young boy.

We can generalize the preceding result by replacing 100 with n. It follows that

$$1 + 2 + 3 + \cdots + n = \frac{n(n + 1)}{2}$$

Written in summation notation, this result is the first of the following formulas.

Summation Formulas

1. $\sum_{k=1}^{n} k = \dfrac{n(n + 1)}{2}$ **2.** $\sum_{k=1}^{n} k^2 = \dfrac{n(n + 1)(2n + 1)}{6}$

A derivation of the second summation formula is outlined in Exercise 62.

EXAMPLE 5   Using the Summation Formulas

Use the summation formulas and properties to evaluate $\sum_{k=1}^{20} (3k^2 - 2k + 5)$.

SOLUTION

$$\sum_{k=1}^{20} (3k^2 - 2k + 5)$$

$$= \sum_{k=1}^{20} 3k^2 - \sum_{k=1}^{20} 2k + \sum_{k=1}^{20} 5 \qquad \text{Use properties 3 and 4.}$$

$$= 3\sum_{k=1}^{20} k^2 - 2\sum_{k=1}^{20} k + 20(5) \qquad \text{Use properties 1 and 2.}$$

$$= 3\,\frac{20(20 + 1)[2(20) + 1]}{6}$$

$$\quad - 2\,\frac{20(20 + 1)}{2} + 100 \qquad \text{Use formulas 1 and 2.}$$

$$= 8610 - 420 + 100$$

$$= 8290$$

Figure 7.11

Applications

The distance that an object falls during equally spaced time intervals can be measured using multiflash photography. Figure 7.11 shows the position of a falling ball at equal time intervals. The distance the ball travels during each successive time interval is a sequence. The next example shows how we use summation properties to determine a formula for the total distance an object falls as a function of time.

⬤ EXAMPLE 6 Application: Distance an Object Falls

If air resistance is negligible, a dropped object falls 16 feet during the first second, 48 feet during the next second, 80 feet during the third second, 112 feet during the fourth second, and so on.

(a) Find an expression for the apparent kth term of the sequence 16, 48, 80, 112, ...

(b) Use the results of part (a) and the summation properties to determine a formula for the total distance the object falls n seconds after it has been dropped.

SOLUTION

(a) If we factor 16 out of each term, then the sequence is

$$16(1), \ 16(3), \ 16(5), \ 16(7), \ \dots$$

From Table 7.1, we know the kth term for the sequence 1, 3, 5, 7, ... is $2k - 1$. So the kth term for the sequence 16(1), 16(3), 16(5), 16(7), ... is apparently

$$a_k = 16(2k - 1)$$

(b) The total distance the object falls after n seconds is

$$16(1) + 16(3) + 16(5) + 16(7) + \cdots + 16(2n - 1)$$

In summation notation, the total distance is

$$\sum_{k=1}^{n} 16(2k - 1)$$

Using the summation properties and formulas, we have

$$\sum_{k=1}^{n} 16(2k - 1) = \sum_{k=1}^{n} (32k - 16)$$

$$= 32 \sum_{k=1}^{n} k - \sum_{k=1}^{n} 16$$

$$= 32 \frac{n(n + 1)}{2} - 16n$$

$$= 16n^2$$

Thus, the total distance d the object falls n seconds after it has been dropped is given by

$$d = 16n^2$$

We close this section with an example that incorporates a model with the properties of summation to solve a problem.

EXAMPLE 7 Application: Court Commitments to State Prisons

The number of people committed to state prisons in the United States for drug offenses from 1981 to 1989 can be approximated by the model

$$a_k = 0.52k^2 - 1.5k + 2.8 \qquad k = 1, 2, 3, \ldots, 8$$

where a_k is the increase in the number of commitments (in thousands) after the kth year, with $k = 1$ corresponding to 1981. The number of commitments in 1981 was approximately 11,500. Use the model to approximate the number of commitments in 1989.

SOLUTION

The number of commitments in 1989 is the sum of the number of commitments in 1981 and the eight subsequent annual increases:

$$11.5 + \sum_{k=1}^{8} a_k = 11.5 + \sum_{k=1}^{8} (0.52k^2 - 1.5k + 2.8)$$

Using the properties of summation, we have

$$11.5 + \sum_{k=1}^{8} (0.52k^2 - 1.5k + 2.8)$$

$$= 11.5 + 0.52 \sum_{k=1}^{8} k^2 - 1.5 \sum_{k=1}^{8} k + \sum_{k=1}^{8} 2.8$$

$$= 11.5 + 0.52 \frac{8(8 + 1)[2(8) + 1]}{6} - 1.5 \frac{8(8 + 1)}{2} + 8(2.8)$$

$$= 85.98$$

Thus, the number of commitments to state prisons for drug offenses in 1989 was approximately 86,000.

The bar graph in Figure 7.12 shows the actual number of commitments for the period from 1981 through 1989. You can use it to compare the model in Example 7 with the actual data.

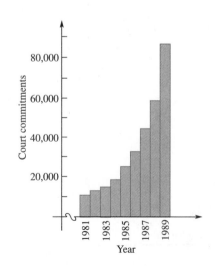

Figure 7.12

EXERCISE SET 7.2

A

In Exercises 1–12, write each expression in expanded form (without summation notation) and simplify.

1. $\displaystyle\sum_{k=1}^{5}(3k-2)$

2. $\displaystyle\sum_{k=1}^{6}(2k-3)$

3. $\displaystyle\sum_{k=0}^{5}(k-1)(k+1)$

4. $\displaystyle\sum_{k=0}^{4}\frac{k(k+1)}{2}$

5. $\displaystyle\sum_{k=1}^{4}\left(\frac{1}{3}\right)^{k-1}$

6. $\displaystyle\sum_{k=1}^{4}\left(\frac{3}{2}\right)^{k}$

7. $\displaystyle\sum_{k=1}^{6}4$

8. $\displaystyle\sum_{k=1}^{7}9$

9. $\displaystyle\sum_{j=0}^{4}(-1)^{j}\,j!$

10. $\displaystyle\sum_{j=0}^{4}\frac{(-2)^{j}}{j!}$

11. $\displaystyle\sum_{k=2}^{9}(14-k)$

12. $\displaystyle\sum_{k=2}^{8}(50-k^{2})$

In Exercises 13–18, given that

$$\sum_{k=1}^{40}a_{k}=75 \quad\text{and}\quad \sum_{k=1}^{40}b_{k}=50$$

evaluate each summation.

13. $\displaystyle\sum_{k=1}^{40}(3a_{k}+b_{k})$

14. $\displaystyle\sum_{k=1}^{40}(a_{k}-2b_{k})$

15. $\displaystyle\sum_{k=1}^{40}(4a_{k}-3)$

16. $\displaystyle\sum_{k=1}^{40}(2b_{k}+5)$

17. $\displaystyle\sum_{k=1}^{40}(2a_{k}+3b_{k}-6)$

18. $\displaystyle\sum_{k=1}^{40}(2a_{k}-5b_{k}+9)$

B

In Exercises 19–30, express each sum in summation notation (answers are not unique).

19. $1+3+5+7+9+11$

20. $1+8+27+64+125$

21. $\frac{1}{2}+\frac{1}{4}+\frac{1}{6}+\frac{1}{8}+\cdots+\frac{1}{30}$

22. $\frac{1}{1}+\frac{1}{4}+\frac{1}{9}+\frac{1}{16}+\cdots+\frac{1}{144}$

23. $1+7+13+19+\cdots+43$

24. $1+6+11+16+\cdots+61$

25. $-3+6-9+12-\cdots+60$

26. $4-8+12-16+\cdots+44$

27. $\frac{1}{2}-\frac{3}{4}+\frac{5}{6}-\frac{7}{8}+\cdots-\frac{39}{40}$

28. $\frac{2}{5}-\frac{4}{10}+\frac{8}{15}-\frac{16}{20}+\cdots-\frac{256}{40}$

29. $100+100(1.1)+100(1.1)^{2}+100(1.1)^{3}+\cdots+100(1.1)^{n-1}$

30. $50(1.06)+50(1.06)^{2}+50(1.06)^{3}+\cdots+50(1.06)^{n}$

In Exercises 31 and 32, refer to the polynomial

$$P_{n}(x)=\sum_{k=0}^{n}\frac{x^{k}}{k!}$$

used to approximate e^{x}.

31. **(a)** Express $P_{4}(x)$ in expanded form and simplify.
(b) Compute $e^{0.5}$ and $P_{4}(0.5)$ to the nearest 0.001.
(c) Graph $y=e^{x}$ and $y=P_{4}(x)$ with a viewing window of $[-3,3]\times[-1,12]$.

32. **(a)** Express $P_{5}(x)$ in expanded form and simplify.
(b) Compute $e^{0.5}$ and $P_{5}(0.5)$ to the nearest 0.001.
(c) Graph $y=e^{x}$ and $y=P_{5}(x)$ with a viewing window of $[-3,3]\times[-1,12]$.

In Exercises 33 and 34, refer to the rational function

$$f_{n}(x)=\sum_{k=1}^{n}\frac{2}{2k-1}\left(\frac{x-1}{x+1}\right)^{2k-1}$$

used to approximate $\ln x$.

33. **(a)** Express $f_{2}(x)$ in expanded form.
(b) Graph $y=\ln x$ and $y=f_{2}(x)$ with a viewing window of $[-1,12]\times[-4,3]$.

34. **(a)** Express $f_{3}(x)$ in expanded form.
(b) Graph $y=\ln x$ and $y=f_{3}(x)$ with a viewing window of $[-1,12]\times[-4,3]$.

In Exercises 35–40, use the summation formulas and properties to evaluate each sum.

35. $\displaystyle\sum_{k=1}^{20}(4k+3)$

36. $\displaystyle\sum_{k=1}^{30}(3k-2)$

37. $\displaystyle\sum_{k=1}^{30} (2k^2 + k)$

38. $\displaystyle\sum_{k=1}^{20} (5k - k^2)$

39. $\displaystyle\sum_{k=1}^{20} \frac{k(k-3)}{2}$

40. $\displaystyle\sum_{k=1}^{20} \frac{k(2k+1)}{3}$

41. Determine a_{12} if $\displaystyle\sum_{k=1}^{12} a_k = 84$ and $\displaystyle\sum_{k=1}^{11} a_k = 70$.

42. Determine $\displaystyle\sum_{k=1}^{14} b_k$ if $\displaystyle\sum_{k=1}^{15} b_k = 91$ and $b_{15} = 8$.

43. A block of ice is released down a ramp with an inclination of 45°. If air and surface resistance is negligible, the block slides $8\sqrt{2}$ feet during the first second, $24\sqrt{2}$ feet during the next second, $40\sqrt{2}$ feet during the third second, $56\sqrt{2}$ feet during the fourth second, and so on.
   (a) Find the apparent kth term of the sequence
   $8\sqrt{2}, 24\sqrt{2}, 40\sqrt{2}, 56\sqrt{2}, \dots$.

   HINT Quick Review: Factor $8\sqrt{2}$ out of each term in the sequence.

   (b) Use the results of part (a) and the summation properties to determine a formula for the total distance the block of ice slides n seconds after it has been released.

44. A skier travels 5 feet down a mountain slope during the first second of descent. During the next second the skier travels 8 feet, and during the third second the skier travels 11 feet.
   (a) Assume that in each successive second, the skier travels 3 feet farther than the previous second. Determine the kth term of the sequence 5, 8, 11, ...
   (b) Use the results of part (a) and the summation properties to determine a formula for the total distance the skier travels during the first n seconds of descent.

45. The annual interest paid on public debt in the United States from 1980 to 1992 can be approximated by the model

$a_k = 18.22k + 59.56 \qquad k = 1, 2, 3, \dots , 13$

where a_k is the interest paid (in billions of dollars) during the kth year, with $k = 1$ corresponding to 1980. Use the model to approximate the total interest paid on public debt from 1980 through 1992.

46. The yearly production of tobacco in the United States from 1986 to 1992 is approximated by the model

$a_k = 0.10k + 1.04 \qquad k = 1, 2, 3, \dots , 7$

where a_k is the amount produced (in millions of pounds) during the kth year, with $k = 1$ corresponding to 1986. Use the model to determine the total amount of tobacco produced from 1986 through 1992.

47. The yearly expenditures for personal health care in the United States from 1985 to 1991 can be approximated by the model

$a_k = 3.66k^2 + 18.71k + 348.34 \qquad \text{for } k = 1, 2, 3, \dots , 7$

where a_k is the annual expenditure (in billions of dollars) during the kth year, with $k = 1$ corresponding to 1985. Use the model to determine the total expenditures from 1985 through 1991. Compare your results with the actual totals using the data given in Figure 7.13.

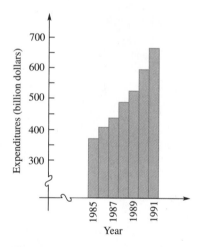

Figure 7.13

Least Squares Line (Exercises 48 and 49)

The most common method of fitting a line $y = mx + b$ to a set of numerical data points $(x_1, y_1), (x_2, y_2), \dots , (x_n, y_n)$, is to select the line that minimizes the *sum of the squares* of the vertical distances $d_1, d_2, \dots , d_n$ from the points to the line (Figure 7.14).

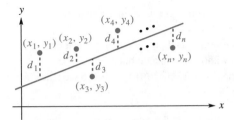

Figure 7.14

The values of m and b that determine this line are given by the formulas

$$m = \frac{\left(\sum_{k=1}^{n} x_k\right)\left(\sum_{k=1}^{n} y_k\right) - n\sum_{k=1}^{n} x_k y_k}{\left(\sum_{k=1}^{n} x_k\right)^2 - n\sum_{k=1}^{n} x_k^2}$$

and

$$b = \frac{1}{n}\left(\sum_{k=1}^{n} y_k - m\sum_{k=1}^{n} x_k\right)$$

The line $y = mx + b$ determined by these values of m and b is called the *least squares line* or *regression line*.

For example, for the points (2, 6), (3, 8), (4, 7), and (4.5, 9):

$$n = 4$$

$$\sum_{k=1}^{4} x_k = 2 + 3 + 4 + 4.5 = 13.5$$

$$\sum_{k=1}^{4} y_k = 6 + 8 + 7 + 9 = 30$$

$$\sum_{k=1}^{4} x_k y_k = 2(6) + 3(8) + 4(7) + 4.5(9) = 104.5$$

$$\sum_{k=1}^{4} x_k^2 = 2^2 + 3^2 + 4^2 + 4.5^2 = 49.25$$

Thus,

$$m = \frac{(13.5)(30) - 4(104.5)}{(13.5)^2 - 4(49.25)} \approx 0.88$$

and

$$b = \frac{1}{4}[30 - 0.88(13.5)] \approx 4.53$$

The regression line is $y = 0.88x + 4.53$.

48. **(a)** Determine the least squares line for the points (−1, 6), (2, 3.5), (3, 2), and (4, 1).
 (b) Plot the data points and graph the least squares line in the same coordinate plane.
 (c) Use the equation for the least squares line to predict the value of y that would correspond to $x = 5$.

49. **(a)** Determine the least squares line for the points (0, 9), (1, 7), (2, 5.5), (5, 0), and (6, −2).
 (b) Plot the data points and graph the least squares line in the same viewing window.
 (c) According to the equation for the least squares line, predict the value of y that would correspond to $x = 8$.

Standard Deviation (Exercises 50 and 51)

One of the most common measures of the dispersion of data $x_1, x_2, x_3, \ldots, x_n$, is the *standard deviation*, s. To determine the standard deviation:

Step 1: Compute the mean (average) of the data: $\bar{x} = \dfrac{1}{n}\sum_{k=1}^{n} x_k$

Step 2: Compute s: $\quad s = \sqrt{\dfrac{1}{n-1}\sum_{k=1}^{n}(x_k - \bar{x})^2}$

For example, consider the data 3, 4, 4, 7, 10:

$$n = 5$$

$$\bar{x} = \frac{1}{5}\sum_{k=1}^{5} x_k = \frac{1}{5}(3 + 4 + 4 + 7 + 10) = 5.6$$

$$s = \sqrt{\frac{1}{n-1}\sum_{k=1}^{n}(x_k - \bar{x})^2}$$

$$= \sqrt{\frac{1}{5-1}[(3 - 5.6)^2 + \cdots + (10 - 5.6)^2]} \approx 2.881$$

The mean is $\bar{x} = 5.6$, and the standard deviation is $s \approx 2.881$.

50. Determine the mean and standard deviation for the data: 1, 3, 4, 6, 7

51. Determine the mean and standard deviation for the data: 12, 15, 16, 17

For Exercises 52–55, refer to the following. In calculus, many important sequences are defined by a summation,

$$\sum_{k=1}^{n} a_k$$

The *sequence of partial sums* is the sequence

$$S_n = \sum_{k=1}^{n} a_k$$

For example, if

$$S_n = \sum_{k=1}^{n} (k^2 + 1)$$

then

$$S_1 = (1^2 + 1) = 2$$
$$S_2 = (1^2 + 1) + (2^2 + 1) = 7$$
$$S_3 = (1^2 + 1) + (2^2 + 1) + (3^2 + 1) = 17$$

and so on. The sequence of partial sums is 2, 7, 17, ...

52. Determine the first four terms of the sequence of partial sums defined by

$$S_n = \sum_{k=1}^{n} k^3$$

53. Determine the first four terms of the sequence of partial sums defined by

$$S_n = \sum_{k=1}^{n} \left(\frac{1}{2}\right)^{k-1}$$

54. Consider the sequence of partial sums defined by

$$S_n = \sum_{k=1}^{n} k(k + 1)$$

    (a) Determine the first four terms of the sequence of partial sums.

    (b) Use the properties of summation to express S_n without summation notation.

    (c) Using the result from part (b), write the sequence in the form $S_1, S_2, S_3, \dots , S_n, \dots$.

55. Consider the sequence of partial sums defined by

$$S_n = \sum_{k=1}^{n} \left(k^2 + 3\right)$$

    (a) Determine the first five terms of the sequence of partial sums.

    (b) Use the formulas and properties of summation to express S_n without summation notation.

    (c) Using the result from part (b), write the sequence in the form $S_1, S_2, S_3, \dots , S_n, \dots$.

56. Use the summation formulas and properties to determine the value of c such that

$$\sum_{k=1}^{10} (ck^2 + k) = 286$$

57. Use the summation formulas and properties to determine the value of c such that

$$\sum_{k=1}^{12} (2k^2 + ck) = 1352$$

C

58. **(a)** Use the formulas and properties of summation to evaluate

$$\frac{1}{100} \sum_{k=1}^{100} \frac{k + 1}{100}$$

    (b) Use the formulas and properties of summation to express

$$\frac{1}{n} \sum_{k=1}^{n} \frac{k + 1}{n}$$

    without summation notation.

    (c) Use the result from part (b) to determine the value

$$\frac{1}{n} \sum_{k=1}^{n} \frac{k + 1}{n}$$

    approaches as $n \to +\infty$.

59. Prove the first summation property: $\displaystyle\sum_{k=1}^{n} c = nc$

60. Prove the fourth summation property:

$$\sum_{k=1}^{n} (a_k - b_k) = \sum_{k=1}^{n} a_k - \sum_{k=1}^{n} b_k$$

61. Write Gauss's solution to $1 + 2 + 3 + \cdots + 100$ using summation notation and the properties of summation.

HINT: $100 + 99 + 98 + \cdots + 1 = \displaystyle\sum_{k=1}^{100}(101 - k)$

62. The second summation formula can be derived as follows.

(a) Show that: $\displaystyle\sum_{k=1}^{n}[(k + 1)^3 - k^3] = (n + 1)^3 - 1$

    HINT: Write out the summation without summation notation; most of the terms will cancel.

(b) Show that

$$\sum_{k=1}^{n}[(k + 1)^3 - k^3] = 3\sum_{k=1}^{n}k^2 + 3\sum_{k=1}^{n}k + \sum_{k=1}^{n}1$$

$$= 3\sum_{k=1}^{n}k^2 + 3\,\frac{n(n + 1)}{2} + n$$

(c) Using the results from parts (a) and (b), show that

$$\sum_{k=1}^{n}k^2 = \frac{n(n + 1)(2n + 1)}{6}$$

ARITHMETIC SEQUENCES AND SERIES

In the next two sections we will focus on two types of sequences: arithmetic and geometric. In this section we will examine arithmetic sequences, one of the most fundamental types of sequences because the formula for the nth term of an arithmetic sequence corresponds to a linear function. In Section 7.4 we will explore geometric sequences, which have many applications, particularly in the mathematics of finance. Both types of sequences have convenient summation formulas.

Arithmetic Sequences

Consider the sequence

$$2, 5, 8, 11, 14, \ldots$$

where each term is 3 more than the previous term. In other words, consecutive terms differ by 3:

$$a_2 - a_1 = 5 - 2 = 3$$
$$a_3 - a_2 = 8 - 5 = 3$$
$$a_4 - a_3 = 11 - 8 = 3$$

and so on.

    The sequence $2, 5, 8, 11, 14, \ldots$ is an example of an *arithmetic sequence*. The feature that qualifies a sequence as arithmetic is that consecutive terms differ by a constant.

Arithmetic Sequence

A sequence in which consecutive terms differ by a constant is called an **arithmetic sequence**. More precisely, a sequence

$$a_1, a_2, a_3, \ldots, a_n, \ldots$$

is arithmetic if there exists a constant d such that

$$a_{n+1} - a_n = d$$

for every positive integer n. The constant d is called the **common difference**.

● EXAMPLE 1   Identifying Arithmetic Sequences

Determine whether the terms given are the first five terms of an arithmetic sequence. If so, specify the common difference and the sixth term.

(a) 1, 2.5, 4, 5.5, 7, …   **(b)** 22, 14, 6, -2, -10, …
(c) 2, 3, 5, 7, 11, …

SOLUTION

(a) The difference of the given consecutive terms is:

$$a_2 - a_1 = 2.5 - 1 = 1.5$$
$$a_3 - a_2 = 4 - 2.5 = 1.5$$
$$a_4 - a_3 = 5.5 - 4 = 1.5$$
$$a_5 - a_4 = 7 - 5.5 = 1.5$$

Since the difference of consecutive terms is constant, the given terms are the first five terms of an arithmetic sequence. The common difference is

$$d = 1.5$$

Since $a_6 - a_5 = 1.5$, the sixth term is 1.5 more than the fifth term:

$$a_6 = a_5 + 1.5 = 7 + 1.5 = 8.5$$

(b) Computing the difference of the given consecutive terms, we have

$$a_2 - a_1 = 14 - 22 \qquad = -8$$
$$a_3 - a_2 = 6 - 14 \qquad = -8$$
$$a_4 - a_3 = -2 - 6 \qquad = -8$$
$$a_5 - a_4 = -10 - (-2) = -8$$

The given terms of the sequence are the first five terms of an arithmetic sequence with a common difference of

$$d = -8$$

Adding the common difference to the fifth term, we get the sixth term:

$$a_6 = -10 + (-8) = -18$$

(c) The sequence is not arithmetic, because

$$a_2 - a_1 = 3 - 2 = 1$$

which is not the same as

$$a_3 - a_2 = 5 - 3 = 2$$

Every term of an arithmetic sequence can be written in terms of the first term a_1 and common difference d as follows:

$$a_1 = a_1$$
$$a_2 = a_1 + d$$
$$a_3 = a_2 + d = (a_1 + d) + d = a_1 + 2d$$
$$a_4 = a_3 + d = (a_1 + 2d) + d = a_1 + 3d$$
$$a_5 = a_4 + d = (a_1 + 3d) + d = a_1 + 4d$$

and so on. Notice the apparent pattern; in each case the coefficient of d is 1 less than the number of the term. For example, the 100th term is

$$a_{100} = a_1 + 99d$$

This pattern leads to the following formula:

The nth Term of an Arithmetic Sequence

> The nth term of an arithmetic sequence with first term a_1 and common difference d is
> $$a_n = a_1 + (n - 1)d$$

The next two examples illustrate some of the uses of this formula.

EXAMPLE 2 Finding the nth Term of an Arithmetic Sequence

Find the nth term of the arithmetic sequence

$$23, 19, 15, 11, \ldots$$

and simplify. Write the sequence in the form $a_1, a_2, a_3, \ldots, a_n, \ldots$.

SOLUTION
The common difference is

$$d = a_2 - a_1 = 19 - 23 = -4$$

Using the formula for the nth term of an arithmetic sequence, we get

$$a_n = a_1 + (n - 1)d$$
$$= 23 + (n - 1)(-4) \quad \text{Replace } a_1 \text{ with 23 and } d \text{ with } -4.$$
$$= 27 - 4n$$

Therefore, $a_n = 27 - 4n$, and the sequence is $23, 19, 15, \ldots, 27 - 4n, \ldots$.

EXAMPLE 3   Finding the Number of Terms of an Arithmetic Sequence

Find the number of terms of the finite arithmetic sequence

$$-2, 1, 4, 7, \ldots, 43$$

SOLUTION

First, we determine that $a_1 = -2$ and $d = 3$. The number of terms is the value of n such that $a_n = 43$. Substituting these values into the formula for the nth term of an arithmetic sequence yields

$$a_n = a_1 + (n - 1)d$$
$$43 = -2 + (n - 1)3$$
$$43 = -5 + 3n \qquad \text{Distribute 3 and add like terms.}$$
$$16 = n \qquad \text{Solve for } n.$$

There are 16 terms in the sequence.

The Graph of an Arithmetic Sequence

Recall from Section 7.1 that the ordered pairs for the sequence $a_1, a_2, a_3, \ldots, a_n, \ldots$ are $(1, a_1), (2, a_2), (3, a_3), \ldots, (n, a_n), \ldots$. The graph of an arithmetic sequence can give us additional insight into its characteristics and properties.

EXAMPLE 4   Graphing Arithmetic Sequences

Plot the first five ordered pairs of each arithmetic sequence.

(a) $-\frac{1}{2}, 0, \frac{1}{2}, 1, 1\frac{1}{2}, \ldots$ **(b)** $11, 8, 5, 2, -1, \ldots$

SOLUTION

(a) The first five ordered pairs are $\left(1, -\frac{1}{2}\right), (2, 0), \left(3, \frac{1}{2}\right), (4, 1)$, and $\left(5, 1\frac{1}{2}\right)$, as shown in Figure 7.15.

(b) The first five ordered pairs, $(1, 11), (2, 8), (3, 5), (4, 2)$, and $(5, -1)$, are shown in Figure 7.16.

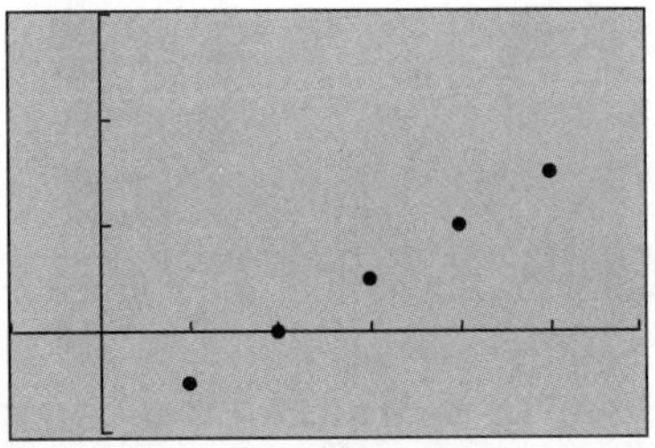

Figure 7.15
$[-1, 6] \times [-1, 3]$

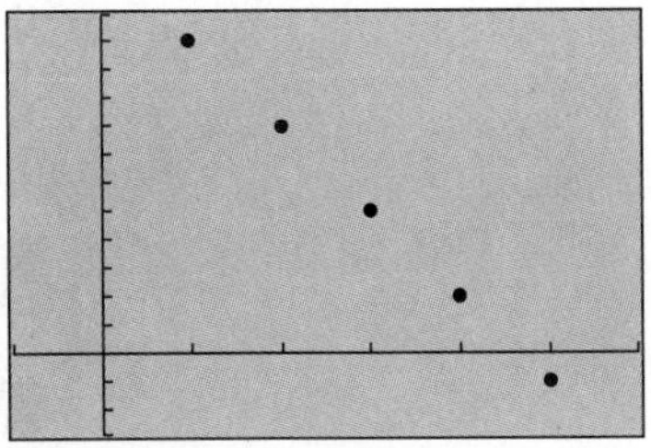

Figure 7.16
$[-1, 6] \times [-3, 12]$

The graphs in Example 4 suggest that the graph of an arithmetic sequence is a set of collinear points. In fact, the expression for the nth term of an arithmetic sequence corresponds to a linear function in point–slope form. For example, the nth term of the sequence in part (a) of Example 4 is

$$a_n = -\tfrac{1}{2} + (n - 1)\tfrac{1}{2} = \tfrac{1}{2}(n - 1) - \tfrac{1}{2}$$

which corresponds to the point–slope form of the linear function

$$f(x) = \tfrac{1}{2}(x - 1) - \tfrac{1}{2}$$

Similarly, the nth term of the sequence in part (b) of Example 4,

$$a_n = 11 + (n - 1)(-3) = -3(n - 1) + 11$$

corresponds to the point–slope form of the linear function

$$f(x) = -3(x - 1) + 11$$

In general, an arithmetic sequence corresponds to a linear function whose graph contains the point (k, a_k) and has slope equal to the common difference of the sequence.

EXAMPLE 5 Finding the nth Term of an Arithmetic Sequence

The common difference of an arithmetic sequence is -4, and the sixth term is 9. Find an expression for the nth term.

SOLUTION

The line corresponding to the arithmetic sequence has a slope that is equal to the common difference of the sequence, -4. The corresponding line also contains the ordered pair $(6, 9)$, since the sixth term of the sequence is 9. Recall the point–slope form of a linear function from Section 2.1:

$$f(x) = m(x - h) + k$$

Substituting $(6, 9)$ for (h, k) and -4 for m, we have

$$f(x) = -4(x - 6) + 9$$
$$= -4x + 33$$

Therefore, the nth term of the arithmetic sequence is $a_n = -4n + 33$.

EXAMPLE 6 Finding the Common Difference and nth Term

The third term of an arithmetic sequence is 10, and the eleventh term is 22. Find the common difference and an expression for the nth term.

SOLUTION

We are given the ordered pairs $(3, 10)$ and $(11, 22)$. Since the sequence is arithmetic, the ordered pairs are points on a line with slope

$$\frac{\Delta y}{\Delta x} = \frac{22 - 10}{11 - 3} = \frac{3}{2}$$

Again, we use the point–slope form of a linear function, $f(x) = m(x - h) + k$. If we replace (h, k) with $(3, 10)$ and m with $\frac{3}{2}$, we get

$$f(x) = \frac{3}{2}(x - 3) + 10$$
$$= \frac{3}{2}x + \frac{11}{2}$$

Thus, the nth term of the arithmetic sequence is $a_n = \frac{3}{2}n + \frac{11}{2}$. The common difference is the slope of the line, so $d = \frac{3}{2}$.

The Sum of an Arithmetic Sequence

We can apply Karl Gauss's method of finding $1 + 2 + 3 + \cdots + 100$ (given in Section 7.2) to determine the sum of the terms of any finite arithmetic sequence. As we discussed earlier, the first n terms of an arithmetic sequence can be written as

$$a_1, a_1 + d, a_1 + 2d, a_1 + 3d, \ldots, a_1 + (n - 1)d$$

Let S_n represent the sum of these n terms:

$$S_n = a_1 + [a_1 + d] + [a_1 + 2d] + \cdots + [a_1 + (n - 1)d]$$

Reversing the order of addition and then adding the equations vertically, we have

$$
\begin{aligned}
S_n &= a_1 &+ [a_1 + d] &+ [a_1 + 2d] &+ \cdots + [a_1 + (n - 1)d] \\
S_n &= [a_1 + (n - 1)d] &+ [a_1 + (n - 2)d] &+ [a_1 + (n - 3)d] &+ \cdots + a_1 \\
\hline
2S_n &= [2a_1 + (n - 1)d] &+ [2a_1 + (n - 1)d] &+ [2a_1 + (n - 1)d] &+ \cdots + [2a_1 + (n - 1)d]
\end{aligned}
$$

Since the right side of the last equation has the term $2a_1 + (n - 1)d$, n times, we have

$$2S_n = n[2a_1 + (n - 1)d]$$

Dividing each side by 2 gives

$$S_n = \frac{n}{2}[2a_1 + (n - 1)d]$$

Using the fact that $a_n = a_1 + (n - 1)d$, we also can write our result as

$$S_n = \frac{n}{2}(a_1 + a_n)$$

Sum of an Arithmetic Sequence

> The sum S_n of the first n terms of an arithmetic sequence with first term a_1 and common difference d is
>
> $$S_n = \frac{n}{2}(a_1 + a_n)$$
>
> An alternative form is
>
> $$S_n = \frac{n}{2}[2a_1 + (n - 1)d]$$

EXAMPLE 7   Finding the Sum of a Given Number of Terms of an Arithmetic Sequence

Find the sum of the first 80 terms of the arithmetic sequence 4, 7, 10, 13, ...

SOLUTION
The first term is $a_1 = 4$, and the common difference is $d = 7 - 4 = 3$. Since we want to know the sum of the first 80 terms, we have $n = 80$. Substituting these values into the alternate formula for the sum of an arithmetic sequence, we get

$$S_n = \frac{n}{2}[2a_1 + (n - 1)d] = \frac{80}{2}[2(4) + (80 - 1)3] = 9800$$

EXAMPLE 8   Finding the Sum of an Arithmetic Sequence

The terms of the sum $1 + 6 + 11 + \cdots + 91$ are the terms of a finite arithmetic sequence. Find the sum.

SOLUTION

For the arithmetic sequence 1, 6, 11, ... , 91, we identify the first term $a_1 = 1$ and common difference $d = 6 - 1 = 5$. We first determine the number of terms, which is the value of n such that $a_n = 91$. Using the formula for the nth term of an arithmetic sequence, we get

$$a_n = a_1 + (n - 1)d$$
$$91 = 1 + (n - 1)5$$
$$91 = -4 + 5n \qquad \text{Distribute 5 and add like terms.}$$
$$19 = n \qquad \text{Solve for } n.$$

By the formula for the sum of an arithmetic sequence, we have

$$S_n = \frac{n}{2}(a_1 + a_n)$$

$$S_{19} = \frac{19}{2}(a_1 + a_{19}) \quad \text{Replace } n \text{ with 19.}$$

$$= \frac{19}{2}(1 + 91) \quad \text{Replace } a_1 \text{ with 1 and } a_{19} \text{ with 91.}$$

$$= 874$$

Applications

In Sections 7.1 and 7.2, we have seen applied problems whose solutions lead to various sequences or series. We finish this section with an application in which we compare two arithmetic series.

EXAMPLE 9   Application: Selecting an Employer

You must decide upon one of two job offers, each having similar benefits and working conditions, but different salary schedules. Company A offers a starting salary of $34,000 per year with an annual raise of $1300. Company B offers $32,000 per year with a raise of $1600 per year. Determine the total salary that would be earned at each company over the next 10 years.

SOLUTION

The annual salary at company A for each year is given by the arithmetic sequence

$$34{,}000, \; 34{,}000 + 1300, \; 34{,}000 + 2(1300), \; 34{,}000 + 3(1300), \ldots$$

The total earned in 10 years at company A is the sum of the first ten terms of the arithmetic sequence. Substituting $a_1 = 34{,}000$, $d = 1300$, and $n = 10$ into the alternative formula for the sum of an arithmetic sequence, we get

$$S_n = \frac{n}{2}[2a_1 + (n-1)d]$$

$$= \frac{10}{2}[2(34{,}000) + (10-1)1300] = 398{,}500$$

Similarly, the total earned in 10 years at company B is the sum of the first ten terms of the arithmetic sequence

$$32{,}000, \ 32{,}000 + 1600, \ 32{,}000 + 2(1600), \ 32{,}000 + 3(1600), \ \ldots$$

Substituting $a_1 = 32{,}000$, $d = 1600$, and $n = 10$ into the alternative formula for the sum of an arithmetic sequence gives

$$S_n = \frac{n}{2}[2a_1 + (n-1)d]$$

$$= \frac{10}{2}[2(32{,}000) + (10-1)1600] = 392{,}000$$

The total salary earned at company A would be \$398,500 and the total salary earned at company B would be \$392,000.

EXERCISE SET 7.3

A

In Exercises 1–14, determine whether the terms given are the first five terms of an arithmetic sequence. If so, specify the common difference and the sixth term.

1. 5, 11, 17, 23, 29, ...
2. 10, 13, 16, 19, 22, ...
3. 2.1, 2.3, 2.5, 2.7, 2.9, ...
4. 4.5, 8, 11.5, 15, 18.5, ...
5. 19, 14, 9, 4, −1, ...
6. 23, 14, 5, −4, −13, ...
7. 1, 4, 9, 16, 25, ...
8. 2, 4, 8, 16, 32, ...
9. 4, 4, 4, 4, 4, ...
10. 7, −7, 7, −7, 7, ...
11. 6, 3, 6, 3, 6, ...
12. 2, −4, 6, −8, 10, ...
13. 1, 3, 5, 7, 10, ...
14. 3, 6, 9, 12, 14, ...

In Exercises 15–20, find the number of terms of each finite arithmetic sequence.

15. −3, −1, 1, 3, ... , 37
16. −10, −6, −2, 2, ... , 46
17. 51, 45, 39, 33, ... , −123
18. 122, 114, 106, 98, ... , −86
19. 9, $13\frac{1}{3}$, $17\frac{2}{3}$, 22, ... , 100
20. 2, $2\frac{5}{6}$, $3\frac{2}{3}$, $4\frac{1}{2}$, ... , 22

In Exercises 21–26, plot the first four ordered pairs of each arithmetic sequence.

21. −3, 0, 3, 6, ...
22. −2, −1, 0, 1, ...
23. 1, $2\frac{1}{3}$, $3\frac{2}{3}$, 5, ...
24. 4, $4\frac{1}{2}$, 5, $5\frac{1}{2}$, ...
25. 6.5, 5, 3.5, 2, ...
26. 4.4, 3.8, 3.2, 2.6, ...

B

In Exercises 27–42, find an expression for the nth term of each arithmetic sequence and simplify. Write the sequence in the form $a_1, a_2, a_3, \ldots, a_n, \ldots$.

27. 4, 7, 10, 13, ...
28. 12, 17, 22, 27, ...
29. 33, 29, 25, 21, ...
30. 28, 22, 16, 10, ...

31. $36, 34\frac{2}{3}, 33\frac{1}{3}, 32, \ldots$

32. $20, 23.5, 27, 30.5, \ldots$

33. The sequence with common difference 8 and fifth term 20

34. The sequence with common difference -3 and seventh term 12

35. The sequence with $d = -\frac{5}{4}$ and $a_9 = 17$

36. The sequence with $d = \frac{3}{5}$ and $a_{16} = 31$

37. The sequence whose seventh term is 16 and thirteenth term is 2

38. The sequence whose third term is 2 and tenth term is 8

39. The sequence with $a_2 = 2$ and $a_{14} = 38$

40. The sequence with $a_3 = 10$ and $a_{17} = 45$

41. $x + 1, -x + 3, -3x + 5, -5x + 7, \ldots$

42. $k + 1, 2k + 4, 3k + 7, 4k + 10, 5k + 13, \ldots$

In Exercises 43–48, find the sum of the first n terms of each arithmetic sequence.

43. $2, 5, 8, 11, \ldots ; n = 30$

44. $7, 9, 11, 13, \ldots ; n = 30$

45. $44, 39, 34, 29, \ldots ; n = 40$

46. $56, 50, 44, 38, \ldots ; n = 40$

47. $3, 3.25, 3.5, 3.75, \ldots ; n = 60$

48. $1.6, 1.9, 2.2, 2.5, \ldots ; n = 60$

49. The terms of the sum $1 + 4 + 7 + \cdots + 58$ are the terms of a finite arithmetic sequence. Find the sum.

50. The terms of the sum $2 + 9 + 16 + \cdots + 72$ are the terms of a finite arithmetic sequence. Find the sum.

51. Find the sum of the even integers from 2 to 300, inclusive.

52. Find the sum of the odd integers from 1 to 199, inclusive.

53. Find the sum

$$\sum_{k=1}^{20} (4k + 3)$$

using one of the formulas for the sum of an arithmetic sequence. (Compare your result to the answer you obtained for Exercise 35, Section 7.2.)

54. Find the sum

$$\sum_{k=1}^{30} (3k - 2)$$

using one of the formulas for the sum of an arithmetic sequence. (Compare your result to the answer you obtained for Exercise 36, Section 7.2.)

55. The life expectancy in the United States is approximated by the arithmetic sequence

$$a_n = 0.2n + 69.4$$

where a_n is the life expectancy of a person born during the nth year after 1960, with $n = 1$ corresponding to 1961.
   (a) Determine a_1 and d.
   (b) Use the sequence to approximate the life expectancy of a person born in the United States in 1980.
   (c) Use the sequence to approximate the life expectancy of a person born in the year 2000.

56. The weekly earnings of production workers in the United States can be approximated by the arithmetic sequence

$$a_n = 11.8n + 112.1$$

where a_n is the average weekly earnings during the nth year after 1970, with $n = 1$ corresponding to 1971.
   (a) Determine a_1 and d.
   (b) Use the sequence to approximate the weekly earnings of a U.S. production worker in 1985.
   (c) Use the sequence to approximate the weekly earnings of a U.S. production worker in the year 2000.

57. According to the *consumer price index*, $3.15 had the same purchasing power in 1993 as $1.00 in 1974. The amount a_n of money n years after 1973 that is equivalent to $1.00 in 1974 can be approximated by the arithmetic sequence

$$a_n = 0.113(n - 1) + 1.00$$

with $n = 1$ corresponding to 1974.
   (a) Determine a_1 and d.
   (b) Use the sequence to determine the amount in 1990 that is equivalent to $1.00 in 1974.
   (c) Use the sequence to predict the amount in 1999 that is equivalent to $1.00 in 1974.

58. A stack of sprinkler pipes has 20 pipes in the first (bottom) row, 19 pipes in the second, 18 in the third row, and so on:

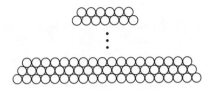

How many pipes are in the stack if there are 6 pipes in the last (top) stack?

59. The seats in auditoriums and theaters are often arranged so that there are more seats per row in the rows toward the back. Suppose a theater has 22 seats in the first row, 24 seats in the second row, 26 in the third row, and so on:

Third row
Second row
First row

Find the seating capacity if there are 34 rows of seats in the theater.

60. A drug store chain pays new managers a starting salary of $36,500, with annual increases of $1750 for the next 6 years.
   (a) What is the annual salary during the fifth year of employment?
   (b) What is the total salary earned during the first 5 years of employment?

61. A firm offers a starting salary of $31,500, with annual increases of $1450 for the next 8 years.
   (a) What is the annual salary during the sixth year of employment?

(b) What is the total salary earned during the first 6 years of employment?

62. The personnel manager has been instructed to design a salary schedule with fixed annual raises for a new position at the company. The personnel manager determines that a competitive starting salary is $28,000 per year. In order to keep an employee that is hired now, he projects that the company must pay that employee $34,600 per year during the fifth year of employment.
   (a) Determine a formula for the annual salary during the nth year of employment.
   (b) Determine the total compensation earned by the employee over the first 8 years of employment.

C

63. The sum of the first ten terms of an arithmetic sequence is 180. The sum of the first twelve terms is 204. Determine a_1 and d.

64. How many numbers are there between 1 and 1001 that are multiples of 2, or multiples of 3, or both?
   HINT: This is not the same as the number of multiples of 6 between 1 and 1001.

65. Let a_n and b_n be two arithmetic sequences.
   (a) Prove or disprove that the sequence $c_n = a_n + b_n$ is an arithmetic sequence.
   (b) Prove or disprove that the sequence $c_n = a_n b_n$ is an arithmetic sequence.

66. Find the sum: $1 + 2 + 4 + 5 + 7 + 8 + 10 + 11 + 13 + \cdots + 100 + 101$, where every third integer is missing.
   HINT: Group the terms in pairs.

SECTION 7.4

- Geometric Sequences
- The Graph of a Geometric Sequence
- The Sum of a Geometric Sequence
- Other Applications

GEOMETRIC SEQUENCES AND SERIES

In this section we will discuss geometric sequences, following an outline similar to that of our discussion of arithmetic sequences. We will define a geometric sequence, derive a formula for the nth term, consider the graph, and develop a convenient formula for the sum of any number of terms of a geometric sequence.

Geometric Sequences

Each term in the sequence

$$1, 5, 25, 125, 625, \ldots$$

is 5 times the previous term. In other words, the ratio of consecutive terms is 5:

$$\frac{a_2}{a_1} = \frac{5}{1} = 5$$

$$\frac{a_3}{a_2} = \frac{25}{5} = 5$$

$$\frac{a_4}{a_3} = \frac{125}{25} = 5$$

and so on.

The sequence 1, 5, 25, 125, 625, … is an example of a *geometric sequence*. This sequence is characterized by the feature that the ratio of consecutive terms is constant.

Geometric Sequence

> A sequence in which consecutive terms have a constant ratio is called a **geometric sequence**. More precisely, a sequence
>
> $$a_1, a_2, a_3, \ldots, a_n, \ldots$$
>
> is geometric if there exists a constant r such that
>
> $$\frac{a_{n+1}}{a_n} = r$$
>
> for every positive integer n. The constant r is called the **common ratio.**

■ EXAMPLE 1   Identifying Geometric Sequences

Determine whether the terms given are the first five terms of a geometric sequence. If so, specify the common ratio and the sixth term.

(a) 3, 6, 12, 24, 48, …   **(b)** 4, 10, 20, 50, 125, …

(c) 432, − 72, 12, − 2, $\frac{1}{3}$, …

SOLUTION

(a) We compute the ratios of consecutive terms:

$$\frac{a_2}{a_1} = \frac{6}{3} = 2$$

$$\frac{a_3}{a_2} = \frac{12}{6} = 2$$

$$\frac{a_4}{a_3} = \frac{24}{12} = 2$$

$$\frac{a_5}{a_4} = \frac{48}{24} = 2$$

Since the ratio of consecutive terms is constant, the given terms are the first five terms of a geometric sequence. The common ratio is 2. Since $a_6 / a_5 = 2$, the sixth term is 2 times the fifth term:

$$a_6 = 2(48) = 96$$

(b) Computing the ratios of consecutive terms, we get

$$\frac{a_2}{a_1} = \frac{10}{4} = \frac{5}{2}$$

$$\frac{a_3}{a_2} = \frac{20}{10} = 2$$

$$\frac{a_4}{a_3} = \frac{50}{20} = \frac{5}{2}$$

$$\frac{a_5}{a_4} = \frac{125}{50} = \frac{5}{2}$$

The ratios are not the same in each case, so the given terms are not the first five terms of a geometric sequence.

(c) The ratios of the given consecutive terms are

$$\frac{a_2}{a_1} = \frac{-72}{432} = -\frac{1}{6}$$

$$\frac{a_3}{a_2} = \frac{12}{-72} = -\frac{1}{6}$$

$$\frac{a_4}{a_3} = \frac{-2}{12} = -\frac{1}{6}$$

$$\frac{a_5}{a_4} = \frac{\frac{1}{3}}{-2} = -\frac{1}{6}$$

The ratio of consecutive terms is constant, so the given terms are the first five terms of a geometric sequence. The common ratio is $-\frac{1}{6}$, so the sixth term is $-\frac{1}{6}$ times the fifth term:

$$a_6 = -\frac{1}{6}\left(\frac{1}{3}\right) = -\frac{1}{18}$$

Each term of a geometric sequence can be expressed in terms of the first term a_1 and common ratio r as follows:

$$a_1 = a_1$$
$$a_2 = a_1 r$$
$$a_3 = a_2 r = (a_1 r)r = a_1 r^2$$
$$a_4 = a_3 r = (a_1 r^2)r = a_1 r^3$$
$$a_5 = a_4 r = (a_1 r^3)r = a_1 r^4$$

and so on. In each case the exponent of r is 1 less than the number of the term. For example,

$$a_{100} = a_1 r^{99}$$

This pattern leads to the following formula:

The nth Term of a Geometric Sequence

> The nth term of a geometric sequence with first term a_1 and common ratio r is
>
> $$a_n = a_1 r^{n-1}$$

The next two examples illustrate some of the uses of this formula.

EXAMPLE 2 Finding the *n*th Term of a Geometric Sequence

Find an expression for the nth term of each geometric sequence. Write the sequence in the form $a_1, a_2, a_3, \ldots, a_n, \ldots$.

(a) 8, 12, 18, 27, ...
(b) The common ratio is 3, and the sixth term is $\frac{81}{2}$.

SOLUTION
(a) We are given that the sequence is geometric, so the common ratio can be determined by taking the ratio of any two consecutive terms:

$$r = \frac{a_2}{a_1} = \frac{12}{8} = \frac{3}{2}$$

Using the formula for the nth term of a geometric sequence, we get

$$a_n = a_1 r^{n-1}$$
$$= 8\left(\tfrac{3}{2}\right)^{n-1} \quad \text{Replace } a_1 \text{ with 8 and } r \text{ with } \tfrac{3}{2}.$$

Thus, $a_n = 8\left(\tfrac{3}{2}\right)^{n-1}$, and the sequence is $8, 12, 18, \ldots, 8\left(\tfrac{3}{2}\right)^{n-1}, \ldots$.

(b) Starting with the formula for the nth term of a geometric sequence, we get

$$a_n = a_1 r^{n-1}$$
$$= a_1 (3)^{n-1} \quad \text{Replace } r \text{ with 3.}$$

We need to determine a_1. Since the sixth term is given, we let $n = 6$ in $a_n = a_1 (3)^{n-1}$ to get

$$a_6 = a_1 (3)^{6-1}$$
$$\frac{81}{2} = a_1 (3)^5 \quad \text{Replace } a_6 \text{ with } \tfrac{81}{2}.$$
$$\frac{81}{2} \cdot \frac{1}{3^5} = a_1 \quad \text{Multiply each side by } 1/3^5.$$
$$\frac{1}{6} = a_1$$

Therefore, an expression for the nth term is $a_n = \tfrac{1}{6}(3)^{n-1}$. The sequence is

$$\tfrac{1}{6}, \tfrac{1}{2}, \tfrac{3}{2}, \ldots, \tfrac{1}{6}(3)^{n-1}, \ldots$$

EXAMPLE 3 Finding the Common Ratio and nth Term

The second term of a geometric sequence is -6, and the fifth term is $\tfrac{3}{4}$. Find the common ratio and an expression for the nth term.

SOLUTION
Since we are given the second and fifth terms, we let $n = 2$ and $n = 5$ in $a_n = a_1 r^{n-1}$:

$$a_2 = a_1 r^{2-1} \quad \text{and} \quad a_5 = a_1 r^{5-1}$$

Replacing a_2 with -6 and a_5 with $\tfrac{3}{4}$ in these equations leads to the system

$$\begin{cases} -6 = a_1 r \\ \tfrac{3}{4} = a_1 r^4 \end{cases}$$

Multiplying the first equation in the system by r^3, we get

$$\begin{cases} -6r^3 = a_1 r^4 \\ \tfrac{3}{4} = a_1 r^4 \end{cases}$$

These two equations imply

$$-6r^3 = \tfrac{3}{4}$$
$$r^3 = \tfrac{3}{4}\left(-\tfrac{1}{6}\right) = -\tfrac{1}{8}$$

Thus,

$$r = \sqrt[3]{-\tfrac{1}{8}} = -\tfrac{1}{2}$$

To determine an expression for the nth term, we need to know a_1. We can determine a_1 by substituting $-\tfrac{1}{2}$ for r in the equation $-6 = a_1 r$ and then solving for a_1:

$$-6 = a_1\left(-\tfrac{1}{2}\right)$$
$$12 = a_1$$

Therefore, an expression for the nth term is

$$a_n = 12\left(-\tfrac{1}{2}\right)^{n-1}$$

The Graph of a Geometric Sequence

In Section 7.3 we noted that an arithmetic sequence corresponds to a linear function. Graphing geometric sequences will help you see that a geometric sequence corresponds to one or two exponential functions.

EXAMPLE 4   Graphing Geometric Sequences

Plot the first four ordered pairs of each geometric sequence in the same viewing window as the indicated function(s) of x.

(a) $a_n = \left(\tfrac{2}{3}\right)^n$; $f(x) = \left(\tfrac{2}{3}\right)^x$

(b) $a_n = \left(-\tfrac{1}{2}\right)^n$; $f(x) = \left(\tfrac{1}{2}\right)^x$, $g(x) = -\left(\tfrac{1}{2}\right)^x$

SOLUTION

(a) Substituting $n = 1, 2, 3,$ and 4 into $a_n = \left(\tfrac{2}{3}\right)^n$, we get

$$a_1 = \left(\tfrac{2}{3}\right)^1 = \tfrac{2}{3}$$
$$a_2 = \left(\tfrac{2}{3}\right)^2 = \tfrac{4}{9}$$
$$a_3 = \left(\tfrac{2}{3}\right)^3 = \tfrac{8}{27}$$
$$a_4 = \left(\tfrac{2}{3}\right)^4 = \tfrac{16}{81}$$

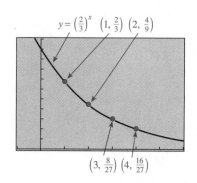

$y = \left(\tfrac{2}{3}\right)^x$ $\left(1, \tfrac{2}{3}\right)$ $\left(2, \tfrac{4}{9}\right)$

$\left(3, \tfrac{8}{27}\right)$ $\left(4, \tfrac{16}{27}\right)$

Figure 7.17
$[-1, 6] \times [0, 1.1]_{0.1}$

The first four points of the sequence are $\left(1, \tfrac{2}{3}\right)$, $\left(2, \tfrac{4}{9}\right)$, $\left(3, \tfrac{8}{27}\right)$, and $\left(4, \tfrac{16}{81}\right)$. These points and the graph of $f(x) = \left(\tfrac{2}{3}\right)^x$ are shown in Figure 7.17.

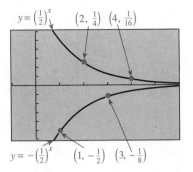

$y = \left(\frac{1}{2}\right)^x$ $\left(2, \frac{1}{4}\right)$ $\left(4, \frac{1}{16}\right)$

$y = -\left(\frac{1}{2}\right)^x$ $\left(1, -\frac{1}{2}\right)$ $\left(3, -\frac{1}{8}\right)$

Figure 7.18
$[-1, 6] \times [-0.6, 0.6]_{0.1}$

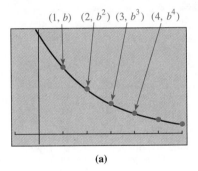

$(1, b)$ $(2, b^2)$ $(3, b^3)$ $(4, b^4)$

(a)

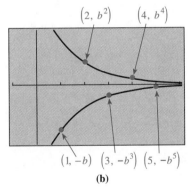

$\left(2, b^2\right)$ $\left(4, b^4\right)$

$\left(1, -b\right)$ $\left(3, -b^3\right)$ $\left(5, -b^5\right)$

(b)

Figure 7.19

(b) Substituting $n = 1, 2, 3$, and 4 into $a_n = \left(-\frac{1}{2}\right)^n$, we get

$$a_1 = \left(-\tfrac{1}{2}\right)^1 = -\tfrac{1}{2}$$
$$a_2 = \left(-\tfrac{1}{2}\right)^2 = \tfrac{1}{4}$$
$$a_3 = \left(-\tfrac{1}{2}\right)^3 = -\tfrac{1}{8}$$
$$a_4 = \left(-\tfrac{1}{2}\right)^4 = \tfrac{1}{16}$$

The first four points of the sequence are $\left(1, -\frac{1}{2}\right)$, $\left(2, \frac{1}{4}\right)$, $\left(3, -\frac{1}{8}\right)$, and $\left(4, \frac{1}{16}\right)$. These points and the graphs of $f(x) = \left(\frac{1}{2}\right)^x$ and $g(x) = -\left(\frac{1}{2}\right)^x$ are shown in Figure 7.18.

The graphs in Example 4 point out the behavior of b^n and $(-b)^n$ for values of b between 0 and 1. From Section 4.1, we know that the exponential functions $f(x) = b^x$ and $g(x) = -b^x$, $0 < b < 1$, approach 0 as x grows without bound. The ordered pairs of the sequence $a_n = b^n$ lie on the graph of $f(x) = b^x$, so the sequence $a_n = b^n$ approaches 0 as n grows without bound (Figure 7.19a).

The ordered pairs of the sequence $a_n = (-b)^n$ lie alternately on the graphs of $f(x) = b^x$ and $g(x) = -b^x$, implying that the sequence $a_n = (-b)^n$ also approaches 0 as n grows without bound (Figure 7.19b).

If we let r represent b or $-b$, then we have the following fact:

$$\text{If } -1 < r < 1, \text{ then } r^n \to 0 \text{ as } n \to +\infty.$$

We will need this observation when we discuss the sum of an infinite number of terms of a geometric sequence.

The Sum of a Geometric Sequence

Just as with arithmetic sequences, there is a convenient formula for adding the first n terms of a geometric sequence. The first n terms of a geometric sequence with first term a_1 and common ratio r are

$$a_1, a_1 r, a_1 r^2, a_1 r^3, \ldots, a_1 r^{n-2}, a_1 r^{n-1}$$

The sum of these terms is

$$S_n = a_1 + a_1 r + a_1 r^2 + a_1 r^3 + \cdots + a_1 r^{n-2} + a_1 r^{n-1}$$

If we multiply each side of this equation by $-r$, we get

$$-rS_n = -a_1 r - a_1 r^2 - a_1 r^3 - a_1 r^4 - \cdots - a_1 r^{n-1} - a_1 r^n$$

Adding these expressions for S_n and $-rS_n$, we get

$$S_n = a_1 + a_1r + a_1r^2 + a_1r^3 + \cdots + a_1r^{n-2} + a_1r^{n-1}$$
$$-rS_n = \quad\quad -a_1r - a_1r^2 - a_1r^3 - \cdots - a_1r^{n-2} - a_1r^{n-1} - a_1r^n$$
$$S_n - rS_n = a_1 + 0 \quad + 0 \quad + 0 \quad + \cdots + 0 \quad\quad + 0 \quad\quad - a_1r^n$$

Therefore,

$$S_n(1 - r) = a_1 - a_1r^n = a_1(1 - r^n)$$

If we divide each side by $1 - r$, we have the following result:

Sum of a Geometric Sequence

> The sum S_n of the first n terms of a geometric sequence with first term a_1 and common ratio r is
>
> $$S_n = \frac{a_1(1 - r^n)}{1 - r} \qquad r \neq 1$$

EXAMPLE 5   Finding the Sum of a Geometric Sequence

Find the sum of the first ten terms of the geometric sequence

$$16, 24, 36, 54, \ldots$$

SOLUTION

The common ratio is $\frac{3}{2}$. Substituting $a_1 = 16$, $r = \frac{3}{2}$, and $n = 10$ into

$$S_n = \frac{a_1(1 - r^n)}{1 - r}$$

we get

$$S_{10} = \frac{16\left[1 - \left(\frac{3}{2}\right)^{10}\right]}{1 - \frac{3}{2}} = \frac{-\frac{58{,}025}{64}}{-\frac{1}{2}} = \frac{58{,}025}{32}$$

EXAMPLE 6   Finding the Sum of a Geometric Sequence

Find each sum.

(a) $\displaystyle\sum_{k=1}^{7} 18\left(-\frac{1}{3}\right)^{k-1}$ **(b)** $\displaystyle\sum_{k=1}^{10} (1.06)^{k-1}$

SOLUTION

(a) Writing out the terms of the summation, we have

$$\sum_{k=1}^{7} 18\left(-\frac{1}{3}\right)^{k-1} = 18 - 6 + 2 - \cdots + \frac{2}{81}$$

The terms of the sum are the first seven terms of the geometric sequence with $a_1 = 18$ and $r = -\frac{1}{3}$. Substituting these values into the formula for the sum of a geometric sequence yields

$$S_7 = \frac{18\left[1 - \left(-\frac{1}{3}\right)^7\right]}{1 - \left(-\frac{1}{3}\right)} = \frac{\frac{4376}{243}}{\frac{4}{3}} = \frac{1094}{81}$$

(b) Expanding the summation, we have

$$\sum_{k=1}^{10} (1.06)^{k-1} = 1 + 1.06 + 1.06^2 + \cdots + 1.06^9$$

The terms are from a geometric sequence with $a_1 = 1$ and $r = 1.06$, so the sum is

$$S_{10} = \frac{1(1 - 1.06^{10})}{1 - 1.06} \approx 13.181$$

Consider the sum

$$S_n = \sum_{k=1}^{n} 3(0.1)^k = 0.3 + 0.03 + 0.003 + \cdots + 3(0.1)^n$$

The first few values of the sequence S_n are

$$S_1 = 0.3$$
$$S_2 = 0.3 + 0.03 = 0.33$$
$$S_3 = 0.3 + 0.03 + 0.003 = 0.333$$

As n grows without bound, it seems that S_n approaches the repeating decimal $0.\overline{3}$. Recognizing that $0.\overline{3}$ is the decimal representation for $\frac{1}{3}$ suggests that

$$S_n \to \frac{1}{3} \qquad \text{as} \qquad n \to +\infty$$

In summation notation, this behavior is denoted by

$$\sum_{k=1}^{\infty} 3(0.1)^k = \frac{1}{3}$$

and we say that this is the *sum* of the *infinite geometric sequence*

$$0.3, 0.03, 0.003, \ldots, 3(0.1)^n, \ldots$$

Given the infinite geometric sequence

$$a_1, a_1r, a_1r^2, a_1r^3, \ldots, a_1r^{n-1}, \ldots$$

we know the sum of the first n terms is

$$S_n = \frac{a_1(1 - r^n)}{1 - r} \qquad r \neq 1$$

We noted earlier that if $-1 < r < 1$, then r^n approaches 0 as n grows without bound. Thus,

$$S_n \rightarrow \frac{a_1(1 - 0)}{1 - r} = \frac{a_1}{1 - r} \qquad \text{as} \qquad n \rightarrow +\infty$$

The following is a summary of our results.

Sum of an Infinite
Geometric Sequence

> If $-1 < r < 1$, then the infinite geometric sequence
>
> $$a_1, a_1r, a_1r^2, a_1r^3 \ldots, a_1r^{n-1}, \ldots$$
>
> has the sum
>
> $$\sum_{k=1}^{\infty} a_1 r^{k-1} = \frac{a_1}{1 - r}$$
>
> If $r \geq 1$ or $r \leq -1$, the infinite geometric sequence does not have a sum.

EXAMPLE 7   The Sum of an Infinite Geometric Sequence

Find the indicated sum of each infinite geometric sequence, if it exists.

(a) $10 + 4 + 1.6 + 0.64 + \cdots$ **(b)** $4 - 6 + 9 - \frac{27}{2} + \cdots$

SOLUTION

(a) The common ratio is $r = \frac{4}{10} = \frac{2}{5}$. Since $-1 < r < 1$, the sum is

$$\frac{a_1}{1 - r} = \frac{10}{1 - \frac{2}{5}} = \frac{50}{3}$$

(b) The common ratio is $r = \frac{-6}{4} = -\frac{3}{2}$. Since $r \leq -1$, the infinite geometric sequence does not have a sum.

EXAMPLE 8   Application: Distance a Pendulum Swings

The length of the initial swing of the tip of a pendulum is 24 centimeters. If the length of each swing is $\frac{7}{8}$ as long as the previous swing, how far does the tip travel before coming to rest?

SOLUTION

The tip travels 24 centimeters through the first swing, $24\left(\frac{7}{8}\right)$ centimeters through the second swing, $24\left(\frac{7}{8}\right)^2$ centimeters through the third swing, and so on. The pendulum will swing an indefinite number of times before eventually coming to rest. The total distance traveled is

$$24 + 24\left(\tfrac{7}{8}\right) + 24\left(\tfrac{7}{8}\right)^2 + \cdots + 24\left(\tfrac{7}{8}\right)^{n-1} + \cdots$$

This series is the sum of an infinite geometric sequence with $r = \frac{7}{8}$ and $a_1 = 24$. Since $-1 < r < 1$, the sum is

$$\frac{a_1}{1-r} = \frac{24}{1-\frac{7}{8}} = 192$$

The pendulum travels 192 centimeters.

The sum of an infinite geometric sequence can be used to express the repeating decimal representation of a number as the ratio of two integers.

EXAMPLE 9   A Repeating Decimal As a Ratio of Integers

Express the repeating decimal as a ratio of two integers.

(a) $0.\overline{18}$ **(b)** $3.1\overline{6}$

SOLUTION

(a) The repeating decimal can expressed as

$$0.\overline{18} = 0.18181818... = 0.18 + 0.0018 + 0.000018 + \cdots$$

Thus, $0.\overline{18}$ is the sum of the infinite geometric sequence with $a_1 = 0.18$ and $r = 0.01$. Since $-1 < r < 1$, the sum is

$$\frac{a_1}{1-r} = \frac{0.18}{1-0.01} = \frac{0.18}{0.99} = \frac{18}{99} = \frac{2}{11}$$

Divide 2 by 11 on a calculator to check this result.

(b) The first two digits are not repeating, so we group them separately from the repeating block:

$$3.1\overline{6} = 3.1 + (0.06 + 0.006 + 0.0006 + \cdots)$$

The expression in parentheses is the sum of an infinite geometric sequence with $a_1 = 0.06$ and $r = 0.1$. Checking first that $-1 < r < 1$, we have

$$0.06 + 0.006 + 0.0006 + \cdots = \frac{a_1}{1-r} = \frac{0.06}{1-0.1} = \frac{0.06}{0.9} = \frac{1}{15}$$

Thus,

$$3.1\overline{6} = 3.1 + (0.06 + 0.006 + 0.0006 + \cdots)$$

$$= \frac{31}{10} + \frac{1}{15}$$

$$= \frac{19}{6}$$

You can check this result by dividing 19 by 6 on a calculator.

Other Applications

Many financial planners recommend making deposits into an interest-bearing account on a regular schedule. An account receiving periodic payments is called an *annuity*. The accumulated amount of an annuity can be determined using the formula for the sum of a (finite) geometric sequence.

EXAMPLE 10   Application: Monthly Payments into Savings

At the end of each month $100 is deposited into an account paying 6% compounded monthly. Determine the balance after:

(a) 4 months   **(b)** 3 years

SOLUTION

(a) Consider each payment as an individual investment. The $100 deposit made at the end of the first month will earn interest for 3 months. Using the formula

$$A = P\left(1 + \frac{r}{n}\right)^{nt}$$

the first deposit of $100 will grow to

$$100\left(1 + \frac{0.06}{12}\right)^{12(3/12)} = 100(1.005)^3$$

The deposit of $100 made at the end of the second month will earn interest for 2 months; therefore, the second $100 deposit will grow to

$$100\left(1 + \frac{0.06}{12}\right)^{12(2/12)} = 100(1.005)^2$$

Similarly, the third deposit of $100 will earn interest for 1 month, amounting to

$$100\left(1 + \frac{0.06}{12}\right)^{12(1/12)} = 100(1.005)$$

The fourth $100 deposit will earn no interest. Adding these amounts (starting with the fourth, then the third, second, and first), the balance after 4 months is

$$100 + 100(1.005) + 100(1.005)^2 + 100(1.005)^3$$

We could use the formula for the sum of a geometric sequence with $a_1 = 100$, $n = 4$, and $r = 1.005$. However, since there are only four terms, we choose to simply find the sum as written:

$$100 + 100(1.005) + 100(1.005)^2 + 100(1.005)^3 \approx 403.01$$

After 4 months, the balance will be $403.01.

(b) Our reasoning is the same as in our solution to part (a). The first $100 deposit earns interest for 35 months, the second $100 deposit earns interest for 34 months, and so on. The balance after 3 years is

$$100 + 100(1.005) + 100(1.005)^2 + \cdots + 100(1.005)^{35}$$

Using the formula for the sum of a geometric sequence with $a_1 = 100$, $n = 36$, and $r = 1.005$, we calculate the balance:

$$100 + 100(1.005) + 100(1.005)^2 + \cdots + 100(1.005)^{35} = \frac{a_1(1 - r^n)}{1 - r}$$

$$= \frac{100(1 - 1.005^{36})}{1 - 1.005}$$

$$\approx 3933.61$$

After 3 years, the balance will be $3933.61.

EXERCISE SET 7.4

A

In Exercises 1–12, determine whether the terms given are the first five terms of a geometric sequence. If so, specify the common ratio and the sixth term.

1. 5, 10, 20, 40, 80, ...

2. 2, 6, 18, 54, 162, ...

3. 16, 24, 36, 54, 81, ...

4. 162, 108, 72, 48, 32, ...

5. $\frac{2}{9}, -\frac{2}{3}, 2, -6, 18, \ldots$

6. $-\frac{5}{16}, \frac{5}{4}, -5, 20, -80, \ldots$

7. $12, 6, 2, 1, \frac{1}{2}, \ldots$

8. 144, 48, 16, 6, 2, ...

9. 5, 5, 5, 5, 5, ...

10. 8, 4, 8, 4, 8, ...

11. $\sqrt{3}, 3 + \sqrt{3}, 6 + 4\sqrt{3}, 18 + 10\sqrt{3}, 48 + 28\sqrt{3}, \ldots$

12. $3 + \sqrt{2}, 2 + 3\sqrt{2}, 6 + 2\sqrt{2}, 4 + 6\sqrt{2}, 12 + 4\sqrt{2}, \ldots$

In Exercises 13–18, plot the first five ordered pairs of each geometric sequence in the same viewing window as the graph of f (or f and g).

13. $a_n = \left(\frac{3}{4}\right)^n; f(x) = \left(\frac{3}{4}\right)^x$

14. $a_n = \left(\frac{2}{5}\right)^n; f(x) = \left(\frac{2}{5}\right)^x$

15. $a_n = 18\left(-\frac{1}{3}\right)^{n-1}; f(x) = 18\left(\frac{1}{3}\right)^{x-1}; g(x) = -18\left(\frac{1}{3}\right)^{x-1}$

16. $a_n = 15\left(-\frac{3}{5}\right)^{n-1}; f(x) = 15\left(\frac{3}{5}\right)^{x-1}, g(x) = -15\left(\frac{3}{5}\right)^{x-1}$

17. $a_n = 4(1.5)^{n-1}; f(x) = 4(1.5)^{x-1}$

18. $a_n = 3(-2)^{n-1}; f(x) = 3(2)^{x-1}, g(x) = -3(2)^{x-1}$

B

In Exercises 19–38, find the nth term of each geometric sequence. Then write the sequence in the form $a_1, a_2, a_3, \ldots , a_n, \ldots$.

19. $18, 6, 2, \frac{2}{3}, \ldots$ **20.** $32, 8, 2, \frac{1}{2}, \ldots$

21. $16, -12, 9, -\frac{27}{4}, \ldots$ **22.** $20, -30, 45, -\frac{135}{2}, \ldots$

23. $200, 220, 242, 266.20, \ldots$ **24.** $500, 600, 720, 864, \ldots$

25. $r = -\frac{5}{4}$ and $a_1 = 8$ **26.** $r = -\frac{5}{6}$ and $a_1 = 12$

27. The sequence with common ratio $\frac{3}{7}$ and third term 9

28. The sequence with common ratio $\frac{2}{5}$ and fourth term 8

29. The sequence with $r = -3$ and $a_5 = 54$

30. The sequence with $r = -2$ and $a_6 = -40$

31. The sequence whose fourth term is $\frac{1}{3}$ and seventh term is $-\frac{1}{81}$

32. The sequence whose third term is $\frac{5}{4}$ and sixth term is 80

33. The sequence with $a_4 = \frac{1}{6}$ and $a_9 = \frac{16}{3}$

34. The sequence with $a_3 = 36$ and $a_8 = \frac{128}{27}$

35. The sequence with $r = -2x$ and $a_1 = 1$

36. The sequence with $r = x/3$ and $a_1 = 9$

37. $\ln 2, \ln 4, \ln 16, \ln 256, \ln 65{,}536, \ldots$

38. $\log \frac{1}{2}, \log 4, \log \frac{1}{16}, \log 256, \log \frac{1}{65{,}536}, \ldots$

In Exercises 39–44, find the sum of the first n terms of each geometric sequence.

39. $1, 3, 9, 27, \ldots ; n = 10$ **40.** $1, 2, 4, 8, \ldots ; n = 12$

41. $44, -22, 11, -\frac{11}{2}, \ldots ; n = 8$

42. $54, -18, 6, -2, \ldots ; n = 8$

43. $200, 220, 242, 266.20, \ldots ; n = 20$

44. $500, 600, 720, 864, \ldots ; n = 20$

In Exercises 45–52, find each sum.

45. $\sum_{k=1}^{10} 36\left(\frac{1}{3}\right)^{k-1}$ **46.** $\sum_{k=1}^{9} 108\left(\frac{5}{6}\right)^{k-1}$

47. $\sum_{k=1}^{10} 2^k$ **48.** $\sum_{k=1}^{8} 3^k$

49. $\sum_{k=1}^{12} 250\left(-\frac{2}{5}\right)^{k-1}$ **50.** $\sum_{k=1}^{11} 162\left(-\frac{2}{3}\right)^{k-1}$

51. $\sum_{k=1}^{18} 100(1.05)^{k-1}$ **52.** $\sum_{k=1}^{16} 400(1.02)^{k-1}$

53. Use the properties of summation and appropriate formulas to find the sum:

$$\sum_{k=1}^{14} (3k + 2^k)$$

54. Use the properties of summation and appropriate formulas to find the sum:

$$\sum_{k=1}^{12} (3^{k-1} - k^2)$$

55. A firm offers a starting salary of \$36,000 with 4% annual increases for 5 years.
   (a) List the annual salaries during each of the 5 years as a sequence.
   (b) Does the annual salary during the third year of employment represent an 8% raise over the original salary? Explain your answer.
   (c) Suppose the 4% annual raise continued for 12 years. Determine the annual salary during the twelfth year of employment.

56. For tax purposes, a real estate investor depreciates the value of a property by 5% every year. The property was originally purchased for \$80,000, so during the second year of ownership the depreciated value is 95% of \$80,000.
   (a) List the depreciated values of the property during each of the first 5 years as a sequence.
   (b) Does the depreciated value during the fourth year represent a 15% decrease? Explain your answer.
   (c) Determine the depreciated value of the property during the tenth year.

57. The recommended weight, a_n, of men with medium frames can be approximated by the geometric sequence

$$a_n = 135(1.017)^n$$

where a_n is in pounds and n is the number of inches in height over 5 feet.

(a) Determine a_1 and r.

(b) Use the sequence to complete the table:

Height	5'4"	5'5"	5'6"	5'7"	5'8"
Recommended weight					

Height	5'9"	5'10"	5'11"	6'0"
Recommended weight				

58. At the end of each month $50 is deposited into an annuity that pays 8% compounded monthly. Determine the balance after:

(a) 3 months   **(b)** 2 years

59. At the end of each year $1000 is deposited into an annuity that pays 7% compounded annually. Determine the balance after:

(a) 3 years   **(b)** 15 years

In Exercises 60–73, find the indicated sum of each infinite geometric sequence, if it exists.

60. $10 + 6 + 3.6 + 2.16 + \cdots$

61. $12 + 9 + 6.75 + 5.0625 + \cdots$

62. $9 - 6 + 4 - \frac{8}{3} + \cdots$

63. $4 - 2 + 1 - \frac{1}{2} + \cdots$

64. $3 - 5 + \frac{25}{3} - \frac{125}{9} + \cdots$

65. $36 - 48 + 64 - \frac{256}{3} + \cdots$

66. $2 + 2\sqrt{3} + 6 + 6\sqrt{3} + \cdots$

67. $\sqrt{5} + 5 + 5\sqrt{5} + 25 + \cdots$

68. $\displaystyle\sum_{k=1}^{\infty} 6(0.2)^{k-1}$ **69.** $\displaystyle\sum_{k=1}^{\infty} 40(0.25)^{k-1}$

70. $\displaystyle\sum_{k=1}^{\infty} (0.2)6^{k-1}$ **71.** $\displaystyle\sum_{k=1}^{\infty} (0.25)40^{k-1}$

72. $\displaystyle\sum_{k=0}^{\infty} \frac{2}{5^k}$ **73.** $\displaystyle\sum_{k=2}^{\infty} \frac{4}{3^{k-1}}$

In Exercises 74–79, express each repeating decimal as a ratio of two integers.

74. $0.\overline{36}$ **75.** $0.\overline{45}$ **76.** $2.\overline{5}$

77. $4.\overline{1}$ **78.** $1.2\overline{3}$ **79.** $1.8\overline{6}$

80. The expression $1 + x + x^2 + x^3 + \cdots$ is the sum of an infinite geometric sequence with common ratio x and first term 1. For $-1 < x < 1$, the sum is

$$\frac{a_1}{1 - r} = \frac{1}{1 - x}$$

(a) Graph

$$y = \frac{1}{1 - x}$$

with a viewing window of $[-1, 1] \times [-1, 5]$.

(b) Graph $y = 1$, $y = 1 + x$, $y = 1 + x + x^2$, and $y = 1 + x + x^2 + x^3$ with a viewing window of $[-1, 1] \times [-1, 5]$. Compare with the graph from part (a).

81. The expression $1 - x + x^2 - x^3 + \cdots$ is the sum of an infinite geometric sequence with common ratio $-x$ and first term 1. For $-1 < x < 1$, the sum is

$$\frac{a_1}{1 - r} = \frac{1}{1 - x}$$

(a) Graph

$$y = \frac{1}{1 + x}$$

with a viewing window of $[-1, 1] \times [-1, 5]$.

(b) Graph $y = 1$, $y = 1 - x$, $y = 1 - x + x^2$, and $y = 1 - x + x^2 - x^3$ with a viewing window of $[-1, 1] \times [-1, 5]$. Compare with the graph from part (a).

82. Suppose a tourist spends $100 in a resort town, and each person in the town spends 70% of their income locally. Thus, the people who receive the $100 from the tourists spend 0.70($100) = $70 in the town, the recipients of the $70 spend 0.70($70) = $49 locally, and so on. If this process, called the *multiplier effect* by economists, continues indefinitely, determine the total amount spent locally.

83. A ball is dropped from a height of 16 feet. Each time it falls, it bounces to 49% of its previous height. Use the sum of an infinite geometric sequence to approximate the total distance traveled by the ball.

84. If wind resistance is negligible, after t seconds the height s of an object falling from an initial height of s_0 feet is given by $s = s_0 - 16t^2$. Thus, the time it takes the ball in Exercise 83 to reach the ground after it is first released can be determined by solving the equation $0 = 16 - 16t^2$. The positive solution to $0 = 16 - 16t^2$, is $t_1 = 1$. The ball then bounces back to 49% of 16 feet and falls again. The time it takes the ball to reach this height is the same as the time it takes to fall to the ground. The time it takes the ball to fall to the ground is the positive solution to $0 = (0.49)16 - 16t^2$, which is $t_2 = 0.7$. Therefore, the total time between the first bounce and the second bounce is $2t_2 = 2(0.7)$ seconds. Continuing, the total time elapsed until the ball comes to rest is

$$t_1 + t_2 + t_3 + t_4 + \cdots$$
$$= 1 + 2(0.7) + 2(0.7)^2 + 2(0.7)^3 + \cdots$$
$$= 1 + \sum_{k=1}^{\infty} 2(0.7)^k$$

Determine the total time.

C

85. At the end of each month m dollars is deposited into an account paying an interest rate r compounded monthly.
(a) Show that the balance A after n months is

$$A = m\left[\frac{\left(1 + \dfrac{r}{12}\right)^n - 1}{\dfrac{r}{12}}\right]$$

(b) Consider a single deposit of P dollars into an account at an annual interest rate r compounded monthly. The balance A after n months is

$$A = P\left(1 + \frac{r}{12}\right)^n$$

Equate this with the result from part (a), and solve for m to show that

$$m = \frac{P\left(\dfrac{r}{12}\right)}{1 - \left(1 + \dfrac{r}{12}\right)^{-n}}$$

(c) Discuss a situation in which the result from part (b) would be useful.

86. Find an infinite geometric series for the rational function

$$f(x) = \frac{6}{2 - x}$$

State the values of x for which the sum equals the rational function.

87. A trail construction crew parks its truck at the same spot every day and walks up a trail it has been building to continue its construction. The crew can walk 4 miles per hour and can build the trail at the rate of 0.2 mile per hour. The crew spends a total of 8 hours either walking or building the trail and must return to the same parking spot every day. On the first day, trail construction began at the parking lot, but each day more time is spent walking and less time is spent building the trail. If n is the number of days spent working on the trail, and L_n is the entire length of the trail at the end of the nth day, find a formula for L_n whose only variable is n.

● QUICK REFERENCE

Topic	Page	Remarks
Sequence	570, 571	A sequence is an ordered list of numbers. A sequence can be defined as a function whose domain is the set of positive integers.
Graph of a sequence	574	The graph of a sequence $a_1, a_2, \ldots, a_n, \ldots$ is the graph of the ordered pairs $(1, a_1), (2, a_2), \ldots, (n, a_n), \ldots$.
Summation notation	581	The expression $\displaystyle\sum_{k=1}^{n} a_k$ represents the sum of $a_1 + a_2 + \cdots + a_n$
Factorial	583	If k is a positive integer, the expression $k!$ represents the product $k \cdot (k - 1) \cdot \cdots \cdot 3 \cdot 2 \cdot 1$. The expression $0!$ is defined to be 1.
Summation properties and formulas	584, 586	See the boxes on pages 584 and 586.
Arithmetic sequence	594	An arithmetic sequence is a sequence in which consecutive terms differ by a constant. More precisely, a sequence is arithmetic if there exists a constant d such that $a_{n+1} - a_n = d$. The constant d is the common difference.
General or nth term of an arithmetic sequence	595	The nth term of an arithmetic sequence with first term a_1 and common difference d is given by $$a_n = a_1 + (n - 1)d$$ The graph of an arithmetic sequence is a set of collinear points; that is, an arithmetic sequence corresponds to a linear function with slope equal to the common difference.
Sum of an arithmetic sequence	599	The sum of the first n terms of an arithmetic sequence with first term a_1 and common difference d is given by $$S_n = \frac{n}{2}(a_1 + a_n)$$ or $$S_n = \frac{n}{2}[2a_1 + (n - 1)d]$$
Geometric sequence	603, 604	A geometric sequence is a sequence in which the ratio of consecutive terms is constant. More precisely, a sequence is geometric if there exists a constant r such that $\dfrac{a_{n+1}}{a_n} = r$. The constant r is the common ratio.
General or nth term of a geometric sequence	606	The nth term of a geometric sequence with first term a_1 and common ratio r is given by $$a_n = a_1 r^{n-1}$$ The graph of a geometric sequence corresponds to the graph of one or two exponential functions.

Continued

Topic	Page	Remarks
Sum of a geometric sequence	610	The sum of the first n terms of a geometric sequence with first term a_1 and common ratio r is given as $$S_n = \frac{a_1(1 - r^n)}{1 - r} \qquad r \neq 1$$
Sum of an infinite geometric sequence	612	If $-1 < r < 1$, the sum of the terms of an infinite geometric sequence $a_1, a_1r, a_1r^2, \ldots, a_1r^{n-1}, \ldots$ is given by $$\sum_{k=1}^{\infty} a_1 r^{k-1} = \frac{a_1}{1 - r}$$

MISCELLANEOUS EXERCISES

In Exercises 1–8, list the first four terms of each sequence with the indicated nth term.

1. $a_n = \dfrac{n^3 - n}{6}$

2. $a_n = \dfrac{n^2 - 1}{4}$

3. $a_n = 2n + 5(-1)^n$

4. $a_n = 4n + 3(-1)^n$

5. $a_n = (x + 1)^n$

6. $a_n = \dfrac{x^n}{n!}$

7. $a_n = \begin{cases} 5 & \text{for } n = 1 \\ 2n + a_{n-1} & \text{for } n = 2, 3, 4, \ldots \end{cases}$

8. $a_n = \begin{cases} 1 & \text{for } n = 1 \\ 3a_{n-1} - n & \text{for } n = 2, 3, 4, \ldots \end{cases}$

In Exercises 9–12, find the most evident expression for the nth term of each sequence. Write the sequence in the form $a_1, a_2, a_3, \ldots, a_n, \ldots$.

9. $5, -7, 9, -11, 13, \ldots$

10. $-4, 9, -14, 19, -24, \ldots$

11. $\dfrac{1}{4}, \dfrac{3}{16}, \dfrac{5}{36}, \dfrac{7}{64}, \dfrac{9}{100}, \ldots$

12. $\dfrac{3}{2}, \dfrac{7}{4}, \dfrac{11}{8}, \dfrac{15}{16}, \dfrac{19}{32}, \ldots$

In Exercises 13–16, graph f and the first five ordered pairs of the sequence a_n in the same viewing window.

13. $f(x) = 9\left(\frac{2}{3}\right)^x + 1; \ a_n = 9\left(\frac{2}{3}\right)^n + 1$

14. $f(x) = 12\left(\frac{3}{4}\right)^x + 2; \ a_n = 12\left(\frac{3}{4}\right)^n + 2$

15. $f(x) = \dfrac{4x}{2x - 1}; \ a_n = \dfrac{4n}{2n - 1}$

16. $f(x) = \dfrac{6x - 5}{2x}; \ a_n = \dfrac{6n - 5}{2n}$

In Exercises 17–20, determine the value that each sequence approaches as $n \to +\infty$.

17. The sequence in Exercise 13

18. The sequence in Exercise 14

19. The sequence in Exercise 15

20. The sequence in Exercise 16

In Exercises 21–26, write each expression in expanded form (without summation notation) and simplify.

21. $\displaystyle\sum_{k=1}^{4} \dfrac{k^3}{k!}$

22. $\displaystyle\sum_{k=1}^{5} \dfrac{(k + 1)!}{2k}$

23. $\displaystyle\sum_{k=1}^{5} (2^k - k^2)$

24. $\displaystyle\sum_{k=1}^{5} \dfrac{(2k + 1)^2 - 1}{8}$

25. $\displaystyle\sum_{k=1}^{5} (-1)^{k+1} \dfrac{x^k}{2k + 1}$

26. $\displaystyle\sum_{k=1}^{5} (-1)^k \dfrac{x^k}{k^2}$

In Exercises 27–32, express each sum in summation notation. (Answers are not unique.)

27. $1 + 7 + 13 + 19 + 25 + 31$

28. $36 + 28 + 20 + 12 + 4 + (-4)$

29. $48 + 24 + 12 + 6 + 3 + \frac{3}{2}$

30. $54 + 18 + 6 + 2 + \frac{2}{3}$

31. $3 + 8 + 15 + 24 + 35 + 48$

32. $4 + 7 + 12 + 19 + 28 + 39$

In Exercises 33 and 34, refer to the polynomial

$$P_n(x) = \sum_{k=1}^{n} (-1)^{k+1} \frac{x^k}{k}$$

used to approximate $\ln(x + 1)$ on the interval $-1 < x \le 1$.

33. (a) Express $P_3(x)$ in expanded form and simplify.
   (b) Compute $\ln 1.5$ and $P_3(0.5)$ to the nearest 0.001.
   (c) Graph $y = \ln(x + 1)$ and $y = P_3(x)$ with a viewing window of $[-1, 1] \times [-2, 2]$.

34. (a) Express $P_4(x)$ in expanded form and simplify.
   (b) Compute $\ln 1.5$ and $P_4(0.5)$ to the nearest 0.001.
   (c) Graph $y = \ln(x + 1)$ and $y = P_4(x)$ with a viewing window of $[-1, 1] \times [-2, 2]$.

In Exercises 35 and 36, use the summation formulas and properties to evaluate each sum.

35. $\displaystyle\sum_{k=1}^{30} 2(k + 3)(k - 3)$ **36.** $\displaystyle\sum_{k=1}^{20} k(3k - 2)$

In Exercises 37–46, each sequence is either arithmetic or geometric. Determine the common difference (if arithmetic) or the common ratio (if geometric). Also, write the sequence in the form $a_1, a_2, a_3, \ldots, a_n, \ldots$.

37. $-25, 15, -9, \frac{27}{5}, -\frac{81}{25}, \ldots$ **38.** $27, -18, 12, -8, \frac{16}{3}, \ldots$

39. $5, \frac{13}{3}, \frac{11}{3}, 3, \frac{7}{3}, \ldots$ **40.** $-2, \frac{3}{2}, 5, \frac{17}{2}, 12, \ldots$

41. $2, 2\sqrt{5}, 10, 10\sqrt{5}, 50, \ldots$

42. $3\sqrt{2}, 6, 6\sqrt{2}, 12, 12\sqrt{2}, \ldots$

43. $4, 4 + 3\pi, 4 + 6\pi, 4 + 9\pi, 4 + 12\pi, \ldots$

44. $11, 11 - 2e, 11 - 4e, 11 - 6e, 11 - 8e, \ldots$

45. $z^b, z^{b+2}, z^{b+4}, z^{b+6}, z^{b+8}, \ldots$

46. $z^b, z^b + 2, z^b + 4, z^b + 6, z^b + 8, \ldots$

47. Find the fifteenth term of the arithmetic sequence whose fourth term is 16 and thirteenth term is 13.

48. Find the twentieth term of the arithmetic sequence whose sixth term is 7 and twelfth term is 12.

49. Find the eighth term of the geometric sequence whose third term is 6 and sixth term is 18.

50. Find the second term of the geometric sequence whose fourth term is 100 and seventh term is $\frac{4}{5}$.

In Exercises 51–54, each sequence is either arithmetic or geometric. Find the sum of the first n terms of each sequence.

51. $-6, -1, 4, 9, \ldots$; $n = 20$

52. $2, 4.5, 7, 9.5, \ldots$; $n = 30$

53. $375, 75, 15, 3, \ldots$; $n = 8$

54. $192, 48, 12, 3, \ldots$; $n = 7$

In Exercises 55–60, find the indicated sum of each infinite geometric sequence, if it exists.

55. $28 - 14 + 7 - \frac{7}{2} + \cdots$ **56.** $45 - 15 + 5 - \frac{5}{3} + \cdots$

57. $\displaystyle\sum_{k=1}^{\infty} 12\left(\frac{5}{8}\right)^k$ **58.** $\displaystyle\sum_{k=1}^{\infty} 15\left(\frac{1}{6}\right)^k$

59. $\displaystyle\sum_{k=1}^{\infty} 9\left(\frac{4}{3}\right)^{k-1}$ **60.** $\displaystyle\sum_{k=1}^{\infty} 6\left(-\frac{10}{9}\right)^{k-1}$

61. Express $\ln x^3 + \ln x^6 + \ln x^9 + \ln x^{12} + \cdots + \ln x^{300}$ in the form $a \ln x$, where a is a constant.

62. For what values of x does the following equation hold?

$$\frac{3}{1 + 2x} = 3 - 6x + 12x^2 - 24x^3 + \cdots$$

63. Let i represent the imaginary number $\sqrt{-1}$. Determine the sum

$$1 + i + i^2 + i^3 + i^4 + \cdots + i^{59}$$

64. Let A_1 represent the area of a square with each side of length 2 units. Form a second square by joining midpoints of adjacent sides of the first square, and let A_2 represent the area of this second square. Continue this process to form a third square with area A_3, a fourth square with area

A_4, and so on (Figure 7.20). Determine the sum

$$A_1 - A_2 + A_3 - A_4 + A_5 - \cdots$$

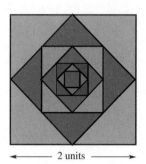

←———— 2 units ————→

Figure 7.20

65. **(a)** Determine the values of $x > 0$ (to the nearest 0.0001) such that

$$1 - x + x^2 - x^3 + x^4 - \cdots + x^{50} = \tfrac{2}{3}$$

HINT: The left side of the equation is a geometric series; use your graphing utility to solve the resulting equation.

(b) Determine the exact value of x such that

$$1 - x + x^2 - x^3 + x^4 - \cdots = \tfrac{2}{3}$$

■ CHAPTER TEST

1. Write the first four terms of the sequence defined by

$$a_n = \begin{cases} 12 & \text{for } n = 1 \\ \dfrac{a_{n-1}}{2} + (-1)^n & \text{for } n = 2, 3, 4, \ldots \end{cases}$$

2. Write

$$\sum_{k=1}^{4} \frac{k!}{(2k)^2}$$

in expanded form (without summation notation) and simplify.

3. Express $1 - 3 + 5 - 7 + 9 - 11 + 13$ in summation notation. (The answer is not unique.)

In Exercises 4 and 5, each sequence is either arithmetic or geometric. Determine the indicated term.

4. $3, 7, 11, 15, \ldots ; a_{24}$ **5.** $\tfrac{8}{9}, \tfrac{4}{3}, 2, 3, \ldots ; a_{10}$

6. For a particular geometric sequence, the fourth term is 6 and the sixth term is 72. If the common ratio is negative, find the seventh term.

7. Find the sum, if it exists, of the infinite geometric series:

$$40 - 10 + \tfrac{5}{2} - \tfrac{5}{8} + \cdots$$

8. Given that

$$\sum_{k=1}^{40} a_k = 75 \quad \text{and} \quad \sum_{k=1}^{40} b_k = 45$$

evaluate

$$\sum_{k=1}^{40} (a_k - 2b_k + 3)$$

9. Determine the sum of the first ten terms of an arithmetic sequence whose first term is 15 and tenth term is $\tfrac{7}{2}$.

10. Express $0.\overline{42}$ as a ratio of two integers.

11. The lengths of the sides of a right triangle represent three consecutive terms of an arithmetic sequence with common difference $\tfrac{3}{2}$. Determine the sides of the right triangle.

12. A ball is dropped from a height of 12 feet. Each time it falls, it bounces to 75% of its previous height. Use the sum of an infinite geometric sequence to approximate the total distance traveled by the ball.

13. Graph the first five points of the sequence

$$a_n = \frac{n^2 - 1}{8}$$

14. Determine the value of $x > 0$ (to the nearest 0.0001) such that

$$1 + x + x^2 + x^3 + x^4 + \cdots + x^{100} = 2$$

HINT: The left side of the equation is a finite geometric series; use a graphing utility to solve the resulting equation.

BINOMIAL
THEOREM
AND
PROBABILITY

This chapter concludes our study of college algebra with three topics that have applications in statistics, engineering, the biological sciences, business, economics, computer science, and the social sciences. In Section 8.1, we discuss ways to expand a binomial raised to any positive integer power. Such expansions are used in calculating probabilities of activities or experiments that have only two outcomes. Binomial expansions are also important in calculus. In Section 8.2, we develop techniques for determining the number of ways in which certain activities can occur or selections can be made. These techniques have applications in many fields, but our primary motivation for discussing counting techniques is to determine probabilities, which we discuss in Section 8.3. We consider applications such as the probability that the Internal Revenue Service will discover an illegal deduction or the probability that a lot of computer chips will pass a quality control inspection. We also discuss probabilities of outcomes associated with "games of chance," such as the probability of being dealt a five-card hand consisting of all spades, or the probability of rolling a total of seven with two dice.

Our treatment of probability is merely an introduction; entire books have been written on the subject. A more extensive discussion of combinations, permutations, and probability can be found in texts on finite mathematics or discrete mathematics.

THE BINOMIAL THEOREM

Certain problems in statistics, calculus, and other branches of mathematics require writing expressions of the form $(a + b)^n$ as a sum, called the *expanded form* of $(a + b)^n$. For example, the expanded form of $(a + b)^2$ is $a^2 + 2ab + b^2$. In this section, we discuss techniques for expanding $(a + b)^n$, where n is a nonnegative integer.

Pascal's Triangle

We begin by considering $(a + b)^n$ for $n = 0, 1, 2, 3, 4,$ and 5. Obviously,

$$(a + b)^0 = 1$$
$$(a + b)^1 = a + b$$

and we know that

$$(a + b)^2 = a^2 + 2ab + b^2$$

Multiplying $(a + b)^3$ and adding like terms, we get

$$
\begin{aligned}
(a + b)^3 &= (a + b)(a + b)^2 \\
&= (a + b)(a^2 + 2ab + b^2) \\
&= a(a^2 + 2ab + b^2) + b(a^2 + 2ab + b^2) \\
&= a^3 + 2a^2b + ab^2 \\
&\quad\ \ \underline{+\ a^2b\ +\ 2ab^2 + b^3} \\
&= a^3 + 3a^2b + 3ab^2 + b^3
\end{aligned}
$$

Distribute and vertically align like terms.

The expressions $(a + b)^4$ and $(a + b)^5$ can be expanded in a similar way, but the process becomes more and more tedious. The results are listed below:

$$
\begin{aligned}
(a + b)^0 &= 1 \\
(a + b)^1 &= a + b \\
(a + b)^2 &= a^2 + 2ab + b^2 \\
(a + b)^3 &= a^3 + 3a^2b + 3ab^2 + b^3 \\
(a + b)^4 &= a^4 + 4a^3b + 6a^2b^2 + 4ab^3 + b^4 \\
(a + b)^5 &= a^5 + 5a^4b + 10a^3b^2 + 10a^2b^3 + 5ab^4 + b^5
\end{aligned}
$$

Fortunately, there is a pattern that makes subsequent expansions easier. Notice the following for $n = 1, 2, 3, 4,$ or 5, in the expansion of $(a + b)^n$:

The first term is a^n and the last term is b^n.

The powers of a decrease by 1 in successive terms from left to right.

The powers of b increase by 1 in successive terms from left to right.

The sum of the exponents of each term (the degree of each term) is n.

Thus, we anticipate that the form for the next case is

$$(a + b)^6 = a^6 + \underline{?}a^5b + \underline{?}a^4b^2 + \underline{?}a^3b^3 + \underline{?}a^2b^4 + \underline{?}ab^5 + b^6$$

where $\underline{?}$ denotes a missing coefficient. To determine these coefficients, reconsider the expansions of $(a + b)^n$ that we have already written for $n = 0, 1, 2, 3, 4,$ and 5. If we list only the coefficients, called *binomial coefficients*, the resulting triangular array of integers is known as *Pascal's triangle*.

$(a + b)^0$:					1					Row 0
$(a + b)^1$:				1		1				Row 1
$(a + b)^2$:			1		2		1			Row 2
$(a + b)^3$:		1		3		3		1		Row 3
$(a + b)^4$:	1		4		6		4		1	Row 4
$(a + b)^5$: 1		5		10		10		5	1	Row 5

To generate additional rows, we start by writing a 1 at the beginning and the end. Each of the remaining entries is the sum of the two entries diagonally above it. For example, the sixth row starts with a 1; the second entry is the sum of the first two entries in the previous row:

$$\begin{array}{cccccc} 1 & 5 & 10 & 10 & 5 & 1 \quad \text{Row 5} \\ 1 & 6 & & & & 1 \quad \text{Row 6} \end{array}$$

Similarly, the third entry in the sixth row is the sum of the second and third entries (5 and 10) of the previous row, and so on:

$$\begin{array}{ccccccc} 1 & 5 & 10 & 10 & 5 & 1 & \quad \text{Row 5} \\ 1 & 6 & 15 & 20 & 15 & 6 & 1 \quad \text{Row 6} \end{array}$$

Recall that the expansion of $(a + b)^6$ has the form

$$(a + b)^6 = a^6 + \underline{?}a^5b + \underline{?}a^4b^2 + \underline{?}a^3b^3 + \underline{?}a^2b^4 + \underline{?}ab^5 + b^6$$

The sixth row of Pascal's triangle supplies the respective coefficients:

$$(a + b)^6 = a^6 + 6a^5b + 15a^4b^2 + 20a^3b^3 + 15a^2b^4 + 6ab^5 + b^6$$

The next example shows how to apply our observations to more complicated expansions.

■ EXAMPLE 1 Using Pascal's Triangle to Expand a Binomial

Use Pascal's triangle to expand each expression, and simplify.

(a) $(2x + 3)^4$ **(b)** $(x - 2y)^7$

SOLUTION

(a) First, we determine the expanded form of $(a + b)^4$ by starting with

$$(a + b)^4 = a^4 + \underline{?}a^3b + \underline{?}a^2b^2 + \underline{?}ab^3 + b^4$$

The coefficients are given by the fourth row of Pascal's triangle:

$$1 \quad 4 \quad 6 \quad 4 \quad 1$$

Thus,

$$(a + b)^4 = a^4 + 4a^3b + 6a^2b^2 + 4ab^3 + b^4$$

To expand $(2x + 3)^4$, we replace a with $2x$, replace b with 3, and simplify:

$$(2x + 3)^4 = (2x)^4 + 4(2x)^3(3) + 6(2x)^2(3)^2 + 4(2x)(3)^3 + (3)^4$$
$$= 16x^4 + 96x^3 + 216x^2 + 216x + 81$$

(b) We first expand $(a + b)^7$ by starting with the form

$$(a + b)^7 = a^7 + \underline{?}a^6b + \underline{?}a^5b^2 + \underline{?}a^4b^3 + \underline{?}a^3b^4 + \underline{?}a^2b^5 + \underline{?}ab^6 + b^7$$

The seventh row of Pascal's triangle can be determined from the sixth row:

$$\begin{array}{ccccccccccccc} 1 & & 6 & & 15 & & 20 & & 15 & & 6 & & 1 \\ & & & & & & & & & & & & \end{array} \quad \text{Row 6}$$
$$\begin{array}{ccccccccccccccc} 1 & & 7 & & 21 & & 35 & & 35 & & 21 & & 7 & & 1 \end{array} \quad \text{Row 7}$$

Therefore,

$$(a + b)^7 = a^7 + 7a^6b + 21a^5b^2 + 35a^4b^3 + 35a^3b^4 + 21a^2b^5 + 7ab^6 + b^7$$

Since

$$(x - 2y)^7 = [x + (-2y)]^7$$

we replace a with x and replace b with $-2y$:

$$(x - 2y)^7 = x^7 + 7x^6(-2y) + 21x^5(-2y)^2 + 35x^4(-2y)^3$$
$$+ 35x^3(-2y)^4 + 21x^2(-2y)^5 + 7x(-2y)^6 + (-2y)^7$$
$$= x^7 - 14x^6y + 84x^5y^2 - 280x^4y^3 + 560x^3y^4$$
$$- 672x^2y^5 + 448xy^6 - 128y^7$$

■ *In Section 8.2, we will show that $_nC_r$ is equal to the number of ways of choosing a subset of r elements from a set of n elements.*

A Formula for Binomial Coefficients

Using Pascal's triangle to determine the expansions of $(a + b)^n$ is not a practical method for large values of n. Instead, we will use a formula that determines the binomial coefficients directly. To do so, we introduce the notation $_nC_r$, which is read as "n choose r." The calculation of $_nC_r$ uses factorials (introduced in Section

7.2), so we restate the definition and notation: If k is a positive integer, then $k!$, called k *factorial*, is defined as

$$k! = k \cdot (k - 1) \cdot \cdots \cdot 3 \cdot 2 \cdot 1$$

The expression $0!$ is defined to be 1.

The binomial coefficient $_nC_r$ is defined as follows:

The Binomial Coefficient $_nC_r$

> Let n and r be nonnegative integers with $r \leq n$. The binomial coefficient $_nC_r$ is defined as
>
> $$_nC_r = \frac{n!}{r!(n - r)!}$$

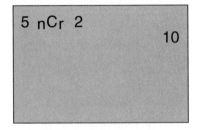

Figure 8.1

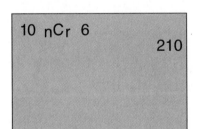

Figure 8.2

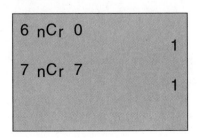

Figure 8.3

EXAMPLE 2 Evaluating $_nC_r$

Evaluate each expression.

(a) $_5C_2$ **(b)** $_{10}C_6$ **(c)** $_6C_0$ **(d)** $_7C_7$

SOLUTION

(a) $_5C_2 = \dfrac{5!}{2!(5 - 2)!} = \dfrac{5!}{2!3!} = \dfrac{5 \cdot 4 \cdot \cancel{3 \cdot 2 \cdot 1}}{(2 \cdot 1)\cancel{(3 \cdot 2 \cdot 1)}} = \dfrac{5 \cdot 4}{2 \cdot 1} = 10$

The calculation also can be determined using the $_nC_r$ feature on a graphing utility (Figure 8.1).

(b) $_{10}C_6 = \dfrac{10!}{6!(10 - 6)!} = \dfrac{10!}{6!4!} = \dfrac{10 \cdot 9 \cdot 8 \cdot 7 \cdot \cancel{6 \cdot 5 \cdot 4 \cdot 3 \cdot 2 \cdot 1}}{\cancel{(6 \cdot 5 \cdot 4 \cdot 3 \cdot 2 \cdot 1)}(4 \cdot 3 \cdot 2 \cdot 1)}$

$$= \dfrac{10 \cdot 9 \cdot 8 \cdot 7}{4 \cdot 3 \cdot 2 \cdot 1} = 210$$

Figure 8.2 shows the calculation with a graphing utility.

(c) $_6C_0 = \dfrac{6!}{0!(6 - 0)!} = \dfrac{6!}{0!6!} = \dfrac{6!}{1 \cdot 6!} = 1$ Recall that $0! = 1$.

(d) $_7C_7 = \dfrac{7!}{7!(7 - 7)!} = \dfrac{7!}{7!0!} = \dfrac{7!}{7! \cdot 1} = 1$

You should verify the results of parts (c) and (d) on a graphing utility (Figure 8.3).

Now we can state a formula for the expansion of $(a + b)^n$, known as the *binomial theorem*.

Binomial Theorem

Let n and r be nonnegative integers with $r \leq n$. Then

$$(a + b)^n = {}_nC_0a^n + {}_nC_1a^{n-1}b + {}_nC_2a^{n-2}b^2 + \cdots + {}_nC_{n-1}ab^{n-1} + {}_nC_nb^n$$

In summation notation,

$$(a + b)^n = \sum_{k=0}^{n} {}_nC_ka^{n-k}b^k$$

EXAMPLE 3   Using the Binomial Theorem

Use the binomial theorem to expand each expression, and simplify the result.

(a) $(a + b)^5$ **(b)** $(x - 2y)^9$

SOLUTION

(a) $(a + b)^5 = {}_5C_0a^5 + {}_5C_1a^4b + {}_5C_2a^3b^2 + {}_5C_3a^2b^3 + {}_5C_4ab^4 + {}_5C_5b^5$
$= a^5 + 5a^4b + 10a^3b^2 + 10a^2b^3 + 5ab^4 + b^5$

(b) First, we determine $(a + b)^9$ using the binomial theorem:

$$(a + b)^9 = {}_9C_0a^9 + {}_9C_1a^8b + {}_9C_2a^7b^2 + {}_9C_3a^6b^3 + {}_9C_4a^5b^4$$
$$+ {}_9C_5a^4b^5 + {}_9C_6a^3b^6 + {}_9C_7a^2b^7 + {}_9C_8ab^8 + {}_9C_9b^9$$
$$= a^9 + 9a^8b + 36a^7b^2 + 84a^6b^3 + 126a^5b^4 + 126a^4b^5$$
$$+ 84a^3b^6 + 36a^2b^7 + 9ab^8 + b^9$$

Substituting x for a and $-2y$ for b, we get

$$(x - 2y)^9 = x^9 + 9x^8(-2y) + 36x^7(-2y)^2 + 84x^6(-2y)^3$$
$$+ 126x^5(-2y)^4 + 126x^4(-2y)^5 + 84x^3(-2y)^6$$
$$+ 36x^2(-2y)^7 + 9x(-2y)^8 + (-2y)^9$$
$$= x^9 - 18x^8y + 144x^7y^2 - 672x^6y^3 + 2016x^5y^4 - 4032x^4y^5$$
$$+ 5376x^3y^6 - 4608x^2y^7 + 2304xy^8 - 512y^9$$

EXAMPLE 4   Finding a Particular Term in an Expansion

Determine the specified term in the expansion of $(5x - 2y)^8$.

(a) x^2y^6 term   **(b)** Fourth term

SOLUTION

(a) Using the binomial theorem, the a^2b^6 term in the expansion of $(a + b)^8$ is

$${}_8C_6a^2b^6 = 28a^2b^6$$

Letting $a = 5x$ and $b = -2y$, the x^2y^6 term in the expansion of $(5x - 2y)^8$ is

$$28(5x)^2(-2y)^6 = 44{,}800x^2y^6$$

(b) The fourth term in the expansion of $(a + b)^8$ is

$$_8C_3a^5b^3 = 56a^5b^3$$

Letting $a = 5x$ and $b = -2y$, we find that the fourth term in the expansion of $(5x - 2y)^8$ is

$$56(5x)^5(-2y)^3 = -1,400,000x^5y^3$$

In Section 1.2, we simplified expressions of the form

$$\frac{f(x + h) - f(x)}{h}$$

for a given function f. This expression, called a *difference quotient* of f, plays an essential role in calculus. The next example illustrates how the binomial theorem is used to simplify a difference quotient of a monomial function.

EXAMPLE 5 Finding a Difference Quotient

For $f(x) = x^5$, find and simplify $\dfrac{f(x + h) - f(x)}{h}$.

SOLUTION

$$
\begin{aligned}
\frac{f(x + h) - f(x)}{h} &= \frac{(x + h)^5 - x^5}{h} \\
&= \frac{(x^5 + 5x^4h + 10x^3h^2 + 10x^2h^3 + 5xh^4 + h^5) - x^5}{h} \\
&= \frac{5x^4h + 10x^3h^2 + 10x^2h^3 + 5xh^4 + h^5}{h} \\
&= \frac{h(5x^4 + 10x^3h + 10x^2h^2 + 5xh^3 + h^4)}{h} \\
&= 5x^4 + 10x^3h + 10x^2h^2 + 5xh^3 + h^4
\end{aligned}
$$

Applications

Many important investigations involve activities or experiments that have only two outcomes. For example, a manufactured item that is tested is either defective or nondefective; a new treatment for a disease is either effective or not effective. One important application of the binomial theorem is calculating probabilities of such experiments.

EXAMPLE 6   Application: Medical Treatment

A new antibiotic was found to be successful in 85% of the patients. Seven patients are treated.

(a) The probability of curing exactly r of them with the antibiotic is

$$_7C_r\,(0.15)^{7-r}\,(0.85)^r$$

Determine (to the nearest 0.001) the probability that exactly four of the patients are cured.

(b) The probability of curing at least r of them with the antibiotic is

$$\sum_{k=r}^{7} {}_7C_k(0.15)^{7-k}(0.85)^k$$

To the nearest 0.001, what is the probability that at least four of the patients are cured?

SOLUTION

(a) Substituting $r = 4$, we get

$$_7C_4(0.15)^{7-4}(0.85)^4 = 35(0.15)^3(0.85)^4 \approx 0.062$$

Thus, the probability that exactly four patients are cured is about 6.2%.

(b) Substituting $r = 4$, we get

$$
\begin{aligned}
\sum_{k=4}^{7} {}_7C_k(0.15)^{7-k}(0.85)^k &= {}_7C_4(0.15)^{7-4}(0.85)^4 + {}_7C_5(0.15)^{7-5}(0.85)^5 \\
&\quad + {}_7C_6(0.15)^{7-6}(0.85)^6 + {}_7C_7(0.15)^{7-7}(0.85)^7 \\
&= 35(0.15)^3(0.85)^4 + 21(0.15)^2(0.85)^5 \\
&\quad + 7(0.15)^1(0.85)^6 + 1(0.15)^0(0.85)^7 \\
&= (0.061661\ldots) + (0.209650\ldots) \\
&\quad + (0.396006\ldots) + (0.320577\ldots) \\
&\approx 0.987896\ldots
\end{aligned}
$$

Thus, the probability that at least four patients are cured is about 98.8%.

EXERCISE SET 8.1

A

In Exercises 1–8, use Pascal's triangle to expand each expression and simplify.

1. $(a + h)^5$ **2.** $(k + b)^6$ **3.** $(x + 2)^4$ **4.** $(y + 3)^4$

5. $(x - 3)^5$ **6.** $(p - 2)^5$ **7.** $(x + y)^6$ **8.** $(x - y)^6$

In Exercises 9–16, evaluate each expression.

9. $_8C_3$ **10.** $_7C_2$ **11.** $_{11}C_4$ **12.** $_9C_6$

13. $_{10}C_{10}$ **14.** $_8C_0$ **15.** $_{14}C_{12}$ **16.** $_{15}C_{12}$

In Exercises 17–20, use Pascal's triangle to expand each expression and simplify.

17. $(2x + y)^5$ **18.** $(x + 3y)^5$ **19.** $(2x - 3y)^6$

20. $(3x - 4y)^6$

In Exercises 21–24, determine the eighth and ninth rows of Pascal's triangle, and use them to expand and simplify each expression.

21. $(x + y)^8$ **22.** $(x - y)^8$ **23.** $(2x - 1)^9$

24. $(2x + 1)^9$

B

In Exercises 25–40, use the binomial theorem to expand each expression and simplify.

25. $(x + 1)^5$ **26.** $(x - 1)^6$ **27.** $(2x - y)^6$

28. $(x + 3y)^5$ **29.** $(3m + 2n)^4$ **30.** $(4r + 3s)^4$

31. $(x^2 + y)^7$ **32.** $(x - y^2)^8$ **33.** $(u^2 - 2v^3)^6$

34. $(2c^3 + d^4)^5$ **35.** $\left(\dfrac{x}{2} + 8\right)^4$ **36.** $\left(9x + \dfrac{1}{3}\right)^4$

37. $\left(z^3 + \dfrac{1}{z}\right)^5$ **38.** $\left(p^2 + \dfrac{1}{p}\right)^6$ **39.** $(\sqrt{h} + 2k)^7$

40. $(a + 2\sqrt{b})^7$

41. Determine the x^4y^8 term in the expansion of $(x + y)^{12}$.

42. Determine the x^6y^5 term in the expansion of $(x + y)^{11}$.

43. Determine the m^5n^9 term in the expansion of $(2m - n)^{14}$.

44. Determine the u^3v^5 term in the expansion of $(u - 3v)^8$.

45. Determine the first four terms in the expansion of $(r^3 + 2s)^{10}$.

46. Determine the first four terms in the expansion of $(f - 3g^2)^{12}$.

47. Determine the sixth term in the expansion of $(3x + 4y)^{11}$.

48. Determine the seventh term in the expansion of $(2a + 5b)^{10}$.

In Exercises 49–52, find and simplify

$$\frac{f(x + h) - f(x)}{h}$$

49. $f(x) = x^6$ **50.** $f(x) = x^7$

51. $f(x) = 5x^4 + 2$ **52.** $f(x) = 3x^5 + 1$

In Exercises 53–56, solve for k.

53. $_4C_k = {}_4C_1$ **54.** $_6C_k = {}_6C_2$

55. $_nC_k = {}_nC_3, n \geq 3$ **56.** $_nC_k = {}_nC_4, n \geq 4$

57. A new drug was found to prevent epileptic seizures in 70% of the patients. The drug is prescribed to 10 patients.

(a) The probability that the drug will prevent seizures for exactly r of the patients is

$$_{10}C_r \, (0.30)^{10-r} \, (0.70)^r$$

Determine (to the nearest 0.001) the probability that the drug will prevent seizures for exactly 7 of the patients.

(b) The probability that the new drug will prevent seizures for at least r of the patients is

$$\sum_{k=r}^{10} {}_{10}C_k \, (0.30)^{10-k} \, (0.70)^k$$

Determine (to the nearest 0.001) the probability that the drug will prevent seizures for at least 7 of the patients.

58. An electrical supplies manufacturer produces dipole switches and packages them in boxes of 20. Quality control studies have shown that 3% of the switches are defective.

(a) The probability that a box of switches contains exactly r defective switches is

$$_{20}C_r \, (0.97)^{20-r} \, (0.03)^r$$

Determine (to the nearest 0.001) the probability that a box of switches contains exactly 3 defective switches.

(b) The probability that a box of switches contains at most r defective switches is

$$\sum_{k=0}^{r} {}_{20}C_k \, (0.97)^{20-k} \, (0.03)^k$$

Determine (to the nearest 0.001) the probability that a box of switches contains at most 3 defective switches.

59. A recent poll determines that 60% of the voters in an election intend to vote for the Republican candidate. Fifteen voters are surveyed.

(a) The probability that exactly r of the 15 voters intend to vote for the Republican candidate is

$$_{15}C_r \, (0.40)^{15-r} \, (0.60)^r$$

Determine (to the nearest 0.001) the probability that exactly 12 of the 15 voters intend to vote for the Republican candidate.

(b) The probability that at least r of the 15 voters intend to vote for the Republican candidate is

$$\sum_{k=r}^{15} {}_{15}C_k \, (0.40)^{15-k} \, (0.60)^k$$

Determine (to the nearest 0.001) the probability that at least 12 of the 15 voters intend to vote for the Republican candidate.

60. Approximate $(1.1)^7$ by adding the first four terms in the expansion of $(1 + 0.1)^7$. Use your calculator to determine the exact value of $(1.1)^7$.

61. Approximate $(0.9)^6$ by adding the first four terms in the expansion of $(1 - 0.1)^6$. Use your calculator to determine the exact value of $(0.9)^6$.

C

62. Let $f(x) = 1/x^4$. Determine and simplify

$$\frac{f(x + h) - f(x)}{h}$$

63. Use the binomial theorem to show that

$${}_nC_0 + {}_nC_1 + {}_nC_2 + \cdots + {}_nC_n = 2^n$$

HINT: $2^n = (1 + 1)^n$

64. Let $f(k) = {}_{10}C_k$, for $k = 0, 1, 2, \ldots, 10$. Is $f(k)$ a one-to-one function? Justify your answer.

65. Use the formula for binomial coefficients to show that

$${}_nC_{r-1} + {}_nC_r = {}_{n+1}C_r$$

66. Use the formula for binomial coefficients to show that

$$\left(\frac{n - r}{r + 1}\right) {}_nC_r = {}_nC_{r+1}$$

PERMUTATIONS AND COMBINATIONS

In this section we develop techniques to determine the number of ways that various activities can occur. The mathematics of enumeration (counting) began long ago. The early Arabs studied counting methods because they believed the arrangements of planets had mystical significance. Modern applications of counting techniques include the study of gene arrangements on a chromosome and coding information for optical readers. In Section 8.3, we will apply methods of counting to determine various probabilities.

Counting Techniques

Often, the most straightforward way to count the number of ways in which an activity can occur is to list all the possibilities.

EXAMPLE 1   Listing the Outcomes

Find the number of ways each activity can occur by listing all the possibilities.

(a) A school election is held for president and secretary. There are 5 candidates for president: Al, Beth, Carmen, Dawn, and Ed. There are 3 candidates for secretary: Flo, Garth, and Hector.

(b) A coin is tossed 3 times and the sequence of heads or tails is recorded.

SOLUTION

(a) Let, for example, BH represent the outcome that Beth is elected president and Hector is elected secretary. The outcomes are listed below:

$$
\begin{array}{ccc}
AF & AG & AH \\
BF & BG & BH \\
CF & CG & CH \\
DF & DG & DH \\
EF & EG & EH
\end{array}
$$

The number of possible outcomes is 15.

(b) Let, for example, HTT represent the first toss is heads, the second is tails, and the third is tails. The possibilities can be represented as

$$
\begin{array}{cccc}
HHH & HHT & HTH & HTT \\
THH & THT & TTH & TTT
\end{array}
$$

So the number of outcomes is 8.

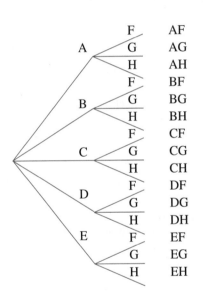

Figure 8.4

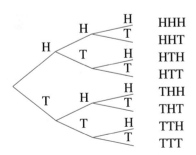

Figure 8.5

In many cases, it is not feasible to list each of the possible outcomes of an activity, yet we need to know the number. If an activity consists of two or more consecutive tasks, then we can often determine the number of outcomes without listing them. For example, the event in part (a) in Example 1 consists of two consecutive tasks—electing a president and electing a secretary. There are 5 possible outcomes for the first activity (Al, Beth, Carmen, Dawn, or Ed), and for each of these outcomes, there are 3 possible outcomes for the second activity (Flo, Garth, or Hector). The *tree diagram* shown in Figure 8.4 illustrates the possibilities. Thus, there are 5 · 3 = 15 possibilities.

Similarly, the activity in part (b) in Example 1 consists of 3 consecutive tasks—the first toss, the second toss, and the third toss. There are 2 possible outcomes for the first toss; for each of these, there are 2 possible outcomes for the second toss; and then, for each of these, there are 2 possible outcomes for the third toss (Figure 8.5). Consequently, there are 2 · 2 · 2 = 8 possibilities.

A summary of the reasoning we used is known as the *fundamental counting principle*.

Fundamental Counting Principle

■ **NOTE:** *The fundamental counting principle can be extended when there are more than two tasks (see Example 2b).*

Suppose that an activity is composed of two tasks. If the first task can occur in m ways, and for each of these, the second task can occur in n ways, then the number of ways the activity can occur is

$$m \cdot n$$

■ EXAMPLE 2   Applying the Fundamental Counting Principle

Find the number of ways each activity can occur.

(a) A school election is held for president and secretary. There are 12 candidates for president and 7 candidates for secretary. Assume no person can run for more than one position.

(b) A coin is tossed 6 times and the sequence of heads or tails is recorded.

SOLUTION

(a) The election consists of 2 consecutive tasks, selecting a president and selecting a secretary. There are 12 ways in which the first task can occur, and there are 7 ways in which the second task can occur. By the fundamental counting principle, the number of outcomes in the election is

$$12 \cdot 7 = 84$$

Number of ways   Number of ways
of selecting president   of selecting secretary

There are 84 possible election results.

(b) Each toss of the coin has 2 outcomes, heads or tails. Extending the fundamental counting principle, we determine the number of outcomes:

$$2 \cdot 2 \cdot 2 \cdot 2 \cdot 2 \cdot 2 = 64$$

Number of	Number of		Number of
outcomes	outcomes	$\cdots$	outcomes
of first toss	of second toss		of sixth toss

Thus, there are 64 outcomes.

■ EXAMPLE 3   Application: Computer Network Security

To gain access to a computer network, a user must enter a password. If a password consists of three capital letters followed by a digit, how many passwords are possible?

SOLUTION

Selecting a password consists of 4 activities. The first, second, and third activities are the selection of the first, second, and third letter, respectively. The fourth activity consists of selecting a digit. There are 26 letters in the English alphabet and 10 digits (0–9). Using the fundamental counting principle, the number of ways to make these selections is

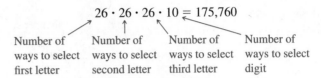

$$26 \cdot 26 \cdot 26 \cdot 10 = 175{,}760$$

Number of ways to select first letter   Number of ways to select second letter   Number of ways to select third letter   Number of ways to select digit

Therefore, the number of possible passwords is 175,760.

EXAMPLE 4 Application: Business Inventory

A car dealer offers a sports car with the following features:

Feature	Choices
Model	SE, LE, LS, DX
Exterior color	Red, black, white
Interior color	Tan, charcoal
Seats	Leather, cloth
Stereo	Basic (AM/FM 4-speaker), basic with cassette, deluxe (AM/FM 6-speaker), deluxe with cassette, deluxe with cassette and CD player

How many sports cars must the dealer have in stock in order to have one of each kind available?

SOLUTION

There are 4 ways to select the model, 3 ways to select the exterior color, 2 ways to select the interior color, 2 ways to select the seats, and 5 ways to select the stereo. By the fundamental counting principle, there are

$$4 \cdot 3 \cdot 2 \cdot 2 \cdot 5 = 240$$

ways to select all the features of the car. Thus, the dealer must have 240 sports cars in stock to have one of each kind available.

Permutations

Four sales representatives—Mort, Pam, Quint, and Rhonda—are competing in a sales contest for a first prize (new car), second prize (steak knives), and third prize (continued employment). How many outcomes are possible? By the fundamental counting principle, there are

$$4 \cdot 3 \cdot 2 = 24 \text{ outcomes}$$

Number of   Number of   Number of
ways to select   ways to select   ways to select
first place   second place   third place

Another way to state our result is to say that there are 24 ways to arrange 3 of the 4 representatives. Each particular arrangement of the representatives is called a *permutation* of the representatives. Thus,

$$\text{Mort, Pam, Rhonda} \qquad \text{Mort, Rhonda, Pam}$$

are two permutations of the 4 representatives if 3 are selected at a time. The number of permutations of the 4 representatives selected 3 at a time is denoted as $_4P_3$, read as "4 pick 3." Thus,

$$_4P_3 = 24$$

The notation for the number of permutations in general follows:

Permutation of n Objects Taken r at a Time

If r objects are selected from a set of n objects ($r \leq n$), a particular arrangement of the r objects is called a **permutation of the n objects taken r at a time**. The number of such arrangements is denoted by

$$_nP_r$$

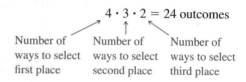 **EXAMPLE 5   Evaluating $_nP_r$ by Listing the Permutations**

Evaluate $_5P_2$ by listing all permutations of the letters a, b, c, d, e taken 2 at a time.

SOLUTION

The permutations of the 5 letters taken 2 at a time are

a, b	*a, c*	*a, d*	*a, e*
b, a	*b, c*	*b, d*	*b, e*
c, a	*c, b*	*c, d*	*c, e*
d, a	*d, b*	*d, c*	*d, e*
e, a	*e, b*	*e, c*	*e, d*

Since there are 20 permutations, $_5P_2 = 20$.

To check the result of Example 5, we can apply the fundamental counting principle to calculate $_5P_2$. Since there are 5 ways to select the first letter, followed by 4 ways to select the second letter, there are $5 \cdot 4 = 20$ possibilities. Thus,

$$_5P_2 = 5 \cdot 4$$

Recall that in our discussion of the permutations of 4 sales representatives (Mort, Pam, Quint, and Rhonda) taken 3 at a time, we also applied the fundamental counting principle to obtain

$$_4P_3 = 4 \cdot 3 \cdot 2$$

These observations can be generalized as follows:

Evaluating $_nP_r$

> Let n and r be nonnegative integers with $r \leq n$. Then
>
> $$_nP_r = n(n - 1)(n - 2) \cdot \cdots \cdot (n - r + 1)$$
>
> or equivalently,
>
> $$_nP_r = \frac{n!}{(n - r)!}$$

◼ EXAMPLE 6 Evaluating $_nP_r$

(a) Evaluate $_{10}P_6$, the number of ways 6 drivers can be assigned to 10 available buses.

(b) Evaluate $_7P_7$, the number of ways 7 managers can be assigned to 7 regions.

SOLUTION

(a) $_{10}P_6 = \dfrac{10!}{(10 - 6)!} = \dfrac{10!}{4!} = \dfrac{10 \cdot 9 \cdot 8 \cdot 7 \cdot 6 \cdot 5 \cdot \cancel{4 \cdot 3 \cdot 2 \cdot 1}}{\cancel{4 \cdot 3 \cdot 2 \cdot 1}} = 151{,}200$

(b) $_7P_7 = \dfrac{7!}{(7 - 7)!} = \dfrac{7!}{0!} = \dfrac{7!}{1} = 7! = 5040$ Recall that $0! = 1$.

You should verify these results using the $_nP_r$ feature on a graphing utility (Figure 8.6). ◼

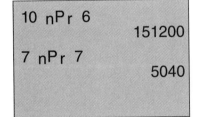

Figure 8.6

```
10 nPr 6
            151200
7 nPr 7
            5040
```

◼ EXAMPLE 7 Application: Selecting Committee Officers

A committee of 12 people must select a chair, vice-chair, and secretary. How many outcomes are possible? Assume no person can hold more than one position.

SOLUTION

We want the number of permutations of 12 people taken 3 at a time, $_{12}P_3$. The $_nP_r$ feature on a graphing utility gives

$$_{12}P_3 = 1320$$

This can also be calculated as follows:

$$_{12}P_3 = \frac{12!}{(12-3)!} = \frac{12!}{9!} = 12 \cdot 11 \cdot 10 = 1320$$

There are 1320 ways to select the officers.

EXAMPLE 8   Application: Portfolio Assignments

In how many ways can 4 different portfolios be assigned to 7 financial advisors if each advisor is to receive no more than one of the portfolios?

SOLUTION

It may be helpful to list a few of the possibilities. Suppose the advisors are Pat, Quint, Ron, Sherry, Tom, Uma, and Vera. Let, for example

Quint, Ron, Pat, Tom

represent the case in which portfolios #1, #2, #3, and #4 are assigned to Quint, Ron, Pat, and Tom, respectively. A few other possible assignments are

Vera, Quint, Ron, Tom
Pat, Quint, Ron, Sherry
Pat, Ron, Quint, Sherry

Each assignment is a permutation of the 7 advisors taken 4 at a time. The number of such assignments is

$$_7P_4 = 840$$

There are 840 ways to assign the 4 portfolios to the 7 advisors.

Combinations

In a problem that calls for permutations, order is important. There are many situations where order is not important. For example, suppose a group of 4 people—Kate, Lou, Marta, and Pham—needs to form a task force of 3. There are 4 possible task forces:

Kate, Lou, and Marta
Kate, Lou, and Pham
Kate, Marta, and Pham
Lou, Marta, and Pham

Notice that the order in which the people are listed does not matter; all that matters is who is on the task force. Thus, the possibilities can be listed as

{Kate, Lou, Marta},   {Kate, Lou, Pham},
{Kate, Marta, Pham},   {Lou, Marta, Pham}

Each possibility is a subset of the set of 4 people selected 3 at a time. This is called a *combination* of the 4 people taken 3 at a time.

■ EXAMPLE 9 Listing the Combinations

List the possible combinations of the letters a, b, c, d, and e taken 2 at a time.

SOLUTION

$\{a, b\}, \{a, c\}, \{a, d\}, \{a, e\}, \{b, c\}, \{b, d\}, \{b, e\}, \{c, d\}, \{c, e\}, \{d, e\}$ ■

The number of combinations of n objects selected r at a time is related to $_nP_r$. Consider forming each of the permutations of n objects selected r at a time as 2 consecutive activities:

Activity 1: Select a combination of r of the n objects.
Activity 2: Arrange the r objects in a particular order.

By the fundamental counting principle,

$$\begin{pmatrix} \text{Number of} \\ \text{permutations of } n \text{ objects} \\ \text{taken } r \text{ at a time} \end{pmatrix} = \begin{pmatrix} \text{Number of ways to} \\ \text{select a combination} \\ \text{of } r \text{ of the } n \text{ objects} \end{pmatrix} \begin{pmatrix} \text{Number of ways} \\ \text{to arrange the} \\ r \text{ objects} \end{pmatrix}$$

Therefore,

$$_nP_r = \begin{pmatrix} \text{Number of ways to} \\ \text{select a combination} \\ \text{of } r \text{ of the } n \text{ objects} \end{pmatrix} (r!)$$

Dividing each side by $r!$, and replacing $_nP_r$ with $n!/(n-r)!$, we have

$$\frac{n!}{r!(n-r)!} = \begin{pmatrix} \text{Number of ways to} \\ \text{select a combination} \\ \text{of } r \text{ of the } n \text{ objects} \end{pmatrix}$$

Notice that the left side of the equation is the formula we used to determine the binomial coefficients, $_nC_r$, in Section 8.1. Therefore, the number of ways to choose (select a combination of) r of the n objects is $_nC_r$.

Combination of n Objects
Taken r at a Time

■ *The formula for $_nC_r$ here is the same as the one used in Section 8.1 for binomial coefficients.*

Suppose r objects are selected from a set of n objects ($r \leq n$). Each subset of the r objects is called a **combination of the n objects taken r at a time**. The number of such subsets is

$$_nC_r = \frac{_nP_r}{r!} = \frac{n!}{r!(n-r)!}$$

In most cases, finding the number of combinations by listing all the possibilities is not practical, and we must calculate $_nC_r$ instead.

EXAMPLE 10   Application: Selecting Job Applicants

A personnel manager must select 5 clerks from a pool of 14 applicants. How many selections are possible?

SOLUTION

The order in which the clerks are selected is not important. Thus, the number of selections is the number of combinations of 14 applicants taken 5 at a time, $_{14}C_5$. Using the $_nC_r$ feature on a graphing utility, we get

$$_{14}C_5 = 2002$$

Alternatively, $_{14}C_5$ can be calculated by the formula

$$_nC_r = \frac{n!}{r!(n-r)!}$$

We have

$$_{14}C_5 = \frac{14!}{5!(14-5)!} = \frac{14!}{5!9!} = \frac{14 \cdot 13 \cdot 12 \cdot 11 \cdot 10 \cdot 9!}{5!9!}$$

$$= \frac{14 \cdot 13 \cdot 12 \cdot 11 \cdot 10}{5 \cdot 4 \cdot 3 \cdot 2 \cdot 1}$$

$$= 2002$$

Therefore, there are 2002 ways in which the personnel manager can select 5 clerks from the 14 applicants.

EXAMPLE 11   Application: The Number of Hands in a Card Game

The game of cribbage is played with a standard deck of 52 cards, and the "crib" hand consists of 4 cards. How many crib hands are possible?

SOLUTION

The number of possibilities is the number of combinations of 52 cards taken 4 at a time, $_{52}C_4$. The $_nC_r$ feature on a graphing utility gives

$$_{52}C_4 = 270{,}725$$

This can also be calculated as follows:

$$_{52}C_4 = \frac{52!}{4!(52-4)!} = \frac{52!}{4!48!} = \frac{52 \cdot 51 \cdot 50 \cdot 49 \cdot \cancel{48!}}{4!\cancel{48!}}$$

$$= \frac{52 \cdot 51 \cdot 50 \cdot 49}{4 \cdot 3 \cdot 2 \cdot 1}$$

$$= 270,725$$

There are 270,725 possible crib hands.

Other Applications

Each of the counting problems we have encountered has called for only one of the techniques we have developed. Many applications require combining several of these techniques.

EXAMPLE 12   Application: Survey Analysis

A reporter decides to survey a sample of 12 of the 48 college football coaches participating in major bowl games to see if they are in favor of a playoff system.

(a) How many samples are possible?

(b) Suppose that 32 of the coaches oppose and 16 favor a playoff system. How many samples of 12 reflect the distribution of the 48 coaches; that is, how many samples are possible with 8 opposing and 4 favoring?

SOLUTION

(a) The number of possible samples is the number of combinations of 48 taken 12 at a time:

$$_{48}C_{12} = 69,668,534,468$$

(b) The selection of such a sample consists of 2 consecutive activities, selecting 8 opposing coaches and selecting 4 favoring coaches. The number of ways to select 8 of the 32 opposing coaches is

$$_{32}C_8 = 10,518,300$$

The number of ways to select 4 of the 16 favoring coaches is

$$_{16}C_4 = 1820$$

By the fundamental counting principle, the number of samples with 8 opposing and 4 favoring is

$$_{32}C_8 \cdot {}_{16}C_4 = 10,518,300 \cdot 1820 = 19,143,306,000$$

▬ EXERCISE SET 8.2

A

In Exercises 1–6, find the number of ways each activity can occur by listing all the possibilities.

1. An appointment is scheduled for lunch hour (noon) or after work (5 PM) on Monday through Friday.

2. Dinner is ordered by selecting exactly one of each choice:
   (1) soup or salad; (2) chicken, steak, or pasta; (3) cake or
   sherbet.

3. A soccer team plays a game in the morning and another
   in the afternoon, and the outcome of each game—win,
   lose, or tie—is recorded.

4. A manager is assigned to one of 5 departments—garden,
   electronics, housewares, furniture, or clothing—at a store
   in one of 3 regions: west, central, or south.

5. In a softball league of 5 teams, 2 teams are selected to
   play a game.

6. From 5 points on a circle, 3 are selected to be vertices of
   a triangle.

In Exercises 7–12, use the fundamental counting principle to
determine the number of ways each activity can occur.

7. A standard die with six sides is rolled, and then a letter is
   selected from the alphabet.

8. A digit (0, 1, 2, ... , 9) is selected, and then a coin is
   flipped.

9. In a photography contest, a blue ribbon is awarded to one
   black-and-white photograph and to one color photograph.
   There are 14 black-and-white entries and 18 color entries.

10. An election is held for president, vice-president, and sec-
    retary. There are 8 candidates for president, 6 candidates
    for vice-president, and 6 candidates for secretary.

11. A market research questionnaire is filled out. The ques-
    tionnaire consists of 5 questions, each of which can be an-
    swered in 4 different ways.

12. (a) A student takes a test consisting of 15 true–false
        questions and guesses the answers to all 15 questions.
    (b) A student takes a test consisting of 15 true–false
        questions and either guesses the answers or leaves
        blanks.

13. Evaluate $_4P_3$ by listing all the permutations of the letters
    a, b, c, d taken 3 at a time.

14. Evaluate $_4P_2$ by listing all the permutations of the letters
    a, b, c, d taken 2 at a time.

15. Evaluate $_6P_2$ by listing all the permutations of the letters
    a, b, c, d, e, f taken 2 at a time.

16. Evaluate $_5C_3$ by listing all the combinations of the letters
    a, b, c, d, e taken 3 at a time.

17. Evaluate $_5C_2$ by listing all the combinations of the letters
    a, b, c, d, e taken 2 at a time.

18. Evaluate $_6C_2$ by listing all the combinations of the letters
    a, b, c, d, e, f taken 2 at a time.

In Exercises 19–24, evaluate each expression.

19. $_5P_4$ 20. $_8P_3$ 21. $_9P_5$

22. $_7P_6$ 23. $_8P_0$ 24. $_{10}P_0$

25. Evaluate $_{12}P_3$, the number of ways 12 runners in a race
    can finish first, second, and third.

26. Evaluate $_{16}P_5$, the number of ways 5 different door prizes
    can be given to 16 participants.

27. Evaluate $_{10}P_{10}$, the number of ways 10 lockers can be as-
    signed to 10 basketball players.

28. Evaluate $_{11}P_{11}$, the number of ways 11 different books can
    be arranged on a shelf.

B

29. In how many ways can a union select a president and
    vice-president from a group of 10 candidates? Assume no
    person can hold more than one position.

30. In how many ways can the chief executive officer of a
    magazine select an editor and an assistant editor from a
    pool of 14 applicants? Assume each applicant applied for
    both positions, and no person can hold both positions.

31. A customer orders a sandwich at a delicatessen by select-
    ing one type of meat, cheese, and bread. He also specifies
    whether or not he wants onions. If there are 4 kinds of

meat, 5 kinds of cheese, and 3 kinds of bread, how many sandwiches are possible?

32. A coffee retailer offers the following options:

Variety: French, Colombian, Kona, Italian, Kenyan
Roast: Dark, medium, light
Grind: Whole bean, coarse, fine
Bag size: $\frac{1}{2}$ pound, 1 pound, 2 pounds

If the customer must specify exactly one of each option for an order, how many orders are possible?

33. A traffic court classifies violators as to whether they have a current license, whether they have a current registration, whether it is a moving violation, and whether they have committed another violation in the previous year. In how many ways can a violator be classified?

34. A builder may select from 3 plumbing subcontractors, 3 electrical subcontractors, 4 flooring subcontractors, and 2 roofing subcontractors. In how many ways can the builder select one of each?

35. An office manager is to distribute 5 different referrals to 6 agents. If each agent is to receive no more than one of the referrals, in how many ways can the referrals be distributed?

36. In how many ways can a congressional committee award 6 different defense contracts to 9 competing companies? Assume a company can win no more than one contract.

37. A network executive must assign 4 different time slots to 4 daytime dramas. In how many ways can this be done?

38. A personnel director will ask exactly one question of each of 6 candidates. In how many ways can the personnel director assign 6 different questions to the 6 candidates?

39. How many different license plates that contain 3 letters (selected from A through Z) followed by 3 digits (selected from 0 through 9) are possible?

40. The phone numbers in a certain region begin with 83 followed by 5 digits (selected from 0 through 9). How many phone numbers are possible in this region?

41. In order to ensure that only purchasers are using their products, many software companies require entering codes to install programs. If a particular code has 7 characters consisting of digits or capital letters, how many codes are possible?

42. Automated bank tellers require entering a personal identification number to make transactions. If each identification number consists of a nonzero digit followed by any 3 digits, how many different identification numbers can the bank issue?

43. In how many ways can a subcommittee of 4 be formed from a committee of 10 people?

44. In how many ways can a 10 person softball team select 3 co-captains?

45. A poker hand consists of 5 cards dealt from a standard deck of 52 cards. How many poker hands are possible?

46. A cribbage player receives 6 cards from a standard deck of 52 cards. In how many ways can this be done?

47. A standard deck of 52 cards consists of 13 spades, 13 diamonds, 13 clubs, and 13 hearts. How many 5 card hands consisting of only hearts are possible?

48. A standard deck of 52 cards contains 12 "face cards" (4 jacks, 4 queens, and 4 kings). How many 6 card hands consisting of only face cards are possible?

49. A club has 10 male and 14 female members. A committee of 3 men and 3 women is to be formed.
    (a) In how many ways can the 3 men be selected?
    (b) In how many ways can the 3 women be selected?
    (c) In how many ways can the committee be selected?
       HINT: Apply the fundamental counting principle, and use the answers to parts (a) and (b).

50. A school board election has 8 candidates from the east district and 7 candidates from the west district. There are 2 positions to be filled with candidates from the east district and 3 positions to be filled with candidates from the west district.
    (a) In how many ways can the 2 east district positions be filled?
    (b) In how many ways can the 3 west district positions be filled?
    (c) How many outcomes are possible for the election?
       HINT: Apply the fundamental counting principle, and use the answers to parts (a) and (b).

51. A network news program surveys a sample of 15 of the 100 U.S. senators to find out if they are in favor of the president's foreign policy.

   (a) How many samples are possible?

   (b) Suppose that 67 of the senators are in favor and 33 oppose the president's foreign policy. How many samples of 15 reflect the distribution; that is, how many samples are possible with 10 in favor and 5 opposing?

52. An editor for a business newsletter knows of 36 stock analysts with enough expertise to comment on the policy of the Federal Reserve. The editor plans to select a panel of 9 of these analysts to comment on whether recent policy will stimulate growth in the stock market.

   (a) How many selections are possible?

   (b) Suppose that 24 of the analysts feel that Federal Reserve policy will stimulate growth and 12 do not. The editor decides to maintain this distribution, so the panel will consist of 6 who believe growth will result and 3 who do not. In how many ways can such a panel be selected?

C

53. In the game of poker, 5 cards are dealt to each player. A "full house" is a hand that consists of 3 of one kind and 2 of another. For example, a hand with 3 tens and 2 kings is a "full house of tens over kings." In how many ways can a full house be dealt?

54. In a softball league, each team plays the other exactly once. If the league plays a total of 91 games, how many teams are in the league?

55. **(a)** If a coin is flipped 2 times, in how many ways can the outcome be all heads? 1 head and 1 tail? All tails?

   (b) If a coin is flipped 3 times, in how many ways can the outcome be all heads? 2 heads and 1 tail? 1 head and 2 tails? All tails?

   (c) If a coin is flipped 4 times, in how many ways can the outcome be all heads? 3 heads and 1 tail? 2 heads and 2 tails? 1 head and 3 tails? All tails?

   (d) Compare the results from parts (a)–(c) with Pascal's triangle. Conjecture the number of ways to get a heads and b tails if a coin is flipped $a + b$ times.

56. Find n such that $n!$ has exactly n digits.

SECTION 8.3

- Sample Spaces and Events
- Probability of an Event
- Calculating Probabilities with Counting Techniques
- Other Applications

PROBABILITY

The theory of *probability* originated in the seventeenth century and was motivated by the study of games of chance. But there are many other aspects of our daily life that involve the likelihood of certain outcomes or consequences. When a meteorologist predicts the weather, a jury decides on guilt or innocence, a public-opinion poll predicts the winner of an election, or a doctor considers the success rate of a particular treatment, uncertainty exists. Probability provides a measure of the amount of certainty or likelihood of an outcome. In this section we introduce the basic notions of probability.

Sample Spaces and Events

In order to discuss probability, we need to introduce some terminology. An *experiment* is an activity or process in which the result or outcome is unknown. For example, rolling a die is an experiment in which there are six possible outcomes: 1, 2, 3, 4, 5, or 6. A *sample space* is the set of all possible outcomes of an experiment. In this section we use S to represent the sample space. Thus, for the experiment of rolling a die, we have

$$S = \{1, 2, 3, 4, 5, 6\}$$

Any subset of the sample space is called an *event*. For example, the subset

$$E = \{5, 6\}$$

is the event of rolling a number greater than 4.

▮ EXAMPLE 1 Specifying the Sample Space and Events

Suppose a coin is tossed 3 times and the sequence of heads or tails is noted.

(a) Determine the sample space, S.
(b) Specify the event A of obtaining exactly 2 heads.
(c) Specify the event B of obtaining at least 2 heads.

SOLUTION

(a) In Example 1b in Section 8.2, we listed the 8 possible outcomes of tossing a coin 3 times. We use the same notation for each outcome; for example, TTH represents the first toss is tails, the second is tails, and the third is heads. The sample space is

$$S = \{HHH, HHT, HTH, HTT, THH, THT, TTH, TTT\}$$

(b) Event A is the event of tossing exactly 2 heads, meaning 2 heads and only 2 heads, so

$$A = \{HHT, HTH, THH\}$$

(c) Event B consists of the outcomes of tossing 2 heads or more; thus,

$$B = \{HHH, HHT, HTH, THH\} \qquad ▮$$

Probability of an Event

The number of outcomes in an event E is denoted as $n(E)$. In Example 1, we have

$$n(A) = n(\{HHT, HTH, THH\}) = 3$$
$$n(B) = n(\{HHH, HHT, HTH, THH\}) = 4$$
$$n(S) = n(\{HHH, HHT, HTH, HTT, THH, THT, TTH, TTT\}) = 8$$

We will use the notation $P(E)$ for the probability of an event E. Since each outcome in the sample space S in Example 1 is equally likely, the probability of event A is the ratio of $n(A)$ and $n(S)$:

$$P(A) = \frac{n(A)}{n(S)} = \frac{3}{8}$$

Similarly, the probability of event B in Example 1 is given by

$$P(B) = \frac{n(B)}{n(S)} = \frac{4}{8} = \frac{1}{2}$$

In general, we have the following:

Probability of an Event

> Let S and E be the sample space and an event of an experiment, respectively. If S consists of $n(S)$ equally likely outcomes and E consists of $n(E)$ outcomes, then the **probability** of event E is given by
>
> $$P(E) = \frac{n(E)}{n(S)}$$

Since an event E is a subset of the sample space S,

$$0 \le n(E) \le n(S)$$

Dividing all three sides by $n(S)$ gives

$$0 \le P(E) \le 1$$

If $P(E) = 0$, then the event E is impossible; if $P(E) = 1$, then E is certain to occur. In practice, $P(E)$ is usually between 0 and 1.

EXAMPLE 2 Selecting a Card from a Deck

A single card is drawn from a standard deck of 52 cards.

(a) Find the probability that the card is a jack, queen, or king. (These cards are called "face" cards.)

(b) Find the probability that the card is a diamond.

SOLUTION

(a) Let F be the event that the card is a face card. Since there are 4 jacks, 4 queens, and 4 kings in a standard deck of 52, there are 12 face cards in the deck. Thus, the probability is

$$P(F) = \frac{n(F)}{n(S)} = \frac{12}{52} = \frac{3}{13} = 0.2307...$$

(b) Let D represent the event of drawing a diamond. There are 13 diamonds in the deck, so the probability is

$$P(D) = \frac{n(D)}{n(S)} = \frac{13}{52} = \frac{1}{4} = 0.25$$

■

■ **EXAMPLE 3   Rolling Two Dice**

Suppose two dice are rolled and the two numbers that turn up are noted. Determine the probability of each event.

(a) The sum of the numbers is 6.
(b) The two numbers that turn up are the same (called "doubles").
(c) The sum of the numbers is 7 or 11.

SOLUTION

It is convenient to picture the dice having two different colors. A representation of the sample space is shown in Figure 8.7. Notice, for example, that the ordered pair (2, 5)—representing a 2 on the red die and 5 on the white die—is not the same as (5, 2). Thus, the sample space S has 36 equally likely outcomes:

$$n(S) = 36$$

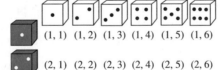

	(1, 1)	(1, 2)	(1, 3)	(1, 4)	(1, 5)	(1, 6)
	(2, 1)	(2, 2)	(2, 3)	(2, 4)	(2, 5)	(2, 6)
	(3, 1)	(3, 2)	(3, 3)	(3, 4)	(3, 5)	(3, 6)
	(4, 1)	(4, 2)	(4, 3)	(4, 4)	(4, 5)	(4, 6)
	(5, 1)	(5, 2)	(5, 3)	(5, 4)	(5, 5)	(5, 6)
	(6, 1)	(6, 2)	(6, 3)	(6, 4)	(6, 5)	(6, 6)

Figure 8.7

(a) Let A be the event of rolling a sum of 6. From Figure 8.7, we list the ways to roll a sum of 6.

$$A = \{(1, 5), (2, 4), (3, 3), (4, 2), (5, 1)\}$$

Event A has 5 equally likely outcomes. So the probability of rolling a sum of 6 is given by:

$$P(A) = \frac{n(A)}{n(S)} = \frac{5}{36} = 0.1388...$$

(b) Let B represent the event of rolling the same number on each of the dice. From Figure 8.7, we have

$$B = \{(1, 1), (2, 2), (3, 3), (4, 4), (5, 5), (6, 6)\}$$

Thus, the probability of rolling doubles is

$$P(B) = \frac{n(B)}{n(S)} = \frac{6}{36} = \frac{1}{6} = 0.1666...$$

(c) If we let C represent the event of rolling a sum of 7 or 11, then

$$C = \{(1, 6), (2, 5), (3, 4), (4, 3), (5, 2), (6, 1), (5, 6), (6, 5)\}$$

and the probability is

$$P(C) = \frac{n(C)}{n(S)} = \frac{8}{36} = \frac{2}{9} = 0.2222...$$

Calculating Probabilities with Counting Techniques

Often, listing all the outcomes of a sample space or event is difficult or impractical. For example, consider the experiment of randomly entering a password that consists of three capital letters followed by a digit. In Example 3 of Section 8.2, we found that the number of possibilities is

$$26 \cdot 26 \cdot 26 \cdot 10 = 175{,}760$$

EXAMPLE 4   Probability of a Hand of Cards

Three cards are dealt from a standard deck of 52 cards. Determine the probability that all 3 cards are hearts.

SOLUTION

The sample space S consists of all possible 3 card hands. The order in which the cards are dealt does not distinguish one hand from another. For example, if the first card is the jack of diamonds, the second is the ten of clubs, and the third is the four of spades, the hand is the same as when the first card is the ten of clubs, the second is the four of spades, and the third is the jack of diamonds. Thus, the number of 3 card hands is

$$n(S) = {}_{52}C_3 = 22{,}100$$

There are 13 hearts in a deck. Letting H represent the event that all 3 cards are hearts, we have

$$n(H) = {}_{13}C_3 = 286$$

Therefore, the probability that all 3 cards are hearts is

$$P(H) = \frac{n(H)}{n(S)} = \frac{286}{22,100} = \frac{11}{850} = 0.0129...$$

EXAMPLE 5 Application: Processing Area Codes

For many years, area codes for phone numbers consisted of 3 digit numbers where the middle digit was a 0 or 1. New area codes, using a middle digit of $2-9$, began appearing in 1995 when cellular phones and facsimile machines created a dramatic increase in demand for phone numbers. At that time, most of the private business switchboards in the United States needed to be upgraded because they could only process the traditional area codes. If a sequence of 3 digits is selected at random, determine the probability that an outmoded private switchboard can process the sequence as an area code.

SOLUTION

Let the sample space S consist of all possible ways to select a sequence of 3 digits. There are 10 ways to select the first digit, 10 ways to select the second digit, and 10 ways to select the third digit. Using the fundamental counting principle,

$$n(S) = 10 \cdot 10 \cdot 10 = 1000$$

Let E represent the event that the outmoded private switchboard can process the area code. Since it can only process middle digits of 0 or 1, the number of area codes it can process is

$$n(E) = 10 \cdot 2 \cdot 10 = 200$$

Thus, the probability that it is able to process the area code is

$$P(E) = \frac{n(E)}{n(S)} = \frac{200}{1000} = 0.2$$

EXAMPLE 6 Application: DNA Sequencing

Since the late 1970's, methods of determining the nucleotide sequence of a DNA molecule have been developed. The sequence essentially consists of a list of four letters—A, T, C, G. Suppose we are given the first 10 terms of a DNA sequence, such as TCCAGCTACT. If the first 10 terms of another DNA sequence are selected at random, determine the probability that the letters match the given sequence.

SOLUTION

Using the fundamental counting principle, the number of ways to form the first 10 terms of a DNA sequence is

$$n(S) = 4 \cdot 4 \cdot 4 \cdot 4 \cdot 4 \cdot 4 \cdot 4 \cdot 4 \cdot 4 \cdot 4 = 1{,}048{,}576$$

If E represents the event of selecting a sequence that matches the given sequence, then E contains only the given sequence, so

$$n(E) = 1$$

Thus, the probability of a match is

$$P(E) = \frac{n(E)}{n(S)} = \frac{1}{1{,}048{,}576} = 0.0000009\ldots$$

Consider the experiment of tossing a coin 3 times and noting the outcome. Using the notation from Example 1, the sample space S is

$$S = \{HHH, HHT, HTH, HTT, THH, THT, TTH, TTT\}$$

Let E be the event that exactly 2 of the flips are heads. We have

$$E = \{HHT, HTH, THH\}$$

and

$$P(E) = \frac{3}{8}$$

Let the event $\overline{E}$ represent the set of all outcomes (in S) other than those in E. Specifically,

$$\overline{E} = \{HHH, HTT, THT, TTH, TTT\}$$

The event $\overline{E}$ is called the *complement of E*. The probability of E is related to the probability of $\overline{E}$ as follows:

$$P(\overline{E}) = \frac{5}{8} = 1 - \frac{3}{8} = 1 - P(E)$$

A generalization of the relationship between the probability of an event and the probability of its complement follows:

Complement of an Event

Let S and E be the sample space and an event of an experiment, respectively. The **complement of E**, denoted $\overline{E}$, is the set of all outcomes in S that are not in E. The probability of E and the probability of $\overline{E}$ are related as follows:

$$P(E) = 1 - P(\overline{E})$$

EXAMPLE 7 Application: Finding the Probability of an Event by Using the Complement

The Internal Revenue Service has identified 20 tax returns that appear to need audits. Budget restrictions permit the IRS to conduct audits on only 4 of the 20. Suppose that 8 of the 20 contain illegal deductions, and the remaining 12 do not. If the IRS selects 4 returns at random, what is the probability that at least 1 of them contains illegal deductions?

SOLUTION

Let E represent the event that at least 1 return with illegal deductions is among the 4 selected. Calculating $P(E)$ directly requires considering several cases, namely, if exactly 1, 2, 3, or 4 returns with illegal deductions are selected. Instead, consider the complement, $\overline{E}$, which is the event that no returns with illegal deductions are selected. Since there are 12 such returns, the probability is given by

$$P(\overline{E}) = \frac{n(\overline{E})}{n(S)} = \frac{{}_{12}C_4}{{}_{20}C_4} = \frac{495}{4845} = \frac{33}{323}$$

Therefore, the probability that at least 1 return with illegal deductions is among the 4 selected is

$$P(E) = 1 - P(\overline{E}) = 1 - \frac{33}{323} = \frac{290}{323} = 0.8978...$$

Other Applications

Probability is used frequently by manufacturers to set standards for the quality of their products and by the government in testing the quality of various products for consumers.

EXAMPLE 8 Application: Quality Control

A computer chip manufacturer requires that 4 chips be randomly selected and tested from each production lot of 25 chips. If 1 or more of the 4 chips tested is defective, the entire lot is rejected. Suppose a lot contains 2 defective chips and 23 satisfactory chips. Determine the probability that the lot will pass inspection.

SOLUTION

The number of ways to select 4 chips from a lot of 25 is the number of outcomes for the sample space S:

$$n(S) = {}_{25}C_4 = 12{,}650$$

Let E represent the event of selecting 4 satisfactory chips. Since there are 23 satisfactory chips in the lot, the number of ways to select 4 of them is

$$n(E) = {}_{23}C_4 = 8855$$

Consequently, the probability that the lot will pass inspection is

$$P(E) = \frac{n(E)}{n(S)} = \frac{8855}{12,650} = \frac{7}{10} = 0.7$$

EXERCISE SET 8.3

A

In Exercises 1–4, determine the sample space and the indicated event for each experiment.

1. Experiment: Flip a coin 4 times.
   Event: Obtain at least 2 heads.

2. Experiment: Roll a die and flip a coin.
   Event: Obtain an even number on the die.

3. Experiment: Select 2 marbles from a bowl containing 1 red, 1 yellow, 1 green, and 1 blue marble.
   Event: Obtain a blue marble.

4. Experiment: Select 3 people as representatives from a committee consisting of Al, Beth, Carmen, Don, and Ed.
   Event: Beth does not get selected.

In Exercises 5–8, refer to the experiment of drawing 1 card from a standard deck of 52 cards described in Example 2. Determine the probability of each event.

5. The card is a nine of hearts or a ten of hearts.

6. The card is a queen of spades or a jack of diamonds.

7. The card is a five, six, seven, or eight.

8. The card is an ace of hearts or diamonds.

In Exercises 9–16, refer to the experiment of rolling two dice described in Example 3. Determine the probability of each event.

9. A sum of 2 turns up.   10. A sum of 4 turns up.

11. A sum of 5 turns up.   12. A sum of 8 turns up.

13. The two numbers that turn up are both even.

14. One or both of the two numbers that turn up is a 1.

15. A sum of at least 10 turns up.

16. A sum of 5 or less turns up.

In Exercises 17–20, consider the experiment of randomly drawing 1 bead from a bowl containing 6 red beads, 4 blue beads, and 2 green beads. Determine the probability of each event.

17. The bead is red.   18. The bead is green.

19. The bead is *not* blue.   20. The bead is *not* green.

B

21. A chair and a secretary are selected at random among the members of a committee that consists of 10 Republicans and 7 Democrats. Determine the probability that:
    (a) The chair is a Republican and the secretary is a Democrat.
    (b) Both the chair and the secretary are Democrats.
    (c) The chair is a Republican.

22. A student's music collection consists of 30 compact discs. Eight of them are classical music, 15 are jazz, and 7 are rap. She randomly selects 1 CD for her roommate and 1 CD for herself. Determine the probability that:
    (a) Each gets a jazz CD.
    (b) She gets a classical music CD and her roommate gets a rap CD.
    (c) Her roommate gets a jazz CD.

In Exercises 23–26, consider the experiment of randomly drawing a pair of beads from a bowl containing 6 red, 4 blue, and 2 green beads. Determine the probability of each event.

23. Both beads are red.　　**24.** Both beads are blue.

25. None of the beads are blue.

26. None of the beads are green.

27. A bottle of vitamins contains 26 potent capsules and 4 capsules that have lost their potency. A pair of capsules are randomly taken. Determine the probability that:
(a) They are both potent.
(b) At least one is potent.

28. Suppose that 60 Peace Corps volunteers return to the United States from Africa, and that 5 of them have unknowingly contracted a parasitic disease. If the government randomly selects 10 of the 60 volunteers to test for the disease, what is the probability that at least 1 of the 10 selected has it?

29. Four cards are dealt from a standard deck of 52 cards. Determine the probability that:
(a) They are all face cards.
(b) At least 1 is a face card.

30. Three cards are dealt from a standard deck of 52 cards. Determine the probability that:
(a) They are all black.
(b) At least 1 is black.

31. Four cards are dealt from a standard deck of 52 cards. Determine the probability that 2 are diamonds and 2 are clubs.

32. Five cards are dealt from a standard deck of 52 cards. Determine the probability that 3 are red and 2 are spades.

33. An exam has 5 true–false questions. Find the probability that a student guessing at all the answers gets:
(a) All the answers correct.
(b) Four right and 1 wrong.

34. A couple decides to have 4 children. Find the probability that:
(a) All children are girls.
(b) Three of the children are girls and 1 is a boy.

35. A briefcase lock has 3 wheels, each with the digits 1–9. Determine the probability of correctly guessing the sequence of 3 numbers to unlock the briefcase.

36. The password for a computer account consists of a sequence of 4 characters. Each character can be a capital letter, a digit, or a blank. Determine the probability of correctly guessing the password.

37. Suppose 2 people are selected at random and their social security numbers are compared. What is the probability that the last 4 digits match?

38. Suppose 3 people compare the day of the week when they were born. What is the probability that at least 2 of them were born on the same day of the week?

39. Suppose 4 people compare the day of the week when they were born. What is the probability that at least 2 of them were born on the same day of the week?

40. A state lottery game is played by choosing 6 different integers from 1 through 49. If the 6 chosen numbers match the 6 numbers randomly drawn by lottery officials, in any order, then the player wins the grand prize. Determine the probability of winning the grand prize.

41. The state lottery game described in Exercise 40 also awards a lesser prize if 5 of the 6 chosen numbers match. Determine the probability of winning this lesser prize.

42. To select employees for a psychic telephone network, an employer tests each applicant's psychic ability by asking them to predict the outcome of 5 tosses of a coin. The applicant is certified if she/he correctly predicts at least 4 of the 5 outcomes. (Assume that none of the applicants has psychic ability.)
(a) Determine the probability that a person guessing randomly will be certified.
(b) If 100 people are tested, approximately how many of them would you expect to be certified?

C

43. Let a, b, and c be three positive digits, not necessarily different, chosen at random.
(a) Determine the probability that the product ab is odd.
(b) Determine the probability that $ab + c$ is odd.

44. Let a, b, and c be three positive digits less than 5, not necessarily different, chosen at random. Determine the probability that the quadratic trinomial $ax^2 + bx + c$ factors as a product of two binomials over the integers.

45. Consider a region with area A that contains a subregion with area a (Figure 8.8). The probability that a randomly selected point from the region is in the subregion is

$$\frac{\text{Area of subregion}}{\text{Area of region}} = \frac{a}{A}$$

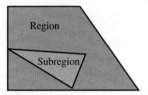

Figure 8.8

Now, suppose a circular disk with a radius of 1 inch is randomly tossed onto a grid consisting of squares that are 3 inches on a side (Figure 8.9). Determine the probability that the disk does not cross any of the grid lines. (In other words, the disk is completely contained in one of the squares.)

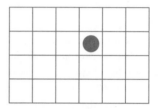

Figure 8.9

QUICK REFERENCE

Topic	Page	Remarks
Pascal's triangle	624	Pascal's triangle is the triangular array of binomial coefficients on page 624. In each row, the first and last entry is 1; each of the remaining entries is the sum of the entries diagonally above.
Binomial coefficients	627	The binomial coefficients are the coefficients of the terms in the expanded form of $(a + b)^n$, for $n = 0, 1, 2, \ldots$.
Binomial coefficient, $_nC_r$	627	Let n and r be nonnegative integers with $r \le n$. The binomial coefficient $_nC_r$ is defined as $$_nC_r = \frac{n!}{r!(n - r)!}$$
Binomial theorem	628	$$(a + b)^n = {_nC_0}a^n + {_nC_1}a^{n-1}b + \cdots + {_nC_r}a^{n-r}b^r + \cdots + {_nC_n}b^n$$ In summation notation, $$(a + b)^n = \sum_{k=0}^{n} {_nC_k}a^{n-k}b^k.$$
Fundamental counting principle	634	If an activity consists of two tasks in which there are m outcomes for the first task and n outcomes for the second task, then the entire activity has $m \cdot n$ outcomes.
Permutation	636	A permutation is an ordered arrangement of objects.

Topic	Page	Remarks
Number of permutations, $_nP_r$	636, 637	Let n and r be nonnegative integers with $r \leq n$. The number of permutations of n objects taken r at a time is $$_nP_r = n(n-1)(n-2) \cdot \cdots \cdot (n-r+1)$$ $$= \frac{n!}{(n-r)!}$$
Combination	639	A combination is a selection of objects without regard to order. More precisely, given a set of n objects, form a subset of r objects ($r \leq n$). The subset of r objects is a combination of the n objects taken r at a time.
Number of combinations, $_nC_r$	639	Let n and r be nonnegative integers with $r \leq n$. The number of combinations of n objects taken r at a time is $$_nC_r = \frac{_nP_r}{r!} = \frac{n!}{r!(n-r)!}$$
Sample space	644	A sample space is the set of all possible outcomes of an experiment.
Event	645	An event is a subset of the sample space.
Probability of an event, $P(E)$	646	For a given experiment, let S be the sample space consisting of equally likely outcomes. Let $n(S)$ and $n(E)$ denote the number of outcomes in S and event E, respectively. The probability of E, denoted $P(E)$, is given by $$P(E) = \frac{n(E)}{n(S)}$$
Complement of an event	650	Let S and E be the sample space and an event, respectively, of an experiment. The complement of E is the set of all outcomes in S that are not in E. The complement of E is denoted $\overline{E}$. The probabilities of E and $\overline{E}$ are related by $$P(E) = 1 - P(\overline{E})$$

MISCELLANEOUS EXERCISES

In Exercises 1–8, evaluate each expression.

1. $_{14}C_3$ **2.** $_{11}C_4$ **3.** $_8C_8$

4. $_7C_0$ **5.** $_6P_3$ **6.** $_9P_5$

7. $_{15}P_0$ **8.** $_6P_6$

In Exercises 9–12, use Pascal's triangle to expand each expression and simplify.

9. $(m+2)^3$ **10.** $(3s+1)^4$ **11.** $(x-2y)^6$

12. $(2u-v)^5$

In Exercises 13–16, use the binomial theorem to expand each expression and simplify.

13. $(p+q^2)^9$ **14.** $(w^3+z)^8$ **15.** $\left(4f - \frac{1}{2}\right)^8$

16. $\left(\frac{x}{3} - 6\right)^6$

17. Determine the p^9q^3 term in the expansion of $(p+2q)^{12}$.

18. Determine the a^2b^7 term in the expansion of $(5a-b)^9$.

19. Determine the first four terms in the expansion of $(x + 3y^2)^{10}$.

20. Determine the first four terms in the expansion of $(x^3 + 2y)^9$.

In Exercises 21 and 22, find and simplify

$$\frac{g(x + h) - g(x)}{h}$$

21. $g(x) = x^4$ **22.** $g(x) = x^3$

23. A certain type of seed is known to germinate 98% of the time. Fifteen of these seeds are planted.
  (a) The probability that exactly r of the seeds germinate is

$$_{15}C_r(0.02)^{15-r}(0.98)^r$$

  Determine (to the nearest 0.001) the probability that exactly 13 of the seeds germinate.
  (b) The probability that at least r of the seeds germinate is

$$\sum_{k=r}^{15} {}_{15}C_k(0.02)^{15-k}(0.98)^k$$

  Determine (to the nearest 0.001) the probability that at least 13 of the seeds germinate.

24. Refer to Exercise 23.
  (a) Determine (to the nearest 0.001) the probability that exactly 12 of the seeds germinate.
  (b) Determine (to the nearest 0.001) the probability that at least 12 of the seeds germinate.

In Exercises 25 and 26, find the number of ways each activity can occur by listing all the possibilities.

25. Three different company logos are arranged horizontally at the top of a letter.

26. Two representatives are selected from a group of 5 people.

27. How many 5 digit numbers can be formed using the digits 1, 2, 3, 4, 5, 6, 7, and 8 if repetitions are not allowed?

28. How many 5 digit numbers can be formed using the digits 1, 2, 3, 4, 5, 6, 7, and 8 if repetitions are allowed?

29. In a tennis tournament, every player plays every other player exactly once. How many matches must be scheduled if 12 players enter the tournament?

30. Each person in a club must share the secret handshake with every other person in the club exactly once. If there are 10 people in the club, how many handshakes occur?

31. Twenty people enter an essay contest in which scholarships for first, second, and third place are awarded. How many outcomes are possible?

32. Many radio stations are named by 4 call letters that begin with a W or K, and end with any 3 letters. Examples are KKSF in San Francisco and WTKS in Orlando. How many 4 letter names are possible?

33. Given 6 points, no 3 of which are collinear, how many triangles can be formed using the points as vertices? (One example is shown.)

34. Given 7 points, no 3 of which are collinear, how many line segments can be formed using the points as endpoints? (One example is shown.)

35. A committee of 3 men and 4 women is to be formed from a group of 10 men and 9 women. How many committees are possible?

36. A standard deck of 52 cards consists of 13 spades, 13 diamonds, 13 clubs, and 13 hearts. How many 6 card hands consisting of 2 spades and 4 hearts are possible?

37. Three teams of 6 players are to be formed from 18 players. In how many ways can this be done?

38. Three construction crews of 4 workers are to be formed from 12 workers. In how many ways can this be done?

39. A contractor wants to build different model homes on 3 available lots. If there are 5 different models to choose from, in how many ways can this be done?

40. A signal consists of 3 flags arranged vertically on a flagpole. If there are 6 different flags, how many signals are possible?

41. A student taking a multiple-choice test omits any 3 of the 10 questions, and responds to the remaining 7 questions by marking a, b, c, d, or e. In how many ways can this be done?

42. An employee must select 3 out of 7 holidays to work. For each working holiday selected, the employee also selects whether she works on the day shift or night shift. How many different selections are possible?

In Exercises 43 and 44, determine the sample space and the indicated event for each experiment.

43. Experiment: A spinner with the numbers 1, 2, and 3 is spun twice.

Event: Obtain at least one 3.

44. Experiment: Flip a coin. If it turns up heads, select a bead from a bowl with a red bead and a green bead. If the coin turns up tails, select a bead from a bag with a red bead and a blue bead.
Event: Obtain a red bead.

In Exercises 45 and 46, 1 card is drawn from a standard deck of 52 cards. Determine the probability of each event.

45. The card is a diamond or an ace (or both).

46. The card is a club or spade, or a king (or both).

In Exercises 47 and 48, two dice are rolled. Determine the probability of each event.

47. The numbers are 1 apart.

48. One number is a multiple of the other.

In Exercises 49–54, 4 cards are dealt from a standard deck of 52 cards. Determine the probability of each event.

49. All 4 cards are spades.

50. None of the cards are hearts.

51. At least 1 of the cards is red.

52. At least 1 of the cards is a five, six, or seven.

53. Each card is a different suit. (In other words, there is 1 club, 1 spade, 1 heart, and 1 diamond.)

54. There is 1 king, 1 queen, 1 jack, and 1 nonface card.

55. Two batteries are randomly chosen from a carton containing 4 dead batteries and 16 good ones. What is the probability that they are both reliable?

56. Three fuses are randomly chosen from a package containing 2 defective fuses and 13 reliable fuses. What is the probability that they are all reliable?

57. Automatic teller machines use personal identification numbers (PIN) consisting of 4 digits. Suppose someone tries to guess another person's PIN. Determine the probability that:
   (a) None of the 4 digits match the PIN.
   (b) At least 1 of the 4 digits match the PIN.

58. The automobile license plates in a certain state consist of 3 letters followed by 3 digits. Determine the probability that:
   (a) The letters are all different.
   (b) At least 2 of the letters are the same.

59. Determine the probability that 4 people seated in a row are in alphabetical order.

60. Determine the probability that the first 5 phone numbers in a phone book are in decreasing order.

CHAPTER TEST

1. **(a)** Write the first six rows of Pascal's triangle.
   (b) Use Pascal's triangle to expand $(z + 1)^5$.

2. Write $(f - 2g)^4$ in expanded form and simplify.

3. Determine the x^3y^6 term in the expansion of $(2x + y)^9$.

4. Determine the first four terms in the expansion of $(\sqrt{a} - 3b^2)^{10}$.

5. Let $f(x) = 7x^4$. Determine and simplify

   $$\frac{f(x + h) - f(x)}{h}$$

6. **(a)** List all the permutations of the letters w, x, y, and z taken 2 at a time.
   (b) List all the combinations of the letters w, x, y, and z taken 2 at a time.
   (c) Determine the number of permutations of 10 objects taken 3 at a time.
   (d) Determine the number of combinations of 10 objects taken 3 at a time.

7. A student may earn a grade of A, B, C, D, or F in each of 4 subjects. How many outcomes are possible?

8. In how many ways can 6 different boxes of cereal be arranged on a display shelf?

9. In how many ways can 2 co-captains be selected from a team of 12 players?

10. In how many ways can 6 people be seated on a bus with 8 empty seats?

11. Suppose two dice are rolled. Determine the probability that the sum of the numbers that turn up is 8 or 9.

12. A hand of 4 cards is dealt from a deck of 52 cards.
    (a) How many hands are possible?
    (b) What is the probability that all the cards are clubs?
    (c) What is the probability that at least 1 of the cards is a club?

13. A committee of 5 people is to be randomly selected from a group consisting of 8 teachers, 5 counselors, and 3 administrators. What is the probability that the chosen committee consists of 3 teachers and 2 counselors?

14. An automobile license plate consists of 3 letters followed by 3 digits. What is the probability that all the letters are different and all the digits are different?

LINEAR PROGRAMMING

In Example 6 of Section 5.8, we analyzed an investment in stocks, bonds, and real estate. We restate the problem here:

A \$15 million pension fund is to be invested in stocks, bonds, and real estate. A regulatory board requires that the amount of each type of investment is at least 20% of the total investment, and the money invested in bonds must be at least half that invested in stocks.

(a) Let x represent the amount (in millions of dollars) invested in stocks, and let y represent the amount (in millions of dollars) invested in bonds. Determine a system of inequalities that models the regulations.

(b) Graph the system of inequalities.

(c) What is the maximum amount that can be invested in stocks?

In the solution to part (a), we found that the regulations are modeled by the system

$$\begin{cases} x \geq 3 \\ y \geq 3 \\ x + y \leq 12 \\ y \geq \frac{1}{2}x \end{cases}$$

Since x represents the amount invested in stocks, parts (a) and (c) can be expressed as follows:

Find the maximum value of x

subject to $\begin{cases} x \geq 3 \\ y \geq 3 \\ x + y \leq 12 \\ y \geq \frac{1}{2}x \end{cases}$

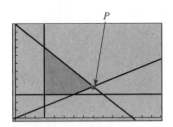

Figure A.1
$[0, 15] \times [0, 12]$

This is an example of a *linear programming problem*. In general, a linear programming problem involves finding the values of variables, say x and y, that maximize (or minimize) a linear expression in those variables, where x and y must satisfy restrictions in the form of linear inequalities. The quantity to be maximized (or minimized) is called the *objective function*, and the inequalities are called the *constraints*. The solutions of the system of inequalities are the *feasible solutions*. The graph of the system of inequalities determines the *feasible region*. In the solution of Example 6 in Section 5.8, we determined the graph of the feasible solutions, and found that the maximum value of x corresponds to the vertex $P(8, 4)$, as shown in Figure A.1. In fact, the maximum value of any linear expression in x and y subject to the same constraints will occur at one of the vertices of the region.

For example, what is the maximum value of $z = 3x + 4y$? If we rewrite this equation in the form

$$y = -\frac{3}{4}x + \frac{z}{4}$$

we see that each value of z corresponds to a line with slope $-\frac{3}{4}$ and y-intercept $z/4$. Figure A.2 shows the feasible region and the graphs of the lines $y = -\frac{3}{4}x + (z/4)$ for various values of z.

Notice the case where $z = 53$. Since the graph of

$$y = -\frac{3}{4}x + \frac{53}{4}$$

does not pass through the feasible region, subject to the given constraints, the objective function $z = 3x + 4y$ cannot be as large as 53.

There is an infinite number of parallel lines, each corresponding to a choice for z. The greatest value of z corresponds to the line intersecting the feasible region with the greatest y-intercept. Clearly, the line we are seeking passes through a vertex. In this case, we can see from Figure A.2 that the vertex $Q(3, 9)$ gives the maximum value of the objective function:

$$z = 3(3) + 4(9) = 45$$

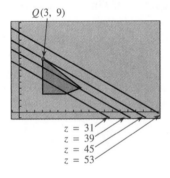

Figure A.2
$[-1, 18] \times [-1, 15]$

The following theorem generalizes our discussion:

Linear Programming Theorem

> Consider a linear objective function $z = ax + by + c$, subject to constraints (linear inequalities). If the objective function has an optimal (maximum or minimum) value, it must occur at a vertex of the feasible region.

The linear programming theorem leads us to a practical way to solve a linear programming problem.

Solving a Linear
Programming Problem

> To solve a linear programming problem:
>
> **Step 1:** Graph the constraints (system of inequalities) to determine the feasible region.
>
> **Step 2:** Determine the vertices of the feasible region.
>
> **Step 3:** Compute the values of the objective function at each vertex. The maximum (minimum) value at the vertex is the desired maximum (minimum).

The next example illustrates this method.

◉ EXAMPLE 1 Solving a Linear Programming Problem

Determine the maximum and minimum value of the objective function

$$z = x - 2y$$

subject to the constraints

$$\begin{cases} x \geq 0 \\ y \geq 0 \\ -2x + 3y \leq 12 \\ 3x + y \leq 15 \end{cases}$$

SOLUTION

The graph of the feasible region is shown in Figure A.3. Clearly, the origin is one of the four vertices. The other vertex on the y-axis is the y-intercept of the associated equation of $-2x + 3y \leq 12$. Letting $x = 0$ in $-2x + 3y = 12$, we get

$$-2(0) + 3y = 12$$
$$y = 4$$

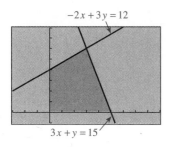

$-2x + 3y = 12$

$3x + y = 15$

Figure A.3
$[-3, 9] \times [-1, 8]$

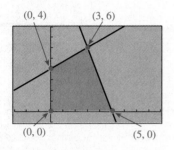

(0, 4) (3, 6)

(0, 0) (5, 0)

Figure A.4

Thus, (0, 4) is a vertex. Similarly, the x-intercept of the associated equation of $3x + y \leq 15$ is a vertex. Substituting $y = 0$ in $3x + y = 15$ leads to

$$3x + 0 = 15$$
$$x = 5$$

Therefore, (5, 0) is a vertex. The remaining vertex is the intersection of the graphs of $-2x + 3y = 12$ and $3x + y = 15$. Using a graphing utility, we find that the lines intersect at (3, 6). The graph of the feasible region, along with the coordinates of the vertices, is shown in Figure A.4.

Computing the value of the objective function at each vertex, we get the following:

Vertex	Objective function, $z = x - 2y$
(0, 0)	$z = 0 - 2(0) = 0$
(0, 4)	$z = 0 - 2(4) = -8$
(5, 0)	$z = 5 - 2(0) = 5$
(3, 6)	$z = 3 - 2(6) = -9$

Therefore, the maximum value of z is 5; this occurs when $x = 5$ and $y = 0$. The minimum value of z is -9; this occurs when $x = 3$ and $y = 6$.

It is possible for the feasible region to be *unbounded*. If so, the maximum (or minimum) value of the objective function may not exist.

● EXAMPLE 2 An Unbounded Feasible Region

Determine the maximum and minimum values (if possible) of the objective function

$$z = 5x + 2y$$

subject to the constraints

$$\begin{cases} x \geq 0 \\ y \geq 0 \\ 3x + 4y \geq 20 \\ x - y \leq 2 \end{cases}$$

SOLUTION

The feasible region and its vertices is shown in Figure A.5.

Testing the value of the objective function at each vertex, we have the following:

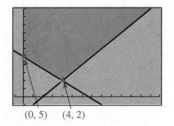

(0, 5) (4, 2)

Figure A.5
$[-1, 14] \times [-1, 11]$

Vertex	Objective function, $z = 5x + 2y$
(0, 5)	$z = 5(0) + 2(5) = 10$
(4, 2)	$z = 5(4) + 2(2) = 24$

In this case, the objective function has no maximum—it can be as large as we want. To see this, notice that any point on the y-axis above (0, 5) is a feasible solution. Thus, for $x = 0$, the value of the objective function is

$$z = 5(0) + 2y = 2y$$

where y can be as large as we like. Therefore, the objective function has no maximum.

The minimum value of the objective function is 10; this occurs when $x = 0$ and $y = 5$.

The maximum (or minimum) value of an objective function may occur at two of the vertices of the feasible region. In this case, the maximum (or minimum) occurs at any point on the line segment containing the two vertices. See Exercises 17, 18, and 32 for examples of this situation.

Our discussion of linear programming problems began by investigating an investment in stocks, bonds, and real estate. Many mathematical models formulated to solve problems in business and economics involve linear programming problems.

EXAMPLE 3 Application: Maximizing Profit

An electronics company makes two types of radar detectors, a standard unit and a deluxe unit. Each unit is assembled and then tested. The standard unit requires 2 hours of assembly and 1 hour of testing. The deluxe unit requires 3 hours of assembly and 2 hours of testing. Each day, the company has 36 hours of labor available for assembly and 22 hours of labor available for testing. The profit on the standard unit is $20, and the profit on the deluxe unit is $28. How many of each unit should be produced each day to maximize the total daily profit?

SOLUTION

The total daily profit is given by

$$\text{Profit} = 20\left(\begin{array}{c}\text{Number of}\\\text{standard units}\end{array}\right) + 28\left(\begin{array}{c}\text{Number of}\\\text{deluxe units}\end{array}\right)$$

The daily restrictions on hours of labor can be expressed as

$$2\left(\begin{array}{c}\text{Number of}\\ \text{standard units}\end{array}\right) + 3\left(\begin{array}{c}\text{Number of}\\ \text{deluxe units}\end{array}\right) \le 36 \quad \text{Assembly}$$

$$1\left(\begin{array}{c}\text{Number of}\\ \text{standard units}\end{array}\right) + 2\left(\begin{array}{c}\text{Number of}\\ \text{deluxe units}\end{array}\right) \le 22 \quad \text{Testing}$$

Let x represent the number of standard units produced and y represent the number of deluxe units produced. Clearly, neither x nor y can be negative, so we have the following linear programming problem:

Determine the maximum value of the objective function

$$P = 20x + 28y$$

subject to the constraints

$$\begin{cases} x \ge 0 \\ y \ge 0 \\ 2x + 3y \le 36 \\ x + 2y \le 22 \end{cases}$$

The feasible region and its vertices is shown in Figure A.6.
    The value of the objective function at each vertex is computed below:

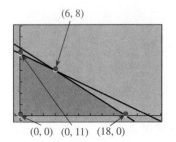

(6, 8)

(0, 0) (0, 11) (18, 0)

Figure A.6
$[-1, 24]_2 \times [-1, 16]_2$

Vertex	Objective function, $P = 20x + 28y$
(0, 0)	$P = 20(0) + 28(0) = 0$
(0, 11)	$P = 20(0) + 28(11) = 308$
(18, 0)	$P = 20(18) + 28(0) = 360$
(6, 8)	$P = 20(6) + 28(8) = 344$

The company should produce 18 of the standard models and none of the deluxe models to achieve a maximum daily profit of $360.

◉ EXAMPLE 4   Application: Minimizing Shipping Costs

A canoe manufacturer stores its canoes in two warehouses, warehouse #1 and warehouse #2. Warehouse #1 has 34 canoes, and warehouse #2 has 30 canoes. The company needs to fill orders for two retail stores, store #1 and store #2. Store #1 has ordered 16 canoes, and store #2 has ordered 24 canoes. The shipping costs for each canoe are as follows:

$12 from warehouse #1 to store #1

$16 from warehouse #1 to store #2

$15 from warehouse #2 to store #1

$14 from warehouse #2 to store #2

Determine the number of canoes that should be shipped from each warehouse to each store to minimize the total shipping cost.

SOLUTION

The total shipping cost can be expressed as

$$C = 12\begin{pmatrix} \text{Number of canoes} \\ \text{shipped from} \\ \text{warehouse \#1} \\ \text{to store \#1} \end{pmatrix} + 16\begin{pmatrix} \text{Number of canoes} \\ \text{shipped from} \\ \text{warehouse \#1} \\ \text{to store \#2} \end{pmatrix} + 15\begin{pmatrix} \text{Number of canoes} \\ \text{shipped from} \\ \text{warehouse \#2} \\ \text{to store \#1} \end{pmatrix} + 14\begin{pmatrix} \text{Number of canoes} \\ \text{shipped from} \\ \text{warehouse \#2} \\ \text{to store \#2} \end{pmatrix}$$

If we let x represent the number of canoes shipped from warehouse #1 to store #1, then the number of canoes shipped from warehouse #2 to store #1 is $16 - x$. Similarly, if y represents the number of canoes shipped from warehouse #1 to store #2, then the number of canoes shipped from warehouse #2 to store #2 is given by $24 - y$. Thus, the total shipping cost can be expressed as

$$C = 12x + 16y + 15(16 - x) + 14(24 - y)$$

$$= 12x + 16y + 240 - 15x + 336 - 14y \qquad \text{Use the distributive property to remove parentheses.}$$

$$= -3x + 2y + 576 \qquad \text{Combine like terms.}$$

The number of canoes shipped cannot be negative, giving

$$x \geq 0, \quad 16 - x \geq 0, \quad y \geq 0, \quad \text{and} \quad 24 - y \geq 0$$

Since warehouse #1 has 34 canoes and warehouse #2 has 30 canoes, we have

$$x + y \leq 34 \qquad \text{and} \qquad (16 - x) + (24 - y) \leq 30$$

Thus, the constraints are given as the following system:

$$\begin{cases} x \geq 0 \\ 16 - x \geq 0 \\ y \geq 0 \\ 24 - y \geq 0 \\ x + y \leq 34 \\ (16 - x) + (24 - y) \leq 30 \end{cases}$$

which can be simplified as

$$\begin{cases} x \geq 0 \\ x \leq 16 \\ y \geq 0 \\ y \leq 24 \\ x + y \leq 34 \\ x + y \geq 10 \end{cases}$$

Figure A.7 shows the feasible region and its vertices.

The value of the objective function at each vertex is computed below:

(0, 24) (10, 24) (16, 18)

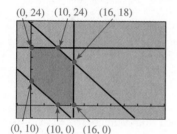

(0, 10) (10, 0) (16, 0)

Figure A.7
$[-5, 50]_5 \times [-5, 35]_5$

Vertex	Objective function, $C = -3x + 2y + 576$
(0, 10)	$C = -3(0) + 2(10) + 576 = 596$
(0, 24)	$C = -3(0) + 2(24) + 576 = 624$
(10, 24)	$C = -3(10) + 2(24) + 576 = 594$
(16, 18)	$C = -3(16) + 2(18) + 576 = 564$
(16, 0)	$C = -3(16) + 2(0) + 576 = 528$
(10, 0)	$C = -3(10) + 2(0) + 576 = 546$

The minimum of the objective function is 528; this occurs when $x = 16$ and $y = 0$. The company will achieve a minimum shipping cost of \$528 by shipping:

16 canoes from warehouse #1 to store #1 $x = 16$
0 canoes from warehouse #1 to store #2 $16 - x = 0$
0 canoes from warehouse #2 to store #1 $y = 0$
24 canoes from warehouse #2 to store #2 $24 - y = 24$

EXERCISE SET

A

In Exercises 1–4, determine the maximum and minimum value of each objective function subject to the constraints

$$\begin{cases} x \geq 0 \\ y \geq 0 \\ 2x - 5y \leq 0 \\ 2x + y \leq 12 \\ 2x + 3y \leq 24 \end{cases}$$

The feasible region is shown in Figure A.8.

1. Objective function: $z = -3x + 5y$

2. Objective function: $z = -x + 4y$

3. Objective function: $z = 7x + 2y$

4. Objective function: $z = x + y$

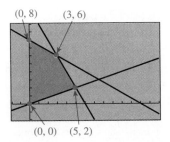

$(0, 8)$ $(3, 6)$

$(0, 0)$ $(5, 2)$

Figure A.8
$[-2, 14] \times [-2, 10]$

In Exercises 5–10, determine the maximum and minimum value of each objective function subject to the indicated constraints.

5. Objective function: $z = 4x + 7y$

Constraints: $\begin{cases} x \geq 0 \\ y \geq 0 \\ 2x + 3y \leq 12 \end{cases}$

6. Objective function: $z = 6x + 11y$

Constraints: $\begin{cases} x \geq 0 \\ y \geq 0 \\ 4x + 5y \leq 20 \end{cases}$

7. Objective function: $z = 6x + 7y$

Constraints: $\begin{cases} x \geq 0 \\ y \geq 0 \\ x + 2y \leq 12 \\ 3x + 2y \leq 24 \end{cases}$

8. Objective function: $z = x + 2y$

Constraints: $\begin{cases} x \geq 0 \\ y \geq 0 \\ 3x + y \leq 15 \\ x + 4y \leq 16 \end{cases}$

9. Objective function: $z = 13x + 9y$

Constraints: $\begin{cases} x \geq 0 \\ y \geq 0 \\ 4x - y \leq 12 \\ x + y \leq 8 \end{cases}$

10. Objective function: $z = 21x + 17y$

Constraints: $\begin{cases} x \geq 0 \\ y \geq 0 \\ 2x + y \leq 8 \\ -x + y \leq 2 \end{cases}$

B

In Exercises 11–18, determine the maximum and minimum value of each objective function subject to the indicated constraints.

11. Objective function: $z = 11x + 3y + 22$

Constraints: $\begin{cases} x \geq 0 \\ y \geq 0 \\ y \leq 10 \\ x + y \leq 16 \\ 3x + y \leq 36 \end{cases}$

12. Objective function: $z = 10x + 11y + 16$

Constraints: $\begin{cases} x \geq 0 \\ y \geq 0 \\ x \leq 6 \\ x + y \leq 12 \\ x + 4y \leq 36 \end{cases}$

13. Objective function: $z = 5x + 2y$

Constraints: $\begin{cases} x \geq 0 \\ y \geq 3 \\ x + y \leq 11 \\ 4x + 5y \geq 35 \end{cases}$

14. Objective function: $z = 8x + 12y$

Constraints: $\begin{cases} x \geq 0 \\ y \geq 2 \\ 2x + 3y \leq 24 \\ x + 2y \geq 10 \end{cases}$

15. Objective function: $z = 4x + 3y$

Constraints: $\begin{cases} x \geq 0 \\ y \geq 0 \\ x \leq 10 \\ 2x + y \geq 14 \end{cases}$

16. Objective function: $z = 10x + 7y$

Constraints:
$$\begin{cases} x \geq 0 \\ y \geq 0 \\ -x + 3y \leq 45 \\ 3x + 2y \geq 18 \end{cases}$$

17. Objective function: $z = 10x + 24y$

Constraints:
$$\begin{cases} x \geq 0 \\ y \geq 0 \\ 5x + 12y \leq 240 \\ 3x + 2y \leq 66 \end{cases}$$

18. Objective function: $z = 15x + 6y$

Constraints:
$$\begin{cases} x \geq 0 \\ y \geq 0 \\ -2x + 3y \leq 33 \\ 5x + 2y \leq 60 \end{cases}$$

19. A company makes two models of oak filing cabinets. The profit on the two-drawer model is $80, and the profit on the four-drawer model is $110. The two-drawer model requires 2 hours of assembly time and 4 hours of finishing time. The four-drawer model requires 3 hours of assembly time and 5 hours of finishing time. Each week, the company has 270 hours of labor available for assembly and 500 hours of labor available for finishing. How many of each model should be produced each week in order to maximize profit?

20. A furniture manufacturer makes two types of chairs—one that reclines and one that does not. The recliner requires 2 hours of assembly time and 1 hour of packing time. The nonrecliner requires 1 hour of assembly and 1 hour of packing time. Each month, the manufacturer has 420 work-hours available for assembly and 300 work-hours available for packing. The profit on each recliner is $140, and the profit on each nonrecliner is $100. How many of each type should be produced per month in order to maximize profit?

21. A farmer has at most 500 acres on which to plant two crops—tomatoes and cotton. Producing tomatoes requires 2 hours of labor per acre, and cotton requires 3 hours of labor per acre. The farmer has 1200 hours of labor avail-

able. If the profit per acre is $80 for tomatoes and $100 for cotton, how many acres of each crop should be planted to maximize profit?

22. A rancher has 1500 acres of grazing pasture to raise sheep and cattle. The ranch has the resources to raise no more than 4800 animals. Each acre of pasture sustains 2 cows or 6 sheep. The profit on each cow is $40, and the profit on each sheep is $15. How many acres should be allotted for cattle and how many acres should be allotted for sheep in order to maximize profit?

23. A refrigerator manufacturer stores its refrigerators in two warehouses, the downtown warehouse and the industrial park warehouse. The downtown warehouse has 45 refrigerators, and the industrial park warehouse has 51 refrigerators. The company needs to fill orders for two retail stores, store A and store B. Store A has ordered 36 refrigerators and store B has ordered 24 refrigerators. The shipping costs for each refrigerator are as follows:

$9 from downtown warehouse to store A

$11 from downtown warehouse to store B

$12 from industrial park warehouse to store A

$10 from industrial park warehouse to store B

Determine the number of refrigerators that should be shipped from each warehouse to each store to minimize the total shipping cost.

24. A paper goods manufacturer has 740 reams of a particular type of paper in their east warehouse and 580 reams of the same type in their west warehouse. An office supply retailer orders 600 reams of the paper, and a stationery store orders 570 reams. The shipping costs per ream are as follows:

$0.60 from the east warehouse to the office supply retailer

$0.80 from the east warehouse to the stationery store

$0.90 from the west warehouse to the office supply retailer

$0.70 from the west warehouse to the stationery store

Determine the number of reams that should be shipped from each warehouse to each store to minimize the total shipping cost.

25. A tea company has 600 pounds of Indonesian tea, 450 pounds of English tea, and 320 pounds of Brazilian tea. The company blends these three teas to produce two blends: premium blend and house blend. One batch of the premium blend requires 6 pounds of Indonesian tea and 4 pounds of Brazilian tea. One batch of the house blend requires 3 pounds of Indonesian tea and 5 pounds of English tea. The profit is $19 per batch of premium blend and $14 per batch of house blend. How many batches of each blend should be produced to maximize the total profit?

26. A company that produces canned mixed fruit has 5000 pounds of peaches, 8000 pounds of melons, and 4000 pounds of pineapples. The cannery produces two mixtures, tropical and domestic, sold in 2 pound cans. One can of the tropical mixture requires 1 pound of melons and 1 pound of pineapples. One can of the domestic mixture requires 1.5 pounds of peaches and 0.5 pound of melons. The tropical mixture gives a profit of $0.35, and the domestic mixture gives a profit of $0.25. How many cans of each should be produced to maximize the total profit?

27. The manager of a bird rescue center uses a blend of two brands of food, brand A and brand B, to supply at least 700 units of calcium and 1660 units of thiamin per week for the birds. Each pound of brand A supplies 5 units of calcium and 8 units of thiamin. Each pound of brand B supplies 4 units of calcium and 10 units of thiamin. If brand A costs $0.50 per pound and brand B costs $0.60 per pound, how many pounds of each brand should be blended to minimize the cost of the blend?

28. A special diet is made from a blend of two foods—food A and food B. Food A consists of 80% carbohydrates and 20% protein, and food B consists of 60% carbohydrates and 40% protein. Food A costs $0.75 per pound and food B costs $1.20 per pound. What blend of these two foods provides at least 2 pounds of carbohydrates and 1 pound of protein at a minimum cost?

29. An investor wants to invest no more than $120,000 in two types of investments—stocks and bonds. The annual return on stocks is predicted to be 10%, and the annual return on bonds is expected to be 8%. The investor wants the amount invested in stocks to be at least twice as much

as the amount invested in bonds. In addition, at least $20,000 is to be invested in bonds. How much should be allocated to each investment to maximize the annual return?

C

30. An oil refinery produces gasoline and diesel. The cost of producing 1 barrel of gasoline is $8, and the production cost of 1 barrel of diesel is $6. The refinery has a maximum daily production of 2400 barrels and a daily budget of $15,000. In order to meet demand for a contractual agreement, the refinery must produce at least 800 barrels of diesel per day. If the profit is $12 per barrel of gasoline and $9 per barrel of diesel, how many barrels of each should be produced daily to maximize profit?

31. A company makes two models of bicycles—the Tourist and the Hillbreaker. A contractual agreement requires that at least 30 Tourists and 25 Hillbreakers are produced each month. Producing one Tourist bike requires 2 hours of assembly, 2 hours of painting, and 2 hours of packaging. Producing a Hillbreaker requires 1 hour of assembly, 2 hours of painting, and 3 hours of packaging. The total number of hours available each month for assembly, painting, and packaging is 800 hours, 960 hours, and 1350 hours, respectively. The profit on each Tourist bike is $180, and the profit on each Hillbreaker is $90. How many of each model should be produced per month in order to maximize profit?

32. (a) Determine the maximum and minimum value of the objective function

$$z = 7x + 14y$$

subject to the constraints

$$\begin{cases} x \geq 0 \\ y \geq 0 \\ x + y \geq 2 \\ 4x - y \leq 12 \\ x + 2y \leq 16 \end{cases}$$

(b) Explain why the maximum value of z occurs at any point on the line segment with endpoints $(0, 6)$ and $(4, 4)$.

A REVIEW: SECTION 1

1. Integers: $-2, 5$ Rational numbers: $-2, 5, 1.34$ Irrational numbers: $-\sqrt{5}$

3. Integers: none   Rational numbers: $\dfrac{13}{7}$ Irrational numbers: $-\sqrt{8}, \dfrac{3\pi}{2}, \sqrt{7}$

5. Integers: 3   Rational numbers: $3, \dfrac{22}{7}, 3.14$ Irrational numbers: π

7. **(a)** Commutative property of addition   **(b)** Distributive property of multiplication over addition
(c) Multiplicative identity

9. **(a)** Distributive property of multiplication over addition   **(b)** Additive inverse   **(c)** Associative property of multiplication

11. $(4, +\infty)$

13. $[\tfrac{11}{2}, +\infty)$

15. $(1, 5)$

17. $[-2, 11]$

19. $5 + \pi$ **21.** $3 - \sqrt{6}$ **23.**

25.   **27.**   **29.** $|x| = 2$

31. $|x - 4| < 5$ **33.** $|x| \geq 1$ **35.** $[-2, 4)$

37. $(-\infty, -3)$ or $[4, +\infty)$ **39.** $(-\infty, -4]$ or $[-1, 3]$

41. $(-4, -2]$ or $[0, 4)$ **43.** $(4.5, 5.5)$

45. $[-3, -1)$ or $(1, 3]$ **47.** $a + 2$ **49.** $17 - c^2$ **51.** $a + c - 2$

53. **(a)** No real numbers   **(b)** No real numbers   **(c)** No real numbers

SECTION 2

1. (a) $4a$ (b) $\dfrac{3}{2}$ **3.** (a) $\dfrac{2}{x}$ (b) $-\dfrac{1}{y^2}$ **5.** (a) $2a^3b$ (b) $\dfrac{3}{xy^4}$ **7.** (a) $\dfrac{y^2}{x^2}$ (b) b^4

9. (a) 1 (b) -2 **11.** $\dfrac{-4y^7}{x^3}$ **13.** $\dfrac{b^2}{3a^2}$ **15.** $\dfrac{4x^2}{y^3}$ **17.** $\dfrac{2a^5}{b^{12}}$ **19.** $48x^4y^{12}$ **21.** 5.13×10^{-4} **23.** 2.2×10^{47}

25. 1.7×10^{37} **27.** $3x^2 - 4x - 2$ **29.** $7x - 6$ **31.** $8x^2 + 4x$ **33.** $7x + 12$ **35.** $-2x - 1$ **37.** $4x - 4$

39. $x^5(x + 1)^8$ **41.** $\dfrac{-40x}{(5x^2 + 3)^5}$ **43.** $12a^{n+2}b^5$ **45.** $3x^{2n+4}$ **47.** $x^3 + 6x^2 + 12x + 8$

49. $-4x^4 + 8x^3 - 14x^2 - 4x + 8$ **51.** $4x^4 - 20x^3 + 49x^2 - 60x + 36$ **53.** $x^3 + x^2 - 6x$

55. (a) $x^3x^4 = x^7$ (b) $3x^{-2} = \dfrac{3}{x^2}$ (c) $(x + y)^{-1} = \dfrac{1}{x + y}$ (d) $(x^3)^4 = x^{12}$ **57.** 3^{2000}

SECTION 3

1. $3x(x^2 + 2)$ **3.** $xy^2(x^2 + 5xy + 8y^2)$ **5.** $(x + 3)(5x - 2)$ **7.** $(a - 1)(a^2 + 3)$ **9.** $(x - 2)(x^5 - 3)$
11. $(n - 3)(5n^2 - 2)$ **13.** $(x + 1)(x - 2)$ **15.** $(x + 1)(3x + 2)$ **17.** $(3x + 2)^2$ **19.** Prime
21. $(2x - 1)(2x + 1)$ **23.** Prime   **25.** $(x - 2)(x^2 + 2x + 4)$ **27.** $(n + m)(n^2 - nm + m^2)$
29. $(2x - yz)(4x^2 + 2xyz + y^2z^2)$ **31.** $6(x - 1)^2$ **33.** $x(x - 2)(x + 2)$ **35.** $xy(x - y)(x^2 + xy + y^2)$
37. $x^4(x + 1)(x - 6)$ **39.** $(x^2 + 4)(x^2 + 2)$ **41.** $(2x - y)(2x + y)(4x^2 + y^2)$ **43.** $(x^2 - y)(x^4 + x^2y + y^2)$
45. $(x - 2)(x + 2)(x^4 + 1)$ **47.** $(x - 5)(x - 6)(x - 4)$ **49.** $2x(7x - 3)(2x - 3)^4$ **51.** $3(x - 3)^2(x^2 + 1)^2(3x^2 - 6x + 1)$
53. $x^{-2}(3x^4 - 4x^3 + x - 2)$ **55.** (a) $2(x^2 - 6)$ (b) $2(x + \sqrt{6})(x - \sqrt{6})$
57. (a) $(x^2 + 5)(x^2 - 5)$ (b) $(x^2 + 5)(x + \sqrt{5})(x - \sqrt{5})$ **61.** $(x^2 - xy + y^2)(x^2 + xy + y^2)$

SECTION 4

1. $\dfrac{2x + 1}{2}$ **3.** $\dfrac{4x}{x + 3}$ **5.** $\dfrac{-5}{x + 2}$ **7.** $\dfrac{x - 3}{2x(x - 2)}$ **9.** $\dfrac{-1}{4x}$ **11.** $\dfrac{x + 2}{3 - 2x}$ **13.** $\dfrac{6x}{5(x + 3)}$ **15.** $\dfrac{2(2x - 5)}{x(x - 2)}$

17. $\dfrac{x + 1}{x - 1}$ **19.** $\dfrac{3(x - 2)}{x}$ **21.** $\dfrac{a - 2}{2a - 1}$ **23.** $\dfrac{5(x + 1)}{(x - 1)(x + 4)}$ **25.** $\dfrac{3x + 7}{2x(x + 1)}$ **27.** $\dfrac{5x + 3}{(2x + 1)(x + 1)}$

29. $\dfrac{x^2 - 17x - 28}{2(x + 5)(x - 7)}$ **31.** $\dfrac{7x^2 - 41x + 42}{2(x + 7)(x - 7)}$ **33.** $\dfrac{x^3 - 3x - 3}{x(x - 1)}$ **35.** $\dfrac{13x^2 - 10x + 5}{2x(x - 1)^2}$ **37.** $\dfrac{8x^2 + 16x + 13}{2x}$

39. $\dfrac{-4x^2 + 11x - 15}{(x + 1)(x - 1)^2}$ **41.** $\dfrac{2x + 1}{3x}$ **43.** $\dfrac{-1}{x(x + 3)}$ **45.** $\dfrac{(2x + 1)(x + 1)}{x(x - 3)}$ **47.** $6x + 9 - \dfrac{5}{x}$ **49.** $3x^3 - \dfrac{7x}{5} + 2 + \dfrac{4}{5x}$

51. $3 + \dfrac{13}{x - 4}$ **53.** $4 + \dfrac{-20}{2x + 5}$ **55.** $\dfrac{3}{2} + \dfrac{1}{2x + 6}$ **57.** $3x - 4 + \dfrac{3}{x + 2}$ **59.** $2x + 2 + \dfrac{9}{2x - 1}$

61. $3x + 4 + \dfrac{-x - 5}{x^2 + 2x}$ **63.** $2x - 1 + \dfrac{-2x + 4}{2x^2 + 1}$ **65.** $4x^3 + 3x^2 - 13x + 3 + \dfrac{5x + 9}{x^2 + 3x + 1}$

67. $x + \dfrac{3}{2} + \dfrac{-\frac{5}{2}}{2x - 3}$ **69.** $k = -25$

SECTION 5

1. $\dfrac{2}{7}$ **3.** 4 **5.** $-\dfrac{1}{2}$ **7.** Not a real number   **9.** -3 **11.** 5 **13.** -3

15. Not a real number   **17.** 2 **19.** 25 **21.** $\dfrac{1}{81}$ **23.** $2xy\sqrt{2x}$ **25.** $3x\sqrt[3]{2x}$ **27.** $2x^2\sqrt{x}$ **29.** $-\dfrac{a^3\sqrt[3]{3a^2}}{2b^3}$

31. $\sqrt[3]{x^2}$ **33.** $15x^2$ **35.** $\dfrac{\sqrt{x}}{3y}$ **37.** $y\sqrt[6]{y}$ **39.** $\dfrac{\sqrt{x^2+y^2}}{a+b}$ **41.** $8\sqrt{3x}+7\sqrt{2x}$ **43.** $5x\sqrt[3]{2x}$ **45.** $7x+2x\sqrt{x}$

47. x^2-3 **49.** $x+5+4\sqrt{x+1}$ **51.** $9x+18$ **53.** $3xy^2$ **55.** $12x^{5/2}y$ **57.** $6x^2+9$ **59.** $x^3-2x^{3/2}y^{1/2}+y$

61. $2x^{4/3}+2x^{2/3}$ **63.** x **65.** $x^{7/12}$ **67.** $-2+\sqrt{6}$ **69.** $\dfrac{3-\sqrt{2}}{4}$ **71.** $-4\pm\sqrt{14-x}$ **73.** $\dfrac{x^2+1}{2x}$

75. $\dfrac{5\sqrt{6}}{6}$ **77.** $\dfrac{2\sqrt{2x}}{3x}$ **79.** $\dfrac{21-7\sqrt{5}}{4}$ **81.** $\dfrac{\sqrt{3}+1}{2}$ **83.** $\dfrac{x-\sqrt{x}}{x-1}$ **85.** $\dfrac{-4x^{4/3}-16x^{1/3}}{(x-2)^3}$

91. (a) $\dfrac{\sqrt[3]{4\sqrt{5}-4}}{2}+\dfrac{\sqrt[3]{-4\sqrt{5}-4}}{2}$ **(b)** $\dfrac{\sqrt[3]{12(27+\sqrt{2229})}}{6}+\dfrac{\sqrt[3]{12(27-\sqrt{2229})}}{6}$

SECTION 6

1. 7 **3.** $-\frac{4}{3}$ **5.** 0 **7.** 3 **9.** 13 **11.** -19 **13.** 8 **15.** -4 **17.** -4 **19.** 2.2 **21.** 2

23. $(-\infty,3]$ **25.** $(-\infty,2]$ **27.** $(-\infty,2]$ **29.** $(\frac{1}{4},+\infty)$ **31.** $(6,+\infty)$ **33.** $(-\infty,5]$ **35.** $[4,8]$ **37.** $(-\frac{7}{5},4)$

39. $[\frac{1}{2},4)$ **41.** $R=\dfrac{PV}{nT}$ **43.** $w=\dfrac{P-2l}{2}$ **45.** $p=\dfrac{5400-x}{800}$ **47.** $R=\dfrac{R_1R_2}{R_1+R_2}$ **49.** $l=\dfrac{A}{w-2}+4$

51. $P=\dfrac{A}{1+rt}$ **53.** $x=zs+\mu$ **55.** $x\geq3$ **57.** $x\leq\frac{9}{2}$ **59.** $x<\frac{5}{2}$ **61.** $2.5\,\text{qt}$ **63.** More than 175 mi

65. No more than 5 **67.** 31.25 min (31 min, 15 sec) **71.** 45 mi

SECTION 7

1. $-6,1$ **3.** $\frac{3}{2},2$ **5.** $0,\frac{10}{3}$ **7.** $-\frac{3}{2},\frac{3}{2}$ **9.** 7 **11.** $\pm\frac{5}{2}$ **13.** $\pm\sqrt{3}$ **15.** $5\pm2\sqrt{3}$ **17.** $\dfrac{-2\pm\sqrt{13}}{2}$

19. $\dfrac{-1\pm\sqrt{6}}{2}$ **21.** No real solution **23.** $-3\pm\sqrt{10}$ **25.** $2\pm\sqrt{2}$ **27.** $\dfrac{-2\pm\sqrt{5}}{2}$ **29.** $\dfrac{3\pm\sqrt{23}}{2}$

31. $\dfrac{-3\pm\sqrt{5}}{2}$ **33.** $\dfrac{5\pm3\sqrt{5}}{2}$ **35.** $\dfrac{3\pm\sqrt{19}}{2}$ **37.** No real solution **39.** $\dfrac{-1\pm\sqrt{5}}{2}$ **41.** $-4,1$ **43.** $\pm\dfrac{\sqrt{5}}{3}$

45. $-\frac{3}{2}$ **47.** $\dfrac{5\pm\sqrt{29}}{2}$ **49.** No real solution **51.** $-\frac{3}{2},6$ **53.** $\dfrac{-3\pm3\sqrt{3}}{2}$ **55.** $-\frac{3}{4},1$ **57.** $\sqrt{3}\pm2$

59. 16 by 10 in., or $\frac{20}{3}$ by 24 in. **61.** $8+8\sqrt{6}$ in. by $8+8\sqrt{6}$ in. (27.60 in. by 27.60 in.) **63.** 18 sides **65.** $\pm\sqrt{6\pm\sqrt{3}}$

MISCELLANEOUS EXERCISES

1. Integers: $-3,9$ Rational numbers: $-3,5.1,9,\frac{7}{13}$ Irrational numbers: $-\sqrt{30},2\pi$

3. (a) Commutative property of addition **(b)** Distributive property of multiplication over addition **(c)** Additive identity

5. (a) $[2,+\infty)$ **(b)** $[1,5)$

7. (a) $(-2,2)$ **(b)** $(-\infty,1.5)$ or $(2.5,+\infty)$

9. (a) $\dfrac{-12}{x}$ **(b)** $\dfrac{x}{2}$ **11. (a)** $\dfrac{1}{6t^2}$ **(b)** $\dfrac{4r^9}{t^7}$ **13. (a)** $2a^2\sqrt[3]{3a^2}$ **(b)** $-\dfrac{x^4\sqrt[3]{2x}}{y^4}$ **15. (a)** x^2 **(b)** $\dfrac{1}{x}$

17. $2x^3+x^2-4x-5$ **19.** $4a^3+19a^2-5a$ **21.** $x^2+7x-15$ **23.** x^5-4x^3+4x **25.** $a^2b^2(7a^2+ab^2-b^3)$

27. $(4s^2-t)(4s^2+t)$ **29.** $t^3(4t+3)(t-1)$ **31.** $(x+3)(x+2)(x+1)$ **33.** $x^2(7x^2+30)(x^2+10)$

35. $4x$ **37.** $\dfrac{4(x + a)}{x - a}$ **39.** $(t - 2)^2$ **41.** $\dfrac{7x + 1}{(2x + 1)(x - 2)}$ **43.** $\dfrac{3x + 1}{3(x + 1)}$ **45.** $\dfrac{x^2 + 1}{x}$ **47.** $4x - 2 + \dfrac{7}{2x^2}$

49. $2 - \dfrac{7}{2x + 1}$ **51.** $2x + 3 + \dfrac{24}{x - 4}$ **53.** $x^3 - 36$ **55.** $5\sqrt{3t}$ **57.** $-3 - 3\sqrt{3}$ **59.** $-x + \sqrt{x^2 - 2}$ **61.** $-\frac{3}{14}$

63. $-\frac{5}{9}$ **65.** $-\frac{5}{2}, 3$ **67.** $-\frac{5}{3}, -\frac{1}{3}$ **69.** $\frac{3}{2} - \frac{1}{2}\sqrt{29}, \frac{3}{2} + \frac{1}{2}\sqrt{29}$ **71.** $-\frac{1}{2}, 1$ **73.** $x \le \frac{7}{3}$ **75.** $-3 < x \le 0$

77. $-1 < x < 2$ **79.** $n = \dfrac{PV}{RT}$ **81.** $r = \dfrac{\sqrt{A\pi}}{\pi}$

CHAPTER TEST

1. Integers: $0, -\sqrt{9}$ ($-\sqrt{9}$ is -3); Rational numbers: $0, \dfrac{4}{7}, -0.3, -\sqrt{9}$; Irrational numbers: $\dfrac{\sqrt{3}}{2}, \dfrac{5}{\pi}$

(see Section 1, Example 1)

2. 2π is approximately 6.28, so $2\pi - 8$ is negative. Thus, $|2\pi - 8| = 8 - 2\pi$ (see Section 1, Example 4)

3. $|x + 1|$ represents $d(x, -1)$, so x is within 3 units of -1. This interval is described by $-4 \le x \le 2$:

x (see Section 1, Example 6).

4. $(-\infty, 1]$ or $(2, 3]$ (see Section 1, Example 3)

5. $(-2x^2y)(3xy^{-4}) = (-2 \cdot 3)(x^2x)\left(y\dfrac{1}{y^4}\right) = \dfrac{-6x^3}{y^3}$ (see Section 2, Example 1)

6. $\left(\dfrac{25a^6}{b^2}\right)^{-1/2} = \left(\dfrac{b^2}{25a^6}\right)^{1/2} = \dfrac{(b^2)^{1/2}}{(25a^6)^{1/2}} = \dfrac{b}{5a^3}$ (see Section 5, Example 5)

7. $\dfrac{p^3q^{-4}}{-5p^{-2}q} = \dfrac{1}{-5} \cdot \dfrac{p^3}{p^{-2}} \cdot \dfrac{q^{-4}}{q} = -\dfrac{1}{5}p^5q^{-5} = -\dfrac{p^5}{5q^5}$ (see Section 2, Example 2)

8. $\dfrac{14x^{3n+1}}{(5x^n)(4x^{n-2})} = \dfrac{14x^{3n+1}}{(5 \cdot 4)(x^n x^{n-2})} = \dfrac{7x^{3n+1}}{10x^{2n-2}} = \dfrac{7x^{(3n+1)-(2n-2)}}{10} = \dfrac{7x^{n+3}}{10}$ (see Section 2, Example 2)

9. $(3.8 \times 10^{17})(1.5 \times 10^{-12}) = (3.8)(1.5) \times (10^{17})(10^{-12}) = 5.7 \times 10^5$ (see Section 2, Example 3)

10. $x^2 - (x + 4)(2x^2 + x - 1) = x^2 - (2x^3 + 9x^2 + 3x - 4) = -2x^3 - 8x^2 - 3x + 4$ (see Section 2, Example 7)

11. $(a^{3/2} + 2a^{5/2})(a^{3/2} - 2a^{5/2}) + 3a^5 = [(a^{3/2})^2 - (2a^{5/2})^2] + 3a^5 = (a^3 - 4a^5) + 3a^5 = a^3 - a^5$ (see Section 5, Example 5)

12. $2a^3b - 18ab^3 = 2ab(a^2 - 9b^2) = 2ab(a - 3b)(a + 3b)$ (see Section 3, Example 6)

13. $2x^2 + 7x - 4 = (2x - 1)(x + 4)$ (see Section 3, Example 3)

14. $x^2(x + 2) + 5(x + 2) = (x^2 + 5)(x + 2)$ (see Section 3, Example 6)

15. $3y^3 + 5y^2 - 2y = y(3y^2 + 5y - 2) = y(3y - 1)(y + 2)$ (see Section 3, Example 6)

16. $\dfrac{2 - x}{3x^2 - 2x - 8} = \dfrac{-(x - 2)}{(3x + 4)(x - 2)} = \dfrac{-1}{3x + 4}$ (see Section 4, Example 1)

17. $\dfrac{(6x^2 + 3x)}{\left(x^2 - \frac{1}{4}\right)} \cdot \dfrac{4}{4} = \dfrac{24x^2 + 12x}{4x^2 - 1} = \dfrac{12x(2x + 1)}{(2x - 1)(2x + 1)} = \dfrac{12x}{2x - 1}$ (see Section 4, Example 2)

18. $\dfrac{7x - 1}{2x} - \dfrac{6x}{x + 1} = \dfrac{7x - 1}{2x} \cdot \dfrac{x + 1}{x + 1} - \dfrac{6x}{x + 1} \cdot \dfrac{2x}{2x} = \dfrac{7x^2 + 6x - 1}{2x(x + 1)} - \dfrac{12x^2}{2x(x + 1)} = \dfrac{-5x^2 + 6x - 1}{2x(x + 1)}$ (see Section 4, Example 4)

19. $\dfrac{x^2 - 4x + 4}{x^2 - 1} \div \dfrac{x^2 - 4}{x^2 - 2x + 1} = \dfrac{x^2 - 4x + 4}{x^2 - 1} \cdot \dfrac{x^2 - 2x + 1}{x^2 - 4} = \dfrac{(x - 2)(x - 2)}{(x - 1)(x + 1)} \cdot \dfrac{(x - 1)(x - 1)}{(x - 2)(x + 2)} = \dfrac{(x - 2)(x - 1)}{(x + 2)(x + 1)}$

(see Section 4, Example 2)

20. $\dfrac{x}{2x+3} + \dfrac{5x+1}{4x^2-9} = \dfrac{x}{2x+3} \cdot \dfrac{2x-3}{2x-3} + \dfrac{5x+1}{(2x+3)(2x-3)} = \dfrac{2x^2-3x}{(2x+3)(2x-3)} + \dfrac{5x+1}{(2x+3)(2x-3)} = \dfrac{2x^2+2x+1}{(2x+3)(2x-3)}$
(see Section 4, Example 4)

21. $\dfrac{6-\sqrt{108}}{12} = \dfrac{6-6\sqrt{3}}{12} = \dfrac{6(1-\sqrt{3})}{12} = \dfrac{1-\sqrt{3}}{2}$ (see Section 5, Example 7)

22. $\frac{1}{3}(-12x + \sqrt{18x^2-45}) = \frac{1}{3}(-12x + 3\sqrt{2x^2-5}) = -4x + \sqrt{2x^2-5}$ (see Section 5, Example 7)

23. $(\sqrt{x}+\sqrt{3})^2 + \sqrt{2}(\sqrt{6x}-\sqrt{2}) = [(\sqrt{x})^2 + 2\sqrt{x}\cdot\sqrt{3} + (\sqrt{3})^2] + (\sqrt{2}\cdot\sqrt{6x} - \sqrt{2}\cdot\sqrt{2}) = $
$(x + 2\sqrt{3x} + 3) + (x + 2\sqrt{3x} - 2) = x + 4\sqrt{3x} + 1$ (see Section 5)

24.
$$\dfrac{3x}{2} + 5 = \dfrac{x-2}{6}$$
$$6\left(\dfrac{3x}{2} + 5\right) = 6\left(\dfrac{x-2}{6}\right)$$
$$9x + 30 = x - 2$$
$$8x + 30 = -2$$
$$8x = -32$$
$$x = -4$$
(see Section 6, Example 2)

25.
$$(2x+1)(x-1) = 2$$
$$2x^2 - x - 1 = 2$$
$$2x^2 - x - 3 = 0$$
$$(2x-3)(x+1) = 0$$
$$2x - 3 = 0 \mid x + 1 = 0$$
$$2x = 3 \qquad\; x = -1$$
$$x = \tfrac{3}{2}$$
(see Section 7, Example 1)

26.
$$2(3x+4)^2 = 81$$
$$3x + 4 = \pm 9$$
$$3x = -4 \pm 9$$
$$x = \dfrac{-4 \pm 9}{3} = -\dfrac{13}{3}, \dfrac{5}{3}$$
(see Section 7, Example 2)

27.
$$\dfrac{x^2}{3} + x = \dfrac{1}{2}$$
$$6\left(\dfrac{x^2}{3} + x\right) = 6\left(\dfrac{1}{2}\right)$$
$$2x^2 + 6x = 3$$
$$2x^2 + 6x - 3 = 0$$
$$x = \dfrac{-6 \pm \sqrt{6^2 - 4(2)(-3)}}{2(2)}$$
$$= \dfrac{-6 \pm 2\sqrt{15}}{4} = \dfrac{-3 \pm \sqrt{15}}{2}$$
(see Section 7, Example 4)

28.
$$S = \dfrac{a}{1-r}$$
$$(1-r)S = (1-r)\dfrac{a}{1-r}$$
$$S - Sr = a$$
$$Sr = S - a$$
$$r = \dfrac{S-a}{S}$$
(see Section 6, Example 3)

29.
$$2(7 - 3x) \le 17$$
$$14 - 6x \le 17$$
$$-6x \le 3$$
$$x \ge -\tfrac{1}{2}$$
(see Section 6, Example 4)

30.
$$0 < \dfrac{2x+11}{3} \le 5$$
$$0 < 2x + 11 \le 15$$
$$-11 < 2x \le 4$$
$$-\tfrac{11}{2} < x \le 2$$
(see Section 6, Example 5)

CHAPTER 1: SECTION 1.1

1. x-intercept: $(6, 0)$; y-intercept: $(0, 4)$ **3.** x-intercepts: $(-3, 0)$, $(3, 0)$; y-intercept: $(0, -9)$
5. x-intercepts: $(-1, 0)$, $(2, 0)$, $(5, 0)$; y-intercept: $(0, 10)$ **7.** $d(A, B) = 5$; $M(\tfrac{7}{2}, 5)$ **9.** $d(A, B) = \sqrt{29}$; $M\left(1, -\tfrac{1}{5}\right)$
11. $d(A, B) = 3$; $M(\sqrt{3}/2, \sqrt{6}/2)$ **13.** $(x+4)^2 + (y-2)^2 = 16$; Center: $(-4, 2)$, Radius: 4
15. $(x-5)^2 + y^2 = 20$; Center: $(5, 0)$, Radius: $2\sqrt{5}$ **17.** $(x-3)^2 + (y+4)^2 = 25$; Center: $(3, -4)$; Radius: 5

19. Origin **21.** Origin, x-axis, y-axis **23.** y-axis **25.** None **27.** Symmetric with respect to the origin

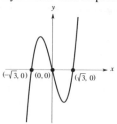

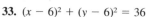

29. Symmetric with respect to the origin **31.** $(x - 3)^2 + (y + 3)^2 = 9$ **33.** $(x - 6)^2 + (y - 6)^2 = 36$

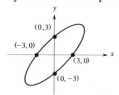

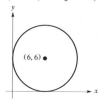

35. $(x - 4)^2 + (y + 3)^2 = 25$ **37.** $x^2 + y^2 = 9$ **39.** $(x - 3)^2 + (y - 2)^2 = 40$ **41.** A, B, and C are collinear.

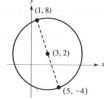

43. Triangle ABC is isosceles. **45.** Triangle ABC is obtuse. **47.** Quadrilateral $ABCD$ is a parallelogram.
49. $(x - 1)^2 + y^2 = 1$ **51.** The graph is the point $(7, 3)$. **53.** $(1, -1)$ **55.** $(1, 7)$ **57.**

59. **61.** $y = -\sqrt{9 - x^2}$ **63.** $y = 2 + \sqrt{4 - x^2}$

SECTION 1.2

1. (a) 6 (b) 0 (c) -12 (d) -15 **3.** (a) $-\frac{3}{8}$ (b) Undefined (c) 0 (d) $-\frac{7}{12}$
5. (a) $(-\infty, +\infty)$ (b) -2 **7.** (a) $(-8, +\infty)$ (b) -8 **9.** (a) $(-\infty, -2)$ or $(-2, 3)$ or $(3, +\infty)$ (b) $\frac{5}{3}$
11. (a) $(-\infty, +\infty)$ (b) None **13.** (a) $t^2 + 8t + 19$ (b) $t^2 + 22$ (c) $t^2 + 7$ (d) $t + 19$
15. (a) $\frac{5}{8}x^3 - 4$ (b) $\frac{5x^3 - 4}{36}$ (c) $\frac{x}{36}$ (d) $\frac{5x^3 - 4}{2}$ **17.** 2 **19.** $a + 5$ **21.** 2 **23.** $2x + 3$

25. 4   **27.** $2x + h - 2$ **29. (a)** $-\frac{3}{2}$ **(b)** 0   **(c)** 0   **(d)** -5 **31. (a)** $\frac{14}{3}$ **(b)** 5   **(c)** 2   **(d)** 6
33. 12   **35.** $4 - \sqrt{11}, 4 + \sqrt{11}$ **37.** $-\sqrt{7}, \sqrt{7}$ **39.** $-4, 4$
41. (a) 50 units; $2750   **(b)** $P(x) = 0.25x^2 + 5x - 1200$
 (c) $x = 60$; when 60 units are manufactured, there is no profit or loss.
43. (a) $6640   **(b)** $d(1980) - d(1976)$
 (c) $79.50, $198.75, $283.25, $663.75, $973.50, $1114.75; these numbers represent the average yearly increase in pcfd during
 each 4 year period.
45. (a) 8.84 sec
 (b) -141.42 ft/sec; -266.84 ft/sec; the second number, which represents the average velocity over the last second.

47. $f(x) = x^3$ **49.** $f(x) = \frac{3}{2}x$ **51.** $f(x) = \frac{1}{2}x^2$ **53.** $f(x) = \frac{1}{2}x + 6$ **55.** $f(x) = \dfrac{1}{x - 9}$ **57.** $\dfrac{1}{(x + h + 1)(x + 1)}$

59. $\dfrac{-2}{x\sqrt{x + h} + \sqrt{x}(x + h)}$

SECTION 1.3

1. $f(-4) = -1, f(-2) = 3, f(2) = 3, f(4) = 1, f(7) = -2$ **3.** Domain: $-5 \le x \le 8$; Range: $-3 \le y \le 4$
5. (a) $(-5, 0), (-3, 0), (5, 0)$ **(b)** Positive: $-3 < x < 5$; Negative: $-5 < x < -3$ or $5 < x \le 8$
7. (a) $(-4, -1), (1, 4)$ **(b)** Increasing: $-4 < x < 1$; Decreasing: $-5 < x < -4$ or $1 < x < 8$
9. Yes   **11.** No   **13.** No   **15.** No   **17.** $f(-x) = -(x^3 + 2x)$; odd function   **19.** $f(-x) = |x - 8|$; neither

21. $f(-x) = \dfrac{12}{x^2 + 4}$; even function   **23.** $f(-x) = -x^{3/5} + 5$; neither   **25. (a)** $(0, -8)$ **(b)** $(4, 0)$

27. (a) $(0, -6)$ **(b)** $(-\sqrt{6}, 0), (\sqrt{6}, 0)$ **29.**

31.

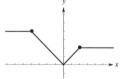

33.

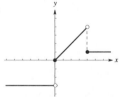

35. (a) $(-2\sqrt{3}, 0), (0, 0), (2\sqrt{3}, 0)$ **(b)** Positive: $x < -2\sqrt{3}$ or $0 < x < 2\sqrt{3}$; Negative: $-2\sqrt{3} < x < 0$ or $x > 2\sqrt{3}$
 (c) Increasing: $-2 < x < 2$; Decreasing: $x < -2$ or $x > 2$
37. (a) None   **(b)** Positive: $x > 0$, Negative: $x < 0$ **(c)** Increasing: $x < -2$ or $x > 2$; Decreasing: $-2 < x < 0$ or $0 < x < 2$
39. Odd function   **41.** Even function   **43.** Odd function   **45.** Even function

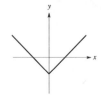
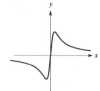

47.

49.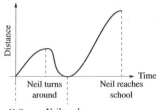

51. (a)

d	0	40	80	120	160	200
F	5	30	27	32	21	15

(b) Increasing: $0 < d < 40$ or $90 < d < 100$; Decreasing: $40 < d < 90$ or $100 < d < 200$

(c) Max: 3500 ft³/sec on day 100; Min: 500 ft³/sec on day 0

SECTION 1.4

1. (a) **(b)** 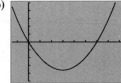 **(c)**

3. (a) 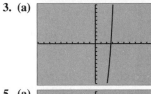 **(b)** 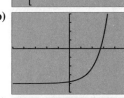 **(c)**

5. (a) 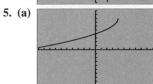 **(b)** 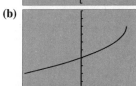 **(c)**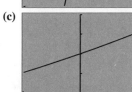

7.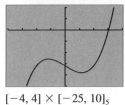

$[-4, 4] \times [-25, 10]_5$

9.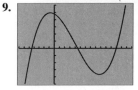

$[-7, 15] \times [-250, 300]_{50}$

11.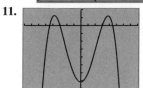

$[-8, 8] \times [-200, 50]_{25}$

13.

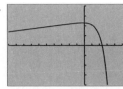

$[-10, 5] \times [-40, 40]_{10}$

15.

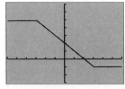

$[-30, 30]_5 \times [-30, 70]_{10}$

17.

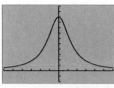

$[-6, 6] \times [-50, 300]_{25}$

19. $\frac{8}{3}$ (2.67)

21. $1 - \sqrt{6}$ (−1.45), $1 + \sqrt{6}$ (3.45)

23. $1 - \sqrt{5}$ (−1.24), 0, $1 + \sqrt{5}$ (3.24)

$[-6, 6] \times [-2, 14]$

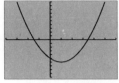

$[-4, 6] \times [-10, 10]$

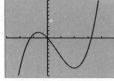

$[-3, 5] \times [-10, 10]$

25. −8 **27.** 2, 4 **29.** −1, 5 **31.** (1.2, 4.9) **33.** (1.0, 1.5)

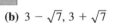

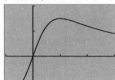

35. (2.1, 0.0) **37.** 2.4101 **39.** −1.8933 **41.** 2.8136

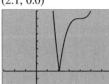

43. (a) 0.354, 5.646 **(b)** $3 - \sqrt{7}$, $3 + \sqrt{7}$ **45. (a)**

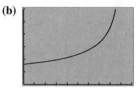

(b)

47. The arch is 630 ft high.

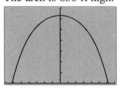

$[-350, 350]_{50} \times [0, 700]_{100}$

SECTION 1.5

1. $A(p) = \dfrac{p^2}{16}$ **3.** $l(w) = \dfrac{72}{w}$ **5.** $V(s) = \dfrac{s\sqrt{6s}}{36}$ **7.** $h(x) = \sqrt{64 - x^2}$; Domain: [0, 8]

9. (a) Yes   **(b)** 1.4; 3.9

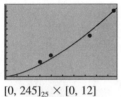

$[0, 245]_{25} \times [0, 12]$

11. (a)   **(b)** 16   **(c)** Yes   **(d)** $S_h(x) = \dfrac{89.25}{\sqrt{x}}$; $S_l(x) = \dfrac{80.75}{\sqrt{x}}$

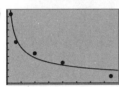

$[0, 80]_{10} \times [0, 55]_5$

13. (a) $C_1(x) = 13x + 15{,}000$; $C_2(x) = 15.5x + 12{,}000$; $C_1(2500) = 47{,}500$; $C_2(2500) = 50{,}750$
    (b)   **(c)** 1200 backpacks; the cost is less from the second supplier.

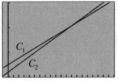

$[-100, 2000]_{100} \times [10{,}000, 40{,}000]_{10{,}000}$

15. (a) $A(w) = -2w^2 + 280w$; Domain: (0, 140)   **(b)** Intercepts: (0, 0), (140, 0); Increasing: (0, 70); Decreasing: (70, 140)

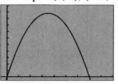

$[-10, 180]_{20} \times [-500, 11{,}000]_{1000}$

(c) 70 yd by 140 yd; maximum area is 9800 yd^2

17. (a) $S(x) = 2x^2 + \dfrac{120}{x}$ **(b)** Domain: $(0, +\infty)$; a base of 3 ft on a side has less surface area.

(c)

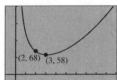

$[-1, 10]_5 \times [-20, 200]_{50}$

The minimum surface area is 58 ft^2 when the length is 3.1 ft.

19. (a) $C(x) = 21x + \dfrac{3840}{x}$ **(b)** Domain: $(0, +\infty)$; 12 yd   **(c)**

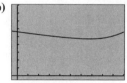

$[-1, 40]_{10} \times [-100, 2000]_{200}$
Minimum cost is $568 when dimensions are 13.5 yd by 17.8 yd.

21. (a) $C(x) = 3\sqrt{50^2 + (180 - x)^2} + 2x$ **(b)**

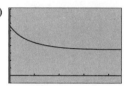

$[-10, 200]_{10} \times [-50, 900]_{100}$
Minimum cost is $471.80 when $x = 135.3$ m. Maximum cost when $x = 0$.

23. (a) $C(s) = 15 + 0.2s + \dfrac{500}{s}$ **(b)**   **(c)** Minimum cost is $35 when speed is 50 mph.

$[10, 55]_{10} \times [-10, 90]_{10}$
No intercepts; decreasing on $(10, 50)$; increasing on $(50, 55)$.

25. $x = \dfrac{25{,}600 - 500p}{3}$ **27. (a)** $x = 13{,}600 - 4000p$ **(b)** $R(x) = x\left(\dfrac{13{,}600 - x}{4000}\right)$

(c)   **(d)** Revenue is maximized when 6800 bags are sold.
(e) $1.70 per bag

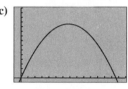

$[-1000, 15{,}000]_{1000} \times [-1000, 15{,}000]_{1000}$

29. (a) $C(x) = 0.30\pi x^2 + \dfrac{10.8}{x}$ **(b)** Minimum cost is $9.05 when radius is 1.8 in.

31. $A(w) = \dfrac{1200w}{w - 8} + 16w$ **33.** $V(w) = 2w\sqrt{64 - w^2}$

SECTION 1.6

1. The graph of $y = |x|$ shifted 2 units to the right   **3.** The graph of $y = \sqrt{4 - x^2}$ shifted 5 units to the left

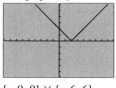

$[-9, 9] \times [-6, 6]$

$[-8, 4] \times [-4, 4]$

5. The graph of $y = x^2$ shifted 4 units down

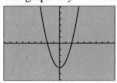

$[-9, 9] \times [-6, 6]$

7. The graph of $y = 1/x$ shifted 2 units to the right and 3 units up

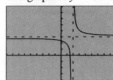

$[-9, 9] \times [-4, 8]$

9. The graph of $y = x^3$ shifted 2 units to the right and 1 unit down

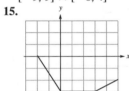

$[-9, 9] \times [-8, 4]$

11.

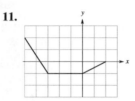

13.

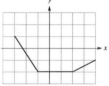

15.

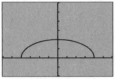

17.

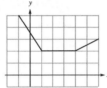

19. The graph of $y = \sqrt{x}$ expanded from the x-axis by a factor of 2

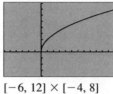

$[-6, 12] \times [-4, 8]$

21. The graph of $y = \sqrt{16 - x^2}$ compressed toward the x-axis by a factor of $\frac{1}{2}$

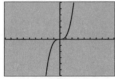

$[-6, 6] \times [-2, 6]$

23. The graph of $y = x^3$ compressed toward the x-axis by a factor of $\frac{1}{2}$

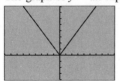

$[-9, 9] \times [-6, 6]$

25. The graph of $y = |x|$ expanded from the x-axis by a factor of $\frac{4}{3}$

$[-9, 9] \times [-4, 8]$

27. The graph of $y = \sqrt{9 - x^2}$ compressed toward the y-axis by a factor of $\frac{1}{3}$

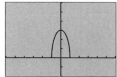

$[-6, 6] \times [-2, 6]$

29. The graph of $y = \sqrt{36 - x^2}$ expanded from the y-axis by a factor of 2

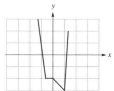

$[-15, 15]_2 \times [-5, 10]_2$

31.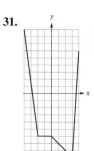

33.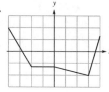

35.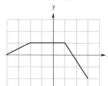

37.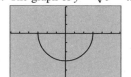

39.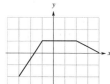

41.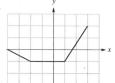

43.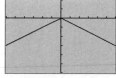

45. The graph of $y = \sqrt{9 - x^2}$ reflected about the x-axis

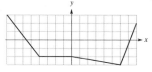

$[-6, 6] \times [-5, 3]$

47. The graph of $y = \frac{1}{2} |x|$ reflected about the x-axis

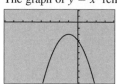

$[-6, 6] \times [-6, 2]$

49. The graph of $y = x^2$ reflected about the x-axis, shifted 1 unit to the left and 2 units down

$[-8, 4] \times [-10, 2]$

51. The graph of $y = \sqrt{x}$ reflected about the y-axis, shifted 4 units to the left   **53.** $y = (x - 2)^2 - 4$

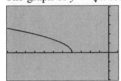

$[-11, 1] \times [-3, 5]$

55. $y = -\sqrt{16 - x^2}$ **57.** $y = -(x + 2)^2$ **59.** $y = \frac{1}{2}x^2$ **61.** $y = 3(1/x)$, or $y = 3/x$

63. The graph of $y = 1/x$ reflected through either the x-axis or y-axis, shifted 2 units to the right and 3 units down

65. The graph of $y = \sqrt{16 - x^2}$ shifted 3 units to the left   **67.** $y = 3S(-x)$

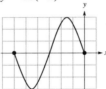

69. (a) Expand the graph of $f(r) = 1/r$ from the r-axis by a factor of 10.   **(b)** $t(r) = \dfrac{10}{r - 5}$

(c) Shift the graph of $T(r) = \dfrac{10}{r}$ to the right 5 units

71. (a) It is the graph of $y = 1/x$ expanded from the x-axis by a factor of 500, then translated up 17 units.

(b)

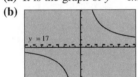

$[-10, 10] \times [-250, 250]_{50}$

73. (a)

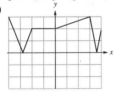

(b)

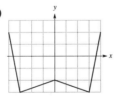

SECTION 1.7

1. (a) $(f + g)(x) = x^2 + 2x - 5$; $(f - g)(x) = x^2 - 6x - 3$ **(b)** $(f + g)(2) = 3$; $(f - g)(2) = -11$

3. (a) $(f + g)(x) = x^2 - 9 + \sqrt{2x}$; $(f - g)(x) = x^2 - 9 - \sqrt{2x}$ **(b)** $(f + g)(2) = -3$; $(f - g)(2) = -7$

5. (a) $(f + g)(x) = \dfrac{3x + 2}{x(x + 1)}$; $(f - g)(x) = \dfrac{-x - 2}{x(x + 1)}$ **(b)** $(f + g)(2) = \dfrac{4}{3}$; $(f - g)(2) = -\dfrac{2}{3}$

7. (a) $(fg)(x) = -2x^3 - 3x^2 + 4x + 6$; $\left(\dfrac{f}{g}\right)(x) = \dfrac{2 - x^2}{2x + 3}$ **(b)** $(fg)(-3) = 21$; $\left(\dfrac{f}{g}\right)(-3) = \dfrac{7}{3}$

9. (a) $(fg)(x) = (x^2 - 9)\sqrt{x + 4}$; $\left(\dfrac{f}{g}\right)(x) = \dfrac{x^2 - 9}{\sqrt{x + 4}}$ **(b)** $(fg)(-3) = 0$; $\left(\dfrac{f}{g}\right)(-3) = 0$

11. (a) $(fg)(x) = 2x^3 - x^2 - 21x$; $\left(\dfrac{f}{g}\right)(x) = \dfrac{2x - 7}{x^2 + 3x}$ **(b)** $(fg)(-3) = 0$; $\left(\dfrac{f}{g}\right)(-3)$ is undefined

13. (a) $[-3, +\infty)$ **(b)** $(-3, +\infty)$ **15. (a)** $[-2, 0)$ or $(0, 2]$ **(b)** $(-2, 0)$ or $(0, 2)$ **17.**

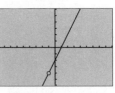

$[-9, 9] \times [-6, 6]$

19. (a) $(f \circ g)(x) = 2x^2 - 2x + 3$ **(b)** $(g \circ f)(x) = 4x^2 + 10x + 6$ **21. (a)** $(f \circ g)(x) = \dfrac{9\sqrt{x}}{x}$ **(b)** $(g \circ f)(x) = \dfrac{3\sqrt{x}}{x}$

23. (a) $(f \circ g)(x) = x^4 - 6x^3 + 9x^2 + 3$ **(b)** $(g \circ f)(x) = -x^4 - 3x^2$ **25. (a)** $(f \circ g)(x) = 3$ **(b)** $(g \circ f)(x) = 23$

27.
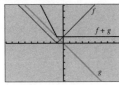

$[-9, 9] \times [-6, 6]$

29.
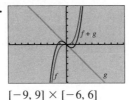

$[-9, 9] \times [-6, 6]$

31.
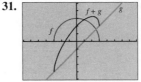

$[-7, 8] \times [-5, 5]$

33. a **35.** b

37.
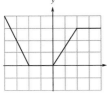

39.
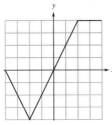

41.
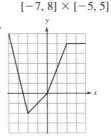

43.
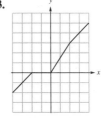

45. $f(x) = \sqrt{x}$, $g(x) = 2x + 1$ **47.** $f(x) = \dfrac{3}{x}$, $g(x) = \sqrt{4 - x}$; or $f(x) = \dfrac{3}{\sqrt{x}}$, $g(x) = 4 - x$

49. $f(x) = \dfrac{x}{x - 2}$, $g(x) = x^4$ **51.** $f(x) = x^5$, $g(x) = x^3 - 4x^2 + x$

53. (a) $(f \circ g)(x) = \dfrac{2}{5}\left[\dfrac{5}{2}(x - 3)\right] + 3 = x$ **(b)** $(g \circ f)(x) = \dfrac{5}{2}\left[\left(\dfrac{2}{5}x + 3\right) - 3\right] = x$

55. (a) $(f \circ g)(x) = \dfrac{\dfrac{3}{x - 1} + 3}{\dfrac{3}{x - 1}} = x$ **(b)** $(g \circ f)(x) = \dfrac{3}{\dfrac{3}{x + 3} - 1} = x$ **57.**

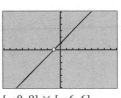

$[-9, 9] \times [-6, 6]$

59. (a) $p = -\dfrac{1}{120}x + \dfrac{61}{12}$ **(b)** $R(x) = -\dfrac{1}{120}x^2 + \dfrac{61}{12}x$ **(c)** $P(x) = -\dfrac{1}{120}x^2 + \dfrac{55}{12}x - 300$ **(d)** 275

61. $5h, 9h, 13h, \ldots$ units   **63. (a)** $g(t) = 40t$ **(b)** $f(t) = 40\sqrt{4 + t^2}$

65. The rate of growth decreases. The rate is greatest when $t = 0$.

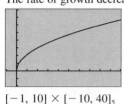

$[-1, 10] \times [-10, 40]_5$

67.

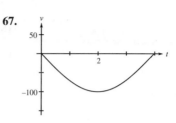

MISCELLANEOUS EXERCISES

1. $d(A, B) = 25$; $M\left(\frac{3}{2}, -1\right)$ **3.** $(x - 2)^2 + (y + 2)^2 = 8$; Center: $(2, -2)$; Radius; $2\sqrt{2}$

5. $(x + 1)^2 + (y - 5)^2 = 18$; Center: $(-1, 5)$; Radius: $3\sqrt{2}$ **7.**

9.

11. $(9, 1)$ **13. (a)** Zero: 13; Domain: $[4, +\infty)$ **(b)** 2   **(c)** $3 - 2\sqrt{t^2 - 1}$ **(d)** $18 - 6r$

15. -3 **17. (a)** 5   **(b)** $\frac{13}{9}$ **(c)** $-\frac{2}{3}$ **(d)** $1 + a^4$ **19.** x-intercepts: $(-2.303, 0)$, $(1, 0)$, $(1.303, 0)$

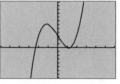

$[-6, 6] \times [-8, 12]$

21. x-intercepts: $(-2.478, 0)$, $(6.105, 0)$ **23.** x-intercepts: $(-4.123, 0)$, $(4.123, 0)$

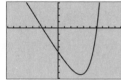

$[-8, 10] \times [-40, 20]_5$

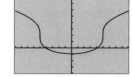

$[-9, 9] \times [-4, 8]$

25. (a)

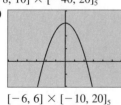

$[-6, 6] \times [-10, 20]_5$

(b) $(\sqrt{5}, 0)$, $(-\sqrt{5}, 0)$

27. (a)

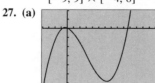

$[-4, 14] \times [-160, 50]_{25}$

(b) $(0, 0)$, $\left(\dfrac{9 + \sqrt{105}}{2}, 0\right)$, $\left(\dfrac{9 - \sqrt{105}}{2}, 0\right)$

29. $-3.73, 0.60, 3.12$ **31.** The graph of $y = |x|$ shifted 4 units to the left

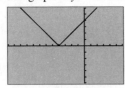

$[-12, 6] \times [-4, 4]$

33. The graph of $y = x^3$ shifted 1 unit to the right and 4 units up

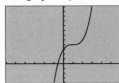

$[-6, 6] \times [-4, 12]$

35. The graph of $y = 1/x$ reflected about the x-axis and shifted 2 units to the right

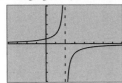

$[-4, 8] \times [-4, 4]$

37. $f(-3) = -1; f(-1) = 0; f(1) = 2; f(4) = -3$

39. x-intercepts: $(-4, 0), (-1, 0), (\frac{5}{2}, 0)$; turning points: $(-3, -1), (1, 2)$

41. 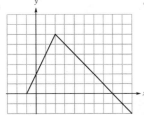   **43.** 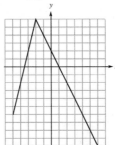   **45.**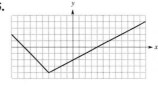

47. 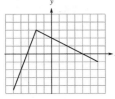   **49.** 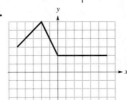   **51.** $y = -x^2 + 3$

53. (a) $x^2 + 4x - 5$ **(b)** $x^2 - 8x - 8$ **55. (a)** $4x^3 + x^2 - 24x - 6$ **(b)** $x^4 - 12x^2 + 36$

57. (a) $16x^2 + 8x - 5$ **(b)** $x^4 - 12x^2 + 30$ **59.** $f(x) = 2^x, g(x) = 4x - 3$

61. (a) Midpoints of AB, BC, CD, and AD are $M_{AB}(-\frac{3}{2}, 4)$, $M_{BC}(-\frac{3}{2}, 6)$, $M_{CD}(2, 4)$, $M_{AD}(2, 2)$; $d(M_{AB}, M_{BC}) = 2 = d(M_{CD}, M_{AD})$, $d(M_{BC}, M_{CD}) = \sqrt{65}/2 = d(M_{AB}, M_{AD})$

(b) Midpoints of AB, BC, CD, and AD are $M_{AB}\left(\dfrac{x_1 + x_2}{2}, \dfrac{y_1 + y_2}{2}\right)$, $M_{BC}\left(\dfrac{x_2 + x_3}{2}, \dfrac{y_2 + y_3}{2}\right)$, $M_{CD}\left(\dfrac{x_3 + x_4}{2}, \dfrac{y_3 + y_4}{2}\right)$,

$M_{AD}\left(\dfrac{x_1 + x_4}{2}, \dfrac{y_1 + y_4}{2}\right)$; $d(M_{AB}, M_{BC}) = \sqrt{\left(\dfrac{x_1 - x_3}{2}\right)^2 + \left(\dfrac{y_1 - y_3}{2}\right)^2} = d(M_{CD}, M_{AD})$;

$d(M_{BC}, M_{CD}) = \sqrt{\left(\dfrac{x_2 - x_4}{2}\right)^2 + \left(\dfrac{y_2 - y_4}{2}\right)^2} = d(M_{AB}, M_{AD})$

63. $A = 4$, $n = 3$ **65.**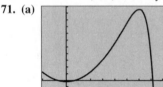

67. (a)

t	0	0.5	1.0	1.5	2.0	2.5
h	-8	0	8	0	-8	0

(b) Moving up over the intervals $(0, 1.0)$ and $(2.0, 2.5)$; moving down over the interval $(1.0, 2.0)$

(c) Maximum height is 8 in. when $t = 1.0$ sec; minimum height is -8 in. when $t = 0$ and 2.0 sec.

69. $C = 35x + 24{,}000$; 2000 keyboards cost \$94,000

71. (a) **(b)** 6860 units; 71,480 units   **(c)** 60,320 units; \$264,600

$[-20, 80]_{10} \times [-25{,}000, 275{,}000]_{25{,}000}$

73. (a) $A(x) = 18x - \frac{1}{2}x^3$; Domain: $(0, 6)$ **(b)** 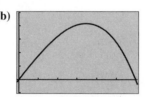   **(c)** 41.57 square units

$[0, 6] \times [-10, 50]_{10}$

75. (a) $C(s) = \dfrac{1440}{s} + 20.8 + 0.56s$ **(b)** 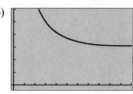   **(c)** \$77.59

$[0, 55]_5 \times [-10, 150]_{25}$

CHAPTER TEST

1.
$$x^2 + y^2 + x - 6y + 7 = 0$$
$$\left(x^2 + x + \tfrac{1}{4}\right) + (y^2 - 6y + 9) = -7 + \tfrac{1}{4} + 9$$
$$\left(x + \tfrac{1}{2}\right)^2 + (y - 3)^2 = \tfrac{9}{4}$$

Center: $\left(-\tfrac{1}{2}, 3\right)$; Radius: $\tfrac{3}{2}$
(see Section 1.1, Example 5)

2.
$$d(A, B) = 10$$
$$\sqrt{(2 - x)^2 + (13 - 5)^2} = 10$$
$$(2 - x)^2 + (13 - 5)^2 = 100$$
$$(2 - x)^2 + 64 = 100$$
$$(2 - x)^2 = 36$$
$$2 - x = \pm 6$$
$$x = 2 \pm 6$$
$$x = -4, 8$$

(see Section 1.1, Example 2)

3. We use the tests from Section 1.1:

x-axis: $x^3 + (-y)^2 = 8$
$\phantom{x\text{-axis:}}\quad x^3 + y^2 = 8$

y-axis: $(-x)^3 + y^2 = 8$
$\phantom{y\text{-axis:}}\quad -x^3 + y^2 = 8$

origin: $(-x)^3 + (-y)^2 = 8$
$\phantom{\text{origin:}}\quad -x^3 + y^2 = 8$

The graph is symmetric with respect to the x-axis.

The graph is not symmetric with respect to the origin.

The graph is not symmetric with respect to the origin.

The graph in quadrant I is reflected across the x-axis to complete the graph.

To find the x-intercept, let $y = 0$:

$$x^3 + y^2 = 8$$
$$x^3 + (0)^2 = 8$$
$$x = 2$$

The x-intercept is $(2, 0)$.
(see Section 1.1, Example 8)

To find the y-intercept, let $x = 0$:

$$x^3 + y^2 = 8$$
$$(0)^3 + y^2 = 8$$
$$y = \pm 2\sqrt{2}$$

The y-intercepts are $(0, \pm 2\sqrt{2})$

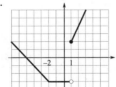

4. (a) Because -7 is in the interval described by $x < -2$, $P(-7) = -(-7) - 5 = 2$.
   (b) Because $t \geq 0$, $t + 2$ is in the interval described by $x \geq 1$. Thus, $P(t + 2) = 2(t + 2) = 2t + 4$.
   (c) The graph of P is constructed from the graph of $y = -x - 5$ over the interval $x < -2$, $y = -3$ over the interval $-2 \leq x < 1$, and $y = 2x$ over the interval $x \geq 1$. \qquad (see Section 1.3, Example 5)

5. (a) $f(9) = \dfrac{4\sqrt{9}}{9 - 2} = \dfrac{4 \cdot 3}{7} = \dfrac{12}{7}$ **(b)** $f(4k + 4) = \dfrac{4\sqrt{4k + 4}}{(4k + 4) - 2} = \dfrac{4\sqrt{4(k + 1)}}{4k + 2} = \dfrac{4 \cdot 2\sqrt{k + 1}}{2(2k + 1)} = \dfrac{4\sqrt{k + 1}}{2k + 1}$

(see Section 1.2, Example 3)

6. Let $f(x) = 0$:

$$2x^3 - 6x = 0$$
$$2x(x^2 - 3) = 0$$

$$2x = 0 \quad | \quad x^2 - 3 = 0$$
$$x = 0 \quad | \quad x^2 = 3$$
$$\quad | \quad x = \pm\sqrt{3}$$

The zeros of f are $-\sqrt{3}, 0, \sqrt{3}$.
(see Section 1.1, Example 1)

8.

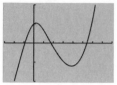

$[-3, 8] \times [-2, 2]$
(see Section 1.4, Example 2)

7. $\dfrac{g(x + h) - g(x)}{h} = \dfrac{\dfrac{3(x + h) - 1}{2} - \dfrac{3x - 1}{2}}{h} \cdot \dfrac{2}{2}$

$$= \frac{(3x + 3h - 1) - (3x - 1)}{h}$$

$$= \frac{3h}{2h} = \frac{3}{2}$$

(see Section 1.2, Example 4)

9. Let $f(x) = 0$:

$$x^3 + x^2 - 5x - 5 = 0$$
$$x^2(x + 1) - 5(x + 1) = 0$$
$$(x^2 - 5)(x + 1) = 0$$

$$x^2 - 5 = 0 \quad | \quad x + 1 = 0$$
$$x^2 = 5 \quad | \quad x = -1$$
$$x = \pm\sqrt{5} \quad |$$

x-intercepts: $(-\sqrt{5}, 0), (-1, 0), (\sqrt{5}, 0)$

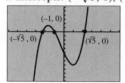

$[-6, 6] \times [-12, 6]$
(see Section 1.4, Example 5)

10. The complete graph of $y = x^4 - 8x^3 + 15$ appears to have x-intercepts close to $x = 1$ and $x = 8$:

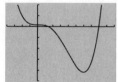

$[-4, 10] \times [-500, 200]_{100}$

Either using the root feature of a graphing utility or the method described in Section 1.4, we find that the x-intercepts are $(1.31, 0)$ and $(7.97, 0)$. Thus, the solutions of the equation are 1.31 and 7.97 (to the nearest 0.01). (see Section 1.4, Example 8)

11. As the tank fills, the depth increases, at first more quickly, then more slowly (since the tank is narrower at the bottom and wider toward the middle). Once the tank is half full, the depth again increases more quickly until the pipe is shut off.

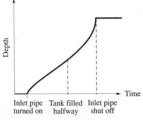

(see Section 1.3, Example 7)

12. (a) The minimum value of C corresponds to the lowest point on the graph for which $700 \le x \le 1200$; this point is (960, 12,068). Thus, the minimum cost in this interval of x is 960 ft.^{2}.

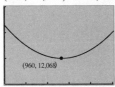

$[700, 1200]_{100} \times [11,000, 14,000]_{1000}$

(b) The lowest point over this interval is (1100, 12,460). Thus, the minimum cost in this interval of x is 1100 ft.^{2}

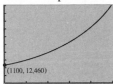

$[1100, 1600]_{100} \times [10,000, 12,000]_{1000}$

(see Section 1.5, Example 7)

13. The capacity of the gutter is maximum when the area of the cross section of the gutter, a rectangle, is maximum. If the height of the gutter is x inches, then the width is $9 - 2x$:   Thus, this area is given by the function $A = x(9 - 2x)$.

Because the highest point on the graph of $y = x(9 - 2x)$ is (2.25, 10.125), the maximum capacity occurs when the height is 2.25 in. and the width is 4.5 in.

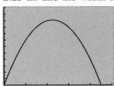

$[0, 5] \times [0, 12]$
(see Section 1.5, Example 5)

14. The graph is a reflection about the x-axis and a vertical translation, 3 units up, of the graph of $y = 1/x$, the reciprocal function in the directory of graphs. Thus, the equation of the graph is $y = (-1/x) + 3$. (see Section 1.6, Examples 4, 12)

15. (a) $f(x) + g(x) = (5x^2 + x - 1) + (3 - 2x^2)$
$$= 3x^2 + x + 2$$

(b) $f(x)g(x) = (5x^2 + x - 1)(3 - 2x^2)$
$$= (5x^2 + x - 1)(3) + (5x^2 + x - 1)(-2x^2)$$
$$= 15x^2 + 3x - 3 - 10x^4 - 2x^3 + 2x^2$$
$$= -10x^4 - 2x^3 + 17x^2 + 3x - 3$$

(c) $(f \circ g)(x) = f[g(x)]$
$$= f[g(3 - 2x^2)]$$
$$= 5(3 - 2x^2)^2 + (3 - 2x^2) - 1$$
$$= 5(9 - 12x^2 + 4x^4) + (3 - 2x^2) - 1$$
$$= 20x^4 - 62x^2 + 47$$

(see Section 1.7, Examples 1, 2)

16. The graph of $y = f(x + 2) + 1$ is a horizontal and vertical translation of the graph of f, 2 units to the left and 1 unit up.

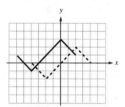

(see Section 1.6, Examples 5, 6)

17. The graph of $y = -2g(x)$ is a vertical expansion of the graph of g by a factor of 2, and a reflection through the x-axis.

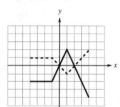

(see Section 1.6, Examples 7, 8)

18. The graph of $y = f(x) + g(x)$ is determined by the technique of adding y-coordinates.

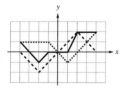

(see Section 1.7, Example 4)

19. Many pairs of functions f and g will work, but the most obvious pair is $f(x) = x^3$ and $g(x) = 2^x + x$.
(see Section 1.7, Example 6)

20. From the table, we see that $f(0)$ is 5. Since $f(0) = m(0) + b = b$, it follows that $b = 5$ and that $f(x) = mx + 5$ for some number m. From the second column of the table, we see that

$$f(2) = 6.5$$
$$m(2) + 5 = 6.5$$
$$2m = 1.5$$
$$m = 0.75$$

The function is $f(x) = 0.75x + 5$. We check by noting that $f(4) = 0.75(4) + 5 = 8$ and $f(10) = 0.75(10) + 5 = 12.5$. (see Section 1.2, page 91)

CHAPTER 2: SECTION 2.1

1. (a) 0   **(b)** 2   **(c)** Undefined   **(d)** $-\frac{1}{3}$ **3.**

5.
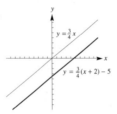

7.

9.
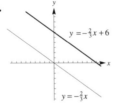
11.
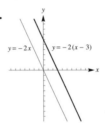
13.

15. $y = \frac{3}{5}x - 2$ **17.** $y = 7$ **19.** $y = -6x + 25$ **21.** $y = -\frac{1}{2}x - 2$ **23.** $y = \frac{1}{2}x - 4$

25. $y = 0x + 4$ **27.** $y = \frac{1}{4}x$ **29.** $y = \frac{8}{3}x - 8$ **31.** $f(x) = -\frac{1}{2}x + 2$

33. $f(x) = -\frac{1}{2}x + 2$ **35.** $f(x) = \frac{4}{3}x + \frac{8}{3}$ **37.** $f(x) = \frac{4}{3}x + \frac{25}{3}$ **39.** $f(x) = -2x + 6$ **41.** $f(x) = \frac{2}{3}x - 2$

43. $f(x) = -\frac{1}{20}x + 8$ **45.** 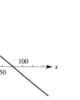   **47.**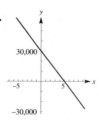

49. $m_{AB} = \frac{4}{3}$, $m_{BC} = \frac{4}{3}$, $m_{CA} = \frac{4}{3}$; the points A, B, and C are collinear   **51.** $m_{AB} = -\frac{1}{3}$, $m_{CA} = 3$, the triangle ABC is a right triangle
53. Quadrilateral $ABCD$ is a rhombus.
55. **(a)** $y = -2250x + 25{,}000$ **(b)** \$20,500   **57.** $n = \frac{1}{4}T + 40$

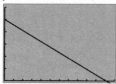

$[0, 12] \times [0, 30{,}000]_{5000}$
59. **(a)** $C = 16x + 4500$ **(b)** $R = 25x$ **(c)** $P = 9x - 4500$ **(d)** 500 hr

SECTION 2.2

1. **(a)** Vertical expansion by a factor of 3   **(b)** Reflection about the x-axis

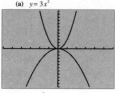

(a) $y = 3x^2$

(b) $y = -x^2$
$[-6, 6] \times [-12, 12]$
3. **(a)** Vertical expansion by a factor of 2, and a reflection about the x-axis   **(b)** Vertical expansion by a factor of 1.5

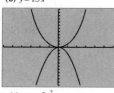

(b) $y = 1.5x^2$

(a) $y = -2x^2$
$[-6, 6] \times [-12, 12]$

5. Horizontal translation 3 units right of the graph of $f(x) = x^2$

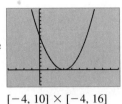

$[-4, 10] \times [-4, 16]$

7. Horizontal translation 2 units right and vertical translation 4 units up of the graph of $f(x) = x^2$

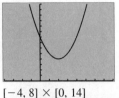

$[-4, 8] \times [0, 14]$

9. Horizontal translation 3 units left and vertical translation 8 units down of the graph of $f(x) = 2x^2$

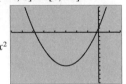

$[-8, 2] \times [-12, 6]$

11. Horizontal translation 4 units left and vertical translation 6 units down of the graph of $f(x) = -x^2$

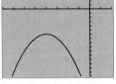

$[-8, 2] \times [-16, 2]$

13. Horizontal translation 3 units left of the graph of $f(x) = -\frac{1}{3}x^2$ **15.** $y = (x - 3)^2 - 2$ **17.** $y = -\frac{1}{2}(x - 2)^2 + 5$

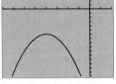

$[-10, 4] \times [-10, 2]$

19. $y = 2(x - 3)^2$ **21.** $y = (x + 2)^2 - 12$ **23.** $y = 2(x - 1)^2 + 3$ **25.** $y = -\frac{1}{2}(x - 2)^2$
27. $y = 3(x + 3)^2 - 27$ **29.** $y = (x + \frac{3}{2})^2 - 2$ **31.** $y = 2(x + 1)^2 + 18$
33. Vertex: $(2, -4)$; x-intercepts: $(0, 0)$, $(4, 0)$; y-intercept: $(0, 0)$
35. Vertex: $(-6, 8)$; x-intercepts: $(-4, 0)$, $(-8, 0)$; y-intercept: $(0, -64)$
37. Vertex: $(1, -6)$; x-intercepts: $(1 \pm \sqrt{6}, 0)$; y-intercept: $(0, -5)$
39. Vertex: $(1, -3)$; x-intercepts: none; y-intercept: $(0, -4)$ **41.** 130 units   **43. (a)** $y = \frac{1}{200}x^2 + 5$ **(b)** 23 ft
45. 3 in.   **47. (a)** The domain is $[0, 9]$.   **(b)** 40.5 square units   **49. (a)** $|k| < 8$ **(b)** $k = \pm 8$ **(c)** $|k| > 8$
53. $(6, 4)$

SECTION 2.3

1. $-2, 0, 6$ **3.** $-6, -1, 1$ **5.** $-3, -\sqrt{5}, \sqrt{5}, 3$ **7.** 4   **9.** 1   **11.** No real solutions   **13.** $-\sqrt[5]{2}$ **15.** 0
17. 4   **19.** 26   **21.** 5   **23.** -45 **25.** $-4, 4$ **27.** $-3, -1, 1, 3$ **29.** 6   **31.** 9   **33.** 62   **35.** $-\sqrt{3}, \sqrt{3}$
37. $-5, -1$ **39.** $2 - 2\sqrt{2}, 2, 2 + 2\sqrt{2}$ **41.** $\frac{1}{8}, 1$ **43.** $-3, 3$ **45.** $1, 5$ **47.** -3 **49.** $0, 9$ **51.** 7.5%
53. 3.1 cm   **55.** 2   **57.** 25   **59.** $-2, 6$ **61.** $-2\sqrt{2}, -3, 2\sqrt{2}$ **63.** $\dfrac{7 - \sqrt{21}}{2}, \dfrac{7 + \sqrt{21}}{2}$
65. $0, 4$ **67.** $\dfrac{-5 - \sqrt{33}}{2}, \dfrac{-5 + \sqrt{33}}{2}$

SECTION 2.4

1. (a) $(-3, 5)$ **(b)** $(-5, -3)$ or $(5, 8]$ **(c)** $[-3, 5]$ **(d)** $[-5, -3]$ or $[5, 8]$ **3.** $x > 5$ **5.** $x \le -3$
7. $x \le 8$ **9.** $x < -2$ or $x > 5$ **11.** $-1 \le x \le 3$ **13.** All real numbers   **15.** $x \le -3$ or $x \ge 7$
17. $2.4 < x < 3.6$ **19.** $x \le -1$ or $x \ge 4$ **21.** $x < -8$ or $x > 4$ **23.** $x = 6$ **25.** $-\sqrt{3} \le x \le \sqrt{3}$
27. $x \le -1$ or $0 \le x \le 1$ **29.** $x \le -\frac{2}{3}$ or $x \ge 4$ **31.** $x = -7$ or $-3 \le x \le 2$ **33.** $-11 < x < 21$
35. $-3 < x < -\sqrt{7}$ or $\sqrt{7} < x < 3$ **37.** $x > -\frac{5}{2}$ **39.** $4 \le x \le 13$ **41.** $x > 2$ **43.** $x \ge 2$
45. $x \le -\sqrt{14}$ or $x \ge \sqrt{14}$ **47.** $x < -1$ or $x > 4$ **49.** Between 1 and 3 sec after toss
51. (a) $0 < x < 118.24$ **(b)** \$57   **53.** $159.9984 < x < 160.0016$
57. (a) No real numbers   **(b)** All real numbers   **(c)** No real numbers

SECTION 2.5

1. The numbers are 15 and 16.   **3.** Hal must score 91.   **5.** Max weighs 32 lb; Lefty weighs 16 lb.
7. \$57,142.86 profit in first quarter   **9.** 8 m $\times$ 16 m   **11.** 60 L of 10% solution   **13.** $\frac{24}{17}$ hr (1 hr, 25 min)
15. \$14,500 in savings; \$10,500 in bonds   **17.** About 7% and 14%   **19.** 9 m $\times$ 16 m
21. $8 + 4\sqrt{15}$ (approx. 23.5) in. square   **23.** 6 m   **25.** 45 mph   **27.** 750,000 standard-sized; 250,000 compacts
29. \$7400   **31.** 0.6 hr   **33.** 4 ft   **35.** $\sqrt{40/\pi}$ (approx. 3.57) in.   **37.** 50 mph

MISCELLANEOUS EXERCISES

1. $y = -\frac{2}{3}x + 7$ **3.** $y = 4x - 6$ **5.** $y = 2x + 20$ **7.** $y = \frac{1}{3}x - 3$

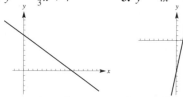

9. $y = -4x + 8$ **11.** $y = 200x - 1000$ **13. (a)** $y = (x + 2)^2 - 25$

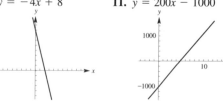

(b)

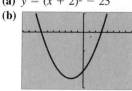

$[-10, 8] \times [-30, 10]_{10}$
(c) Vertex: $(-2, -25)$; x-intercepts: $(-7, 0), (3, 0)$;
y-intercept: $(0, -21)$

15. (a) $y = \left(x + \frac{5}{2}\right)^2 - \frac{15}{4}$

(b)

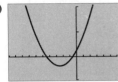

$[-10, 6] \times [-10, 20]_{10}$

(c) Vertex: $\left(-\frac{5}{2}, -\frac{15}{4}\right)$; x-intercepts: $\left(\dfrac{-5 \pm \sqrt{15}}{2}, 0\right)$; y-intercept: $\left(0, \frac{5}{2}\right)$

17. (a) $y = \frac{1}{4}(x + 2)^2 - 3$

(b)

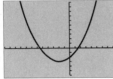

$[-12, 8] \times [-6, 10]$

(c) Vertex: $(-2, -3)$; x-intercepts: $(-2 \pm 2\sqrt{3}, 0)$; y-intercept: $(0, -2)$

19. $-2 - \sqrt{13}, -2 + \sqrt{13}$ **21.** $2 - 2\sqrt{3}, \frac{5}{2}, 2 + 2\sqrt{3}$ **23.** $8 - 2\sqrt[3]{2}$ **25.** 40   **27.** $-6, 12$ **29.** 4   **31.** 1

33. $\dfrac{11 - \sqrt{17}}{2}, \dfrac{11 + \sqrt{17}}{2}$ **35.** $[-3, 1]$ **37.** $[-3, 2]$ **39.** $(-\infty, -2]$ **41.** $(-4, 3)$

43. $f(x) = 0.03x + 2800$; 40,000 ft (7.6 mi)   **45.** 130   **47. (a)** 3 sec   **(b)** 0.75–2.25 sec   **(c)** 1.5 sec; 36 ft

49. 41, 43, 45   **51.** 9.6 oz   **53.** \$14,400 at 4%; \$9600 at 6%   **55.** 30 and 40 mph

CHAPTER TEST

1. Since $f(3) = -2$ and $f(-6) = 10$, the graph of f passes through $(3, -2)$ and $(-6, 10)$:

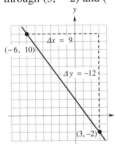

The figure shows that $\Delta x = 9$ and $\Delta y = -12$, so the slope is

$$m = \frac{\Delta y}{\Delta x} = \frac{-12}{9} = -\frac{4}{3}$$

The point–slope form for f is $f(x) = m(x - h) + k$; using $(h, k) = (3, -2)$ and $m = -\frac{4}{3}$, we have

$$f(x) = -\frac{4}{3}(x - 3) - 2 = -\frac{4}{3}x + 2$$

So the slope–intercept form of f is $f(x) = -\frac{4}{3}x + 2$.
(see Section 2.1, Example 6)

2. We solve the equation for y:

$$2x - 3y = 6$$
$$-3y = -2x + 6$$
$$y = \frac{2}{3}x - 2$$

So the slope is $\frac{2}{3}$ and the y-intercept is -2. To graph the line, we start at the y-intercept, $(0, -2)$, and locate another point on the line, $(3, 0)$, by moving 3 units horizontally and 2 units vertically from $(0, -2)$:

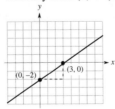

(see Section 2.1, Example 8)

3. We find the x-intercept by letting $y = 0$:

$$3x + 22(0) = 33$$
$$3x = 33$$
$$x = 11$$

The x-intercept is $(11, 0)$.
To find the y-intercept, we let $x = 0$:

$$3(0) + 22y = 33$$
$$22y = 33$$
$$y = \tfrac{3}{2}$$

The y-intercept is $\left(0, \tfrac{3}{2}\right)$.

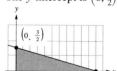

The triangle described is a right triangle with legs of lengths $\tfrac{3}{2}$ and 11. The area A is $A = \tfrac{1}{2}\left(\tfrac{3}{2}\right)(11) = \tfrac{33}{4}$.
(see Section 2.1, Example 8)

4. By completing the square, we find that

$$f(x) = x^2 - 2x - 1$$
$$= (x^2 - 2x + \mathbf{1}) - 1 - \mathbf{1}$$
$$= (x - 1)^2 - 2$$

The vertex is $(1, -2)$. Since $f(0) = 0^2 - 2(0) - 1 = -1$, the y-intercept is $(0, -1)$. Similarly, solving $f(x) = 0$ gives us the x-intercepts:

$$f(x) = 0$$
$$(x - 1)^2 - 2 = 0$$
$$(x - 1)^2 = 2$$
$$x - 1 = \pm\sqrt{2}$$
$$x = 1 \pm \sqrt{2}$$

The x-intercepts are $(1 - \sqrt{2}, 0)$ and $(1 + \sqrt{2}, 0)$.

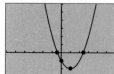

$[-6, 6] \times [-3, 6]$
(see Section 2.2, Example 6)

5. Because the vertex is $(h, k) = (-7, 12)$, we get $f(x) = a(x + 7)^2 + 12$. Because the graph passes through $(-1, -6)$, it follows that

$$f(-1) = 6$$
$$a[(-1) + 7]^2 + 12 = -6$$
$$36a + 12 = -6$$
$$36a = -18$$
$$a = -\tfrac{1}{2}$$

The function is $f(x) = -\tfrac{1}{2}(x + 7)^2 + 12$.
(see Section 2.2, Example 3)

6. The graph of C is shown in the figure:

$[0, 800]_{100} \times [0, 16{,}000]_{1000}$
The minimum value of the function corresponds to the vertex (h, k) of the graph. It follows that

$$h = -\frac{b}{2a} = -\frac{-24}{2(0.03)} = 400$$

Thus, the minimum cost occurs when 400 units are manufactured.
(see Section 2.2, Example 7)

7. The graph of $y = |2x^3 - 1| - 15$ suggests that the equation has two solutions:

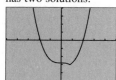

$[-5, 5] \times [-25, 20]_5$
A solution to the equation is a solution to either $2x^3 - 1 = \pm 15$.

$2x^3 - 1 = 15$	$2x^3 - 1 = -15$
$2x^3 = 16$	$2x^3 = -14$
$x^3 = 8$	$x^3 = -7$
$x = 2$	$x = -\sqrt[3]{7}$

The solutions are 2 and $-\sqrt[3]{7}$; these values agree with the figure.
(see Section 2.3, Example 6)

8. We are given that $w = 3438$ and $s = 91$. The equation becomes

$$91 = 234 \sqrt[3]{\frac{h}{3438}}$$

$$\frac{91}{234} = \sqrt[3]{\frac{h}{3438}}$$

$$\left(\frac{91}{234}\right)^3 = \frac{h}{3438}$$

$$3438\left(\frac{91}{234}\right)^3 = h$$

or $h = 202$ horsepower (to the nearest horsepower).
(see Section 2.3, Example 4)

9. The graph of $y = x + 2\sqrt{x - 1} - 9$ appears to have an x-intercept at $x = 5$:

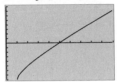

$[0, 10] \times [-8, 8]$
We check:

$$(5) + 2\sqrt{(5) - 1} = 9$$

$$5 + 2\sqrt{4} = 9$$

$$5 + 4 = 9$$

$$9 = 9$$

The solution to the equation is 5.
(see Section 2.3, Example 4)

10. The inequality is equivalent to $|2x - 6| - |x| \leq 0$. The graph of $f(x) = |2x - 6| - |x|$ appears to have x-intercepts at $x = 2$ and $x = 6$:

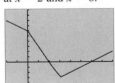

$[-2, 8] \times [-4, 10]$
This conjecture is verified by noting that

$$f(2) = |2(2) - 6| - |2| = |-2| - |2| = 0$$

$$f(2) = |2(6) - 6| - |6| = |6| - |6| = 0$$

The graph is on or below the x-axis over the interval $2 \leq x \leq 6$. The solution to the inequality is $2 \leq x \leq 6$.
(see Section 2.4, Example 7)

11. The concentration of fluoride solution is the ratio of the amount of fluoride to the total amount of mixture:

$$\frac{\left(\begin{array}{c}\text{Amount of} \\ \text{fluoride in mixture}\end{array}\right)}{\left(\begin{array}{c}\text{Total amount} \\ \text{of mixure}\end{array}\right)} = \left(\begin{array}{c}\text{Concentration of} \\ \text{fluoride in mixture}\end{array}\right)$$

This verbal equation is equivalent to

$$\frac{\left(\begin{array}{c}\text{Amount of} \\ \text{fluoride in} \\ \text{6\% solution}\end{array}\right) + \left(\begin{array}{c}\text{Amount of} \\ \text{fluoride in} \\ \text{11\% solution}\end{array}\right)}{\left(\begin{array}{c}\text{Total amount} \\ \text{of 6\%} \\ \text{solution}\end{array}\right) + \left(\begin{array}{c}\text{Total amount} \\ \text{of 11\%} \\ \text{solution}\end{array}\right)} = \left(\begin{array}{c}\text{Concentration} \\ \text{of fluoride} \\ \text{in mixture}\end{array}\right)$$

We let $x =$ Amount of 6% solution and get

$$\frac{0.06x + 0.11(12)}{x + 12} = 0.08$$

$$\frac{0.06x + 1.32}{x + 12} = 0.08$$

$$0.06x + 1.32 = 0.08(x + 12)$$

$$0.06x + 1.32 = 0.08x + 0.96$$

$$-0.02x = -0.36$$

$$x = 18$$

Thus, 18 oz of 6% solution of fluoride must be added. (see Section 2.5, Example 5)

12. The boat travels the same distance upstream as it does downstream. So,

$$\left(\begin{array}{c}\text{Distance} \\ \text{traveled} \\ \text{upstream}\end{array}\right) = \left(\begin{array}{c}\text{Distance} \\ \text{traveled} \\ \text{downstream}\end{array}\right)$$

or

$$\left(\begin{array}{c}\text{Time} \\ \text{upstream}\end{array}\right)\left(\begin{array}{c}\text{Rate} \\ \text{upstream}\end{array}\right) = \left(\begin{array}{c}\text{Time} \\ \text{downstream}\end{array}\right)\left(\begin{array}{c}\text{Rate} \\ \text{downstream}\end{array}\right)$$

We let $r =$ Rate of the stream, and get

$$2(4 - r) = 1(4 + r)$$

$$8 - 2r = 4 + r$$

$$-3r = -4$$

$$r = \frac{4}{3}$$

The rate of the stream is $\frac{4}{3}$ mph.
(see Section 2.1, Example 2)

CHAPTER 3: SECTION 3.1

1.
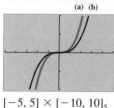
$[-5, 5] \times [-10, 10]_5$

3.
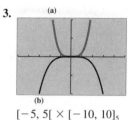
$[-5, 5[\times [-10, 10]_5$

5.
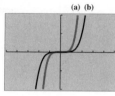
$[-5, 5] \times [-10, 10]_5$

7. (a)
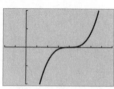
$[-2, 8] \times [-10, 10]_5$

(b) $y \to -\infty$ as $x \to -\infty$; $y \to +\infty$ as $x \to +\infty$ **(c)** $(4, 0)$

9. (a)
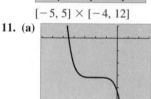
$[-5, 5] \times [-4, 12]$

(b) $y \to -\infty$ as $x \to -\infty$; $y \to -\infty$ as $x \to +\infty$ **(c)** $(-2\sqrt[4]{3}, 0), (2\sqrt[4]{3}, 0)$

11. (a)
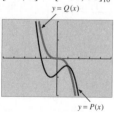
$[-7, 3] \times [-25, 5]_5$

(b) $y \to +\infty$ as $x \to -\infty$; $y \to -\infty$ as $x \to +\infty$ **(c)** $(-2 - 2\sqrt[5]{3}, 0)$

13.
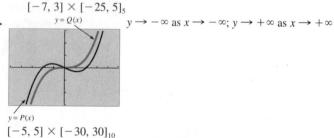
$[-5, 5] \times [-30, 30]_{10}$

$y \to -\infty$ as $x \to -\infty$; $y \to +\infty$ as $x \to +\infty$

15.
$[-5, 5] \times [-10, 10]_5$

$y \to +\infty$ as $x \to -\infty$; $y \to -\infty$ as $x \to +\infty$

17. 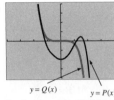 $y \to +\infty$ as $x \to -\infty$; $y \to -\infty$ as $x \to +\infty$

$y = Q(x)$ $y = P(x)$

$[-4, 4] \times [-40, 40]_{10}$

19. No, y does not grow without bound as $x \to -\infty$ or as $x \to +\infty$ **21.** No, not continuous

23. Yes, leading coefficient is positive; degree is even and at least 4.   **25.**

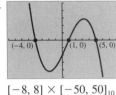

$(-4, 0)$ $(1, 0)$ $(5, 0)$

$[-8, 8] \times [-50, 50]_{10}$

27. 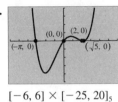   **29.**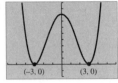

$(0, 0)$ $(2, 0)$

$(-\pi, 0)$ $(\sqrt{5}, 0)$

$(-3, 0)$ $(3, 0)$

$[-6, 6] \times [-25, 20]_5$ $[-6, 6] \times [-20, 100]_{10}$

31. **(a)** $y \to -\infty$ as $x \to -\infty$; $y \to +\infty$ as $x \to +\infty$ **(b)** $(-5, 0), (0, 0), (5, 0)$ **(c)**

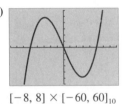

$[-8, 8] \times [-60, 60]_{10}$

33. **(a)** $y \to +\infty$ as $x \to -\infty$; $y \to -\infty$ as $x \to +\infty$ **(b)** $(0, 0), (7, 0)$ **(c)**

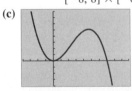

$[-4, 10] \times [-40, 80]_{10}$

35. **(a)** $y \to -\infty$ as $x \to -\infty$; $y \to +\infty$ as $x \to +\infty$ **(b)** $(-2, 0), (0, 0), (4, 0)$ **(c)**

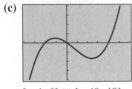

$[-4, 6] \times [-40, 40]_{10}$

37. (a) $y \to -\infty$ as $x \to -\infty$; $y \to -\infty$ as $x \to +\infty$ **(b)** $(-3, 0), (0, 0), (7, 0)$ **(c)**

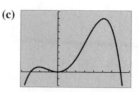

$[-4, 2] \times [-100, 450]_{50}$

39. (a) $y \to -\infty$ as $x \to -\infty$; $y \to +\infty$ as $x \to +\infty$ **(b)** $(-1, 0), (1, 0), (4, 0)$ **(c)**

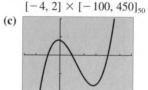

$[-3, 6] \times [-10, 10]_5$

41. (a) $y \to +\infty$ as $x \to -\infty$; $y \to +\infty$ as $x \to +\infty$ **(b)** $(-2\sqrt{2}, 0), (-1, 0), (1, 0), (2\sqrt{2}, 0)$ **(c)**

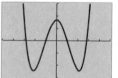

$[-5, 5] \times [-15, 15]_5$

43. (a) $y \to +\infty$ as $x \to -\infty$; $y \to +\infty$ as $x \to +\infty$ **(b)** $(-7, 0), (-2\sqrt{3}, 0), (0, 0), (2\sqrt{3}, 0)$ **(c)**

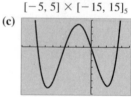

$[-9, 5] \times [-175, 100]_{25}$

45. (a) $y \to -\infty$ as $x \to -\infty$; $y \to +\infty$ as $x \to +\infty$ **(b)** $(0, 0), (2, 0), (4, 0)$ **(c)**

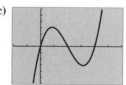

$[-2, 6] \times [-6, 6]$

47. (a) $V(h) = 3h^3 - 195h^2 + 3000h$; Domain: $0 < h < 25$ **(b)** Approx. 4.62 cm

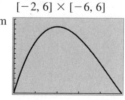

$[0, 25]_5 \times [0, 15{,}000]_{1000}$

SECTION 3.2

1. $(2x + 1)(x^2 - 5) + 7$ **3.** $(x^2 + 2x - 1)(x^2 - 3x - 1) + 5x - 5$ **5.** $(x^2 - 3)(x^2 - x + 2) + 5x + 10$ **7.** $-4, 3, 8$
9. $-2, -1, 3$ **11.** $-7, -2, 0, 2$ **13.** $-1, 1, 2$ **15.** $-3, -2\sqrt{2}, 2\sqrt{2}$ **17.** $-3, 4$
19. $P(x) = x^3 - 6x^2 - 4x + 24$ **21.** $P(x) = 2x^4 + 3x^3 - 32x^2 - 48x$ **23.** $P(x) = x^4 - 7x^2 + 10$
25. $1 - \sqrt{7}, 3, 1 + \sqrt{7}$ **27.** $-\sqrt{5}, -1, \sqrt{5}, 8$ **29.** $-4, 4$ **31.** $2, \sqrt[3]{10}, 4$
33. $-2\sqrt{5}, -4, -1 - \sqrt{3}, -1 + \sqrt{3}, 2\sqrt{5}$ **35.** $-2\sqrt{2}, -2, 2\sqrt{2}, 6$ **37.** $-\frac{5}{2}, \frac{7}{4}, 4$

39. $P(x) = -4x^3 + 10x^2 + 28x - 16$ **41.** $P(x) = 2x^4 + 2x^3 - 10x^2 - 6x + 12$ **43.** $P(x) = -\frac{1}{2}x^3 + \frac{1}{2}x^2 + 8x - 8$
45. $(+\infty, 0]$ or $[2, 3]$ **47.** $[-3, 3]$ **49.** $(2 - \sqrt{6}, 0), (2 + \sqrt{6}, 0)$ **51.** $(x + 5)(x - 3)(x^2 - 8)$ **53. (b)** 5 sec
55. 3 m

SECTION 3.3

1. (a) $3i$ **(b)** $\sqrt{13}\,i$ **3. (a)** $2\sqrt{3}\,i$ **(b)** $-2\sqrt{2}\,i$ **5. (a)** $7 - \sqrt{3}\,i$ **(b)** $-2 + 2i$
7. (a) $2 + i$ **(b)** $7 - \sqrt{3}\,i$ **9. (a)** $6 - 6i$ **(b)** $-8 + 12i$ **11. (a)** $-4 + 8i$ **(b)** $17 - 11i$
13. (a) $-3 + \frac{37}{4}i$ **(b)** $-5 + 12i$ **15. (a)** $\frac{1}{2} - \frac{3}{2}i$ **(b)** $\frac{17}{25} + \frac{19}{25}i$ **17.** $-\sqrt{3}\,i, \sqrt{3}\,i$ **19.** $4 + 4i, 4 - 4i$
21. $5 - \frac{4}{3}i, 5 + \frac{4}{3}i$ **23.** $1 - \sqrt{5}\,i, 1 + \sqrt{5}\,i$ **25.** $-6, 1$ **27.** $\dfrac{1 - \sqrt{11}\,i}{6}, \dfrac{1 + \sqrt{11}\,i}{6}$ **29.** $38 + 8i$ **31.** $2 - 11i$
33. 5   **35.** $5 - 31i$ **37.** $\frac{11}{58} - \frac{13}{58}i$ **39.** $\frac{3}{25} + \frac{4}{25}i$ **41.** $2, 4, 2\sqrt{3}\,i, -2\sqrt{3}\,i$ **43.** $4, -2 - 2\sqrt{3}\,i, -2 + 2\sqrt{3}\,i$
45. $-3i, -2i, 2i, 3i$ **47.** $-3, -2\sqrt{2}\,i, 2\sqrt{2}\,i$ **49.** $-5, -3, 2$ **51.** $-5, 7, -\sqrt{3}\,i, \sqrt{3}\,i$ **53.** $f(x) = 2(x + 5)(x - 3)$
55. $f(x) = \frac{1}{2}(x - 4i)(x + 4i)$ **57.** $f(x) = (x - 2)(x - 5i)(x + 5i)$ **59.** $f(x) = 2(x - 2)(x + 2)(x + 2 + 2i)(x + 2 - 2i)$
61. $P(x) = -2x^3 - 2x^2 + 10x - 6$ **63.** $P(x) = x^4 + 8x^3 + 19x^2 + 24x + 48$ **65.** $P(x) = \frac{1}{4}x^3 + \frac{3}{4}x^2 - 1$ **67.** $i, -3i$

SECTION 3.4

1. Domain: all real numbers except 2 and -2; zeros: $\frac{5}{2}$ **3.** Domain: all real numbers except $-\frac{1}{2}$; zeros: none
5. Domain: all real numbers except 0; zeros: 2   **7.** $x = -\frac{4}{3}$ **9.** $x = 3, x = -3$ **11.** No vertical asymptotes

13. $f(x) = 3 + \dfrac{-7}{x + 2}$ 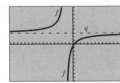   **15.** $f(x) = 0 + \dfrac{5x - 1}{x^2 + 4}$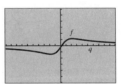

17. $f(x) = 2x - 2 + \dfrac{7}{x + 1}$ 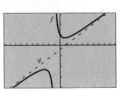   **19.** Intercept: $\left(0, \frac{2}{7}\right)$; asymptotes: $x = -\frac{7}{2}, y = 0$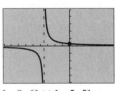

$[-9, 6] \times [-5, 5]$

21. Intercept: $(0, 0)$; asymptotes: $x = 4, y = 2$ **23.** Intercepts: $\left(0, -\frac{5}{2}\right), \left(\frac{5}{2}, 0\right)$; asymptotes: $x = 2, y = -2$

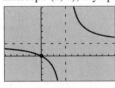   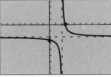

$[-6, 12] \times [-4, 8]$ $[-8, 10] \times [-8, 4]$

25. Intercepts: $\left(0, -\frac{3}{4}\right), (3, 0)$; asymptotes: $x = -2, y = 0$ **27.** Intercepts: $(0, 0)$; asymptotes: $x = -2, x = 2, y = 0$

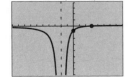

$[-10, 8] \times [-8, 4]$ $[-9, 9] \times [-6, 6]$

29. Intercepts: $(0, -2)$, $(\frac{3}{2}, 0)$; asymptote: $y = 0$

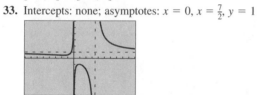

$[-9, 9] \times [-6, 6]$

31. Intercept: $(-1, 0)$; asymptotes: $x = 0$, $y = \frac{1}{2}x + 1$

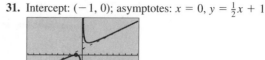

$[-9, 9] \times [-6, 6]$

33. Intercepts: none; asymptotes: $x = 0$, $x = \frac{7}{2}$, $y = 1$

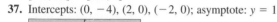

$[-8, 10] \times [-6, 6]$

35. Intercepts: $\left(0, \frac{1}{5}\right)$, $\left(-\frac{1}{3}, 0\right)$, $(1, 0)$; asymptotes: $x = \frac{5}{3}$, $y = x + 1$

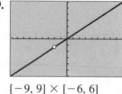

$[-9, 9] \times [-6, 10]$

37. Intercepts: $(0, -4)$, $(2, 0)$, $(-2, 0)$; asymptote: $y = 1$

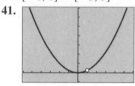

$[-6, 6] \times [-5, 3]$

39.

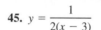

$[-9, 9] \times [-6, 6]$

41.

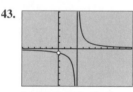

$[-6, 6] \times [-1, 7]$

43.

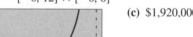

$[-6, 12] \times [-6, 6]$

45. $y = \dfrac{1}{2(x - 3)}$

47. (a) $k = 213{,}333\frac{1}{3}$ **(b)**

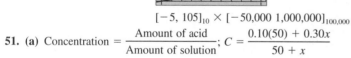

$[-5, 105]_{10} \times [-50{,}000 \ 1{,}000{,}000]_{100{,}000}$

(c) $\$1{,}920{,}000$

49. (a) $y = \dfrac{6t}{t + 1.2}$ **(b)** 6 yr

51. (a) Concentration $= \dfrac{\text{Amount of acid}}{\text{Amount of solution}}$; $C = \dfrac{0.10(50) + 0.30x}{50 + x}$ **(b)**

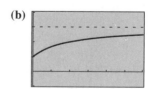

$[-1, 150]_{25} \times [-0.1, 0.4]_{0.1}$

(c) y-intercept $(0, 0.1)$ means that adding no 30% solution results in a 10% solution; horizontal asymptote $y = 0.3$ means that the upper bound of concentration is 30%

53. **(a)** $4.80   **(b)**

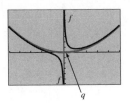

$[-1000, 20{,}000]_{1000} \times [-1, 10]$

Horizontal asymptote indicates that average cost approaches $4.20 as more units are produced.

55. $f(x) = \frac{1}{5}x^2 + \frac{1}{2x}$

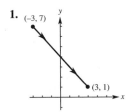

SECTION 3.5

1. $-\frac{3}{10}$ **3.** -5 **5.** $\frac{5}{4}$ **7.** $-\frac{19}{6}$ **9.** $-\frac{1}{3}$ **11.** $\frac{1}{6}, 2$ **13.** 7   **15.** No solution   **17.** $\frac{1}{2}, 1$ **19.** 3   **21.** 75

23. $x > 2$ **25.** $x \le -3$ or $x > 0$ **27.** $x < -4$ or $x \ge \frac{7}{2}$ **29.** $x \le 0$ or $1 < x \le 4$ **31.** $x \le -\frac{3}{2}$ or $0 < x \le 2$

33. $x < -2$ or $0 < x < 2$ or $x > 3$ **35.** $-3, 1$ **37.** $-1, \dfrac{1 + \sqrt{5}}{4}, \dfrac{1 - \sqrt{5}}{4}$ **39.** **(a)** $\frac{9}{2}\,\Omega$ **(b)** $r_1 = \dfrac{Rr_2}{r_2 - R}$

41. **(a)** 2 or 458   **(b)** $2 < x < 458$ **43.** **(a)** $A(l) = 72 + 4l + \dfrac{288}{l}$

(b) 4 in. vertical length by 12 in. wide, or 18 in. vertical length by $2\frac{2}{3}$ in. wide

(c) $4 < l < 18$; one example is 8 in. vertical length by 6 in. wide

45. 10 min, 15 min   **47.** 32 mph   **49.** 6 hr

SECTION 3.6

1.

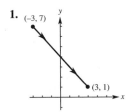

3.

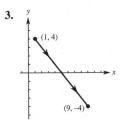

5.

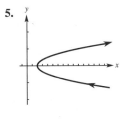

7.

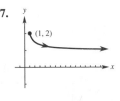

9. $y = (x + 2)^2$

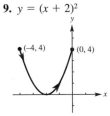

11. $y = 2x + 1$

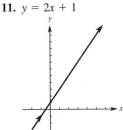

13. $y = \sqrt{x + 2}$

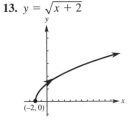

15.

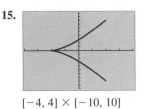

$[-4, 4] \times [-10, 10]$

17.

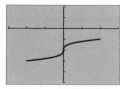

$[-3, 3] \times [-5, 2]$

19.

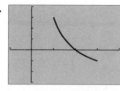

$[-1, 4] \times [-3, 4]$

21.

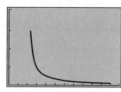

$[0, 12] \times [0, 4]$

23. $x = t, y = t^2 + 2t - 4, -4 \le t \le 3$ **25.** $x = 4 - t^2, y = t, -3 \le t \le 3$ **27.** $x = \frac{1}{2} \cos t, y = \frac{1}{2} \sin t, 0 \le t \le 2\pi$

29. $x = 5 - 8t, y = 5 - 4t, 0 \le t \le 1$ **31.** $x^2 + y^2 = 4$

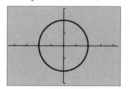

$[-4.5, 4.5] \times [-3, 3]$

33. $x^2 + y^2 = 1$ **35.** $x^2 + y^2 = 25$

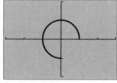

$[-3, 3] \times [-2, 2]$

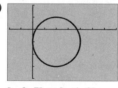

$[-9, 9] \times [-6, 6]$

37. $x = 2 - 5t, y = 4 + 16t, 0 \le t \le 1$ **39.** $x = t, y = t^2 - 4, -2 \le t \le 2$ **41.** $x = 2t, y = 4 - 4t, 0 \le t \le 1$

43. $x = 4 \cos t, y = 4 \sin t, 0 \le t \le \pi$ **45. (a)** **(b)** $(x - 2)^2 + (y + 1)^2 = 4$

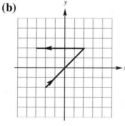

$[-2, 7] \times [-4, 2]$

47. (a)

t	x	y
-4	-2	-2
-2	0	0
0	2	2
2	0	2
4	-2	2

(b)

49. $x = 2t, y = \frac{1}{2}t + 1, -2 \le t \le 2$ **51. (a)** **(b)** 3.75 sec, 390 ft

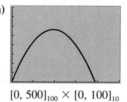

$[0, 500]_{100} \times [0, 100]_{10}$

53. (a) 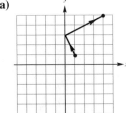 (b) 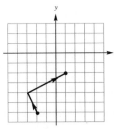 (c)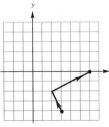

55. (a) 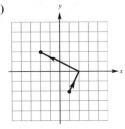 (b) 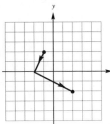 (c)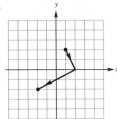

MISCELLANEOUS EXERCISES

1. (a) $y \to -\infty$ as $x \to -\infty$; $y \to +\infty$ as $x \to +\infty$ (b) $(2\sqrt[5]{3} - 4, 0)$ (c)

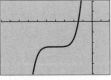
$[-10, 2] \times [-50, 20]_{10}$

3. (a) $y \to +\infty$ as $x \to -\infty$; $y \to +\infty$ as $x \to +\infty$ (b) $(1, 0), (4, 0)$ (c)

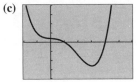
$[-2, 6] \times [-40, 40]_{10}$

5. (a) $y \to +\infty$ as $x \to -\infty$; $y \to +\infty$ as $x \to +\infty$ (b) $(-7, 0), (-2\sqrt{2}, 0),$ $(0, 0), (2\sqrt{2}, 0)$ (c)

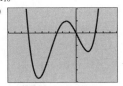
$[-10, 6] \times [-200, 100]_{50}$

7. $y \to 0$ as $x \to \pm\infty$; intercepts: none; vertical asymptote: $x = 0$; horizontal asymptote: $y = 0$

$[-6, 6] \times [-2, 2]$

9.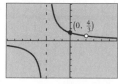
$[-8, 6] \times [-6, 6]$

$y \rightarrow 0$ as $x \rightarrow \pm\infty$; vertical asymptote: $x = -3$; horizontal asymptote: $y = 0$

11.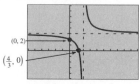
$[-6, 9] \times [-5, 8]$

$y \rightarrow 3$ as $x \rightarrow \pm\infty$; vertical asymptote: $x = 2$; horizontal asymptote: $y = 3$

13.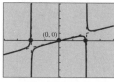
$[-8, 8]_2 \times [-10, 10]_2$

$y \rightarrow \frac{1}{2}x$ as $x \rightarrow \pm\infty$; vertical asymptotes: $x = -4$, $x = 4$; slant asymptote: $y = \frac{1}{2}x$

15. $-5, 1 - \sqrt{13}, 1 + \sqrt{13}$ **17.** $-\frac{7}{5}$ **19.** $-4, -\frac{7}{3}$ **21.** $P(x) = 2x^4 + 3x^3 - 12x^2 - 7x + 6$
23. $(x - 4)(x + 4)(x + 2 + \sqrt{2}\,i)(x + 2 - \sqrt{2}\,i)$ **25. (a)** $1 + 6i$ **(b)** $5 + 14i$ **27. (a)** $\frac{2}{17} + \frac{26}{17}i$ **(b)** $\frac{4}{25} - \frac{3}{25}i$
29. (a) $16 + 18i$ **(b)** $\frac{2}{5} + \frac{1}{5}i$ **31. (a)** $-27 + 53i$ **(b)** $\frac{3}{58} - \frac{7}{58}i$ **33.** $-1 - \sqrt{3}\,i, -1 + \sqrt{3}\,i$
35. $-\sqrt{3}, \sqrt{3}, -\sqrt{3}\,i, \sqrt{3}\,i,$ **37.** $-2\sqrt{2}, -2, 2\sqrt{2}$ **39.** $(-\infty, -\sqrt{5})$ or $(\sqrt{5}, +\infty)$ **41.** $(2, 4]$
43. $[-\frac{3}{2}, 0)$ or $(0, 2]$ **45.** $[-2, 2)$ or $[6, +\infty)$

47.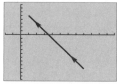
$[-2, 12] \times [-12, 8]$
$y = 7 - 2x$

49.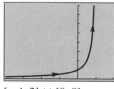
$[-4, 2] \times [0, 8]$
$y = \dfrac{1}{1 - x}$

51.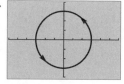
$[-6, 6] \times [-4, 4]$
$x^2 + y^2 = 9$

53. $x = 2 - 2t, y = -5t, 0 \leq t \leq 1$ **55.** $x = t^3 - 4t, y = t, 0 \leq t \leq 4$
57. (a)

t	x	y
-4	-2	-3
-2	0	-1
0	2	1
2	0	3
4	-2	5

(b)
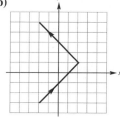

59. 40 mph and 60 mph

CHAPTER TEST

1. (a) The leading term of f is $5x^3$. Therefore, the behavior of the graph of f is the same as that of $y = 5x^3$: $y \to -\infty$ as $x \to -\infty$ and $y \to +\infty$ as $x \to +\infty$

(b) The complete graph of f shows three x-intercepts:

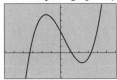

$[-6, 6] \times [-80, 150]_{20}$
To find the x-intercepts, we solve $f(x) = 0$:

$$(5x - 6)(x^2 - 11) = 0$$

$5x - 6 = 0$	$x^2 - 11 = 0$
$5x = 6$	$x^2 = 11$
$x = \frac{6}{5}$	$x = \pm\sqrt{11}$

The x-intercepts are $(-\sqrt{11}, 0)$, $(\frac{6}{5}, 0)$, $(\sqrt{11}, 0)$.
(see Section 3.1, Example 5)

2. (a) To find the x-intercepts, we solve $f(x) = 0$:

$$-x^3 + 6x = 0$$
$$x(-x^2 + 6) = 0$$

$x = 0$	$-x^2 + 6 = 0$
	$x^2 = 6$
	$x = \pm\sqrt{6}$

The x-intercepts are $(-\sqrt{6}, 0)$, $(0, 0)$, $(\sqrt{6}, 0)$.

(b)

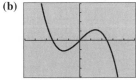

$[-5, 5] \times [-20, 20]_5$
(see Section 3.1, Example 6)

3.
$$
\require{enclose}
\begin{array}{r}
3x - 7 \\
x^2 - x + 3 \enclose{longdiv}{3x^3 - 10x^2 + 17x - 19} \\
\underline{3x^3 - 3x^2 + 9x} \\
-7x^2 + 8x - 19 \\
\underline{-7x^2 + 7x - 21} \\
x + 2
\end{array}
$$

$P(x) = (x^2 - x + 3)(3x - 7) + (x + 2)$
(see Section 3.2, Example 1)

4. The graph of the corresponding function, $f(x) = 7x^3 + 3x^2 - 98x - 42$, has three x-intercepts.

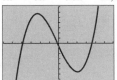

$[-6, 6] \times [-180, 180]_{25}$
The zeros of f are found by solving $f(x) = 0$:

$$7x^3 + 3x^2 - 98x - 42 = 0$$
$$x^2(7x + 3) - 14(7x + 3) = 0$$
$$(x^2 - 14)(7x + 3) = 0$$

$x^2 - 14 = 0$	$7x + 3 = 0$
$x^2 = 14$	$7x = -3$
$x = \pm\sqrt{14}$	$x = -\frac{3}{7}$

The solutions are $-\sqrt{14}$, $-\frac{3}{7}$, and $\sqrt{14}$.
(see Section 3.1, Example 6)

5.
$$2x^3 + 3x^2 \le 23x + 12$$
$$2x^3 + 3x^2 - 23x - 12 \le 0$$

The graph of $y = 2x^3 + 3x^2 - 23x - 12$ is shown.

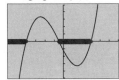

$[-6, 6] \times [-50, 50]_{10}$
The solution to the inequality corresponds to all values of x on the graph that lie on or below the x-axis: $x \le -4$ or $-\frac{1}{2} \le x \le 3$.
(see Section 3.5, Example 2, and Section 3.2, Exercise 45)

6. We start with $P(x) = a(x + 1)(x - \frac{1}{2})(x - 3)$, where a is a constant. In expanded form, $a(x + 1)(x - \frac{1}{2})(x - 3)$ has a leading coefficient of ax^3, and since the leading term must be $4x^3$, $P(x) = 4(x + 1)(x - \frac{1}{2})(x - 3) = 4x^3 - 10x^2 - 8x + 6$.
(see Section 3.1, Example 5, and Section 3.2, Example 3)

7. (a) $\overline{w}z = \overline{(4 - 3i)}(2 + i) = (4 + 3i)(2 + i)$
$= 8 + 10i + 3i^2 = 5 + 10i$

(b) $\dfrac{10}{z} = \dfrac{10}{2 + i} = \dfrac{10}{2 + i} \cdot \dfrac{2 - i}{2 - i} = \dfrac{20 - 10i}{5} = 4 - 2i$
(see Section 3.3, Example 3)

8.
$$8x^3 - 3x^2 + 40x - 15 = 0$$
$$x^2(8x - 3) + 5(8x - 3) = 0$$
$$(x^2 + 5)(8x - 3) = 0$$

$x^2 + 5 = 0$ | $8x - 3 = 0$
$x = \pm\sqrt{5}\,i$ | $x = \frac{3}{8}$

The solutions are $\frac{3}{8}, \sqrt{5}\,i, -\sqrt{5}\,i$.
(see Section 3.3, Example 5)

9. Because

$$f(0) = \frac{-3(0) + 12}{2(0) - 4} = -3$$

the y-intercept is $(0, -3)$. The x-intercepts correspond to the zeros of the numerator:

$$-3x + 12 = 0$$
$$x = 4$$

The x-intercept is $(4, 0)$. The vertical asymptotes correspond to the zeros of the denominator:

$$2x - 4 = 0$$
$$x = 2$$

The vertical asymptote is $x = 2$. To express $f(x)$ in the form $f(x) = q(x) + \dfrac{r(x)}{d(x)}$, we divide $2x - 4$ into $-3x + 12$:

$$\frac{-3x + 12}{2x - 4} = -\frac{3}{2} + \frac{6}{2x - 4} = -\frac{3}{2} + \frac{3}{x - 2}$$

Thus, $y = -\frac{3}{2}$ is a horizontal asymptote.
(see Section 3.4, Example 4)

10. The y-intercept is $\left(0, \frac{1}{9}\right)$ since

$$f(0) = \frac{0 + 1}{0^2 - 9} = -\frac{1}{9}$$

The x-intercepts correspond to the zeros of the numerator:

$$x + 1 = 0$$
$$x = -1$$

The x-intercept is $(-1, 0)$. The vertical asymptotes correspond to the zeros of the denominator:

$$x^2 - 9 = 0$$
$$x = \pm 3$$

The vertical asymptotes are $x = 3$ and $x = -3$. Expressing $f(x)$ in the form $f(x) = q(x) + \dfrac{r(x)}{d(x)}$, we have $\dfrac{x + 1}{x^2 - 9} = 0 + \dfrac{x + 1}{x^2 - 9}$. The horizontal asymptote is $y = 0$. The graph is shown below:

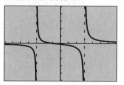

$[-7, 7] \times [-5, 5]$
(see Section 3.4, Example 5)

11. Note that the values 0 and -2 are not possible solutions since they cause division by 0.

$$\frac{x + 1}{x + 2} = \frac{2}{x^2 + 2x}$$
$$\frac{x + 1}{x + 2} = \frac{2}{x(x + 2)}$$
$$\frac{x + 1}{x + 2} \cdot x(x + 2) = \frac{2}{x(x + 2)} \cdot x(x + 2)$$
$$x(x + 1) = 2$$
$$x^2 + x - 2 = 0$$
$$(x + 2)(x - 1) = 0$$

$x + 2 = 0$ | $x - 1 = 0$
$x = -2$ | $x = 1$

Since -2 is not a possible solution, $x = 1$.
(see Section 3.5, Example 3)

12.
$$\frac{x}{2} - 1 + \frac{72}{x} < 19$$
$$\frac{x}{2} - 20 + \frac{72}{x} < 0$$
$$\frac{x^2}{2x} - \frac{40x}{2x} + \frac{144}{2x} < 0$$
$$\frac{(x - 4)(x - 36)}{2x} < 0$$

The graph of $f(x) = \dfrac{(x-4)(x-36)}{2x} < 0$ is shown below:

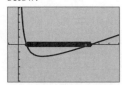

$[-5, 50]_5 \times [-25, 25]_5$

The solutions to the inequality correspond to the values of x for which the graph is below the x-axis. This occurs for $4 < x < 36$.
(see Section 3.5, Example 5)

13. Set the tMin $= -1$ and tMax $= 2$. The graph is shown below:

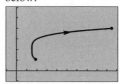

$[-1, 10] \times [-1, 6]$
(see Section 3.6, Example 3)

14. From Table 3.2, the line segment from $P(a, b)$ to $Q(c, d)$ is given parametrically as $x = a(c - a)t$, $y = b + (d - b)t$, with $0 \le t \le 1$: $x = 6 + (-3 - 6)t$, $y = 5 + (-1 - 5)t$. Simplifying, we get $x = 6 - 9t$, $y = 5 - 6t$, with $0 \le t \le 1$.
(see Section 3.6, Example 6)

CHAPTER 4: SECTION 4.1

1. (a) 2.3784   **(b)** 4.7111   **(c)** 1.4646   **(d)** 1.2469
3. (a) 1003.7291   **(b)** 1061.3636   **(c)** 1030.2250   **(d)** 1563.0802
5. (a) 7.0000   **(b)** 5.3050   **(c)** 13.9226   **(d)** 0.4375   **7.** Vertical translation 5 units down

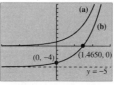

$[-3, 3] \times [-8, 10]$

9. Vertical translation 3 units up   **11.** Horizontal translation 1 unit left; no x-intercept

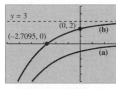

$[-6, 3] \times [-5, 5]$

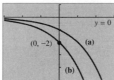

$[-3, 3] \times [-5, 1]$

13. Vertical expansion by a factor of 2; no x-intercept   **15.** Horizontal compression by a factor of 6; no x-intercept

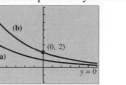

$[-6, 6] \times [-2, 8]$

$[-3, 3] \times [-1, 4]$

17. Vertical and horizontal reflection; no x-intercept   **19.** \$504.99   **21.** \$1915.75   **23.** \$2198.46   **25.** \$6719.84

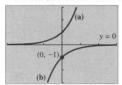

$[-3, 3] \times [-3, 3]$

27. \$3892.93   **29.** 9.54%   **31.** $y = \left(\frac{1}{3}\right)^x \left(k = 1, b = \frac{1}{3}\right)$ **33.** $y = 4(3)^x (k = 4, b = 3)$

35. $y = -2(2)^x (k = -2, b = 2)$ **37. (a)** 11.6 billion, 23.2 billion   **(b)** $A(t) = 5.8(2)^{t/38}$ **(c)** 6.4 billion

39. (a) $A(t) = 25\left(\frac{1}{2}\right)^{t/20}$ **(b)** 21.0 mg, 3.1 mg   **41. (a)** $A(t) = 18(2)^{t/8.4}$ **(b)** 62 million

43. $n = 1$: \$12,624.77; $n = 4$: \$12,689.86; $n = 12$: \$12,704.89; $n = 500$: \$12,712.31; $n = 10{,}000$: \$12,712.48

45. (a) $W = \dfrac{Pr}{1 - (1 + r)^{-n}}$ **(b)** \$30,888.83

47. (a) Monthly payment is \$955.65; total payment is \$172,017   **(b)** \$104,640.59

49. (a) $I(x) = I_0\left(\frac{4}{5}\right)^x$ **(b)** 211.1 W/m²

SECTION 4.2

1. (a) $5e^{-1}$; 1.8394   **(b)** $e^{1/3}$; 1.3956   **3. (a)** $3e^6$; 1210.2864   **(b)** $2e^3$; 40.1711

5. Vertical translation 2 units up; no x-intercept   **7.** Vertical translation 6 units down

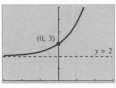

$[-3, 3] \times [0, 6]$

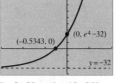

$[-3, 3] \times [-8, 2]$

9. Horizontal translation 3 units left; no x-intercept   **11.** Vertical translation 32 units down and horizontal translation 4 units left

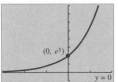

$[-3, 2] \times [-40, 80]_{10}$

$[-3, 2] \times [-40, 80]_{10}$

13. Vertical compression by a factor of $\frac{1}{2}$; no x-intercept

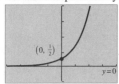

$[-4, 4] \times [-1, 4]$

15. Horizontal expansion by a factor of 3 and reflection about the *y*-axis; no *x*-intercept

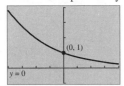

$[-4, 4] \times [-1, 4]$

17. Vertical expansion by a factor of 200, horizontal expansion by a factor of 2, and reflection about the *y*-axis; no *x*-intercept

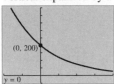

$[-2, 4] \times [-50, 450]_{50}$

19. \$508.50   **21.** \$1916.43   **23.** \$2198.52   **25.**

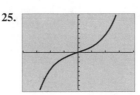

$[-4, 4] \times [-8, 8]$

$y \approx \frac{1}{2} e^x$ as $x \to +\infty$, $y \approx -\frac{1}{2} e^x$ as $x \to -\infty$

27.

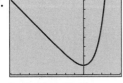

$[-8, 4] \times [0, 8]$

$y \approx e^x$ as $x \to +\infty$, $y \approx -x$ as $x \to -\infty$

29. $g(x) = 4x - 1$, $f(x) = \frac{1}{3} e^x$ **31.** $g(x) = e^x$, $f(x) = x^2 - 4x + 7$

33. (a) 4.35 psi   **(b)** 14.7 psi   **(c)** 8.6 psi   **35. (a)** 0.5522   **(b)** 0.7768   **(c)** 1.000

37. (a) $A(t) = 41.67e^{0.0139t}$ **(b)** 55.02 million, 72.66 million

39. (a) $A(t) = 22,000e^{0.015t}$ **(b)** 24,072   **(c)** 2015   **41. (a)** $A(t) = 54,000e^{-0.2t}$ **(b)** \$29,636   **(c)** \$10,902

43. (a) $T(t) = 85e^{-0.045t} + 70$ **(b)** 113°F   **(c)** 11.8 min   **45. (a)** $T(t) = 70e^{-0.18t} + 20$ **(b)** 5.9 min

47. (a) 778   **(b)** 22 wk; the number of words learned approaches 3000   **49. (a)** 23 persons   **(b)** 44 days   **53.** 108°F

SECTION 4.3

1.

$[-6, 6] \times [-4, 4]$

3.

$[-6, 6] \times [-4, 4]$

5.

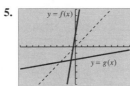

$[-9, 9] \times [-6, 6]$

7.

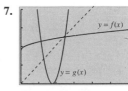

$[0, 6] \times [0, 4]$

9. One-to-one   **11.** One-to-one   **13.** One-to-one   **15.** One-to-one

17. $f^{-1}(x) = \frac{1}{3}x + \frac{4}{3}$ **19.** $R^{-1}(x) = \sqrt[3]{x - 1}$ **21.** $f^{-1}(x) = \dfrac{1}{x} + 5$ **23.** $q^{-1}(x) = x^3 + 5$

25. No   **27.** Yes   **29.** Yes   **31.** $f^{-1}(x) = \dfrac{4x + 1}{2 - x}$ **33.** $f^{-1}(x) = \dfrac{16}{x^2},\ x \geq 0$ **35.** $f^{-1}(x) = (x - 3)^2 + 4,\ x \geq 3$

37. 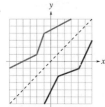   **39.** 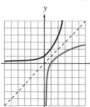   **41.** $f^{-1}(x) = -\sqrt{x + 4}$ **43.** $f^{-1}(x) = -\sqrt{36 - x^2},\ x \geq 0$

45. $f(x) = (x - 2)^2 + 1,\ x \geq 2;\ g(x) = \sqrt{x - 1} + 2$ **47.** $f(x) = -\frac{1}{2}(x + 3)^2 + \frac{17}{2},\ x \geq -3;\ g(x) = -3 + \sqrt{17 - 2x}$

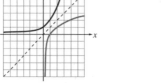

$[0, 6] \times [0, 6]$

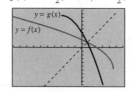

$[-10, 6] \times [-10, 10]$

49. $I^{-1}(x) = \dfrac{13x}{100}$ **51.** $F^{-1}(c) = \frac{5}{9}(c - 32)$ **53.** (a)

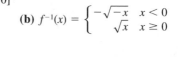

(b) $f^{-1}(x) = \begin{cases} -\sqrt{-x} & x < 0 \\ \sqrt{x} & x \geq 0 \end{cases}$

55. 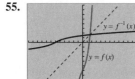   **57.** 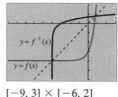   **59.**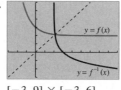

$[-12, 12] \times [-8, 8]$ $[-9, 3] \times [-6, 2]$ $[-3, 9] \times [-3, 6]$

65. $f^{-1}(x) = \dfrac{3x}{x - 2};\ f(x^{-1}) = \dfrac{2}{1 - 3x};\ [f(x)]^{-1} = \dfrac{x - 3}{2x}$

SECTION 4.4

1. **(a)** 3   **(b)** -2 **(c)** 0   **(d)** 1   **3.** **(a)** 1.5   **(b)** -2 **(c)** 1   **(d)** Undefined

5. **(a)** $\log_4(2P) = t$ **(b)** $\log_{xy} 8 = t$ **7.** **(a)** $5^M = 2t$ **(b)** $(x/2)^5 = 4t$

9. **(a)** 1.1271   **(b)** -1.8697 **(c)** 2.1282   **(d)** -2.5903 **11.** **(a)** $\log_3 11 + \log_3 x$ **(b)** $\log_9 13 + \log_9 s - \log_9 t$

13. **(a)** $\log x + \log y + \log z$ **(b)** $\log_7 3 + 2 \log_7 P - \log_7 Q$ **15.** **(a)** $\frac{1}{2} \ln(1 - x)$ **(b)** $-\frac{1}{2} \log_2(x - 3) - \frac{1}{2} \log_2(x + 3)$

17. **(a)** $\log(3y)$ **(b)** $\ln\left(\dfrac{5a}{2b}\right)$ **19.** **(a)** $\log\left[\dfrac{(a - 3)^2}{a + 4}\right]$ **(b)** $\log_2\left[\dfrac{x^2 - 3}{yz}\right]$ **21.** **(a)** $\log(x\sqrt{x^2 - 6})$ **(b)** $\ln(3\sqrt{x})$

23. **(a)** $5y$ **(b)** $10x$ **(c)** $\dfrac{1}{4x}$ **25.** **(a)** $2.54w$ **(b)** $-4x$ **(c)** $-x$ **27.** $(-\infty, -4)$ or $(4, +\infty)$

29. $(-\infty, 3)$ or $(3, +\infty)$

31. $(0, +\infty)$ **33.**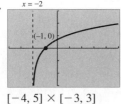

$[-4, 5] \times [-3, 3]$

35.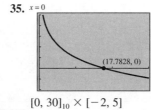

$[0, 30]_{10} \times [-2, 5]$

37.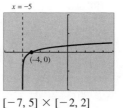

$[-7, 5] \times [-2, 2]$

39. (a) False   **(b)** False   **(c)** True   **(d)** False

41. (a) False   **(b)** False   **(c)** False   **(d)** False   **43.** $f(x) = \ln x$, $g(x) = 2x^3 - 4$

45. 3.5850   **47.** -2.5850 **49. (a)** About 5700 yr ago   **(b)** About 77%

51. 7.4 and 6.7   **53. (a)** 20 dB   **(b)** 150 dB   **55.** 0   **57. (a)** 0.11   **(b)** 11.1

61.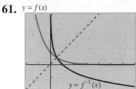

$[-2, 7] \times [-2, 4]$

SECTION 4.5

1. 3.5850   **3.** -2.5850 **5.** -2.8074 **7.**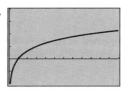

$[0, 12] \times [-2, 4]$

9.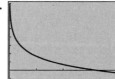

$[0, 40]_{10} \times [-1, 8]$

11.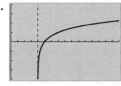

$[0, 16] \times [-4, 4]$

13. 3   **15.** -2 **17.** $-\frac{5}{2}$ **19.** 81   **21.** $\frac{1}{36}$ **23.** $\frac{1}{16}$ **25.** $\frac{1}{2} \log_5 74$; 1.3371

27. $\frac{1}{2} \ln 123 + \frac{1}{2}$; 2.9061   **29.** 24   **31.** $\frac{50}{3} \ln \frac{100}{27}$; 21.8222   **33.** 4   **35.** $\ln 4$; 1.3863   **37.** $\frac{32}{3}$; 10.6667

39. $\frac{1}{4} e^{8/3}$; 3.5980   **41.** 20   **43.** $\frac{3}{2}$ **45.** $-2, 3$ **47.** No solution   **49. (c)** 11.7699   **51.** $\dfrac{\ln 4}{\ln 5 - \ln 2}$; 1.5129

53. $f^{-1}(x) = \log_2(x + 5)$ **55.** $f^{-1}(x) = \ln(\frac{1}{4} x + \frac{11}{4}) + 2$ **57.** 1966   **59.** 56 quarters

61. 4.00×10^{21} ergs   **63.** $t = -\dfrac{L}{R} \ln\left(1 - \dfrac{RI}{E}\right)$ **65.** $\ln(3 + \sqrt{10})$

MISCELLANEOUS EXERCISES

1.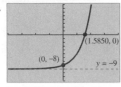

$[-4, 4] \times [-12, 8]$

3.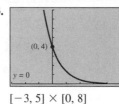

$[-3, 5] \times [0, 8]$

5.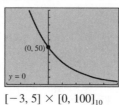

$[-3, 5] \times [0, 100]_{10}$

7.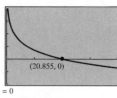

$[0, 40]_{10} \times [-2, 4]$

9.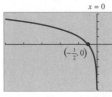

$[-5, 1] \times [-6, 4]$

11. (a) 4   **(b)** -4 **(c)** $\frac{1}{2}$ **13.** $(-2, 2)$ **15.** \$696.78   **17.** \$21,311.31

19. $f^{-1}(x) = \frac{1}{2}x + 2$ **21.** $f^{-1}(x) = 8 - x^3$ **23.** $f^{-1}(x) = \dfrac{x}{x-2}$ **25.** $f^{-1}(x) = \frac{1}{2}\ln(2x) + 4$

27. (a) $2t$ **(b)** $w/10$ **(c)** t **29. (a)** $\log 7 + \log x - 1$ **(b)** $\ln 2 + \ln x + \ln y$ **(c)** $\frac{1}{2}\ln(1-x)$

31. (a) $\log[(m-3)^3(m+2)]$ **(b)** $\ln(x^2 + 2x)$ **33.** $\frac{1}{3}\log_7 24$; 0.5444   **35.** $-\frac{5}{2}\ln 20$; -0.8345 **37.** 0, $\ln 2$ (0.6931)

39. 12   **41.** No solution   **43.** $\dfrac{1}{1+e^2}$; 0.1192   **45.** $t = \dfrac{1}{r}\ln\left(\dfrac{A}{P}\right)$ **47.** $I = I_0(10)^{d/10}$ **49.** $k = 1$, $b = \frac{1}{3}$, $c = 2$

51. (a) 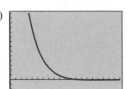   **(b)** $[0, 0.70]$ **53. (a)** $A(t) = 1.75(\frac{1}{2})^{t/4}$ **(b)** 0.62 million; 0.31 million

$[0.0, 2.0]_{0.1} \times [-0.2, 1.4]_{0.1}$

55. 51 quarters   **57.** About 18,500 yr old   **59.** 2.30 yr

CHAPTER TEST

1.

$[0, 80]_{10} \times [-5, 1]$
(see Section 4.4, Example 4)

2. Let $y = f^{-1}(x)$. Then

$$f(y) = x$$
$$2e^{y-1} = x$$
$$e^{y-1} = \frac{x}{2}$$

The equivalent logarithmic equation is

$$y - 1 = \ln\left(\frac{x}{2}\right)$$
$$y = \ln\left(\frac{x}{2}\right) + 1$$

Thus,

$$f^{-1}(x) = \ln\left(\frac{x}{2}\right) + 1$$

(see Section 4.3, Examples 4, 5)

3. $3\log(x+2) - \log(x^2-4)$

$= \log[(x+2)^3] - \log(x^2-4)$

$= \log\left[\dfrac{(x+2)^3}{x^2-4}\right]$

$= \log\left[\dfrac{(x+2)^3}{(x+2)(x-2)}\right]$

$= \log\left[\dfrac{(x+2)^2}{x-2}\right]$

(see Section 4.4, Example 7)

4. By exponent identities 4 and 5 of Section 4.4,

$$e^{\ln b} = b$$

$$\log\left(\dfrac{1}{10^a}\right) = \log 10^{-a} = -a$$

so

$$e^{\ln b} + \log\left(\dfrac{1}{10^a}\right) = b - a$$

(see Section 4.4, Example 8)

5. $17(3^{2x}) = 250$

$3^{2x} = \frac{250}{17}$

The equivalent logarithmic equation is

$2x = \log_3\left(\frac{250}{17}\right)$

$x = \frac{1}{2}\log_3\left(\frac{250}{17}\right)$

To approximate this expression with a calculator, we use the change-of-base formula:

$$x = \frac{1}{2}\log_3\left(\frac{250}{17}\right) = \frac{\ln\left(\frac{250}{17}\right)}{2\ln 3} \approx 1.2235$$

(see Section 4.5, Example 5)

6. $\ln(5x) = 3 + \ln 2$

$\ln(5x) - \ln 2 = 3$

$\ln\left(\dfrac{5x}{2}\right) = 3$

Writing this last equation in exponential form gives

$\dfrac{5x}{2} = e^3$

$x = \frac{2}{5}e^3$

To the nearest 0.0001, this is 8.0342.
(see Section 4.5, Example 7)

7. In general, the horizontal asymptote of $y = k \cdot b^x + c$ is $y = c$. Thus, from the figure, c is -4, and $y = k \cdot b^x - 4$. Because the graph passes through the point $(0, -2)$,

$-2 = k \cdot b^0 - 4$

$-2 = k - 4$

$k = 2$

Thus, $y = 2b^x - 4$. Finally, the graph passes through the point $(1, 2)$, which tells us that

$2 = 2b^1 - 4$

$6 = 2b$

$b = 3$

The graph is that of $y = 2 \cdot 3^x - 4$.
(see Section 4.1, Example 4)

8. Using the compound interest formula, we get

$$A = P\left(1 + \frac{r}{n}\right)^{nt}$$

$$= 2000\left(1 + \frac{0.065}{12}\right)^{12\cdot6} = 2950.854323...$$

The accumulated amount is $2950.85.
(see Section 4.1, Example 4)

9. Using the half-life formula, we get a function for the amount $A(t)$ of thorium-228 left after t years:

$$A(t) = 50\left(\tfrac{1}{2}\right)^{t/1.9}$$

Thus,

$$A(4) = 50\left(\tfrac{1}{2}\right)^{4/1.9} = 11.6244194... \ .$$

The amount left after 4 yr is approximately 11.6 g.
(see Section 4.1, Example 6)

10. First, we solve this equation for e^{rt}:

$$A = C + Be^{rt}$$

$$A - C = Be^{rt}$$

$$\frac{A-C}{B} = e^{rt}$$

The corresponding logarithmic equation is

$$rt = \ln\left(\frac{A-C}{B}\right)$$

so

$$r = \frac{1}{t}\ln\left(\frac{A-C}{B}\right)$$

(see Section 4.4, Example 2)

11. If P dollars are invested, then the accumulated amount after it doubles is $2P$. Thus,

$$2P = P\left(1 + \frac{0.07}{52}\right)^{52t}$$

$$2 = \left(1 + \frac{0.07}{52}\right)^{52t}$$

Rewriting this as a logarithmic equation gives

$$52t = \log_{(1+0.07/52)} 2$$
$$t = \tfrac{1}{52}\log_{(1+0.07/52)} 2$$
$$= \frac{\ln 2}{52\ln(1+0.07/52)} = 9.908\ 765\ 962...$$

The investment doubles in approximately 9.9 yr.
(see Section 4.5, Example 9)

12. $8.0 = \dfrac{\log E - 11.4}{1.5}$

$$12.0 = \log E - 11.4$$
$$23.4 = \log E$$
$$E = 10^{23.4}$$

This is approximately 2.5×10^{23} ergs.
(see Section 4.5, Example 10)

13. (a) $N(6) = \dfrac{500}{1 + 49e^{-0.2(6)}} = 31.728\ 875\ 22...$

After 6 days, approx. 32 people will be infected.

(b) We solve $N(t) = 100$:

$$\frac{500}{1 + 49e^{-0.2t}} = 100$$
$$500 = 100(1 + 49e^{-0.2t})$$
$$5 = 1 + 49e^{-0.2t}$$
$$4 = 49e^{-0.2t}$$
$$\tfrac{4}{49} = e^{-0.2t}$$
$$-0.2t = \ln\left(\tfrac{4}{49}\right)$$
$$t = -\tfrac{1}{0.2}\ln\left(\tfrac{4}{49}\right) = 12.527\ 629\ 68...$$

Thus, 100 people will be infected in approx. 13 days.

(c)

$[0, 60]_{10} \times [0, 600]_{100}$
Note that $y \to 500$ as $x \to +\infty$. Thus, as time goes on, nearly 500 people will be infected.
(see Section 4.2, Exercises 49, 50)

14. $A(10) = 68 + (150 - 68)e^{-0.06(10)}$
$$= 68 + 82e^{-0.6}$$
$$= 113.002\ 554\ 2...$$

After 10 min, the soup is approx. 113°F.
(see Section 4.2, Example 7)

CHAPTER 5: SECTION 5.1

1. $(3, -2)$ **3.** $(-2, 4), (1, 1)$ **5.** $(3, 4), (-3, 4)$ **7.** $(1.296, 1.679), (-2, 4)$ **9.** $(1, 0), (2, 1), (3, 2)$
11. $(-0.426, 0.653)$ **13.** $(1.857, 6.406), (4.536, 93.354)$ **15.** $\left(\frac{5}{2}, -\frac{1}{2}\right)$ **17.** $\left(\frac{3}{2}, 2\right)$ **19.** No solution (inconsistent)
21. Dependent   **23.** Dependent   **25.** $(-3.2, -2.4), (2.4, 3.2)$ **27.** $(3.394, 2.606), (10.606, -4.606)$
29. $(-2.701, -4.207), (-1.903, 4.624), (-0.645, 4.958), (2.443, 4.363)$ **31.** $(1.397, 3.045)$ **33.** No solution
35. $(a, b) = (11, 7.5)$ **37.** $(u, v) = (-3.450, 7.080), (4.641, 5.794)$ **39.** $t = 13.327, A = 2173.913$
41. $(4, -6), (8, -15), (12, -24)$, corresponding to $t = 1, 2, 3$ **43.** \$60.67   **45.** \$49.55
47. (a) Supply: $p = 8x - 90$; Demand: $p = -5x + 350$ **(b)** \$180.80   **49. (a)** 16.651 yr   **(b)** Simple interest
51. Wind: 45 mph; Jet: 495 mph   **53.** $r = 1.3$ in., $h = 11.0$ in., or $r = 3.2$ in., $h = 1.8$ in.   **55. (a)** $e^{1/e}$ **(b)** (e, e)

SECTION 5.2

1. $(6, 7)$ **3.** $(2, -2)$ **5.** $(0, 5)$ **7.** $\left(2, \frac{2}{3}\right)$ **9.** Inconsistent **11.** Dependent; $(x, y) = \left(\frac{5}{3}t - \frac{2}{3}, t\right)$ **13.** $(-4, 3)$
15. $(-2, -5)$ **17.** $\left(\frac{1}{2}, 6\right)$ **19.** Inconsistent **21.** Dependent; $(x, y) = \left(-\frac{3}{4}t + \frac{3}{4}, t\right)$ **23.** Dependent; $(x, y) = \left(\frac{5}{2}t, t\right)$
25. $\left(\frac{1}{2}, \frac{1}{3}\right)$ **27.** $\left(-3, \frac{2}{5}\right)$ **29.** $(14, 8)$ **31.** Dependent; $(x, y) = (-3t + 23, t)$
33. (a) $(-3, 0), (1, 1), (5, 2)$ **(b)** $(-3, 0), (1, 1), (5, 2)$ **35.** 840 gal 87% isooctane, 560 gal 92% isooctane
37. 25 L 20% acid, 15 L 60% acid **39.** 2 qt **41.** 1600 phones **43.** 0.6 cup crispy rice, 0.4 cup wheat flakes
45. 112.5 oz 20% protein, 4% fat; 50 oz 15% protein, 2% fat **47.** $y = \frac{1}{2}x^2 + 3x$ **49.** $\left(\frac{5}{2}, 2\right)$

SECTION 5.3

1. $(-3, 3)$ **3.** $(4, 5, 3)$ **5.** $\left(\frac{20}{3}, \frac{2}{3}, \frac{1}{3}\right)$ **7.** $(0, -3, 2)$ **9.** $(4, 2, 1)$ **11.** $(2, 2, 3)$ **13.** $(1, -2, 1)$
15. $(4, 1, -1)$ **17.** $(2, 1, -1)$ **19.** $(5, -4, 2)$ **21.** $(1, 1, 2)$ **23.** $(1, 0, 1)$ **25.** Dependent; $(x, y, z) = (-2t, -t, t)$
27. Dependent; $(x, y, z) = (1 + 2t, 3 + t, t)$ **29.** Inconsistent **31.** Dependent; $(x, y, z) = (12.4 + -4.6t, -3.4 - 1.6t, t)$
33. $\left(\frac{1}{4}, -\frac{1}{2}, \frac{3}{8}\right)$ **35.** $(2.5, 7, 0.5)$ **37.** $\left(\frac{1}{2}, \frac{2}{3}, \frac{1}{2}\right)$ **39.** $(x, y, z, w) = \left(-1, \frac{1}{2}, 3, 4\right)$
41. (a) $(2, 1, 0), (3, -2, 1)$ **(b)** $(2, 1, 0)(3, -2, 1)$; identical to part (a)
(c) $(4, -5, 2), (5, -8, 3), (6, -11, 4)$ using $t = 2, 3, 4$ in first set of solutions or $T = 4, 5, 6$ in the second set
43. $y = -x^2 + 0.5x + 1.5$ **45.** $x = y^2 - 5y + 2$ **47.** $x^2 + y^2 - 2x + 4y - 20 = 0$
49. $6000 in CD, $3000 in stocks, $6000 in municipal bonds **51.** 10 lb almonds, 5 lb peanuts, 5 lb walnuts
53. There is more than one way; three examples: 1000 to L.A., 200 to N.Y., 0 to Denver; 600 to L.A., 400 to N.Y., 200 to Denver;
200 to L.A., 600 to N.Y., 400 to Denver
55. There is more than one way; three examples: 8 lb French roast, 1 lb Kona, 31 lb Colombian; 12 lb French roast, 4 lb Kona,
24 lb Colombian; 16 lb French roast, 7 lb Kona, 17 lb Colombian
57. $\left(\frac{1}{4}, 2, \frac{2}{3}\right), (a, 0, 0), (0, a, 0),$ or $(0, 0, a),$ where a is any real number

SECTION 5.4

1. $\begin{bmatrix} 5 & 1 & | & 6 \\ -1 & 3 & | & 1 \end{bmatrix}$; 2×3 **3.** $\begin{bmatrix} 2 & 1 & 0 & | & 6 \\ 1 & -2 & 4 & | & 5 \\ 6 & 0 & -2 & | & -1 \end{bmatrix}$; 3×4 **5.** $\begin{bmatrix} -1 & 1 & 4 & 7 \\ 0 & 12 & \frac{1}{2} & -5 \\ 4 & 1 & -9 & 6 \end{bmatrix}$

7. $\begin{bmatrix} 1 & -8 & 2 & 10 \\ 0 & 1 & \frac{2}{3} & -2 \\ -1 & 1 & 5 & 14 \end{bmatrix}$ **9.** $\begin{bmatrix} 1 & -1 & -5 & 2 \\ 2 & 3 & 5 & 6 \\ 0 & 4 & 15 & -10 \end{bmatrix}$ **11.** Already reduced **13.** $\begin{bmatrix} 1 & 0 & 0 & | & -17 \\ 0 & 1 & 0 & | & 2 \\ 0 & 0 & 1 & | & -3 \end{bmatrix}$

15. Already reduced **17.** $\frac{1}{2}R_1 \rightarrow R_1$ **19.** $5R_1 + R_3 \rightarrow R_3$ **21.** $R_3 \rightarrow R_2$ and $R_2 \rightarrow R_3$
23. $(-4, 1, 7)$ **25.** Inconsistent **27.** $(-3 - 7t, 2 - 5t, t)$

29. (a) $\begin{bmatrix} 1 & 1 & 1 & | & 1 \\ 2 & 3 & 1 & | & 0 \\ -1 & 1 & 1 & | & 11 \end{bmatrix}$ **(b)** $\begin{bmatrix} 1 & 1 & 1 & | & 1 \\ 0 & 1 & -1 & | & -2 \\ 0 & 2 & 2 & | & 12 \end{bmatrix}$ **(c)** $\begin{bmatrix} 1 & 0 & 2 & | & 3 \\ 0 & 1 & -1 & | & -2 \\ 0 & 0 & 4 & | & 16 \end{bmatrix}$ **(d)** $\begin{bmatrix} 1 & 0 & 2 & | & 3 \\ 0 & 1 & -1 & | & -2 \\ 0 & 0 & 1 & | & 4 \end{bmatrix}$

(e) $\begin{bmatrix} 1 & 0 & 0 & | & -5 \\ 0 & 1 & 0 & | & 2 \\ 0 & 0 & 1 & | & 4 \end{bmatrix}$

31. $(4, 1, -1)$ **33.** $(2, 1, -1)$ **35.** $(5, -4, 2)$ **37.** $(1, 1, 2)$ **39.** $(-1, 3, -2)$ **41.** $\left(-\frac{8}{3}, \frac{16}{3}, \frac{23}{3}\right)$
43. $\left(\frac{2}{3}, \frac{1}{3}, -1\right)$ **45.** $(-3, 1, 0)$ **47.** $\left(-\frac{5}{7}, \frac{18}{7}, \frac{4}{7}\right)$ **49.** $(2, 1, 0)$ **51.** Dependent; $(x, y, z) = (40 - 13t, 24 - 9t, t)$
53. Dependent; $(x, y, z) = (3 - 5t, 2 + 3t, t)$ **55.** $(x, y, z, w) = (-17, 14, 15, -5)$ **57.** $f(x) = -x^2 + 7x + 3$
59. $f(x) = -2x^3 + 5x^2 + 3x + 1$ **61.** $15,000 at 6%, $17,500 at 8%, $10,500 at 10%
63. 26 $220 bags, 52 $170 bags, 78 $140 bags **65.** $I_1 = \frac{3}{4}$ A, $I_2 = \frac{9}{4}$ A, $I_3 = 3$ A **67.** $t_1 = \frac{89}{56}°$F, $t_2 = \frac{19}{14}°$F, $t_3 = \frac{47}{56}°$F

SECTION 5.5

1. -3 **3.** -1 **5.** $r^2 - 7$ **7.** $\frac{27}{2}$ sq units   **9.** 26 sq units   **11.** -20 **13.** 90

15. (a) $8i + 11j + 9k$ **(b)** $8i + 11j + 9k$ **17.** $(94, 26)$ **19.** $(-10, 9)$ **21.** $\left(\frac{3}{13}, \frac{12}{13}\right)$ **23.** -37.41 **25.** 1210

27. $(1, -2, 1)$ **29.** $(2, -3, 1)$ **31.** $\left(\frac{1}{2}, \frac{3}{2}, -3\right)$ **33.** $\left(\frac{2}{9} - \frac{k}{9}, \frac{4}{9} - \frac{2k}{9}, -\frac{1}{3} + \frac{2k}{3}\right)$ **35.** $(5.83, -0.62)$

37. $(0.88, 0.17, 0.53)$ **39.** $x = -\dfrac{f}{5u}, y = \dfrac{f}{10v}$ **41.** $x = \frac{11}{6}$ **43.** $y = 0$ **45.** $z = \frac{5}{8}$ **47.** $r = -5, 5$

49. $r = -3, 5$

SECTION 5.6

1. $\begin{bmatrix} 1 & 11 & 7 \\ 17 & -11 & 1 \end{bmatrix}$ **3.** $\begin{bmatrix} -5 & -11 & 21 \\ 1 & -9 & -11 \end{bmatrix}$ **5.** Not possible   **7.** $\begin{bmatrix} -4 & 0 & 28 \\ 18 & -20 & -10 \end{bmatrix}$ **9.** $\begin{bmatrix} 8 & -24 \\ -32 & -2 \end{bmatrix}$

11. $\begin{bmatrix} 16 & 22 & -84 \\ -29 & 48 & 37 \end{bmatrix}$ **13.** Not possible   **15.** $\frac{1}{3}\begin{bmatrix} 6 & -3 \\ 9 & 0 \end{bmatrix}$ **17.** $2\begin{bmatrix} -4 & \frac{3}{2} \\ 2 & 0 \end{bmatrix}$ **19.** $\begin{bmatrix} 2 & 28 \\ 3 & -18 \\ -2 & 32 \end{bmatrix}$

21. $MN = \begin{bmatrix} -22 & 48 \\ 3 & -46 \end{bmatrix}$; $NM = \begin{bmatrix} -13 & -51 \\ 3 & -55 \end{bmatrix}$ **23.** $\begin{bmatrix} 57 & 36 \\ 0 & -26 \end{bmatrix}$ **25.** $\begin{bmatrix} -9 & 0 & -18 \\ 11 & -20 & 10 \\ 21 & 30 & 60 \end{bmatrix}$ **27.** Not possible

29. $\begin{bmatrix} 28 & 4 \\ -13 & -14 \end{bmatrix}$ **31.** $\begin{bmatrix} 52 & 196 \\ -12 & -91 \end{bmatrix}$ **33.** $\begin{bmatrix} -3 & 14 \\ -2 & -7 \end{bmatrix}$ **35.** Not possible   **37.** $\begin{bmatrix} 470 & 125 & 160 \\ 724 & 194 & 232 \\ 134 & 19 & 212 \end{bmatrix}$

39. (a) $A^2 = \begin{bmatrix} 0.56 & 0.16 & 0.28 \\ 0.33 & 0.31 & 0.36 \\ 0.37 & 0.28 & 0.35 \end{bmatrix}$ **(b)** $A^4 = \begin{bmatrix} 0.470 & 0.218 & 0.312 \\ 0.420 & 0.250 & 0.330 \\ 0.429 & 0.244 & 0.327 \end{bmatrix}$ **(c)** $A^8 = \begin{bmatrix} 0.446 & 0.233 & 0.321 \\ 0.444 & 0.234 & 0.322 \\ 0.445 & 0.234 & 0.321 \end{bmatrix}$

41. (a) $AX = \begin{bmatrix} 3x + 2y + 5z \\ 5x + 9y + 8z \\ -x - 5y + 4z \end{bmatrix}$ **(b)** $AX = B$ represents the system of equations $\begin{cases} 3x + 2y + 5z = 16 \\ 5x + 9y + 8z = 21 \\ -x - 5y + 4z = 15 \end{cases}$

43.

	Lecturer	Assist. prof.	Assoc. prof.	Prof.	
	29,469	37,098	43,890	53,713	0–3 yr
	33,963	40,860	48,802	59,774	4–7 yr
	37,620	43,681	54,967	65,104	8 or more yr

45. Portland: $54 standard, $82 deluxe; Atlanta: $50.50 standard, $76.50 deluxe; Boston: $58 standard, $87.75 deluxe

47. $\begin{bmatrix} \frac{1}{3} & \frac{11}{30} & \frac{3}{10} \end{bmatrix}$ (Sharp T.I. Kasia)   **49.** $\begin{bmatrix} 1 & 0 \\ 0 & 1 \end{bmatrix}$

51. (a) $AB = \begin{bmatrix} ax + bz & bw + ay \\ cx + dz & dw + cy \end{bmatrix}$ **(b)** $AB + AC = \begin{bmatrix} ap + br + ax + bz & aq + bs + bw + ay \\ cp + dr + cx + dz & cq + ds + dw + cy \end{bmatrix}$

(c) $B + C = \begin{bmatrix} p + x & q + y \\ r + z & s + w \end{bmatrix}$ **(d)** $A(B + C) = \begin{bmatrix} a(p + x) + b(r + z) & b(s + w) + a(q + y) \\ c(p + x) + d(r + z) & d(s + w) + c(q + y) \end{bmatrix}$

(e) The distributive property holds for matrix algebra.

SECTION 5.7

1. $MM^{-1} = M^{-1}M = I_2$ **3.** $MM^{-1} = M^{-1}M = I_3$ **5.** $\begin{bmatrix} -5 & 2 \\ 3 & -1 \end{bmatrix}$ **7.** $\begin{bmatrix} -2 & \frac{1}{2} \\ -1 & \frac{1}{2} \end{bmatrix}$ **9.** Does not exist

11. $\begin{bmatrix} -7 & 2 & -4 \\ -1 & 1 & -1 \\ 2 & 0 & 1 \end{bmatrix}$ **13.** $\begin{bmatrix} 2 & -1 & 0 \\ -2 & 1 & -\frac{1}{2} \\ -1 & 0 & -\frac{1}{2} \end{bmatrix}$ **15.** Does not exist **17.** $\begin{bmatrix} 5 & -2 \\ 4 & 3 \end{bmatrix}\begin{bmatrix} x \\ y \end{bmatrix} = \begin{bmatrix} 7 \\ -1 \end{bmatrix}$

19. $\begin{bmatrix} 1 & -2 & 3 \\ 5 & 4 & -4 \\ 2 & 0 & 9 \end{bmatrix}\begin{bmatrix} x \\ y \\ z \end{bmatrix} = \begin{bmatrix} 14 \\ -1 \\ 13 \end{bmatrix}$ **21.** $x = -27, y = -14$ **23.** $x = 21, y = 25, z = 37$ **25.** $(33, 26)$

27. $(-50, -27)$ **29.** $(-33, -6, 13)$ **31. (a)** $(-28, 18)$ **(b)** $(10, -3)$ **(c)** $(-276, 174)$

33. (a) $\left(-\frac{39}{2}, 25\right)$ **(b)** $\left(-\frac{15}{2}, 8\right)$ **(c)** $(-24, 33)$ **35. (a)** $(-9, 7, -5)$ **(b)** $(-5, 7, -5)$ **(c)** $(53, -18, 3)$

37. (a) $(7, -6, 1)$ **(b)** $(4, -10, -12)$ **(c)** $(77, -72, -11)$

39. (a) $(-11, 4, 5)$ **(b)** $\left(\frac{5}{2}, 11, \frac{9}{2}\right)$ **(c)** $\left(-\frac{165}{4}, 40, \frac{131}{4}\right)$

41. (a) $I_1 = \frac{12}{5}$ A, $I_2 = \frac{6}{5}$ A, $I_3 = \frac{6}{5}$ A **(b)** $I_1 = \frac{18}{5}$ A, $I_2 = \frac{9}{5}$ A, $I_3 = \frac{9}{5}$ A **(c)** $I_1 = \frac{24}{5}$ A, $I_2 = \frac{12}{5}$ A, $I_3 = \frac{12}{5}$ A

43. Group #1: 4 g food A, 2 g food B; Group #2: 2.7 g food A, 6.9 g food B

45. (a) Model A: 40; model B: 200; model C: 20 **(b)** Model A: 260; model B: 80; model C: 0

(c) Model A: 316; model B: 158; model C: 16

47. (a) $M = A^{-1}CB^{-1}$ **(b)** $C^{-1}B^{-1}A^{-1}$

SECTION 5.8

1. 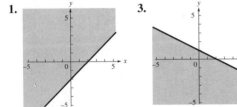 **3.** **5.** **7.**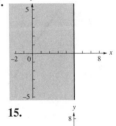

9. 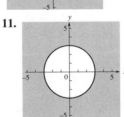 **11.** 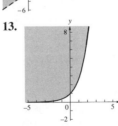 **13.** 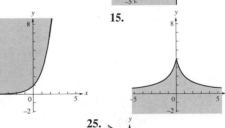 **15.**

17. IV **19.** I **21.** 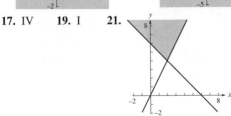 **23.** 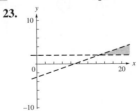 **25.**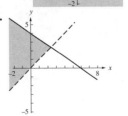

27.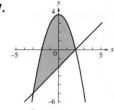

29.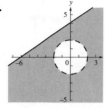

31.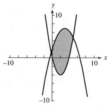

33.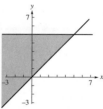

35.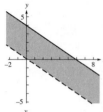

37.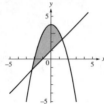

39.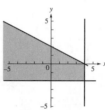

41.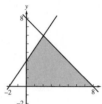

43.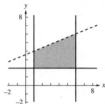

45.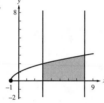

47.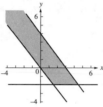

49. $(0, 0), (8, 0), (0, 3), (2, 6)$ **51.** $(1, 2), (6, 2), (1, 4), (6, 6)$ **53.** $9 million

55. (a) $\begin{cases} 20 \le x + y \le 30 \\ x \le 3y \\ x \ge 0 \\ y \ge 0 \end{cases}$

(b)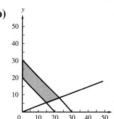

(c) 22; 5   **57.** $\begin{cases} x \ge 2 \\ y \ge 0 \\ x + 2y \le 8 \end{cases}$ **59.** $x \le y \le e^x$

MISCELLANEOUS EXERCISES

1. $(3, 4), (-1.203, 4.853)$ **3.** $(2, 3), (-2.408, -2.491)$ **5.** $(65, 4)$ **7.** $\left(\frac{5}{2}, -3\right)$ **9.** Dependent; $(3t + 3, t)$
11. $59.17   **13.** 93.3 ft   **15.** $\left(0, \frac{1}{2}\right)$ **17.** $\left(10\frac{2}{3}, 5\frac{1}{3}\right)$ **19.** Inconsistent   **21.** $(12, 0, 7)$ **23.** $(1, 3, -2)$
25. $(-1, 6, 10)$ **27.** $\left(0, -\frac{3}{7}, \frac{18}{7}\right)$ **29.** Dependent; $\left(\dfrac{25 - 5t}{7}, \dfrac{4t - 6}{7}, t\right)$ **31.** Inconsistent
33. $(x, y, z, w) = (3, -4, 3, 2)$ **35.** $f(x) = -x^2 + 1.5x + 0.5$ **37.** $I_1 = \frac{7}{17}$ A, $I_2 = \frac{53}{17}$ A, $I_3 = \frac{60}{17}$ A   **39.** $\frac{27}{2}$ sq units
41. $(86, 66)$ **43.** $x = -\dfrac{2q}{3p^2 + 2pq}, y = \dfrac{3}{3p + 2q}$ **45.** $(-3, 4, 9)$ **47.** $(0.1, 4.6, 2.7)$
49. $\begin{bmatrix} -6 & 11 \\ 5 & 15 \end{bmatrix}$ **51.** $\begin{bmatrix} 0 & 1 & 6 \\ 11 & -29 & 24 \end{bmatrix}$ **53.** Not possible
55. [8.34 8.45 8.41]; this represents the total cost of buying all four products at each market.

57. $\begin{bmatrix} 1 & 8 \\ 4 & -3 \end{bmatrix} \begin{bmatrix} x \\ y \end{bmatrix} = \begin{bmatrix} 12 \\ 6 \end{bmatrix}$ **59.** $\begin{bmatrix} 5 & 3 & -4 \\ 1 & 2 & -4 \\ 2 & 0 & -5 \end{bmatrix} \begin{bmatrix} x \\ y \\ z \end{bmatrix} = \begin{bmatrix} -14 \\ 1 \\ 10 \end{bmatrix}$ **61. (a)** $(29, -16)$ **(b)** $(-186, 110)$

63. (a) $\left(6, -\frac{3}{2}, -\frac{9}{2}\right)$ **(b)** $(408, -72, -222)$ **65.**

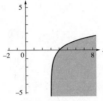

67. 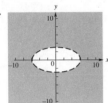   **69.** 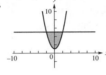   **71.** 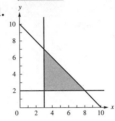   **73.**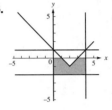

CHAPTER TEST

1. The graph of the system is shown below:

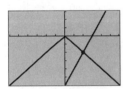

$[-6, 6] \times [-6, 3]$

The solution to the system is $(2, -2)$.
(see Section 5.1, Example 1)

2. To graph $x^2 + y^2 = 4$ on a graphing utility, we graph $y = \sqrt{4 - x^2}$ and $y = -\sqrt{4 - x^2}$. The graph of the system is shown below:

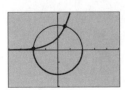

$[-4, 5] \times [-3, 3]$

Using the zoom and trace features on a graphing utility, we find that the graphs intersect at $(0.590, 1.911)$ and $(-1.997, 0.111)$, to the nearest 0.001. The solutions are $(0.590, 1.911)$ and $(-1.997, 0.111)$.
(see Section 5.1, Example 3)

3. Writing the system in slope–intercept form, we get

$$\begin{cases} y = -\frac{1}{2}x + 5 \\ y = \frac{1}{2}x \end{cases}$$

The graph of the system is shown below:

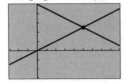

$[-3, 9] \times [-3, 5]$
Since the graphs intersect at $(5, 2.5)$, the solution to the system is $(5, 2.5)$.
(see Section 5.1, Example 4a)

4. Writing the system in slope–intercept form, we get

$$\begin{cases} y = \frac{5}{8}x - 2 \\ y = \frac{5}{8}x + \frac{5}{6} \end{cases}$$

The graph of the system is shown below:

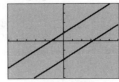

$[-6, 6] \times [-4, 4]$

Since the slope of each graph is $\frac{5}{8}$ but the y-intercepts are different, the graphs are parallel and do not intersect. The system is inconsistent.
(see Section 5.1, Example 4b)

5. *Substitution method:* Isolate y in the first equation:

$$y = 1 - 2x$$

Substitute into the second equation:

$$8x + 3(1 - 2x) = 10$$

Solve for y:

$$8x + 3 - 6x = 10$$
$$2x = 7$$
$$x = \tfrac{7}{2}$$

Substitute $x = \frac{7}{2}$ in $y = 1 - 2x$:

$$y = 1 - 2\left(\tfrac{7}{2}\right) = -6$$

The solution is $\left(\tfrac{7}{2}, -6\right)$.
Elimination method: Multiply the first equation in the system by -4 and add the equations:

$$\begin{cases} -8x - 4y = -4 \\ 8x + 3y = 10 \end{cases}$$
$$-y = 6$$

Hence, $y = -6$. Substituting into the first equation of the original system, we get

$$2x + (-6) = 1$$
$$2x = 7$$
$$x = \tfrac{7}{2}$$

The solution is $\left(\tfrac{7}{2}, -6\right)$.
(see Section 5.2, Examples 1, 3)

6. Starting with the augmented matrix for the system, we have

$$\begin{bmatrix} 1 & 1 & 1 & | & -1 \\ 1 & -2 & -1 & | & 3 \\ 3 & 0 & -1 & | & 3 \end{bmatrix} \begin{matrix} \\ -1R_1 + R_2 \rightarrow R_2 \\ -3R_1 + R_3 \rightarrow R_3 \end{matrix} \begin{bmatrix} 1 & 1 & 1 & | & -1 \\ 0 & -3 & -2 & | & 4 \\ 0 & -3 & -4 & | & 6 \end{bmatrix} \begin{matrix} \\ -\frac{1}{3}R_2 \rightarrow R_2 \\ \\ \end{matrix} \begin{bmatrix} 1 & 1 & 1 & | & -1 \\ 0 & 1 & \frac{2}{3} & | & -\frac{4}{3} \\ 0 & -3 & -4 & | & 6 \end{bmatrix}$$

$$\begin{matrix} -1R_2 + R_1 \rightarrow R_1 \\ \\ 3R_2 + R_3 \rightarrow R_3 \end{matrix} \begin{bmatrix} 1 & 0 & \frac{1}{3} & | & \frac{1}{3} \\ 0 & 1 & \frac{2}{3} & | & -\frac{4}{3} \\ 0 & 0 & -2 & | & 2 \end{bmatrix} \qquad \begin{bmatrix} 1 & 0 & \frac{1}{3} & | & \frac{1}{3} \\ 0 & 1 & \frac{2}{3} & | & -\frac{4}{3} \\ 0 & 0 & 1 & | & -1 \end{bmatrix} \begin{matrix} \\ \\ -\frac{1}{2}R_3 \rightarrow R_3 \end{matrix} \begin{matrix} -\frac{1}{3}R_3 + R_1 \rightarrow R_1 \\ -\frac{2}{3}R_3 + R_2 \rightarrow R_2 \\ \\ \end{matrix} \begin{bmatrix} 1 & 0 & 0 & | & \frac{2}{3} \\ 0 & 1 & 0 & | & -\frac{2}{3} \\ 0 & 0 & 1 & | & -1 \end{bmatrix}$$

The solution is $\left(\tfrac{2}{3}, -\tfrac{2}{3}, -1\right)$.
(see Section 5.4, Example 5)

7. Starting with the augmented matrix for the system, we have

$$\begin{bmatrix} 7 & 4 & -1 & | & 32 \\ 3 & 2 & -3 & | & 18 \\ 2 & 1 & 1 & | & 7 \end{bmatrix} \begin{matrix} \\ \frac{1}{3}R_2 \rightarrow R_2 \\ \\ \end{matrix} \begin{bmatrix} 7 & 4 & -1 & | & 32 \\ 1 & \frac{2}{3} & -1 & | & 6 \\ 2 & 1 & 1 & | & 7 \end{bmatrix} \begin{matrix} R_2 \rightarrow R_1 \\ R_1 \rightarrow R_2 \\ \\ \end{matrix} \begin{bmatrix} 1 & \frac{2}{3} & -1 & | & 6 \\ 7 & 4 & -1 & | & 32 \\ 2 & 1 & 1 & | & 7 \end{bmatrix}$$

$$\begin{matrix} \\ -7R_1 + R_2 \rightarrow R_2 \\ -2R_1 + R_3 \rightarrow R_3 \end{matrix} \begin{bmatrix} 1 & \frac{2}{3} & -1 & | & 6 \\ 0 & -\frac{2}{3} & 6 & | & -10 \\ 0 & -\frac{1}{3} & 3 & | & -5 \end{bmatrix} \begin{matrix} \\ -\frac{3}{2}R_2 \rightarrow R_2 \\ \\ \end{matrix} \begin{bmatrix} 1 & \frac{2}{3} & -1 & | & 6 \\ 0 & 1 & -9 & | & 15 \\ 0 & -\frac{1}{3} & 3 & | & -5 \end{bmatrix}$$

$$\begin{matrix} -\frac{2}{3}R_2 + R_1 \rightarrow R_1 \\ \\ \frac{1}{3}R_2 + R_3 \rightarrow R_3 \end{matrix} \begin{bmatrix} 1 & 0 & 5 & | & -4 \\ 0 & 1 & -9 & | & 15 \\ 0 & 0 & 0 & | & 0 \end{bmatrix}$$

The matrix is in reduced row-echelon form. The first and second rows represent the equation $x + 5z = -4$ and $y - 9z = 15$, respectively. Solving for x and y, we get $x = -4 - 5z$ and $y = 15 + 9z$. Letting $z = t$, we get the parametric solution $(-4 - 5t, 15 + 9t, t)$. The system is a dependent system.
(see Section 5.4, Example 6)

8. We are given that $f(2) = a(2)^2 + b(2) + c = 29$, $f(4) = a(4)^2 + b(4) + c = 22$, and $f(6) = a(6)^2 + b(6) + c = 9$. We have the following system:

$$\begin{cases} 4a + 2b + c = 29 \\ 16a + 4b + c = 22 \\ 36a + 6b + c = 9 \end{cases}$$

Solving this system, we get $a = -\frac{3}{4}$, $b = 1$, $c = 30$. Thus, $f(t) = -\frac{3}{4}t^2 + t + 30$. The height after 5 sec is given by

$$f(5) = -\frac{3}{4}(5)^2 + 5 + 30 = 16\frac{1}{4}$$

After 5 sec the height is $16\frac{1}{4}$ ft.
(see Section 5.3, Example 7)

9. The area is given by

$$\frac{1}{2}\left(\begin{vmatrix} 2 & 3 \\ 4 & 1 \end{vmatrix} + \begin{vmatrix} 4 & 1 \\ 6 & 3 \end{vmatrix} + \begin{vmatrix} 6 & 3 \\ 3 & 7 \end{vmatrix} + \begin{vmatrix} 3 & 7 \\ 2 & 3 \end{vmatrix}\right)$$
$$= \frac{1}{2}(-10 + 6 + 33 - 5) = 12$$

The area is 12 sq units.
(see Section 5.5, Example 8)

10. $x = \dfrac{D_x}{D} = \dfrac{\begin{vmatrix} 9 & -4 \\ 7 & 2 \end{vmatrix}}{\begin{vmatrix} 5 & -4 \\ -2 & 2 \end{vmatrix}} = \dfrac{46}{2} = 23$

$y = \dfrac{D_y}{D} = \dfrac{\begin{vmatrix} 5 & 9 \\ -2 & 7 \end{vmatrix}}{\begin{vmatrix} 5 & -4 \\ -2 & 2 \end{vmatrix}} = \dfrac{53}{2}$

The solution is $\left(23, \frac{53}{2}\right)$.
(see Section 5.5, Example 5)

11. $x = \dfrac{D_x}{D} = \dfrac{\begin{vmatrix} 1 & L+1 \\ 0 & 2L \end{vmatrix}}{\begin{vmatrix} 3 & L+1 \\ 4 & 2L \end{vmatrix}} = \dfrac{2L}{2L-4} = \dfrac{L}{L-2}$

$y = \dfrac{D_y}{D} = \dfrac{\begin{vmatrix} 3 & 1 \\ 4 & 0 \end{vmatrix}}{\begin{vmatrix} 3 & L+1 \\ 4 & 2L \end{vmatrix}} = \dfrac{-4}{2L-4} = \dfrac{-2}{L-2}$

The solution is $\left(\dfrac{L}{L-2}, -\dfrac{2}{L-2}\right)$.
(see Section 5.5, Example 6)

12. $\begin{bmatrix} 7 & 3 \\ 4 & 2 \end{bmatrix}\begin{bmatrix} 1 & 6 \\ -2 & 3 \end{bmatrix}$
$$= \begin{bmatrix} 7(1) + 3(-2) & 7(6) + 3(3) \\ 4(1) + 2(-2) & 4(6) + 2(3) \end{bmatrix} = \begin{bmatrix} 1 & 51 \\ 0 & 30 \end{bmatrix}$$

(see Section 5.6, Examples 4 and 5)

13. Since the number of columns in B (2) does not equal the number of rows in C (3), the multiplication is not possible.
(see Section 5.6, Example 5)

14. Expanding along the first row of C, we have

$$\det C = -2\begin{vmatrix} 5 & 6 \\ -1 & -3 \end{vmatrix} - 7\begin{vmatrix} 0 & 6 \\ 4 & -3 \end{vmatrix} + 7\begin{vmatrix} 0 & 5 \\ 4 & -1 \end{vmatrix}$$
$$= -2(-9) - 7(-24) + 7(-20)$$
$$= 46$$

(see Section 5.5, Example 2)

15. $A^{-1} = \dfrac{1}{\det A}\begin{bmatrix} 2 & -3 \\ -4 & 7 \end{bmatrix}$
$$= \frac{1}{2}\begin{bmatrix} 2 & -3 \\ -4 & 7 \end{bmatrix} = \begin{bmatrix} 1 & -\frac{3}{2} \\ -2 & \frac{7}{2} \end{bmatrix}$$

(see Section 5.7, Example 2)

16. (a) $\begin{bmatrix} -3 & 1 & 1 \\ 1 & -1 & 2 \\ 0 & 3 & -10 \end{bmatrix}\begin{bmatrix} x \\ y \\ z \end{bmatrix} = \begin{bmatrix} 6 \\ 10 \\ 1 \end{bmatrix}$

(b) Using a graphing utility,

$$\begin{bmatrix} -3 & 1 & 1 \\ 1 & -1 & 2 \\ 0 & 3 & -10 \end{bmatrix}^{-1} = \begin{bmatrix} 4 & 13 & 3 \\ 10 & 30 & 7 \\ 3 & 9 & 2 \end{bmatrix}$$

(c) $\begin{bmatrix} x \\ y \\ z \end{bmatrix} = \begin{bmatrix} -3 & 1 & 1 \\ 1 & -1 & 2 \\ 0 & 3 & -10 \end{bmatrix}^{-1}\begin{bmatrix} 6 \\ 10 \\ 1 \end{bmatrix}$
$$= \begin{bmatrix} 4 & 13 & 3 \\ 10 & 30 & 7 \\ 3 & 9 & 2 \end{bmatrix}\begin{bmatrix} 6 \\ 10 \\ 1 \end{bmatrix} = \begin{bmatrix} 157 \\ 367 \\ 110 \end{bmatrix}$$

The solution of the system is (157, 367, 110).
(see Section 5.7, Example 3)

17. The graph of $y \le 8$ is the set of all points on or below the graph of $y = 8$. The graph of $x \ge 3$ is the set of all points on or to the right of the graph of $x = 3$. To graph $y \ge 2x$, we graph the associated equation, $y = 2x$, and select a test point not on the graph, say (0, 1). Since $1 \ge 2(0)$ is true, the graph of $y \ge 2x$ consists of all points on or above the graph of $y = 2x$. The graph of the system is the set of all points common to all three graphs:

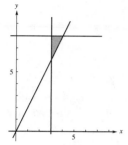

(see Section 5.8, Example 3)

18. The compound inequality $-2 \leq x \leq 4$ is equivalent to the system

$$\begin{cases} x \geq -2 \\ x \leq 4 \end{cases}$$

The graph of $x \geq -2$ is the set of all points on or to the right of the graph of $x = -2$, and the graph of $x \leq 4$ is all points on or to the left of the graph of $x = 4$. Thus, the graph

of $-2 \leq x \leq 4$ is the set of all points between the vertical lines $x = -2$ and $x = 4$. To graph $x^2 + y^2 \leq 25$, we first graph the associated equation, $x^2 + y^2 = 25$, which is a circle centered at the origin with radius 5. Selecting a test point not on the graph, say $(0, 0)$, we get $0^2 + 0^2 \leq 25$, which is true. Therefore, the graph of $x^2 + y^2 \leq 25$ is the set of all points inside the graph of $x^2 + y^2 = 25$. The graph of the system is the set of all points common to all the graphs:

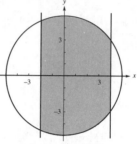

(see Section 5.8, Examples 4, 5)

CHAPTER 6: SECTION 6.1

5.

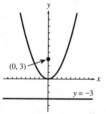

7.

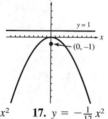

9.

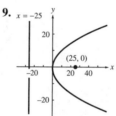

11.

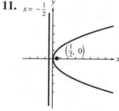

13. $y = -\frac{1}{20}x^2$ **15.** $y = \frac{1}{2}x^2$ **17.** $y = -\frac{1}{12}x^2$ **19.** $x = \frac{1}{8}y^2$

21.

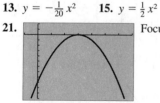

$[-1, 7] \times [-12, 2]$

Focus: $\left(3, -\frac{1}{4}\right)$; directrix: $y = \frac{1}{4}$

23.

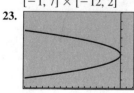

$[-16, 2] \times [0, 9]$

Focus: $\left(-\frac{1}{8}, 4\right)$; directrix: $x = \frac{1}{8}$

25. 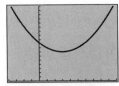 Focus: $(3, 8)$; directrix: $y = 6$

$[-4, 10] \times [0, 18]$

27. 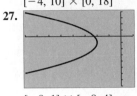 Focus: $(-3, -1)$; directrix: $x = -1$

$[-8, 1] \times [-8, 4]$

29. $y = \frac{1}{4}(x - 2)^2 - 1$; vertex: $(2, -1)$; focus: $(2, 0)$; directrix: $y = -2$

31. $x = \frac{1}{8}(y - 2)^2 - 5$; vertex: $(-5, 2)$; focus: $(-3, 2)$; directrix: $x = -7$

33. $y = \frac{3}{8}(x - 2)^2 - 2$; vertex: $(2, -2)$; focus: $\left(2, -\frac{4}{3}\right)$; directrix: $y = -\frac{8}{3}$

35. $y = \frac{1}{8}(x - 3)^2 - 2$; vertex: $(3, -2)$; focus: $(3, 0)$; directrix: $y = -4$ **37.** $y = \frac{1}{4}(x - 1)^2 - 4$

39. $y = -\frac{1}{12}(x - 2)^2 + 4$ **41.** $y = -\frac{1}{8}(x - 2)^2 + 3$ **43.** $y = -\frac{1}{8}x^2 + 2$ **45.** 6 in.

49. $x = -\frac{1}{4}(y - 7)^2 - 3$; $x = \frac{1}{4}(y - 7)^2 - 5$

SECTION 6.2

5. 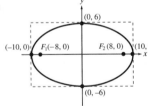   **7.** 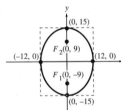   **9.**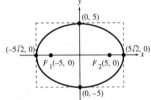

11. 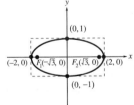   **13.** 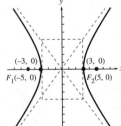   **15.**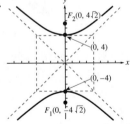

17.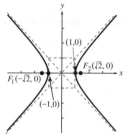

19.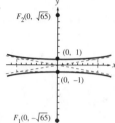

21.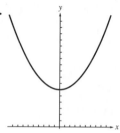

23.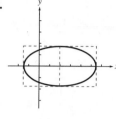

25.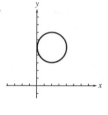

27.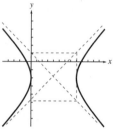

29.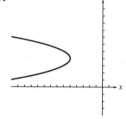

31.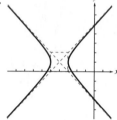

33. Vertices: $(2 \pm 2\sqrt{3}, 0)$; foci: $(-1, 0)$, $(5, 0)$

35. Vertices: $(0, -2)$, $(8, -2)$; foci: $(-1, -2)$, $(9, -2)$; asymptote: $y = \pm\frac{3}{4}(x - 4) - 2$

37. $\dfrac{x^2}{36} + \dfrac{y^2}{4} = 1$ **39.** $\dfrac{x^2}{4} - \dfrac{y^2}{5} = 1$ **41.** $\dfrac{(x + 2)^2}{9} + \dfrac{(y - 2)^2}{4} = 1$ **43.** $\dfrac{(y - 1)^2}{4} - \dfrac{(x + 3)^2}{5} = 1$

45. $\dfrac{x^2}{72} + \dfrac{y^2}{8} = 1$ **47.** $(2, 3)$, $\left(-\frac{11}{5}, -\frac{6}{5}\right)$ **49.** $(2, \sqrt{3})$, $(2, -\sqrt{3})$, $(-2, \sqrt{3})$, $(-2, -\sqrt{3})$

51. $(2, 0)$, $(5, \sqrt{3})$, $(5, -\sqrt{3})$ **53.** 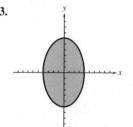   **55.** 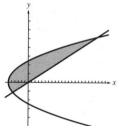   **57.**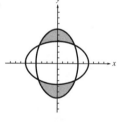

59. 151.7 km   **61.** $\dfrac{20}{\sqrt{3}} \approx 11.5$ ft

SECTION 6.3

1. $\dfrac{x^2}{16} + \dfrac{y^2}{4} = 1$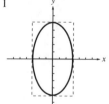

3. $\dfrac{x^2}{9} + \dfrac{y^2}{25} = 1$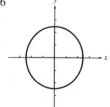

5. $x^2 + y^2 = 6$

7. $y = -\frac{1}{2}(x - 2)^2 + 4$

9. $y = 2 \pm \sqrt{x - 3}$

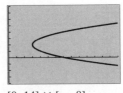

$[0, 14] \times [-, 8]$

11. $y = \pm \sqrt{x^2 - 9}$

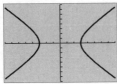

$[-8, 8] \times [-8, 8]$

13. $y = \pm\frac{1}{3}\sqrt{36 - 3x^2}$

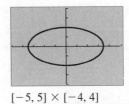

$[-5, 5] \times [-4, 4]$

15. $y = \pm \sqrt{4 - x^2}$

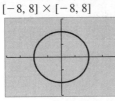

$[-4, 4] \times [-3, 3]$

17. $x = (t - 2)^2 + 3,\ y = t$

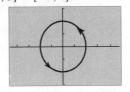

$[-4.5, 4.5] \times [-3, 3]$

19. $x = 3 \cosh t,\ y = 3 \sinh t$

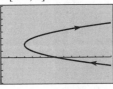

$[0, 14] \times [-4, 8]$

21. $x = 2\sqrt{3} \cos t,\ y = 2 \sin t,\ 0 \le t \le 2\pi$

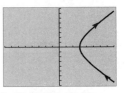

$[-8, 8] \times [-8, 8]$

23. $x = 2 \cos t,\ y = 2 \sin t,\ 0 \le t \le 2\pi$

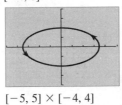

$[-5, 5] \times [-4, 4]$

25. $\dfrac{x^2}{4} + \dfrac{(y + 3)^2}{1} = 1$

27. $(x + 4)^2 + (y - 2)^2 = 1$

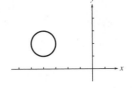

29. $\dfrac{(y - 2)^2}{11} - \dfrac{(x - 1)^2}{11} = 1$

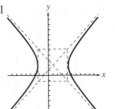

31. Empty graph

33. $y^2 - x^2 = 1$

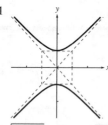

35. $y = -x \pm \sqrt{x}$; parabola

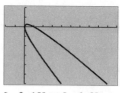

$[-2, 10] \times [-6, 2]$

37. $y = \frac{1}{2}x \pm 2\sqrt{4 - x^2}$; ellipse

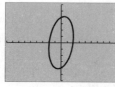

$[-9, 9] \times [-6, 6]$

39. $y = x \pm 2\sqrt{x^2 - 1}$; hyperbola

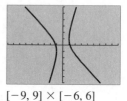

$[-9, 9] \times [-6, 6]$

41. $y = -\frac{1}{2}x$, $y = x - 1$; degenerate conic

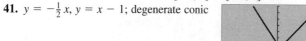

$[-9, 9] \times [-6, 6]$

43.

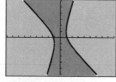

$[-9, 9] \times [-6, 6]$

45.

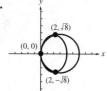

47. $x = -3 \cosh t$, $y = 3 \sinh t$

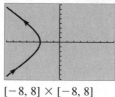

$[-8, 8] \times [-8, 8]$

49. (a) Parabola   **(b)** Circle   **(c)** Hyperbola   **(d)** Ellipse

51. (a) $\frac{5}{4} = 1.20$ **(b)** $\frac{\sqrt{35}}{6} \approx 0.99$ **(c)** $\frac{\sqrt{2}}{2} \approx 0.71$ **(d)** $\sqrt{2} \approx 1.41$

53. Maximum distance is 10,673.6 million km; minimum distance is 150.8 million km   **59.** $2xy + 1 = 0$

MISCELLANEOUS EXERCISES

1. $y = \frac{1}{10}x^2$ **3.** $x = \frac{1}{12}y^2$ **5.** $y = \frac{1}{12}(x - 3)^2 - 2$; focus: $(3, 1)$; directrix: $y = -5$
7. $x = -\frac{2}{3}(y + 2)^2$; focus: $(-\frac{3}{8}, -2)$; directrix: $x = \frac{3}{8}$ **9.** $x = \frac{1}{6}(y - 1)^2 - \frac{9}{2}$
11. $y = -\frac{1}{8}(x - 1)^2 + 4$, $y = \frac{1}{8}(x - 1)^2$ **13.** 4 ft
15.

17.

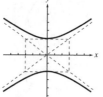

19.

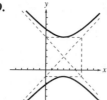

21.

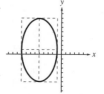

23. Vertices: $(\pm\sqrt{11}, 4)$, $(0, 10)(0, -2)$; foci: $(0, 9)(0, -1)$ **25.** Vertices: $(2, -4)$, $(2, 0)$; foci: $(2, -5)$, $(2, 1)$

27. $\dfrac{(x-1)^2}{5} + \dfrac{(y-3)^2}{9} = 1$ **29.** $\dfrac{x^2}{8} - \dfrac{y^2}{1} = 1$

31. 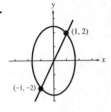   **33.** 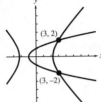   **35.** 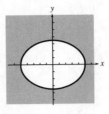   **37.** Empty graph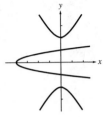

39. $y = -\frac{1}{4}(x-1)^2$ 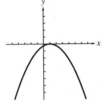   **41.** $\dfrac{(x-5)^2}{4} + \dfrac{y^2}{1} = 1$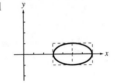

43. $\dfrac{(x-6)^2}{4} - \dfrac{y^2}{4} = 1$ 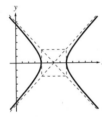   **45.** $\dfrac{(x-3)^2}{4} - \dfrac{(y+3)^2}{1} = 1$

47. $y = \frac{1}{12}(x+4)^2$ 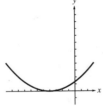   **49.** $\dfrac{x^2}{16} + \dfrac{(y+3)^2}{4} = 1$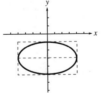

51. $\dfrac{x^2}{16} - \dfrac{(y+1)^2}{16} = 1$ 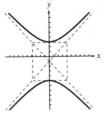   **53.** $y = \pm\frac{1}{2}\sqrt{16 - (x+4)^2}$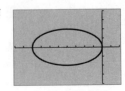

$[-10, 2] \times [-4, 4]$

55. Parabola; $y = 2x \pm \sqrt{x}$

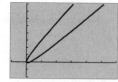

$[-1, 6] \times [-2, 8]$

57. Ellipse; $y = -\frac{1}{2}x \pm \sqrt{16 - x^2}$

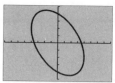

$[-8, 8] \times [-5, 5]$

59. $x = 3 \cosh t, y = 2 \sinh t$

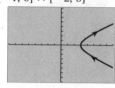

$[-8, 8] \times [-8, 8]$

61. $x = 7 \cos t, y = 5 \sin t, 0 \le t \le 2\pi$

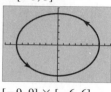

$[-9, 9] \times [-6, 6]$

63. Hyperbola with rectangular equation $x^2 - y^2 = 4$

CHAPTER TEST

1. The parabola opens to the left, so the standard equation is in the form

$$x = \frac{1}{4p}(y - k)^2 + h$$

The vertex is $V(1, 3)$ and $p = -2$. The equation is $x = -\frac{1}{8}(y - 3)^2 + 1$.

(see Section 6.1, Example 6)

2. Solving the equation $x^2 = 10y$ for y gives $y = \frac{1}{10}x^2$. From

$$\frac{1}{4p} = \frac{1}{10}$$

we get $p = \frac{5}{2}$.

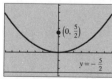

$[-7, 7] \times [-3, 6]$
(see Section 6.1, Example 2)

3. We write this equation in standard form:

$$8x = y^2 - 8y - 8$$
$$x = \tfrac{1}{8}(y^2 - 8y + \mathbf{16}) - 1 - \mathbf{2}$$
$$x = \tfrac{1}{8}(y - 4) - 3$$

The vertex is $(-3, 4)$. Since $p = 2$, the focus is $(-1, 4)$.
(see Section 6.1, Example 5)

4. Comparing

$$\frac{x^2}{36} + \frac{y^2}{20} = 1 \quad \text{with} \quad \frac{x^2}{a^2} + \frac{y^2}{b^2} = 1$$

tells us that a is 6 and b is $2\sqrt{5}$. Because

$$c = \sqrt{a^2 - b^2} = \sqrt{36 - 20} = 4$$

the foci are $(-4, 0)$ and $(4, 0)$.

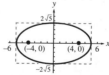

(see Section 6.2, Example 2)

5. The standard form is

$$\frac{(x - 1)^2}{4} + \frac{(y + 2)^2}{9} = 1$$

The center is $(1, -2)$, $a = 2$, and $b = 3$.

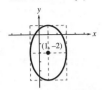

(see Section 6.2, Example 6)

6. We rewrite the equation in standard form:

$$4x^2 - 9y^2 - 24x - 18y + 63 = 0$$
$$4(x^2 - 6x + 9) - 9(y^2 + 2y + 1) = -63 + 36 - 9$$
$$4(x - 3)^2 - 9(y + 1)^2 = -36$$
$$\frac{4(x - 3)^2}{-36} - \frac{9(y + 1)^2}{-36} = 1$$
$$\frac{(y + 1)^2}{4} - \frac{(x - 3)^2}{9} = 1$$

The graph is a hyperbola with center $(3, -1)$. Also, $a = 3$ and $b = 2$.

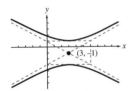

(see Section 6.3, Example 1)

7. Because the hyperbola has vertices $(0, -2)$ and $(0, 2)$, it follows that the center is the origin, and that $a = 2$:

$$\frac{y^2}{4} - \frac{x^2}{b^2} = 1$$

Since the hyperbola passes through the point $(3, 4)$, we can let $x = 3$ and $y = 4$ in the equation and solve to determine b^2:

$$\frac{(4)^2}{4} - \frac{(3)^2}{b^2} = 1$$
$$4 - \frac{9}{b^2} = 1$$
$$\frac{9}{b^2} = 3$$
$$9 = 3b^2$$
$$b^2 = 3$$

The equation is

$$\frac{y^2}{4} - \frac{x^2}{3} = 1$$

(see Section 6.2, Example 5)

8. The equation of the ellipse, which is shown on a coordinate plane below, is of the form

$$\frac{x^2}{26^2} + \frac{y^2}{b^2} = 1$$

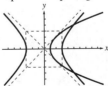

Letting $x = 24$ and $y = 5$, we get

$$\frac{(24)^2}{26^2} + \frac{(5)^2}{b^2} = 1$$

which can be solved to show that $b^2 = 169$. Thus, the equation is

$$\frac{x^2}{676} + \frac{y^2}{169} = 1$$

(see Section 6.2, Example 8)

9. The standard form of $x^2 - y^2 = 9$ is

$$\frac{x^2}{9} - \frac{y^2}{9} = 1$$

The graph of this equation is a hyperbola centered at the origin, with $a = 3$ and $b = 3$. The standard form of the graph of $4(x - 1) = y^2$ is $x = \frac{1}{4}y^2 + 1$. The graph of this equation is a parabola opening to the right with vertex $(1, 0)$.

This system appears to have two solutions. The system is most easily solved by substitution:

$$x^2 - y^2 = 9$$
$$x^2 - [4(x - 1)] = 9$$
$$x^2 - 4x - 5 = 0$$
$$(x - 5)(x + 1) = 0$$
$$x - 5 = 0 \quad | \quad x + 1 = 0$$
$$x = 5 \quad | \quad x = -1$$

If $x = 5$, then

$$(5)^2 - y^2 = 9$$
$$y^2 = 16$$
$$y = \pm 4$$

There are no solutions to the system when $x = -1$.

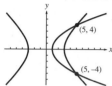

The points of intersection are $(5, -4)$ and $(5, 4)$.
(see Section 6.2, Exercises 47–52)

10. We start by solving the equation for y:

$$7x^2 + 4y^2 = 60$$
$$4y^2 = 60 - 7x^2$$
$$y^2 = \tfrac{1}{4}(60 - 7x^2)$$
$$y = \pm\tfrac{1}{2}\sqrt{60 - 7x^2}$$

The graph of $7x^2 + 4y^2 = 60$ is determined by graphing both $y = -\tfrac{1}{2}\sqrt{60 - 7x^2}$ and $y = \tfrac{1}{2}\sqrt{60 - 7x^2}$:

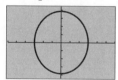

$[-6, 6] \times [-4.5, 4.5]$
(see Section 6.3, Example 3)

11. The equation is quadratic in y:

$$3y^2 + (3x)y + (x^2 + 2x) = 0$$

Using the quadratic formula, we get

$$y = \frac{-(3x) \pm \sqrt{(3x)^2 - 4(3)(x^2 + 2x)}}{2(3)}$$
$$y = \frac{-3x \pm \sqrt{-3x^2 - 24x}}{6}$$

The graphs of these two equations are shown below. The graph is an ellipse.

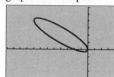

$[-12, 4] \times [-4, 6]$
(see Section 6.3, Example 4)

12. In standard form, the equation becomes

$$\frac{x^2}{36} - \frac{y^2}{4} = 1$$

Thus, $a = 6$ and $b = 2$. By Table 6.1, we get

$$x = 6 \cosh t, \quad y = 2 \sinh t$$

(see Section 6.3, Example 5)

CHAPTER 7: SECTION 7.1

1. 1, 4, 7, 10 **3.** $0, \frac{1}{4}, \frac{1}{3}, \frac{3}{8}$ **5.** 19, 16, 11, 4 **7.** $0, -3, 8, -15$ **9.** $-5, 5, -5, 5$ **11.** $1, -\frac{1}{2}, \frac{1}{4}, -\frac{1}{8}$
13. 3, 0, 3, 0 **15.** 5, 10, 20, 40 **17.** $\frac{1}{2}, 4, \frac{1}{16}, 256$ **19.** 1, 1, 3, 7 **21.** $a_n = 7n;\ 7, 14, 21, \ldots, 7n, \ldots$
23. $a_n = 4n - 1;\ 3, 7, 11, \ldots, 4n - 1, \ldots$ **25.** $a_n = (-1)^{n+1}(2^n);\ 2, -4, 8, -16, \ldots, (-1)^{n+1}(2^n), \ldots$
27. $a_n = 2n - 2;\ 0, 2, 4, \ldots, 2n - 2, \ldots$ **29.** $a_n = \dfrac{n^3}{n+1};\ \dfrac{1}{2}, \dfrac{8}{3}, \dfrac{27}{4}, \ldots, \dfrac{n^3}{n+1}, \ldots$

31. $a_n = \dfrac{(-1)^n}{n(n+1)};\ \dfrac{-1}{1 \cdot 2}, \dfrac{1}{2 \cdot 3}, \dfrac{-1}{3 \cdot 4}, \ldots, \dfrac{(-1)^n}{n(n+1)}, \ldots$

33.
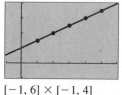

$[-1, 6] \times [-1, 4]$

35.
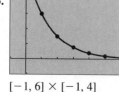

$[-1, 6] \times [-1, 4]$

37.
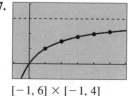

$[-1, 6] \times [-1, 4]$

39. 0 **41.** 3 **43.** 31

45. 55   **47.** 46   **49.** $1 + 2x + 3x^2 + 4x^3 + 5x^4 + 6x^5$ **51.** $2x, 4x, 6x, 8x$ **53.** $P, P(1 + r), P(1 + r)^2, P(1 + r)^3$
55. (a) 1004.17, 1008.35, 1012.55, 1016.77   **(b)** \$1104.94.
57. (a) 14 additional connections; $a_5 = 14$ **(b)** 35   **(c)** $a_n = \begin{cases} 0 & \text{for } n = 1 \\ a_{n-1} + n & \text{for } n = 2, 3, 4, \ldots \end{cases}$ or $a_n = \dfrac{(n-3)n}{2}$
59. (a) 66.817, 67.465, 69.465, 72.529, 76.369, 80.697   **(b)** \$98,329   **61.** 3, 2.5, 2.45, 2.449 489 796   **63.** 1, 1, 2, 3, 5, 8

SECTION 7.2

1. $1 + 4 + 7 + 10 + 13 = 35$ **3.** $-1 + 0 + 3 + 8 + 15 + 24 = 49$ **5.** $1 + \frac{1}{3} + \frac{1}{9} + \frac{1}{27} = \frac{40}{27}$
7. $4 + 4 + 4 + 4 + 4 + 4 = 24$ **9.** $1 - 1 + 2 - 6 + 24 = 20$ **11.** $12 + 11 + 10 + 9 + 8 + 7 + 6 + 5 = 68$

13. 275   **15.** 180   **17.** 60   **19.** $\displaystyle\sum_{k=1}^{6} (2k - 1)$ **21.** $\displaystyle\sum_{k=1}^{15} \frac{1}{2k}$ **23.** $\displaystyle\sum_{k=1}^{8} (6k - 5)$

25. $\displaystyle\sum_{k=1}^{20} (-1)^k (3k)$ **27.** $\displaystyle\sum_{k=1}^{20} (-1)^{k+1}\left(\frac{2k-1}{2k}\right)$ **29.** $\displaystyle\sum_{k=1}^{n} 100(1.1)^{k-1}$

31. (a) $P_4(x) = 1 + x + \dfrac{x^2}{2} + \dfrac{x^3}{6} + \dfrac{x^4}{24}$ **(b)** $e^{0.5} = 1.649; P_4(0.5) = 1.648$ **(c)**

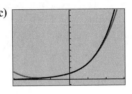

33. (a) $f_2(x) = 2\left(\dfrac{x-1}{x+1}\right) + \dfrac{2}{3}\left(\dfrac{x-1}{x+1}\right)^3$ **(b)**

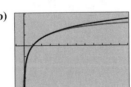

35. 900   **37.** 19,375   **39.** 1120

41. 14   **43. (a)** $a_k = 8\sqrt{2}(2k - 1)$ **(b)** After n sec it slides $8\sqrt{2}\,n^2$ ft.   **45.** \$2432.3 billion   **47.** \$3474.66 billion
49. (a) $y = -1.813x + 8.978$ **(b)**

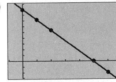

$[-1, 7] \times [-3, 10]$

(c) -5.526 **51.** $\bar{x} = 15; s \approx 2.160$

53. $1, \frac{3}{2}, \frac{7}{4}, \frac{15}{8}$ **55. (a)** 4, 11, 23, 42, 70   **(b)** $S_n = \dfrac{2n^3 + 3n^2 + 19n}{6}$ **(c)** $4, 11, 23, \ldots, \dfrac{2n^3 + 3n^2 + 19n}{6}, \ldots$ **57.** $\frac{2}{3}$

SECTION 7.3

1. Yes; $d = 6, a_6 = 35$ **3.** Yes; $d = 0.2, a_6 = 3.1$ **5.** Yes; $d = -5, a_6 = -6$ **7.** Not arithmetic
9. Yes; $d = 0, a_6 = 4$ **11.** Not arithmetic   **13.** Not arithmetic   **15.** 21   **17.** 30   **19.** 22
21.

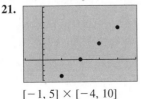

$[-1, 5] \times [-4, 10]$

23.

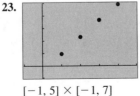

$[-1, 5] \times [-1, 7]$

25.

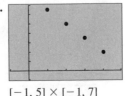

$[-1, 5] \times [-1, 7]$

27. $a_n = 4 + (n - 1)3 = 3n + 1$; 4, 7, 10, ... , $3n + 1$, ...
29. $a_n = 33 + (n - 1)(-4) = -4n + 37$; 33, 29, 25, ... , $-4n + 37$, ...
31. $a_n = 36 + (n - 1)\left(-\frac{4}{3}\right) = -\frac{4}{3}n + \frac{112}{3}$; 36, $34\frac{2}{3}$, $33\frac{1}{3}$, ... , $-\frac{4}{3}n + \frac{112}{3}$, ...
33. $a_n = -12 + (n - 1)8 = 8n - 20$; $-12, -4, 4$, ... , $8n - 20$, ...
35. $a_n = 27 + (n - 1)\left(-\frac{5}{4}\right) = -\frac{5}{4}n + \frac{113}{4}$; 27, $\frac{103}{4}$, $\frac{49}{2}$, ... , $-\frac{5}{4}n + \frac{113}{4}$, ...
37. $a_n = -\frac{7}{3}n + \frac{97}{3}$; 30, $\frac{83}{3}$, $\frac{76}{3}$, ... , $-\frac{7}{3}n + \frac{97}{3}$, ... **39.** $a_n = 3n - 4$; $-1, 2, 5$, ... , $3n - 4$, ...
41. $a_n = (3 - 2n)x + 2n - 1$; $x + 1, -x + 3$, ... , $(3 - 2n)x + 2n - 1$, ... **43.** 1365 **45.** -2140 **47.** 622.5
49. 590 **51.** 22,650 **53.** 900 **55.** (a) $a_1 = 69.6$; $d = 0.2$ (b) 73.4 yr (c) 77.4 yr
57. (a) $a_1 = 1.00$; $d = 0.113$ (b) \$2.70 (c) \$3.71 **59.** 1870 **61.** (a) \$38,750 (b) \$210,750
63. $a_1 = \frac{45}{2}$; $d = -1$

SECTION 7.4

1. Yes; $r = 2$, $a_6 = 160$ **3.** Yes; $r = \frac{3}{2}$, $a_6 = \frac{243}{2}$ **5.** Yes; $r = -3$, $a_6 = -54$ **7.** Not geometric
9. Yes; $r = 1$, $a_6 = 5$ **11.** Yes; $r = 1 + \sqrt{3}$, $a_6 = 132 + 76\sqrt{3}$

13.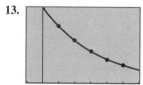

$[-1, 6] \times [0, 1]$

15.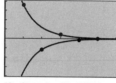

$[0, 6] \times [-20, 20]_5$

17.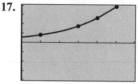

$[0, 6] \times [-20, 20]_5$

19. $a_n = 18\left(\frac{1}{3}\right)^{n-1}$; 18, 6, 2, ... , $18\left(\frac{1}{3}\right)^{n-1}$, ...
21. $a_n = 16\left(-\frac{3}{4}\right)^{n-1}$; 16, -12, 9, ... , $16\left(-\frac{3}{4}\right)^{n-1}$, ... **23.** $a_n = 200\left(\frac{11}{10}\right)^{n-1}$; 200, 220, 242, ... , $200\left(\frac{11}{10}\right)^{n-1}$, ...
25. $a_n = 8\left(-\frac{5}{4}\right)^{n-1}$; 8, -10, $\frac{25}{2}$, ... , $8\left(-\frac{5}{4}\right)^{n-1}$, ... **27.** $a_n = 49\left(\frac{3}{7}\right)^{n-1}$; 49, 21, 9, ... , $49\left(\frac{3}{7}\right)^{n-1}$, ...
29. $a_n = \frac{2}{3}(-3)^{n-1}$; $\frac{2}{3}$, -2, 6, ... , $\frac{2}{3}(-3)^{n-1}$, ...
31. $a_n = -9\left(-\frac{1}{3}\right)^{n-1} = -(-3)^{3-n}$; -9, 3, -1, ... , $-(-3)^{3-n}$, ...
33. $a_n = \frac{1}{48}(2)^{n-1} = \frac{1}{3}(2)^{n-5}$; $\frac{1}{48}$, $\frac{1}{24}$, $\frac{1}{12}$, ... , $\frac{1}{3}(2)^{n-5}$, ... **35.** $a_n = (-2x)^{n-1}$, 1, $-2x$, $4x^2$, ... , $(-2x)^{n-1}$, ...
37. $a_n = 2^{n-1} \ln 2$; $\ln 2$, $\ln 4$, $\ln 16$, ... , $2^{n-1} \ln 2$, ... **39.** 29,524 **41.** $\frac{935}{32}$
43. $2000(1.12^{20} - 1) \approx 11{,}454.9998987$
45. $\frac{118{,}096}{2187}$ **47.** 2046 **49.** $\frac{69{,}753{,}294}{390{,}625}$ **51.** $\frac{25{,}918{,}366}{9213} \approx 2813.2384674$ **53.** 33,081
55. (a) \$36,000, \$37,440, \$38,937.60, \$40,495.10, \$42,114.91 (b) No (c) \$55,420.35
57. (a) $a_1 = 137.3$; $r = 1.017$
(b)

Height	5'4"	5'5"	5'6"	5'7"	5'8"	5'9"	5'10"	5'11"	6'0"
Recommended weight	137.3	139.6	142.0	144.4	146.9	149.4	151.9	154.5	157.1

59. (a) \$3214.90 (b) \$25,129.02 **61.** 48 **63.** $\frac{8}{3}$ **65.** Does not have a sum **67.** Does not have a sum **69.** $\frac{160}{3}$
71. Does not have a sum **73.** 2 **75.** $\frac{5}{11}$ **77.** $\frac{37}{9}$ **79.** $\frac{28}{15}$
81. (a) 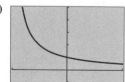 (b) 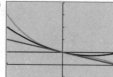 **83.** $\frac{2384}{51}$ ft **87.** $L_n = \sum_{k=1}^{n} \frac{32}{21}\left(\frac{19}{21}\right)^{k-1}$

MISCELLANEOUS EXERCISES

1. $0, 1, 4, 10$ **3.** $-3, 9, 1, 13$ **5.** $x + 1, (x + 1)^2, (x + 1)^3, (x + 1)^4$ **7.** $5, 9, 15, 23$

9. $a_n = (-1)^{n-1}(2n + 3); 5, -7, 9, \ldots, (-1)^{n-1}(2n + 3), \ldots$ **11.** $a_n = \dfrac{2n - 1}{(2n)^2}; \dfrac{1}{4}, \dfrac{3}{16}, \dfrac{5}{36}, \ldots, \dfrac{2n - 1}{(2n)^2}, \ldots$

13.

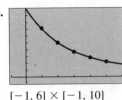

$[-1, 6] \times [-1, 10]$

15.

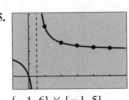

$[-1, 6] \times [-1, 5]$

17. 1 **19.** 2 **21.** $\dfrac{1}{1} + \dfrac{8}{2} + \dfrac{27}{6} + \dfrac{64}{24} = \dfrac{73}{6}$

23. $1 + 0 - 1 + 0 + 7 = 7$ **25.** $\dfrac{x}{3} - \dfrac{x^2}{5} + \dfrac{x^3}{7} - \dfrac{x^4}{9} + \dfrac{x^5}{11}$ **27.** $\displaystyle\sum_{k=1}^{6} 6k - 5$ **29.** $\displaystyle\sum_{k=1}^{6} 48\left(\dfrac{1}{2}\right)^{k-1}$

31. $\displaystyle\sum_{k=1}^{6} k^2 - 1$ **33. (a)** $x - \dfrac{x^2}{2} + \dfrac{x^3}{3}$ **(b)** $\ln 1.5 = 0.405; P_3(0.5) = 0.417$ **(c)**

35. $18{,}370$ **37.** Geometric; $r = -\dfrac{3}{5}; -25, 15, -9, \ldots, -25\left(-\dfrac{3}{5}\right)^{n-1}, \ldots$

39. Arithmetic; $d = -\dfrac{2}{3}; 5, \dfrac{13}{3}, \dfrac{11}{3}, \ldots, 5 + (n - 1)\left(-\dfrac{2}{3}\right), \ldots$

41. Geometric; $r = \sqrt{5}; 2, 2\sqrt{5}, 10, \ldots, 2(\sqrt{5})^{n-1}, \ldots$

43. Arithmetic; $d = 3\pi; 4, 4 + 3\pi, 4 + 6\pi, \ldots, 4 + (n - 1)3\pi, \ldots$

45. Geometric; $r = z^2; z^b, z^{b+2}, z^{b+4}, \ldots, z^b(z^2)^{n-1}, \ldots$

47. $\dfrac{37}{3}$ **49.** $18\sqrt[3]{9}$ **51.** 830 **53.** $\dfrac{292{,}968}{625}$ **55.** $\dfrac{56}{3}$ **57.** 20 **59.** Does not have a sum.   **61.** $15{,}150 \ln x$

63. 0 **65. (a)** $0.5000, 0.9778$ **(b)** $\dfrac{1}{2}$

CHAPTER TEST

1. The first term is $a_1 = 12$. To find the second term, let

$n = 2$ in $a_n = \dfrac{a_{n-1}}{2} + (-1)^n$.

$a_2 = \dfrac{a_1}{2} + (-1)^2 = \dfrac{12}{2} + 1 = 7$

Similarly, the third and fourth terms are determined by letting $n = 3$ and then letting $n = 4$ in

$a_n = \dfrac{a_{n-1}}{2} + (-1)^n$:

$a_3 = \dfrac{a_2}{2} + (-1)^3 = \dfrac{7}{2} - 1 = \dfrac{5}{2}$

$a_4 = \dfrac{a_3}{2} + (-1)^4 = \dfrac{\frac{5}{2}}{2} + 1 = \dfrac{9}{4}$

The first four terms are $12, 7, \dfrac{5}{2}, \dfrac{9}{4}$.
(see Example 2, Section 7.1)

2. $\displaystyle\sum_{k=1}^{4} \dfrac{k!}{(2k)^2} = \dfrac{1!}{(2 \cdot 1)^2} + \dfrac{2!}{(2 \cdot 2)^2} + \dfrac{3!}{(2 \cdot 3)^2} + \dfrac{4!}{(2 \cdot 4)^2}$

$= \dfrac{1}{4} + \dfrac{2}{16} + \dfrac{6}{36} + \dfrac{24}{64} = \dfrac{11}{12}$

(see Example 1, Section 7.2)

3. Referring to Table 7.1 in Section 7.1, we see the terms 1, 3, 5, $\ldots$, 13 are given by $a_k = 2k - 1$. We can make the sign alternate with the factor $(-1)^{k+1}$. Thus, the sum can be expressed as $\displaystyle\sum_{k=1}^{7}(-1)^{k-1}(2k - 1)$.

(see Example 2, Section 7.2)

4. The difference of consecutive terms is always 4, so the sequence is arithmetic with common difference $d = 4$. Using the formula $a_n = a_1 + (n - 1)d$, we get
$a_{24} = 3 + (24 - 1)4 = 95$.
(see Example 2, Section 7.3)

5. The ratio of consecutive terms is always $\frac{3}{2}$, so the sequence is geometric with common ratio $r = \frac{3}{2}$. Using the formula $a_n = a_1 r^{n-1}$, we get $a_{10} = \frac{8}{9}(\frac{3}{2})^{10-1} = \frac{2187}{64}$.
(see Example 2, Section 7.4)

6. Since the fourth term is 6, we have $6 = a_4 = a_1 r^3$. Since the sixth term is 72, we have $72 = a_6 = a_1 r^5$. Given that r is negative, the solution of the system

$$\begin{cases} 6 = a_1 r^3 \\ 72 = a_1 r^5 \end{cases}$$

is $r = -2\sqrt{3}$ and $a_1 = -\sqrt{3}/12$. Therefore,

$$a_7 = a_1 r^{n-1} = -\frac{\sqrt{3}}{12}(-2\sqrt{3})^6 = -144\sqrt{3}$$

(see Example 3, Section 7.4)

7. The common ratio is $r = -\frac{1}{4}$. Since $-1 < r < 1$, we use the formula $S = a_1/(1 - r)$:

$$S = \frac{40}{1 - \left(-\frac{1}{4}\right)} = 32$$

(see Example 7, Section 7.4)

8. $\displaystyle\sum_{k=1}^{40}(a_k - 2b_k + 3) = \sum_{k=1}^{40} a_k - 2\sum_{k=1}^{40} b_k + \sum_{k=1}^{40} 3$
$$= 75 - 2(45) + 40(3) = 105$$

(see Example 4, Section 7.2)

9. Using the formula

$$S_n = \frac{n}{2}(a_1 + a_n)$$

we have

$$S_{10} = \frac{10}{2}(a_1 + a_{10}) = \frac{10}{2}\left(15 + \frac{7}{2}\right) = \frac{185}{2}$$

(see Example 8, Section 7.3)

10. $0.\overline{42} = 0.42 + 0.0042 + 0.000042 + \cdots$
This is an infinite geometric series with $r = \frac{1}{100}$ and $a_1 = \frac{42}{100}$. The sum is

$$S = \frac{a_1}{1 - r} = \frac{\frac{42}{100}}{1 - \frac{1}{100}} = \frac{\frac{42}{100}}{\frac{99}{100}} = \frac{14}{33}$$

(see Example 9, Section 7.4)

11. The first three terms of an arithmetic sequence with common difference $\frac{3}{2}$ can be expressed as $a_1, a_1 + \frac{3}{2}, a_1 + 3$. By the Pythagorean theorem,

$$a_1^2 + \left(a_1 + \frac{3}{2}\right)^2 = (a_1 + 3)^2$$

Expanding and solving, we get

$$a_1^2 + a_1^2 + 3a_1 + \frac{9}{4} = a_1^2 + 6a_1 + 9$$
$$a_1^2 - 3a_1 - \frac{27}{4} = 0$$
$$4a_1^2 - 12a_1 - 27 = 0$$
$$(2a_1 - 9)(2a_1 + 3) = 0$$

$$2a_1 - 9 = 0 \quad \bigg| \quad 2a_1 + 3 = 0$$
$$a_1 = \frac{9}{2} \quad \bigg| \quad a_1 = -\frac{3}{2}$$

Since the length of the side of a triangle must be positive, $a_1 = \frac{9}{2}$. The sides of the triangle are $\frac{9}{2}, 6, \frac{15}{2}$.
(see the nth term of an arithmetic sequence, Section 7.3)

12. After the ball falls 12 ft, it bounces 12(0.75) ft and falls 12(0.75) ft. Then it bounces [12(0.75)](0.75) ft and falls the same distance, and so on. The total distance is

$$12 + 24(0.75) + 24(0.75)^2 + \cdots = 12 + \sum_{k=1}^{\infty} 24(0.75)^k$$

The summation is an infinite geometric series with $a_1 = 24(0.75)$ and $r = 0.75$. Thus, its sum is given by

$$\sum_{k=1}^{\infty} 24(0.75)^k = \frac{a_1}{1 - r} = \frac{24(0.75)}{1 - 0.75} = 72$$

Therefore,

$$12 + \sum_{k=1}^{\infty} 24(0.75)^k = 12 + 72 = 84$$

The ball travels 84 ft.
(see Example 7, Section 7.4)

13. The first five terms of the sequence are

$$a_1 = \frac{1^2 - 1}{8} = 0, \quad a_2 = \frac{2^2 - 1}{8} = \frac{3}{8},$$
$$a_3 = \frac{3^2 - 1}{8} = 1, \quad a_4 = \frac{4^2 - 1}{8} = \frac{15}{8}, \quad a_5 = \frac{5^2 - 1}{8} = 3$$

Plotting the points $(1, 0)$, $(2, \frac{3}{8})$, $(3, 1)$, $(4, \frac{15}{8})$, and $(5, 3)$, we get the graph shown below.

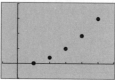

$[-1, 6] \times [-1, 4]$
(see Example 4, Section 7.1)

14. $1 + x + x^2 + x^3 + \cdots + x^{100} = \dfrac{a_1(1 - r^n)}{1 - r} = \dfrac{1(1 - x^{101})}{1 - x}$

Thus, the given equation is equivalent to

$$\dfrac{1 - x^{101}}{1 - x} = 2$$

The graph of

$$y = \dfrac{1 - x^{101}}{1 - x} - 2 \quad \text{for } x > 0$$

is shown at the top of the next column.

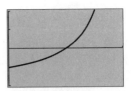

$[0, 1] \times [-2, 2]$

Using the zoom and trace features on a graphing utility, the x-intercept (to the nearest 0.0001) is (0.5000,0). The solution to the equation is $x = 0.5$.
(see the sum of a geometric sequence, Section 7.4)

CHAPTER 8: SECTION 8.1

1. $a^5 + 5a^4h + 10a^3h^2 + 10a^2h^3 + 5ah^4 + h^5$ **3.** $x^4 + 8x^3 + 24x^2 + 32x + 16$
5. $x^5 - 15x^4 + 90x^3 - 270x^2 + 405x - 243$ **7.** $x^6 + 6x^5y + 15x^4y^2 + 20x^3y^3 + 15x^2y^4 + 6xy^5 + y^6$ **9.** 56   **11.** 330
13. 1   **15.** 91   **17.** $32x^5 + 80x^4y + 80x^3y^2 + 40x^2y^3 + 10xy^4 + y^5$
19. $64x^6 - 576x^5y + 2160x^4y^2 - 4320x^3y^3 + 4860x^2y^4 - 2916xy^5 + 729y^6$
21. 1 8 28 56 70 56 28 8 1; $x^8 + 8x^7y + 28x^6y^2 + 56x^5y^3 + 70x^4y^4 + 56x^3y^5 + 28x^2y^6 + 8xy^7 + y^8$
23. 1 9 36 84 126 126 84 36 9 1; $512x^9 - 2304x^8 + 4608x^7 - 5376x^6 + 4032x^5 - 2016x^4 + 672x^3 - 144x^2 + 18x - 1$
25. $x^5 + 5x^4 + 10x^3 + 10x^2 + 5x + 1$ **27.** $64x^6 - 192x^5y + 240x^4y^2 - 160x^3y^3 + 60x^2y^4 - 12xy^5 + y^6$
29. $81m^4 + 216m^3n + 216m^2n^2 + 96mn^3 + 16n^4$ **31.** $x^{14} + 7x^{12}y + 21x^{10}y^2 + 35x^8y^3 + 35x^6y^4 + 21x^4y^5 + 7x^2y^6 + y^7$
33. $u^{12} - 12u^{10}v^3 + 60u^8v^6 - 160u^6v^9 + 240u^4v^{12} - 192u^2v^{15} + 64v^{18}$ **35.** $\frac{1}{16}x^4 + 4x^3 + 96x^2 + 1024x + 4096$
37. $z^{15} + 5z^{11} + 10z^7 + 10z^3 + \dfrac{5}{z} + \dfrac{1}{z^5}$
39. $h^3\sqrt{h} + 14h^3k + 84h^2\sqrt{h}\,k^2 + 280h^2k^3 + 560h\sqrt{h}\,k^4 + 672hk^5 + 448\sqrt{h}\,k^6 + 128k^7$
41. $_{12}C_8x^4y^8 = 495x^4y^8$ **43.** $_{14}C_9(2m)^5(-n)^9 = -64,064m^5n^9$ **45.** $r^{30} + 20r^{27}s + 180r^{24}s^2 + 960r^{21}s^3$
47. $_{11}C_5(3x)^6(4y)^5 = 344,881,152x^6y^5$ **49.** $6x^5 + 15x^4h + 20x^3h^2 + 15x^2h^3 + 6xh^4 + h^5$ **51.** $20x^3 + 30x^2h + 20xh^2 + 5h^3$
53. $k = 1$ or 3   **55.** $k = 3$ or $n - 3$ **57.** **(a)** 0.267   **(b)** 0.650   **59.** **(a)** 0.063   **(b)** 0.091
61. $(1 - 0.1)^6 \approx 0.53$; $(0.9)^6 = 0.531441$

SECTION 8.2

1. Noon Mon., noon Tues., noon Wed., noon Thurs., noon Fri., 5 PM Mon., 5 PM Tues., 5 PM Wed., 5 PM Thurs., 5 PM Fri.
3. Morning win, morning lose, morning tie, afternoon win, afternoon lose, afternoon tie
5. {Team #1, Team #2}, {Team #1, Team #3}, {Team #1, Team #4}, {Team #1, Team #5}, {Team #2, Team #3},
   {Team #2, Team #4}, {Team #2, Team #5}, {Team #3, Team #4}, {Team #3, Team #5}, {Team #4, Team #5}
7. 156   **9.** 252   **11.** 1024
13. abc, acb, bac, bca, cab, cba, abd, adb, bad, bda, dab, dba, dbc, dcb, bdc, bcd, cdb, cbd, adc, acd, dac, dca, cad, cda
15. ab, ba, ac, ca, ad, da, ae, ea, af, fa, bc, cb, bd, db, be, eb, bf, fb, cd, dc, ce, ec, cf, fc, de, ed, df, fd, fe, ef
17. {a, b}, {a, c}, {a, d}, {a, e}, {b, c}, {b, d}, {b, e}, {c, d}, {c, e}, {d, e}   **19.** 120   **21.** 15,120   **23.** 1   **25.** 1320
27. 3,268,800   **29.** 90   **31.** 120   **33.** 16   **35.** 720   **37.** 24   **39.** 17,576,000   **41.** 78,364,164,096   **43.** 120
45. 2,598,960   **47.** 1287   **49.** **(a)** 120   **(b)** 364   **(c)** 43,680
51. **(a)** $_{100}C_{15} \approx 2.533 \times 10^{17}$ **(b)** $(_{67}C_{10})(_{33}C_5) \approx 5.886 \times 10^{16}$ **53.** 3744
55. **(a)** 1 way for all heads; 2 ways for 1 head and 1 tail; 1 way for all tails
   (b) 1 way for all heads; 3 ways for 2 heads and 1 tail; 3 ways for 1 head and 2 tails; 1 way for all tails

55. (c) 1 way for all heads; 4 ways for 3 heads and 1 tail; 6 ways for 2 heads and 2 tails; 4 ways for 1 head and 3 tails; 1 way for all tails

(d) $_{a+b}C_a$, or the $(a + 1)$st entry in row $a + b$ of Pascal's triangle

SECTION 8.3

1. Sample space:
{HHHH, HHHT, HHTH, HTHH, HHTT, HTHT, HTTH, HTTT, THHH, THHT, THTH, TTHH, THTT, TTHT, TTTH, TTTT};
Event: {HHHH, HHHT, HHTH, HTHH, HHTT, HTHT, HTTH, THHH, THHT, THTH, TTHH}

3. Sample space: {{R, Y}, {R, G}, {R, B}, {Y, G}, {Y, B}, {G, B}}; Event: {{R, B}, {Y, B}, {G, B}}   **5.** $\frac{2}{52} \approx 0.038$

7. $\frac{4}{52} \approx 0.077$ **9.** $\frac{1}{36} \approx 0.028$ **11.** $\frac{4}{36} \approx 0.111$ **13.** $\frac{9}{36} = 0.25$ **15.** $\frac{6}{36} \approx 0.167$ **17.** $\frac{6}{12} = 0.5$ **19.** $\frac{8}{12} \approx 0.667$

21. (a) $\frac{35}{136} \approx 0.257$ **(b)** $\frac{21}{136} \approx 0.154$ **(c)** $\frac{10}{17} = 0.588$ **23.** $\frac{5}{22} \approx 0.227$ **25.** $\frac{14}{33} \approx 0.424$

27. (a) $\frac{65}{87} \approx 0.747$ **(b)** $\frac{143}{145} \approx 0.986$ **29. (a)** $\frac{99}{54,145} \approx 0.002$ **(b)** $\frac{2759}{8165} \approx 0.338$ **31.** $\frac{468}{20,825} \approx 0.022$

33. (a) $\frac{1}{32} \approx 0.031$ **(b)** $\frac{5}{32} \approx 0.156$ **35.** $\frac{1}{729} \approx 0.001$ **37.** $\frac{1}{10,000} = 0.0001$ **39.** $\frac{223}{343} \approx 0.650$

41. $\frac{129}{6,991,908} \approx 0.00002$ **43. (a)** $\frac{1}{4} = 0.25$ **(b)** $\frac{1}{2} = 0.5$ **45.** $\frac{1}{9} \approx 0.111$

MISCELLANEOUS EXERCISES

1. 364   **3.** 1   **5.** 120   **7.** 1   **9.** $m^3 + 6m^2 + 12m + 8$

11. $x^6 - 12x^5y + 60x^4y^2 - 160x^3y^3 + 240x^2y^4 - 192xy^5 + 64y^6$

13. $p^9 + 9p^8q^2 + 36p^7q^4 + 84p^6q^6 + 126p^5q^8 + 126p^4q^{10} + 84p^3q^{12} + 36p^2q^{14} + 9pq^{16} + 9q^{18}$

15. $65,536f^8 - 65,536f^7 + 28,672f^6 - 7168f^5 + 1120f^4 - 112f^3 + 7f^2 - \frac{1}{4}f + \frac{1}{256}$ **17.** $_{12}C_3 p^9(2q)^3 = 1760p^9q^3$

19. $x^{10} + 30x^9y^2 + 405x^8y^4 + 3240x^7y^6$ **21.** $4x^3 + 6x^2h + 4xh^2 + h^3$ **23. (a)** 0.032299   **(b)** 0.996961

25. Logo #1, Logo #2, Logo #3; Logo #1, Logo #3, Logo #2; Logo #2, Logo #1, Logo #3; Logo #2, Logo #3, Logo #1; Logo #3, Logo #2, Logo #1; Logo #3, Logo #1, Logo #2

27. 6720   **29.** 66   **31.** 6840   **33.** 20   **35.** 15,120   **37.** 17,153,136   **39.** 60   **41.** 9,375,000

43. Sample space: {(1, 1), (1, 2), (1, 3), (2, 1), (2, 2), (2, 3), (3, 1), (3, 2), (3, 3)}; Event: {(1, 3), (2, 3), (3, 1), (3, 2), (3, 3)}

45. $\frac{4}{13} \approx 0.308$ **47.** $\frac{5}{18} \approx 0.278$ **49.** $\frac{11}{4165} \approx 0.003$ **51.** $\frac{787}{833} \approx 0.945$ **53.** $\frac{2197}{20,825} \approx 0.105$ **55.** $\frac{1}{120} \approx 0.008$

57. (a) 0.6561   **(b)** 0.3439   **59.** $\frac{1}{24} \approx 0.042$

CHAPTER TEST

1. (a)

```
            1
          1   1
        1   2   1
      1   3   3   1
    1   4   6   4   1
  1   5  10  10   5   1
1   6  15  20  15   6   1
```

(b) $(z + 1)^5 = \underline{?}z^5 + \underline{?}z^4 \cdot 1 + \underline{?}z^3 \cdot 1^2$
$$+ \underline{?}z^2 \cdot 1^3 + \underline{?}z \cdot 1^4 + \underline{?} \cdot 1^5$$

We determine the coefficient of each term using the fifth row of Pascal's triangle:

$(z + 1)^5 = 1z^5 + 5z^4 \cdot 1 + 10z^3 \cdot 1^2 + 10z^2 \cdot 1^3$
$$+ 5z \cdot 1^4 + 1 \cdot 1^5$$
$$= z^5 + 5z^4 + 10z^3 + 10z^2 + 5z + 1$$

(see Example 1, Section 8.1)

2. $(f - 2g)^4 = {}_4C_0 f^4 + {}_4C_1 f^3(-2g) + {}_4C_2 f^2(-2g)^2$
$$+ {}_4C_3 f(-2g)^3 + {}_4C_4(-2g)^4$$
$$= 1 \cdot f^4 + 4f^3(-2g) + 6f^2(-2g)^2$$
$$+ 4f(-2g)^3 + 1(-2g)^4$$
$$= f^4 - 8f^3g + 24f^2g^2 - 32fg^3 + 16g^4$$

(see Example 3, Section 8.1)

3. The x^3y^6 term is given by
$${}_9C_6(2x)^3y^6 = 84(8x^3)y^6 = 672x^3y^6$$
(see Example 4, Section 8.1)

4. The first four terms in the expansion are

${}_{10}C_0(\sqrt{a})^{10} + {}_{10}C_1(\sqrt{a})^9(-3b^2)$
$$+ {}_{10}C_2(\sqrt{a})^8(-3b^2)^2$$
$$+ {}_{10}C_3(\sqrt{a})^7(-3b^2)^3$$
$$= a^5 - 30a^4\sqrt{a}\,b^2 + 405a^4b^4 - 3240a^3\sqrt{a}\,b^6$$

(see Example 3, Section 8.1)

5. $\dfrac{f(x+h)-f(x)}{h} = \dfrac{7(x+h)^4 - 7x^4}{h}$

$$= \dfrac{7(x^4 + 4x^3h + 6x^2h^2 + 4xh^3 + h^4) - 7x^4}{h}$$

$$= \dfrac{28x^3h + 42x^2h^2 + 28xh^3 + 7h^4}{h}$$

$$= 28x^3 + 42x^2h + 28h^2 + 7h^3$$

(see Example 5, Section 8.1)

6. **(a)** $wx, xw, wy, yw, wz, zw, xy, yx, xz, zx, yz, zy$

(b) $\{w, x\}, \{w, y\}, \{w, z\}, \{x, y\}, \{x, z\}, \{y, z\}$

(c) $_{10}P_3 = 720$

(d) $_{10}C_3 = 120$

(see Examples 1, 6, 10, Section 8.2)

7. There are 5 possibilities for the first subject, and for each of these there are 5 possibilities for the second subject, and for each of these there are 5 possibilities for the third subject, followed by 5 possibilities for the fourth subject. By the fundamental counting principle, there are $5 \cdot 5 \cdot 5 \cdot 5 = 625$ outcomes.

(see Example 2, Section 8.2)

8. $_6P_6 = 720$

(see Example 7, Section 8.2)

9. $_{12}C_2 = 66$

(see Example 10, Section 8.2)

10. $_8P_6 = 20{,}160$

(see Example 8, Section 8.2)

11. See Figure 8.7. Let E be the event of rolling 8 or 9. There are 5 ways to roll an 8 and 4 ways to roll a 9:

$E = \{(2, 6), (6, 2), (3, 5), (5, 3), (4, 4), (3, 6), (6, 3), (4, 5), (5, 4)\}$.
Therefore, $P(E) = \frac{9}{36} = \frac{1}{4}$.

(see Example 3, Section 8.3)

12. **(a)** $_{52}C_4 = 270{,}725$

(b) $\dfrac{_{13}C_4}{_{52}C_4} = \dfrac{715}{270{,}725} = \dfrac{11}{4165} \approx 0.003$

(c) Letting A represent the event that at least one of the cards is a club, $\overline{A}$ represents the event of getting no clubs.

$$P(\overline{A}) = \dfrac{_{39}C_4}{_{52}C_4} = \dfrac{82{,}251}{270{,}725} = \dfrac{6327}{20{,}825}$$

Thus,

$$P(A) = 1 - P(\overline{A}) = 1 - \dfrac{6327}{20{,}825}$$

$$= \dfrac{14{,}498}{20{,}825} \approx 0.696$$

(see Examples 4, 7, Section 8.3)

13. $\dfrac{(_8C_3)(_5C_2)}{_{16}C_5} = \dfrac{5}{39} \approx 0.128$

(see Example 12, Section 8.2, and probability of an event, Section 8.3)

14. $\dfrac{26 \cdot 25 \cdot 24 \cdot 10 \cdot 9 \cdot 8}{26^3 \cdot 10^3} = \dfrac{108}{169} \approx 0.639$

(see fundamental counting principle, Section 8.2, and probability of an event, Section 8.3)

APPENDIX A

1. Minimum 5 occurs at (5, 2); Maximum 40 occurs at (0, 8).

3. Minimum 0 occurs at (0, 0); Maximum 39 occurs at (5, 2).

5. Minimum 0 occurs at (0, 0); Maximum 28 occurs at (0, 4).

7. Minimum 0 occurs at (0, 0); Maximum 57 occurs at (6, 3).

9. Minimum 0 occurs at (0, 0); Maximum 88 occurs at (4, 4).

11. Minimum 22 occurs at (0, 0); Maximum 154 occurs at (12, 0).

13. Minimum 14 occurs at (0, 7); Maximum 46 occurs at (5, 3).

15. Minimum 42 occurs at (7, 0); no maximum exists.

17. Minimum 0 occurs at (0, 0); Maximum 480 occurs at any point on the line segment with endpoints (0, 20) and (12, 15).

19. Maximum profit is $10,400; this occurs when 75 two-drawer and 40 four-drawer cabinets are produced.

21. Maximum profit is $60,000; this occurs when 500 acres of tomatoes are planted and no cotton is planted.

23. Minimum cost is $564; this occurs when 36 units are shipped from downtown to store A, 0 units are shipped from downtown to store B, 0 units are shipped from industrial park to store A, and 24 units are shipped from industrial park to store B.

25. Maximum profit is $2305; this occurs when 55 batches of premium blend and 90 batches of house blend are produced.

27. Minimum cost is $100; this occurs when 20 lb of brand A and 150 lb of brand B are blended.

29. Maximum return is $11,600; this occurs when $100,000 is invested in stocks and $20,000 is invested in bonds.

31. Maximum profit is $72,000; the number of each model to produce may be any of the following combinations:

Tourist	Hillbreaker
320	160
321	158
322	156
⋮	⋮
387	26